Springer Proceedings in Mathematics

Volume 1

For further volumes:
http://www.springer.com/series/8806

Springer Proceedings in Mathematics

The book series will feature volumes of selected contributions from workshops and conferences in all areas of current research activity in mathematics. Besides an overall evaluation, at the hands of the publisher, of the interest, scientific quality, and timeliness of each proposal, every individual contribution will be refereed to standards comparable to those of leading mathematics journals. It is hoped that this series will thus propose to the research community well-edited and authoritative reports on newest developments in the most interesting and promising areas of mathematical research today.

Mauricio Matos Peixoto · Alberto Adrego Pinto
David A. Rand
Editors

Dynamics, Games and Science I

DYNA 2008, in Honor of Mauricio Peixoto and David Rand, University of Minho, Braga, Portugal, September 8-12, 2008

Editors

Mauricio Matos Peixoto
Instituto de Matemática
Pura e Aplicada (IMPA)
Estrada Dona Castorina 110
22460-320 Rio de Janeiro
Brazil
peixoto@impa.br

David A. Rand
University of Warwick
Warwick Systems Biology
Coventry House
Coventry CV4 7AL
United Kingdom
d.a.rand@warwick.ac.uk

Alberto Adrego Pinto
Universidade do Porto
Faculdade de Ciências
Departamento de
Matemática
Rua do Campo Alegre 687
4169-007 Porto
Portugal
aapinto@fc.up.pt

ISSN 2190-5614
ISBN 978-3-642-11455-7   e-ISBN 978-3-642-11456-4
DOI 10.1007/978-3-642-11456-4
Springer Heidelberg Dordrecht London New York

Library of Congress Control Number: 2011925151

Mathematical Subject Classification (2010): 34; 37; 49; 60; 83; 91; 92; 93; 94

© Springer-Verlag Berlin Heidelberg 2011
This work is subject to copyright. All rights are reserved, whether the whole or part of the material is
concerned, specifically the rights of translation, reprinting, reuse of illustrations, recitation, broadcasting,
reproduction on microfilm or in any other way, and storage in data banks. Duplication of this publication
or parts thereof is permitted only under the provisions of the German Copyright Law of September 9,
1965, in its current version, and permission for use must always be obtained from Springer. Violations
are liable to prosecution under the German Copyright Law.
The use of general descriptive names, registered names, trademarks, etc. in this publication does not
imply, even in the absence of a specific statement, that such names are exempt from the relevant protective
laws and regulations and therefore free for general use.

Cover design: deblik, Berlin

Printed on acid-free paper

Springer is part of Springer Science+Business Media (www.springer.com)

*To my wife Alcilea Augusto and my four
children and eight grandchildren:*
   *Marta: Daniel, Thomas and Mariana;
   Ricardo: Gabriel;
   Marcos: Andre and Livia; and
   Elisa: Clara and Bruno.*

To António Adrego Pinto

*To Barbel Finkenstädt and the Rand Kids:
   Ben,
   Tamsin,
   Rupert and
   Charlotte.*

Mauricio Peixoto

Alberto Pinto has asked me to write about Mauricio Peixoto in this book that honors him as well as David Rand. I am happy to do so. Mauricio is among my oldest friends in mathematics, having met him more than fifty years ago. Moreover he was instrumental in my entry into the field of dynamical systems. So important is this part of my life that my collected works contain four articles that bear on Mauricio in one way or another. That is fortunate since I wrote that material when events were fresher in my mind than they are now. Thus I will borrow freely from these references.

A most important period in my relationship with Mauricio is the summer of 1958 to June of 1960. This is discussed in an article titled "On how I got started in dynamical systems" appearing in the "Mathematics of Time", based on a talk given at a Berkeley seminar circa 1976. There I wrote how I met Mauricio in the summer of 1958 through a mutual friend, Elon Lima, who was a student from Brazil finishing his PhD at Chicago in topology. Through Lefshetz, Peixoto had become interested in structural stability and he explained to me that subject and described his own work in that area. I became immediately enthusiastic, and started making some early conjectures on how to pass from two to higher dimension. Shortly thereafter, Peixoto and Lima invited me and Clara to Rio for a visit to IMPA, or Instituto de Matemática, Pura e Aplicada.

It was during the next six months (January–June, 1960) that I did some of my most well known work, firstly the introduction of the horseshoe dynamical system and its consequences and secondly the proof of Poincare's conjecture in dimensions five or more. I sometimes described these works as having been done on the beaches of Rio; this part of the story is told in two articles in the Mathematics Intelligencer in the 1980s.

Thus we may see here what a big influence Mauricio had on my career. Another impact was his "sending" me a student to write a PhD thesis at Berkeley. That student in fact finished such a thesis and went on to become a world leader in dynamical systems. Jacob Palis' contributions in science go well beyond that. He is a main figure in developing third world science, and mathematics in Brazil in particular.

In the article "What is Global Analysis", based on a talk I gave before the Mathematical Association of America, 1968, I gave a focus to one result as an excellent theorem in global analysis. That result was Peixoto's theorem that structurally stable differential equations on a two dimensional manifold form an open and dense set. Another example of the influence of Mauricio!

I will end on a final note that reinforces all that I have said here. Over the last fifty years I have made fifteen visits to IMPA, the institute founded by Mauricio Peixoto (and Leopoldo Nachbin).

Steve Smale

Alberto Adrego Pinto

I met Alberto a few years ago, in the office of Mauricio Peixoto at IMPA, the Brazilian Institute of Pure and Applied Mathematics. Alberto was on a summer visit, and he wanted to discuss results he had obtained with his former student Diogo Pinheiro on the focal decomposition proposed years earlier by Mauricio.

By sheer accident, I had come across an application of the focal decomposition in finite temperature quantum mechanics. In fact, semi-classical approximations to the problem practically forced one to make use of the focal decomposition, although it was only much later that I became aware of its existence. That was why I was part of the meeting: my mathematician friends were curious about possible applications, and we were eager to collaborate.

Alberto immediately impressed me by his enthusiasm, his genuine interest in science, and by his easy-going style, much appreciated by a "carioca" like myself. Besides, our discussions were lively, and touched upon various conceptual points that seemed quite natural to a physicist, and eventually proved very useful from a mathematical point of view. Our collaboration has been going on ever since, and has already led to a couple of articles.

Alberto has also offered us all with a wonderful event back in 2008, when he organized a conference in honor of David Rand and Mauricio Peixoto in the precious city of Braga. The conference made me appreciate, even more, the versatility and scientific depth of Alberto, as he and his PhD. students and postdocs presented seminars that covered a wide variety of subjects.

As a final word about Alberto, it must be said that he is a marvelous host. He showed us the finest of the region of Minho, using a well balanced combination of science, art, good food, good wine, and above all, good humor. That is the reason I always look forward to our next meeting: whether in Brazil or in Portugal, I am sure we will have a pleasant and productive time.

Carlos Alberto Aragão de Carvalho

David Rand

David Rand has had a world-leading influence in dynamical systems theory, in transferring dynamical systems ideas into the sciences, particularly physical and life sciences but also economics, and in developing relevant new mathematics for these areas. Highlights are his theories of the two-frequency route to chaos, invasion exponents in evolutionary dynamics, and robustness of circadian rhythms. He is widely appreciated for his leadership and for his highly pertinent and generous insights into research projects of others.

He was one of the first to bring ideas on dynamical systems with symmetry into fluid mechanics, predicting modulated wave states in circular Couette flow, subsequently confirmed experimentally by Swinney and Gorman.

A major advance was his proposal of a renormalization explanation for observations of asymptotic self-similarity in the transition from quasiperiodic to chaotic dynamics for circle maps. He extended the theory to dissipative annulus maps, providing a complete picture of the breakup of invariant circles in this scenario. Similar analysis of his has been important in understanding the spectrum of quasiperiodic Schrodinger operators.

He put the theory of multifractal scaling for chaotic attractors on a firm footing, including theory for the distribution of Lyapunov exponents.

He contributed significantly to the dynamical theory of evolutionary stability and co-evolution, including the fundamental concept of invasion exponents. He developed pair approximations for spatial ecologies and epidemics, which are now widely used.

With Alberto Pinto, he developed an extensive theory of the smooth conjugacy classes of hyperbolic dynamics in one and two dimensions, surveyed in a recent Springer Monograph in Mathematics.

He made one of the earliest analyses of nonlinear dynamics in an economics context, showing that a duopoly game has chaotic trajectories. Game theory has been a recurrent interest of his, particularly in the contexts of ecology and evolution.

Much of his recent work falls under "systems biology". He has proposed a theory of the immune system, based on large deviation theory. He has developed theory of the robustness of circadian rhythms, which has generated much interest with experimental collaborators. The work is part of a larger project to develop mathematical tools to aid in the understanding of biological regulatory and signaling networks.

He has played a leading role in establishing Nonlinear Dynamics in the UK, co-founding the Nonlinear Systems Laboratory in Warwick and the journal Nonlinearity. He is doing the same now for Systems Biology, creating the Warwick Systems Biology Centre.

He exudes energy and enthusiasm. So it was a pleasure for me when he attracted me to Warwick. We had great fun setting up and running the Nonlinear Systems Laboratory, building up the applied side to Warwick's Mathematics department and its curriculum, and setting up the Interdisciplinary Mathematics Research Programme which he rebranded as Mathematical Interdisciplinary Research at Warwick and of which I took over directorship from David in 2000. He is a great friend and I have greatly appreciated his insightful comments, suggestions and support for my own work.

Robert S. MacKay

Preface

A couple of years ago Alberto Pinto informed me that he was planning to organize an international conference on dynamical systems and game theory in honor of Mauricio Peixoto and David Rand. I told him that I wholeheartedly support the idea and will ask the International Society of Difference Equations (ISDE) to support the proposed conference which it did later. Through my frequent visits to Portugal, I became aware of the significant contributions in dynamical systems and game theory made by Portuguese mathematicians and have subsequently been involved in fruitful discussions or joint research with a number of them. The growth of dynamical systems and game theory research in Portugal has placed Portuguese mathematicians at the forefront of these emerging fields, bringing worldwide recognition to their contributions. Indeed, in addition to DYNA2008, Portuguese researchers organized two of the last three International conferences on difference equations and applications (ICDEA), which included important talks on dynamical systems and game theory.

The work in this area has unveiled beautiful and deep mathematical theories that capture universal characteristics observed in many apparently unrelated natural phenomena and complex social behavior. Mauricio Peixoto has made lasting contributions in classifying and understanding a variety of behaviors of dynamical systems. Today these problems are the main research focus in diverse yet complementary areas at distinguished research institutions like IMPA, the institute founded by Mauricio Peixoto and Leopoldo Nachbin, and University of Warwick. Alberto Pinto has made notable contributions through his studies on rigidity properties of infinitely renormalizable dynamical systems. In addition, he discovered stochastic universalities in complex natural and social phenomena, e.g. rivers, sunspots and stock market indices, and is developing a theory with Peixoto in semi-classics physics using Peixoto's focal decomposition. David Rand, a world-leading authority, has contributed deeply and broadly to this area by developing theoretical aspects of these two fields, and identifying properties of infinitely renormalizable, universal and chaotic phenomena throughout the sciences – especially in biology, economics and physics. In collaboration with Alberto Pinto, he constructed a fine classification of dynamical systems. Moreover, the research groups led by David Rand and Alberto Pinto have, independently, developed new schools of inquiry using

xiii

game theoretical and dynamic models applied to biology, economics, finances, psychology and sociology.

The research and survey papers in these volumes, written by leading researchers in their scientific areas, focus on these and many other relevant aspects of dynamical systems, game theory and their applications to science and engineering. The papers in these volumes are based on talks given at the International Conference DYNA2008, in honor of Mauricio Peixoto and David Rand. This conference, held at the University of Minho, was organized by Alberto Pinto and his colleagues and brought together influential researchers from around the world. It is worthwhile to note the warmth and hospitality of the organizers who made sure we enjoyed the beautiful region of Minho with its rich culture and fine cuisine.

Saber Elaydi

Acknowledgments

We thank all the authors for their contributions to this book and the anonymous referees. We are grateful to Steve Smale, Carlos Aragão and Robert MacKay for contributing the forwards for the book. We thank Saber Elaydi, President of the International Society of Difference Equations (ISDE), for contributing the preface for the book and for supporting and dedicating two special issues of the Journal of Difference Equations and Applications to DYNA2008, in honor of Mauricio Peixoto and David Rand, held at University of Minho, Portugal. We thank the Executive Editor for Mathematics, Computational Science and Engineering, Martin Peters, at Springer-Verlag, for his invaluable suggestions and advice throughout this project. We thank João Paulo Almeida, Isabel Duarte, Fernanda Ferreira, Flávio Ferreira, Helena Ferreira, Miguel Ferreira, Patrícia Gonçalves, Rui Gonçalves, Susan Jenkins, José Tenreiro Machado, José Martins, Filipe Mena, Abdelrahim Mousa, Bruno Oliveira, Telmo Parreira, Diogo Pinheiro, Ana Jacinta Soares and Nico Stollenwerk for their invaluable help in assembling these volumes and, especially, to João Paulo Almeida, Helena Ferreira, Miguel Ferreira, Susan Jenkins, and Abdelrahim Mousa for editorial assistance.

At DYNA2008, Mauricio Peixoto and David Rand were honored with the medal of Universidade do Minho, given by the Vice-Chancellor of Universidade do Minho, Leandro Almeida, and with a commemorative plaque given by the Vice-President of Instituto de Matemática Pura e Aplicada, Claudio Landim. We thank the members of the Scientific Committee of DYNA2008 for their invaluable help: Aragão de Carvalho (FRJ, Brazil), Edson de Faria (USP, Brazil), Pablo Ferrari (IME USP, Brazil), Livio Flaminio (U. Lille, France), Claudio Landim (IMPA, Brazil), Marco Martens (SUNYSB, USA), Welington de Melo (IMPA, Brazil), Mauricio Peixoto (IMPA, Brazil), Alberto Adrego Pinto (U. Minho, Portugal), Enrique Pujals (IMPA, Brazil), David Rand (Warwick, UK), Marcelo Viana (IMPA, Brazil) and Athanasios Yannacopoulos (Athens, Greece). We gratefully acknowledge the members of the Organizing Committee of DYNA2008 for their help: Isabel Caiado, Salvatore Cosentino, Wolfram Erlhagen, Miguel Ferreira, Patrícia Gonçalves, Filipe Mena, Alberto Adrego Pinto, Ana Jacinta Soares and Joana Torres.

We thank the Chancellor of Universidade do Minho António Guimarães Rodrigues, the Vice-Chancellors Leandro Almeida and Manuel Mota, the President of Escola de Ciências Graciete Dias, and the President of Escola de Enfermagem

Beatriz Araújo for supporting DYNA2008. We thank Fundação para a Ciência e Tecnologia, Fundação Calouste Gulbenkian, Universidade do Minho, Universidade do Porto, Departamento de Matemática e Aplicações da Escola de Ciências da Universidade do Minho, Departamento de Matemática da Faculdade de Ciências da Universidade do Porto, Centro de Matemática da Universidade do Minho, Laboratório de Inteligência Artificial e Apoio à Decisão and Laboratório Associado Instituto de Engenharia de Sistemas e Computadores for their support.

Contents

1 A Brief Survey of Focal Decomposition 1
Mauricio M. Peixoto

2 Anosov and Circle Diffeomorphisms 11
João P. Almeida, Albert M. Fisher, Alberto A. Pinto,
and David A. Rand

**3 Evolutionarily Stable Strategies and Replicator Dynamics
in Asymmetric Two-Population Games** 25
Elvio Accinelli and Edgar J. Sánchez Carrera

4 Poverty Traps, Rationality and Evolution 37
Elvio Accinelli, Silvia London, Lionello F. Punzo,
and Edgar J. Sanchez Carrera

5 Leadership Model 53
Leandro Almeida, José Cruz, Helena Ferreira,
and Alberto A. Pinto

6 Lorenz-Like Chaotic Attractors Revised 61
Vítor Araújo and Maria José Pacifico

**7 A Dynamical Point of View of Quantum Information:
Entropy and Pressure** 81
A.T. Baraviera, C.F. Lardizabal, A.O. Lopes,
and M. Terra Cunha

8 Generic Hamiltonian Dynamical Systems: An Overview 123
Mário Bessa and João Lopes Dias

**9 Microeconomic Model Based on MAS Framework:
Modeling an Adaptive Producer** 139
Pavel Brazdil and Frederico Teixeira

xvii

10 A Tourist's Choice Model 159
J. Brida, M.J. Defesa, M. Faias, and Alberto A. Pinto

11 Computability and Dynamical Systems 169
J. Buescu, D.S. Graça, and N. Zhong

12 Dynamics and Biological Thresholds 183
N.J. Burroughs, M. Ferreira, J. Martins, B.M.P.M. Oliveira,
Alberto A. Pinto, and N. Stollenwerk

13 Global Convergence in Difference Equations 193
Elias Camouzis and Gerasimos Ladas

**14 Networks Synchronizability, Local Dynamics
and Some Graph Invariants** 221
Acilina Caneco, Sara Fernandes, Clara Grácio,
and J. Leonel Rocha

**15 Continuous Models for Genetic Evolution in Large
Populations** 239
Fabio A.C.C. Chalub and Max O. Souza

**16 Forecasting of Yield Curves Using Local State Space
Reconstruction** 243
Eurico O. Covas and Filipe C. Mena

17 KAM Theory as a Limit of Renormalization 253
João Lopes Dias

**18 An Overview of Optimal Life Insurance Purchase,
Consumption and Investment Problems** 271
Isabel Duarte, Diogo Pinheiro, Alberto A. Pinto, and Stanley
R. Pliska

**19 Towards a Theory of Periodic Difference Equations
and Its Application to Population Dynamics** 287
Saber N. Elaydi, Rafael Luís, and Henrique Oliveira

**20 Thompson's Group, Teichmüller Spaces,
and Dual Riemann Surfaces** 323
Edson de Faria

21 Bargaining Skills in an Edgeworthian Economy 339
M. Ferreira, B. Finkenstädt, B.M.P.M. Oliveira,
Alberto A. Pinto, and A.N. Yannacopoulos

22	**Fractional Analysis of Traffic Dynamics**	353

Lino Figueiredo and J.A. Tenreiro Machado

23 The Set of Planar Orbits of Second Species in the RTBP 359
Joaquim Font, Ana Nunes, and Carles Simó

24 Statistical Properties of the Maximum for Non-Uniformly Hyperbolic Dynamics 365
Ana Cristina Moreira Freitas, Jorge Milhazes Freitas, and Mike Todd

25 Adaptive Learning and Central Bank Inattentiveness in Optimal Monetary Policy 375
Orlando Gomes, Vivaldo M. Mendes, and Diana A. Mendes

26 Discrete Time, Finite State Space Mean Field Games 385
Diogo A. Gomes, Joana Mohr, and Rafael Rigão Souza

27 Simple Exclusion Process: From Randomness to Determinism 391
Patrícia Gonçalves

28 Universality in PSI20 fluctuations 405
Rui Gonçalves, Helena Ferreira, and Alberto A. Pinto

29 Dynamical Systems with Nontrivially Recurrent Invariant Manifolds 421
Viacheslav Grines and Evgeny Zhuzhoma

30 Some Recent Results on the Stability of Endomorphisms 471
J. Iglesias, A. Portela, and A. Rovella

31 Differential Rigidity and Applications in One-Dimensional Dynamics 487
Yunping Jiang

32 Minimum Regret Pricing of Contingent Claims in Incomplete Markets 503
C. Kountzakis, S.Z. Xanthopoulos, and A.N. Yannacopoulos

33 A Class of Infinite Dimensional Replicator Dynamics 529
D. Kravvaritis, V. Papanicolaou, T. Xepapadeas, and A.N. Yannacopoulos

34 Kinetic Theory for Chemical Reactions Without a Barrier 533
Gilberto M. Kremer and Ana Jacinta Soares

**35 Dynamical Gene-Environment Networks
Under Ellipsoidal Uncertainty: Set-Theoretic
Regression Analysis Based on Ellipsoidal OR** 545
Erik Kropat, Gerhard-Wilhelm Weber, and Selma Belen

**36 Strategic Interaction in Macroeconomic Policies:
An Outline of a New Differential Game Approach** 573
T. Krishna Kumar

37 Renormalization of Hénon Maps 597
M. Lyubich and M. Martens

38 Application of Fractional Calculus in Engineering 619
J.A. Tenreiro Machado, Isabel S. Jesus, Ramiro Barbosa,
Manuel Silva, and Cecilia Reis

**39 Existence of Invariant Circles for Infinitely
Renormalisable Area-Preserving Maps** 631
R.S. MacKay

40 The Dynamics of Expectations 637
Wilfredo L. Maldonado and Isabel M.F. Marques

41 Dynamics on the Circle 651
W. de Melo

42 Rolling Ball Problems 661
Waldyr M. Oliva and Gláucio Terra

**43 On the Dynamics of Certain Models Describing the HIV
Infection** 671
Dayse H. Pastore and Jorge P. Zubelli

44 Tilings and Bussola for Making Decisions 689
Alberto A. Pinto, Abdelrahim S. Mousa,
Mohammad S. Mousa, and Rasha M. Samarah

45 A Hotelling-Type Network 709
Alberto A. Pinto and Telmo Parreira

46 The Closing Lemma in Retrospect 721
Charles Pugh

Contents

47 From Peixoto's Theorem to Palis's Conjecture 743
Enrique R. Pujals

**48 Dynamics Associated to Games (Fictitious Play)
with Chaotic Behavior** 747
Colin Sparrow and Sebastian van Strien

49 A Finite Time Blowup Result for Quadratic ODE's 761
Dennis Sullivan

**50 Relating Material and Space-Time Metrics Within
Relativistic Elasticity: A Dynamical Example** 763
E.G.L.R. Vaz, Irene Brito, and J. Carot

**51 Strategic Information Revelation Through Real Options
in Investment Games** 769
Takahiro Watanabe

**52 On Consumption Indivisibilities, the Demand
for Durables, and Income Distribution** 785
David Zilberman and Jenny Hsing-I Liu

Contributors

Elvio Accinelli Facultad de Economía, UASLP, San Luis Potosí, México, elvio.accinell@eco.uaslp.mx

João P. Almeida LIAAD-INESC Porto LA, Porto, Portugal
and
Centro de Matemática da Universidade do Minho, Campus de Gualtar, 4710-057 Braga, Portugal
and
Departamento de Matemática, Escola Superior de Tecnologia e Gestão, Instituto Politécnico de Bagança, Campus de Santa Apolónia, Ap. 1134, 5301-857 Bragança, Portugal, jpa@ipb.pt

Leandro Almeida IEP, Campus de Gualtar, 4710-057 Braga, Portugal, leandro@iep.uminho.pt

Vítor Araújo Instituto de Matemática, Universidade Federal do Rio de Janeiro, C. P. 68.530, 21.945-970, Rio de Janeiro, RJ, Brazil
and
Centro de Matemática da, Universidade do Porto, Rua do Campo Alegre 687, 4169-007 Porto, Portugal, vitor.araujo@im.ufrj.br, vdaraujo@fc.up.pt

A.T. Baraviera I.M. – UFRGS, 91500-000, Porto Alegre, Brazil, atbaraviera@gmail.com

Ramiro Barbosa Institute of Engineering of Porto, Rua Dr. António Bernardino de Almeida, 4200-072 Porto, Portugal, rsb@isep.ipp.pt

Selma Belen CAG University, Yenice-Tarsus, 33800 Mersin, Turkey, sbelen@cag.edu.tr

Mário Bessa Departamento de Matemática da, Universidade do Porto, Rua do Campo Alegre 687, 4169-007 Porto, Portugal
and
ESTGOH-Instituto Politécnico de Coimbra, Rua General Santos Costa, 3400-124 Oliveira do Hospital, Portugal, bessa@fc.up.pt

Pavel Brazdil LIAAD-INESC Porto LA, Porto, Portugal

and

FEP, University of Porto, Porto, Portugal, pbrazdil@liaad.up.pt

J. Brida Free University of Bolzano, Bolzano, Italy, JuanGabriel.Brida@unibz.it

Irene Brito Departamento de Matemática para a Ciência e Tecnologia, Universidade do Minho, 4800 058 Guimarães, Portugal, ireneb@mct.uminho.pt

J. Buescu DM/FCUL, University of Lisbon, Lisbon, Portugal

and

CMAF, Lisbon, Portugal, jbuescu@ptmat.fc.ul.pt

N.J. Burroughs University of Warwick, Coventry CV4 7AL, UK, njb@maths.warwick.ac.uk

Elias Camouzis Department of Mathematics and Natural Sciences, The American College of Greece, Gravias 6 Str, 15342 Aghia Paraskevi, Athens, Greece, camouzis@acgmail.gr

Acilina Caneco Mathematics Unit, Instituto Superior de Engenharia de Lisboa, Lisbon, Portugal

and

CIMA-UE, Universidade de Évora, Évora, Portugal, acilina@deetc.isel.ipl.pt

J. Carot Departament de Fsica, Universitat de les Illes Balears, Cra Valldemossa pk 7.5, 07122 Palma de Mallorca, Spain, jcarot@uib.es

Edgar J. Sánchez Carrera Department of Economics, University of Siena, Siena, Italy, sanchezcarre@unisi.it

Fabio A.C.C. Chalub Departamento de Matemática and Centro de Matemática e Aplicações, Universidade Nova de Lisboa, Quinta da Torre, 2829-516, Caparica, Portugal, chalub@fct.unl.pt

Eurico O. Covas HSBC Bank Plc, Hedging Software Development, 8 Canada Square, Canary Wharf, London E14 5HQ, UK, eurico.covas@hsbcgroup.com

José Cruz IEP, Campus de Gualtar, 4710-057 Braga, Portugal, jcruz@iep.uminho.pt

M. Terra Cunha D.M – UFMG, Belo Horizonte 30161-970, Brazil, marcelo.terra.cunha@gmail.com

M.J. Defesa Universidad de Alcalá, Alcalá de Henares, Spain, mjesus.such@fct.uah.es

W. de Melo Instituto de Matemática Pura e Aplicada (IMPA), Estrada Dona Castorina, 110, 22460-320 Rio de Janeiro, RJ, Brazil, demelo@impa.br

João Lopes Dias Departamento de Matemática and Cemapre, ISEG, Universidade Técnica de Lisboa, Rua do Quelhas 6, 1200-781 Lisbon, Portugal, jldias@iseg.utl.pt

I. Duarte Department of Mathematics, University of Minho, Braga, Portugal, isabelduarte@math.uminho.pt

Saber N. Elaydi Department of Mathematics, Trinity University, San Antonio, TX, USA, selaydi@trinity.edu

M. Faias Universidade Nova de Lisboa, Lisbon, Portugal, mcm@fct.unl.pt

Edson de Faria IME-USP, Rua do Matão, 1010 Butantã, 05508-090 – Sao Paulo, SP, Brazil, edson@ime.usp.br

Sara Fernandes Department of Mathematics, Universidade de Évora, Évora, Portugal

and

CIMA-UE, Universidade de Évora, Évora, Portugal, saf@uevora.pt

Helena Ferreira LIAAD-INESC Porto LA, Porto, Portugal

and

Research Center of Mathematics, School of Sciences, University of Minho, Campus of Gualtar, Braga, Portugal, helenaisafer@gmail.com

M. Ferreira Escola Superior de Estudos Industriais e de Gestão do Instituto Politécnico do Porto (IPP), LIAAD-INESC Porto LA, Escola de Ciências, Universidade do Minho, Braga, Portugal, migferreira2@gmail.com

Lino Figueiredo Department of Electrical Engineering, ISEP-Institute of Engineering of Porto, Rua Dr. Antonio Bernardino de Almeida, 4200-072 Porto, Portugal, lbf@isep.ipp.pt

B. Finkenstädt Department of Statistics, University of Warwick, Coventry CV4 7AL, UK, b.f.finkenstadt@warwick.ac.uk

Albert M. Fisher Departamento de Matemática, IME-USP, Caixa Postal 66281, CEP 05315-970, São Paulo, Brazil, afisher@ime.usp.br

Joaquim Font Departament de Matemàtica Aplicada i Anàlisi, Universitat de Barcelona, Gran Via 587, 08007 Barcelona, Spain, quim@maia.ub.es

Ana Cristina Moreira Freitas Centro de Matemática & Faculdade de Economia da, Universidade do Porto, Rua Dr. Roberto Frias, 4200-464 Porto, Portugal, amoreira@fep.pt

Jorge Milhazes Freitas Centro de Matemática da, Universidade do Porto, Rua do Campo Alegre 687, 4169-007 Porto, Portugal, jmfreita@fc.up.pt

Diogo A. Gomes Instituto Superior Técnico, Av. Rovisco Pais, 1049-001 Lisbon, Portugal, dgomes@math.ist.utl.pt

Orlando Gomes Escola Superior de Comunicao Social, Campus de Benfica do IPL, 1549-014 Lisbon, Portugal, ESCS/IPL and UNIDE/ERC, ogomes@escs.ipl.pt

Patrícia Gonçalves Centro de Matemática da, Universidade do Minho, Campus de Gualtar, 4710-057 Braga, Portugal, patg@math.uminho.pt

Rui Gonçalves LIAAD-INESC Porto LA e Secção de Matemática, Faculdade de Engenharia, Universidade do Porto, R. Dr. Roberto Frias s/n, 4200-465 Porto, Portugal, rjasg@fe.up.pt

D.S. Graça DM/Faculdade de Ciências e Tecnologia, Universidade do Algarve, Faro, Portugal

and

SQIG/Instituto de Telecomunicações, Lisbon, Portugal, dgraca@ualg.pt

Clara Grácio Department of Mathematics, Universidade de Évora, Évora, Portugal

and

CIMA-UE, Universidade de Évora, Évora, Portugal, mgracio@uevora.pt

Viacheslav Grines Nizhny Novgorod State University, 23 Gagarin Ave, Nizhny Novgorod, 603950, Russia, vgrines@yandex.ru

J. Iglesias IMERL, Facultad de Ingeniería, Universidad de la República, J. Herrera y Reissig 565, Montevideo 11300, Uruguay, jorgei@fing.edu.uy

Isabel S. Jesus Institute of Engineering of Porto, Rua Dr. António Bernardino de Almeida, 4200-072 Porto, Portugal, isj@isep.ipp.pt

Yunping Jiang Department of Mathematics, Queens College of the City, University of New York, New York, NY, USA

and

Department of Mathematics, The Graduate Center, CUNY, New York, NY, USA, yunping.jiang@qc.cuny.edu

Christos Kountzakis Department of Statistic and Actuarial-Financial Mathematics, University of the Aegean, Karlovassi, 83200 Samos, Greece, chrkoun@aegean.gr

D. Kravvaritis Department of Mathematics, National Technical University, Athens, Greece, dkrav@math.ntua.gr

Gilberto M. Kremer Department of Physics, University of Paraná, Curitiba, Brazil, kremer@fisica.ufpr.br

Erik Kropat Institute for Theoretical Computer Science, Mathematics and Operations Research, Universität der Bundeswehr München, Werner-Heisenberg-Weg 39, 85577 Neubiberg, Germany, erik.kropat@unibw.de

T. Krishna Kumar Samkhya Analytica India Pvt. Ltd., Bangalore, India
and
Economic Analysis Unit, Indian Statistical Institute, Bangalore, India
and
Indian Institute of Management, Bangalore, India
and
3 Graystone Court, Naperville, IL 60565, USA, tkkumar@gmail.com

Gerasimos Ladas Department of Mathematics, University of Rhode Island, Kingston, RI 02881-0816, USA, gladas@math.uri.edu

C.F. Lardizabal I.M. – UFRGS, 91500-000, Porto Alegre, Brazil, carlos.lardizabal@gmail.com

Jenny Liu Department of Agricultural and Resource Economics, UC Berkeley, Berkeley, CA, USA, jliu@berkeley.edu

Silvia London Facultad de Economía, Universidad Nacional del Sur – CONICET, Bahía Blanca, Argentina, slondon@uns.edu.ar

A.O. Lopes I.M. – UFRGS, 91500-000, Porto Alegre, Brazil, arturoscar.lopes@gmail.com

Rafael Luís Center for Mathematical Analysis, Geometry, and Dynamical Systems, Instituto Superior Tecnico, Technical University of Lisbon, Lisbon, Portugal, rafael@ebsaas.com

M. Lyubich SUNY, Stony Brook, NY, USA, mlyubich@math.sunysb.edu

J.A. Tenreiro Machado Department of Electrical Engineering, ISEP-Institute of Engineering of Porto, Rua Dr. Antonio Bernardino de Almeida, 4200-072 Porto, Portugal, jtm@isep.ipp.pt

R.S. MacKay Mathematics Institute, University of Warwick, Coventry CV4 7AL, UK, R.S.MacKay@warwick.ac.uk

Wilfredo L. Maldonado Graduate School in Economics, Catholic University of Brasília, SGAN 916, Módulo B, Brasília DF, CEP 70790-160, Brazil, wilfredo@pos.ucb.br

Isabel M.F. Marques Graduate School in Economics, Catholic University of Brasília, SGAN 916, Módulo, Brasília DF, CEP 70790-160, Brazil, isabel.marques@catolica.edu.br

M. Martens SUNY, Stony Brook, NY, USA, marco@math.sunysb.edu

J. Martins LIAAD-INESC Porto LA, Porto, Portugal
and
Department of Mathematics, School of Technology and Management, Polytechnic Institute of Leiria, Campus 2, Morro do Lena – Alto do Vieiro, 2411-901 Leiria, Portugal, jmmartins@estg.ipleiria.pt

Filipe C. Mena Departamento de Matemática, Universidade do Minho, Campus de Gualtar, 4710 Braga, Portugal, fmena@math.uminho.pt

Diana A. Mendes Department of Quantitative Methods, IBS – ISCTE Business School, ISCTE, Avenida das Forças Armadas, 1649-026 Lisbon, Portugal, diana.mendes@iscte.pt

Vivaldo M. Mendes Department of Economics, Instituto Superior de Ciências do Trabalho e da Empresa, Avenida das Forças Armadas, 1649-026 Lisbon, Portugal, vivaldo.mendes@iscte.pt

Joana Mohr Instituto de Matemática, UFRGS, 91509-900, Porto Alegre, Brazil, joanamohr@yahoo.com.br

Abdelrahim S. Mousa LIAAD-INESC Porto LA, Porto, Portugal

and

Department of Mathematics, University of Minho, Braga, Portugal, abed11@ritaj.ps

Mohammad S. Mousa Department of Mathematics, Birzeit University, Birzeit, Palestine, moha_mousa@yahoo.com

Ana Nunes CFTC/Departamento de Física, Faculdade de Ciências, Universidade de Lisboa, Av. Prof. Gama Pinto 2, 1649-003 Lisbon, Portugal, anunes@ptmat.fc.ul.pt

Waldyr M. Oliva Instituto Superior Técnico, ISR and Departamento de Matemática, Centro de Análise Matemática, Geometria e Sistemas Dinâmicos, Av. Rovisco Pais, 1, 1049-001 Lisbon, Portugal, wamoliva@math.ist.utl.pt

B.M.P.M. Oliveira Faculdade de Ciências da Nutrição e Alimentação da, Universidade do Porto, LIAAD-INESC Porto LA, Porto, Portugal, bmpmo@fcna.up.pt

Henrique Oliveira Department of Mathematics, Instituto Superior Tecnico, Technical University of Lisbon, Lisbon, Portugal, holiv@math.ist.utl.pt

Maria José Pacifico Instituto de Matemática, Universidade Federal do Rio de Janeiro, C. P. 68.530, 21.945-970 Rio de Janeiro, RJ, Brazil, pacifico@im.ufrj.br, pacifico@impa.br

V. Papanicolaou Department of Mathematics, National Technical University, Athens, Greece, papanico@math.ntua.gr

Telmo Parreira Universidade de Aveiro, Aveiro, Portugal, telmoparreira@ua.pt

D.H. Pastore CEFET-RJ, Av. Maracanã 229, Rio de Janeiro, RJ 20271-110, Brazil, dayse@impa.br

Mauricio M. Peixoto Instituto de Matemática Pura e Aplicada (IMPA), Estrada Dona Castorina, 110, 22460-320 Rio de Janeiro, RJ, Brazil, peixoto@impa.br

Contributors xxix

D. Pinheiro CEMAPRE, ISEG, Technical University of Lisbon, Lisbon, Portugal, dpinheiro@iseg.utl.pt

Alberto A. Pinto LIAAD-INESC Porto LA e Departamento de Matemática, Faculdade de Ciências, Universidade do Porto, Rua do Campo Alegre, 687, 4169-007, Portugal
and
Centro de Matemática e Departamento de Matemática e Aplicaçöes, Escola de Ciências, Universidade do Minho, Campus de Gualtar, 4710-057 Braga, Portugal, aapinto@fc.up.pt

S.R. Pliska Department of Finance, University of Illinois at Chicago, Chicago, IL 60607, USA, srpliska@uic.edu

A. Portela IMERL, Facultad de Ingeniería, Universidad de la República, J. Herrera y Reissig 565, Montevideo 11300, Uruguay, aldo@fing.edu.uy

Charles Pugh Department of Mathematics, University of Toronto, 40 St. George Street, Toronto, ON, Canada M5S 2E4, cpugh@math.utoronto.ca

Enrique R. Pujals Instituto de Matemática Pura e Aplicada (IMPA), Estrada Dona Castorina, 110, 22460-320 Rio de Janeiro, RJ, Brazil, enrique@impa.br

Lionello F. Punzo Department of Economics, University of Siena, Piazza S. Francesco, 7, 53100 Siena, Italy
and
PPED-INCT, UFRJ, Rio de Janeiro, RJ, Brazil, punzo@unisi.it

D.A. Rand Warwick Systems Biology & Mathematics Institute, University of Warwick, Coventry CV4 7AL, UK, dar@maths.warwick.ac.uk

Cecilia Reis Institute of Engineering of Porto, Rua Dr. António Bernardino de Almeida, 4200-072 Porto, Portugal, cmr@isep.ipp.pt

J. Leonel Rocha Mathematics Unit, Instituto Superior de Engenharia de Lisboa, Lisbon, Portugal, jrocha@deq.isel.ipl.pt

A. Rovella Centro de Matemática Facultad de Ciencias, Universidad de la República, Iguá 4225, Montevideo 11400, Uruguay, leva@cmat.edu.uy

Rasha M. Samarah Department of Mathematics, Birzeit University, Birzeit, Palestine, rasha.abed1@yahoo.com

Manuel Silva Institute of Engineering of Porto, Rua Dr. António Bernardino de Almeida, 4200-072 Porto, Portugal, mss@isep.ipp.pt

Carles Simó Departament de Matemàtica Aplicada i Anàlisi, Universitat de Barcelona, Gran Via 587, 08007 Barcelona, Spain, carles@maia.ub.es

Ana Jacinta Soares Department of Mathematics, University of Minho, 4710-057 Braga, Portugal, ajsoares@math.uminho.pt

Max O. Souza Departamento de Matemática Aplicada, Universidade Federal Fluminense, R.Mário Santos Braga, s/n, 22240-920, Niterói, RJ, Brazil, msouza@mat.uff.br

Rafael Rigão Souza Instituto de Matemática, UFRGS, 91509-900, Porto Alegre, Brazil, rafars@mat.ufrgs.br

Colin Sparrow Maths Department, University of Warwick, Coventry CV4 7AL, UK, C.Sparrow@warwick.ac.uk

N. Stollenwerk Centro de Matemática e Aplicações Fundamentais, Universidade de Lisboa, Lisbon, Portugal, nico@ptmat.fc.ul.pt

Dennis Sullivan Mathematics Department, CUNY Graduate Center, New York, NY, USA

and

Mathematics Department, SUNY, Stony Brook, NY, USA, dennis@math.sunysb.edu

Frederico Teixeira McKinsey, Lisbon, Portugal, jfredericoteixeira@gmail.com

Gláucio Terra Departamento de Matemática, Instituto de Matemática e Estatística, Universidade de São Paulo, Rua do Matão, 1010, 05508-090 São Paulo, Brazil, glaucio.terra@gmail.com, glaucio@ime.usp.br

Mike Todd Centro de Matemática da, Universidade do Porto, Rua do Campo Alegre 687, 4169-007 Porto, Portugal, mtodd@fc.up.pt

Sebastian van Strien Maths Department, University of Warwick, Coventry CV4 7AL, UK, S.J.van-Strien@warwick.ac.uk

E.G.L.R. Vaz Departamento de Matemática para a Ciência e Tecnologia, Universidade do Minho, 4800 058 Guimarães, Portugal, evaz@mct.uminho.pt

Takahiro Watanabe Department of Business Administration, Tokyo Metropolitan University, Minamiosawa 1-1, Hachiouji, Tokyo, Japan, contact_nabe08@nabenavi.net

Gerhard-Wilhelm Weber Institute of Applied Mathematics, Middle East Technical University, 06531 Ankara, Turkey

and

Faculty of Economics, Business and Law, University of Siegen, Siegen, Germany

and

Center for Research on Optimization and Control, University of Aveiro, Aveiro, Portugal

and

Faculty of Science, Universiti Teknologi Malaysia (UTM), Skudai, Malaysia, gweber@metu.edu.tr

Stylianos Z. Xanthopoulos Department of Statistic and Actuarial-Financial Mathematics, University of the Aegean, Karlovassi, 83200 Samos, Greece, sxantho@aegean.gr

T. Xepapadeas Department of International and European Economic Studies, Athens University of Economics and Business, Athens, Greece, xepapad@aueb.gr

A.N. Yannacopoulos Department of Statistics, Athens University of Economics and Business, Patission 76, 10434 Athens, Greece, ayannaco@aueb.gr

N. Zhong DMS, University of Cincinnati, Cincinnati, OH 45221-0025, USA, ning.zhong@uc.edu

Evgeny Zhuzhoma Nizhny Novgorod State Pedagogical University, 1 Ulyanova Str,, Nizhny Novgorod, 603600, Russia, zhuzhoma@mail.ru

David Zilberman Department of Agricultural and Resource Economics, UC Berkeley, Berkeley, CA, USA, zilber11@berkeley.edu

J.P. Zubelli IMPA, Est. D. Castorina 110, Rio de Janeiro, RJ 22460-320, Brazil, zubelli@impa.br

Chapter 1
A Brief Survey of Focal Decomposition

Mauricio M. Peixoto

Abstract We present a brief survey of focal decomposition stressing how this subject relates naturally to some a priori unrelated mathematical and physical subjects.

1.1 Introduction

The concept of focal decomposition is a formalization of the rough geometric idea of focalization. Given that we have a number of "trajectories" passing through a point P will they ever meet again, and how at some subsequent point(s)? The trajectories we will consider will be the trajectories of a second order ordinary differential equation in Euclidean spaces, or else the geodesics of a Riemannian manifold M. Focal decomposition was then introduced in [10], in the context of the simplest instance of focalization namely the 2-point boundary value problem for a second order ordinary differential equation. The concept of focal decomposition was then naturally extended to Riemannian manifolds by Kupka and Peixoto in [6]. Originally and for some time later focal decomposition was called "σ-decomposition", a misnomer. We present here a brief survey of focal decomposition stressing how this subject relates naturally to some a priori unrelated mathematical and physical subjects. As shown by the examples below.

1. The semi-classical quantization of the pendulum equation $\ddot{x} + \sin x = 0$ via the Feynman path integral method.
2. The arithmetic of binary quadratic forms.
3. The Brillouin zones of a crystal.
4. The Landau–Ramanujan function associated to binary quadratic forms.

M.M. Peixoto
Instituto de Matemática Pura e Aplicada (IMPA), Estrada Dona Castorina, 110,
22460-320 Rio de Janeiro, RJ, Brazil
e-mail: peixoto@impa.br

M.M. Peixoto et al. (eds.), *Dynamics, Games and Science I*, Springer Proceedings
in Mathematics 1, DOI 10.1007/978-3-642-11456-4_1,
© Springer-Verlag Berlin Heidelberg 2011

2   M.M. Peixoto

5. Inspired by the Mostow Rigidity Theorem, F. Kwakkel in his Groningen thesis [8] introduced, essentially, the concept of Focal Rigidity opening up what is likely to be a fruitful field of research.

1.2 Ordinary Differential Equations

Our starting point is the 2-point boundary value problem for a second order ordinary differential equation

$$\ddot{x} = f(t, x, \dot{x}) \quad x(t_1) = x_1, \quad x(t_2) = x_2, \quad t, x, \dot{x} \in \mathbb{R}. \tag{1.1}$$

This is the simplest and oldest of all boundary value problems introduced by Euler in his work on the foundations of the calculus of variations in the eighteenth century. There is a vast literature on this problem, mostly in the context of applied mathematics and functional analysis. Here we are primarily interested in the number of solutions of the problem (1.1). Let $\mathbb{R}^4(t_1, x_1, t_2, x_2) = \mathbb{R}^2(t_1, x_1) \times \mathbb{R}^2(t_2, x_2)$ be the totality of pairs of points of the (t, x) plane and to each point $(t_1, x_1, t_2, x_2) \in \mathbb{R}^4$ associate its index $i(t_1, x_1, t_2, x_2)$ the number of solutions of the problem (1.1), a non negative number or ∞. So the possible values of i are $0, 1, 2, \ldots, \infty$. When $t_1 = t_2$, the index i is defined to be 0, if $x_1 \neq x_2$, and ∞, if $x_1 = x_2$. Call $\sum_i \subset \mathbb{R}^4$ the set of points to which the index i has been assigned. Then

$$\mathbb{R}^4 = \Sigma_0 \cup \Sigma_1 \cup \ldots \cup \Sigma_\infty \tag{1.2}$$

and (1.2) is called Focal Decomposition associated to the 2-point problem (1.1). Clearly the sets $\sum_i$ are disjoint. Fixing the point (t_1, x_1) and calling

$$\sigma_i = \Sigma_i \cap \{(t_1, x_1)\} \times \mathbb{R}^2(t_2, x_2)$$

we get

$$\mathbb{R}^2 = \sigma_0 \cup \sigma_1 \cup \ldots \cup \sigma_\infty \tag{1.3}$$

with the σ_i disjoint sets. Then (1.3) is called the Focal Decomposition relative to the 2-point boundary value problem (1.1) with base point (t_1, x_1). This is called the restricted problem with base point (t_1, x_1). So far there is no indication whatsoever of what the sets σ_i and $\sum_i$ might be, and how they decompose $\mathbb{R}^2$ or $\mathbb{R}^4$. Consider the case of the restricted problem (1.3). The equation

$$\ddot{x} = f(t, x, \dot{x})$$

can be written

$$\dot{t} = 1, \dot{x} = u, \dot{u} = f(t, x, u) \tag{1.4}$$

the base point being (t_1, x_1). Call $\mathscr{F}$ the foliation of $\mathbb{R}^3(t, x, u)$ defined by the trajectories of (1.4).

1 A Brief Survey of Focal Decomposition

Definition 1.1. The star of the base point (t_1, x_1), $\Omega(t_1, x_1)$ is the union of all the leaves $\mathscr{F}(t_1, x_1, u)$, $-\infty < u < \infty$.

Now consider the projection

$$\prod : \mathbb{R}^3 \to \mathbb{R}^2 \tag{1.5}$$

defined by $\prod(t, x, u) = (t, x)$. It is straightforward that

$$(t_2, x_2) \in \sigma_i \Leftrightarrow card\left\{ \left(\prod |\Omega\right)^{-1} (t_2, x_2) \right\} = i.$$

This shows then that the whole focal decomposition (1.3) can be obtained by applying the projection $\prod$ on the star. Thanks to a theorem by Hironaka [5], pp. 40–43, we have the

Theorem 1.1 (*Existence Theorem (Peixoto - Thom)*). *If f in (1.1) is analytic,*

$$\left(\prod |\Omega\right) (t_1, x_1)$$

is proper on $\mathbb{R}^2 \setminus \delta$, where $\delta = \{(t, x)|t = t_1\}$ then there exists an analytical Whitney stratification of $\mathbb{R}^2 \setminus \delta$ such that each $\sigma_i \setminus \delta$ is the locally finite union of strata of this stratification.

For a proof of the above see [13]. For a more general situation see [12].

1.2.1 *Focal Decomposition of the Pendulum Equation*

The pendulum equation $\ddot{x} + \sin x = 0$ can be formally integrated by Jacobi elliptic functions so that we have the equation of the star at the origin, $\Omega(0, 0)$. To begin with we determine the focal decomposition restricted to the t-axis, which follows from the very definitions of the Jacobian elliptic functions involved. Local sections of $\Omega(0, 0)$ by the planes $t = \pm\pi, \pm 2\pi, \pm 3\pi, \dots$ give local sections which are then glued together and projected on the plane $u = 0$ producing the focal decomposition below. See [13] for the justification of Fig. 1.1.

1.3 Semiclassical Quantization and the Pendulum Equation

Consider a second order equation (1.6) which is the Euler equation of a certain action integral

$$S = \int_{P_1}^{P_2} L(t, x, \dot{x}) dt, \quad \frac{d}{dt}\left(\frac{\partial L}{\partial \dot{x}}\right) - \frac{\partial L}{\partial x} = 0. \tag{1.6}$$

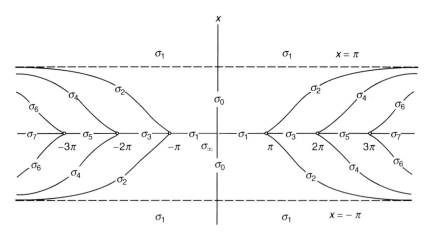

Fig. 1.1 Focal decomposition of the pendulum

Then there is a well defined mathematical operation called the quantization of (1.6) which corresponds to find a certain number of eigenfunctions and the corresponding eigenvalues. One of the more accepted method of quantization is Feynman path integral method. In this method a crucial step is the calculation of a certain amplitude (angle, real number) called the propagator $K(P_1, P_2)$ relative to a pair of points $P_1(t_1, x_1)$ and $P_2(t_2, x_2)$. By Feynman method the calculation of $K(P_1, P_2)$ is a complicated thing: you consider all continuous curves joining P_1 to P_2. On each such curve one does some operation the result which is some kind of "infinitesimal" dK. Integrating over all continuous curves we get the propagator K. Of course this may be in some cases a good numerical procedure, without a mathematical justification, far from it. Suppose now that we are dealing with the semiclassical quantization. This is the case where the data of the problem such as length, mass, time, etc are such that the action S in (1.6) makes $S \setminus \bar{h}$ very big. Most important, in the semiclassical case Feynman–Hibbs [4], p. 29, show that in the calculation of the propagator $K(P_1, P_2)$ we need to consider only the curves which are solutions of the Euler equation. Now given a differential equation of the second order such as (1.6) and two points P_1 and P_2 the problem of knowing the number of solutions of this equation passing through P_1 to P_2 is exactly what the focal decomposition is about. True the focal decomposition gives just the number of solutions. But this is just the first step which helps getting the actual solutions through P_1 to P_2. All this points out to the interest of doing the semiclassical quantization of the pendulum equation. With this and some related problems in mind I joined the physicist C.A.A. de Carvalho and the mathematicians D. Pinheiro and A. A. Pinto. The first paper of a projected series has been accepted for publication [3], another was submitted to a journal.

1 A Brief Survey of Focal Decomposition

1.4 Focal Decomposition on Riemannian Manifolds

Let (M, g) be a complete smooth Riemannian manifold with metric g, $p \in M$ and $T_p M$ the tangent plane at p. Since M is complete the exponential map $\exp_p : T_p M \to M$ is defined everywhere on $T_p M$. Recall that for a vector $v \in T_p M$, $\exp_p v$ is obtained by considering the geodesic passing through p, tangent to v, and mark on it the length $||v||$ obtaining the point $\exp_p v$ on that geodesic. Now define the index $I(v)$ as the cardinality of all vector $w \in T_p M$ with the same length and same exponential as v

$$I(v) = \mathrm{card}\{w \in T_p M \mid ||w|| = ||v|| \text{ and } \exp_p w = \exp_p v\}.$$

To every point $p \in M$ we call $\sigma_i(p)$ the totality of the vectors $v \in T_p M$ with index $i = I(v)$. Since every vector of $T_p M$ belongs to one and only one of the $\sigma_i(p)$, we have

$$T_p(M) = \bigcup_i \sigma_i(p) \tag{1.7}$$

and (1.7) is called the focal decomposition of $T_p M$. The tangent bundle has a corresponding focal decomposition

$$TM = \bigcup_i \Sigma_i \tag{1.8}$$

where

$$\Sigma_i = \bigcup_{p \in M} \sigma_i(p). \tag{1.9}$$

All this for the given Riemannian metric g on M. The expression in (1.8) is called the focal decomposition of the tangent bundle. It is clear that the focal decomposition of TM depends only on the metric g. It is also a global concept, all geodesics passing through p have a role in the construction of the sets σ_i and all geodesics of M have a role in the construction of the sets Σ_i.

Theorem 1.2 (Existence Theorem). *If M is analytic then there is an analytical Whitney stratification of TM such that each Σ_i is the locally finite union of strata of this stratification.*

Similarly for the restricted problem with base point $p \in M$: there is an analytical Whitney stratification of $T_p M$ such that each $\sigma_i(p)$ is the locally finite union of strata of this stratification. See [6]. In the case of ordinary differential equations the stratification is a consequence of analyticity plus a properness condition. In the case of a complete Riemannian manifold M, analyticity is enough to imply the stratification. As in the case of o.d.e. no σ_i or Σ_i can be too complicated, say a Cantor set. In any case the natural place to study focal decomposition is a complete real analytical Riemannian manifold. Relating to the variation of the metric, we have in [7] the genericity theorems

Theorem 1.3 (Pointwise Index Theorem). *Given a point p on a manifold M with dimension m, the generic metric g has no more than $m+1$ geodesics of equal length joining p to some point $q \in M$.*

Theorem 1.4 (Uniform Index Theorem). *The generic metric of a compact manifold M of dimension m has no more than $2m+2$ geodesics of equal length joining distinct points of M.*

A final comment about focal decomposition on a Riemannian manifold is that as a consequence of the angle lemma in [6] we get that in the focal decomposition of $T_p M$, (1.7), only σ_1 has non empty interior, i.e. contains some open set of $T_p M$. So σ_1 has full measure. This is in complete disagreement with Fig. 1.1 corresponding to the focal decomposition of the pendulum equation. There σ_i has positive measure if and only if i is odd. The theories are different.

1.4.1 Focal Decomposition on the Flat Torus

Consider the flat torus $T^2 = \mathbb{R}^2/\mathbb{Z}^2$ with the usual metric determined by the quadratic form $x^2 + y^2$. The exponential map coincides with the covering map in which the coordinates (x, y) are reduced mod 1. Take $(0,0)$ as the base point. To determine the index $I(x, y)$ of $p = (x, y)$ on the tangent space T_p, draw a circle centred at $(0,0)$ and then $I(x, y)$ equals the cardinality of the pair of integers (m, n) such that

$$x^2 + y^2 = (x + m)^2 + (y + n)^2. \tag{1.10}$$

If $(x, y) \in \mathbb{Z}^2$ then $I(x, y)$ is exactly the number of integer solutions of the equation

$$X^2 + Y^2 = N, \quad N = x^2 + y^2, \tag{1.11}$$

then $I(x, y) = R(N)$ and there is a formula of Gauss expressing $R(N)$ in terms of the factorization of N in prime factors. If $(x, y) \notin \mathbb{Z}^2$ then the determination of the index $I(x, y)$ becomes more interesting. In fact we invert the problem, assuming that we know (m, n) and want (x, y). Then (1.10) is

$$2mx + 2ny + m^2 + n^2 = 0. \tag{1.12}$$

But this is the equation of the perpendicular bisector of the lattice point $(-m, -n)$ call it $L(m, n)$. Now we have proved the

Proposition 1.1. *The index $I(x, y)$ equals 1 plus the number of lines $L(m, n)$ that pass through the point (x, y). The 1 corresponds to the fact that $m = 0$, $n = 0$ defines no line $L(m, n)$.*

The family of lines $L(m, n)$, $\mathscr{L}$, (Fig. 1.2) determines the focal decomposition associated to the quadratic form $x^2 + y^2$. So for N real and positive, consider the circle

1 A Brief Survey of Focal Decomposition

Fig. 1.2 Lines $L(m,n)$

$x^2 + y^2 = N$ and let P be one of its points. Then the number of lines $L(m,n)$ passing through P is exactly the index i for which $P \in \sigma_i$. In this sense the focal decomposition incorporates an extension of all Diophantine equations

$$x^2 + y^2 = N \qquad (1.13)$$

for every N and to the whole plane. And one may well say that the index $I(a,b)$ defined for every N is an extension to the whole $\mathbb{R}^2(x,y)$ of the arithmetic function R defined above only for the natural integers.

1.4.2 The Landau–Ramanujan Function

All that was said above about the positive definite quadratic form $x^2 + y^2$ can be translated in a straightforward and appropriate way to any other positive definite quadratic form

$$ax^2 + 2bxy + cy^2, \quad a,b,c \in \mathbb{Z}. \qquad (1.14)$$

Going back to the original form $x^2 + y^2$, Figure 1.2 is just the totality of all lines $L(m,n)$. Now every line $L(m,n)$, by definition, is tangent to the circle centred at the origin and passing through the point $(-m/2, -n/2)$. So it is natural to "organize" the lines $L(m,n)$ by the circles to which they are tangent. This is motivation to the following

Definition 1.2. Consider (1.13), $x^2 + y^2 = N$, and call $R(N)$ the number of its integer solutions. The Landau–Ramanujan function associated to it is defined by

$$L(N) = card\{0 < v \le N \,|\, R(v) \ne 0\}$$

$R(v) \ne 0$ means that v is the sum of 2 squares $v = x^2 + y^2$. Then

Theorem 1.5. $R(N) < 2L(4N)$

The proof of this is quite simple. The $4N$ comes from the fact that if $L(m,n)$ is tangent to $x^2 + y^2 = N$, we have $m^2 + n^2 = 4N$. But the most important fact is that if instead of $x^2 + y^2 = N$ we had $ax^2 + 2bxy + cy^2 = N$, the above theorem remains true. The above on the Landau–Ramanujan function is contained on a forthcoming joint work with C. Pugh.

1.5 The Brillouin Zones

The Brillouin zones is a way of organizing the complement of the lines $L(m,n)$ on the plane. Each connected component C of this complement receives an integer number as follows. Join the origin to a point of C by a straight line and count the number i of the lines $L(m,n)$ that this segment meets. All the sets C that have received the same number i constitute the ith zone of Brillouin. See Fig. 1.3.

1.6 Other Works on Focal Decomposition

In [1,2] and [15] the authors study focal decomposition of linear second order differential equations in $\mathbb{R}^n$. In [1], to such differential equation one can associate a sequence of eigenvalues of the linear operator called the *resonance sequence*. It is shown that two such differential equations are focally equivalent, in the natural topological sense, if and only if the two equations generate the same resonance sequence. In [9] we consider the pair of Diophantine equations

$$Ax^2 + Cy^2 = N \tag{i}$$
$$ACx^2 + y^2 = N \tag{ii}$$

1 A Brief Survey of Focal Decomposition

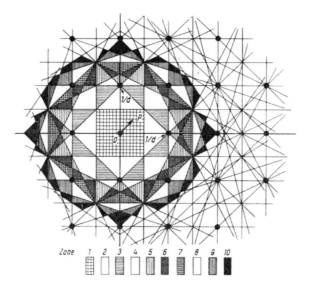

Fig. 1.3 Brillouin zones

with A, C odd, square free, $A \equiv C \pmod{4}$. If $R(N)$ is the number of solutions of (i), $r(N)$ is the number of primitive solutions of (ii) and $\Delta = 4AC$, then we have the convolution formula

$$R(N)^2 = \frac{1}{2} \sum_{d \mid \Delta N} r(d) R\left(\frac{\Delta N}{d}\right). \tag{1.15}$$

In [14] the same formula is obtained with different conditions on A and C.

References

1. Alvarez, S., Berend, D., Birbrair, L., Girão, D.: Resonance sequences and focal decomposition. Isr. J. Math. **170**, 269–284 (2009)
2. Birbrair, L., Sobolevsky, M., Sobolevskii, P.: Focal decompositions for linear differential equations of the second order. Abstr. Appl. Anal., Estados Unidos, **14**, 813–821 (2003)
3. Carvalho, C.A.A., Peixoto, M.M., Pinheiro, D., Pinto, A.A.: Focal decomposition, renormalization and semiclassical physics. J. Differ. Equ. Appl. (2011)
4. Feynman, R., Hibbs, A.: Quantum Mechanics and Path Integrals. McGraw-Hill (1965)
5. Goresky, M., MacPherson, R.: Stratified Morse Theory Ergebnisse der Mathematik und ihrer Grenzgebiete 3 Folge 14. Springer (1989)
6. Kupka, I., Peixoto, M.M.: On the enumerative geometry of geodesics. From Topology to Computation, Proceedings of the Smalefest, pp. 243–253. Springer (1993)
7. Kupka, I., Peixoto, M.M., Pugh, C.C.: Focal stability of Riemannian Metrics. J. für die Reine Angewandte Mathe. (Crelle's Journal) **593**, 31–72 (2006)
8. Kwakkel, F.: Rigidity of Brillouin zones, Master Thesis, University of Groningen, Department of Mathematics (2006)

9. Peixoto, M.M.: Sigma décomposition et arithmétique de quelques formes quadratiques définies positives. In: Porte, M. (ed.) Thom's Festschrift, Passion des Formes, ENS Editions, pp. 455–479. Fontenay (1994)
10. Peixoto, M.M.: On end point boundary values problem. J. Differ. Equ. **44**, 273–280 (1982)
11. Peixoto, M.M.: On a generic theory of end point boundary value problems. An. Acad. Bras. Cienc. **41**, 1–6 (1969)
12. Peixoto, M.M., Da Silva, A.R.: Focal decomposition and some results of S.Bernstein on the 2-point boundary value problem. J. Lond. Math. Soc. **60**(2), 517–547 (1999)
13. Peixoto, M.M., Thom, R.: Le point de vue énumeratif dans les problémes aux limites pour les équations differentielles ordinaires I, II. C.R. Acad. Sci. **303** (1986); Série I, **13**, 629–632; **14**, 693–698; Erratum 307 pp. 197–198
14. Pombo Júnior, D.P.: A convolution formula for certain Diophantine equations. Acta Sci. Math. (Szeged) **62**(3-4), 381–390 (1996)
15. Sobolevsky, M.: Decomposição Focal. Seminário Iberoamericano de Matemáticas. Singularidades en Tordesillas **3**, 21–27 (2003)

Chapter 2
Anosov and Circle Diffeomorphisms

João P. Almeida, Albert M. Fisher, Alberto A. Pinto, and David A. Rand

Abstract We present an infinite dimensional space of C^{1+} smooth conjugacy classes of circle diffeomorphisms that are C^{1+} fixed points of renormalization. We exhibit a one-to-one correspondence between these C^{1+} fixed points of renormalization and C^{1+} conjugacy classes of Anosov diffeomorphisms.

2.1 Introduction

The link between Anosov diffeomorphisms and diffeomorphisms of the circle is due to D. Sullivan and E. Ghys through the observation that the holonomies of Anosov diffeomorphisms give rise to C^{1+} circle diffeomorphisms that are C^{1+}

J.P. Almeida (✉)
LIAAD-INESC Porto LA, Porto, Portugal
and
Centro de Matemática da Universidade do Minho, Campus de Gualtar, 4710-057 Braga, Portugal
and
Departamento de Matemática, Escola Superior de Tecnologia e Gestão, Instituto Politécnico de Bagança, Campus de Santa Apolónia, Ap. 1134, 5301-857 Bragança, Portugal
e-mail: jpa@ipb.pt

A.M. Fisher
Departamento de Matemática, IME-USP, Caixa Postal 66281, CEP 05315-970, São Paulo, Brazil
e-mail: afisher@ime.usp.br

A.A. Pinto
LIAAD-INESC Porto LA e Departamento de Matemática, Faculdade de Ciências, Universidade do Porto, Rua do Campo Alegre, 687, 4169-007, Portugal
and
Centro de Matemática e Departamento de Matemática e Aplicações, Escola de Ciências, Universidade do Minho, Campus de Gualtar, 4710-057 Braga, Portugal
e-mail: aapinto@fc.up.pt

D.A. Rand
Warwick Systems Biology & Mathematics Institute, University of Warwick, Coventry CV4 7AL, UK
e-mail: dar@maths.warwick.ac.uk

M.M. Peixoto et al. (eds.), *Dynamics, Games and Science I*, Springer Proceedings in Mathematics 1, DOI 10.1007/978-3-642-11456-4_2,
© Springer-Verlag Berlin Heidelberg 2011

fixed points of renormalization (see also [2]). A. Pinto and D. Rand [23] proved that this observation gives one-to-one correspondence between the corresponding smooth conjugacy classes. After the works of Thurston [41] and Williams [43], a key object in this link is the smooth horocycle equipped with a hyperbolic Markov map.

2.2 Circle Difeomorphisms

Fix a natural number $a \in \mathbb{N}$ and let $\mathbb{S}$ be a *counterclockwise oriented circle* homeomorphic to the circle $\mathbb{S}^1 = \mathbb{R}/(1 + \gamma)\mathbb{Z}$, where $\gamma = (-a + \sqrt{a^2 + 4})/2 = 1/(a + 1/(a + \cdots))$. We note that if $a = 1$ then γ is the inverse of the golden number $(1 + \sqrt{5})/2$. A key feature of γ is that it satisfies the relation $a\gamma + \gamma^2 = 1$.

An *arc* in $\mathbb{S}$ is the image of a non trivial interval I in $\mathbb{R}$ by a homeomorphism $\alpha : I \to \mathbb{S}$. If I is closed (resp. open) we say that $\alpha(I)$ is a *closed* (resp. *open*) *arc* in $\mathbb{S}$. We denote by (a, b) (resp. $[a, b]$) the positively oriented open (resp. closed) arc in $\mathbb{S}$ starting at the point $a \in \mathbb{S}$ and ending at the point $b \in \mathbb{S}$. A C^{1+} atlas $\mathscr{A}$ of $\mathbb{S}$ is a set of charts such that (i) every small arc of $\mathbb{S}$ is contained in the domain of some chart in $\mathscr{A}$, and (ii) the overlap maps are $C^{1+\alpha}$ compatible, for some $\alpha > 0$.

A C^{1+} *circle diffeomorphism* is a triple $(g, \mathbb{S}, \mathscr{A})$ where $g : \mathbb{S} \to \mathbb{S}$ is a $C^{1+\alpha}$ diffeomorphism, with respect to the $C^{1+\alpha}$ atlas $\mathscr{A}$, for some $\alpha > 0$, and g is quasi-symmetric conjugate to the rigid rotation $r_\gamma : \mathbb{S}^1 \to \mathbb{S}^1$, with rotation number equal to $\gamma/(1 + \gamma)$. We denote by $\mathscr{F}$ the set of all C^{1+} circle diffeomorphisms $(g, \mathbb{S}, \mathscr{A})$, with respect to a C^{1+} atlas $\mathscr{A}$ in $\mathbb{S}$.

In order to simplify the notation, we will denote the C^{1+} circle diffeomorphism $(g, \mathbb{S}, \mathscr{A})$ only by g.

2.2.1 The Horocycle and Renormalization

Let us mark a point in $\mathbb{S}$ that we will denote by $0 \in \mathbb{S}$, from now on. Let $S_0 = [0, g(0)]$ be the oriented closed arc in $\mathbb{S}$, with endpoints 0 and $g(0)$. For $k = 0, \ldots, a$ let $S_k = [g^k(0), g^{k+1}(0)]$ be the oriented closed arc in $\mathbb{S}$, with endpoints $g^k(0)$ and $g^{k+1}(0)$ and such that $S_k \cap S_{k-1} = \{g^k(0)\}$. Let $S_{a+1} = [g^{a+1}(0), 0]$ be the oriented closed arc in $\mathbb{S}$, with endpoints $g^{a+1}(0)$ and 0. We introduce an *equivalence relation* $\sim$ in $\mathbb{S}$ by identifying the $a + 1$ points $g(0), \ldots, g^{a+1}(0)$ and form the topological space $H(\mathbb{S}, g) = \mathbb{S}/\sim$ with the orientation induced by $\mathscr{S}$. We call this oriented topological space the *horocycle* (see Fig. 2.1) and we denote it by $H = H(\mathbb{S}, g)$. We consider the quotient topology in H. Let $\pi_g : \mathbb{S} \to H$ be the natural projection. The point $\xi = \pi_g(g(0)) = \cdots = \pi_g(g^{a+1}(0)) \in H$ is called the *junction* of the horocycle H. For every $k \in \{0, \ldots, a\}$, let $S_k^H = S_k^H(\mathbb{S}, g) \subset H$ be the projection by π_g of the closed arc S_k. Let $R\mathbb{S} = S_0^H \cup S_{a+1}^H$ be the *renormalized circle*. The horocycle H is the union of the renormalized circle $R\mathbb{S}$ with the circles S_k^H for every $k \in \{1, \ldots, a\}$. A *parametrization* in H is the image of a non trivial interval I in $\mathbb{R}$ by a homeomorphism $\alpha : I \to H$. If I is closed (resp.

2 Anosov and Circle Diffeomorphisms

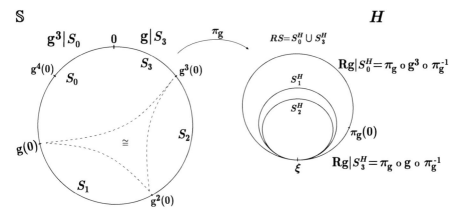

Fig. 2.1 The horocycle H and the renormalization Rg for the case $a = 2$. The junction ξ of the horocycle is equal to $\xi = \pi_g(g(0)) = \pi_g(g^2(0)) = \pi_g(g^3(0))$

open) we say that $\alpha(I)$ is a *closed* (resp. *open*) arc in H. A *chart* in H is the inverse of a parametrization. A *topological atlas* $\mathscr{B}$ on the horocycle H is a set of charts $\{(j, J)\}$, on the horocycle, with the property that every small arc is contained in the domain of a chart in $\mathscr{B}$, i.e. for any open arc K in H and any $x \in K$ there exists a chart $\{(j, J)\} \in \mathscr{B}$ such that $J \cap K$ is a non trivial open arc in H and $x \in J \cap K$. A C^{1+} *atlas* $\mathscr{B}$ in H is a topological atlas $\mathscr{B}$ such that the overlap maps are $C^{1+\alpha}$ and have $C^{1+\alpha}$ uniformly bounded norms, for some $\alpha > 0$.

Let $\mathscr{A}$ be a C^{1+} atlas in $\mathbb{S}$ for which g is C^{1+}. We are going to construct a C^{1+} atlas $\mathscr{A}^H$ in the horocycle that is the *extended pushforward* $\mathscr{A}^H = (\pi_g)_* \mathscr{A}$ of the atlas $\mathscr{A}$ in $\mathbb{S}$.

If $x \in H \setminus \{\xi\}$ then there exists a sufficiently small open arc J in H, containing x, such that $\pi_g^{-1}(J)$ is contained in the domain of some chart (I, i) in $\mathscr{A}$. In this case, we define $(J, i \circ \pi_g^{-1})$ as a chart in $\mathscr{A}^H$. If $x = \xi$ and J is a small arc containing ξ, then either (i) $\pi_g^{-1}(J)$ is an arc in $\mathbb{S}$ or (ii) $\pi_g^{-1}(J)$ is a disconnected set that consists of a union of two connected components. In case (i), $\pi_g^{-1}(J)$ is connected and we define $(J, i \circ \pi_g^{-1})$ as a chart in $\mathscr{A}^H$. In case (ii), $\pi_g^{-1}(J)$ is a disconnected set that is the union of two connected arcs J_k^L and J_l^R of the form $J_l^R = [g^l(0), d)$ and $J_k^L = (c, g^k(0)]$, respectively, for some $k, l \in \{0, \ldots, a+1\}$ with $k \neq l$ (see Fig. 2.2). Let $(I, i) \in \mathscr{A}$ be a chart such that $I \supset (c, d)$. We define $j : J \to \mathbb{R}$ as follows,

$$j(x) = \begin{cases} i \circ \pi_g^{-1}(x), & \text{if } x \in \pi_g([g^l(0), d)) \\ i \circ g^{l-k} \circ \pi_g^{-1}(x), & \text{if } x \in \pi_g((c, g^k(0)]) \end{cases}.$$

We call the atlas determined by these charts, the *extended pushforward atlas of* $\mathscr{A}$ and, by abuse of notation, we will denote it by $\mathscr{A}^H = (\pi_g)_* \mathscr{A}$.

Let $g = (g, \mathbb{S}, \mathscr{A})$ be a C^{1+} circle diffeomorphism with respect to a C^{1+} atlas $\mathscr{A}$ in $\mathbb{S}$. Let $R\mathscr{A}$ be the restriction $\mathscr{A}|R\mathbb{S}$, of the C^{1+} atlas $\mathscr{A}$ to $R\mathbb{S}$. The

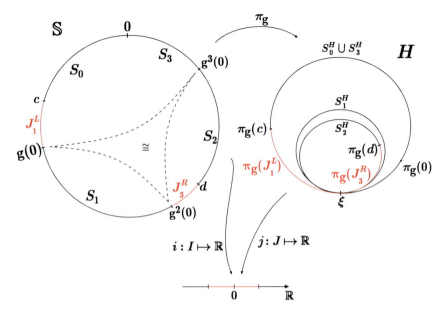

Fig. 2.2 The chart $j : J \to \mathbb{R}$ in case (ii)

renormalization of $g = (g, \mathbb{S}, \mathscr{A})$ is the triple $(Rg, R\mathbb{S}, R\mathscr{A})$, where (i) $R\mathbb{S}$ has the reversed orientation of the horocycle H; and (ii) $Rg : R\mathbb{S} \to R\mathbb{S}$ is the continuous map given by

$$Rg(x) = \begin{cases} \pi_g \circ g^{a+1} \circ \pi_g^{-1}(x) & if \ x \in S_0^H \setminus \{\xi\} \\ \pi_g \circ g \circ \pi_g^{-1}(x) & if \ x \in S_{a+1}^H \setminus \{\xi\} \\ \pi_g \circ g^{a+1}(x) & if \ x \in \{\xi\} \end{cases}.$$

For simplicity of notation, we will denote the renormalization $(Rg, R\mathbb{S}, R\mathscr{A})$ of a C^{1+} circle diffeomorphism only by Rg.

We recall that $\mathscr{F}$ denotes the set of all C^{1+} circle diffeomorphisms $(g, \mathbb{S}, \mathscr{A})$ with respect to a C^{1+} atlas $\mathscr{A}$ in $\mathbb{S}$.

Lemma 2.1. *The renormalization Rg of a C^{1+} circle diffeomorphism $g \in \mathscr{F}$ is a C^{1+} circle diffeomorphism, i.e. the map $R : \mathscr{F} \to \mathscr{F}$ given by $R(g) = Rg$ is well defined. In particular, the renormalization Rr_γ of the rigid rotation is the rigid rotation r_γ.*

The proof of Lemma 2.1 is in [23].

The marked point $0 \in \mathbb{S}$ determines the marked point 0 in the circle $R\mathbb{S}$. Since Rg is homeomorphic to a rigid rotation, there exists $h : \mathbb{S} \to R\mathbb{S}$, with $h(0) = 0$, such that h conjugates g with Rg.

Definition 2.1. We call g a C^{1+} *fixed point of renormalization* if $h : \mathbb{S} \to R\mathbb{S}$ is C^{1+}. We will denote by $\mathscr{R}$ the set of all C^{1+} circle diffeomorphisms $g \in \mathscr{F}$ that are C^{1+} fixed points of renormalization.

Fig. 2.3 The Markov map M_{r_γ} with respect to the atlas $\mathcal{A}_{iso}^H$

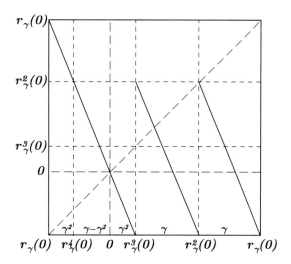

We note that the rigid rotation r_γ, with respect to the atlas $\mathcal{A}_{iso}$, whose charts are isometries with respect to the usual norm in $\mathbb{S}$, is an affine fixed point of renormalization. Hence $r_\gamma \in \mathcal{R}$.

2.2.2 Markov Maps

Let $g = (g, \mathbb{S}, \mathcal{A})$ be a C^{1+} circle diffeomorphism, with respect to a C^{1+} atlas $\mathcal{A}$, and let $H = H(\mathbb{S}, g)$ be the horocycle determined by the C^{1+} circle diffeomorphism. Let $\mathcal{A}^H$ be the atlas in the horocycle H, that is the extended pushforward of the atlas $\mathcal{A}$. Let $\pi_g : \mathbb{S} \to H$ be the natural projection. Let $h : \mathbb{S} \to R\mathbb{S}$ be the homeomorphism that conjugates g and Rg sending the marked point 0 of $\mathbb{S}$ in the marked point 0 of $R\mathbb{S}$.

Definition 2.2. The *Markov map* M_g, associated to the $C^{1+\alpha}$ circle diffeomorphism $g \in \mathcal{F}$, is the map $M_g : H \to H$ defined by

$$M_g(x) = \begin{cases} \pi_g \circ h^{-1}(x) & \text{if } x \in R\mathbb{S} \\ \pi_g \circ h^{-1} \circ \pi_g \circ g^k \circ \pi_g^{-1}(x) & \text{if } x \in S_k^H, \text{ for } k = 1, \ldots, a \end{cases}.$$

We observe that, in particular, the *rigid Markov map* M_{r_γ} is an affine map with respect to the atlas $\mathcal{A}_{iso}^H$. Noting that $M_g(\pi_g \circ g^{k+2}(0)) = \pi_g \circ g^2(0)$, for every $k \in \{0, \ldots, a+1\}$, we represent M_g in Figure 2.3. We observe that the identification in H of $\pi_g \circ g(0)$ with $\pi_g \circ g^2(0)$ makes the Markov map M_g a local homeomorphism.

Lemma 2.2. *Let g be a C^{1+} circle diffeomorphism. The Markov map M_g associated to g is a C^{1+} local diffeomorphism with respect to the atlas $\mathcal{A}^H = (\pi_g)_* \mathcal{A}$ if, and only if, the diffeomorphism g is a C^{1+} fixed point of renormalization.*

The proof of Lemma 2.2 is in [23].

2.3 Anosov Diffeomorphisms

Fix a positive integer $a \in \mathbb{N}$ and consider the *Anosov automorphism* $G_a : \mathbb{T} \to \mathbb{T}$ given by $G_a(x, y) = (ax+y, x)$, where $\mathbb{T}$ is equal to $\mathbb{R}^2/(v\mathbb{Z} \times w\mathbb{Z})$ with $v = (\gamma, 1)$ and $w = (-1, \gamma)$. Let $\pi : \mathbb{R}^2 \to \mathbb{T}$ be the natural projection. Let A_0 and B_0 be the rectangles $[0, 1] \times [0, 1]$ and $[-\gamma, 0] \times [0, \gamma]$ respectively. A Markov partition $\mathscr{M}_{G_a}$ of G_a is given by $\mathbf{A} = \pi(A_0)$ and $\mathbf{B} = \pi(B_0)$ (see Fig. 2.4). The unstable manifolds of G_a are the projection by π of the vertical lines in the plane, and the stable manifolds of G_a are the projection by π of the horizontal lines in the plane.

A C^{1+} *Anosov diffeomorphism* $G : \mathbb{T} \to \mathbb{T}$ is a $C^{1+\alpha}$ diffeomorphism, with $\alpha > 0$, such that (i) G is topologically conjugate to G_a; (ii) the tangent bundle has a $C^{1+\alpha}$ uniformly hyperbolic splitting into a stable direction and an unstable direction (see [40]). We denote by $\mathscr{G}$ the set of all such C^{1+} Anosov diffeomorphisms with an invariant measure absolutely continuous with respect to the Lebesgue measure.

If h is the topological conjugacy between G_a and G, then a Markov partition $\mathscr{M}_G$ of G is given by $h(\mathbf{A})$ and $h(\mathbf{B})$. Let $d = d_\rho$ be the distance on the torus $\mathbb{T}$, determined by a Riemannian metric ρ. We define the map $G_\iota = G$ if $\iota = u$, or $G_\iota = G^{-1}$ if $\iota = s$. For $\iota \in \{s, u\}$ and $x \in \mathbb{T}$, we denote the local ι-manifolds through x by

$$W^\iota(x, \varepsilon) = \{y \in \mathbb{T} : d(G_\iota^{-n}(x), G_\iota^{-n}(y)) \leq \varepsilon, \text{ for all } n \geq 0\}.$$

By the Stable Manifold Theorem (see [40]), these sets are respectively contained in the stable and unstable immersed manifolds

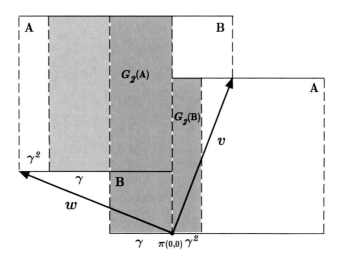

Fig. 2.4 The Anosov automorphism $G_2 : \mathbb{T} \to \mathbb{T}$

$$W^\iota(x) = \bigcup_{n \geq 0} G_\iota^n \left(W^\iota \left(G_\iota^{-n}(x), \varepsilon_0 \right) \right)$$

which are the image of $C^{1+\alpha}$ immersions $\kappa_{\iota,x} : \mathbb{R} \to \mathbb{T}$, for some $0 < \alpha \leq 1$ and some small $\varepsilon_0 > 0$. An *open* (resp. *closed*) *ι-leaf segment I* is defined as a subset of $W^\iota(x)$ of the form $\kappa_{\iota,x}(I_1)$ where I_1 is an open (resp. closed) subinterval (non-empty) in $\mathbb{R}$. An *ι-leaf segment* is either an open or closed ι-leaf segment. The *endpoints* of an ι-leaf segment $I = \kappa_{\iota,x}(I_1)$ are the points $\kappa_{\iota,x}(u)$ and $\kappa_{\iota,x}(v)$ where u and v are the endpoints of I_1. The *interior* of an ι-leaf segment I is the complement of its boundary. A map $c : I \to \mathbb{R}$ is an *ι-leaf chart* of an ι-leaf segment I if c is a homeomorphism onto its image.

2.3.1 Spanning Leaf Segments

One can find a small enough $\varepsilon_0 > 0$, such that for every $0 < \varepsilon < \varepsilon_0$ there is $\delta = \delta(\varepsilon) > 0$ with the property that, for all points $w, z \in \mathbb{T}$ with $d(w, z) < \delta$, $W^u(w, \varepsilon)$ and $W^s(z, \varepsilon)$ intersect in a unique point that we denote by

$$[w, z] = W^u(w, \varepsilon) \cap W^s(z, \epsilon).$$

A *rectangle R* is a subset of $\mathbb{T}$ which is (i) closed under the bracket, i.e. $x, y \in R \Rightarrow [x, y] \in R$, and (ii) proper, i.e. it is the closure of its interior in $\mathbb{T}$. If ℓ^u and ℓ^s are respectively unstable and stable closed leaf segments intersecting in a single point then we denote by $[\ell^u, \ell^s]$ the set consisting of all points of the form $[w, z]$ with $w \in \ell^u$ and $z \in \ell^s$. We note that $[\ell^u, \ell^s]$ is a rectangle. Conversely, given a rectangle R, for each $x \in R$ there are closed unstable and stable leaf segments of $\mathbb{T}$, $\ell^u(x, R) \subset W^u(x)$ and $\ell^s(x, R) \subset W^s(x)$ such that $R = [\ell^u(x, R), \ell^s(x, R)]$. The leaf segments $\ell^u(x, R)$ and $\ell^s(x, R)$ are called, respectively, *unstable* and *stable spanning leaf segments*.

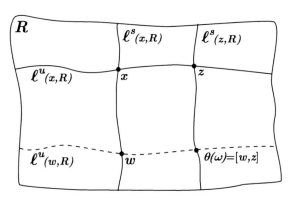

Fig. 2.5 A basic stable holonomy $\theta : \ell^s(x, R) \to \ell^s(z, R)$

2.3.2 Basic Holonomies

Suppose that x and z are two points inside any rectangle R of $\mathbb{T}$. Let $\ell^s(x, R)$ and $\ell^s(z, R)$ be two stable spanning leaf segments of R containing, respectively, x and z. We define the map $\theta : \ell^s(x, R) \to \ell^s(z, R)$ by $\theta(w) = [w, z]$ (see Fig. 2.5). Such maps are called the *basic stable holonomies*. They generate the pseudo-group of all stable holonomies. Similarly, we can define the *basic unstable holonomies*.

2.3.3 Lamination Atlas

The *stable lamination atlas* $\mathscr{L}^s = \mathscr{L}^s(G, \rho)$, *determined by a Riemannian metric* ρ, is the set of all maps $e : I \to \mathbb{R}$, where e is an isometry between the induced Riemannian metric on the stable leaf segment I and the Euclidean metric on the reals. We call the maps $e \in \mathscr{L}^s$ the *stable lamination charts*. Similarly, we can define the *unstable lamination atlas* $\mathscr{L}^u = \mathscr{L}^u(G, \rho)$. By Theorem 2.1 in [28], the basic unstable and stable holonomies are C^{1+} with respect to the lamination atlas $\mathscr{L}^s$.

2.3.4 Circle Diffeomorphisms

Let G be a C^{1+} Anosov diffeomorphism topologically conjugate to the Anosov automorphism G_a by the homeomorphism h. For each Markov rectangle R, let t_R^s be the set of all unstable spanning leaf segments of R. Thus, by the local product structure, one can identify t_R^s with any stable spanning leaf segment $\ell^s(x, R)$ of R. We form the space $\mathbb{S}_G$ by taking the disjoint union $t_{h(\mathbf{A})}^s \bigsqcup t_{h(\mathbf{B})}^s$, where $h(\mathbf{A})$ and $h(\mathbf{B})$ are the Markov rectangles of the Markov partition $\mathscr{M}_G$, and identifying two points $I \in t_R^s$ and $J \in t_{R'}^s$ if (i) $R \neq R'$, (ii) the unstable leaf segments I and J are unstable boundaries of Markov rectangles, and (iii) $int(I \cap J) \neq \emptyset$. Topologically, the space $\mathbb{S}_G$ is a *counterclockwise oriented circle*. Let $\pi_{\mathbb{S}_G} : \bigsqcup_{R \in \mathscr{M}_G} R \to \mathbb{S}_G$ be the natural projection sending $x \in R$ to the point $\ell^u(x, R)$ in $\mathbb{S}_G$.

Let $I_{\mathbb{S}}$ be an arc of $\mathbb{S}_G$ and I a leaf segment such that $\pi_{\mathbb{S}_G}(I) = I_{\mathbb{S}}$. The chart $i : I \to \mathbb{R}$ in $\mathscr{L} = \mathscr{L}^s(G, \rho)$ determines a *circle chart* $i_{\mathbb{S}} : I_{\mathbb{S}} \to \mathbb{R}$ for $I_{\mathbb{S}}$ given by $i_{\mathbb{S}} \circ \pi_{\mathbb{S}_G} = i$. We denote by $\mathscr{A}_G = \mathscr{A}(G, \rho)$ the set of all circle charts $i_{\mathbb{S}}$ determined by charts i in $\mathscr{L} = \mathscr{L}^s(G, \rho)$. Given any circle charts $i_{\mathbb{S}} : I_{\mathbb{S}} \to \mathbb{R}$ and $j_{\mathbb{S}} : J_{\mathbb{S}} \to \mathbb{R}$, the overlap map $j_{\mathbb{S}} \circ i_{\mathbb{S}}^{-1} : i_{\mathbb{S}}(I_{\mathbb{S}} \cap J_{\mathbb{S}}) \to j_{\mathbb{S}}(I_{\mathbb{S}} \cap J_{\mathbb{S}})$ is equal to $j_{\mathbb{S}} \circ i_{\mathbb{S}}^{-1} = j \circ \theta \circ i^{-1}$, where $i = i_{\mathbb{S}} \circ \pi_{\mathbb{S}_G} : I \to \mathbb{R}$ and $j = j_{\mathbb{S}} \circ \pi_{\mathbb{S}_G} : J \to \mathbb{R}$ are charts in $\mathscr{L}$, and

$$\theta : i^{-1}(i_{\mathbb{S}}(I_{\mathbb{S}} \cap J_{\mathbb{S}})) \to j^{-1}(j_{\mathbb{S}}(I_{\mathbb{S}} \cap J_{\mathbb{S}}))$$

is a basic stable holonomy. By Theorem 2.1 in [28], there exists $\alpha > 0$ such that, for all circle charts $i_{\mathbb{S}}$ and $j_{\mathbb{S}}$ in $\mathscr{A}_G$, the overlap maps $j_{\mathbb{S}} \circ i_{\mathbb{S}}^{-1} = j \circ \theta \circ i^{-1}$ are $C^{1+\alpha}$

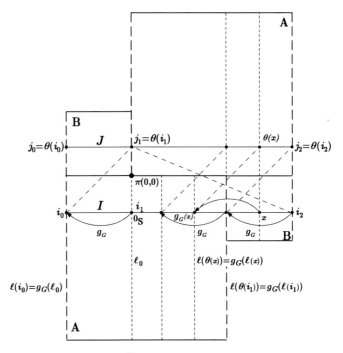

Fig. 2.6 The arc rotation map $g_G = \tilde{\theta}_G : \pi_{\mathbb{S}_G}(I) \to \pi_{\mathbb{S}_G}(J)$. We note that $\mathbb{S} = \pi_{\mathbb{S}_G}(I) = \pi_{\mathbb{S}_G}(J)$ and $\ell(x) = \pi_{\mathbb{S}_G}(x)$ is the unstable spanning leaf segment containing x

diffeomorphisms with a uniform bound in the $C^{1+\alpha}$ norm. Hence, $\mathscr{A}_G = \mathscr{A}(G, \rho)$ is a C^{1+} atlas.

Suppose that I and J are stable leaf segments and $\theta : I \to J$ is a holonomy map such that, for every $x \in I$, the unstable leaf segments with endpoints x and $\theta(x)$ cross once, and only once, a stable boundary of a Markov rectangle. We define the *arc rotation map* $\tilde{\theta}_G : \pi_{\mathbb{S}_G}(I) \to \pi_{\mathbb{S}_G}(J)$, associated to θ, by $\tilde{\theta}_G(\pi_{\mathbb{S}_G}(x)) = \pi_{\mathbb{S}_G}(\theta(x))$ (see Fig. 2.6). By Theorem 2.1 in [28] there exists $\alpha > 0$ such that the holonomy $\theta : I \to J$ is a $C^{1+\alpha}$ diffeomorphism, with respect to the C^{1+} lamination atlas $\mathscr{L}^s(G, \rho)$. Hence, the arc rotation maps $\tilde{\theta}_G$ are C^{1+} diffeomorphisms, with respect to the C^{1+} atlas $\mathscr{A}(G, \rho)$.

Lemma 2.3. *There is a well-defined C^{1+} circle diffeomorphism g_G, with respect to the C^{1+} atlas $\mathscr{A}_G = \mathscr{A}(G, \rho)$, such that $g_G | \pi_{\mathbb{S}_G}(I) = \tilde{\theta}_G$, for every arc rotation map $\tilde{\theta}_G$. In particular, if G_a is the Anosov automorphism, then g_{G_a} is the rigid rotation r_γ, with respect to the isometric atlas $\mathscr{A}_{iso} = \mathscr{A}(G_a, E)$, where E corresponds to the Euclidean metric in the plane.*

The proof of Lemma 2.3 is in [23].

2.3.5 Train-Tracks and Markov Maps

Roughly speaking, train-tracks are the optimal leaf-quotient spaces on which the stable and unstable Markov maps induced by the action of G on leaf segments are local homeomorphisms.

Let G be a C^{1+} Anosov diffeomorphism topologically conjugate to G_a by a homeomorpishm h. We recall that, for each Markov rectangle R, t_R^s denotes the set of all unstable spanning leaf segments of R and, by the local product structure, one can identify t_R^s with any stable spanning leaf segment $\ell^s(x, R)$ of R. We form the space $\mathbf{T}_G$ by taking the disjoint union $t_{h(\mathbf{A})}^s \bigsqcup t_{h(\mathbf{B})}^s$, where $h(\mathbf{A})$ and $h(\mathbf{B})$ are the Markov rectangles of the Markov partition $\mathcal{M}_G$ and identifying two points $I \in t_R^s$ and $J \in t_{R'}^s$ if (i) the unstable leaf segments I and J are unstable boundaries of Markov rectangles and (ii) $int(I \cap J) = \emptyset$. This space is called the *stable train-track* and it is denoted by $\mathbf{T}_G$.

Let $\pi_{\mathbf{T}_G} : \bigsqcup_{R \in \mathcal{M}_G} R \to \mathbf{T}_G$ be the natural projection sending the point $x \in R$ to the point $\ell^u(x, R)$ in $\mathbf{T}_G$. A *topologically regular point* I in $\mathbf{T}_G$ is a point with a unique preimage under $\pi_{\mathbf{T}_G}$ (i.e. the preimage of I is not a union of distinct unstable boundaries of Markov rectangles). If a point has more than one preimage by $\pi_{\mathbf{T}_G}$, then we call it a *junction*. Hence, there is only one junction.

A chart $i : I \to \mathbb{R}$ in $\mathcal{L} = \mathcal{L}^s(G, \rho)$ determines a *train-track chart* $i_T : I_T \to \mathbb{R}$ for I_T given by $i_T \circ \pi_{\mathbf{T}_G} = i$. We denote by $\mathcal{A}^{\mathbf{T}_G} = \mathcal{A}^{\mathbf{T}_G}(G, \rho)$ the set of all train-track charts i_T determined by charts i in $\mathcal{L} = \mathcal{L}^s(G, \rho)$. Given any train-track charts $i_T : I_T \to \mathbb{R}$ and $j_T : J_T \to \mathbb{R}$ in $\mathcal{A}^{\mathbf{T}_G}$, the overlap map $j_T \circ i_T^{-1} : i_T(I_T \cap J_T) \to j_T(I_T \cap J_T)$ is equal to $j_T \circ i_T^{-1} = j \circ \theta \circ i^{-1}$, where $i = i_T \circ \pi_{\mathbf{T}_G} : I \to \mathbb{R}$ and $j = j_T \circ \pi_{\mathbf{T}_G} : J \to \mathbb{R}$ are charts in $\mathcal{L}$, and

$$\theta : i^{-1}(i_T(I_T \cap J_T)) \to j^{-1}(j_T(I_T \cap J_T))$$

is a basic stable holonomy. By Theorem 2.1 in [28] there exists $\alpha > 0$ such that, for all train-track charts i_T and j_T in $\mathcal{A}^{\mathbf{T}_G}(G, \rho)$, the overlap maps $j_T \circ i_T^{-1} = j \circ \theta \circ i^{-1}$ have $C^{1+\alpha}$ diffeomorphic extensions with a uniform bound in the $C^{1+\alpha}$ norm. Hence, $\mathcal{A}^{\mathbf{T}_G}(G, \rho)$ is a $C^{1+\alpha}$ atlas in $\mathbf{T}_G$.

The *(stable) Markov map* $M_G : \mathbf{T}_G \to \mathbf{T}_G$ is the mapping induced by the action of G on unstable spanning leaf segments, that it is defined as follows: if $I \in \mathbf{T}_G$, $M_G(I) = \pi_{\mathbf{T}_G}(G(I))$ is the unstable spanning leaf segment containing $G(I)$. This map M_G is a local homeomorphism because G sends short stable leaf segments homeomorphically onto short stable leaf segments.

A *stable leaf primary cylinder of a Markov rectangle* R is a stable spanning leaf segment of R. For $n \geq 1$, a *stable leaf n-cylinder of* R is a stable leaf segment I such that (i) $G^n I$ is a stable leaf primary cylinder of a Markov rectangle $R'(I) \in \mathcal{M}_G$; (ii) $G^n(\ell^u(x, R)) \subset R'(I)$ for every $x \in I$, where $\ell^u(x, R)$ is an unstable spanning leaf segment of R. For $n \geq 1$, an *n-cylinder* is the projection into $\mathbf{T}_G$ of a stable leaf n-cylinder segment. Thus, each Markov rectangle in $\mathbb{T}$ projects in a unique primary stable leaf segment in $\mathbf{T}_G$.

2 Anosov and Circle Diffeomorphisms

Given a topological chart (e, U) on the train-track $\mathbf{T}_G$ and a train-track segment $C \subset U$, we denote by $|C|_e$ the length of $e(C)$. We say that M_G has *bounded geometry* in a C^{1+} atlas $\mathscr{A}$, if there is $\kappa_1 > 0$ such that, for every n-cylinder C_1 and n-cylinder C_2 with a common endpoint with C_1, we have $\kappa_1^{-1} < |C_1|_e/|C_2|_e < \kappa_1$, where the lengths are measured in any chart (e, U) of the atlas such that $C_1 \cup C_2 \subset U$. We note that M_G has bounded geometry, with respect to a C^{1+} atlas $\mathscr{A}$, if, and only if, there are $\kappa_2 > 0$ and $0 < \nu < 1$ such that $|C|_e \le \kappa_2 \nu^n$, for every n-cylinder and every $e \in \mathscr{B}$.

By Sect. 4.3 in Pinto-Rand [26], we obtain that M_G is C^{1+} and has *bounded geometry* in $\mathscr{A}^{\mathbf{T}_G}(G, \rho)$. In [23] it is proved that M_G corresponds to the Markov map M_{g_G} (see Definition 2.2). Hence, g_G is a C^{1+} fixed point of renormalization.

Theorem 2.1. *The map $G \to g_G$ induces a one-to-one correspondence between C^{1+} conjugacy classes of Anosov diffeomorphisms, with an invariant measure absolutely continuous with respect to the Lebesgue measure, and C^{1+} conjugacy classes of circle diffeomorphisms that are fixed points of renormalization.*

The proof of Theorem 2.1 is in [23].

Since the eigenvalues of G are invariants of its smooth conjugacy class, there is an infinite dimensional space of C^{1+} conjugacy classes for circle diffeomorphisms g_G that are invariant under renormalization. Our result contrasts with the theory of Arnol'd, Herman and Yoccoz [1, 11, 44] that proves the uniqueness of the smooth conjugacy class of circle diffeomorphisms with bounded rotation number and with a degree of smoothness higher than 2.

Acknowledgements We are grateful to Dennis Sullivan by sharing with us his ideas leading to the construction of the smooth horocycle with a hyperbolic Markov map. Previous versions of this work were presented in the International Congresses of Mathematicians ICM 2006 and 2010, EURO 2010, ICDEA 2009 and in the celebration of David Rand's 60th birthday, achievements and influence, University of Warwick. We thank LIAAD-INESC Porto LA, Calouste Gulbenkian Foundation, PRODYN-ESF, FEDER, POFC, POCTI and POSI by FCT and Ministério da Ciência e da Tecnologia, and the FCT Pluriannual Funding Program of LIAAD-INESC Porto LA and of Research Centre of Mathematics of the University of Minho for their financial support. A. Fisher would like to thank FAPESP, CNPq and CNRS for their financial support.

References

1. Arnol'd, V.I.: Small denominators I: on the mapping of a circle into itself. *Investijia Akad. Nauk. Math.*, **25**(1), 21–96 (1961). Transl. A.M.S., 2nd series, 46, 213–284
2. Cawley, E.: The Teichmüller space of an Anosov diffeomorphism of T^2. Inventiones Math. **112**, 351–376 (1993)
3. Coullet, P., Tresser, C.: Itération d'endomorphismes et groupe de renormalisation. J. de Physique **C5**, 25 (1978)
4. de Faria, E., de Melo, W., Pinto, A.A.: Global hyperbolicity of renormalization for C^r unimodal mappings. Ann. Math. **164**, 731–824 (2006)
5. Feigenbaum, M.: Quantitative universality for a class of nonlinear transformations. J. Stat. Phys. **19**, 25–52 (1978)

6. Feigenbaum, M.: The universal metric properties of nonlinear transformations. J. Stat. Phys. **21**, 669–706 (1979)
7. Ferreira, F., Pinto, A.A.: Explosion of smoothness from a point to everywhere for conjugacies between diffeomorphisms on surfaces. Ergod. Theory Dyn. Syst. **23**, 509–517 (2003)
8. Ferreira, F., Pinto, A.A., Rand, D.A.: Hausdorff dimension versus smoothness. In: Vasile Staicu (ed.) Differential Equations Chaos and Variational Problems. Nonlinear Differ. Equ. Appl. **75**, 195–209 (2007)
9. Franks, J.: Anosov diffeomorphisms. In: Smale, S. (ed) Global Analysis, vol. 14, pp. 61–93. AMS Providence (1970)
10. Ghys, E.: Rigidité différentiable des groupes Fuchsiens. Publ. IHES **78**, 163–185 (1993)
11. Herman, M.R.: Sur la conjugaison différentiable des difféomorphismes du cercle à des rotations. Publ. IHES **49**, 5–233 (1979)
12. Jiang, Y.: Metric invariants in dynamical systems. J. Dyn. Differ. Equ. **17**(1), 51–71 (2005)
13. Lanford, O.: Renormalization group methods for critical circle mappings with general rotation number. VIIIth International Congress on Mathematical Physics, pp. 532–536. World Scientific Publishing, Singapore (1987)
14. de la Llave, R.: Invariants for Smooth conjugacy of hyperbolic dynamical systems II. Commun. Math. Phys. **109**(3), 369–378 (1987)
15. Manning, A.: There are no new Anosov diffeomorphisms on tori. Am. J. Math. **96**, 422 (1974)
16. Marco, J.M., Moriyon, R.: Invariants for Smooth conjugacy of hyperbolic dynamical systems I. Commun. Math. Phys. **109**, 681–689 (1987)
17. Marco, J.M., Moriyon, R.: Invariants for Smooth conjugacy of hyperbolic dynamical systems III. Commun. Math. Phys. **112**, 317–333 (1987)
18. de Melo, W.: Review of the book *Fine structure of hyperbolic diffeomorphisms*. In: Pinto, A.A., Rand, D., Ferreira, F. (eds.) Monographs in Mathematics. Bulletin of the AMS, S 0273-0979 01284-2. Springer (2010)
19. de Melo, W., Pinto, A.: Rigidity of C^2 infinitely renormalization for unimodal maps. Commun. Math. Phys. **208**, 91–105 (1999)
20. Penner, R.C., Harer, J.L.: Combinatorics of Train-Tracks. Princeton University Press, Princeton, New Jersey (1992)
21. Pinto, A.A.: Hyperbolic and minimal sets, Proceedings of the 12th International Conference on Difference Equations and Applications, Lisbon, World Scientific, 1–29 (2007)
22. Pinto, A.A., Almeida, J.P., Portela, A.: Golden tilings, Transactions of the American Mathematical Society, 1–21 (accepted)
23. Pinto, A.A., Rand, D.A.: Renormalisation gives all surface Anosov diffeomorphisms with a smooth invariant measure (submitted)
24. Pinto, A.A., Rand, D.A.: Train-tracks with C^{1+} self-renormalisable structures. Journal of Difference Equations and Applications, **16**(8), 945–962 (2010)
25. Pinto, A.A., Rand, D.A.: Solenoid functions for hyperbolic sets on surfaces. In: Boris Hasselblat (ed.) Dynamics, Ergodic Theory and Geometry, vol. 54, pp. 145–178. MSRI Publications (2007)
26. Pinto, A.A., Rand, D.A.: Geometric measures for hyperbolic sets on surfaces. Stony Brook, pp. 1–59. preprint (2006)
27. Pinto, A.A., Rand, D.A.: Rigidity of hyperbolic sets on surfaces. J. Lond. Math. Soc. **71**(2), 481–502 (2004)
28. Pinto, A.A., Rand, D.A.: Smoothness of holonomies for codimension 1 hyperbolic dynamics. Bull. Lond. Math. Soc. **34**, 341–352 (2002)
29. Pinto, A.A., Rand, D.A.: Teichmüller spaces and HR structures for hyperbolic surface dynamics. Ergod. Theory Dyn. Syst. **22**, 1905–1931 (2002)
30. Pinto, A.A., Rand, D.A.: Existence, uniqueness and ratio decomposition for Gibbs states via duality. Ergod. Theory Dyn. Syst. **21**, 533–543 (2001)
31. Pinto, A.A., Rand, D.A.: Classifying C^{1+} structures on dynamical fractals, 1. The moduli space of solenoid functions for Markov maps on train tracks. Ergod. Theory Dyn. Syst. **15**, 697–734 (1995)

2 Anosov and Circle Diffeomorphisms

32. Pinto, A.A., Rand, D.A.: Classifying C^{1+} structures on dynamical fractals, 2. Embedded trees. Ergod. Theory Dyn. Syst. **15**, 969–992 (1995)
33. Pinto, A.A., Rand, D.A., Ferreira, F.: C^{1+} self-renormalizable structures. In: Ruffing, A., Suhrer, A., Suhrer, J. (eds.) Communications of the Laufen Colloquium on Science 2007, pp. 1–17 (2007)
34. Pinto, A.A., Rand, D.A., Ferreira, F.: Arc exchange systems and renormalization. J. Differ. Equ. Appl. **16**(4), 347–371 (2010)
35. Pinto, A.A., Rand, D.A., Ferreira, F.: Fine structures of hyperbolic diffeomorphisms. Springer Monograph in Mathematics, Springer (2009)
36. Pinto, A.A., Rand, D.A., Ferreira, F.: Hausdorff dimension bounds for smoothness of holonomies for codimension 1 hyperbolic dynamics. J. Differe. Equ. **243**, 168–178 (2007)
37. Pinto, A.A., Rand, D.A., Ferreira, F.: Cantor exchange systems and renormalization. J. Differ. Equ. **243**, 593–616 (2007)
38. Pinto, A.A., Sullivan, D.: The circle and the solenoid. Dedicated to Anatole Katok On the Occasion of his 60th Birthday. DCDS A, **16**(2), 463–504 (2006)
39. Rand, D.A.: Global phase space universality, smooth conjugacies and renormalisation, 1. The $C^{1+\alpha}$ case. Nonlinearity **1**, 181–202 (1988)
40. Schub, M.: Global Stability of Dynamical Systems. Springer (1987)
41. Thurston, W.: On the geometry and dynamics of diffeomorphisms of surfaces. Bull. Am. Math. Soc. **19**, 417–431 (1988)
42. Veech, W.: Gauss measures for transformations on the space of interval exchange maps. Ann. Math., 2nd Ser. **115**(2), 201–242 (1982)
43. Williams, R.F.: Expanding attractors. Publ. I.H.E.S. **43**, 169–203 (1974)
44. Yoccoz J. C.: Conjugaison différentiable des difféomorphismes du cercle dont le nombre de rotation vérifie une condition diophantienne. Ann. Scient. Éc. Norm. Sup., 4 série, t, **17**, 333–359 (1984)

Chapter 3
Evolutionarily Stable Strategies and Replicator Dynamics in Asymmetric Two-Population Games

Elvio Accinelli and Edgar J. Sánchez Carrera

Abstract We analyze the main dynamical properties of the evolutionarily stable strategy ($\mathcal{ESS}$) for asymmetric two-population games of finite size and its corresponding replicator dynamics. We introduce a definition of $\mathcal{ESS}$ for two-population asymmetric games and a method of symmetrizing such an asymmetric game. We show that every strategy profile of the asymmetric game corresponds to a strategy in the symmetric game, and that every Nash equilibrium ($\mathcal{NE}$) of the asymmetric game corresponds to a (symmetric) $\mathcal{NE}$ of the symmetric version game. We study the (standard) replicator dynamics for the asymmetric game and we define the corresponding (non-standard) dynamics of the symmetric game. We claim that the relationship between $\mathcal{NE}$, $\mathcal{ESS}$ and the stationary states ($\mathcal{SS}$) of the dynamical system for the asymmetric game can be studied by analyzing the dynamics of the symmetric game.

3.1 Introduction

Evolutionary dynamics originally appeared in biology and then started to be used in economics. Evolutionary stability, introduced by Maynard Smith and Price [10], is a criterion for the robustness of an incumbent strategy against the entry of individuals or mutants using a different strategy. The framework considered is a conflict within a homogenous population. This game is symmetric since all players have the same strategy set and the payoff for a given strategy depends only on the strategies being played and not on who is playing them.

E. Accinelli (✉)
Facultad de Economía, UASLP, San Luis Potosí, México
e-mail: elvio.accinell@eco.uaslp.mx

E.J.S. Carrera
Facultad de Economía, UASLP, México and Department of Economics, University of Siena, Italy
e-mail: sanchezcarre@unisi.it

M.M. Peixoto et al. (eds.), *Dynamics, Games and Science I*, Springer Proceedings in Mathematics 1, DOI 10.1007/978-3-642-11456-4_3,
© Springer-Verlag Berlin Heidelberg 2011

Nevertheless, many economic applications come from multi-population rather than single-population dynamics on asymmetric environments. So, in most applications, the game is not symmetric and involves at least two players with different strategies and each player's role is represented by a different population. In the spirit of Nash's [12] "mass action interpretation", each type of player is drawn from his or her "player-role population". For instance, the players may play the role of buyers or sellers, incumbents or entrants in oligopolistic markets, workers or firms, or the social relationships between migrants and residents; all of them with non-homogeneous behaviors on the state of the economy and different attitudes towards – and perceptions about – development efforts or environmental quality of the state of the economy and so forth.

Recall that, from the framework of symmetric games, there is a seminal refinement of the Nash equilibrium ($\mathcal{NE}$) concept that is the notion of Evolutionarily Stable Strategy ($\mathcal{ESS}$) (see Maynard Smith and Price [10] and Maynard Smith [11]). We known that every ESS is robust against mutant strategies from the postentry population (so-called equilibrium entrants), and asymptotically stable steady state of the associated replicator dynamics. The relationship between $\mathcal{NE}$, $\mathcal{ESS}$ and the steady states ($\mathcal{SS}$) of the replicator dynamics are well known (see Weibull [18]).

In this paper, we consider the evolution of two populations facing a conflictive situation modeled by an asymmetric normal form game. The main purpose of this work is to analyze the evolution and stability of the behaviors of the populations, involved in asymmetric games. Our approach is to symmetrize the asymmetric game because it give us the possibility to characterize the $\mathcal{ESS}$, using the well known properties of these strategies for the case of symmetric games. We introduce an approach to symmetrize a game that differs from the usual ones of symmetrizing a bimatrix game (see Hofbauer and Sigmund [9] and Cressman [6]).

We extend the concept of $\mathcal{ESS}$ for asymmetric two-population games, following the definition of Selten [16] and Samuelson [14], but in those papers it was not analyzed the evolutionary dynamics of such a population. We exhibit connections between $\mathcal{ESS}$, $\mathcal{NE}$ and $\mathcal{SS}$ for these two dynamics. Close to our argument is the one by Fishman [8], nevertheless our approach is quite different while we do not allow for invader's frequencies – the fundamental approach (invasion dynamics analysis, IDA) used in that paper is due to Cressman [4, 5] – we benefit by gaining a through familiarity with IDA. In particular, we consider the necessity of requiring independence in the invader's frequencies that precludes "symmetrization". By symmetrizing the game, we get the advantage of generalizing the standard definition of $\mathcal{ESS}$ and its relationship with the stability of the dynamical equilibria of the replicator dynamics and with the strategic stability for asymmetric games. We note that, much of the topic of this paper can be generalized for cases of finite ($n > 2$) asymmetric populations. However, to simplify the notation, we shall consider the case of two asymmetric populations.

Following this approach it is straightforward to see that a strategic profile is an $\mathcal{ESS}$ if and only if it is a strict Nash equilibrium (see Balkenborg and Schlag [1,2]; Cressman [4–6]; Samuelson [14]; Selten [16]; Weibull [18]) and that every $\mathcal{ESS}$ is an asymptotically stable steady state of the replicator dynamics (see Retchkiman (2007); Samuelson and Zhang [15]).

3 Evolutionarily Stable Strategies and Replicator Dynamics

The paper is organized as follows. Section 3.2 draws the notation and basic definitions to set up the baseline model, namely a two-player asymmetric normal-form game. Section 3.3 defines the $\mathscr{ESS}$ for our model. In Sect. 3.4, we introduce the symmetric version of an asymmetric two population game. Section 3.5 studies the dynamics of our model. Section 3.6 states the relationships between $\mathscr{ESS}$, $\mathscr{NE}$ and $\mathscr{SS}$. Section 3.7 draws some concluding remarks.

3.2 The Model

Consider a normal-form (strategic) game with a player set composed by individuals that comprise τ populations, namely residents R and migrants M i.e. $\tau = \{R, M\}$. Each population splits in different clubs denoted by n_i^τ with $i \in \{1, \ldots, k_\tau\}$, i.e. $(n_1^R, \ldots, n_{k_R}^R)$ and $(n_1^M, \ldots, n_{k_M}^M)$. The split depends on the strategy agents play or the behavior that agents follow. Strategies are in correspondence with the clubs. Individuals belonging to the n_i^τ club are called i-strategists. Thus, the set S^τ of pure strategies are $S^R = \{n_1^R, \ldots, n_{k_R}^R\}$ and $S^M = \{n_1^M, \ldots, n_{k_M}^M\}$. For each population $\tau \in \{M, R\}$ we represent the set of mixed strategies by

$$\Delta^\tau = \left\{ x \in R^{k_\tau} : \sum_{j=1}^{k_\tau} x_j = 1, \ x_j \geq 0, \ j = 1, \ldots, n_i \right\}$$

A profile distribution $x = (x_1, \ldots, x_{k_\tau}) \in \Delta^\tau$ can bee seen as the individual behavior of a player spending a part of his time x_j in the n_j^τ-club. Hence, the population state represents the vector of individuals' share belonging to each club $i \in \{1, \ldots, k_\tau\}$, for all $\tau \in \{R, M\}$. The normal form representation of our described game is given by the next matrix payoff

$R \setminus M$	y_1	$\cdots$	y_{k_M}
x_1	a_{11}, b_{11}	$\cdots$	a_{1k_M}, b_{1k_M}
$\vdots$	$\vdots$	$\cdots$	$\vdots$
x_{k_R}	a_{k_R1}, b_{k_R1}	$\cdots$	$a_{k_Rk_M}, b_{k_Rk_M}$

$$(3.1)$$

where a_{ij} denotes the payoff of an i-strategist from population R playing against a j-strategist from population M. Similarly, we define b_{ij} by replacing M by R and vice-versa.

The matching between individuals from different populations is random. The i-strategist's expected payoff, supposing that the i-strategist's belongs to the n_i^R-club from population R is given by

$$E^R(n_i^R \mid y) = \sum_{j=1}^{k_M} a_{ij} y_j, \ \forall \ n_i^R \in S^R$$

where x is the clubs' distribution for the other population M.

Similarly, the expected payoff of the i-strategist belonging to n_i^M-club from population M is given by

$$E^M(n_i^M/x) = \sum_{j=1}^{k_R} b_{ij} x_j, \ \forall n_i^M \in S^M$$

where x is the clubs' distribution for the other population R. Rational individuals follow the strategic profile that maximizes their expected payoffs.

3.3 The Asymmetric Game and the Definition of $\mathcal{ESS}$

Consider the two-population normal form game

$$G = \{(\tau = \{R, M\}), S^\tau, (A = (a_{ij}), B = (b_{ij}))\} \tag{3.2}$$

where each population splits into clubs denoted by n_i^τ with $i \in \{1, \ldots, k_\tau\}$ and $\tau = \{R, M\}$. Hence:

- The population of residents is the set: $R = \bigcup_{i=1}^{k_R} n_i^R$, and $\forall h \neq j \ n_h^R \cap n_j^R = \emptyset$.
- The population of migrants is the set: $M = \bigcup_{i=1}^{k_M} n_i^M$, and $\forall h \neq j \ n_h^M \cap n_j^M = \emptyset$.

Let $p \in \Delta^R$ be the profile distribution of individuals' behavior from population R and let $q \in \Delta^M$ be the profile distribution of individuals' behavior in population M is at time t_0.

Let us postulate that an invasion occurs like a post-entry population at a post-period of time $t_1 > t_0$, by a small number of individuals of both types associated with an alternative strategy profile $(\bar{q}, \bar{p})$. The profile distribution from population R after suffering a small mutation is

$$q_\epsilon = (1 - \epsilon)q + \epsilon \bar{q},$$

which is called the fitness of the post-entry population in M. Similarly, the profile distribution from population R after suffering a small mutation is

$$p_\epsilon = (1 - \epsilon)p + \epsilon \bar{p}.$$

Definition 3.1. Let $(p^*, q^*) \in \Delta^R \times \Delta^M$ be a profile of mixed strategies. We say that the profile (p^*, q^*) is an $\mathcal{ESS}$ for an asymmetric two-population normal form game G, if for each pair $(\bar{p}, \bar{q}) \neq (p^*, q^*) \in \Delta^R \times \Delta^M$ there exists $\bar{\epsilon}$ such that:

3 Evolutionarily Stable Strategies and Replicator Dynamics

$$1) \ E^R(p^*/q_\epsilon^*) > E^R(\bar{p}/q_\epsilon^*)$$

$$2) \ E^M(q^*/p_\epsilon^*) > E^M(\bar{q}/p_\epsilon^*),$$

(3.3)

for all ϵ, $0 < \epsilon \leq \bar{\epsilon}$, where $p_\epsilon^* = (1 - \epsilon)p^* + \epsilon\bar{p}$ and $q_\epsilon^* = (1 - \epsilon)q^* + \epsilon\bar{q}$ are the respective post-entry populations.

Hence, individuals' behavior who adopt an $\mathscr{ESS}$ brings more offspring (with higher fitness) than the mutant individuals' behavior from the post-entry population. It has already been noticed by Selten [16] that an evolutionary stable strategy pair is not only stable when mutants appear in one of the populations but also if mutants appear in both populations. Definition 3.1 can be extended to the case of multipopulation models.

A well known result (see Cressman [6] and Weibull [18]) that characterizes the ESS in terms of NE is:

Proposition 3.1. *A profile x is $\mathscr{ESS}$ if and only if x is a strict Nash equilibrium.*

The evolutive properties of the $\mathscr{ESS}$ and its relationship with the set of Nash equilibria and the stationary states ($\mathscr{SS}$) of the replicator dynamics for the case of symmetric games are well known (see Hofbauer and Sigmund [9]; Weibull [18]). Then, with the purpose of analyzing the dynamical properties of $\mathscr{ESS}$, we introduce the symmetric (one-population) version of the asymmetric two-population game G.

3.4 The Symmetrized Game

Consider the asymmetric two-population normal form game G (see 3.2), where each population splits into clubs $n_1^R, \ldots, n_{k_R}^R$ and $n_1^M, \ldots, n_{k_M}^M$ and the payoff matrixes are A and B, respectively. Now, instead of pairwise matching, we consider the case that all players are interacting together, i.e. all players are "playing the field". Thus, the payoff of a player is determined by his own strategy and the strategies of all other players. So, the corresponding symmetrized one-population game is defined by as follows:

Let G be an asymmetric game defined by (3.2). Let $P = R \cup M$ be the big population. Let $N = \left\{ n_1^R, \ldots, n_{k_R}^R, n_1^M, \ldots, m_{k_M}^M \right\}$ be the set of pure strategy for P. The matrix payoff for the big population P is given by:

$$\Pi = \begin{bmatrix} 0 & A \\ B^T & 0 \end{bmatrix}$$

(3.4)

where we assume that the elements of $A(\cdot)$ and $B(\cdot)$ are "well behaved" in the sense of being continuously differentiable.

The symmetrized game version of the asymmetric game G is $G^s = \{P, N, \Pi\}$. For each asymmetric two-population game G, there exists a corresponding symmetric version G^s. It is worth to note that, these two versions are not equivalent in several aspects but every Nash equilibrium of the asymmetric game is a Nash equilibrium of its symmetric version. Our purpose is to characterize the main dynamics properties of the $\mathcal{ESS}$.

Let us consider the strategic profile $(p, q) \in \Delta^R \times \Delta^M$ and the profile distribution $x = (x_1, \ldots, x_{k_R + k_M})$ verifying the following identities:

$$
x_i = \begin{cases} p_i \frac{|R|}{|R|+|M|} & \text{if } 1 \le i \le k_R \\[2ex] q_i \frac{|M|}{|R|+|M|} & \text{if } k_R < i \le k_R + k_M \end{cases}
\tag{3.5}
$$

where $|\cdot|$ denotes the cardinality on the sets R and M defining the corresponding mixed strategy for the symmetric version G^s.

Proposition 3.2. *For each strategic profile* $(p, q) \in \Delta^R \times \Delta^M$, *there exists a mixed strategy* $x \in \Delta^P$ *of the corresponding one-population game, and vice-versa.*

Proof. Let $(p, q) \in \Delta^R \times \Delta^M$ be a strategic profile for the asymmetric game. Consider $x \in \Delta^P$ given by the expression (3.5), i.e. $x = \left(\frac{|R|}{|M|+|R|} p, \frac{|M|}{|M|+|R|} q \right)$. Thus, x is a mixed strategy for the symmetric game. To see the reciprocal, suppose that $x \in \Delta^P$. Since $x_i = \frac{|R_i|}{|R|+|M|}$ if $1 \le i \le k_R$ and $x_i = \frac{|M_i|}{|R|+|M|}$ if $k_R < i \le k_R + k_M$, we get $p_i = \frac{|R|+|M|}{|R|} x_i$, and $q_i = \frac{|R|+|M|}{|M|} x_i$. $\qquad\square$

Let us denote by $B_\tau(z)$ the set of best replies for the population $\tau = \{M, R\}$, where the profile distribution over the clubs in the opposite population $\tau' \ne \tau$ is given by z.

The following propositions offer an insight about the relationship between the set of $\mathcal{NE}$ and the set of $\mathcal{ESS}$ for asymmetric games and their respective symmetric versions.

Proposition 3.3. *If the strategic profile* (p^*, q^*) *is a* $\mathcal{NE}$ *of the original asymmetric two-population game, then the corresponding* x^* *defined by the expression (3.5) is the symmetric* $\mathcal{NE}$ *in the corresponding symmetric version.*

Proof. Suppose that the profile (p^*, q^*) is a $\mathcal{NE}$ of the asymmetric two-population game. Let $x^* = (x_1^*, \ldots, x_{k_M + k_R}^*)$ be the corresponding strategy in the corresponding symmetrized one-population game. Then, $p^* \in B_R(q^*)$ and $q^* \in B_M(p^*)$ implies that $x^* P x^* \ge y P x^*$, for all $y \in \Delta^P$ because

$$
\begin{aligned}
y P x^* &= \frac{|M||R|}{(|M|+|R|)^2} \left(q B^T p^* + p^* A q \right) \\
&\le \frac{|M||R|}{(|M|+|R|)^2} \left(q^* B^T p^* + p^* A q^* \right) = x^* P x^*.
\end{aligned}
$$

$\qquad\square$

3 Evolutionarily Stable Strategies and Replicator Dynamics   31

Proposition 3.4. *If the profile (p^*, q^*) is a strict Nash equilibrium for the asymmetric two population game, then the corresponding x^* is a strict Nash equilibrium for the symmetric version.*

Proof. Let (p^*, q^*) be a strict Nash equilibrium for the asymmetric two population game and let x^* be the corresponding profile for the symmetric version. Assume that there exist $y \neq x^* \in \Delta^P$, such that $y \Pi x^* = x^* \Pi x^*$. Using Proposition 3.2, there exist $p \neq p^*$ such that $pAq^* \geq p^*Aq^*$ or, there exist $q \neq q^*$ such that $p^*Bq \geq p^*Bq$, which is in contradiction with our assumption. $\square$

Proposition 3.5. *If the profile (p^*, q^*) is an $\mathscr{ESS}$ for the asymmetric two-population game, then the corresponding x^* is an $\mathscr{ESS}$ for the symmetric version.*

Proof. Let (p^*, q^*) be an $\mathscr{ESS}$. By Proposition 3.1, (p^*, q^*) is a strict Nash equilibrium. From Proposition 3.4, the corresponding strategy x^* is a strict Nash equilibrium for the symmetric version and it is straightforward to see that the reciprocal of this Proposition does not hold. $\square$

3.5 The Dynamics of the Model

The symmetric version of the asymmetric game allows us to characterize the main dynamical properties of the asymmetric game, because these properties are well known in the symmetric case.

Consider the asymmetric two-population normal form game G (see 3.2).

Let $n_i^\tau(t)$ be the number of individuals at time t belonging to the i-club in the population τ. Let $p_i(t)$ be the share of individuals in the i-club from the population R and, similarly, let $q_i(t)$ the share of individuals in the i-club from the population M, at time t. Hence,

$$p_i(t) = \frac{n_i^R}{|R|}$$

and

$$q_i(t) = \frac{n_i^M}{|M|}.$$

The vector $(p(t), q(t))$ is the profile distribution (or population state) at time t. Furthermore, $p(t) \in \Delta^R$ and $q(t) \in \Delta^M$.

Recall that the members of the i-club from population τ are called i-strategists from the population $\tau \in \{R, M\}$. Rational individuals choose strategies to maximize their expected payoffs. Let $z_0 = (p_0, q_0)$ be the strategic profile at time $t = 0$ for the asymmetric two-population game G. According to the rationality assumption, we define:

$$\begin{aligned}
\dot{p}_i &= ((e_i^R - p)Aq)p_i, \ i = 1, \ldots, k_R \\
\dot{q}_i &= ((e_i^M - q)B^T p))q_i, \ i = 1, \ldots, k_M,
\end{aligned} \tag{3.6}$$

where e_i^R is the i-canonical vector in $\mathbb{R}^R$ and e_i^M is the canonical i-th vector in the $\mathbb{R}^M$. The differential equation (3.6) represents the clubs' evolution for each population. For the system (3.6), a solution of the form $\xi(t, z_0) = (\xi_1(t, z_0), \xi_2(t, z_0))$ represents the evolution of the population states with initial state given by z_0.

From system (3.6), each time t the club of the i-strategists in each population increases if and only if the expected payoff of the i-strategy is greater than the average payoff, and reciprocally.

For each pair $(p(t), q(t))$ in G, there exists a corresponding mixed strategy $x(t)$ in the symmetric version G^s given by the expression (3.5).

The dynamical system (3.6) has a corresponding dynamical system, namely the replicator dynamics, (see Taylor and Jonker [17]) of the symmetric one-population game given by

$$\dot{x}_i = ((e_i - x)Px)x_i \tag{3.7}$$

where $x = (x_1, \ldots, x_{k_R + k_M})$, x_i is given by the expression (3.5), and e_i is the i-canonical vector in $\mathbb{R}^{k_R + k_M}$.

We analyse the relationship between $\mathcal{NE}$, $\mathcal{ESS}$ and $\mathcal{SS}$ of the system (3.6) of the symmetric version game G^s.

If a pair $(\bar{p}, \bar{q})$ is a stationary state of the system (3.6) then the corresponding $\bar{x}$ is a stationary state for the dynamical system (3.7). Furthermore, every strictly positive stationary state of the dynamical system (3.6) is a $\mathcal{NE}$ for the corresponding asymmetric two-population game. Every $\mathcal{NE}$ of an asymmetric two-population game is a stationary state for its corresponding dynamical system given by (3.6). Hence, we can conclude that the set of $\mathcal{NE}$ of an asymmetric two-population game is a subset of the set $\mathcal{SS}$ corresponding to the dynamical system (3.6). Every $\mathcal{NE}$ of a two-population game is a stationary state for the corresponding dynamical system (3.7).

3.6 Evolutionarily Stable Strategies and Liapunov's Stability

Denote by $\mathcal{AS}$ the set of asymptotically stable steady states. From the well known relations between $\mathcal{ESS}$, $\mathcal{NE}$ and $\mathcal{SS}$ for the symmetric cases (see Weibull [18]), the following relationship holds for every asymmetric two-population game

$$\mathcal{ESS} \subseteq \mathcal{AS}, \tag{3.8}$$

and

$$\mathcal{NE} \subseteq \mathcal{SS}. \tag{3.9}$$

Proposition 3.6. *For an asymmetric two-population game, if (p^*, q^*) is an asymptotically stable steady state of the dynamical system (3.6), then (p^*, q^*) is a $\mathcal{NE}$.*

Proof. If $(p^*, q^*) \in AS$ for the dynamical system (3.6) then it is stationary state. If $p^* > 0$ and $q^* > 0$ then (p^*, q^*) is a $\mathcal{NE}$ for the asymmetric game. Now

3 Evolutionarily Stable Strategies and Replicator Dynamics

we consider the case where some strategy is absent in p^* or in q^*. Without loss of generality we assume that $p_j^* = 0$. Suppose that (p^*, q^*) is not a $\mathcal{NE}$. Then, there exists some pure strategy $j \notin supp(p^*)$ such that $E^R(e_j^R/q^*) = e_j^R A q^* > p^* A q^* = E^R(p^*/q^*)$. Assume that a perturbation affects the distribution p^* and that in the population R some j-strategist appears. The post-entry population at time t, is $p_\epsilon(t) = (1 - \epsilon(t))p^* + \epsilon(t)e_j^R$. Substituting in the j-differential equation of (3.6), we obtain

$$\dot{p}_{\epsilon j} = \dot{\epsilon} = [(e_j^R - p_\epsilon)Aq^*]\epsilon. \tag{3.10}$$

Define $F(\epsilon) = (e_j^R - p_\epsilon)Aq^*$. Note that $F(0) = (e_j^R - p^*)Aq^*$ and $F'(0) = (p^* - e_j^R)Aq^*$. The Taylor polynomial is $F(\epsilon) = F(0) + F'(0)\epsilon + 0(\epsilon^2)$. Considering the first order approximation equation (3.10) gives

$$\dot{\epsilon} = [(e_j^R - p^*)Aq^*]\epsilon.$$

In the population R, the members in the n_j^R club increase, contradicting our claim that (p^*, q^*) is an asymptotically stable steady state with $n_j^R = 0$. $\square$

We now study the connection between $\mathcal{ESS}$ and the replicator dynamics in an asymmetric game. We will use the following Proposition (see Taylor and Jonker [17]).

Proposition 3.7. *For symmetric homogeneous population game every $\mathcal{ESS}$ is an asymptotically stable steady state of the replicator dynamics.*

Theorem 3.1. *For the asymmetric two-population game, we obtain the following chain of inclusions:*

$$\mathcal{ESS} \subseteq \mathcal{AS} \subseteq \mathcal{NE} \subseteq \mathcal{SS}.$$

Proof. Let (p^*, q^*) be an $\mathcal{ESS}$ for an asymmetric game and let x^* be the corresponding strategic profile in its symmetric version. So, from Proposition 3.1, it follows that (p^*, q^*) is a strict Nash equilibrium. By Proposition 3.4, it follows that the symmetric strategic profile of every strict Nash equilibrium of an asymmetric game is a strict $\mathcal{NE}$. Then x^* is a strict Nash equilibrium for the symmetric version, and then x^* is a $\mathcal{ESS}$. By Proposition 3.7, it follows that x^* is an asymptotically stable steady state of the replicator dynamics. Then, (p^*q^*) is an asymptotically stable steady state for the asymmetric version and is a $\mathcal{NE}$. $\square$

Bomze [3] shows that every asymptotically stable steady state in the homogeneous population replicator dynamics corresponds to a Nash equilibrium that is trembling hand. However, using the symmetric version of a non-homogeneous asymmetric n-population the following Proposition holds:

Corollary 3.1. *Every $\mathcal{ESS}$ of a non-homogeneous asymmetric n-population game is trembling hand and isolate.*

Proof. Let (p^*, q^*) be an $\mathscr{ESS}$ for an asymmetric game and let x^* be the corresponding strategic profile in its symmetric version. By Theorem 3.1, it follows that every $\mathscr{ESS}$ is asymptotically stable for the symmetric version. Hence, x^* is asymptotically stable steady state for the symmetric version. By Bomze [3], it follows that x^* is trembling hand and isolate equilibrium, and so (p^*, q^*) verifies this property in the original asymmetric game. $\qquad\square$

3.7 Concluding Remarks

We extended the definition of evolutionarily stable strategies ($\mathscr{ESS}$) of symmetric games to asymmetric two-population games. We did it by taking as the strategy space for the symmetrized game the union of strategies from the two-population asymmetric game and assigning zero payoffs to all strategy combinations that belong to the same player position in the asymmetric game. Hence, evolutionary dynamics in a two-population asymmetric game can be analyzed using the well known properties of the replicator dynamics corresponding to the symmetric version of this game. This fact may have interest for economic theory and social analysis, where asymmetric games are useful to analyze the behavior of two populations engaged in non-cooperative games such as, buyers and suppliers, firms and workers or residents and migrant populations interacting in a given country or economy.

Acknowledgements The author is grateful to CONACYT-Mexico (Project 46209) and UASLP Secretaría de Posgrado (Grant C07-FAI-11-46.82) for the financial support. Facultad de Economía, UASLP México. We thank Dan Friedmand and Karl Schlag for their helpful comments to improve this research. The usual disclaimer applies. The author wishes to thank the support given by Programa Integral de Fortalecimiento Insititucional de la UASLP (PIFI) 2010.

References

1. Balkenborg, D., Schlag, K.: Evolutionary Stability in Asymmetric Population Games. Discussion Paper Serie B 314, University of Bonn, Germany (1995)
2. Balkenborg, D., Schlag, K.: On the evolutionary selection of sets of Nash equilibria. J. Econ. Theory **133**, 295–315 (2007)
3. Bomze, I.: Non-cooperative two-person games in biology: A classification. Int. J. Game Theory **15**(1), 31–57 (1986)
4. Cressman, R.: The Stability Concept of Evolutionary Game Theory, Chaps. 2, 3. Springer, Berlin (1992)
5. Cressman, R.: Frequency-dependent stability for two species interactions. Theor. Popul. Biol. **49**, 189–210 (1996)
6. Cressman, R.: Evolutionary Dynamics and Extensive Form Games, Chap. 4. MIT, Cambridge (2003)
7. Cressman, R.: Uninvadability in N-species frequency models for resident-mutant systems with discrete or continuous time. Theor. Popul. Biol. **69**, 253–262 (2006)
8. Fishman, M.A.: Asymmetric evolutionary games with non-linear pure strategy payoffs. Games Econ. Behav. **63**, 77–90 (2008)

9. Hofbauer, J., Sigmund, K.: Evolutionary games and population dynamics. Cambridge University Press, Cambridge (1998)
10. Smith, M., Price, G.R.: The logic of animal conflict. Nature **246**, 15–18 (1973)
11. Smith, M.: The theory of the games and evolution of animal conflicts. J. Theor. Biol. **47**, 209–221 (1974)
12. Nash, J.: Non-cooperative games, unpublished Ph.D. thesis, Mathematics Department, Princeton University (1950)
13. Retchkiman Königsberg, Z.: A Vector Lyapunov Approach to the Stability Problem for the n-Population Continuous Time Replicator Dynamics. Int. Math. Forum **52**, 2587–2591 (2007)
14. Samuelson, L.: Evolutionary Games and Equilibrium Selection. MIT (1998)
15. Samuelson, L., Zhang, J.: Evolutionary stability in asymmetric games. J. Econ. Theory **57**(2), 363–391 (1992)
16. Selten, R.: A note on evolutionary stable strategies in asymmetric contests. J. Theor. Biol. **84**, 93–101 (1980)
17. Taylor, P.D., Jonker, L.: Evolutionarily stable strategies and game dynamics. Math. Biosci. **40**, 145–156 (1978)
18. Weibull, W.J.: Evolutionary Game Theory. MIT (1995)

Chapter 4
Poverty Traps, Rationality and Evolution

**Elvio Accinelli, Silvia London, Lionello F. Punzo,
and Edgar J. Sanchez Carrera**

Abstract We study an economy with heterogenous workers and firms as a two population game, in normal form, and its evolutionary dynamics implied by strategic complementarities. The population of firms is distributed in two groups, innovative and non innovative, while workers need to choose between two strategies, acquiring skills or remaining unskilled. Without having knowledge of the firms' distribution, a worker reviews her strategy by asking herself whether it is worth it to change behavior or not. Rational choice on her part is taken, hereafter, to imply that she will choose the strategy which she expect to yield the greatest payoff, on the basis of her beliefs and the current state of the economy. By imitating successful agents, if the initial shares of innovative firms and skilled agents are "too small", an economy eventually lead into a poverty trap. Hence, when an economy is close to a poverty trap, rationality may act as an actual obstacle to a take-off.

E. Accinelli (✉)
Facultad de Economía, UASLP, San Luis Potosí, México
e-mail: elvio.accinell@eco.uaslp.mx

S. London
Facultad de Economía, Universidad Nacional del Sur – CONICET, Bahía Blanca, Argentina
e-mail: slondon@uns.edu.ar

L.F. Punzo
Department of Economics, University of Siena, Piazza S. Francesco, 7, 53100 Siena, Italy
and
PPED-INCT, UFRJ, Rio de Janeiro, RJ, Brazil
e-mail: punzo@unisi.it

E.J.S. Carrera
Facultad de Economía, UASLP, México and Department of Economics, University of Siena, Italy
e-mail: sanchezcarre@unisi.it

M.M. Peixoto et al. (eds.), *Dynamics, Games and Science I*, Springer Proceedings
in Mathematics 1, DOI 10.1007/978-3-642-11456-4_4,
© Springer-Verlag Berlin Heidelberg 2011

4.1 Introduction

This paper is based on the model by Accinelli et al. [1] where workers' decisions are driven by imitative behavior and firms' decisions depend on the number of skilled workers. Such a model analyzes the dynamic complementarities between innovative firms and skilled workers. When firms invest in R&D to become an innovative firm, they are successful only in the presence of sufficiently high number of skilled workers (see Redding [4]). At the same time workers are encouraged to increase their skills when a large number of firms make investments in high-technology. On the contrary, firms that do not invest in R&D do not look for skilled workers and, so, make the accumulation of skills unprofitable. We show that there exists a threshold number of innovative firms above which it becomes advantageous to accumulate human capital and to become a skilled worker. If the percentage of innovative firms is under a certain threshold value, the economy will evolve to a poverty trap where the number of skilled workers decreases to zero. Thus it will be better for firms not to invest in R&D. On the other hand, if the initial percentage of innovative firms is higher than the threshold value, then by an imitation behavior workers will push the economy to evolve to a higher level equilibrium, a steady state characterized by the coexistence of non innovative firms and innovative firms, and skilled workers and unskilled workers. This is the mechanism that allows an economy to get out from the poverty trap.

Our result may account for the experience of many developing countries in which there is a mismatch between investment in R&D and Human Capital accumulation, the engine of sustained economic growth.

This paper, therefore, focuses on the relationship between evolution, rationality and poverty traps, in an economy with different firms and workers. We assume that, in a game theoretic setting, agents follow a rational behavior, in other words, when faced with the need to choose a behavior or a strategy under uncertainty, they choose the one with greatest expected value, given their own beliefs. Each agent is reviewing his strategy choice when in need to choose a strategy for the next period. Furthermore, we assume that a reviewer imitates the most successful agent. Then, the evolution of the economy is represented by a particular dynamical system where, depending on the initial conditions, the economy will approach to local attractors, one in a set of steady states. Given the structure of the economy, it is the rationality of the agents that will determine the dynamical system and the evolution of the economy.

Under our hypotheses there exists a steady state that is a Pareto inefficient Nash equilibrium of the game, and is an evolutionarily stable strategy against the field. Such inefficient equilibrium is a poverty trap, whereby workers are unskilled and firms do not innovate. In that case, the rational rule followed by agents makes the economy to evolve over time along to a low growth trajectory. We also show that there exist steady state attractors, where a percentage of innovative firms and of skilled workers live together with a percentage of non innovative ones and unskilled workers. The relative weight of these percentages are shown to depend on the initial distribution of firms.

4 Poverty Traps, Rationality and Evolution

Implementing an imitative behavior, workers can choose to increase their competence being trained at a school, and trying to attain the knowledge frontier. Hence, being a skilled worker implies human capital accumulation, and this would be the driving engine of economic growth (see Lucas [3] and Sen [6]). The way agents imitate the others follows the well known theory whereby they question their own performance of their current behavior (or strategy). The way agents imitate the others is opting by strategies whose performances are better than the other agents (see Schlag [5] and Weibull [7]).

The paper is organized as follow. Section 4.2 describes the basic, two-population normal form game characterizing strategies and payoffs for firms and workers. Section 4.3 introduces the dynamic imitation mechanism to analyze the evolution of worker's population. In Sect. 4.4, we analyze the evolutive behavior of an economy as depending upon its initial conditions. In Sect. 4.5, the relationships between Nash and dynamic equilibria are analyzed and the definition of an evolutionarily stable strategy is introduced. In Sect. 4.6, we present the market dynamics for firms, while Sect. 4.8 draws some concluding remarks.

4.2 The Model

Consider that the economy is composed by two population: workers W and firms, F. Each population splits into two clubs according with the strategy followed by individuals. These strategies are denoted by $\{S, NS\}$ for workers, and by $\{I, NI\}$ for firms. By $E(i/Y_F)$ we represent the conditional expected payoff associated with the strategy $i \in \{S, NS\}$ given the initial distribution of the population of firms over their clubs Y_F. By Δ^F we denote the set of distributions of probabilities of the firms, i.e., $Y_F \in \Delta^F$. Hence, workers and firms are engaged in a repeated normal form game. To choose their club or strategy each worker look for the expected payoff $E(i/Y_F)$ and choose according with the strategy having associated the maximum expected payoff. Firms look for cost and profit, and choose workers. Technology is free and each firm can choose without cost to be innovative or not. The costs of the innovative firms decrease with the possibilities to obtain in the labor market skilled workers.

This game between workers and firms is characterized by:

1. *Asymmetric information.* At the beginning of each contractual period, workers do not know the type of firm that is going to hire them. However, workers have to certify their skilled levels so that firms know their profile.
2. *Gross income.* Let $B_i(j)$ be the gross-benefit of the i-firm hiring the j-worker, for all $i \in \{I, NI\}$ and $j \in \{S, NS\}$. The S-type worker gets a salary s, while the NS-type gets $\bar{s} < s$.
3. *Skill premia.* Innovative firms I give a plus (or premia) to their workers, while NI-firms do not do it, at the end of the contractual period. Only at the end of the period, workers will know the type of the firms that they are engaged. Thus, the skilled worker S receives a premium $\bar{p}$ while the unskilled workers NS receives

a premium p when both are engaged with an innovative firm I, such that $0 < p < \bar{p}$.

4. *Training cost.* To become a skilled worker has a cost CS. We assume (only for simplicity) that the firms have no cost to face in order to become innovative. Thus, $CS > \bar{s}$, i.e. there are no incentives to be skilled worker if there are no prizes.

5. *Short run and long run.* We assume that innovative firms face a high level of fixed costs. Non-innovative firms need to do a big investment to become innovative. Hence, the distribution of the firms in the short run remains constant and only in the long run this distribution evolves.

We will show, that in the long run, there are *strategic complementarities* between types of firms as well as between types of workers. Thus:

- If the firm is innovative, the payoff of the skilled worker is greater than the payoff of the unskilled one, i.e., $\bar{s} + \bar{p} - CS > s + p$.
- If the firm is non-innovative, the payoff of a skilled worker is at least as good as the payoff of an unskilled one, i.e., $s \geq \bar{s} - CS$.
- For a skilled worker, the payoffs obtained by the innovative firm are greater than those obtained by the non-innovative firm, i.e., $B_I(S) - \bar{p} > B_{NI}(S)$.
- For an unskilled worker, the benefits of the non innovative firm are greater than those obtained by the innovative one, i.e., $B_I(NS) - p < B_{NI}(NS)$.

In summary, for our two population normal form game, the payoff matrix is represented by,

$W \backslash F$	I	NI
S	$\bar{s} + \bar{p} - CS,\ B_I(S) - (\bar{s} + \bar{p})$	$\bar{s} - CS,\ B_{NI}(S) - \bar{s}$
NS	$s + p,\ B_I(NS) - (s + p)$	$s,\ B_{NI}(NS) - s$

$$(4.1)$$

The expected payoff of the S-type worker, given the chances of being hired either by the I or NI firm, is:

$$E(S/Y_F) = \text{prob}(I)\,[\bar{s} + \bar{p}] + \text{prob}(NI)(\bar{s}) - CS, \qquad (4.2)$$

where $\text{prob}(I)$ represents the probability of being hired by the innovative firm and $\text{prob}(NI)$ the probability of being hired by the non innovative firm. Similarly,

$$E(NS/Y_F) = \text{prob}(I)\,[s + p] + \text{prob}(NI)s. \qquad (4.3)$$

Therefore, workers prefer to be S-type strategists if $E(S) > E(NS)$ and viceversa. The latter happens if and only if $\text{prob}(I)$ is large enough, i.e., when:

$$\text{prob}(I) > \frac{CS - (\bar{s} - s)}{(\bar{p} - p)}. \qquad (4.4)$$

4 Poverty Traps, Rationality and Evolution 41

Workers are indifferent between being skilled or unskilled if and only if

$$\text{prob} : (I) = \frac{CS - (\bar{s} - s)}{(\bar{p} - p)}. \tag{4.5}$$

Note that, $0 < (CS - (\bar{s} - s))/((\bar{p} - p)) < 1$ holds.

Let us label $\text{prob}(I) = P_u = \frac{CS-(\bar{s}-s)}{(\bar{p}-p)}$, and denote the probability to employ a skilled worker for an innovative firm by $\text{prob}(S)$. Hence, a firm gets innovative if and only if its expected payoff is greater than the expected payoff of being non innovative, that is $E(I) > E(NI)$ or

$$\text{prob}(S) > \frac{B_I(NS) - B_{NI}(NS) - p}{B_I(NS) - B_I(S) + B_{NI}(S) - B_{NI}(NS) + (\bar{p} - p)}. \tag{4.6}$$

Let us label $\text{prob}(S) = \bar{x}_s$. The threshold level where economic agents, firms and workers, prefer to be of high-profiles is $(\bar{x}_s, P_u)$. We find three Nash equilibria, two of them in pure strategies: $A = \{S, I\}$ and $B = \{NS, NI\}$, and a mixed strategy Nash equilibrium given by

$$NE = \left(\bar{X}_S, (1 - \bar{X}_{NS}); \ P_u, (1 - P_u) \right). \tag{4.7}$$

We conclude that the A equilibrium Pareto-dominates equilibrium B while the latter is the risk dominant equilibrium.

4.3 Dynamic Imitation by Workers

Assume that in the short run the firms' distribution remain fixed. This assumption is natural because innovative firms have high fixed costs associated. In the short run the number of firms in the market is constant. Let $Y_I = \text{prob}(I) = PI = QI/Q$ be the share (percentage) of innovative firms, where QI is the number of innovative firms and Q is the total number of them. Then, $Y_{NI} = \text{prob}(NI) = PNI = 1 - PI$. Equivalently the firms distribution on the set of pure strategy is given by $Y_F = (Y_I, Y_{NI})$. Let us denote by Δ^W the set of distributions over the possible behaviors of the workers population.

We call an individual reviewer to whom ask to himself whether she needs or not to change her current strategy. The rate at which agents in the population review their strategic choice depends on the current performance of their strategies and on the current distribution of the population state, denoted here by $Z = (X_W, Y_F)$, where $X_W = (X_S, X_{NS}) \in \Delta^W$ and $Y_F \in (Y_I, Y_{NI})$. Let $R_i(Z)$ be the probability that the i-strategist, $i \in \{S, NS\}$ is reviewing. This probability at which agents, in a given population, review their strategies choice represents the rate of the arrival time of a Poisson process. Let $P_{ij}(Z)$ be the probability that such reviewing worker really switches to the strategy $j \neq i$. Then,

$$P(i \rightarrow j)(Z) = R_i(X_W)P_{ij}(Z) \tag{4.8}$$

is the probability that a worker of the i club changes to the j one, $i \neq j \in \{S, NS\}$. In the sequel, $e_S = (1,0)$ and $e_{NS} = (0,1)$ indicate vectors of pure strategies S or NS. Considering that all agent's Poisson process is statistically independent, the aggregate of reviewing times in the subpopulation of i strategists is itself a Poisson process with arrival time $X_i P_{ij}(Z)$. So, for large populations, we may invoke the law of large numbers and model these aggregate stochastic processes as deterministic flows, each flow being set equal to the expected rate of the corresponding Poisson arrival process. Hence, the expected flow share of skilled workers $\dot{X}_S$ will be equal to the percentage of unskilled workers changing to be skilled workers minus the percentage of skilled workers changing to be unskilled workers.

Rearranging terms, we get the differential equation system characterizing the dynamic flow of workers:

$$\begin{cases} \dot{X}_S = R_{NS}(Z)P_{USS}(Z)X_{NS} - R_S(X_W)P_{SNS}(Z)X_S \\ \dot{X}_{NS} = -\dot{X}_S, \end{cases} \tag{4.9}$$

where X_S is the fraction of skilled (X_{NS} of unskilled, respectively) workers.

We assume that reviewers take their decisions under an imitative rule, and this will be reflected in the characteristics of the probability $P_{ij}(Z)$. To simplify the model, we will consider only imitative behavior processes, supported by imitative rules. An imitative rule makes sense if there are at least two distinct behaviors, one of them currently adopted and the other being a candidate behavior to imitate (in our model, if one of the two populations disappears the incentive to change vanishes with it).

Assume that reviewing workers evaluate their current strategy and decide to imitate only the successful one. The problem is how to decide what is a successful behavior. Let us suppose that each reviewing agent, cannot know the true expected payoff of each possible behavior. She randomly samples in the neighboring population, computing average payoffs and imitating the behavior with the highest sampled average value. Let $\bar{E}(S)$ and $\bar{E}(NS)$ be the estimators of the true payoffs or values $E(S/Y_F)$ and $E(NS/Y_F)$.

Hence, an i-worker changes her current strategy if and only if $\bar{E}(i) < \bar{E}(j)$, $i \neq j \in \{S, US\}$.

Thus, the probability for an i-type become a j-type strategist depends on the probability that $\bar{E}(j) - \bar{E}(i) > 0$. Then, (4.9) can be written as:

$$\begin{aligned} \dot{X}_S &= R_{NS}P[\bar{E}(S) - \bar{E}(NS) > 0]X_{NS} - R_S P[\bar{E}(NS) - \bar{E}(S) > 0]X_S, \\ \dot{X}_{NS} &= -\dot{X}_S. \end{aligned} \tag{4.10}$$

4 Poverty Traps, Rationality and Evolution 43

where $P[\bar{E}(j) - \bar{E}(i) > 0]$ denotes the conditional probability distribution that an i reviewer become a j strategist.

We assume that this probability is an increasing (continuously differentiable) function $\phi : R \to [0, 1]$ of the true difference $E(S/Y_F) - E(NS/Y_F)$ if $E(S/Y_F) - E(NS/Y_F) > 0$ and 0 in other case. When we are considering the case $E(S/Y_F) - E(NS/Y_F) > 0$, the system (4.10) becomes:

$$\dot{X}_S = R_{NS}\phi(E(S/Y_F) - E(NS/Y_F))X_{NS}$$
$$\qquad - R_S[1 - \phi(E(S/Y_F) - E(NS/Y_F))]X_S \qquad (4.11)$$
$$\dot{X}_{NS} = -\dot{X}_S.$$

To simplify, we assume that $R_{NS} = R_S = 1$. Then (4.11) results in

$$\dot{X}_S = \phi(E(S/Y_F) - E(NS/Y_F)) - X_S$$
$$\dot{X}_{NS} = -\dot{X}_S \qquad (4.12)$$

Intuitively, this process is supported on the rule: *to switch if the other strategy brings a higher payoff.* The process depends on the ordinal ranking of the expected values, but the speed of adjustment depends on the size of expected value differences.

Let $\xi(t, t_0, X_{S0})$ be the solution of (4.12) for the initial conditions $X_S(t_0) = (X_{S0}, 1 - X_{S0})$. Hence, the solution is on a trajectory converging to

$$(\phi(E(S/Y_F) - E(NS/Y_F)), 1 - \phi(E(S/Y_F) - E(NS/Y_F)),)$$

i.e.

$$\lim_{t \to \infty} \xi(t, t_0, X_{S0})$$
$$= (\phi(E(S/Y_F) - E(NS/Y_F)), 1 - \phi(E(S/Y_F) - E(NS/Y_F))).$$

It follows that the percentage of skilled workers in the steady state increase with the true difference $E(S) - E(NS)$. In the stationary state, coexist a percentage of skilled workers with a percentage of unskilled, the percentage difference in favor of first, will be greater whatever major is the difference between the true expected values $E(S/Y_F) - E(NS/Y_F)$.

Considering now the case where $E(S/Y_F) - E(NS/Y_F) < 0$, the system (4.10) becomes

$$\dot{X}_S = -X_S$$
$$\dot{X}_{NS} = -\dot{X}_S \qquad (4.13)$$

In this case, the solution is given by:

$$\xi(t, t_0, X_{S0}) = \left(X_{S0}e^{-t}, 1 - X_{S0}e^{-t}\right)$$

and then, $\lim_{t\to\infty} \xi(t, t_0, X_{S0}) = (0, 1)$. Note that the club of skilled workers disappears with time and all workers become unskilled workers.

4.4 Initial Conditions Matter

Does the initial number of innovative firms contribute to explain the path of the economy? Consider two countries, 1 and 2 and assume that $PI_1 > PI_2$ when $t = t_0$. From the solution of (4.11), the share of skilled workers in country 1 is larger than the corresponding share workers in country 2, for each $t > t_0$, i.e.

$$X_{1S}(t) > X_{2S}(t), \quad \forall t > t_0 \tag{4.14}$$

Hence, the equilibrium state is higher in country 1 than in country 2.

Figure 4.1 shows the evolution of the dynamical system when the initial percentage of the innovative firms is above or below such threshold value:

1. If $PI > \pi$ then $E(S/Y_F) > E(NS/Y_F)$:

   - If $X_S(0) > \phi(E(S/Y_F) - E(NS/Y_F))$, the percentage of skilled workers in the total population decrease and its share converge to $\phi(E(S/Y_F) - E(NS/Y_F))$.
   - If $X_S(0) < \phi(E(S/Y_F) - E(NS/Y_F))$, the percentage of skilled workers increases.
   - In both cases the economy converges to the high level and diversified equilibrium.
   - This high steady state is not a Nash equilibria.

2. If $PI \leq \pi$, then $E(S/Y_F) \leq E(NS/Y_F)$:

   - The share of skilled workers is decreasing to zero $X_S(0) \to 0$. In this case, the economy is in a poverty trap and the rational workers will choose to be unskilled. This shows that rationality may imply inefficiency. This is the only asymptotically stable Nash equilibrium for the game above.

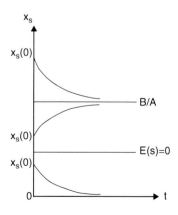

Fig. 4.1 Evolution and steady states, initial condition matter

4 Poverty Traps, Rationality and Evolution 45

- The profile of mixed strategy of inefficient Nash equilibrium for the game is (e_{NS}, e_I) and the steady state of the dynamical system in this case is $e_{NS} = (0, 1)$. This means that in the inefficient (or low) steady state firms prefer to be non-innovative.

The next theorem summarizes our results.

Theorem 4.1. *Consider the dynamic flow of workers, given by the system (4.9). There exists a threshold value $\pi = \frac{CS-\bar{s}}{\bar{p}}$ such that:*

1. *If the initial number of innovative firms PI is larger than the value, i.e., $PI > \pi$ then the percentage of skilled workers $X_S(t)$ will converge to $\phi(E(S/Y_F) - E(NS/Y_F))$.*
2. *If the initial number of innovative firms verifies $PI \leq \pi$, then, the percentage of skilled workers $X_S(t)$ will converge to 0.*

4.5 Dynamic Equilibria, Nash Equilibria and the Evolutionarily Stable Strategy

Note that there is no possibility of observing the high Nash equilibrium $(S, I) = (1, 0; 1, 0) = (e_S, e_I)$ –in pure strategies– because it is not a dynamic equilibrium. On the contrary, the low Nash equilibrium $(NS, NI) = (0, 1; 0, 1) = (e_{NS}, e_{NI})$ is asymptotically stable, and no the poverty trap arises as a result of the rational conduct of economic agents. On the other hand the only mixed Nash equilibrium is not observable, because it is an unstable steady state.

$$NE = (\bar{x}_s, (1 - \bar{x}_s); \pi, (1 - \pi)),$$

We observe that $\pi = (CS - (\bar{s} - s))/(\bar{p} - p)$ and the equality $E(S/\pi) = E(NS/\pi)$ is satisfied. Furthermore,
$\bar{x}_s$ verify the equality $E(NI/\bar{x}_s) = E(I/\bar{x}_s)$, equivalently,

$$\bar{x}_s = \frac{B_{NI}(S) - B_I(NS) + pr'}{B_I(S) - B_I(NS) + B_{NI}(NS) - B_{NI}(S) + pr + pr'},$$

After a perturbation on the distribution of firms, the population of skilled workers either converges to $B/A \neq \bar{x}_s$, or converges to 0.

Let us introduce the concept of an *evolutionarily stable strategy against the field* given a profile distribution of the firms' population.

Let Δ^W be the set of distributions on the workers' population and Δ^F be the set of distributions on the firms. Let us consider a distribution on the workers' population $X_W = (X_S, X_{NS}) \in \Delta^W$ and an initial distribution $Y_F \in \Delta^F$. Let Y_ϵ be a perturbed distribution of the initial distribution of the firms with $\epsilon > 0$ small enough, such that in the Euclidean distance $|Y_F - Y_\epsilon| < \epsilon$. Assume that X_W is a best response against Y_F.

Definition 4.1. We say that the distribution on the population of workers X_W is an *evolutionarily stable strategy against the field* given by Y_F if there exist $\epsilon > 0$ such that X_W is the best response against all distribution Y_ϵ in a neighborhood V_ϵ of radium ϵ, centered at Y_F.

Intuitively, this means that the best response X_W against Y_F continues being a best response against perturbations (in the distributions of the *field*). Notice that, when $Y_F \leq \pi$, the degenerate distribution $e_{NS} = (0, 1)$ (i.e. all workers are unskilled) is an ESS against the field given by Y_F.

Hence, a rational worker choses to be unskilled even in the case when the initial conditions were to change as long as such change is not "too large".

4.6 On the Dynamics of Firms

The following assertion from [2] gives the empirical support to the main results of this section:"Technological and scientific innovation is the engine of U.S. economic growth and human talent is the main input that generates this growth."

In the long period, with changes in demand and with the reduction of costs the distribution of the firms in each branch of the production can change. Non-innovative firms can become innovative because the necessary input costs to do an innovative production decrease. We will see in this section, the relationships between the evolution of the population of workers and the evolution of the population of firms.

Consider an industry producing under a competitive market. To focus on strategic complementarities, let us suppose that the production function of the innovative firms is given by

$$G = f(\mathcal{Z}, X_S, X_{NS}) \tag{4.15}$$

where $\mathcal{Z}$ is the technology, X_S and X_{NS} are respectively, the number of skilled and unskilled workers employed by the firm, and G is the total output. We suppose that technology is a complementary input to skilled labor.[1] Hence, the marginal product of technology is an increasing function of the number of skilled workers.

Suppose that innovative firms face a demand G for its product. Let $X_S(t)$ be the percentage of skilled workers existing in the economy in the time t. Assume that the percentage of skilled worker is an increasing function of time, i.e., $X_S(t_0) < X_S(t_1)$ if $t_0 < t_1$.

Using the usual hypothesis on the technology, it follows that

$$\frac{\partial C(G, X_S(t_1))}{\partial G} \leq \frac{\partial C(G, X_S(t_0))}{\partial G}, \tag{4.16}$$

---

[1] For instance $G = \mathcal{Z}^\alpha X_S^\beta + X_{NS}$ where $0 < \alpha, \beta < 1$. See example in Sect. 4.6.1.

4 Poverty Traps, Rationality and Evolution

where $C(G, X_S)$ stands for the short run cost function. Hence, there exists $\bar{G}$ such that

$$C(G, X_S(t_0)) > C(G, X_S(t_1)) \, ; \, \forall G > \bar{G}.$$

This means that the short run cost decrease when the upper bound on the supply of skilled workers increase. The long run cost $C(G)$ is the envelope of the short run costs curves $C(G, X_S(t))$, where $X_S(t)$ is considered as given in each time t. Figure 4.2 offers a graphic representation. Assume that supply of a skilled worker, at $t = t_0$, $X_S(t_0)$ is lower than the optimal level required by an innovative firm facing a demand of G units of its product $X_S(t_0, G)$, i.e.

$$X_s(t_0) \leq X_S^*(t_0, G).$$

The rate of convergence of the short run cost $C^I(G, X_S(t))$ of the innovative firm to the long run cost $C^I(q)$ of the innovative firm, when the supply of skilled workers is increasing with time, is greater than this convergence process for the non-innovative firms, formally

$$|C^I(G, X_S(t)) - C^I(q)| \leq \epsilon_I(t) \quad (A)$$
$$|C^{NI}(G, X_S(t)) - C^{NI}(q)| \leq \epsilon_{NI}(t) \quad (B)$$
(4.17)

where the infinitesimal order $\epsilon_I(t)$ is greater than the infinitesimal order of $\epsilon_{NI}(t)$, i.e.,

$$o(\epsilon_I(t)) > o(\epsilon_{NI}(t)).$$

If the supply of skilled workers is increasing, short run costs for innovative firms will decrease toward long run cost levels. Innovative firms can obtain positive profits and fixe a price $p' < p$ to obtain positive profits. So, there are incentives for non innovative firms to change their behavior, because the investments in technology can

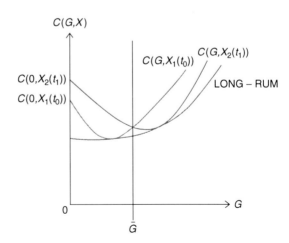

Fig. 4.2 Short run costs Vs. long run costs

48   E. Accinelli et al.

be recovered in a more convenient time. In this process a share of the non-innovative firms will leave the market or become innovative if at the $2x$:

1. The cost to become innovative $C(h, H)$ is greater than the difference between the profits of the innovative firms $\Pi^I(p', q_I^*, t)$ at price p' and he profits of the non-innovative firms $\Pi^{NI}(p, q_I^*, t)$ at price p, and
2. The difference between the market demand $D(p')$ and the aggregate supply $S^I(p')$ of the innovative firms is positive.

Hence, in the long run period a share of non-innovative firms can co-exist with a share of innovative firms, even if the share of innovative firms is increasing.

There is a further reason to reinforce the above argument on the evolution of firms. *Innovative firms require skilled workers whereas non innovative firms prefer unskilled ones.*

4.6.1 Example

To illustrate this situation, consider firms characterized by the production function:

$$f(Z, X_S, X_{NS}) = k \mathscr{Z}^\alpha X_S^\beta + X_{NS}^\lambda, \tag{4.18}$$

$$k = \begin{cases} H & \text{if the firm is innovative} \\ h & \text{if the firm is not innovative,} \end{cases}$$

When $H > h > 0$ and α, β and λ are positive constants, such that $\alpha + \beta = 1$ and $\lambda < 1$. Note that the marginal output of the human capital is greater in the case of an innovative firm than in a non-innovative firm.

Assuming the technology $\mathscr{Z} = \bar{\mathscr{Z}}$ to be a positive constant, (the same for both kind of firms), the salary of a skilled worker is w_s and of an unskilled worker is w_{ns}. The innovative firms give a skill premia (the bonus for skilled workers) pr and give the premia (the bonus for non-skilled workers) $\bar{p}r$. Hence, for a innovative firm the short run cost function $C_I(\cdot, X_S) : R \to R$ is given by:

$$C_I(G, \bar{X}_S) = (w_s + pr)X_S + (w_{ns} + \bar{p}r)\left[G - H\bar{Z}^\alpha X_S^\beta\right]^{\frac{1}{\lambda}} \tag{4.19}$$

For a non-innovative firm the short run cost $C_{NI}(\cdot, X_S) : R \to R$ is given by

$$C_{NI}(G, \bar{X}_S) = w_s X_S + w_{ns}\left[G - h\bar{Z}^\alpha X_S^\beta\right]^{\frac{1}{\lambda}} \tag{4.20}$$

It follows that

4 Poverty Traps, Rationality and Evolution

$$\frac{\partial}{\partial G} C_I(G, \bar{X}_S) = (w_{ns} + \bar{p}r)\frac{1}{\lambda}\left[G - H\bar{Z}^\alpha X_S^\beta\right]^{\frac{1}{\lambda}-1},$$

$$\frac{\partial^2}{\partial X_S \partial G} C_I(G, \bar{X}_S) = -(w_{ns} + \bar{p}r)\left(\frac{1}{\lambda} - 1\right)\frac{1}{\lambda}[Y - H\bar{Z}^\alpha X_S^\beta]^{\frac{1}{\lambda}-2} H\bar{Z}^\alpha X_S^{\beta-1} < 0.$$

So we can conclude that the speed of the adjustment process of the short run cost to the long run cost, considering as given, for each time $t = t_0$, the skilled workers' supply $X_S(t_0)$, is faster for innovative firms than for non-innovative ones. If, at $t = t_0$, the fraction of innovative firms is higher than the threshold value π, then the supply of skilled workers increases with time. If the supply of skilled workers $X_S(t_0)$, at $t = t_0$ is below the optimal demand $X_S(t_0, G)$ of a innovative firm facing a level G of demand. Then, the short run cost of this firm decreases with time.

At the same time, the firms with more intensive utilization of the skilled workers increase the supply of its product and they can reduce the variable short run costs more quickly than the non innovative ones. Hence, this firms can obtain positive profits in the short run.

Assume, that the market price for the final product is given by p. If the market is competitive, the optimal supply for each kind of firms are, respectively, for innovative and non-innovative firms, equal to

$$\begin{aligned} Y_I^* &= pH\mathcal{Z}^\alpha X_{IS}^* + X_{INS}^* \\ Y_{NI}^* &= ph\mathcal{Z}^\alpha X_{NIS}^* + X_{NINS}^* \end{aligned} \tag{4.21}$$

where X_{iS}^* and X_{iNS}^*, $i \in \{I, NI\}$ stand for the long run demand for inputs from innovative and non innovative firms

$$X_{INS}^* = \left(\frac{w_{ns} + \bar{p}r}{\lambda p}\right)^{\frac{1}{\beta-1}}, \quad X_{NINS}^* = \left(\frac{w_{ns}}{\lambda p}\right)^{\frac{1}{\beta-1}},$$

$$X_{IS}^* = \left(\frac{w_s + pr}{\lambda pHZ^\alpha\beta}\right)^{\frac{1}{\beta-1}}, \quad X_{NIS}^* = \left(\frac{w_s}{\lambda phZ^\alpha\beta}\right)^{\frac{1}{\beta-1}}. \tag{4.22}$$

Let $PI > \pi$ be the number of innovative firms at $t = t_0$. By $D(p)$. We denote the total demand for the product, and by $S(p)$ the total supply. $S_I(p)$ is total supply of innovative firms and $S_{NI}(p)$ is the total supply of the non-innovative firms. So

$$\begin{aligned} S_I(p) &= (PI)Y_I^* \\ S_{NI}(p) &= S(p) - S_I(p) \end{aligned} \tag{4.23}$$

In equilibrium $S(p) = D(p)$, so the number of non innovative firms, at the same time, is equal to

$$\max\left\{\frac{D(p) - S_I(p)}{Y_{NI}^*}, 0\right\}.$$

Assume that there is a cost $C(h, H)$ to become innovative corresponding to transforming the technology h in a technology H. Thus, at time $t = t_0$ a non innovative firm has an incentive to change and become innovative if, and only if, the difference between the benefits of the innovative firm and the benefits of a non-innovative firm, verify the following inequality:

$$\Pi(I, t_0) - \Pi(NI, t_0) > C(h, H).$$

Then, in the long run, innovative and non-innovative firms may coexist if

$$\lim_{t_0 \to \infty} \Pi(I, t_0) - \Pi(NI, t_0) = \Pi(I) - \Pi(NI) > C(h, H).$$

This transformation process depends on the possible size of the reduction of the short run cost that an innovative firm can obtain, increasing the percentage of skilled worker engaged, but this size decrease at the same time that the short run cost converges to the long run one. So, even in the long run innovative and non-innovative firms may coexist.

4.7 Defining a Poverty Trap

In the context of this paper, we now introduce a definition of an economy in a poverty trap and a definition of poverty trap.

Consider a two populations normal form game Γ, where populations are workers and firms. Let Δ^F be the set of the distributions of firms over the set of pure strategies (or behaviors)$\{I, NI\}$ and let Δ^W be the set of the distributions of the workers population's over the set of their pure strategies $\{S, NS\}$. Suppose this game has a Pareto inefficient Nash equilibrium.

At the end of any period, workers have to decide their strategic behavior for the following one. Assume that they choose according to the rule of maximizing expected payoffs. Let $Y_F = (Y_I, Y_{NI}) \in \Delta^F$ be the distribution of the firms during the current period. Let Y_I stand for the percentage of innovative and Y_{NI} for the percentage of non innovative firms.

Definition 4.2. *(The economy is in a poverty trap)* Consider a two populations normal form game Γ of firms an workers. Assume that the initial distribution of the firms (the field) is given by $Y_F \in \Delta^F$. We say that the economy is in a poverty trap, if for every initial distribution (or mixed strategy) over the population of workers $X_W(t_0) = (X_S(t_0), X_{NS}(t_0)) \in \Delta^W$ the solution $\xi(t, X_W(t_0))$ of the dynamical system (4.9) define a trajectory $X_W(t) = (X_S(t), X_{NS}(t)) \in \Delta^W$, evolving as time $t \to \infty$, to a steady state X_W of this system. Furthermore, the steady state is an evolutionarily stable strategy against a field Y_F, and the pair (X_W, Y_F) is an inefficient (in the Pareto sense) Nash equilibrium.

4 Poverty Traps, Rationality and Evolution

Definition 4.3. *(poverty traps)* Let Γ be a two population normal form game. We denote the respective populations by W and F. We say that a Pareto inefficient Nash equilibrium $Z = (X_W, Y_F) \in \Delta^W \times \Delta^F$ is a poverty trap for the economy if

1. X_W is an evolutionary stable strategy against the field Y_F, and it is at the same time,
2. X_W is a local attractor for a dynamical system that represents the evolution of the economy, as the system (4.9) does.

So, an economy is in a poverty trap, if for any initial distribution of workers, every solution of the dynamical system representing the evolution of the economy, define a trajectory converging to a poverty trap. The fact that an economy is or not in a trap of poverty, depends exclusively of the distribution of the firms.

In our case, $X_W = e_{NS}$ is a local attractor and $(X_W, Y_F) = (e_{NS}, e_{NI}) = (0, 1; 0, 1)$ represents the poverty trap, where there are neither skilled workers nor innovative firms. The economy is in a poverty trap, if $PI < \pi$.

Therefore, the possibility to be in a poverty trap, for an economy of firms and workers, depends on the field and on such structural characteristics, as costs of education and incentives (or *prizes*) for skilled workers. Rationality alone, on the part of individual agents, may not only prove insufficient to avoid it, but may actually drive towards it. Thus, the only possibility for an economy to escape its fate, the poverty trap where it is heading, and to jump onto a trajectory towards a high equilibrium is to have its structural characteristics altered by intervention from outside. This can be the task of a benevolent central planner.

4.8 Conclusion

Our main conclusion is that poverty trap is the result of structural conditions that render rational for a worker not to acquire any skills. As the economy evolves in a trajectory leading to the poverty trap, more firms tend, progressively, to be non innovative and the absence of skilled workers induces to replace them with unskilled ones, who however perform better when employed in non innovative firms. A proactive policy maker wishing to help a less developed country to exit a poverty trap, can implement policies to reduce the key threshold value π that we identified, in such a way that the economy's trajectory falls into the basin of attraction of a high equilibrium. This policy's aim may be realized by reducing educational costs, or else by introducing incentives to innovative firms to increase prizes for skills. The closer a country gets to that threshold, the more growth-enhancing becomes any investment in education. On the other hand, policy differences are known to be among the determinants of the differences in the degrees of development across countries and over time. To avoid getting into a poverty trap, it is useless to appeal to rationality. Rational behavior may be a source of high growth if and only if initial conditions happen to lie outside the "wrong" basin of attraction.

Acknowledgements The author wishes to thank the support given by Programa Integral de Fortalecimiento Insititucional de la UASLP (PIFI) 2010.

References

1. Accinelli, E., London, S., Punzo, L., Sánchez Carrera, E.: Complementariedades, Dinámicas, Eficiencia y Equilibrio de Nash en un modelo de firmas y trabajadores. EconoQuantum **6**(1) (2010). (Forthcoming)
2. Ezell, Stephen J., Atkinson, Robert D.: RAND's Rose-Colored Glasses: How RAND's Report on U.S. Competitiveness in Science and Technology Gets it Wrong. The information technology & innovation foundation, pp. 1–23 (2008)
3. Lucas, R.: On the mechanics of economic development. J. Monet. Econ. **22**(1), 3–42 (1988)
4. Redding, S.: Low-skill, low quality trap: strategic complementarities between human capital and R&D. Econ. J. **106**, 458–470 (1996)
5. Schlag, K.: Why imitate, and if so, how? A boundedly rational approach to multi-armed bandit. J. Econ. Theory **78**, 130–156 (1998)
6. Sen, A.: Development as Freedom. Anchor Books, New York, US (1999)
7. Weibull, J.: Evolutionary Game Theory. MIT, Cambridge, USA (1995)

Chapter 5
Leadership Model

Leandro Almeida, José Cruz, Helena Ferreira, and Alberto A. Pinto

Abstract The Theory of Planned Behavior studies the decision-making mechanisms of individuals. We construct a game theoretical model to understand the role of leaders in decision-making of individuals or groups. We study the characteristics of the leaders that can have a positive or negative influence over others' behavioral decisions.

5.1 Introduction

The main goal in Planned Behavior or Reasoned Action Theories (see Ajzen [1], Baker [6]) is to understand and forecast how individuals turn intentions into behaviors. Almeida, Cruz, Ferreira and Pinto [5] created a game theoretical model, inspired by the works of Cownley and Wooders [7,8] where specific individual characteristics of the individuals, defined as taste type and crowding type are considered.

L. Almeida
Instituto de Educação, Universidade do Minho, Campus de Gualtar, 4710-057 Braga, Portugal
e-mail: leandro@iep.uminho.pt

J. Cruz
Escola de Psicologia, Universidade do Minho, Campus de Gualtar, 4710-057 Braga, Portugal
e-mail: jcruz@iep.uminho.pt

H. Ferreira (✉)
LIAAD-INESC Porto LA, Porto, Portugal
and
Centro de Matemática da Universidade do Minho, Campus de Gualtar, 4710-057 Braga, Portugal
e-mail: helenaisafer@gmail.com

A.A. Pinto
LIAAD-INESC Porto LA e Departamento de Matemática, Faculdade de Ciências, Universidade do Porto, Rua do Campo Alegre, 687, 4169-007, Porto, Portugal
and
Centro de Matemática e Departamento de Matemática e Aplicações, Escola de Ciências, Universidade do Minho, Campus de Gualtar, 4710-057 Braga, Portugal
e-mail: aapinto@fc.up.pt

M.M. Peixoto et al. (eds.), *Dynamics, Games and Science I*, Springer Proceedings in Mathematics 1, DOI 10.1007/978-3-642-11456-4_5,
© Springer-Verlag Berlin Heidelberg 2011

The taste type determines the inner characteristics of an individual specifying his welfare function. The crowding type of an individual determines his influence in the welfare function of the other individuals. Following the works of Driskel, Salas and Sternberg [9, 10, 13] on leadership, Almeida, Cruz, Ferreira and Pinto [2] presented a new game theoretical model to study the influence of the leaders. This chapter examines the theory of Planned Behavior from a game theoretical point of view and the leaders impact in individual/group decision-making [2, 3, 5]).

5.2 Theory of Planned Behavior or Reasoned Action

The Theory of Planned Behavior or Reasoned Action can be summarized in Fig. 5.1 (see Ajzen [1]), where we can observe that external variables are divided in three categories: intrapersonal associated to individual actions; interpersonal associated to the interaction of the individual with others and; sociocultural associated to social values. These external variables influence, especially, the intermediate variables which are also subdivided in three major categories: social norms, attitude, and self-efficacy. The social norms can be the opinions, conceptions and judgments that others have about a certain behavior attitudes are personal opinions in favor or against a specific behavior and self-efficacy is the extent of ability to control a certain behavior. These external and intermediate variables lead to a consequent intention to adopt a certain behavior.

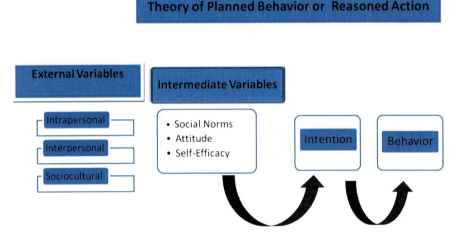

Fig. 5.1 Theory of Planned Behavior

5.3 Game Theoretical Model

Almeida et al. [5] constructed a game theoretical model, that we will describe. Let S denote the set of all individuals. For each individual $s \in S$, we distinguish two types of characteristics: *taste type* and *crowding type*.

We associate to each individual $s \in S$ one taste type $\mathscr{T}(s) = t \in T$ that describes the individual's inner characteristics, which are not always observable by the other individuals. We also associate to each individual $s \in S$ one crowding type $\mathscr{C}(s) = c \in C$ that describes the individual's characteristics observed by the others and that can influence the welfare of the others. In accordance with the Theory of Planned Behavior or Reasoned Action, we associate the intrapersonal external variables and the attitude and self-efficacy intermediate variables with the taste type and the interpersonal and sociocultural external variables and the social norms intermediate variable with the crowding type.

The individuals, with their own characteristics, can define a strategy $\mathscr{G} : S \to G$, i.e., each individual $s \in S$ chooses the behavior/group to which he would like to belong $\mathscr{G}(s)$ taking into account his taste type and the others crowding type. Each strategy $\mathscr{G}$ corresponds to an intention in the Theory of Planned Behavior. Given a behavior/group strategy $\mathscr{G} : S \to G$, the *crowding vector* $m(\mathscr{G}) \in (\mathbb{N}^C)^G$ is the vector whose components $m_c^g = m_c^g(\mathscr{G})$ are the number of individuals in g that have crowding type $c \in C$, i.e.

$$m_c^g = \#\{s \in S : \mathscr{G}(s) = g \wedge \mathscr{C}(s) = c\}.$$

We denote by $s_{t,c}$ the individual s with taste type t and crowding type c. The level of welfare, or personal satisfaction, that an individual $s_{t,c}$ acquires by belonging to a group/behavior $g \in G$ with *crowding vector* $m(\mathscr{G})$, is measured by the utility function $u_{t,c} : G \times (\mathbb{N}^C)^G \to \mathbb{R}$ defined by

$$u_{t,c}(g, m) = V_{t,c}^g + \sum_{c' \in \mathscr{C}} A_{t,c}^{g,c'} m_{c'}^g$$

where (i) $V_{t,c}^g$ measures the satisfaction level that each individual $s_{t,c}$ has in belonging to a group/behavior $g \in G$, (ii) $A_{t,c}^{g,c'}$ evaluates the satisfaction that each individual $s_{t,c}$ has with the presence of an individual with crowding type c' in g.

The group/behavior strategy $\mathscr{G}^* : S \to G$ is a *Nash Equilibrium group/behavior*, if given the choice options of all individuals, no individual feels motivated to change his behavior/group, i.e its utility does not increase by changing his behavior/group decision (see Pinto [12]).

The dictionary between the Game Theory and the Theory of Planned Behavior is summarized in Figure 5.2 (see Almeida [4]).

We denote by $S_{(t,c)}$ the group of all individuals $s_{t,c}$ with the same taste type $t \in T$ and the same crowding type $c \in C$. Let $n(t,c)$ be the number of individuals in $S_{(t,c)}$. An interesting way to interpret $S_{(t,c)}$ is to consider that $n(t,c)$ is

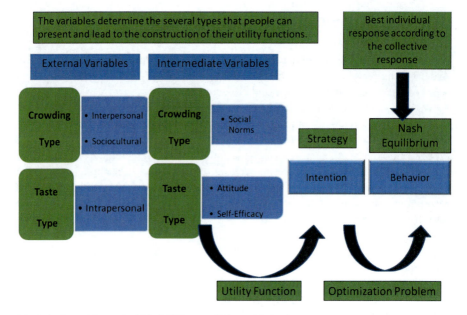

Fig. 5.2 Game Theoretical Model/Theory of Planned Behavior

the number of times that a single individual $s_{t,c}$ has to take an action. In this case, $A_{t,c}^{g,c} > 0$ can be interpreted as the individual positive reward by repeating the same group/behavior choice $c \in C$, i.e., the individual $s_{t,c}$ does not feel a saturation effect by repeating the same choice. On the other hand, $A_{t,c}^{g,c} < 0$ can be interpreted as the individual negative reward by repeating the same group/behavior choice $c \in C$, i.e., the individual $s_{t,c}$ feels a saturation, boredom or frustration effect by repeating the same choice.

5.4 Leadership in a Game Theoretical Model

A leader is an individual who can influence the others to choose a certain behavior/group. We consider that the leader makes his behavior/group decision before the others, and the others already know the leader's decision before taking their behavior/group decision. We study how the choice of the leader s_{t^l,c^l} can influence the followers s_{t^f,c^f} to choose the same behavior/group g as the leader, see [2, 3].

The leaders and the followers are characterized by the parameters (α, P, L) and we distinguish the following types:

- *Altruistic and individualist leaders.* The leader s_{t^l,c^l} values the behavior/group g and can donate a part P to the followers. The *altruistic leader* is the one who distributes a valuation to the followers of the behavior/group g, i.e. $P > 0$ and the *individualist leader* is the one who gives a devaluation or debt to the followers of the behavior/group g, i.e. $P < 0$.

5 Leadership Model

- *Consumption or wealth creation by the followers.* We define α as the parameter of the consumption or wealth creation on the valuation of the good distributed by the leader to the followers. Therefore, the new valuation of the followers s_{t^f,c^f} to choose the behavior/group g is given by

$$V^g_{t^f,c^f} = \bar{V}^g_{t^f,c^f} + \frac{\alpha P}{n(t^f,c^f)},$$

where $\bar{V}^g_{t^f,c^f}$ corresponds to the previous valuation of the followers to choose behavior/group g. There is *wealth creation by the followers* when $P > 0$ and $\alpha > 1$ or when $P < 0$ and $0 < \alpha < 1$. There is *wealth consumption by the followers* when $P > 0$ and $0 < \alpha < 1$ or when $P < 0$ and $\alpha > 1$.

- *Influential and persuasive leaders.* The influence or persuasiveness of the leaders s_{t^l,c^l} on the followers (t^f,c^f) is measured by the parameter L. We consider that

$$A^{g,c^l}_{t^f,c^f} = L$$

corresponds to the satisfaction that the followers have by choosing the same behavior/group as the leader. Alternatively, we consider that $A^{g,c^l}_{t^f,c^f} = 0$ and that the followers have a new valuation $V^{g'}_{t^f,c^f} = V^{g'}_{t^f,c^f} - L$ when they choose the behavior/group $g' \in G \setminus \{g\}$ under the influence of the leader. If $L < 0$, the followers do not like to choose the same behavior/group as the leader, but if $L > 0$, the followers like to choose the same behavior/group as the leader.

We define the *leader worst neighbors* $LWN_g(t^f,c^f)$ of the individual s_{t^f,c^f} in choosing the behavior/group g by:

$$LWN_g(t^f,c^f) = \begin{cases} A^{g,c^f}_{t^f,c^f} + \displaystyle\sum_{c' \in C, A^{g,c'}_{t^f,c^f} < 0} A^{g,c'}_{t^f,c^f} \sum_{t' \in T} n(t',c') & \text{if } A^{g,c^f}_{t^f,c^f} \geq 0 \\[2em] \displaystyle\sum_{c' \in C, A^{g,c'}_{t^f,c^f} < 0} A^{g,c'}_{t^f,c^f} \sum_{t' \in T} n(t',c') & \text{if } A^{g,c^f}_{t^f,c^f} < 0 \end{cases}$$

We define the *leader best neighbors* $LBN_g(t^f,c^f)$ of the individual s_{t^f,c^f} by:

$$LBN_g(t^f,c^f) = \begin{cases} \displaystyle\sum_{c' \in C, A^{g,c'}_{t^f,c^f} > 0} A^{g,c'}_{t^f,c^f} \sum_{t' \in T} n(t',c') & \text{if } A^{g,c^f}_{t^f,c^f} \geq 0 \\[2em] A^{g,c^f}_{t^f,c^f} + \displaystyle\sum_{c' \in C, A^{g,c'}_{t^f,c^f} > 0} A^{g,c'}_{t^f,c^f} \sum_{t' \in T} n(t',c') & \text{if } A^{g,c^f}_{t^f,c^f} < 0 \end{cases}$$

Let

$$gw = \arg \max_{\{g \in G\}} LWN_g(t^f,c^f)$$

and

$$LBN(t^f, c^f) = \max_{\{g \in G: g \neq g_W\}} LBN_g(t^f, c^f)$$

Lemma 5.1. *Let the leader s_{t^l, c^l} choose the behavior/group $g \in G$. If*

$$\frac{\alpha P}{n(t^f, c^f)} + L > LBN(t^f, c^f) - LWN_{g_W}(t^f, c^f)$$

then $\mathscr{G}^*(s_{t^f, c^f}) = g_W$, for every Nash equilibrium $\mathscr{G}^*$.

Inequality above gives a sufficient condition, in the value of the donated part P, in the influence and persuasiveness L of the leader and, also, in the creation or consumption of wealth α by the followers, implying that the followers choose the same behavior/group as the leader.

Lemma 5.1 is proved in [2].

5.5 Conclusion

We defined a dictionary between Game Theory and the Theory of Planned Behavior and we proposed the Nash equilibria as one of many, possible mechanisms of transforming individual intentions in decisions. In this game theoretical model, we studied how the characteristics of the leaders, can have a positive or negative influence over other individual's behavioral decisions. In particular, we show that an individualist leader might have to be more persuasive than an altruistic leader to convince the followers to choose his behavior/group.

Acknowledgements Previous versions of this work were presented in the International Congress of Mathematicians ICM 2010, the Second Brazilian Workshop of the Game Theory Society in honor of John Nash and EURO 2010, ICDEA 2009 and LAMES 2008. This work was highlighted in the article [11] after being presented in ICM 2010. We thank LIAAD-INESC Porto LA, Calouste Gulbenkian Foundation, PRODYN-ESF, POCTI and POSI by FCT and Ministério da Ciência e da Tecnologia, and the FCT Pluriannual Funding Program of the LIAAD-INESC Porto LA and of the Research Center of Mathematics of University of Minho, for their financial support.

References

1. Ajzen, I.: Perceived behavioral control, self-efficacy, locus of control, and the theory of planned behavior. J. Appl. Soc. Psychol. **32**, 665–683 (2002)
2. Almeida, L., Cruz, J., Ferreira, H., Pinto, A.A.: Experts and Leaders effects in the Theory of Planned Behavior (submitted)
3. Almeida, L., Cruz, J., Ferreira, H. and Pinto, A.A., Leaders and decision-making. Gazeta de Matemática (submitted).
4. Almeida, L., Cruz, J., Ferreira, H., Pinto, A.A.: Nash Equilibria in Theory of Reasoned Action. *AIP Conference Proceedings of the 6th International Conference of Numerical Analysis and Applied Mathematics*, Greece (2008)

5. Almeida, L., Cruz, J., Ferreira, H., Pinto, A.A.: Bayesian-Nash equilibria in theory of planned behavior. J. Differ. Equ. Appl., **17**(6), 61–69 (2011)
6. Baker, S., Beadnell, B., Gillmore, M., Morrison, D., Huang, B., Stielstra, S.: The theory of reasoned action and the role of external factors on heterosexual mens monogamy and condom use. J. Appl. Soc. Psychol. **38**(1), 97–134 (2008)
7. Cartwright, E., Wooders, M.: *On Equilibrium in Pure Strategies in Games with Many Players*, FEEM Working Paper 122 (2003)
8. Conley, J., Wooders, M.: Anonymous lindahl pricing in a tiebout economy with crowding types. Can. J. Econ. **31**(4), 952–974 (1998)
9. Daya, D., Gronnb, P., Salas, E.: Leadership capacity in teams. Leadership Q. **15**, 857–880 (2004)
10. Driskell, J., Salas, E.: The effect of content and demeanor on reactions to dominance behavior, group dynamics: theory, research and practice. Educ. Publishing Found., **9**(1), 3–14 (2005)
11. Mudur, G.S.: Maths for movies, medicine & markets. The Telegraph Calcutta, India, 20/09/2010
12. Pinto, A.A.: *Game Theory and Duopoly Models*. Interdisciplinary Applied Mathematics Series. Springer (to appear)
13. Sternberg, R., WICS: A model of leadership. The Psychologist-Manager J. **8**(1), 29–43 (2005)

Chapter 6
Lorenz-Like Chaotic Attractors Revised

Vítor Araújo and Maria José Pacifico

Abstract We describe some recent results on the dynamics of singular-hyperbolic (Lorenz-like) attractors Λ introduced in [26]: (1) there exists an invariant foliation whose leaves are forward contracted by the flow; (2) there exists a positive Lyapunov exponent at every orbit; (3) attractors in this class are expansive and so sensitive with respect to initial data; (4) they have zero volume if the flow is C^2, or else the flow is globally hyperbolic; (5) there is a unique physical measure whose support is the whole attractor and which is the equilibrium state with respect to the center-unstable Jacobian; (6) the hitting time associated to a geometric Lorenz attractor satisfies a logarithm law; (7) the rate of large deviations for the physical measure on the ergodic basin of a geometric Lorenz attractor is exponential.

6.1 Introduction

In this note M is a compact boundaryless 3-manifold and $\mathscr{X}^1(M)$ denotes the set of C^1 vector fields on M endowed with the C^1 topology. Moreover Leb denotes *volume* or *Lebesgue measure*: a normalized volume form given by some Riemannian metric on M. We also denote by dist the Riemannian distance on M.

V. Araújo

Instituto de Matemática, Universidade Federal do Rio de Janeiro, C. P. 68.530, 21.945-970, Rio de Janeiro, RJ, Brazil

and

Centro de Matemática da, Universidade do Porto, Rua do Campo Alegre 687, 4169-007 Porto, Portugal
e-mail: vitor.araujo@im.ufrj.br, vdaraujo@fc.up.pt

M.J. Pacifico
Instituto de Matemática, Universidade Federal do Rio de Janeiro, C. P. 68.530, 21.945-970 Rio de Janeiro, RJ, Brazil
e-mail: pacifico@im.ufrj.br, pacifico@impa.br

M.M. Peixoto et al. (eds.), *Dynamics, Games and Science I*, Springer Proceedings in Mathematics 1, DOI 10.1007/978-3-642-11456-4_6,
© Springer-Verlag Berlin Heidelberg 2011

The notion of singular hyperbolicity was introduced in [24, 26] where it was proved that any C^1 robustly transitive set for a 3-flow is either a singular hyperbolic attractor or repeller.

A compact invariant set Λ of a 3-flow $X \in \mathscr{X}^1(M)$ is an *attractor* if there exists a neighborhood U of Λ (its isolating neighborhood) such that

$$\Lambda = \bigcap_{t>0} X^t(U)$$

and there exists $x \in \Lambda$ such that $X(x) \neq \mathbf{0}$ and whose positive orbit $\{X^t(x) : t > 0\}$ is dense in Λ.

We say that a compact invariant subset is *singular hyperbolic* if all the singularities in Λ are hyperbolic, and the tangent bundle $T\Lambda$ decomposes in two complementary DX^t-invariant bundles $E^s \oplus E^{cu}$, where: E^s is one-dimensional and uniformly contracted by DX^t; E^{cu} is bidimensional, contains the flow direction, DX^t expands area along E^{cu} and $DX^t \mid E^{cu}$ dominates $DX^t \mid E^s$ (i.e., any eventual contraction in E^s is stronger than any possible contraction in E^{cu}), for all $t > 0$.

We note that the presence of an equilibrium together with regular orbits accumulating on it prevents any invariant set from being uniformly hyperbolic, see e.g. [13]. Indeed, in our 3-dimensional setting a compact invariant subset Λ is uniformly hyperbolic if the tangent bundle $T\Lambda$ decomposes in *three* complementary DX^t-invariant bundles $E^s \oplus E^X \oplus E^u$, each one-dimensional, E^X is the flow direction, E^s is uniformly contracted and E^u uniformly expanded by DX^t, $t > 0$. This implies the continuity of the splitting and the presence of a non-isolated equilibrium point in Λ leads to a discontinuity in the splitting dimensions.

In the study of the asymptotic behavior of orbits of a flow $X \in \mathscr{X}^1(M)$, a fundamental problem is to understand how the behavior of the tangent map DX determines the dynamics of the flow X_t. The main achievement along this line is the uniform hyperbolic theory: we have a complete description of the dynamics assuming that the tangent map has a uniformly hyperbolic structure since [13].

In the same vein, under the assumption of singular hyperbolicity, one can show that at each point there exists a strong stable manifold and that the whole set is foliated by leaves that are contracted by forward iteration. In particular this shows that any robust transitive attractor with singularities displays similar properties to those of the geometrical Lorenz model. It is also possible to show the existence of local central manifolds tangent to the central unstable direction. Although these central manifolds do not behave as unstable ones, in the sense that points on them are not necessarily asymptotic in the past, the expansion of volume along the central unstable two-dimensional direction enables us to deduce some remarkable properties.

We shall list some of these properties that give us a nice description of the dynamics of a singular hyperbolic attractor.

6.2 The Geometric Lorenz Attractor

Here we briefly recall the construction of the geometric Lorenz attractor [1, 16], that is the more representative example of a singular-hyperbolic attractor.

In 1963 the meteorologist Edward Lorenz published in the Journal of Atmospheric Sciences [27] an example of a parametrized polynomial system of differential equations

$$
\begin{aligned}
\dot{x} &= a(y - x) & a &= 10 \\
\dot{y} &= rx - y - xz & r &= 28 \\
\dot{z} &= xy - bz & b &= 8/3
\end{aligned}
\tag{6.1}
$$

as a very simplified model for thermal fluid convection, motivated by an attempt to understand the foundations of weather forecast.

The origin $\sigma = (0, 0, 0)$ is an equilibrium of saddle type for the vector field defined by (6.1) with real eigenvalues $\lambda_i, i \leq 3$ satisfying

$$
\lambda_2 < \lambda_3 < 0 < -\lambda_3 < \lambda_1.
\tag{6.2}
$$

(in this case $\lambda_1 \approx 11.83$, $\lambda_2 \approx -22.83$, $\lambda_3 = -8/3$).

Numerical simulations performed by Lorenz for an open neighborhood of the chosen parameters suggested that almost all points in phase space tend to a *chaotic attractor*, whose well known picture is presented in Fig. 6.1. The *chaotic feature* is the fact that trajectories converging to the attractor are *sensitive with respect to initial data*: trajectories of any two nearby points are driven apart under time evolution.

Lorenz's equations proved to be very resistant to rigorous mathematical analysis, and also presented serious difficulties to rigorous numerical study. Indeed, these two main difficulties are:

conceptual: The presence of an equilibrium point at the origin accumulated by regular orbits of the flow prevents this attractor from being hyperbolic [8].

numerical: The presence of an equilibrium point at the origin, implying that solutions slow down as they pass near the origin, which means unbounded return times and, thus, unbounded integration errors.

Moreover the attractor is *robust*, that is, the features of the limit set persist for all nearby vector fields. More precisely, if U is an isolating neighborhood of the attractor Λ for a vector field X, then Λ is *robustly transitive* if, for all vector fields Y which are C^1 close to X, the corresponding Y-invariant set

$$
\Lambda_Y(U) = \bigcup_{t > 0} Y^t(U)
$$

also admits a dense positive Y-orbit. We remark that *the persistence of transitivity*, that is, the fact that, for all nearby vector fields, the corresponding limit set is transitive, *implies a dynamical characterization of the attractor*, as we shall see.

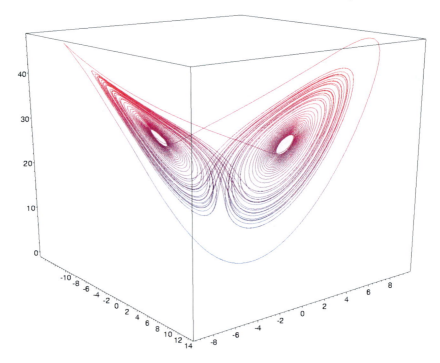

Fig. 6.1 A view of the Lorenz attractor calculated numerically

These difficulties led, in the seventies, to the construction of a geometric flow presenting a similar behavior as the one generated by (6.1). Nowadays this model is known as *geometric Lorenz flow*. Next we briefly describe this construction, see [1, 16] for full details.

We start by observing that under some non-resonance conditions, by the results of Sternberg [31], in a neighborhood of the origin, which we assume to contain the cube $[-1, 1]^3 \subset \mathbb{R}^3$, the Lorenz equations are equivalent to the linear system $(\dot{x}, \dot{y}, \dot{z}) = (\lambda_1 x, \lambda_2 y, \lambda_3 z)$ through smooth conjugation, thus

$$X^t(x_0, y_0, z_0) = (x_0 e^{\lambda_1 t}, y_0 e^{\lambda_2 t}, z_0 e^{\lambda_3 t}), \tag{6.3}$$

where $\lambda_1 \approx 11.83$, $\lambda_2 \approx -22.83$, $\lambda_3 = -8/3$ and $(x_0, y_0, z_0) \in \mathbb{R}^3$ is an arbitrary initial point near $(0, 0, 0)$.

Consider $S = \{(x, y, 1) : |x| \leq 1/2, \quad |y| \leq 1/2\}$ and

$$S^- = \{(x, y, 1) \in S : x < 0\}, \qquad S^+ = \{(x, y, 1) \in S : x > 0\} \quad \text{and}$$
$$S^* = S^- \cup S^+ = S \setminus \ell, \qquad \text{where} \quad \ell = \{(x, y, 1) \in S : x = 0\}.$$

Assume that S is a global transverse section to the flow so that every trajectory eventually crosses S in the direction of the negative z axis.

Consider also $\Sigma = \{(x, y, z) : |x| = 1\} = \Sigma^- \cup \Sigma^+$ with $\Sigma^\pm = \{(x, y, z) : x = \pm 1\}$.

For each $(x_0, y_0, 1) \in S^*$ the time τ such that $X^\tau(x_0, y_0, 1) \in \Sigma$ is given by

$$\tau(x_0) = -\frac{1}{\lambda_1} \log |x_0|,$$

which depends on $x_0 \in S^*$ only and is such that $\tau(x_0) \to +\infty$ when $x_0 \to 0$. This is one of the reasons many standard numerical algorithms were unsuited to tackle the Lorenz system of equations. Hence we get (where $\mathrm{sgn}(x) = x/|x|$ for $x \neq 0$)

$$X^\tau(x_0, y_0, 1) = \left(\mathrm{sgn}(x_0), y_0 e^{\lambda_2 \tau}, e^{\lambda_3 \tau}\right) = \left(\mathrm{sgn}(x_0), y_0|x_0|^{-\frac{\lambda_2}{\lambda_1}}, |x_0|^{-\frac{\lambda_3}{\lambda_1}}\right). \tag{6.4}$$

Since $0 < -\lambda_3 < \lambda_1 < -\lambda_2$, we have $0 < \alpha = -\frac{\lambda_3}{\lambda_1} < 1 < \beta = -\frac{\lambda_2}{\lambda_1}$. Let $L : S^* \to \Sigma$ be such that $L(x, y) = (y|x|^\beta, |x|^\alpha)$ with the convention that $L(x, y) \in \Sigma^+$ if $x > 0$ and $L(x, y) \in \Sigma^-$ if $x < 0$. It is easy to see that $L(S^\pm)$ has the shape of a triangle without the vertex $(\pm 1, 0, 0)$. In fact the vertex $(\pm 1, 0, 0)$ are cusp points at the boundary of each of these sets. The fact that $0 < \alpha < 1 < \beta$ together with (6.4) imply that $L(\Sigma^\pm)$ are uniformly compressed in the y-direction.

From now on we denote by $\Sigma^\pm$ the closure of $L(S^\pm)$. Clearly each line segment $S^* \cap \{x = x_0\}$ is taken to another line segment $\Sigma \cap \{z = z_0\}$ as sketched in Fig. 6.2.

The sets $\Sigma^\pm$ should return to the cross section S through a composition of a translation T, an expansion E only along the x-direction and a rotation R around $W^s(\sigma_1)$ and $W^s(\sigma_2)$, where σ_i are saddle-type singularities of X^t that are outside the cube $[-1, 1]^3$, see [8]. We assume that this composition takes line segments $\Sigma \cap \{z = z_0\}$ into line segments $S \cap \{x = x_1\}$ as sketched in Fig. 6.2. The composition $T \circ E \circ R$ of linear maps describes a vector field V in a region outside $[-1, 1]^3$. The geometric Lorenz flow X^t is then defined in the following way: for each $t \in \mathbb{R}$ and

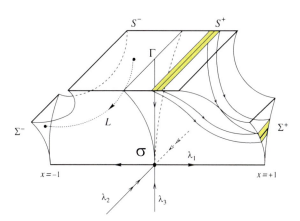

Fig. 6.2 Behavior near the origin

each point $x \in S$, the orbit $X^t(x)$ will start following the linear field until $\tilde{\Sigma}^{\pm}$ and then it will follow V coming back to S and so on. Let us write $\mathscr{B} = \{X^t(x), x \in S, t \in \mathbb{R}^+\}$ the set where this flow acts. The geometric Lorenz flow is then the pair $(\mathscr{B}, X^t)$ defined in this way. The set

$$\Lambda = \cap_{t \geq 0} X^t(S)$$

is the *geometric Lorenz attractor*.

We remark that the existence of a chaotic attractor for the original Lorenz system was established by Tucker with the help of a computer aided proof (see [32]).

The combined effects of $T \circ E \circ R$ and the linear flow given by (6.4) on lines implies that the foliation $\mathscr{F}^s$ of S given by the lines $S \cap \{x = x_0\}$ is invariant under the first return map $F : S \to S$. In another words, we have

($\star$) *for any given leaf γ of $\mathscr{F}^s$, its image $F(\gamma)$ is contained in a leaf of $\mathscr{F}^s$.*

The main features of the geometric Lorenz flow and its first return map can be seen at Figs. 6.3 and 6.4.

The one-dimensional map f is obtained quotienting over the leaves of the stable foliation $\mathscr{F}^s$ defined before (Fig. 6.5).

For a detailed construction of a geometric Lorenz flow see [8, 15].

As mentioned above, a geometric Lorenz attractor is the most representative example of a singular-hyperbolic attractor [24].

6.3 The Dynamical Results

The study of robust attractors is inspired by the Lorenz flow example. Next we list the main dynamical properties of a robust attractor.

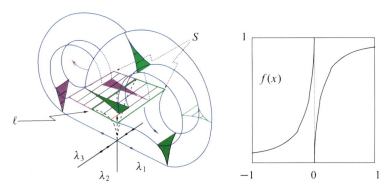

Fig. 6.3 The global cross-section for the geometric Lorenz flow and the associated 1d quotient map, the Lorenz transformation

Fig. 6.4 The image $F(S^*)$

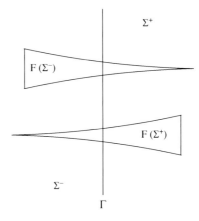

Fig. 6.5 Projection on I

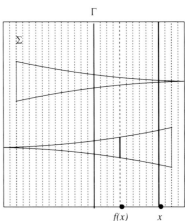

6.3.1 Robustness and Singular-Hyperbolicity

Inspired by the Lorenz flow example we define an equilibrium σ of a flow X^t to be *Lorenz-like* if the eigenvalues $\lambda_1, \lambda_2, \lambda_3$ of $DX(\sigma)$ are real and satisfy the relation at (6.2):
$$\lambda_2 < \lambda_3 < 0 < -\lambda_3 < \lambda_1.$$

These are the equilibria contained in robust attractors naturally, since they are the only kind of equilibria in a 3-flow which cannot be perturbed into saddle-connections which generate sinks or sources when unfolded.

Theorem 6.1. *Let Λ be a robustly transitive set of $X \in \mathscr{X}^1(M)$. Then, either for $Y = X$ or $Y = -X$, every singularity $\sigma \in \Lambda$ is Lorenz-like for Y and satisfies $W_Y^{ss}(\sigma) \cap \Lambda = \{\sigma\}$.*

The fact that a robust attractor does not admit sinks or sources for all nearby vector fields in its isolating neighborhood has several other strong consequences (whose study was pioneered by R. Mañé in its path to solve the Stability Conjecture in [21]) which enable us to show the following, see [26].

Theorem 6.2. *A robustly transitive set for $X \in \mathcal{X}^1(M)$ is a singular-hyperbolic attractor for X or for $-X$.*

The following shows in particular that the notion of singular hyperbolicity is an extension of the notion of uniform hyperbolicity.

Theorem 6.3. *Let Λ be a singular hyperbolic compact set of $X \in \mathcal{X}^1(M)$. Then any invariant compact set $\Gamma \subset \Lambda$ without singularities is uniformly hyperbolic.*

A consequence of Theorem 6.3 is that every periodic orbit of a singular hyperbolic set is hyperbolic. The existence of a periodic orbit in every singular-hyperbolic attractor was proved recently in [12] and also a more general result was obtained in [10].

Theorem 6.4. *Every singular hyperbolic attractor Λ has a dense subset of periodic orbits.*

In the same work [10] it was announced that every singular hyperbolic attractor is the homoclinic class associated to one of its periodic orbits. Recall that the *homoclinic class* of a periodic orbit $\mathcal{O}$ for X is the closure of the set of transversal intersection points of it stable and unstable manifold: $H(\mathcal{O}) = \overline{W^u(\mathcal{O}) \pitchfork W^s(\mathcal{O})}$. This result is well known for the elementary dynamical pieces of uniformly hyperbolic attractors. Moreover, in particular, the geometric Lorenz attractor is a homoclinic class as proved in [11].

6.3.2 Singular-Hyperbolicity and Chaotic Behavior

Using the area expansion along the bidimensional central direction, which contains the direction of the flow, one can show

Theorem 6.5. *Every orbit in any singular-hyperbolic attractor has a direction of exponential divergence from nearby orbits (positive Lyapunov exponent).*

Denote by $S(\mathbb{R})$ the set of surjective increasing continuous real functions $h : \mathbb{R} \to \mathbb{R}$ endowed with the C^0 topology. The flow X_t is *expansive* on an invariant compact set Λ if for every $\epsilon > 0$ there is $\delta > 0$ such that if for some $h \in S(\mathbb{R})$ and $x, y \in \Lambda$

$$\text{dist}(X_t(x), X_{h(t)}y) \leq \delta \quad \text{for all} \quad t \in \mathbb{R},$$

then $X_{h(t_0)}(y) \in X_{[t_0-\epsilon, t_0+\epsilon]}(x)$, for some $t_0 \in \mathbb{R}$. A stronger notion of expansiveness was introduced by Bowen–Ruelle [13] for uniformly hyperbolic attractors, but equilibria in expansive sets under this strong notion must be isolated, see e.g. [8].

6 Lorenz-Like Chaotic Attractors Revised

Komuro proved in [17] that a geometrical Lorenz attractor Λ is expansive. In particular, this implies that this kind of attractor is *sensitive with respect to initial data*, i.e., there is $\delta > 0$ such that for any pair of distinct points $x, y \in \Lambda$, if $\mathrm{dist}(X_t(x), X_t(y)) < \delta$ for all $t \in \mathbb{R}$, then x is in the orbit of y. In [9] this was fully extended to the singular-hyperbolic setting.

Theorem 6.6. *Let Λ be a singular hyperbolic attractor of $X \in \mathscr{X}^1(M)$. Then Λ is expansive.*

Corollary 6.1. *Singular hyperbolic attractors are sensitive with respect to initial data.*

6.3.3 Singular-Hyperbolicity, Positive Volume and Global Hyperbolicity

Recently a generalization of the results of Bowen–Ruelle [13] was obtained in [2] showing that a uniformly hyperbolic transitive subset of saddle-type for a $C^{1+\alpha}$ flow has zero volume, for any $\alpha > 0$. We denote the family of all flows whose differentiability class is at least Hölder-C^1 by C^{1+}.

Theorem 6.7. *A C^{1+} singular-hyperbolic attractor has zero volume.*

This can be extended to the following dichotomy. Recall that a *transitive Anosov vector field X* is a vector field without singularities such that the entire manifold M is a uniformly hyperbolic set of saddle-type.

Theorem 6.8. *Let Λ be a singular hyperbolic attractor for a C^{1+} 3-dimensional vector field X. Then either Λ has zero volume or X is a transitive Anosov vector field.*

6.4 The Ergodic Theory of Singular-Hyperbolic Attractors

The ergodic theory of singular-hyperbolic attractors is incomplete. Most results still are proved only in the particular case of the geometric Lorenz flow, which automatically extends to the original Lorenz flow after the work of Tucker [32], but demand an extra effort to encompass the full singular-hyperbolic setting.

6.4.1 Existence of a Physical Measure

Another main result obtained in [9] is that typical orbits in the basin of every singular-hyperbolic attractor, for a C^2 flow X on a 3-manifold, have well-defined statistical behavior, i.e., for Lebesgue almost every point the forward Birkhoff time average converges, and it is given by a certain physical probability measure. It was also obtained that this measure admits absolutely continuous conditional measures along the center-unstable directions on the attractor. As a consequence, it is a u-Gibbs state and an equilibrium state for the flow.

Theorem 6.9. *A C^2 singular-hyperbolic attractor Λ admits a unique ergodic physical hyperbolic invariant probability measure μ whose basin covers Lebesgue almost every point of a full neighborhood of Λ.*

Recall that an invariant probability measure μ for a flow X is *physical* (or *SRB*) if its *basin*

$$B(\mu) = \left\{ x \in M : \lim_{T \to \infty} \frac{1}{T} \int_0^T \psi(X_t(x)) \, dt = \int \psi \, d\mu, \forall \psi \in C^0(M, \mathbb{R}) \right\}$$

has positive volume in M.

Here hyperbolicity means *non-uniform hyperbolicity* of the probability measure μ: the tangent bundle over Λ splits into a sum $T_z M = E_z^s \oplus E_z^X \oplus F_z$ of three one-dimensional invariant subspaces defined for μ-a.e. $z \in \Lambda$ and depending measurably on the base point z, where μ is the physical measure in the statement of Theorem 6.9, E_z^X is the flow direction (with zero Lyapunov exponent) and F_z is the direction with positive Lyapunov exponent, that is, for every non-zero vector $v \in F_z$ we have

$$\lim_{t \to +\infty} \frac{1}{t} \log \|DX_t(z) \cdot v\| > 0.$$

We note that the invariance of the splitting implies that $E_z^{cu} = E_z^X \oplus F_z$ whenever F_z is defined.

Theorem 6.9 is another statement of sensitiveness, this time applying to the whole essentially open set $B(\Lambda)$. Indeed, since non-zero Lyapunov exponents express that the orbits of infinitesimally close-by points tend to move apart from each other, this theorem means that most orbits in the basin of attraction separate under forward iteration. See Kifer [18], and Metzger [23], and references therein, for previous results about invariant measures and stochastic stability of the geometric Lorenz models.

The u-Gibbs property of μ is stated as follows.

Theorem 6.10. *Let Λ be a singular-hyperbolic attractor for a C^2 three-dimensional flow. Then the physical measure μ supported in Λ has a disintegration into absolutely continuous conditional measures μ_γ along center-unstable surfaces γ such that $\frac{d\mu_\gamma}{dm_\gamma}$ is uniformly bounded from above. Moreover $\mathrm{supp}(\mu) = \Lambda$.*

Here the existence of unstable manifolds is guaranteed by the hyperbolicity of the physical measure: the strong-unstable manifolds $W^{uu}(z)$ are the "integral manifolds" in the direction of the one-dimensional sub-bundle F, tangent to F_z at almost

6 Lorenz-Like Chaotic Attractors Revised

every $z \in \Lambda$. The sets $W^{uu}(z)$ are embedded sub-manifolds in a neighborhood of z which, in general, depend only measurably (including its size) on the base point $z \in \Lambda$. The *strong-unstable manifold* is defined by

$$W^{uu}(z) = \{y \in M : \lim_{t \to -\infty} \operatorname{dist}(X_t(y), X_t(z)) = 0\}$$

and exists for almost every $z \in \Lambda$ with respect to the physical and hyperbolic measure obtained in Theorem 6.9. We remark that since Λ is an attracting set, then $W^{uu}(z) \subset \Lambda$ whenever defined. The central unstable surfaces mentioned in the statement of Theorem 6.10 are just small strong-unstable manifolds carried by the flow, which are tangent to the central-unstable direction E^{cu}.

The absolute continuity property along the center-unstable sub-bundle given by Theorem 6.10 ensures that

$$h_\mu(X^1) = \int \log \left| \det(DX^1 \mid E^{cu}) \right| d\mu,$$

by the characterization of probability measures satisfying the Entropy Formula, obtained in [20]. The above integral is the sum of the positive Lyapunov exponents along the sub-bundle E^{cu} by Oseledets Theorem [22,33]. Since in the direction E^{cu} there is only one positive Lyapunov exponent along the one-dimensional direction F_z, μ-a.e. z, the ergodicity of μ then shows that the following is true.

Corollary 6.2. *If Λ is a singular-hyperbolic attractor for a C^2 three-dimensional flow X^t, then the physical measure μ supported in Λ satisfies the Entropy Formula*

$$h_\mu(X^1) = \int \log \|DX^1 \mid F_z\| d\mu(z).$$

Again by the characterization of measures satisfying the Entropy Formula, we get that μ *has absolutely continuous disintegration along the strong-unstable direction*, along which the Lyapunov exponent is positive, thus μ *is a u-Gibbs state* [29]. This also shows that μ *is an equilibrium state for the potential* $- \log \|DX^1 \mid F_z\|$ with respect to the diffeomorphism X^1. We note that the entropy $h_\mu(X^1)$ of X^1 is the entropy of the flow X^t with respect to the measure μ [33].

Hence we are able to extend most of the basic results on the ergodic theory of hyperbolic attractors to the setting of singular-hyperbolic attractors.

6.4.2 Hitting and Recurrence Time Versus Local Dimension for Geometric Lorenz Flows

Given $x \in M$, let $B_r(x) = \{y \in M; d(x, y) \le r\}$ be the ball centered at x with radius r. The *local dimension* of μ at $x \in M$ is defined by

$$d_\mu(x) = \lim_{r \to \infty} \frac{\log \mu(B_r(x))}{\log r}$$

if this limit exists. In this case $\mu(B_r(x)) \sim r^{d_\mu(x)}$.

This notion characterizes the local geometric structure of an invariant measure with respect to the metric in the phase space of the system, see [35] and [28].

The existence of the local dimension for a Borel probability measure μ on M implies the crucial fact that virtually *all* the known characteristics of dimension type of the measure coincide. The common value is a fundamental characteristic of the fractal structure of μ, see [28].

Let $x_0 \in \mathbb{R}^3$ and

$$\tau_r^{X^t}(x, x_0) = \inf\{t \geq 0 \mid X^t(x) \in B_r(x_0)\}$$

be the time needed for the X-orbit of a point x to enter for the *first time* in a ball $B_r(x_0)$. The number $\tau_r^{X^t}(x, x_0)$ is the *hitting time associated to* the flow X^t and $B_r(x_0)$. If the orbit X^t starts at x_0 itself and we consider the second entrance time in the ball

$$\tau_r'(x_0) = \inf\{t \in \mathbb{R}^+ : X^t(x_0) \in B_r(x_0), \exists i, s.t. X^i(x_0) \notin B_r(x_0)\}$$

we have a quantitative recurrence indicator, and the number $\tau_r'(x_0)$ is called the *recurrence time associated to* the flow X^t and $B_r(x_0)$.

Now let X^t be a geometric Lorenz flow, and μ its X^t-invariant SBR measure. The main result in [15] establishes the following.

Theorem 6.11. *For μ-almost every x,*

$$\lim_{r \to 0} \frac{\log \tau_r(\mathbf{x}, \mathbf{x_0})}{-\log \mathbf{r}} = \mathbf{d}_\mu(\mathbf{x_0}) - \mathbf{1}.$$

Observe that the result above indicates once more the chaoticity of a Lorenz-like attractor: it shows that asymptotically, such attractors behave as an i.d. system.

We can always define the *upper* and the *lower* local dimension at x as

$$d_\mu^+(x) = \limsup_{r \to \infty} \frac{\log \mu(B_r(x))}{\log r}, \quad d_\mu^-(x) = \liminf_{r \to \infty} \frac{\log \mu(B_r(x))}{\log r}.$$

If $d^+(x) = d^-(x) = d$ almost everywhere the system is called *exact dimensional*. In this case many properties of dimension of a measure coincide. In particular, d is equal to the dimension of the measure μ: $d = \inf\{\dim_H Z; \mu(Z) = 1\}$. This happens in a large class of systems, for example, in C^2 diffeomorphisms having non zero Lyapunov exponents almost everywhere, [28].

Using a general result proved in [30] it is also proved in [15] a quantitative recurrence bound for the Lorenz geometric flow:

6 Lorenz-Like Chaotic Attractors Revised

Theorem 6.12. *For a geometric Lorenz flow it holds*

$$\liminf_{r \to 0} \frac{\log \tau_r'(x)}{-\log r} = d_\mu^- - 1, \qquad \limsup_{r \to 0} \frac{\log \tau_r'(x)}{-\log r} = d_\mu^+ - 1, \qquad \mu - a.e..$$

where τ' is the recurrence time for the flow, as defined above.

The proof of Theorem 6.11 is based on the following results, proved in [15].
Let $F : S \to S$ be the first return map to S, a global cross section to X^t through $W^s(p)$, p the singularity at the origin, as indicated at Fig. 6.3. It follows that F has a physical measure μ_F, see e.g. [34]. Recall that we say the system (S, F, μ_F) has exponential decay of correlation for Lipschitz observables if there are constants $C > 0$ and $\lambda > 0$, depending only on the system such that for each n it holds

$$\left| \int g(F^n(x)) f(x) d\mu - \int g(x) d\mu \int f(x) d\mu \right| \leq C \cdot e^{-\lambda n}$$

for any Lipschitz observable g and f with bounded variation,

Theorem 6.13. *Let μ_F an invariant physical measure for F. The system (S, F, μ_F) has exponential decay of correlations with respect to Lipschitz observables.*

We remark that a sub-exponential bound for the decay of correlation for a two dimensional Lorenz like map was given in [14] and [1].

Theorem 6.14. *μ_F is exact, that is, $d_{\mu_F}(x)$ exist almost every $x \in S$.*

Let $x_0 \in S$ and $\tau_r^S(x, x_0)$ be the time needed to $\mathcal{O}_x$ enter for the first time in $B_r(x_0) \cap S = B_{r,S}$.

Theorem 6.15. $\lim_{r \to 0} \frac{\log \tau_r(x,x_0)}{-\log r} = \lim_{r \to 0} \frac{\log \tau_r^S(x,x_0)}{-\log r} = d_{\mu_F}(x_0).$

From the fact that the attractor is a suspension of the support of μ_F we easily deduce the following.

Theorem 6.16. $d_\mu(x) = d_{\mu_F}(x) + 1.$

We remark that the results in this section can be extended to a more general class of flows described in [15]. The interested reader can find the detailed proofs in this article.

6.4.3 Large Deviations for the Physical Measure on a Geometric Lorenz Flow

Having shown that physical probability measures exist, it is natural to consider the rate of convergence of the time averages to the space average, measured by the volume of the subset of points whose time averages stay away from the space average by a prescribed amount up to some evolution time. We extend part of the results on large deviation rates of Kifer [19] from the uniformly hyperbolic setting to semi-flows over non-uniformly expanding base dynamics and unbounded roof function. These special flows model non-uniformly hyperbolic flows like the Lorenz flow, exhibiting equilibria accumulated by regular orbits.

6.4.3.1 Suspension Semiflows

We first present these flows and then state the main assumptions related to the modelling of the geometric Lorenz attractor.

Given a Hölder-C^1 local diffeomorphism $f : M \setminus \mathscr{S} \to M$ outside a volume zero non-flat[1] singular set $\mathscr{S}$, let $X^t : M_r \to M_r$ be a *semiflow with roof function* $r : M \setminus \mathscr{S} \to \mathbb{R}$ *over the base transformation* f, as follows.

Set $M_r = \{(x, y) \in M \times [0, +\infty) : 0 \leq y < r(x)\}$ and X^0 the identity on M_r, where M is a compact Riemannian manifold. For $x = x_0 \in M$ denote by x_n the nth iterate $f^n(x_0)$ for $n \geq 0$. Denote $S_n^f \varphi(x_0) = \sum_{j=0}^{n-1} \varphi(x_j)$ for $n \geq 1$ and for any given real function φ. Then for each pair $(x_0, s_0) \in X_r$ and $t > 0$ there exists a unique $n \geq 1$ such that $S_n r(x_0) \leq s_0 + t < S_{n+1} r(x_0)$ and define (see Fig. 6.6)

$$X^t(x_0, s_0) = (x_n, s_0 + t - S_n r(x_0)).$$

The study of suspension (or special) flows is motivated by modeling a flow admitting a cross-section. Such flow is equivalent to a suspension semiflow over the Poincaré return map to the cross-section with roof function given by the return time function on the cross-section. This is a main tool in the ergodic theory of uniformly hyperbolic flows developed by Bowen and Ruelle [13].

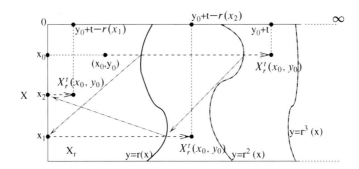

Fig. 6.6 The equivalence relation defining the suspension flow of f over the roof function r

---

[1] f behaves like a power of the distance to $\mathscr{S}$: $\|Df(x)\| \approx \text{dist}(x, \mathscr{S})^{-\beta}$ for some $\beta > 0$ (see Alves–Araújo [3,4] for a precise statement).

6.4.3.2 Conditions on the Base Dynamics

We assume that the singular set $\mathscr{S}$ (containing the points where f is either *not defined, discontinuous* or *not differentiable*) is regular, e.g., a submanifold of M, and that f is *non-uniformly expanding*: there exists $c > 0$ such that for Lebesgue almost every $x \in M$

$$\limsup_{n \to +\infty} \frac{1}{n} S_n \psi(x) \leq -c \quad \text{where} \quad \psi(x) = \log \left\| Df(x)^{-1} \right\|.$$

Moreover we assume that f has *exponentially slow recurrence to the singular set $\mathscr{S}$* i.e. for all $\epsilon > 0$ there is $\delta > 0$ s.t.

$$\limsup_{n \to +\infty} \frac{1}{n} \log \text{Leb} \left\{ x \in M : \frac{1}{n} S_n \left| \log d_\delta(x, \mathscr{S}) \right| > \epsilon \right\} < 0,$$

where $d_\delta(x, y) = \text{dist}(x, y)$ if $\text{dist}(x, y) < \delta$ and $d_\delta(x, y) = 1$ otherwise.

These conditions ensure [5] in particular the existence of finitely many ergodic absolutely continuous (in particular *physical*) f-invariant probability measures $\mu_1, \ldots, \mu_k$ whose basins cover the manifold Lebesgue almost everywhere.

We say that an f-invariant measure μ is an *equilibrium state* with respect to the potential $\log J$, where $J = |\det Df|$, if $h_\mu(f) = \mu(\log J)$, that is if μ *satisfies the Entropy Formula*. Denote by $\mathbb{E}$ the family of all such equilibrium states. It is not difficult to see that each physical measure in our setting belongs to $\mathbb{E}$.

We assume that $\mathbb{E}$ *is formed by a unique absolutely continuous probability measure*.

6.4.3.3 Conditions on the Roof Function

We assume that $r : M \setminus \mathscr{S} \to \mathbb{R}^+$ has *logarithmic growth near $\mathscr{S}$*: there exists $K = K(\varphi) > 0$ such that[2] $r \cdot \chi_{B(\mathscr{S}, \delta)} \leq K \cdot \left| \log d_\delta(x, \mathscr{S}) \right|$ for all small enough $\delta > 0$. We also assume that r is bounded from below by some $r_0 > 0$.

Now we can state the result on large deviations.

Theorem 6.17. *Let X^t be a suspension semiflow over a non-uniformly expanding transformation f on the base M, with roof function r, satisfying all the previously stated conditions.*

Let $\psi : M_r \to \mathbb{R}$ be continuous, $\nu = \mu \times \text{Leb}^1$ be the induced invariant measure[3] for the semiflow X^t and $\lambda = \text{Leb} \times \text{Leb}^1$ be the natural extension of volume to the space M_r. Then

---

[2] $B(\mathscr{S}, \delta)$ is the δ-neighborhood of $\mathscr{S}$.

[3] for any $A \subset M_r$ set $\nu(A) = \mu(r)^{-1} \int d\mu(x) \int_0^{r(x)} ds \, \chi_A(x, s)$.

$$\limsup_{T \to \infty} \frac{1}{T} \log \lambda \left\{ z \in M_r : \left| \frac{1}{T} \int_0^T \psi \left(X^t(z) \right) dt - \nu(\psi) \right| > \epsilon \right\} < 0.$$

6.4.3.4 Consequences for the Lorenz flow

Now consider a Lorenz geometric flow as constructed in Sect. 6.2 and let f be the one-dimensional map associated, obtained quotienting over the leaves of the stable foliation, see Fig. 6.3. This map has all the properties stated previously for the base transformation. The Poincaré return time gives also a roof function with logarithmic growth near the singularity line.

The uniform contraction along the stable leaves implies that the *time averages of two orbits on the same stable leaf under the first return map are uniformly close* for all big enough iterates. If $P : S \to [-1, 1]$ is the projection along stable leaves

Lemma 6.1. *For* $\varphi : U \supset \Lambda \to \mathbb{R}$ *continuous and bounded,* $\epsilon > 0$ *and* $\varphi(x) = \int_0^{r(x)} \psi(x, t) \, dt$, *there exists* $\zeta : [-1, 1] \setminus \mathscr{S} \to \mathbb{R}$ *with logarithmic growth near* $\mathscr{S}$ *such that* $\left\{ \left| \frac{1}{n} S_n^R \varphi - \mu(\varphi) \right| > 2\epsilon \right\}$ *is contained in*

$$P^{-1}\left(\left\{ \left| \frac{1}{n} S_n^f \zeta - \mu(\zeta) \right| > \epsilon \right\} \cup \left\{ \frac{1}{n} S_n^f | \log \mathrm{dist}_\delta(y, \mathscr{S})| > \epsilon \right\} \right).$$

Hence in this setting it is enough to study the quotient map f to get information about deviations for the Poincaré return map. Coupled with the main result we are then able to deduce

Corollary 6.3. *Let* X^t *be a flow on* $\mathbb{R}^3$ *exhibiting a Lorenz or a geometric Lorenz attractor with trapping region* U. *Denoting by* Leb *the normalized restriction of the Lebesgue volume measure to* U, $\psi : U \to \mathbb{R}$ *a bounded continuous function and* μ *the unique physical measure for the attractor, then for any given* $\epsilon > 0$

$$\limsup_{T \to \infty} \frac{1}{T} \log \mathrm{Leb} \left\{ z \in U : \left| \frac{1}{T} \int_0^T \psi \left(X^t(z) \right) dt - \mu(\psi) \right| > \epsilon \right\} < 0.$$

Moreover for any compact $K \subset U$ *such that* $\mu(K) < 1$ *we have*

$$\limsup_{T \to +\infty} \frac{1}{T} \log \mathrm{Leb} \left(\{ x \in K : X^t(x) \in K, 0 < t < T \} \right) < 0.$$

6.4.3.5 Idea of the Proof

We use properties of non-uniformly expanding transformations, especially a large deviation bound recently obtained [6], to deduce a large deviation bound for the

6 Lorenz-Like Chaotic Attractors Revised

suspension semiflow reducing the estimate of the volume of the deviation set to the volume of a certain deviation set for the base transformation.

The initial step of the reduction is as follows. For a continuous and bounded $\psi : M_r \to \mathbb{R}$, $T > 0$ and $z = (x, s)$ with $x \in M$ and $0 \le s < r(x) < \infty$, there exists the *lap number* $n = n(x, s, T) \in \mathbb{N}$ such that $S_n r(x) \le s + T < S_{n+1} r(x)$, and we can write

$$\int_0^T \psi\big(X^t(z)\big)\, dt = \int_s^{r(x)} \psi\big(X^t(x, 0)\big)\, dt + \int_0^{T+s-S_n r(x)} \psi\big(X^t(f^n(x), 0)\big)\, dt$$
$$+ \sum_{j=1}^{n-1} \int_0^{r(f^j(x))} \psi\big(X^t(f^j(x), 0)\big)\, dt.$$

Setting $\varphi(x) = \int_0^{r(x)} \psi(x, 0)\, dt$ we can rewrite the last summation above as $S_n \varphi(x)$. We get the following expression for the time average

$$\frac{1}{T} \int_0^T \psi\big(X^t(z)\big)\, dt = \frac{1}{T} S_n \varphi(x) - \frac{1}{T} \int_0^s \psi\big(X^t(x, 0)\big)\, dt$$
$$+ \frac{1}{T} \int_0^{T+s-S_n r(x)} \psi\big(X^t(f^n(x), 0)\big)\, dt.$$

Writing $I = I(x, s, T)$ for the sum of the last two integral terms above, observe that for $\omega > 0$, $0 \le s < r(x)$ and $n = n(x, s, T)$

$$\left\{ (x, s) \in M_r : \left| \frac{1}{T} S_n \varphi(x) + I(x, s, T) - \frac{\mu(\varphi)}{\mu(r)} \right| > \omega \right\}$$

is contained in

$$\left\{ (x, s) \in M_r : \left| \frac{1}{T} S_n \varphi(x) - \frac{\mu(\varphi)}{\mu(r)} \right| > \frac{\omega}{2} \right\} \cup \left\{ (x, s) \in M_r : I(x, s, T) > \frac{\omega}{2} \right\}.$$

The left hand side above is a *deviation set for the observable φ over the base transformation*, while the right hand side will be *bounded by the geometric conditions on* $\mathscr{S}$ and by a *deviations bound for the observable r over the base transformation*.

Analysing each set using the conditions on f and r and noting that for μ- and Leb-almost every $x \in M$ and every $0 \le s < r(x)$

$$\frac{S_n r(x)}{n} \le \frac{T+s}{n} \le \frac{S_{n+1} r(x)}{n} \quad \text{so} \quad \frac{n(x, s, T)}{T} \xrightarrow[T \to \infty]{} \frac{1}{\mu(r)},$$

we are able to obtain the asymptotic bound of the Main Theorem.

Full details of the proof are presented in [7].

The interested reader can find the proofs of the results mentioned above in the papers listed below, the references therein, and also in one of IMPA's texts [8] for the XXV Brazilian Mathematical Colloquium.

Acknowledgements Vítor Araújo was partially supported by FCT-CMUP (Portugal) and CNPq, FAPERJ and PRONEX-Dyn. Systems (Brazil). Maria José Pacifico was partially supported by CNPq, FAPERJ and PRONEX-Dyn.Systems (Brazil). This work was done while Maria José Pacifico was enjoying a post-doc leave at CRM Ennio De Giorgi-Scuola Normale Superiore di Pisa with a CNPq fellowship and also partially supported by CRM Ennio De Giorgi.

References

1. Afraimovich, V.S., Chernov, N.I., Sataev, E.A.: Statistical properties of 2-D generalized hyperbolic attractors. Chaos 5 **1**, 238–252 (1995)
2. Alves, J.F., Araújo, V., Pacifico, M.J., Pinheiro, V.: On the volume of singular-hyperbolic sets. Dyn. Syst. Int. J. **22**(3), 249–267 (2007)
3. Alves, J.F., Araújo, V.: Hyperbolic times: frequency versus integrability. Ergo. Theory Dyn. Syst. **24**, 1–18 (2004)
4. Alves, J.F., Araújo, V., Pacifico, M.J., Pinheiro, V.: On the volume of singular-hyperbolic sets. Dyn. Syst. Int. J. **22**(3), 249–267 (2007)
5. Alves, J.F., Bonatti, C., Viana, M.: SRB measures for partially hyperbolic systems whose central direction is mostly expanding. Invent. Math. **140**(2), 351–398 (2000)
6. Araújo, V., Pacifico, M.J.: Large deviations for non-uniformly expanding maps. J. Stat. Phys. **125**(2), 415–457 (2006)
7. Araújo, V.: Large deviations bound for semiflows over a non-uniformly expanding base. Bull. Brazil. Math. Soc. **38**(3), 335–376 (2007)
8. Araujo, V., Pacifico, M.J.: Three Dimensional Flows. XXV Brazillian Mathematical Colloquium. IMPA, Rio de Janeiro (2007)
9. Araújo, V., Pujals, E.R., Pacifico, M.J., Viana, M.: Singular-hyperbolic attractors are chaotic. Trans. A.M.S. **361**, 2431–2485 (2009)
10. Arroyo, A., Pujals, E.R.: Dynamical Properties of Singular Hyperbolic Attractors. Discrete Continuous Dyn. Syst. **19**(1), 67–87 (2007)
11. Bautista, S.: The geometric Lorenz attractor is a homoclinic class. Boletin de Matemáticas – Nueva Serie (Dpto. de Matemáticas - Facultad de Ciencias - Universidad Nacional de Colombia) **XI**(1), 69–78 (2004)
12. Bautista, S., Morales, C.: Existence of periodic orbits for singular-hyperbolic sets. Mosc. Math. J. **6**(2), 265–297 (2006)
13. Bowen, R., Ruelle, D.: The ergodic theory of Axiom A flows. Invent. Math. **29**, 181–202 (1975)
14. Bunimovich, L.A.: Statistical properties of Lorenz attractors, Nonlinear Dynamics and turbulence, pp. 71–92. Pitman (1983)
15. Galatolo, S., Pacifico, M.J.: Lorenz-like flows: exponential decay of correlations for the Poincaré map, logarithm law, quantitative recurrence. Ergodic Theory and Dynamical Systems, **30**, 1703–1737 (2010)
16. Guckenheimer, J., Williams, R.F.: Structural stability of Lorenz attractors. Publ. Math. IHES **50**, 59–72 (1979)
17. Komuro, M.: Expansive properties of Lorenz attractors. In The theory of dynamical systems and its applications to nonlinear problems , pp. 4–26. World Scientific Publishing, Kyoto (1984)
18. Kifer, Y.: Random perturbations of dynamical systems, Progress in Probability and Statistics, vol. 16. Birkhäuser Boston, Boston, MA (1988)
19. Kifer, Y.: Large deviations in dynamical systems and stochastic processes. Trans. Am. Math. Soc. **321**(2), 505–524 (1990)

20. Ledrappier, F., Young, L.S.: The metric entropy of diffeomorphisms I. Characterization of measures satisfying Pesin's entropy formula. Ann. Math. **122**, 509–539 (1985)
21. Mañé, R.: An ergodic closing lemma. Ann. Math. **116**, 503–540 (1982)
22. Mañé, R.: Ergodic Theory and Differentiable Dynamics. Springer, New York (1987)
23. Metzger, R.J.: Sinai-Ruelle-Bowen measures for contracting Lorenz maps and flows. Ann. Inst. H. Poincaré Anal. Non Linéaire **17**(2), 247–276 (2000)
24. Morales, C., Pacifico, M.J., Pujals, E.: On C^1 robust singular transitive sets for three-dimensional flows. C. R. Acad. Sci. Paris **326**, Série I, 81–86 (1998)
25. Morales, C.A., Pacifico, M.J., Pujals, E.R.: Singular hyperbolic systems. Proc. Am. Math. Soc. **127**(11), 3393–3401 (1999)
26. Morales, C.A., Pacifico, M.J., Pujals, E.R.: Robust transitive singular sets for 3-flows are partially hyperbolic attractors or repellers. Ann. Math. 2nd Ser. **160**(2), 375–432, (2004)
27. Lorenz, E.N.: Deterministic nonperiodic flow. J. Atmosph. Sci. **20**, 130–141 (1963)
28. Pesin, Y.: Dimension theory in dynamical systems, Chicago Lectures in Mathematics (1997)
29. Pesin, Ya., Sinai, Ya.: Gibbs measures for partially hyperbolic attractors. Ergod. Theory Dyn. Syst. **2**, 417–438 (1982)
30. Saussol, B.: Recurrence rate in rapidly mixing dynamical systems. Discrete Continuous Dyn. Syst. A **15**, 259–267 (2006)
31. Sternberg, E.: On the structure of local homeomorphisms of euclidean n-space - II Am. J. Math. **80**, 623–631 (1958)
32. Tucker, W.: A rigorous ode solver and smale's 14th problem. Found. Comput. Math. **2**(1), 53–117 (2002)
33. Walters, P.: An Introduction to Ergodic Theory. Springer (1982)
34. Viana, M.: Stochastic dynamics of deterministic systems. Brazil. Math. Colloquium, Publicações do IMPA (1997)
35. Young, L-S.: Dimension, entropy and Liapunov exponents. Ergod. Theory Dyn. Syst. **2**, 1230–1237 (1982)

Chapter 7
A Dynamical Point of View of Quantum Information: Entropy and Pressure

A.T. Baraviera, C.F. Lardizabal, A.O. Lopes, and M. Terra Cunha

Abstract Quantum Information is a new area of research which has been growing rapidly since last decade. This topic is very close to potential applications to the so called Quantum Computer. In our point of view it makes sense to develop a more "dynamical point of view" of this theory. We want to consider the concepts of entropy and pressure for "stationary systems" acting on density matrices which generalize the usual ones in Ergodic Theory (in the sense of the Thermodynamic Formalism of R. Bowen, Y. Sinai and D. Ruelle). We consider the operator $\mathscr{L}$ acting on density matrices $\rho \in \mathscr{M}_N$ over a finite N-dimensional complex Hilbert space $\mathscr{L}(\rho) := \sum_{i=1}^{k} tr(W_i \rho W_i^*) V_i \rho V_i^*$, where W_i and V_i, $i = 1, 2, \ldots k$ are operators in this Hilbert space. $\mathscr{L}$ is not a linear operator. In some sense this operator is a version of an Iterated Function System (IFS). Namely, the $V_i(.)V_i^* =: F_i(.)$, $i = 1, 2, \ldots, k$, play the role of the inverse branches (acting on the configuration space of density matrices ρ) and the W_i play the role of the weights one can consider on the IFS. We suppose that for all ρ we have that $\sum_{i=1}^{k} tr(W_i \rho W_i^*) = 1$. A family $W := \{W_i\}_{i=1,\ldots,k}$ determines a Quantum Iterated Function System (QIFS) $\mathscr{F}_W$, $\mathscr{F}_W = \{\mathscr{M}_N, F_i, W_i\}_{i=1,\ldots,k}$.

7.1 Introduction

We will present a survey, and also some new results, of certain topics in Quantum Information from a strictly mathematical point of view. This area is very close to potential applications to the so called Quantum Computer [26]. In our point of view

---

A.T. Baraviera, C.F. Lardizabal (✉), and A.O. Lopes
I.M. – UFRGS, 91500-000, Porto Alegre, Brazil
e-mail: atbaraviera@gmail.com, carlos.lardizabal@gmail.com, arturoscar.lopes@gmail.com

M. Terra Cunha
D.M – UFMG, Belo Horizonte 30161-970, Brazil
e-mail: marcelo.terra.cunha@gmail.com

M.M. Peixoto et al. (eds.), *Dynamics, Games and Science I*, Springer Proceedings in Mathematics 1, DOI 10.1007/978-3-642-11456-4_7,
© Springer-Verlag Berlin Heidelberg 2011

it makes sense to develop a more "dynamical point of view" of this theory. For instance, Von Neumann entropy is a very nice and useful concept, but, in our point of view, it is not a dynamical entropy. A nice exposition about this theory from an Ergodic Theory point of view is presented in [3] (see also [4]). Our setting is different. Part of our work is to justify why the concepts we present here are natural generalizations of the usual ones in Thermodynamic Formalism.

We have to analyze first the fundamental concepts in both theories. It is well-known that the so called Quantum Stochastic Processes have some special features which present a quite different nature than the usual classical Stochastic Processes. A main issue on QSP is the possibility of interference (see [1, 2, 8, 28, 31]). We will analyze carefully Quantum Iterated Function Systems, which were described previously by [22] and [29].

We refer the reader to [1] for the proofs of the results presented in the first part of this exposition.

Density matrices play the role of probabilities on Quantum Mechanics. In this work we investigate a generalization of the classical Thermodynamic Formalism (in the sense of Bowen, Sinai and Ruelle) for the setting of density matrices. We consider the operator $\mathscr{L}$ acting on density matrices $\rho \in \mathscr{M}_N$ over a finite N-dimensional complex Hilbert space

$$\mathscr{L}(\rho) := \sum_{i=1}^{k} tr(W_i \rho W_i^*) V_i \rho V_i^*,$$

where W_i and $V_i, i = 1, 2, \ldots, k$ are operators in this Hilbert space. Note that $\mathscr{L}$ is not a linear operator.

In some sense this operator is a version of an Iterated Function System (IFS). Namely, the $V_i(.)V_i^* =: F_i(.), i = 1, 2, \ldots, k$, play the role of the inverse branches (acting on the configuration space of density matrices ρ) and the W_i play the role of the weights one can consider on the IFS. We suppose that for all ρ we have that $\sum_{i=1}^{k} tr(W_i \rho W_i^*) = 1$. This means that $\mathscr{L}_{\mathscr{F}_W}$ is a normalized operator.

A family $W := \{W_i\}_{i=1,\ldots,k}$ determines a Quantum Iterated Function System (QIFS) $\mathscr{F}_W$,

$$\mathscr{F}_W = \{\mathscr{M}_N, F_i, W_i\}_{i=1,\ldots,k}$$

We want to consider a new concept of entropy for stationary systems acting on density matrices which generalizes the usual one in Ergodic Theory. In our setting the $V_i, i = 1, 2, \ldots, k$ are fixed (i.e. the dynamics of the inverse branches is fixed in the beginning) and we consider the different families $W_i, i = 1, 2, \ldots, k$, (also with the attached corresponding eigendensity matrix ρ_W) as possible Jacobians (of "stationary probabilities").

It is appropriate to make here a remark about the meaning of "stationarity" for us. In Ergodic Theory the action of the shift σ in the Bernoulli space $\Omega = \{1, 2, \ldots, k\}^{\mathbb{N}}$ with k symbols is well understood. The concept of stationarity for a

7 A Dynamical Point of View of Quantum Information

Stochastic Process (where the space of states is $S = \{1, 2, \ldots, k\}$) is defined by the shift-invariance for the associated probability P in the Bernoulli space (the space of paths). Shannon–Kolmogorov entropy is a concept designed for stationary probabilities. When the probability P is associated to a Markov chain, this entropy is given by

$$H(P) := - \sum_{i,j=1}^{N} p_i \, p_{ij} \log p_{ij},$$

where $P = (p_{ij})$ describes the transition matrix, and p_i the invariant probability vector, $i, j = 1, 2, \ldots, k$. This is the key idea for our definition of stationary entropy.

Thermodynamic Formalism and the Ruelle operator for a potential $A : \Omega \to \mathbb{R}$ are natural generalizations of the theory associated to the Perron theorem for positive matrices (see [30]) (this occurs when the potential depends on only the first two symbols of $w = (w_1, w_2, w_3, \ldots) \in \Omega$). We will analyze the Pressure problem for density matrices under this last perspective.

The main point here (and also in [1, 2, 18, 20]) is that in order to define Kolmogorov entropy one can avoid the use of partitions, etc. We just need to look the problem at the level of Ruelle operators (which in some sense captures the underlying dynamics).

Given a normalized family W_i, $i = 1, 2, \ldots, k$, a natural definition of entropy, denoted by $h_V(W)$, is given by

$$- \sum_{i=1}^{k} \frac{tr(W_i \rho_W W_i^*)}{tr(V_i \rho_W V_i^*)} \sum_{j=1}^{k} tr\left(W_j V_i \rho_W V_i^* W_j^*\right) \log \left(\frac{tr(W_j V_i \rho_W V_i^* W_j^*)}{tr(V_i \rho_W V_i^*)}\right),$$

where, ρ_W denotes the barycenter of the unique invariant, attractive measure for the Markov operator $\mathcal{V}$ associated to $\mathscr{F}_W$. We show that this generalizes the entropy of a Markov System. This will be described later on this work.

A different definition of entropy for density operators is presented in [2, 7]. There are examples where the values one gets from these two concepts are different (see [2]).

We also want to present here a concept of pressure for stationary systems acting on density matrices which generalizes the usual one in Ergodic Theory.

In addition to the dynamics obtained by the V_i, which are fixed, a family of potentials H_i, $i = 1, 2, \ldots, k$ induces a kind of Ruelle operator given by

$$\mathscr{L}_H(\rho) := \sum_{i=1}^{k} tr(H_i \rho H_i^*) V_i \rho V_i^* \tag{7.1}$$

We show that such operator admits an eigenvalue β and an associated eigenstate ρ_β, that is, one satisfying $\mathscr{L}_H(\rho_\beta) = \beta \rho_\beta$.

The natural generalization of the concept of pressure for a family H_i, $i = 1, 2, \ldots, k$ is the problem of finding the maximization on the possible normalized families W_i, $i = 1, 2, \ldots, k$, of the expression

$$h_V(W) + \sum_{j=1}^{k} \log \left(tr(H_j \rho_H H_j^*) tr(V_j \rho_H V_j^*) \right) tr(W_j \rho_W W_j^*)$$

We show a relation between the eigendensity matrix ρ_H for the Ruelle operator and the set of W_i, $i = 1, 2, \ldots, k$, which maximizes pressure. In the case each V_i, $i = 1, 2, \ldots, k$, is unitary, then the maximum value is $\log \beta$.

Our work is inspired by the results presented in [22] and [29]. We would like to thank these authors for supplying us with the corresponding references.

We point out that completely positive mappings (operators) acting on density matrices are of great importance in Quantum Computing. These operators can be written in the Stinespring–Kraus form. This motivates the study of operators in the class we will assume here, which are a generalization of such Stinespring–Kraus transformations.

The initial part of our work is dedicated to present all the definitions and concepts that are not well-known (at least for the general audience of people in Dynamical Systems), in a systematic and well organized way. We present many examples and all the basic main definitions which are necessary to understand the theory. However, we do not have the intention to exhaust what is already known. We believe that the theoretical results presented here can be useful as a general tool to understand problems in Quantum Computing.

Several examples are presented with all details in the text. We believe that this will help the reader to understand the main issues of the theory.

In order to simplify the notation we will present most of our results for the case of two by two matrices.

In Sects. 7.2 and 7.3 we present some basic definitions, examples and we show some preliminary relations of our setting to the classical Thermodynamic Formalism. In Sect. 7.4 we present an eigenvalue problem for non-normalized Ruelle operators which will be required later. Some properties and concepts about density matrices and Ruelle operators are presented in Sects. 7.6 and 7.7. In Sect. 7.10 we introduce the concept of stationary entropy for *measures* defined on the set of density matrices. In Sect. 7.11 we compare this definition with the usual one for Markov Chains. Section 7.12 aims to motivate the interest on pressure and the capacity-cost function. The Sects. 7.13, 7.14 and 7.15 are dedicated to the presentation of our main results on pressure, important inequalities, examples and its relation with the classical theory of Thermodynamic Formalism.

In the paper "A dynamical point of view of Quantum Information: Discrete Wigner measures", published in this volume, we consider the discrete Wigner measure, in which part of the analysis described is related to QIFS.

This work is part of the thesis dissertation of C. F. Lardizabal in Prog. Pos-Grad. Mat. UFRGS (Brazil) [16].

7 A Dynamical Point of View of Quantum Information

7.2 Basic Definitions

Let $M_N(\mathbb{C})$ the set of complex matrices of order N. If $\rho \in M_N(\mathbb{C})$ then ρ^* denotes the transpose conjugate of ρ. We consider in $\mathbb{C}^N$ the $\mathscr{L}^2$ norm. A state (or vector) in $\mathbb{C}^N$ will be denoted by ψ or $|\psi\rangle$, and the associated projection will be written $|\psi\rangle\langle\psi|$. Define

$$\mathscr{H}_N := \{\rho \in M_N(\mathbb{C}) : \rho^* = \rho\}$$
$$\mathscr{P}\mathscr{H}_N := \{\rho \in \mathscr{H}_N : \langle\rho\psi,\psi\rangle \geq 0, \forall \psi \in \mathbb{C}^N\}$$
$$\mathscr{M}_N := \{\rho \in \mathscr{P}\mathscr{H}_N : tr(\rho) = 1\}$$
$$\mathscr{P}_N := \{\rho \in \mathscr{H}_N : \rho = |\psi\rangle\langle\psi|, \psi \in \mathbb{C}^N, \langle\psi|\psi\rangle = 1\},$$

the space of hermitian, positive, density operators and pure states, respectively. Density operators are also called mixed states. Any state ρ, by the spectral theorem, can be written as

$$\rho = \sum_{i=1}^{k} p_i |\psi_i\rangle\langle\psi_i|, \tag{7.2}$$

for some choice of p_i, which are positive numbers with $\sum_i p_i = 1$, and ψ_i, which have norm one and are orthogonal.

The set $\mathscr{P}_N$ is the set of extremal points of $\mathscr{M}_N$, that is, the set of points which can not be decomposed as a nontrivial convex combination of elements in $\mathscr{M}_N$.

Definition 7.1. Let $G_i : \mathscr{M}_N \to \mathscr{M}_N$, $p_i : \mathscr{M}_N \to [0,1]$, $i = 1,\ldots,k$ and such that $\sum_i p_i(\rho) = 1$. We call

$$\mathscr{F}_N = \{\mathscr{M}_N, G_i, p_i : i = 1,\ldots,k\} \tag{7.3}$$

a Quantum Iterated Function System (QIFS).

Definition 7.2. A QIFS is *homogeneous* if p_i and $G_i p_i$ are affine mappings, $i = 1,\ldots,k$.

Suppose that the QIFS considered is such that there are V_i and W_i linear maps, $i = 1,\ldots,k$, with $\sum_{i=1}^{k} W_i^* W_i = I$ such that

$$G_i(\rho) = \frac{V_i \rho V_i^*}{tr(V_i \rho V_i^*)} \tag{7.4}$$

and

$$p_i(\rho) = tr(W_i \rho W_i^*) \tag{7.5}$$

Then we have that a QIFS is homogeneous if $V_i = W_i$, $i = 1,\ldots,k$.

Now we can define a Markov operator $\mathcal{V} : \mathcal{M}(\mathcal{M}_N) \to \mathcal{M}(\mathcal{M}_N)$,

$$(\mathcal{V}\mu)(B) = \sum_{i=1}^{k} \int_{G_i^{-1}(B)} p_i(\rho)d\mu(\rho),$$

where $\mathcal{M}(\mathcal{M}_N)$ denotes the space of probability measure over $\mathcal{M}_N$. We also define $\Lambda : \mathcal{M}_N \to \mathcal{M}_N$,

$$\Lambda(\rho) := \sum_{i=1}^{k} p_i(\rho)G_i(\rho)$$

The operator defined above has no counterpart in the classical Thermodynamic Formalism. We will also consider the operator acting on density matrices ρ.

$$\mathcal{L}(\rho) = \sum_{i=k}^{k} q_i(\rho)V_i\rho V_i^*.$$

If for all ρ we have $\sum_{i=k}^{k} q_i(\rho) = 1$, we say the operator is normalized.

In the normalized case, the different possible choices of $q_i, i = 1, 2, \ldots, k$, (which means different choices of $W_i, i = 1, 2, \ldots, k$) play here the role of the different Jacobians of possible invariant probabilities (see [23] II. 1, and [20]) in Thermodynamic Formalism. In some sense the probabilities can be identified with the Jacobians (this is true at least for Gibbs probabilities of Hölder potentials [25]). The set of Gibbs probabilities for Hölder potentials is dense in the set of invariant probabilities [19].

We are also interested on the non-normalized case. If the QIFS is homogeneous, then

$$\Lambda(\rho) = \sum_{i} V_i\rho V_i^* \tag{7.6}$$

Theorem 7.1. *[29] A mixed state ρ_0 is Λ-invariant if and only if*

$$\rho_0 = \int_{\mathcal{M}_N} \rho d\mu(\rho), \tag{7.7}$$

for some $\mathcal{V}$-invariant measure μ.

In order to define hyperbolic QIFS, one has to define a distance on the space of mixed states. For instance, we could choose one of the following:

$$D(\rho_1, \rho_2) = \sqrt{tr[(\rho_1 - \rho_2)^2]}$$
$$D(\rho_1, \rho_2) = tr\sqrt{(\rho_1 - \rho_2)^2}$$
$$D(\rho_1, \rho_2) = \sqrt{2\{1 - tr[(\rho_1^{1/2}\rho_2\rho_1^{1/2})^{1/2}]\}}$$

7 A Dynamical Point of View of Quantum Information

Such metrics generate the same topology on $\mathcal{M}$. Considering the space of mixed states with one of those metrics we can make the following definition. We say that a QIFS is *hyperbolic* if the quantum maps G_i are contractions with respect to one of the distances on $\mathcal{M}_N$ and if the maps p_i are Hölder-continuous and positive, see for instance, [22].

Proposition 7.1. *If a QIFS (7.3) is homogeneous and hyperbolic the associated Markov operator admits a unique invariant measure μ. Such invariant measure determines a unique Λ-invariant state $\rho \in \mathcal{M}_N$, given by (7.7).*

See [22, 29] for the proof.

7.3 Examples of QIFS

Example 7.1. $\Omega = \mathcal{M}_N$, $k = 2$, $p_1 = p_2 = 1/2$, $G_1(\rho) = U_1 \rho U_1^*$, $G_2(\rho) = U_2 \rho U_2^*$. The normalized identity matrix $\rho_* = I/N$ is Λ-invariant, for any choice of unitary U_1 and U_2. Note that we can write

$$\rho_* = \int_{\mathcal{M}_N} \rho \, d\mu(\rho)$$

where the measure μ, uniformly distributed over $\mathcal{P}_N$, is $\mathcal{V}$-invariant. $\diamond$

In the example described below we use Dirac notation for the projections.

Example 7.2. We are interested in finding the fixed point $\hat{\rho}$ for Λ in an example for the case $N = 2$ and $k = 3$.

Consider the bits $|0> = (0, 1)$ and $|1> = (1, 0)$ (the canonical basis). The states ρ are generated by $|0><0|$, $|0><1|$, $|1><0|$ and $|1><1|$. Take $V_1 = I$ and V_2 such that $|0> \to |0>$ and $|1> \to |0>$. Consider V_3 such that $|0> \to |1>$ and $|1> \to |1>$. That is, $V_2 = |0><0| + |0><1|$ and $V_3 = |1><0| + |1><1|$. Therefore, $V_2^* = |0><0| + |1><0|$ and $V_3^* = |0><1| + |1><1|$. Suppose $p_i = \hat{p}_i$, $i = 1, 2, 3$, are such that $\sum_i p_i = 1$ (in this case, each p_i is independent of ρ). Therefore, we consider the operator $\mathcal{L}$ and look for fixed points ρ. Suppose

$$\rho = \rho_{00}|0><0| + \rho_{01}|0><1| + \rho_{10}|1><0| + \rho_{11}|1><1|$$

Then

$$\Lambda(\rho) = \sum_{i=1}^{3} p_i(\rho) \frac{(V_i \rho V_i^*)}{\mathrm{tr}(V_i \rho V_i^*)}$$

$$= \sum_{i=1}^{3} p_i \left[\frac{V_i((\rho_{00}|0><0| + \rho_{01}|0><1| + \rho_{10}|1><0| + \rho_{11}|1><1|))V_i^*}{\mathrm{tr}(V_i \rho V_i^*)} \right]$$

Let us compute first the action of the operator $V_2|0><0|V_2^*$.

Note that $(V_2|0 >< 0|V_2^*)|0 >= V_2|0 >< 0|(|0 > +|1 >) = V_2|0 >= |0 >$ and $(V_2|0 >< 0|V_2^*)|1 >= V_2(0) = 0$. More generally

$$\rho V_2^* = (\rho_{00}|0 >< 0| + \rho_{01}|0 >< 1| + \rho_{10}|1 >< 0| + \rho_{11}|1 >< 1|)$$

$$(|0 >< 0|+|1 >< 0|) = \rho_{00}|0 >< 0|+\rho_{01}|0 >< 0|+\rho_{10}|1 >< 0|+\rho_{11}|1 >< 0|.$$

Therefore,

$$V_2\rho V_2^* = (|0 >< 0| + |0 >< 1|)(\rho_{00}|0 >< 0| + \rho_{01}|0 >< 0| + \rho_{10}|1 >< 0|$$
$$+\rho_{11}|1 >< 0|) = (\rho_{00} + \rho_{01} + \rho_{10} + \rho_{11})|0 >< 0|$$
$$= (1 + 2Re(\rho_{01}))|0 >< 0|,$$

because ρ has trace $1 = \rho_{00} + \rho_{11}$. Note that $tr(V_2\rho V_2^*) = (1 + 2Re(\rho_{01}))$. A similar result can be obtained for V_3. Proceeding in the same way we get that

$$\Lambda(\rho) = p_1(\rho_{00}|0 >< 0| + \rho_{01}|0 >< 1| + \rho_{10}|1 >< 0| + \rho_{11}|1 >< 1|)$$
$$+p_2|0 >< 0| + p_3|1 >< 1|.$$

The equation

$$\Lambda(\rho) = \rho = \rho_{00}|0 >< 0| + \rho_{01}|0 >< 1| + \rho_{10}|1 >< 0| + \rho_{11}|1 >< 1|$$

means

$$p_1\rho_{00} + p_2 = \rho_{00},$$
$$p_1\rho_{01} = \rho_{01},$$
$$p_1\rho_{10} = \rho_{10},$$
$$p_1\rho_{11} + p_3 = \rho_{11}.$$

If $p_1 \neq 0$, then $\rho_{01} = \rho_{10} = 0$. Finally, if $p_1 \neq 1$, then $\rho_{00} = \frac{p_2}{1-p_1}$ and $\rho_{11} = \frac{p_3}{1-p_1}$ and the fixed point is

$$\hat{\rho} = \frac{p_2}{1 - p_1}|0 >< 0| + \frac{p_3}{1 - p_1}|1 >< 1|.$$

$\diamond$

We recall that a mapping Λ is *completely positive* (CP) if $\Lambda \otimes I$ is positive for any extension of the Hilbert space considered $\mathcal{H}_N \to \mathcal{H}_N \otimes \mathcal{H}_E$. We know that every CP mapping which is trace-preserving can be represented (in a nonunique way) in the Stinespring–Kraus form

7 A Dynamical Point of View of Quantum Information

$$\Lambda(\rho) = \sum_{i=1}^{k} V_i \rho V_i^*, \quad \sum_{i=1}^{k} V_i^* V_i = I,$$

where the V_i are linear operators. Moreover if we have $\sum_{j=1}^{k} V_j V_j^* = I$, then $\Lambda(I/N) = I/N$. This is the case if each of the V_i are normal.

We call a unitary trace-preserving CP map a *bistochastic map*. An example of such a mapping is

$$\Lambda_U(\rho) = \sum_{i=1}^{k} p_i U_i \rho U_i^*,$$

where the U_i are unitary operators and $\sum_i p_i = 1$. Note that if we write $G_i(\rho) = U_i \rho U_i^*$, then example 7.1 is part of this class of operators. For such operators we have that ρ_* is an invariant state for Λ_U and also that δ_{ρ_*} is invariant for the Markov operator P_U induced by this QIFS.

We will present a simple example of the kind of problems we are interested here, namely eigenvalues and eigendensity matrices. Let $\mathcal{H}_N$ be a Hilbert space of dimension N. As before, let $\mathcal{M}_N$ be the space of density operators on $\mathcal{H}_N$. A natural problem is to find fixed points for $\Lambda : \mathcal{M}_N \to \mathcal{M}_N$,

$$\Lambda(\rho) = \sum_{i=1}^{k} V_i \rho V_i^*.$$

In order to simplify our reasoning we fix $N = 2$ and $k = 2$. Let

$$V_1 = \begin{pmatrix} v_1 & v_2 \\ v_3 & v_4 \end{pmatrix}, V_2 = \begin{pmatrix} w_1 & w_2 \\ w_3 & w_4 \end{pmatrix}, \rho = \begin{pmatrix} \rho_1 & \rho_2 \\ \overline{\rho_2} & \rho_4 \end{pmatrix},$$

where V_1 and V_2 are invertible and ρ is a density operator. We would like to find ρ such that

$$V_1 \rho V_1^* + V_2 \rho V_2^* = \rho. \tag{7.8}$$

Below we have an example where the matrices V_i are not real.

Example 7.3. Let

$$V_1 = e^{ik} \begin{pmatrix} \sqrt{p} & 0 \\ 0 & -\sqrt{p} \end{pmatrix}, V_2 = e^{il} \begin{pmatrix} \sqrt{1-p} & 0 \\ 0 & -\sqrt{1-p} \end{pmatrix},$$

where $k, l \in \mathbb{R}$, $p \in (0, 1)$. Then $V_1^* V_1 + V_2^* V_2 = I$. A simple calculation shows that $\rho_2 = 0$, and then

$$\rho = \begin{pmatrix} q & 0 \\ 0 & 1-q \end{pmatrix}$$

is invariant to $\Lambda(\rho) = V_1 \rho V_1^* + V_2 \rho V_2^*$, for $q \in (0, 1)$. $\diamond$

Now we make a few considerations about the Ruelle operator $\mathscr{L}$ defined before. In particular, we show that Perron's classic eigenvalue problem is a particular case of the problem for the operator $\mathscr{L}$ acting on matrices. Let

$$V_1 = \begin{pmatrix} p_{00} & 0 \\ 0 & 0 \end{pmatrix}, V_2 = \begin{pmatrix} 0 & p_{01} \\ 0 & 0 \end{pmatrix}$$

$$V_3 = \begin{pmatrix} 0 & 0 \\ p_{10} & 0 \end{pmatrix}, V_4 = \begin{pmatrix} 0 & 0 \\ 0 & p_{11} \end{pmatrix}, \rho = \begin{pmatrix} \rho_1 & \rho_2 \\ \rho_3 & \rho_4 \end{pmatrix}$$

Define

$$\mathscr{L}(\rho) = \sum_{i=1}^{4} q_i(\rho) V_i \rho V_i^*$$

We have that $\mathscr{L}(\rho) = \rho$ implies $\rho_2 = 0$ and

$$a\rho_1 + b\rho_4 = \rho_1 \tag{7.9}$$

$$c\rho_1 + d\rho_4 = \rho_4 \tag{7.10}$$

where

$$a = q_1 p_{00}^2, b = q_2 p_{01}^2, c = q_3 p_{10}^2, d = q_4 p_{11}^2$$

Solving (7.9) and (7.10) in terms of ρ_1 gives

$$\rho_1 = \frac{b}{1-a}\rho_4, \rho_1 = \frac{1-d}{c}\rho_4$$

that is,

$$\frac{b}{1-a} = \frac{1-d}{c} \tag{7.11}$$

which is a restriction over the q_i. For simplicity we assume here that the q_i are constant. One can show that

$$\rho = \begin{pmatrix} \dfrac{q_2 p_{01}^2}{q_2 p_{01}^2 - q_1 p_{00}^2 + 1} & 0 \\ 0 & \dfrac{1 - q_1 p_{00}^2}{q_2 p_{01}^2 - q_1 p_{00}^2 + 1} \end{pmatrix} = \begin{pmatrix} \dfrac{1 - q_4 p_{11}^2}{1 - q_4 p_{11}^2 + q_3 p_{10}^2} & 0 \\ 0 & \dfrac{q_3 p_{10}^2}{1 - q_4 p_{11}^2 + q_3 p_{10}^2} \end{pmatrix} \tag{7.12}$$

Now let

$$P = \sum_i V_i = \begin{pmatrix} p_{00} & p_{01} \\ p_{10} & p_{11} \end{pmatrix},$$

be a column-stochastic matrix. Let $\pi = (\pi_1, \pi_2)$ such that $P\pi = \pi$. Then

$$\pi = \left(\frac{p_{01}}{p_{01} - p_{00} + 1}, \frac{1 - p_{00}}{p_{01} - p_{00} + 1} \right) \tag{7.13}$$

7 A Dynamical Point of View of Quantum Information   91

Comparing (7.13) and (7.12) suggests that we should fix

$$q_1 = \frac{1}{p_{00}}, q_2 = \frac{1}{p_{01}}, q_3 = \frac{1}{p_{10}}, q_4 = \frac{1}{p_{11}} \tag{7.14}$$

Then the nonzero entries of ρ are equal to the entries of π and therefore we associate the fixed point of P to the fixed point of some $\mathscr{L}$ in a natural way. But note that such a choice of q_i is not unique, because

$$q_2 = \frac{1 - q_1 p_{00}^2}{p_{01} p_{10}}, q_4 = \frac{1 - q_3 p_{10} p_{01}}{p_{11}^2}, \tag{7.15}$$

for any q_1, q_3 also produces ρ with nonzero coordinates equal to the coordinates of π.

Now we consider the following problem. Let

$$V_1 = \begin{pmatrix} h_{00} & 0 \\ 0 & 0 \end{pmatrix}, V_2 = \begin{pmatrix} 0 & h_{01} \\ 0 & 0 \end{pmatrix}, V_3 = \begin{pmatrix} 0 & 0 \\ h_{10} & 0 \end{pmatrix}$$

$$V_4 = \begin{pmatrix} 0 & 0 \\ 0 & h_{11} \end{pmatrix}, H = \sum_i V_i, \rho = \begin{pmatrix} \rho_1 & \rho_2 \\ \rho_3 & \rho_4 \end{pmatrix}$$

Define

$$\mathscr{L}(\rho) = \sum_{i=1}^{4} q_i V_i \rho V_i^*,$$

where $q_i \in \mathbb{R}$. Assume that $h_{ij} \in \mathbb{R}$, so we want to obtain λ such that $\mathscr{L}(\rho) = \lambda \rho$, $\lambda \neq 0$, and λ is the largest eigenvalue. With a few calculations we obtain $\rho_2 = \rho_3 = 0$,

$$q_1 h_{00}^2 \rho_1 + q_2 h_{01}^2 \rho_4 = \lambda \rho_1$$

$$q_3 h_{10}^2 \rho_1 + q_4 h_{11}^2 \rho_4 = \lambda \rho_4$$

that is,

$$a\rho_1 + b\rho_4 = \lambda \rho_1 \tag{7.16}$$

$$c\rho_1 + d\rho_4 = \lambda \rho_4, \tag{7.17}$$

with

$$a = q_1 h_{00}^2, b = q_2 h_{01}^2, c = q_3 h_{10}^2, d = q_4 h_{11}^2$$

Therefore

$$\rho = \begin{pmatrix} \frac{\lambda - d}{c} \rho_4 & 0 \\ 0 & \rho_4 \end{pmatrix} = \begin{pmatrix} \frac{b}{\lambda - a} \rho_4 & 0 \\ 0 & \rho_4 \end{pmatrix}$$

and

$$\frac{\lambda - d}{c} = \frac{b}{\lambda - a}$$

Solving for λ, we obtain the eigenvalues

$$\lambda = \frac{a+d}{2} \pm \frac{\zeta}{2} = \frac{a+d}{2} \pm \frac{\sqrt{(d-a)^2 + 4bc}}{2}$$

$$= \frac{1}{2}\left(q_1 h_{00}^2 + q_4 h_{11}^2 \pm \sqrt{(q_4 h_{11}^2 - q_1 h_{00}^2)^2 + 4q_2 q_3 h_{01}^2 h_{10}^2}\right),$$

where

$$\zeta = \sqrt{(d-a)^2 + 4bc} = \sqrt{(q_4 h_{11}^2 - q_1 h_{00}^2)^2 + 4q_2 q_3 h_{01}^2 h_{10}^2}$$

and the associated eigenfunctions

$$\rho = \begin{pmatrix} \frac{a-d\pm\zeta}{2c}\rho_4 & 0 \\ 0 & \rho_4 \end{pmatrix} = \begin{pmatrix} \frac{2b}{d-a\pm\zeta}\rho_4 & 0 \\ 0 & \rho_4 \end{pmatrix}$$

But $\rho_1 + \rho_4 = 1$ so we obtain

$$\rho = \begin{pmatrix} \frac{a-d\pm\zeta}{a-d\pm\zeta+2c} & 0 \\ 0 & \frac{2c}{a-d\pm\zeta+2c} \end{pmatrix}$$

$$= \begin{pmatrix} \frac{q_1 h_{00}^2 - q_4 h_{11}^2 \pm \zeta}{q_1 h_{00}^2 - q_4 h_{11}^2 \pm \zeta + 2q_3 h_{10}^2} & 0 \\ 0 & \frac{2q_3 h_{10}^2}{q_1 h_{00}^2 - q_4 h_{11}^2 \pm \zeta + 2q_3 h_{10}^2} \end{pmatrix} \qquad (7.18)$$

that is,

$$\rho = \begin{pmatrix} \frac{-2b}{a-2b-d\mp\zeta} & 0 \\ 0 & \frac{a-d\mp\zeta}{a-2b-d\mp\zeta} \end{pmatrix}$$

$$= \begin{pmatrix} \frac{-2q_2 h_{01}^2}{q_1 h_{00}^2 - 2q_2 h_{01}^2 - q_4 h_{11}^2 \mp \zeta} & 0 \\ 0 & \frac{q_1 h_{00}^2 - q_4 h_{11}^2 \mp \zeta}{q_1 h_{00}^2 - 2q_2 h_{01}^2 - q_4 h_{11}^2 \mp \zeta} \end{pmatrix} \qquad (7.19)$$

Therefore we obtained that $\rho_1, \rho_4, q_1, \ldots, q_4, \lambda$ are implicit solutions for the set of (7.16)–(7.17). Recall that in this case we obtained $\rho_2 = \rho_3 = 0$.

Now we consider the problem of finding the eigenvector associated to the dominant eigenvalue of H. The eigenvalues are

$$\lambda = \frac{1}{2}\left(h_{00} + h_{11} \pm \sqrt{(h_{00} - h_{11})^2 + 4h_{01} h_{10}}\right)$$

7 A Dynamical Point of View of Quantum Information   93

Then we can find v such that $Hv = \lambda v$ from the set of equations

$$h_{00}v_1 + h_{01}v_2 = \lambda v_1 \tag{7.20}$$

$$h_{10}v_1 + h_{11}v_2 = \lambda v_2 \tag{7.21}$$

which determine v_1, v_2, λ implicitly. Note that if we set

$$q_1 = \frac{1}{p_{00}}, q_2 = \frac{1}{p_{01}}, q_3 = \frac{1}{p_{10}}, q_4 = \frac{1}{p_{11}}$$

we have that the set of (7.16)–(7.17) and (7.20)–(7.21) are the same. Hence we conclude that Perron's classic eigenvalue problem is a particular case of the problem for $\mathscr{L}$ acting on matrices. $\diamond$

7.4 A Theorem on Eigenvalues for the Ruelle Operator

The following proposition is inspired in [25]. We say that a hermitian operator $P :$ $V \to V$ on a Hilbert space $(V, \langle \cdot \rangle)$ is *positive* if $\langle Pv, v \rangle \geq 0$, for all $v \in V$, denoted $P \geq 0$. Consider the positive operator $\mathscr{L}_{W,V} : \mathscr{P}\mathscr{H}_N \to \mathscr{P}\mathscr{H}_N$,

$$\mathscr{L}_{W,V}(\rho) := \sum_{i=1}^{k} tr(W_i \rho W_i^*) V_i \rho V_i^* \tag{7.22}$$

We have the following result:

Proposition 7.2. *[1] There is $\rho \in \mathscr{M}_N$ and $\beta > 0$ such that $\mathscr{L}_{W,V}(\rho) = \beta\rho$.*

7.5 Vector Integrals and Barycenters

We recall here a few basic definitions. For more details, see [22] and [29]. Let X be a metric space. Let $(V, +, \cdot)$ be a real vector space, and τ a topology on V. We say that $(V, +, \cdot; \tau)$ is a topologic vector space if it is Hausdorff and if the operations $+$ and $\cdot$ are continuous. For instance, in the context of density matrices, we will consider V as the Hilbert space $\mathscr{H}_N$ and X will be the space of density matrices $\mathscr{M}_N$.

Definition 7.3. Let (X, Σ) be a measurable space, let $\mu \in M(X)$, let $(V, +, \cdot; \tau)$ be a locally convex space and let $f : X \to V$. we say that $x \in V$ is the *integral* of f in X, denoted by

$$x := \int_X f d\mu$$

if

$$\Psi(x) = \int_X \Psi \circ f d\mu,$$

for all $\Psi \in V^*$.

It is known that if we have a compact metric space X, V is a locally convex space and $f : X \to V$ is a continuous function such that $\overline{co} f(X)$ is compact then the integral of f in X exists and belongs to $\overline{co} f(X)$. We will also use the following well-known result, the barycentric formula:

Proposition 7.3. *[32] Let V be a locally convex space, let $E \subset V$ be a complete, convex and bounded set, and $\mu \in M^1(E)$. Then there is a unique $x \in E$ such that*

$$l(x) = \int_E l d\mu,$$

for all $l \in V^*$.

7.6 Example: Density Matrices

In this section we briefly review how the constructions of the previous section adjust to the case of density matrices.

Define $V := \mathcal{H}_N$, $V^+ := \mathcal{P}\mathcal{H}_N$ (note that such space is a convex cone), and let the partial order $\leq$ on $\mathcal{P}\mathcal{H}_N$ be $\rho \leq \psi$ if and only if $\psi - \rho \geq 0$, i.e., if $\psi - \rho$ is positive. Then

$$(V, V^+, e) = (\mathcal{H}_N, \mathcal{P}\mathcal{H}_N, tr),$$

is a regular state space [29]. Also, the set B of unity trace in V^+ is, of course, the space of density matrices. Hence, $B = \mathcal{M}_N$.

Let $Z \subset V^*$ be a nonempty vector subspace of V^*. The smallest topology in V such that every functional defined in Z is continuous on that topology, denoted by $\sigma(V, Z)$, turns V into a locally convex space. In particular, $\sigma(V, V^*)$ is the weak topology in V. If $(V, \| \cdot \|)$ is a normed space, then $\sigma(V^*, V)$ is called a weak* topology in V^* (we identify V with a subspace of V^{**}). We also have that $(C, \tau) = (\mathcal{P}\mathcal{H}_N, \tau)$, where τ is the weak* topology (and which is equal to the Euclidean, see [29]) is a metrizable compact structure. In this case we have that $B_C = B \cap C = \mathcal{M}_N$.

Definition 7.4. A *Markov operator* for probability measures is an operator $P : M^1(X) \to M^1(X)$ such that

$$P(\lambda\mu_1 + (1 - \lambda)\mu_2) = \lambda P\mu_1 + (1 - \lambda)P\mu_2,$$

for $\mu_1, \mu_2 \in M^1(X)$, $\lambda \in (0, 1)$.

7 A Dynamical Point of View of Quantum Information

An example of such an operator is one which we have defined before and we denote it $\mathscr{V} : M^1(X) \to M^1(X)$,

$$(\mathscr{V}v)(B) = \sum_{i=1}^{k} \int_{F_i^{-1}(B)} p_i \, dv, \tag{7.23}$$

and we call it the Markov operator induced by the IFS $\mathscr{F}$. We will be interested in fixed points for $\mathscr{V}$.

Define

$$m_b(X) := \{f : X \to \mathbb{R} : \text{f is bounded, measurable}\}$$

and also $\mathscr{U} : m_b(X) \to m_b(X)$,

$$(\mathscr{U}f)(x) := \sum_{i=1}^{k} p_i(x) f(F_i(x))$$

Proposition 7.4. *[29] Let $f \in m_b(X)$ and $\mu \in M^1(X)$, then*

$$\langle f, \mathscr{V}\mu \rangle = \langle \mathscr{U}f, \mu \rangle = \sum_{i=1}^{k} \int p_i(f \circ F_i) d\mu,$$

where $\langle f, \mu \rangle$ denotes the integral of f with respect to μ.

Definition 7.5. An operator $Q : V^+ \to V^+$ is *submarkovian* if

1. $Q(x + y) = Q(x) + Q(y)$
2. $Q(\alpha x) = \alpha Q(x)$
3. $\|Q(x)\| \leq \|x\|$,

for all $x, y \in V^+, \alpha > 0$.

Every submarkovian operator $Q : V^+ \to V^+$ can be extended in a unique way to a positive linear contraction on V.

Definition 7.6. Let $P : V^+ \to V^+$ a Markov operator and let $P_i : V^+ \to V^+, i = 1, \ldots, k$ be submarkovian operators such that $P = \sum_i P_i$. We say that $(P, \{P_i\}_{i=1}^k)$ is a *Markov pair*.

From [29], we know that there is a 1–1 correspondence between homogeneous IFS and Markov pairs.

Example 7.4. In this example we want to obtain a probability η such that $\mathscr{V}(\eta) = \eta$. Suppose a QIFS, such that

$$p_i(\rho) = tr(W_i \rho W_i^*), \sum_i W_i^* W_i = I, F_i(\rho) = \frac{V_i \rho V_i^*}{tr(V_i \rho V_i^*)}$$

for $i = 1, \ldots, k$. Denote $m_b(\mathcal{M}_N)$ the space of bounded and measurable functions in $\mathcal{M}_N$. Consider $\Lambda : \mathcal{M}_N \to \mathcal{M}_N$,

$$\Lambda(\rho) = \sum_i p_i(\rho) F_i(\rho) = \sum_i tr(W_i \rho W_i^*) \frac{V_i \rho V_i^*}{tr(V_i \rho V_i^*)}$$

Suppose there exists a density matrix ρ which Λ-invariant. As we know, such state is the barycenter of μ which is $\mathcal{V}$-invariant. Suppose $\mathcal{V}\mu = \mu$, then we can write

$$\int f d\mu = \int f d\mathcal{V}\mu = \sum_{i=1}^{k} \int p_i(\rho) f(F_i(\rho)) d\mu(\rho)$$

$$= \sum_i \int p_i(\rho) f\left(\frac{V_i \rho V_i^*}{tr(V_i \rho V_i^*)}\right) d\mu$$

$$= \sum_i \int tr(W_i \rho W_i^*) f\left(\frac{V_i \rho V_i^*}{tr(V_i \rho V_i^*)}\right) d\mu$$

Therefore, for any $f \in m_b(\mathcal{M}_N)$, we got the condition

$$\int f d\mu = \sum_i \int tr(W_i \rho W_i^*) f\left(\frac{V_i \rho V_i^*}{tr(V_i \rho V_i^*)}\right) d\mu \tag{7.24}$$

Let us consider a particular example where $N = 2, k = 4$, and

$$V_1 = \begin{pmatrix} \sqrt{p_{11}} & 0 \\ 0 & 0 \end{pmatrix}, V_2 = \begin{pmatrix} 0 & \sqrt{p_{12}} \\ 0 & 0 \end{pmatrix},$$

$$V_3 = \begin{pmatrix} 0 & 0 \\ \sqrt{p_{21}} & 0 \end{pmatrix}, V_4 = \begin{pmatrix} 0 & 0 \\ 0 & \sqrt{p_{22}} \end{pmatrix},$$

in such way that the p_{ij} are the entries of a column stochastic matrix P. Let $\pi = (\pi_1, \pi_2)$ be a vector such that $P\pi = \pi$. A simple calculation shows that for ρ, the density matrix such that has entries ρ_{ij}, we have

$$V_1 \rho V_1^* = \begin{pmatrix} p_{11}\rho_{11} & 0 \\ 0 & 0 \end{pmatrix}, V_2 \rho V_2^* = \begin{pmatrix} p_{12}\rho_{22} & 0 \\ 0 & 0 \end{pmatrix} \tag{7.25}$$

$$V_3 \rho V_3^* = \begin{pmatrix} 0 & 0 \\ 0 & p_{21}\rho_{11} \end{pmatrix}, V_4 \rho V_4^* = \begin{pmatrix} 0 & 0 \\ 0 & p_{22}\rho_{22} \end{pmatrix}, \tag{7.26}$$

and therefore

$$\frac{V_1 \rho V_1^*}{tr(V_1 \rho V_1^*)} = \begin{pmatrix} 1 & 0 \\ 0 & 0 \end{pmatrix}, \frac{V_2 \rho V_2^*}{tr(V_2 \rho V_2^*)} = \begin{pmatrix} 1 & 0 \\ 0 & 0 \end{pmatrix} \tag{7.27}$$

7 A Dynamical Point of View of Quantum Information

$$\frac{V_3 \rho V_3^*}{tr(V_3 \rho V_3^*)} = \begin{pmatrix} 0 & 0 \\ 0 & 1 \end{pmatrix}, \quad \frac{V_4 \rho V_4^*}{tr(V_4 \rho V_4^*)} = \begin{pmatrix} 0 & 0 \\ 0 & 1 \end{pmatrix} \tag{7.28}$$

that is, the above values do not depend on ρ.

Define

$$\rho_x = \begin{pmatrix} 1 & 0 \\ 0 & 0 \end{pmatrix}, \rho_y = \begin{pmatrix} 0 & 0 \\ 0 & 1 \end{pmatrix} \tag{7.29}$$

and

$$\eta = \pi_1 \delta_{\rho_x} + \pi_2 \delta_{\rho_y} \tag{7.30}$$

Note that the barycenter of η is

$$\rho_\eta = \pi_1 \rho_x + \pi_2 \rho_y = \pi_1 \begin{pmatrix} 1 & 0 \\ 0 & 0 \end{pmatrix} + \pi_2 \begin{pmatrix} 0 & 0 \\ 0 & 1 \end{pmatrix} = \begin{pmatrix} \pi_1 & 0 \\ 0 & \pi_2 \end{pmatrix}$$

For any mensurable set B we have

$$\mathscr{V}\eta(B) = \sum_{i=1}^{4} \int 1_B(F_i(\rho)) p_i(\rho) d\eta = \sum_{i=1}^{4} \int 1_B\left(\frac{V_i \rho V_i^*}{tr(V_i \rho V_i^*)}\right) tr(V_i \rho V_i^*) d\eta \tag{7.31}$$

We can now consider the following cases:

1. Suppose first that $\rho_x, \rho_y \in B$. The using (7.25) and (7.26), one can show that

$$\begin{aligned} \mathscr{V}\eta(B) &= \sum_{i=1}^{4} p_{11} tr(V_i \rho_x V_i^*) + p_{22} tr(V_i \rho_y V_i^*) \\ &= (\pi_1 p_{11} + 0) + (0 + \pi_2 p_{12}) + (\pi_1 p_{21} + 0) + (0 + \pi_2 p_{22}) \\ &= (\pi_1 + \pi_2) = 1, \end{aligned}$$

because $P\pi = \pi$.

2. Suppose now that $\rho_x \in B, \rho_y \notin B$

$$\mathscr{V}\eta(B) = \sum_{i=1}^{4} \pi_1 tr(V_i \rho_x V_i^*) = \pi_1(p_{11} + 0 + p_{21} + 0) = \pi_1$$

3. Finally, suppose that $\rho_x \notin B, \rho_y \in B$

$$\mathscr{V}\eta(B) = \sum_{i=1}^{4} \pi_2 tr(V_i \rho_y V_i^*) = \pi_2(0 + p_{12} + 0 + p_{22}) = \pi_2$$

4. It is easy to see that if $\rho_x, \rho_y \notin B$ then $\mathscr{V}\eta(B) = 0$.

   The conclusion is that, $\mathscr{V}\eta(B) = \eta(B)$ for any measurable set B.
   Therefore, $\mathscr{V}(\eta) = \eta$. ◇

7.7 Some Lemmas for IFS

We want to understand the structure of $\Lambda : \mathcal{M}_N \to \mathcal{M}_N$,

$$\Lambda(\rho) := \sum_{i=1}^{k} p_i F_i = \sum_{i=1}^{k} tr(W_i \rho W_i^*) \frac{V_i \rho V_i^*}{tr(V_i \rho V_i^*)},$$

where V_i, W_i are linear, $\sum_i W_i^* W_i = I$. Such operator is associated in a natural way to a IFS which is not homogeneous. In this section we state a few useful properties which are relevant for our study. The following lemmas hold for any IFS, except for lemma 7.3, for which a proof is known for homogeneous IFS only.

Lemma 7.1. *Let $\{X, F_i, p_i\}_{i=1,\dots,k}$ be a IFS, Ψ a linear functional on X. Then $\mathcal{U} \circ \Psi = \Psi \circ \Lambda$.*

Corollary 7.1. *Let $\mathcal{F} = (X, F_i, p_i)_{i=1,\dots,k}$ be a IFS and let $\rho_0 \in X$. Then $\Lambda(\rho_0) = \rho_0$ if and only if $\mathcal{U}(\Psi(\rho_0)) = \Psi(\rho_0)$, for all Ψ linear functional.*

Lemma 7.2. *Let $\mathcal{F} = \{X, F_i, p_i\}_{i=1,\dots,k}$ be a IFS.*

1. *Let $\rho_0 \in X$ such that $F_i(\rho_0) = \rho_0$, $i = 1, \dots, k$. Then $\mathcal{V}\delta_{\rho_0} = \delta_{\rho_0}$.*
2. *Let $\rho_0 \in X$ such that $\mathcal{V}\delta_{\rho_0} = \delta_{\rho_0}$, then $\Lambda(\rho_0) = \rho_0$.*

Lemma 7.3. *Let $\{X, F_i, p_i\}_{i=1,\dots,k}$ be a homogeneous IFS, $\Lambda = \sum_i p_i F_i$.*

1. *Let ρ_v be the barycenter of a probability measure v. Then $\Lambda(\rho_v)$ is the barycenter of $\mathcal{V}v$, where $\mathcal{V}$ is the associated Markov operator.*
2. *Let μ be an invariant probability measure for $\mathcal{V}$. Then the barycenter of μ, denoted by ρ_μ, is a fixed point of Λ.*

Example 7.5. Let $k = N = 2$,

$$V_1 = \begin{pmatrix} -1 & 0 \\ 0 & 1 \end{pmatrix}, V_2 = \begin{pmatrix} 0 & -\frac{3\sqrt{2}}{4} \\ -\frac{3\sqrt{2}}{2} & 0 \end{pmatrix},$$

$W_1 = (1/2)I$, $W_2 = (\sqrt{3}/2)I$. Then

$$\Lambda(\rho) = \sum_i p_i(\rho) F_i(\rho) = \sum_i tr(W_i \rho W_i^*) \frac{V_i \rho V_i^*}{tr(V_i \rho V_i^*)}$$

$$= \frac{1}{4} V_1 \rho V_1^* + \frac{3}{4} \frac{V_2 \rho V_2^*}{tr(V_2 \rho V_2^*)} = \frac{1}{4} V_1 \rho V_1^* + \frac{3}{4} \frac{V_2 \rho V_2^*}{(\frac{9}{8} + \frac{27}{8}\rho_1)}$$

induces a IFS and it is such that $\rho_0 = \frac{1}{3}|0\rangle\langle 0| + \frac{2}{3}|1\rangle\langle 1|$ is a fixed point, with $F_1(\rho_0) = F_2(\rho_0) = \rho_0$. We can apply lemma 7.2 and conclude that δ_{ρ_0} is an invariant measure for the Markov operator $\mathcal{V}$ associated to the IFS determined by p_i and F_i. $\diamond$

7 A Dynamical Point of View of Quantum Information

The following lemma, a simple variation from results seen in [29], determines reasonable conditions that we will need in order to obtain a fixed point for $\mathscr{L}$ from a certain measure which is invariant for the Markov operator $\mathscr{V}$.

Lemma 7.4. *Let* $\{\mathscr{M}_N, F_i, p_i\}_{i=1,\dots,k}$ *be an IFS which admits an attractive invariant measure* μ *for* $\mathscr{V}$. *Then* $\lim_{n\to\infty} \Lambda^n(\rho_0) = \rho_\mu$, *for every* $\rho_0 \in \mathscr{M}_N$, *where* ρ_μ *is the barycenter of* μ.

7.8 Integral Formulae for the Entropy of IFS

Part of the results we present here in this section are variations of the results presented in [29]. Let (X, d) be a complete separable metric space. Let (V, V^+, e) be a complete state space, $B = \{x \in V^+ : e(x) = 1\}$ and $\mathscr{F} = (X, F_i, p_i)_{i=1,\dots,k}$ the homogeneous IFS induced by the Markov pair $(\Lambda, \{\Lambda_i\}_{i=1}^k)$. Let $I_k = \{1,\dots,k\}$ Let $n \in \mathbb{N}, \iota \in I_k^n, i \in I_k$. Define $F_{\iota i} := F_i \circ F_\iota$ and

$$p_{\iota i}(x) = \begin{cases} p_i(F_\iota x)p_\iota(x) & \text{if } p_\iota(x) \neq 0 \\ 0 & \text{otherwise} \end{cases} \tag{7.32}$$

Proposition 7.5. *Let* $n \in \mathbb{N}$, $f \in m_b(X)$, $x \in X$. *Then*

$$(\mathscr{U}^n f)(x) = \sum_{\iota \in I_k^n} p_\iota(x) f(F_\iota(x))$$

Proposition 7.6. *Let* $x \in B$, $n \in \mathbb{N}$. *Then*

$$\Lambda^n(x) = \sum_{\iota \in I_k^n} p_\iota(x) F_\iota(x).$$

Proposition 7.7. *Let* $\mathscr{F}$ *be a IFS and let* $g : B \to \mathbb{R}$. *Then for* $n \in \mathbb{N}$,

1. *If g is concave (resp. convex, affine) then $\mathscr{U}^n g \leq g \circ \Lambda^n$ (resp. $\mathscr{U}^n g \geq g \circ \Lambda^n$, $\mathscr{U}^n g = g \circ \Lambda^n$).*
2. *If $\bar{x}$ is a fixed point for Λ then the sequence $(\mathscr{U}^n g)(\bar{x}))_{n\in\mathbb{N}}$ is decreasing (resp. increasing, constant) if g is concave (resp. convex, affine).*
   Also suppose that $\mathscr{F}$ is homogeneous. Then
3. *If g is concave (resp. convex, affine), then $\mathscr{U} g$ is concave (resp. convex, affine).*

Define $\eta : \mathbb{R}^+ \to \mathbb{R}$ as

$$\eta(x) = \begin{cases} -x \log x & \text{if } x \neq 0 \\ 0 & \text{if } x = 0 \end{cases}$$

Define the *Shannon–Boltzmann entropy function* as $h : X \to \mathbb{R}^+$,

$$h(x) := \sum_{i=1}^{k} \eta(p_i(x))$$

Let $n \in \mathbb{N}$. Define the *partial entropy* $H_n : X \to \mathbb{R}^+$ as

$$H_n(x) := \sum_{\iota \in I_k^n} \eta(p_\iota(x)),$$

for $n \geq 1$ and $H_0(x) := 0$, $x \in X$. Define, for $x \in X$,

$$\overline{\mathscr{H}}(x) := \limsup_{n \to \infty} \frac{1}{n} H_n(x),$$

the *upper entropy on* x, and

$$\underline{\mathscr{H}}(x) := \liminf_{n \to \infty} \frac{1}{n} H_n(x),$$

the *lower entropy on* x. If such limits are equal, we call its common value the *entropy on* x, denoted by $\mathscr{H}(x)$.

Denote by $M^{\mathscr{V}}(X)$ the set of $\mathscr{V}$-invariant probability measures on X. Let $\mu \in M^{\mathscr{V}}(X)$. The *partial entropy of the measure* μ is defined by

$$H_n(\mu) := \sum_{\iota \in I_k^n} \eta(\langle p_\iota, \mu \rangle),$$

for $n \geq 1$ and $H_0(\mu) := 0$.

Proposition 7.8. *Let* $\mu \in M^{\mathscr{V}}(X)$. *Then the sequences* $(\frac{1}{n} H_n(\mu))_{n \in \mathbb{N}}$ *and* $(H_{n+1}(\mu) - H_n(\mu))_{n \in \mathbb{N}}$ *are nonnegative, decreasing, and have the same limit.*

We denote the common limit of the sequences mentioned in the proposition above as $\mathscr{H}(\mu)$ and we call it the *entropy of the measure* μ, i.e.,

$$\mathscr{H}(\mu) := \lim_{n \to \infty} \frac{1}{n} H_n(\mu) = \lim_{n \to \infty} (H_{n+1}(\mu) - H_n(\mu))$$

The following result gives us an integral formula for entropy, and also a relation between the entropies defined before. We write $S(\mu) := M^{\mathscr{V}}(X) \cap \mathrm{Lim}(\mathscr{V}^n \mu)_{n \in \mathbb{N}}$, where $\mathrm{Lim}(\mathscr{V}^n \mu)_{n \in \mathbb{N}}$ is the convex hull of the set of accumulation points of $(\mathscr{V}^n \mu)_{n \in \mathbb{N}}$, and $S_{\mathscr{F}}(\mu)$ is the set $S(\mu)$ associated to the Markov operator induced by the IFS $\mathscr{F}$. For the definition of compact structure and (C, τ)-continuity, see [29].

7 A Dynamical Point of View of Quantum Information

Theorem 7.2. *[29] (Integral formula for entropy of homogeneous IFS, compact case). Let (C, τ) be a metrizable compact structure (V, V^+, e) such that $(\Lambda, \{\Lambda_i\}_{i=1}^k)$ is (C, τ)-continuous. Assume that $\rho_0 \in B_C := B \cap C$ is such that $\Lambda(\rho_0) = \rho_0$. Then*

$$\mathcal{H}(\rho_0) = \mathcal{H}(\nu) = \int_X h \, d\nu$$

for each $\nu \in S_{\mathcal{F}_C}(\delta_{\rho_0})$, where $\mathcal{F}_C$ is the IFS $\mathcal{F}$ restricted to (B_C, τ).

The analogous result for hyperbolic IFS is the following.

Theorem 7.3. *[29] Let $\mathcal{F} = (X, F_i, p_i)_{i=1,\ldots,k}$ be a hyperbolic IFS, $x \in X$, $\mu \in M^1(X)$ an invariant attractive measure for $\mathcal{F}$. Then*

$$\mathcal{H}(x) = \lim_{n \to \infty} (H_{n+1}(x) - H_n(x))$$

and

$$\mathcal{H}(x) = \mathcal{H}(\mu) = \int_X h \, d\mu.$$

7.9 Some Calculations on Entropy

Let U be a unitary matrix of order mn acting on $\mathcal{H}_m \otimes \mathcal{H}_n$. Its Schmidt decomposition is

$$U = \sum_{i=1}^K \sqrt{q_i} V_i^A \otimes V_i^B, \; K = min\{m^2, n^2\}$$

The operators V_i^A and V_i^B act on certain Hilbert spaces $\mathcal{H}_m$ and $\mathcal{H}_n$, respectively. We also have that $\sum_{i=1}^K q_i = 1$. Let $\sigma = \rho_A \otimes \rho_*^B = \rho_A \otimes I_n/n$ and define

$$\Lambda(\rho_A) := tr_B(U\sigma U^*) = \sum_{i=1}^K q_i V_i^A \rho_A V_i^{A*}$$

Recall that

$$tr_B(|a_1\rangle\langle a_2| \otimes |b_1\rangle\langle b_2|) := |a_1\rangle\langle a_2| tr(|b_1\rangle\langle b_2|)$$

where $|a_1\rangle$ and $|a_2\rangle$ are vectors on the state space of A and $|b_1\rangle$ and $|b_2\rangle$ are vectors on the state space of B. The trace on the right side is the usual trace on B. A calculation shows that if $\rho_*^A = I_m/m$, then $\Lambda(\rho_*^A) = \rho_*^A$ and so Λ is such that $\Lambda(I_m/m) = I_m/m$ and Λ is trace preserving.

Let $\mathcal{F}$ be the homogeneous IFS associated to the V_i^A, that is, $p_i(\rho) = tr(q_i V_i^A \rho V_i^{A*})$, $F_i(\rho) = (q_i V_i^A \rho V_i^{A*})/tr(q_i V_i^A \rho V_i^{A*})$ and let ρ_0 be a fixed point of $\Lambda = \sum_i p_i F_i$. Following [29], we have that ρ_0 is the barycenter of $\mathcal{V}^n \delta_{\rho_0}$, $n \in \mathbb{N}$. By theorem 7.2, we can calculate the entropy of such IFS. In this case we have

$$\mathcal{H}(\rho_0) = \mathcal{H}(\nu) = \int_{\mathcal{M}_N} h \, d\nu, \qquad (7.33)$$

where $\nu \in M^{\mathcal{V}}(X) \cap \mathrm{Lim}(\mathcal{V}^n \delta_{\rho_0})_{n \in \mathbb{N}}$. $\qquad \Diamond$

Let $\mathcal{F} = (\mathcal{M}_N, F_i, p_i)_{i=1,\dots,k}$ be an IFS, $\Lambda(\rho) = \sum_i p_i F_i$. Let $\mathcal{U}$ be the conjugate of $\mathcal{V}$. By proposition 7.5,

$$(\mathcal{U}^n h)(\rho) = \sum_{\iota \in I_k^n(\rho)} p_\iota(\rho) h(F_\iota(\rho))$$

and since $h(\rho) = \sum_{j=1}^k \eta(p_j(\rho))$, we have, for $\iota = (i_1, \dots, i_n)$, and every $\rho_0 \in \mathcal{M}_N$,

$$\int_{\mathcal{M}_N} h \, d\mathcal{V}^n \delta_{\rho_0} = \int_{\mathcal{M}_N} \mathcal{U}^n h \, d\delta_{\rho_0} \qquad (7.34)$$

$$= -\int_{\mathcal{M}_N} \sum_{\iota \in I_k^n(\rho)} p_\iota(\rho) \sum_{j=1}^k p_j(F_\iota(\rho)) \log p_j(F_\iota(\rho)) d\delta_{\rho_0} \quad (7.35)$$

$$= -\sum_{\iota \in I_k^n(\rho_0)} p_\iota(\rho_0) \sum_{j=1}^k p_j(F_\iota(\rho_0)) \log p_j(F_\iota(\rho_0)) \qquad (7.36)$$

$$= -\sum_{\iota \in I_k^n(\rho_0)} p_{i_1}(\rho_0) p_{i_2}(F_{i_1}\rho_0) \cdots p_{i_n}(F_{i_{n-1}}(F_{i_{n-2}}(\cdots(F_{i_1}\rho_0))))$$

$$\qquad (7.37)$$

$$\times \sum_{j=1}^k p_j(F_{i_n}(F_{i_{n-1}}(\cdots(F_{i_1}\rho_0))))$$

$$\log p_j(F_{i_n}(F_{i_{n-1}}(\cdots(F_{i_1}\rho_0)))) = (\mathcal{U}^n h)(\rho_0) \qquad (7.38)$$

Suppose $\Lambda(\rho_0) = \rho_0$. We have by proposition 7.7, since h is concave, that $(\mathcal{U}^n h)_{n \in \mathbb{N}}$ is decreasing, $\mathcal{U}^n h \le h \circ \Lambda^n$ and so

$$\int_{\mathcal{M}_N} h \, d\mathcal{V}^n \delta_{\rho_0} \le h(\Lambda^n(\rho_0)) = h(\rho_0), \qquad (7.39)$$

for every n.

7 A Dynamical Point of View of Quantum Information

7.10 An Expression for a Stationary Entropy

In this section we present a definition of entropy which captures a stationary behavior.

Let H be a hermitian operator and V_i, $i = 1, \ldots, k$ linear operators. We can define the dynamics $F_i : \mathcal{M}_N \to \mathcal{M}_N$:

$$F_i(\rho) := \frac{V_i \rho V_i^*}{tr(V_i \rho V_i^*)} \tag{7.40}$$

Let W_i, $i = 1, \ldots, k$ be linear and such that $\sum_{i=1}^{k} W_i^* W_i = I$. This determines functions $p_i : \mathcal{M}_N \to \mathbb{R}$,

$$p_i(\rho) := tr(W_i \rho W_i^*) \tag{7.41}$$

Then we have $\sum_{i=1}^{k} p_i(\rho) = 1$, for every ρ. Therefore a family $W := \{W_i\}_{i=1,\ldots,k}$ determines a QIFS $\mathscr{F}_W$,

$$\mathscr{F}_W = \{\mathcal{M}_N, F_i, p_i\}_{i=1,\ldots,k}$$

with F_i, p_i given by (7.40) and (7.41).

Different choices of W_i, $i = 1, 2 \ldots, k$, as above, determine different invariant probabilities.

We introduce the following definition of entropy

Definition 7.7. Suppose that we have a QIFS such that there is a unique attractive invariant measure for the Markov operator $\mathcal{V}$ associated to $\mathscr{F}_W$. Let ρ_W be the barycenter of such measure. Define

$$h_V(W) := -\sum_{i=1}^{k} tr(W_i \rho_W W_i^*) \sum_{j=1}^{k} tr\left(\frac{W_j V_i \rho_W V_i^* W_j^*}{tr(V_i \rho_W V_i^*)}\right) \log tr\left(\frac{W_j V_i \rho_W V_i^* W_j^*}{tr(V_i \rho_W V_i^*)}\right) \tag{7.42}$$

Remember that by lemma 7.4, we have that ρ_W is a fixed point for

$$\widehat{\mathscr{L}}_{\mathscr{F}_W}(\rho) := \sum_{i=1}^{k} p_i(\rho) F_i(\rho) = \sum_{i=1}^{k} tr(W_i \rho W_i^*) \frac{V_i \rho V_i^*}{tr(V_i \rho V_i^*)} \tag{7.43}$$

Lemma 7.5. We have that $0 \leq h_V(W) \leq \log k$, for every family W_i of linear operators satisfying $\sum_{i=1}^{k} W_i^* W_i = I$. Also, for any given dynamics V the maximum can be reached.

We also define

$$\mathscr{L}_{\mathscr{F}_W}(\rho) := \sum_{i=1}^{k} tr(W_i \rho W_i^*) V_i \rho V_i^* \tag{7.44}$$

Note that by the construction made on Sect.7.10, we have $h_V(W) = \mathscr{U}h(\rho_W)$, where $\mathscr{U}h(\rho) = \sum_i p_i(\rho)h(F_i(\rho))$. $\diamond$

Lemma 7.6. *Let $\mathscr{F} = (\mathscr{M}_N, F_i, p_i)$ be a QIFS, with F_i, p_i in the form (7.40) and (7.41). Suppose there is $\rho_0 \in \mathscr{M}_N$ such that δ_{ρ_0} is the unique $\mathscr{V}$-invariant measure. Then $\widehat{\mathscr{L}}_{\mathscr{F}}(\rho_0) = \rho_0$ (7.43) and*

$$\int \mathscr{U}^n h \, d\delta_{\rho_0} = \mathscr{U}^n h(\rho_0) = h(\rho_0),$$

for all $n \in \mathbb{N}$. Besides, $\mathscr{U}^n h(\rho_0) = \mathscr{U}h(\rho_0)$ and so

$$h_V(W) = \mathscr{U}^n h(\rho_0),$$

for all $n \in \mathbb{N}$.

Lemma 7.7. *Let μ be a $\mathscr{V}$-invariant attractive measure. Then if ρ_μ is the barycenter of μ we have, for any ρ,*

$$\lim_{n \to \infty} \mathscr{U}^n h(\rho) = \int \mathscr{U}h \, d\mu = \int h \, d\mu \leq h(\rho_\mu) \tag{7.45}$$

Lemma 7.8. *Let $\mathscr{F} = (\mathscr{M}_N, F_i, p_i)$ be a QIFS, with F_i, p_i in the form (7.40) and (7.41). Suppose that ρ is the unique point such that $\widehat{\mathscr{L}}_{\mathscr{F}}(\rho) = \rho$. Suppose that $F_i(\rho) = \rho, i = 1, \ldots, k$. Then*

$$\mathscr{U}^n h(\rho) = h(\rho),$$

$n = 1, 2, \ldots$, and therefore $h_V(W)$ does not depend on n.

7.11 Entropy and Markov Chains

Let V_i, W_i be linear operators, $i = 1, \ldots, k$, $\sum_{i=1}^{k} W_i^* W_i = I$. Suppose the V_i are fixed and determine a dynamics given by $F_i : \mathscr{M}_N \to \mathscr{M}_N, i = 1, \ldots, k$. Define

$$P := \{(p_1, \ldots, p_k) : p_i : \mathscr{M}_N \to \mathbb{R}^+, i = 1, \ldots, k, \sum_{i=1}^{k} p_i(\rho) = 1, \forall \rho \in \mathscr{M}_N\}$$

$$P' := P \cap \{(p_1, \ldots, p_k) : \exists W_i, i = 1, \ldots, k : p_i(\rho) = tr(W_i \rho W_i^*),$$

$$W_i \text{ linear}, \sum_i W_i^* W_i = I\}$$

$$\mathscr{M}_F := \{\mu \in M^1(\mathscr{M}_N) : \exists p \in P' \text{ such that } \mathscr{V}_p \mu = \mu\},$$

7 A Dynamical Point of View of Quantum Information

where $\mathcal{V}_p : M^1(\mathcal{M}_N) \to M^1(\mathcal{M}_N)$,

$$\mathcal{V}_p(\mu)(B) := \sum_{i=1}^{k} \int_{F_i^{-1}(B)} p_i \, d\mu$$

Note that a family $W := \{W_i\}_{i=1,\dots,k}$ determines a QIFS $\mathcal{F}_W$,

$$\mathcal{F}_W = \{\mathcal{M}_N, F_i, p_i\}_{i=1,\dots,k}$$

As done in the previous section we introduce the following definition (which is in some sense stationary)

$$h_V(W) := -\sum_{i=1}^{k} \frac{tr(W_i \rho_W W_i^*)}{tr(V_i \rho_W V_i^*)} \sum_{j=1}^{k} tr\left(W_j V_i \rho_W V_i^* W_j^*\right) \log\left(\frac{tr(W_j V_i \rho_W V_i^* W_j^*)}{tr(V_i \rho_W V_i^*)}\right)$$

(7.46)

where as before, ρ_W denotes the barycenter of the unique attractive invariant measure for the Markov operator $\mathcal{V}$ associated to $\mathcal{F}_W$.

Let $P = (p_{ij})_{i,j=1,\dots,N}$ be a stochastic, irreducible matrix. Let p be the stationary vector of P. The entropy of P is defined as

$$H(P) := -\sum_{i,j=1}^{N} p_i \, p_{ij} \log p_{ij}$$

(7.47)

We consider an example which shows that the usual Markov chain entropy can be realized as the entropy associated to a certain QIFS.

Example 7.6. (Homogeneous case, 4 matrices). Let $N = 2, k = 4$ and

$$V_1 = \begin{pmatrix} \sqrt{p_{00}} & 0 \\ 0 & 0 \end{pmatrix}, V_2 = \begin{pmatrix} 0 & \sqrt{p_{01}} \\ 0 & 0 \end{pmatrix},$$

$$V_3 = \begin{pmatrix} 0 & 0 \\ \sqrt{p_{10}} & 0 \end{pmatrix}, V_4 = \begin{pmatrix} 0 & 0 \\ 0 & \sqrt{p_{11}} \end{pmatrix}$$

Note that

$$\sum_i V_i^* V_i = \begin{pmatrix} p_{00} + p_{10} & 0 \\ 0 & p_{01} + p_{11} \end{pmatrix}$$

and so $\sum_i V_i^* V_i = I$ if we suppose that

$$P := \begin{pmatrix} p_{00} & p_{01} \\ p_{10} & p_{11} \end{pmatrix}$$

is column-stochastic. We have

$$V_1 \rho V_1^* = \begin{pmatrix} p_{00}p_1 & 0 \\ 0 & 0 \end{pmatrix}, \ V_2 \rho V_2^* = \begin{pmatrix} p_{01}p_4 & 0 \\ 0 & 0 \end{pmatrix}$$

$$V_3 \rho V_3^* = \begin{pmatrix} 0 & 0 \\ 0 & p_{10}p_1 \end{pmatrix}, \ V_4 \rho V_4^* = \begin{pmatrix} 0 & 0 \\ 0 & p_{11}p_4 \end{pmatrix}$$

so

$$tr(V_1 \rho V_1^*) = p_{00}p_1, tr(V_2 \rho V_2^*) = p_{01}p_4$$
$$tr(V_3 \rho V_3^*) = p_{10}p_1, tr(V_4 \rho V_4^*) = p_{11}p_4$$

The fixed point of $\Lambda(\rho) = \sum_i V_i \rho V_i^*$ is

$$\rho_V = \begin{pmatrix} \frac{p_{01}}{1-p_{00}+p_{01}} & 0 \\ 0 & \frac{1-p_{00}}{1-p_{00}+p_{01}} \end{pmatrix}$$

Let $\pi = (\pi_1, \pi_2)$ such that $P\pi = \pi$. We know that

$$\pi = \left(\frac{p_{01}}{1 - p_{00} + p_{01}}, \frac{1 - p_{00}}{1 - p_{00} + p_{01}} \right) \tag{7.48}$$

Then the nonzero entries of ρ_V are the entries of π and so we associate the fixed point of P to the fixed point of a certain Λ in a natural way. Let us calculate $h_V(W)$. Note that Λ defined above is associated to a homogeneous IFS. Then $W_i = V_i$, $i = 1, \ldots, k$ and

$$h_V(W) = h_V(V)$$

$$= -\sum_{i=1}^{k} \frac{tr(W_i \rho_V W_i^*)}{tr(V_i \rho_V V_i^*)} \sum_{j=1}^{k} tr\left(W_j V_i \rho_V V_i^* W_j^* \right) \log \left(\frac{tr(W_j V_i \rho_V V_i^* W_j^*)}{tr(V_i \rho_V V_i^*)} \right)$$

$$= -\sum_{i,j} tr\left(V_j V_i \rho_V V_i^* V_j^* \right) \log \left(\frac{tr(V_j V_i \rho_V V_i^* V_j^*)}{tr(V_i \rho_V V_i^*)} \right) \tag{7.49}$$

A simple calculation yields $H(P) = h_V(V)$, where $H(P)$ is the entropy of P, given by (7.47). This shows that the entropy of Markov chains is a particular case of the entropy for QIFS defined before. $\diamond$

In a similar way, we can reach the same conclusion for the nonhomogeneous case, 4 matrices, and also for 2 matrices [1]. $\diamond$

Lemma 7.9. *Let V_{ij} be matrices of order n,*

$$V_{ij} = \sqrt{p_{ij}} |i\rangle \langle j|$$

for $i, j = 1, \ldots, n$. Let

$$\Lambda_P(\rho) := \sum_{i,j} V_{ij}\rho V_{ij}^*$$

where $P = (p_{ij})_{i,j=1,\dots,n}$. Then for all n, $\Lambda_P^n(\rho) = \Lambda_{P^n}(\rho)$.

Corollary 7.2. *Under the lemma hypothesis, we have* $\lim_{n\to\infty}\Lambda_P^n(\rho) = \Lambda_\pi(\rho)$, *where* $\pi = \lim_{n\to\infty}P^n$ *is the stochastic matrix which has all columns equal to the stationary vector for* P.

7.12 Capacity-Cost Function and Pressure

Recall that every trace preserving, completely positive (CP) mapping can be written in the Stinespring–Kraus form,

$$\Lambda(\rho) = \sum_{i=1}^{k} V_i\rho V_i^*, \quad \sum_{i=1}^{k} V_i^* V_i = I,$$

for V_i linear operators. These mappings are also called *quantum channels*.

This is one of the main motivations for considering the class of operators (a generalization of the above ones) described in the present work. These are natural objets in the study of Quantum Computing.

Definition 7.8. The *Holevo capacity* for sending classic information via a quantum channel Λ is defined as

$$C_\Lambda := \max_{\substack{p_i \in [0,1] \\ \rho_i \in \mathcal{M}_N}} S\left(\sum_{i=1}^{n} p_i \Lambda(\rho_i)\right) - \sum_{i=1}^{n} p_i S\left(\Lambda(\rho_i)\right) \tag{7.50}$$

where $S(\rho) = -tr(\rho\log\rho)$ is the von Neumann entropy. The maximum is, therefore, over all choices of p_i, $i = 1,\dots,n$ and density operators ρ_i, for some $n \in \mathbb{N}$. The Holevo capacity establishes an upper bound on the amount of information that a quantum system contains [24].

Definition 7.9. Let Λ be a quantum channel. Define the *minimum output entropy* as

$$H^{min}(\Lambda) := \min_{|\psi\rangle} S(\Lambda(|\psi\rangle\langle\psi|))$$

Additivity conjecture We have that

$$C_{\Lambda_1 \otimes \Lambda_2} = C_{\Lambda_1} + C_{\Lambda_2}$$

Minimum output entropy conjecture For any channels Λ_1 and Λ_2,

$$H^{min}(\Lambda_1 \otimes \Lambda_2) = H^{min}(\Lambda_1) + H^{min}(\Lambda_2)$$

In [27], is it shown that the additivity conjecture is equivalent to the minimum output entropy conjecture, and in [12] we obtain a counterexample for this last conjecture.
$\diamond$

We will be interested here in a different class of problem which concern maximization (and not minimization) of entropy plus a given potential (a cost) [9,13,14].

Definition 7.10. Let M_F be the set of invariant measures defined in the Sect. 7.11 and let H be a hermitian operator. For $\mu \in \mathcal{M}_F$ let ρ_μ be its barycenter. Define the capacity-cost function $C : \mathbb{R}^+ \to \mathbb{R}^+$ as

$$C(a) := \max_{\mu \in \mathcal{M}_F} \{h_{W,V}(\rho_\mu) : tr(H\rho_\mu) \le a\} \tag{7.51}$$

The following analysis is inspired in [21]. There is a relation between the cost-capacity function and the variational problem for pressure. In fact, let $F : \mathbb{R}^+ \to \mathbb{R}^+$ be the function given by

$$F(\lambda) := \sup_{\mu \in \mathcal{M}_F} \{h_{W,V}(\rho_\mu) - \lambda tr(H\rho_\mu)\} \tag{7.52}$$

We have the following fact. There is a unique probability measure $v_0 \in \mathcal{M}_F$ such that

$$F(\lambda) = h_{W,V}(\rho_{v_0}) - \lambda tr(H\rho_{v_0})$$

Also, we have the following lemma:

Lemma 7.10. Let $\lambda \le 0$, and $\hat{a} = tr(H\rho_{v_0})$. Then

$$C(\hat{a}) = h_{W,V}(\rho_{v_0}) \tag{7.53}$$

7.13   Analysis of the Pressure Problem

Let V_i, W_i be linear operators, $i = 1, \ldots, k$, with $\sum_i W_i^* W_i = I$ and let

$$H\rho := \sum_{i=1}^{k} H_i \rho H_i^* \tag{7.54}$$

a hermitian operator. We are interested in obtaining a version of the variational principle of pressure for our context. We will see that the pressure will be maximum whenever we have a certain relation between the potential H and the probability distribution considered (and represented here by the W_i). Initially we consider that the V_i are fixed. From the reasoning described below, it will be natural to consider as definition of pressure the maximization among the possible stationary W_i of the expression

$$h_V(W) + \sum_{j=1}^{k} \log \left(tr(H_j \rho_\beta H_j^*) tr(V_j \rho_\beta V_j^*) \right) tr(W_j \rho_W W_j^*)$$

Remember that different choices of $W_i, i = 1, 2, \ldots, k$, represent different choices of invariant probabilities.

Our analysis uses the following important lemma.

Lemma 7.11. *If $r_1, \ldots, r_k$ and $q_1, \ldots, q_k$ are two probability distributions over $1, \ldots, k$, such that $r_j > 0$, $j = 1, \ldots, k$, then*

$$-\sum_{j=1}^{k} q_j \log q_j + \sum_{j=1}^{k} q_j \log r_j \le 0$$

and equality holds if and only if $r_j = q_j$, $j = 1, \ldots, k$.

For the proof, see [25].

The potential given by (7.54) together with the V_i induces an operator, given by

$$\mathcal{L}_H(\rho) := \sum_{i=1}^{k} tr(H_i \rho H_i^*) V_i \rho V_i^* \qquad (7.55)$$

We know that such operator admits an eigenvalue β with its associate eigenstate ρ_β. Then $\mathcal{L}_H(\rho_\beta) = \beta \rho_\beta$ implies

$$\sum_{i=1}^{k} tr(H_i \rho_\beta H_i^*) V_i \rho_\beta V_i^* = \beta \rho_\beta \qquad (7.56)$$

In coordinates, (7.56) can be written as

$$\sum_{i=1}^{k} tr(H_i \rho_\beta H_i^*)(V_i \rho_\beta V_i^*)_{jl} = \beta(\rho_\beta)_{jl} \qquad (7.57)$$

Remark. Comparing the above calculation with the problem of finding an eigenvalue λ of a matrix $A = (a_{ij})$, we have that (7.56) can be seen as the analogous of the expression

$$l E^A = \lambda l \qquad (7.58)$$

Above, the matrix A plays the role of a potential, E^A denotes the matrix with entries $e^{a_{ij}}$ and l_j denotes the j-th coordinate of the left eigenvector l associated to the eigenvalue λ. In coordinates,

$$\sum_{i} l_i e^{a_{ij}} = \lambda l_j, i, j = 1, \ldots, k \qquad (7.59)$$

From this point we can perform two calculations. First, considering (7.56) we will take the trace of such equation in order to obtain a scalar equation. In spite of the fact that taking the trace makes us lose part of the information given by the eigenvector equation, we are still able to obtain a version of what we will call a *basic inequality*, which can be seen as a quantum IFS version of the variational principle of pressure. However, there is an algebraic drawback to this approach, namely, that we will not be able to have the classic variational problem as a particular case of such inequality (such disadvantage is a consequence of taking the trace, clearly). The second calculation will consider (7.57), the coordinate equations associated to the matrix equation for the eigenvectors. In this case we also obtain a basic inequality, but now we will have the classic variational problem of pressure as a particular case.

An important question which is of our interest, regarding both calculations mentioned above, is the question of whether it is possible for a given system to attain its maximum pressure. It is not clear that given any dynamics, we can obtain a measure reaching such a maximum. With respect to our context, we will state sufficient conditions on the dynamics which allows us to determine expressions for the measure which maximizes the pressure. We now perform the calculations mentioned above.

Based on (7.56), define

$$r_j = \frac{1}{\beta} tr(H_j \rho_\beta H_j^*) tr(V_j \rho_\beta V_j^*) \tag{7.60}$$

So we have $\sum_j r_j = 1$. Let

$$q_j^i := tr\left(\frac{W_j V_i \rho_W V_i^* W_j^*}{tr(V_i \rho_W V_i^*)} \right) \tag{7.61}$$

where, as before, ρ_W is the fixed point associated to the renormalized operator $\Lambda_{\mathscr{F}_W}$,

$$\Lambda_{\mathscr{F}_W}(\rho) := \sum_{i=1}^k p_i(\rho) F_i(\rho) \tag{7.62}$$

induced by the QIFS $(\mathscr{M}_N, F_i, p_i)_{i=1,\dots,k}$,

$$F_i(\rho) = \frac{V_i \rho V_i^*}{tr(V_i \rho V_i^*)}$$

and

$$p_i(\rho) = tr(W_i \rho W_i^*)$$

7 A Dynamical Point of View of Quantum Information

Note that we have

$$\sum_{j=1}^{k} q_j^i = \frac{1}{tr(V_i \rho w V_i^*)} \sum_{j=1}^{k} tr(W_j^* W_j V_i \rho w V_i^*)$$

$$= \frac{1}{tr(V_i \rho w V_i^*)} tr(\sum_{j=1}^{k} W_j^* W_j V_i \rho w V_i^*) = 1$$

Then we can apply lemma 7.11 for $r_j, q_j^i, j = 1, \ldots k$, with i fixed, to obtain

$$- \sum_j tr\left(\frac{W_j V_i \rho w V_i^* W_j^*}{tr(V_i \rho w V_i^*)}\right) \log tr\left(\frac{W_j V_i \rho w V_i^* W_j^*}{tr(V_i \rho w V_i^*)}\right)$$

$$+ \sum_j tr\left(\frac{W_j V_i \rho w V_i^* W_j^*}{tr(V_i \rho w V_i^*)}\right) \log \left(\frac{1}{\beta} tr(H_j \rho_\beta H_j^*) tr(V_j \rho_\beta V_j^*)\right) \leq 0 \qquad (7.63)$$

and equality holds if and only if for all i, j,

$$\frac{1}{\beta} tr(H_j \rho_\beta H_j^*) tr(V_j \rho_\beta V_j^*) = \frac{tr(W_j V_i \rho w V_i^* W_j^*)}{tr(V_i \rho w V_i^*)} \qquad (7.64)$$

Then

$$- \sum_j tr\left(\frac{W_j V_i \rho w V_i^* W_j^*}{tr(V_i \rho w V_i^*)}\right) \log tr\left(\frac{W_j V_i \rho w V_i^* W_j^*}{tr(V_i \rho w V_i^*)}\right)$$

$$+ \sum_j tr\left(\frac{W_j V_i \rho w V_i^* W_j^*}{tr(V_i \rho w V_i^*)}\right) \log \left(tr(H_j \rho_\beta H_j^*) tr(V_j \rho_\beta V_j^*)\right)$$

$$\leq \sum_j tr\left(\frac{W_j V_i \rho w V_i^* W_j^*}{tr(V_i \rho w V_i^*)}\right) \log \beta$$

which is equivalent to

$$- \sum_j tr\left(\frac{W_j V_i \rho w V_i^* W_j^*}{tr(V_i \rho w V_i^*)}\right) \log tr\left(\frac{W_j V_i \rho w V_i^* W_j^*}{tr(V_i \rho w V_i^*)}\right)$$

$$+ \sum_j \frac{tr(W_j V_i \rho w V_i^* W_j^*)}{tr(V_i \rho w V_i^*)} \log \left(tr(H_j \rho_\beta H_j^*) tr(V_j \rho_\beta V_j^*)\right) \leq \log \beta \qquad (7.65)$$

Multiplying by $tr(W_i \rho_W W_i^*)$ and summing over the i index, we have

$$h_V(W) + \sum_j \log \left(tr(H_j \rho_\beta H_j^*) tr(V_j \rho_\beta V_j^*) \right)$$

$$\sum_i \frac{tr(W_i \rho_W W_i^*)}{tr(V_i \rho_W V_i^*)} tr(W_j V_i \rho_W V_i^* W_j^*)$$

$$\leq \sum_i tr(W_i \rho_W W_i^*) \log \beta = \log \beta \qquad (7.66)$$

and equality holds if and only if for all i, j,

$$\frac{1}{\beta} tr(H_j \rho_\beta H_j^*) tr(V_j \rho_\beta V_j^*) = \frac{tr(W_j V_i \rho_W V_i^* W_j^*)}{tr(V_i \rho_W V_i^*)} \qquad (7.67)$$

Let us rewrite inequality (7.66). First we use the fact that ρ_W is a fixed point of $\Lambda_{\mathscr{F}_W}$,

$$\sum_{i=1}^k tr(W_i \rho_W W_i^*) \frac{V_i \rho_W V_i^*}{tr(V_i \rho_W V_i^*)} = \rho_W \qquad (7.68)$$

Now we compose both sides of the equality above with the operator

$$\sum_{j=1}^k \log \left(tr(H_j \rho_\beta H_j^*) tr(V_j \rho_\beta V_j^*) \right) W_j^* W_j \qquad (7.69)$$

and then we obtain

$$\sum_{i=1}^k tr(W_i \rho_W W_i^*) \frac{V_i \rho_W V_i^*}{tr(V_i \rho_W V_i^*)} \sum_{j=1}^k \log \left(tr(H_j \rho_\beta H_j^*) tr(V_j \rho_\beta V_j^*) \right) W_j^* W_j$$

$$= \rho_W \sum_{j=1}^k \log \left(tr(H_j \rho_\beta H_j^*) tr(V_j \rho_\beta V_j^*) \right) W_j^* W_j \qquad (7.70)$$

Reordering terms we get

$$\sum_{j=1}^k \log \left(tr(H_j \rho_\beta H_j^*) tr(V_j \rho_\beta V_j^*) \right) \sum_{i=1}^k \frac{tr(W_i \rho_W W_i^*)}{tr(V_i \rho_W V_i^*)} V_i \rho_W V_i^* W_j^* W_j$$

$$= \rho_W \sum_{j=1}^k \log \left(tr(H_j \rho_\beta H_j^*) tr(V_j \rho_\beta V_j^*) \right) W_j^* W_j \qquad (7.71)$$

7 A Dynamical Point of View of Quantum Information   113

Taking the trace on both sides we get

$$\sum_{j=1}^{k} \log\left(tr(H_j \rho_\beta H_j^*)tr(V_j \rho_\beta V_j^*)\right)\sum_{i=1}^{k}\frac{tr(W_i \rho_W W_i^*)}{tr(V_i \rho_W V_i^*)}tr(W_j V_i \rho_W V_i^* W_j^*)$$

$$= \sum_{j=1}^{k}\log\left(tr(H_j \rho_\beta H_j^*)tr(V_j \rho_\beta V_j^*)\right)tr(\rho_W W_j^* W_j) \qquad (7.72)$$

Note that the left hand side of (7.72) is one of the sums appearing in (7.66). Therefore replacing (7.72) into (7.66) gives us the following inequality:

$$h_V(W) + \sum_{j=1}^{k}\log\left(tr(H_j \rho_\beta H_j^*)tr(V_j \rho_\beta V_j^*)\right)tr(W_j \rho_W W_j^*) \leq \log \beta \qquad (7.73)$$

and equality holds if and only if for all i, j,

$$\frac{1}{\beta}tr(H_j \rho_\beta H_j^*)tr(V_j \rho_\beta V_j^*) = \frac{tr(W_j V_i \rho_W V_i^* W_j^*)}{tr(V_i \rho_W V_i^*)} \qquad (7.74)$$

So we have the following result.

Theorem 7.4. *Let $\mathscr{F}_W$ be a QIFS such that there is a unique attractive invariant measure for the associated Markov operator $\mathscr{V}$. Let ρ_W be the barycenter of such measure and let ρ_β be an eigenstate of $\mathscr{L}_H(\rho)$ with eigenvalue β. Then*

$$h_V(W) + \sum_{j=1}^{k}\log\left(tr(H_j \rho_\beta H_j^*)tr(V_j \rho_\beta V_j^*)\right)tr(W_j \rho_W W_j^*) \leq \log \beta \qquad (7.75)$$

and equality holds if and only if for all i, j,

$$\frac{1}{\beta}tr(H_j \rho_\beta H_j^*)tr(V_j \rho_\beta V_j^*) = \frac{tr(W_j V_i \rho_W V_i^* W_j^*)}{tr(V_i \rho_W V_i^*)} \qquad (7.76)$$

In Sect. 7.15 we make some considerations about certain cases in which we can reach an equality in (7.75). $\diamond$

For the calculations regarding expression (7.57), define

$$r_{jlm} = \frac{1}{\beta}tr(H_j \rho_\beta H_j^*)\frac{(V_j \rho_\beta V_j^*)_{lm}}{(\rho_\beta)_{lm}} \qquad (7.77)$$

Then we have $\sum_j r_{jlm} = 1$. Let

$$q_{ij} := tr\left(\frac{W_j V_i \rho_W V_i^* W_j^*}{tr(V_i \rho_W V_i^*)}\right) \qquad (7.78)$$

A calculation similar to the one we have made for (7.75) gives us

$$h_V(W) + \sum_{j=1}^{k} tr(W_j \rho_W W_j^*) \log tr(H_j \rho_\beta H_j^*)$$

$$+ \sum_{j=1}^{k} tr(W_j \rho_W W_j^*) \log\left(\frac{(V_j \rho_\beta V_j^*)_{lm}}{(\rho_\beta)_{lm}}\right) \leq \log \beta \qquad (7.79)$$

and equality holds if and only if for all $i, j, l, m,$

$$\frac{1}{\beta} tr(H_j \rho_\beta H_j^*) \frac{(V_j \rho_\beta V_j^*)_{lm}}{(\rho_\beta)_{lm}} = \frac{tr(W_j V_i \rho_W V_i^* W_j^*)}{tr(V_i \rho_W V_i^*)} \qquad (7.80)$$

$$\diamond$$

7.14 Some Classic Inequality Calculations

A natural question is to ask whether the maximum among normalized W_i, $i = 1, \ldots, k$, for the pressure problem associated to a given potential is realized as the logarithm of the main eigenvalue of a certain Ruelle operator associated to the potential H_i, $i = 1, \ldots, k$. This problem will be considered in this section and also in the next one.

We begin by recalling a classic inequality. Consider

$$-\sum_{j=1}^{k} q_j \log q_j + \sum_{j=1}^{k} q_j \log r_j \leq 0 \qquad (7.81)$$

given by lemma 7.11. Let A be a matrix. If v denotes the left eigenvector of matrix E^A (such that each entry is $e^{a_{ij}}$), then $vE^A = \beta v$ can be written as

$$\sum_{i} v_i e^{a_{ij}} = \beta v_j, \forall j \qquad (7.82)$$

Define

$$r_{ij} := \frac{e^{a_{ij}} v_i}{\beta v_j} \qquad (7.83)$$

So $\sum_i r_{ij} = 1$. Let $q_{ij} > 0$ such that $\sum_i q_{ij} = 1$. By (7.81), we have

$$-\sum_{i=1}^{k} q_{ij} \log q_{ij} + \sum_{i=1}^{k} q_{ij} \log \frac{e^{a_{ij}} v_i}{\beta v_j} \leq 0 \qquad (7.84)$$

7 A Dynamical Point of View of Quantum Information

That is,

$$-\sum_{i=1}^{k} q_{ij} \log q_{ij} + \sum_{i=1}^{k} q_{ij} a_{ij} + \sum_{i=1}^{k} q_{ij} (\log v_i - \log v_j) \le \log \beta \qquad (7.85)$$

Let Q be a matrix with entries q_{ij}, let $\pi = (\pi_1, \dots, \pi_k)$ be the stationary vector associated to Q. Since $\sum_i q_{ij} = 1$, Q is column-stochastic so we write $Q\pi = \pi$. Multiplying the above inequality by π_j and summing the j index, we get

$$-\sum_{j} \pi_j \sum_{i} q_{ij} \log q_{ij} + \sum_{j} \pi_j \sum_{i} q_{ij} a_{ij} + \sum_{j} \pi_j \sum_{i} q_{ij} (\log v_i - \log v_j) \le \log \beta$$

$$(7.86)$$

In coordinates, $Q\pi = \pi$ is $\sum_j q_{ij}\pi_j = \pi_i$, for all i. Then

$$-\sum_{j} \pi_j \sum_{i} q_{ij} \log q_{ij} + \sum_{j} \pi_j \sum_{i} q_{ij} a_{ij}$$

$$+ \sum_{j} \pi_j \sum_{i} q_{ij} \log v_i - \sum_{j} \pi_j \sum_{i} q_{ij} \log v_j \le \log \beta \qquad (7.87)$$

These calculations are well-known and give the following inequality:

$$-\sum_{j} \pi_j \sum_{i} q_{ij} \log q_{ij} + \sum_{j} \pi_j \sum_{i} q_{ij} a_{ij} \le \log \beta \qquad (7.88)$$

Definition 7.11. We call inequality (7.88) the *classic inequality* associated to the matrix A with positive entries, and stochastic matrix Q.

Definition 7.12. For fixed k, and $l, m = 1, \dots, k$ we call the inequality

$$h_V(W) + \sum_{j=1}^{k} tr(W_j \rho_W W_j^*) \log tr(H_j \rho_\beta H_j^*)$$

$$+ \sum_{j=1}^{k} tr(W_j \rho_W W_j^*) \log \left(\frac{(V_j \rho_\beta V_j^*)_{lm}}{(\rho_\beta)_{lm}} \right) \le \log \beta, \qquad (7.89)$$

the *basic inequality* associated to the potential $H\rho = \sum_i H_i \rho H_i^*$ and to the QIFS determined by $V_i, W_i, i = 1, \dots, k$. Equality holds if for all i, j, l, m,

$$\frac{1}{\beta} tr(H_j \rho_\beta H_j^*) \frac{(V_j \rho_\beta V_j^*)_{lm}}{(\rho_\beta)_{lm}} = \frac{tr(W_j V_i \rho_W V_i^* W_j^*)}{tr(V_i \rho_W V_i^*)} \qquad (7.90)$$

$\diamond$

As before ρ_β is an eigenstate of $\mathscr{L}_H(\rho)$ and ρ_W is the barycenter of the unique attractive, invariant measure for the Markov operator $\mathscr{V}$ associated to the QIFS $\mathscr{F}_W$. Given the classic inequality (7.88) we want to compare it to the basic inequality (7.89). More precisely, we would like to obtain operators V_i that satisfy the following: given a matrix A with positive entries and a stochastic matrix Q, there are H_i and W_i such that inequality (7.89) becomes inequality (7.88). We have the following proposition.

Proposition 7.9. *[1] Define*

$$V_1 = \begin{pmatrix} 1 & 0 \\ 0 & 0 \end{pmatrix}, V_2 = \begin{pmatrix} 0 & 1 \\ 0 & 0 \end{pmatrix} \tag{7.91}$$

$$V_3 = \begin{pmatrix} 0 & 0 \\ 1 & 0 \end{pmatrix}, V_4 = \begin{pmatrix} 0 & 0 \\ 0 & 1 \end{pmatrix} \tag{7.92}$$

Let $A = (a_{ij})$ be a matrix with positive entries and $Q = (q_{ij})$ a two-dimensional column-stochastic matrix. Define

$$H_{11} = \begin{pmatrix} \sqrt{e^{a_{11}}} & \sqrt{e^{a_{11}}} \\ 0 & 0 \end{pmatrix}, H_{12} = \begin{pmatrix} \sqrt{e^{a_{12}}} & \sqrt{e^{a_{12}}} \\ 0 & 0 \end{pmatrix} \tag{7.93}$$

$$H_{21} = \begin{pmatrix} 0 & 0 \\ \sqrt{e^{a_{21}}} & \sqrt{e^{a_{21}}} \end{pmatrix}, H_{22} = \begin{pmatrix} 0 & 0 \\ \sqrt{e^{a_{22}}} & \sqrt{e^{a_{22}}} \end{pmatrix} \tag{7.94}$$

and also

$$W_1 = \begin{pmatrix} \sqrt{q_{11}} & 0 \\ 0 & 0 \end{pmatrix}, W_2 = \begin{pmatrix} 0 & \sqrt{q_{12}} \\ 0 & 0 \end{pmatrix} \tag{7.95}$$

$$W_3 = \begin{pmatrix} 0 & 0 \\ \sqrt{q_{21}} & 0 \end{pmatrix}, W_4 = \begin{pmatrix} 0 & 0 \\ 0 & \sqrt{q_{22}} \end{pmatrix} \tag{7.96}$$

Then the basic inequality associated to $W_i, V_i, H_i, i = 1, \ldots, 4, l = m = 1$ or $l = m = 2$, is equivalent to the classic inequality associated to A and Q.

Example 7.7. Let

$$H_1 = \begin{pmatrix} 2i & 2i \\ 0 & 0 \end{pmatrix}, \quad H_2 = I, \quad H_3 = \begin{pmatrix} i\sqrt{2} & i\sqrt{2} \\ 0 & 0 \end{pmatrix}, \quad H_4 = I$$

Then

$$H_1^* = \begin{pmatrix} -2i & 0 \\ -2i & 0 \end{pmatrix}, \quad H_2^* = I, \quad H_3^* = \begin{pmatrix} -i\sqrt{2} & 0 \\ -i\sqrt{2} & 0 \end{pmatrix}, \quad H_4^* = I$$

If we suppose the V_i are the same as from proposition 7.9, we have that ρ_β is diagonal, so

$$tr(H_1\rho_\beta H_1^*) = 4, \quad tr(H_2\rho_\beta H_2^*) = 1, \quad tr(H_3\rho_\beta H_3^*) = 2, \quad tr(H_4\rho_\beta H_4^*) = 1$$

7 A Dynamical Point of View of Quantum Information 117

Then $\mathscr{L}_H(\rho) = \beta\rho$ leads us to

$$4\rho_{11} + \rho_{22} = \beta\rho_{11}$$

$$2\rho_{11} + \rho_{22} = \beta\rho_{22}$$

A simples calculation gives

$$\beta = \frac{5 + \sqrt{17}}{2}$$

with eigenstate

$$\rho_\beta = \frac{4}{7 + \sqrt{17}} \begin{pmatrix} \frac{3+\sqrt{17}}{4} & 0 \\ 0 & 1 \end{pmatrix}$$

We want to calculate the W_i which maximize the basic inequality (7.89). Recall that from proposition 7.9, the choice of V_i we made is such that

$$\frac{(V_j \rho_\beta V_j^*)_{lm}}{(\rho_\beta)_{lm}} = 1,$$

So

$$h_V(W) + \sum_{j=1}^{k} tr(W_j \rho_W W_j^*) \log tr(H_j \rho_\beta H_j^*) \leq \log \beta \tag{7.97}$$

and equality holds if and only if, for all i, j, l, m,

$$\frac{1}{\beta} tr(H_j \rho_\beta H_j^*) \frac{(V_j \rho_\beta V_j^*)_{lm}}{(\rho_\beta)_{lm}} = \frac{tr(W_j V_i \rho_W V_i^* W_j^*)}{tr(V_i \rho_W V_i^*)} \tag{7.98}$$

Choose, for instance, $l = m = 1$. Then condition (7.98) becomes

$$\frac{1}{\beta} tr(H_j \rho_\beta H_j^*) = \frac{tr(W_j V_i \rho_W V_i^* W_j^*)}{tr(V_i \rho_W V_i^*)} \tag{7.99}$$

To simplify calculations, write $\widehat{W}_i = W_i^* W_i$ and $\widehat{W}_i = (w_{ij}^i)$. Then we get

$$\frac{tr(H_i \rho_\beta H_i^*)}{\beta} = w_{11}^i = w_{22}^i, i = 1, \ldots, 4 \tag{7.100}$$

So we conclude

$$W_i = \frac{1}{\sqrt{\beta}} \begin{pmatrix} \sqrt{tr(H_i \rho_\beta H_i^*)} & 0 \\ 0 & \sqrt{tr(H_i \rho_\beta H_i^*)} \end{pmatrix}, i = 1, \ldots, 4 \tag{7.101}$$

That is,

$$W_1 = \frac{2}{\sqrt{\beta}} I, W_2 = \frac{1}{\sqrt{\beta}} I, W_3 = \frac{\sqrt{2}}{\sqrt{\beta}} I, W_4 = \frac{1}{\sqrt{\beta}} I \tag{7.102}$$

Note that

$$\sum_i W_i^* W_i = \frac{4 + \sqrt{2}}{\sqrt{\beta}} I \neq I$$

To solve that, we renormalize the potential. Define

$$\tilde{H}_i := \sqrt{\alpha} H_i \tag{7.103}$$

where

$$\alpha := \frac{\sqrt{\beta}}{4 + \sqrt{2}} \tag{7.104}$$

Then a calculation shows that $\mathcal{L}_{\tilde{H}}(\rho) = \tilde{\beta}\rho$ gives us the same eigenstate as before, that is $\rho_{\tilde{\beta}} = \rho_{\beta}$. But note that the associated eigenvalue becomes $\tilde{\beta} = \alpha\beta$. Now, note that it is possible to renormalize the W_i in such a way that we obtain $\tilde{W}_i$ with $\sum_i \tilde{W}_i^* \tilde{W}_i = I$, and that these maximize the basic inequality for the H_i initially fixed. In fact, given the renormalized $\tilde{H}_i$, define

$$\tilde{W}_i = \sqrt{\alpha} W_i, i = 1, \ldots, 4 \tag{7.105}$$

Note that $\sum_i \tilde{W}_i^* \tilde{W}_i = I$. Also we obtain

$$h_V(\tilde{W}) + \sum_{j=1}^{k} tr(\tilde{W}_j \rho_{\tilde{W}} \tilde{W}_j^*) \log tr(\sqrt{\alpha} H_j \rho_\beta \sqrt{\alpha} H_j^*) \leq \log \alpha\beta \tag{7.106}$$

which is equivalent to

$$h_V(\tilde{W}) + \sum_{j=1}^{k} tr(\tilde{W}_j \rho_{\tilde{W}} \tilde{W}_j^*) \log(\alpha tr(H_j \rho_\beta H_j^*)) \leq \log \alpha + \log \beta \tag{7.107}$$

That is

$$h_V(\tilde{W}) + \sum_{j=1}^{k} tr(\tilde{W}_j \rho_{\tilde{W}} \tilde{W}_j^*) \log \alpha$$

$$+ \sum_{j=1}^{k} tr(\tilde{W}_j \rho_{\tilde{W}} \tilde{W}_j^*) \log tr(H_j \rho_\beta H_j^*) \leq \log \alpha + \log \beta, \tag{7.108}$$

7 A Dynamical Point of View of Quantum Information   119

and cancelling $\log \alpha$, we get the same inequality as for the nonrenormalized H_i. As we have seen before, such $\tilde{W}_i$ gives us equality. Hence

$$h_V(\tilde{W}) + \sum_{j=1}^{k} tr(\tilde{W}_j \rho_{\tilde{W}} \tilde{W}_j^*) \log tr(H_j \rho_\beta H_j^*) = \log \beta \qquad (7.109)$$

$\diamond$

7.15 Remarks on the Problem of Pressure and Quantum Mechanics

One of the questions we are interested in is to understand how to formulate a variational principle for pressure in the context of quantum information theory. An appropriate combination of such theories could have as a starting point a relation between the inequality for positive numbers

$$-\sum_i q_i \log q_i + \sum_i q_i \log p_i \leq 0,$$

(seen in certain proofs of the variational principle of pressure), and the entropy for QIFS we defined before. We have carried out such a plan and then we have obtained the basic inequality, which can be written as

$$h_V(W) + \sum_{j=1}^{k} \log\left(tr(H_j \rho_\beta H_j^*) tr(V_j \rho_\beta V_j^*)\right) tr(W_j \rho_W W_j^*) \leq \log \beta \qquad (7.110)$$

where equality holds if and only if for all i, j,

$$\frac{1}{\beta} tr(H_j \rho_\beta H_j^*) tr(V_j \rho_\beta V_j^*) = \frac{tr(W_j V_i \rho_W V_i^* W_j^*)}{tr(V_i \rho_W V_i^*)} \qquad (7.111)$$

As we have discussed before, it is not clear that given any dynamics, we can obtain a measure such that we can reach the maximum value $\log \beta$. Considering particular cases, we can suppose, for instance, that the V_i are unitary. In this way, we combine in a natural way a problem of classic thermodynamics, with an evolution which has a quantum character. In this particular setting, we have for each i that $V_i V_i^* = V_i^* V_i = I$ and then the basic inequality becomes

$$h_V(W) + \sum_{j=1}^{k} tr(W_j \rho_W W_j^*) \log tr(H_j \rho_\beta H_j^*) \leq \log \beta \qquad (7.112)$$

and equality holds if and only if for all i, j,

$$\frac{1}{\beta} tr(H_j \rho_\beta H_j^*) = tr(W_j V_i \rho_W V_i^* W_j^*) \tag{7.113}$$

We have the following:

Lemma 7.12. *Given a QIFS with a unitary dynamics (i.e., V_i is unitary for each i), there are $\hat{W}_i$ which maximize (7.110), i.e., such that*

$$h_V(\hat{W}) + \sum_{j=1}^{k} tr(\hat{W}_j \rho_{\hat{W}} \hat{W}_j^*) \log tr(H_j \rho_\beta H_j^*) = \log \beta \tag{7.114}$$

The above lemma also holds for the basic inequality in coordinates, given by (7.89). Also, it is immediate to obtain a similar version of the above lemma for any QIFS such that the V_i are multiples of the identity, and also for QIFS such that ρ_W fixes each branch of the QIFS, that is, satisfying

$$\frac{V_i \rho_W V_i^*}{tr(V_i \rho_W V_i^*)} = \rho_W$$

References

1. Baraviera, A., Lardizabal, C.F., Lopes, A.O., Terra Cunha, M.: A Thermodynamic Formalism for density matrices in Quantum Information. Appl. Math. Res. Express **1**, 63–118 (2010)
2. Baraviera, A., Lardizabal, C.F., Lopes, A.O., Terra Cunha, M.: Quantum Stochastic Processes, Quantum Iterated Function Systems and Entropy. São Paulo J. Math. Sci. (2010)
3. Benatti, F.: Dynamics, Information and Complexity in Quantum Systems. Springer (2009)
4. Benenti, G., Casati, G. Strini, G.: Principles of Quantum Computation and Information, Vol I and II. World scientific (2007)
5. Bengtsson, I., Życzkowski, K.: Geometry of Quantum States. Cambridge University Press, Cambridge (2006)
6. Busch, P., Ruch, E.: The measure cone: irreversibility as a geometrical phenomenon. Int. J. Quantum Chem. **41**, 163–185 (1992)
7. Castro, G., Lopes, A.O.: KMS states, entropy and a variational principle for pressure. Real Anal. Exch. (2009)
8. Gardiner, C.W., Zoller, P.: Quantum Noise. Springer (2004)
9. Gray, R.M.: Entropy and Information Theory. Springer, New York (1990)
10. de Gosson, M.: Symplectic Geometry and Quantum Mechanics. Birkhauser (2006)
11. Gustafson, S., Sigal, I.: Mathematical Concepts of Quantum Mechanics. Springer (2003)
12. Hastings, M.B.: A counterexample to additivity of minimum output entropy. arXiv:0809.3972v3 [quant-ph] (2008)
13. Hayashi, M.: Capacity with energy constraint in coherent state channel. arXiv:0904.0307v1 [quant-ph] (2009)
14. Hayashi, M., Nagaoka, H.: General formulas for capacity of classical-quantum channels. IEEE Trans. Inf. Theory **7**, 49 (2003)
15. Jordan, T.: Affine maps of density matrices. Phys. Rev. A **71**, 034101 (2005)

7 A Dynamical Point of View of Quantum Information

16. Lardizabal, C.F.: Processos Estocásticos Quânticos, PhD. thesis. Programa de Pos-Graduação em Matemática - Universidade Federal do Rio Grande do Sul (UFRGS), Brazil (2010).
17. Lasota, A., Mackey, M.: Chaos, Fractals and Noise. Springer, New York (1994)
18. Lopes, A.O., Oliveira, E.: Entropy and variational principles for holonomic probabilities of IFS. Discrete Contin. Dyn. Syst. A. **23**(3), 937–955 (2009)
19. Lopes, A.O.: Entropy and large deviation. NonLinearity **3**(2), 527–546 (1990)
20. Lopes, A.O.: An analogy of the charge distribution on Julia sets with the Brownian motion. J. Math. Phys. **30**(9) (1989)
21. Lopes, A.0., Craizer, M.: The capacity-cost function of a hard-constrained channel. Int. J. Appl. Math. **2**(10), 1165–1180 (2000)
22. Lozinski, A., Życzkowski, K., Słomczyński, W.: Quantum iterated function systems. Phys. Rev. E **68**, 04610 (2003)
23. Mañé, R.: Ergodic Theory. Springer (1986)
24. Nielsen, M., Chuang, I.: Quantum Computation and Quantum Information. Cambridge University Press, Cambridge (2000)
25. Parry, W., Pollicott, M.: Zeta Functions and the Periodic Orbit Structure of Hyperbolic Dynamics. Société Mathématique de France, pp. 187–188, Astérisque (1990)
26. Rieffel, E., Polak, W.: An introduction to quantum computing for non-physicists. ACM Comput. Surv. **32**(3), 300–335 (2000)
27. Shor, P.W.: Equivalence of additivity question in quantum information theory. Comm. Math. Phys. **246**, 453–472 (2004)
28. Słomczyński, W., Życzkowski, K.: Quantum Chaos: an entropy approach. J. Math. Phys. **32**(1), 5674–5700 (1994)
29. Słomczyński, W.: Dynamical Entropy, Markov Operators and Iterated Function Systems. Jagiellonian University Press (2003)
30. Spitzer, F.: A Variational characterization of finite Markov chains. Ann. Math. Stat. **43**(1), 303–307 (1972)
31. Srinivas, M.D.: Foundations of a quantum probability theory. J. Math. Phys. **16**(8) (1975)
32. Winkler, G.: Choquet Order and Simplices, Lecture notes in Mathematics, vol. 1145. Springer, Berlin (1985)

Chapter 8
Generic Hamiltonian Dynamical Systems: An Overview

Mário Bessa and João Lopes Dias

Abstract We present for a general audience the state of the art on the generic properties of C^2 Hamiltonian dynamical systems.

8.1 Introduction and Main Definitions

Hamiltonian systems form a fundamental subclass of dynamical systems. Their importance follows from the vast range of applications throughout different branches of science. Generic properties of such systems are thus of great interest since they give us the "typical" behaviour (in some appropriate sense) that one could expect from the class of models at hand (cf. [38]). There are, of course, considerable limitations to the amount of information one can extract from a specific system by looking at generic cases. Nevertheless, it is of great utility to learn that a selected model can be slightly perturbed in order to obtain dynamics we understand in a reasonable way.

M. Bessa (✉)
Departamento de Matemática da, Universidade do Porto, Rua do Campo Alegre 687, 4169-007 Porto, Portugal

and

ESTGOH-Instituto Politécnico de Coimbra, Rua General Santos Costa, 3400-124 Oliveira do Hospital, Portugal
e-mail: bessa@fc.up.pt

J.L. Dias
Departamento de Matemática and Cemapre, ISEG, Universidade Técnica de Lisboa, Rua do Quelhas 6, 1200-781 Lisbon, Portugal
e-mail: jldias@iseg.utl.pt

M.M. Peixoto et al. (eds.), *Dynamics, Games and Science I*, Springer Proceedings in Mathematics 1, DOI 10.1007/978-3-642-11456-4_8,
© Springer-Verlag Berlin Heidelberg 2011

8.1.1 Residual Sets and Generic Properties

A residual set is a countable intersection of dense open sets. The elements of a residual set are called generic. A property that holds within a residual set is also referred as generic.

A Baire space is a topological space with the property that residual sets are dense. The space of C^s, $s \in \mathbb{N} \cup \{0\}$, functions on a manifold is Baire.

8.1.2 Hamiltonian Dynamics

Let M be a $2d$-dimensional smooth manifold endowed with a symplectic structure, i.e. a closed and nondegenerate 2-form ω. The pair (M, ω) is called a symplectic manifold which is also a volume manifold by Liouville's theorem. Let μ be the so-called Lebesgue measure associated to the volume form $\omega^d = \omega \wedge \cdots \wedge \omega$.

A diffeomorphism $g: (M, \omega) \to (N, \omega')$ between two symplectic manifolds is called a symplectomorphism if $g^*\omega' = \omega$. The action of a diffeomorphism on a 2-form is given by the pull-back $(g^*\omega')(X, Y) = \omega'(g_*X, g_*Y)$. Here X and Y are vector fields on M and the push-forward $g_*X = Dg\,X$ is a vector field on N. Notice that a symplectomorphism $g: M \to M$ preserves the Lebesgue measure μ since $g^*\omega^d = \omega^d$.

For any smooth Hamiltonian function $H: M \to \mathbb{R}$ there is a corresponding Hamiltonian vector field $X_H: M \to TM$ determined by $\iota_{X_H}\omega = dH$ being exact, where $\iota_v\omega = \omega(v, \cdot)$ is a 1-form. Notice that H is C^s iff X_H is C^{s-1}. The Hamiltonian vector field generates the Hamiltonian flow, a smooth 1-parameter group of symplectomorphisms φ_H^t on M satisfying $\frac{d}{dt}\varphi_H^t = X_H \circ \varphi_H^t$ and $\varphi_H^0 = $ id. Since $dH(X_H) = \omega(X_H, X_H) = 0$, X_H is tangent to the energy level sets $H^{-1}(\{e\})$, for some energy value $e \in H(M)$.

If $v \in T_x H^{-1}(\{e\})$, i.e. $dH(v)(x) = \omega(X_H, v)(x) = 0$, then its push-forward by φ_H^t is again tangent to $H^{-1}(\{e\})$ on $\varphi_H^t(x)$ since

$$dH(D\varphi_H^t\,v)(\varphi_H^t(x)) = \omega(X_H, D\varphi_H^t\,v)(\varphi_H^t(x)) = \varphi_H^{t*}\omega(X_H, v)(x) = 0.$$

We consider also the tangent flow $D\varphi_H^t: TM \to TM$ that satisfies the linear variational equation (the linearized differential equation)

$$\frac{d}{dt}D\varphi_H^t = DX_H(\varphi_H^t)\,D\varphi_H^t$$

with $DX_H: M \to TTM$.

We say that x is a *regular* point if $dH(x) \neq 0$ (x is not critical). We denote the set of regular points by $\mathscr{R}(H)$ and the set of critical points by $\mathrm{Crit}(H)$. We call

8 Generic Hamiltonian Dynamical Systems: An Overview

$H^{-1}(\{e\})$ a regular energy level of H if $H^{-1}(\{e\}) \cap \text{Crit}(H) = \emptyset$. A regular energy surface is a connected component of a regular energy level.

Given any regular energy level or surface $\mathcal{E}$, we induce a volume form $\omega_{\mathcal{E}}$ on the $(2d - 1)$-dimensional manifold $\mathcal{E}$ in the following way. For each $x \in \mathcal{E}$,

$$\omega_{\mathcal{E}}(x) = \iota_Y \omega^d(x) \quad \text{on } T_x \mathcal{E}$$

defines a $(2d - 1)$ non-degenerate form if $Y \in T_x M$ satisfies $dH(Y)(x) = 1$. Notice that this definition does not depend on Y (up to normalization) as long as it is transversal to $\mathcal{E}$ at x. Moreover,

$$dH(D\varphi_H^t \, Y)(\varphi_H^t(x)) = d(H \circ \varphi_H^t)(Y)(x) = 1.$$

Thus, $\omega_{\mathcal{E}}$ is φ_H^t-invariant, and the measure $\mu_{\mathcal{E}}$ induced by $\omega_{\mathcal{E}}$ is again invariant. In order to obtain finite measures, we need to consider compact energy levels.

On the manifold M we also fix any Riemannian structure which induces a norm $\| \cdot \|$ on the fibers $T_x M$. We will use the standard norm of a bounded linear map A given by $\|A\| = \sup_{\|v\|=1} \|A v\|$ and also the co-norm defined by $\mathbf{m}(A) = \|A^{-1}\|^{-1}$.

The symplectic structure guarantees by Darboux theorem the existence of an atlas $\{h_j : U_j \to \mathbb{R}^{2d}\}$ satisfying $h_j^* \omega_0 = \omega$ with

$$\omega_0 = \sum_{i=1}^{d} dy_i \wedge dy_{d+i}. \tag{8.1}$$

On the other hand, when dealing with volume manifolds (N, Ω) of dimension p, Moser's theorem [30] gives an atlas $\{h_j : U_j \to \mathbb{R}^p\}$ such that $h_j^*(dy_1 \wedge \cdots \wedge dy_p) = \Omega$.

For more on the general symplectic and Hamiltonian theories, see e.g. [1].

8.1.3 Our Setting

In the following we will always assume that M is a $2d$-dimensional compact smooth symplectic manifold with a smooth boundary ∂M (including the case $\partial M = \emptyset$) and $d \geq 2$. Furthermore, C^s Hamiltonians are real-valued functions on M that are constant on each connected component of ∂M. We denote by $C^s(M)$ the set of C^s Hamiltonians. This set is endowed with the C^2-topology.

Under these conditions, the Hamiltonian flow is globally defined with respect to time because H is constant on the components of ∂M or, equivalently, X_H is tangent to ∂M.

8.1.4 Transversal Linear Poincaré Flow

Given any regular point x we take the orthogonal splitting $T_x M = \mathbb{R} X_H(x) \oplus N_x$, where $N_x = (\mathbb{R} X_H(x))^\perp$ is the normal fiber at x. Consider the automorphism of vector bundles

$$
\begin{aligned}
D\varphi_H^t : T_{\mathcal{R}} M &\to T_{\mathcal{R}} M \\
(x, v) &\mapsto (\varphi_H^t(x), D\varphi_H^t(x) v).
\end{aligned}
\tag{8.2}
$$

Of course, in general, the subbundle $N_{\mathcal{R}}$ is not $D\varphi_H^t$-invariant. So we relate to the $D\varphi_H^t$-invariant quotient space $\widetilde{N}_{\mathcal{R}} = T_{\mathcal{R}} M / \mathbb{R} X_H(\mathcal{R})$ with an isomorphism $\phi_1 : N_{\mathcal{R}} \to \widetilde{N}_{\mathcal{R}}$. The unique map

$$
P_H^t : N_{\mathcal{R}} \to N_{\mathcal{R}}
$$

such that $\phi_1 \circ P_H^t = D\varphi_H^t \circ \phi_1$ is called the *linear Poincaré flow* for H. Denoting by $\Pi_x : T_x M \to N_x$ the canonical orthogonal projection, the linear map $P_H^t(x) : N_x \to N_{\varphi_H^t(x)}$ is

$$
P_H^t(x) v = \Pi_{\varphi_H^t(x)} \circ D\varphi_H^t(x) v.
$$

We now consider

$$
\mathcal{N}_x = N_x \cap T_x H^{-1}(e),
$$

where $T_x H^{-1}(e) = \ker dH(x)$ is the tangent space to the energy level set with $e = H(x)$. Thus, $\mathcal{N}_{\mathcal{R}}$ is invariant under P_H^t. So we define the map

$$
\Phi_H^t : \mathcal{N}_{\mathcal{R}} \to \mathcal{N}_{\mathcal{R}}, \qquad \Phi_H^t = P_H^t |_{\mathcal{N}_{\mathcal{R}}},
$$

called the *transversal linear Poincaré flow* for H such that

$$
\Phi_H^t(x) : \mathcal{N}_x \to \mathcal{N}_{\varphi_H^t(x)}, \qquad \Phi_H^t(x) v = \Pi_{\varphi_H^t(x)} \circ D\varphi_H^t(x) v
$$

is a linear symplectomorphism for the symplectic form induced on $\mathcal{N}_{\mathcal{R}}$ by ω.

8.1.5 Oseledets Theorem

Take $H \in C^2(M)$. Since the time-1 map of any tangent flow derived from a Hamiltonian vector field is measure preserving, we obtain a version of Oseledets theorem for Hamiltonian systems. Given a point $x \in M$ we say that x is *Oseledets regular* if there exists a splitting $T_x M = E_x^1 \oplus \dots E_x^{k(x)}$ and numbers $\lambda^1(x) \geq \dots \geq \lambda^{k(x)}(x)$ such that for any (non-zero) vector $v \in E_x^i$ we have

$$
\lim_{t \to \pm\infty} \frac{1}{t} \log \| D\varphi_H^t(x) v \| = \lambda^i(x).
$$

8 Generic Hamiltonian Dynamical Systems: An Overview 127

The Oseledets theorem [32] asserts that Oseledets regular points form a η-full measure set for any φ_H^t-invariant probability measure η.

Moreover,

$$\lim_{t \to \pm\infty} \frac{1}{t} \log \sin \alpha_t = 0, \tag{8.3}$$

where α_t is the angle at time t between any subspaces of the splitting.

The splitting of the tangent bundle is called *Oseledets splitting* and the real numbers $\lambda^i(H, x)$ are called the *Lyapunov exponents*. The full measure set of the *Oseledets points* is denoted by $\mathcal{O}(H) = \mathcal{O}$.

The vector field direction $\mathbb{R} X_H(x)$ is trivially an Oseledets's direction with zero Lyapunov exponent.

If $x \in \mathcal{R} \cap \mathcal{O}$ and $\lambda^i(x) \neq 0$, the Oseledets splitting on $T_x M$ induces a $\Phi_H^t(x)$-invariant splitting on $\mathcal{N}_x$ where $\mathcal{N}_x^i = \Pi_x(E_x^i)$.

The next lemma makes explicit that the dynamics of $D\varphi_H^t$ and Φ_H^t are coherent so that the Lyapunov exponents for both cases are related. The proof uses (8.3).

Lemma 8.1 ([8]). *Given $x \in \mathcal{R} \cap \mathcal{O}$, the Lyapunov exponents of the Φ_H^t-invariant decomposition are equal to the ones of the $D\varphi_H^t$-invariant decomposition.*

We now restate the Oseledets theorem for the dynamic cocycle Φ_H^t: For μ-a.e. $x \in M$ there exists a splitting of the normal bundle $\mathcal{N}_x = \mathcal{N}_x^1 \oplus \cdots \oplus \mathcal{N}_x^{k(x)}$ and numbers $\lambda^1(x) \geq \cdots \geq \lambda^{k(x)}(x)$ such that for any (non-zero) vector $v \in \mathcal{N}_x^i$ we have

$$\lim_{t \to \pm\infty} \frac{1}{t} \log \|\Phi_H^t(x) v\| = \lambda^i(x).$$

Observe that there exist at most $2d - 2$ different exponents for Φ_H^t. Moreover, the Lyapunov exponents of Φ_H^t are symmetric (i.e. if λ is one the exponents, then $-\lambda$ is also one of the exponents and their multiplicity is the same). Finally, $\dim(\mathcal{N}_x^+) = \dim(\mathcal{N}_x^-)$ and since $\dim(\mathcal{N}_x^+)$ is even we obtain that $\dim(\mathcal{N}_x^0)$ is also even.

8.1.6 Hyperbolicity and Dominated Splitting

Let $H \in C^2(M)$. Given any compact and φ_H^t-invariant set $\Lambda \subset H^{-1}(e)$, we say that Λ is a *hyperbolic set* for φ_H^t if there exist $m \in \mathbb{N}$ and a $D\varphi_H^t$-invariant splitting $T_\Lambda H^{-1}(e) = E_\Lambda^+ \oplus E_\Lambda^- \oplus E_\Lambda$ such that for all $x \in \Lambda$ we have:

- $\|D\varphi_H^m(x)|_{E_x^-}\| \leq \frac{1}{2}$ (uniform contraction).
- $\|D\varphi_H^{-m}(x)|_{E_x^+}\| \leq \frac{1}{2}$ (uniform expansion).
- E includes the directions of the vector field and of the gradient of H.

Similarly, we can define a hyperbolic structure for the transversal linear Poincaré flow Φ_H^t. We say that Λ is hyperbolic for Φ_H^t on Λ if $\Phi_H^t|_\Lambda$ is a hyperbolic vector bundle automorphism. The next lemma relates the hyperbolicity for Φ_H^t with the hyperbolicity for φ_H^t. It is an immediate consequence of a result by Doering [22] for the linear Poincaré flow extended to our Hamiltonian setting and the transversal linear Poincaré flow.

Lemma 8.2. *Let Λ be an φ_H^t-invariant and compact set. Then Λ is hyperbolic for φ_H^t iff Λ is hyperbolic for Φ_H^t.*

We now consider a weaker form of hyperbolicity. Let $\Lambda \subset M$ be an φ_H^t-invariant set and $m \in \mathbb{N}$. A splitting of the bundle $\mathcal{N}_\Lambda = \mathcal{N}_\Lambda^1 \oplus \mathcal{N}_\Lambda^2$ is an *m-dominated splitting* for the transversal linear Poincaré flow if it is Φ_H^t-invariant and continuous such that

$$\frac{\|\Phi_H^m(x)|\mathcal{N}_x^2\|}{\mathbf{m}(\Phi_H^m(x)|\mathcal{N}_x^1)} \leq \frac{1}{2}, \qquad \text{for all } x \in \Lambda. \tag{8.4}$$

We call $\mathcal{N}_\Lambda = \mathcal{N}_\Lambda^1 \oplus \mathcal{N}_\Lambda^2$ a *dominated splitting* if it is m-dominated for some $m \in \mathbb{N}$.

If Λ has a dominated splitting, then we may extend the splitting to its closure, except to critical points. Moreover, the angle between $\mathcal{N}^1$ and $\mathcal{N}^2$ is bounded away from zero on Λ. Under the four-dimensional assumption the decomposition is unique. For more details about dominated splitting see [20].

We say that a dominated splitting $\mathcal{N}_\Lambda = \mathcal{N}_\Lambda^- \oplus \mathcal{N}_\Lambda^0 \oplus \mathcal{N}_\Lambda^+$ over the set Λ is *partially hyperbolic* if the bundle $\mathcal{N}_\Lambda^-$ is uniformly contractive and the bundle $\mathcal{N}_\Lambda^+$ is uniformly expanding.

The proof of the next lemma (see [8, Lemma 2.6]) hints to the fact that the four-dimensional setting is crucial in obtaining hyperbolicity from the dominated splitting structure.

Lemma 8.3. *Let $H \in C^2(M)$ and a regular energy surface $\mathcal{E}$. If $\Lambda \subset \mathcal{E}$ has a dominated splitting for Φ_H^t, then $\overline{\Lambda}$ is hyperbolic.*

Actually, the previous lemma is a version of the following general fact proved in [18, Theorem 11] which we trivially adapt for Hamiltonians.

Theorem 8.1. *Let $H \in C^2(M)$ and let $\mathcal{N}_\Lambda = \mathcal{N}_\Lambda^1 \oplus \mathcal{N}_\Lambda^2$ be a dominated splitting over a φ_H^t-invariant set Λ. Assume that $\dim \mathcal{N}_\Lambda^1 \leq \dim \mathcal{N}_\Lambda^2$ and let $\mathcal{N}_\Lambda^+ = \mathcal{N}_\Lambda^1$. Then $\mathcal{N}_\Lambda^2$ splits invariantly as $\mathcal{N}_\Lambda^0 \oplus \mathcal{N}_\Lambda^-$ with $\dim \mathcal{N}_\Lambda^+ = \mathcal{N}_\Lambda^-$, and the splitting $\mathcal{N}_\Lambda = \mathcal{N}_\Lambda^+ \oplus \mathcal{N}_\Lambda^0 \oplus \mathcal{N}_\Lambda^-$ is partially hyperbolic.*

8.1.7 Elliptic, Parabolic and Hyperbolic Closed Orbits

Let $\Gamma \subset M$ be a closed orbit of least period τ. The characteristic multipliers of Γ are the eigenvalues of $\Phi_H^\tau(p)$, which are independent of the point $p \in \Gamma$. We say that Γ is

- *k-elliptic* iff $2k$ characteristic multipliers are simple, non-real and of modulus 1.
- *parabolic* iff the characteristic multipliers are real and of modulus 1.
- *hyperbolic* iff the characteristic multipliers have modulus different from 1.

8 Generic Hamiltonian Dynamical Systems: An Overview   129

We call $d - 1$-elliptic orbits total elliptic. In case $d = 2$ we have that 1-elliptic are total.

It is clear that under small perturbations, d-elliptic and hyperbolic orbits are stable whilst parabolic ones are unstable.

We refer to a point in a closed orbit as periodic. Periodic points are classified in the same way as the respective closed orbit.

8.1.8 Perturbation Lemmas

We include here several perturbation results in our setting. The first is the celebrated Pugh's closing lemma [37, Sect. 9]:

Theorem 8.2 (Pugh's closing lemma). *If $\epsilon > 0$ and $x \in M$ is a recurrent point for the flow φ_H^t associated to $H \in C^2(M)$, then there exists $\widetilde{H} \in C^2(M)$ ϵ-C^2-close to H such that x is a periodic point for $\varphi_{\widetilde{H}}^t$.*

An important upgrade is the Arnaud's closing lemma [4]. It states that the orbit of a non-wandering point can be approximated for a very long time by a closed orbit of a nearby Hamiltonian.

Theorem 8.3 (Arnaud's closing lemma). *Let $H \in C^s(M)$, $2 \leq s \leq \infty$, a non-wandering point $x \in M$ and $\epsilon, r, \tau > 0$. Then, we can find $\widetilde{H} \in C^s(M)$ ϵ-C^2-close to H, a closed orbit Γ of $\widetilde{H}$ with least period ℓ, $p \in \Gamma$ and a map $g : [0, \tau] \to [0, \ell]$ close to the identity such that:*

- dist $\left(\varphi_H^t(x), \varphi_{\widetilde{H}}^{g(t)}(p) \right) < r$, $0 \leq t \leq \tau$, *and*
- $H = \widetilde{H}$ *on* $M \setminus A$, *where* $A = \bigcup_{0 \leq t \leq \ell} \left(B(p, r) \cap B(\varphi_{\widetilde{H}}^t(p), r) \right)$.

The next theorem is a version of Franks' lemma for Hamiltonians proved by Vivier [41]. Roughly, it says that we can realize a Hamiltonian corresponding to a given perturbation of the transversal linear Poincaré flow. It is proved for $2d$-dimensional manifolds with $d \geq 2$.

Theorem 8.4 (Vivier's lemma). *Let $H \in C^s(M)$, $2 \leq s \leq \infty$, $\epsilon, \tau > 0$ and $x \in M$. There exists $\delta > 0$ such that for any flowbox V of an injective arc of orbit $\Sigma = \varphi_H^{[0,t]}(x)$, $t \geq \tau$, and a transversal symplectic δ-perturbation F of $\Phi_H^t(x)$, there is $\widetilde{H} \in C^{\max\{2,s-1\}}(M)$ ϵ-C^2-close to H satisfying:*

- $\Phi_{\widetilde{H}}^t(x) = F$,
- $H = \widetilde{H}$ *on* $\Sigma \cup (M \setminus V)$.

In order to perform local perturbations to our original Hamiltonians, we need an improved version of a lemma by Robinson [39] that provides us with symplectic flowbox coordinates. Consider the canonical symplectic form on $\mathbb{R}^{2d}$ given by ω_0 as in (8.1). The Hamiltonian vector field of any smooth $H : \mathbb{R}^{2d} \to \mathbb{R}$ is then

$$X_H = \begin{bmatrix} 0 & I \\ -I & 0 \end{bmatrix} \nabla H,$$

where I is the $d \times d$ identity matrix. Let the Hamiltonian function $H_0 \colon \mathbb{R}^{2d} \to \mathbb{R}$ be given by $y \mapsto y_{d+1}$, so that

$$X_{H_0} = \frac{\partial}{\partial y_1}.$$

Theorem 8.5 (Symplectic flowbox coordinates [8]). *Let $H \in C^s(M)$, $2 \le s \le \infty$, and $x \in M$. If $x \notin \mathrm{Crit}(M)$, there exists a neighborhood U of x and a local C^{s-1}-symplectomorphism $g \colon (U, \omega) \to (\mathbb{R}^{2d}, \omega_0)$ such that $H = H_0 \circ g$ on U.*

8.2 Abundance of Zero Lyapunov Exponents Away from Hyperbolicity

The computation of Lyapunov exponents is one of the main problems in the modern theory of dynamical systems. They give us fundamental information on the asymptotic exponential behaviour of the linearized system. It is therefore important to understand these objects in order to study the time evolution of orbits. In particular, Pesin's theory deals with non-vanishing Lyapunov exponents systems (non-uniformly hyperbolic). This setting jointly with a C^α regularity, $\alpha > 0$, of the tangent map allows us to derive a very complete geometric picture of the dynamics (stable/unstable invariant manifolds). On the other hand, if we aim at understanding both local and global dynamics, the presence of zero Lyapunov exponents creates lots of obstacles. An example is the case of conservative systems: using enough differentiability, the celebrated KAM theory guarantees persistence of invariant quasiperiodic motion on tori yielding zero Lyapunov exponents.

In this section we study the dependence of the Lyapunov exponents on the dynamics of Hamiltonian flows. For a survey of the theory see [18] and references therein. In Theorem 8.6 we state that zero Lyapunov exponents for four-dimensional Hamiltonian systems are very common, at least for a C^2-residual subset. This picture changes radically for the C^∞ topology, the setting of most common Hamiltonian systems coming from applications. In this case Markus and Meyer showed that there exists a residual of C^∞ Hamiltonians neither integrable nor ergodic [28].

Theorem 8.6 ([8]). *Let $d = 2$. For a C^2-generic Hamiltonian $H \in C^2(M)$, the union of the regular energy surfaces $\mathcal{E}$ that are either Anosov or have zero Lyapunov exponents $\mu_{\mathcal{E}}$-a.e. for the Hamiltonian flow, forms an open μ-mod 0 and dense subset of M.*

Geodesic flows on negative curvature surfaces are well-known systems yielding Anosov energy levels. An example of a mechanical system which is Anosov on each positive energy level was obtained by Hunt and MacKay [25].

8   Generic Hamiltonian Dynamical Systems: An Overview   131

Another dichotomy result for the transversal linear Poincaré flow on the tangent bundle is the following:

Theorem 8.7 ([8]). *Let $d = 2$. There exists a C^2-dense subset $\mathfrak{D}$ of $C^2(M)$ such that, if $H \in \mathfrak{D}$, there exists an invariant decomposition $M = D \cup Z$ (mod 0) satisfying:*

- $D = \bigcup_{n \in \mathbb{N}} D_{m_n}$, *where D_{m_n} is a set with m_n-dominated splitting for the transversal linear Poincaré flow of H.*
- *The Hamiltonian flow of H has zero Lyapunov exponents for $x \in Z$.*

The proof of the above theorems is based on a result that allows us to decay the Lyapunov exponents of points without dominated splitting. This is possible by first constructing a local perturbation in the coordinates given by Lemma 8.5, that mixes the transversal directions of non-zero Lyapunov exponents along an orbit segment. Thus the effects of contraction and expansion average out.

The following problem is the generalization of the recent result by Bochi [15] to our context.

Open problem 1. Show that Theorem 8.7 holds for $d > 2$.

8.3   Denseness of Elliptic Points away from Hyperbolicity

In this section we recall a related C^2-generic dichotomy by Newhouse [31]: for a C^2-generic Hamiltonian, an energy surface through any $p \in M$ is Anosov or is in the closure of 1-elliptical periodic orbits.

The Newhouse dichotomy was first proved for C^1-generic symplectomorphisms in [31], and extensions have appeared afterwards [3,24,40]. Those were all done for discrete-time dynamics.

Theorem 8.8 ([9]). *Let $d = 2$. Given $\epsilon > 0$ and an open subset $U \subset M$, if $H \in C^2(M)$ has a far from Anosov regular energy surface intersecting U, then there is $\widetilde{H} \in C^\infty(M)$ ϵ-C^2-close to H having a closed elliptic orbit through U.*

The above theorem is proved in [9] (see [10] for divergence-free 3-flows) by looking first at the case of hyperbolic closed orbits with a small angle between the stable and unstable directions. Those are then showed to become elliptic by a small perturbation. On the other hand, for hyperbolic closed orbits with large angles and without dominated splitting, an adaptation of Mañé's perturbation techniques [10] leads again to elliptic orbits by a perturbation. The remaining case of hyperbolic closed orbits with dominated splitting and large angle is not true generically (as the case of parabolic ones).

As an almost direct consequence we arrive at the Newhouse dichotomy for four-dimensional Hamiltonians. Recall that for a C^2-generic Hamiltonian all but finitely many points are regular.

Theorem 8.9 ([9]). *Let $d = 2$. For a C^2-generic $H \in C^2(M)$, the union of the Anosov regular energy surfaces and the closed elliptic orbits, forms a dense subset of M.*

Open problem 2. Prove the related result for $d > 2$: For a C^2-generic Hamiltonian, the union of the partially hyperbolic regular energy surfaces and the closed elliptic orbits, forms a dense subset of M.

8.4 Star Energy Surfaces

Consider the set $\mathcal{M} = M \times C^2(M)$ endowed with the standard product topology. Given $(p, H) \in \mathcal{M}$, we denote by $\mathscr{E}_{p,H}$ the energy surface in $H^{-1}(H(p))$ containing p. We say that $\mathscr{E}_{p,H}$ is a *star energy surface* if it is regular and there exists a neighbourhood $\mathscr{U}$ of (p, H) such that all energy surfaces $\mathscr{E}_{\tilde{p},\tilde{H}}$, with $(\tilde{p}, \tilde{H}) \in \mathscr{U}$, are regular and have all closed orbits hyperbolic.

Denote by $\mathscr{G}$ the set of $(p, H) \in \mathcal{M}$ such that $\mathscr{E}_{p,H}$ is star, and by $\mathscr{A}$ if $\mathscr{E}_{p,H}$ is Anosov. If there exists a homeomorphism between $\mathscr{E}_{p,H}$ and any nearby $\mathscr{E}_{\tilde{p},\tilde{H}}$ preserving orbits and their orientations, we say that (p, H) is *structurally stable*, i.e. $(p, H) \in \mathscr{S}$.

The next theorem is classical in the theory of dynamical systems, namely Anosov systems are open and structurally stable (see e.g. [13]).

Theorem 8.10. *Let $d \geq 2$. $\mathscr{A}$ is open and $\mathscr{A} \subset \mathscr{S}$.*

In the $d = 2$ case, there is already a good characterization of Anosov energy surfaces.

Theorem 8.11 ([13]). *$\mathscr{G} = \mathscr{A} = \mathscr{S}$ for $d = 2$.*

In rough terms the proof of the previous theorem goes as follows. By Lemma 8.3, in the four-dimensional context, dominated splitting is tantamount to hyperbolicity. So, we are left to show that in the absence of domination it is possible to create a non-hyperbolic closed orbit by an arbitrary small C^2 perturbation of the Hamiltonian.

Assume that we do not have dominated splitting (cannot be Anosov) and we still have the star property. We claim that we must be far from systems exhibiting elliptic closed orbits, and moreover we must have good uniform constants of hyperbolicity over closed orbits. Since we do not have domination, we use the ideas from the proof of Theorem 8.6 to obtain an Oseledets regular point with (almost) zero exponents. Then, the closing lemma (Theorem 8.3) produce a closed orbit without good constants of hyperbolicity, contradicting our assumption.

We say that (p, H) is *isolated in the boundary of $\mathscr{A}$* if $\mathscr{E}_{p,H}$ is not Anosov but any nearby $\mathscr{E}_{\tilde{p},\tilde{H}}$ such that $H \neq \tilde{H}$ or $\tilde{p} \notin \mathscr{E}_{p,H}$ is Anosov. As a consequence of Theorem 8.11, we obtain the following.

8 Generic Hamiltonian Dynamical Systems: An Overview

Corollary 8.1. *Let $d = 2$. The boundary of $\mathscr{A}$ has no isolated points.*

Open problem 3. Show that Theorem 8.11 holds for $d > 2$.

8.5 Robust Transitivity

We say that a dynamical system is *transitive* if it has a dense orbit. Moreover, it is C^r-*robustly transitive* if in addition any arbitrarily C^r-close system is transitive.

Theorem 8.12 (Horita–Tahzibi [24]). *Any robustly transitive symplectomorphism defined in a compact symplectic manifold is partially hyperbolic.*

Working in the Hamiltonian context, we have that a regular energy surface is *transitive* if it has a dense orbit, and it is *robustly transitive* if the restriction of any sufficiently C^2-close Hamiltonian to a nearby regular energy surface is still transitive.

Theorem 8.13 (Vivier [41]). *Let $d = 2$. Any Hamiltonian admitting a robustly transitive regular energy surface is Anosov on that surface.*

We observe that the proof of this theorem uses the Hamiltonian version of Franks' lemma (Lemma 8.4).

It is easy to see that Theorem 8.8 also implies Theorem 8.13. In fact, if a regular energy surface $\mathscr{E}$ of $H \in C^2(M)$ is far from Anosov, then by Theorem 8.8 there exists a C^2-close C^∞-Hamiltonian with an elliptic closed orbit on a nearby regular energy surface. This invalidates the chance of robust transitivity for H according to a KAM-type criterium (see [41, Corollary 9]).

Taking into account Theorem 8.1 we get the following question.

Open problem 4. Let $d > 2$. Show that if a Hamiltonian admits a robustly transitive regular energy surface, then it is partially hyperbolic there.

8.6 Genericity of Dense Orbits

It follows from Poincaré's recurrence theorem that, in the volume-preserving context, almost any point is recurrent. However, the points can be restricted to some region of the manifold both for the past and for the future. The problem of knowing if a given dynamical system exhibits only one "piece" or, in other words, if there is any dense orbit, is a central problem in the modern theory of dynamical systems. A partial answer to this problem was given by Bonatti and Crovisier in [19] for the volume-preserving discrete-time case and by the same authors and Arnaud in the symplectomorphism framework [5]. They proved that for some C^1-residual subset any map has a dense orbit.

In the continuous-time case the first author proved in [7] the corresponding version for divergence-free flows, and recently Ferreira announced the following result.

Theorem 8.14 ([23]). *For a C^2-generic Hamiltonian H and $e \in H(M)$, we have that $H^{-1}(\{e\})$ has a transitive energy surface.*

Theorem 8.14 is a central tool in order to obtain important results in the generic theory of Hamiltonians (e.g. Open Problems 2, 3 and 4).

The main tool to conclude the proof of the previous result is the next theorem, a version for Hamiltonians of the connecting lemma for pseudo-orbits.

We say that the numbers $\sigma_1, \ldots, \sigma_{2d}$ satisfy a *trivial resonance relation* if

$$\sigma_i = \prod_{j=1}^{2d} \sigma_j^{k_j}, \quad i = 1, \ldots, 2d,$$

where $k_i \in \mathbb{N}$ such that either $k_i \neq 1$ or there exists $j \neq i$ verifying $k_j \neq 0$.

Theorem 8.15. *Let $(p, H) \in \mathcal{M}$ such that $\mathcal{E}_{p,H} \subset H^{-1}(\{p\})$ is a regular surface. Suppose that every closed orbit there has a trivial resonance relation between the Floquet exponents. Then, for any $x, y \in \mathcal{E}_{p,H}$ connected by a pseudo-orbit, there is a C^2-nearby $\widetilde{H}$ and $t > 0$ such that $\varphi^t_{\widetilde{H}}(x) = y$.*

8.7 On Palis' Conjecture

It is known from Peixoto's work [35, 36] that structurally stable flows on surfaces form a dense open set. A few years later Palis formulated the following conjecture for general dynamical systems defined on a closed manifold (flows, diffeomorphisms, or even more general transformations). Any system can always be C^1 approximated by another one which is uniformly hyperbolic or else it exhibits either a homoclinic tangency or a heterodimensional cycle [34].

In the conservative setting a more accurate result holds. In fact, Bessa and Rocha recently proved that any volume-preserving diffeomorphism of dimension $d \geq 3$ (or symplectomorphism of dimension $d \geq 4$) can be C^1 approximated by a volume-preserving (symplectic) diffeomorphism which is Anosov or else it exhibits a heterodimensional cycle [12].

In respect to the two-dimensional area-preserving discrete-time case, we have the following.

Theorem 8.16. *Any area-preserving diffeomorphism in a compact surface can always be C^1 approximated by another area-preserving diffeomorphism which is either Anosov or it exhibits a homoclinic tangency.*

8 Generic Hamiltonian Dynamical Systems: An Overview 135

Proof. By Newhouse's dichotomy [31] for a C^1-dense subset $\mathscr{D}$ of the Baire space of area-preserving diffeomorphisms endowed with the C^1-topology, we have that: if $f \in \mathscr{D}$, then f is Anosov or the elliptic points of f are dense in the manifold. It is sufficient to show that if f is in the C^1-interior of the complementary set of Anosov maps, we can C^1-approximate f by an area-preserving diffeomorphism g displaying a homoclinic tangency.

Now, we choose one elliptic point p for f. Since the C^2 area-preserving diffeomorphisms are C^1-dense in the C^1 area-preserving diffeomorphisms [42] and the elliptic points are stable, we can C^1-approximate f by $f_0 \in C^2$ such that the analytic continuation p_0 of p is elliptic. Now, since f_0 is of class C^2, we use the weak pasting lemma for diffeomorphisms [2] to create an invariant curve for some area-preserving diffeomorphism f_1 arbitrarily close to f_0. Finally, [29] is used to obtain persistence of homoclinic tangencies for g arbitrarily close to f_1. $\square$

Taking into account the previous result, we believe that the following result should hold.

Open problem 5. Let $d = 2$. Given $H \in C^2(M)$, $e \in H(M)$ and $\epsilon > 0$, then there exists $\widetilde{H}$ ϵ-C^2-close to H such that some regular energy surface in $\widetilde{H}^{-1}(\{e\})$ is Anosov or else it contains a homoclinic tangency associated to some hyperbolic closed orbit.

Open problem 6. Let $d > 2$. Given $H \in C^2(M)$, $e \in H(M)$ and $\epsilon > 0$, then there exists $\widetilde{H}$ ϵ-C^2-close to H such that some regular energy surface in $\widetilde{H}^{-1}(\{e\})$ is Anosov or else it contains a heterodimensional cycle.

8.8 Subclasses of Hamiltonian Systems

There are many subclasses of $C^2(M)$ for which it would be very interesting to find generic properties. We will only briefly mention below two of them, because of their high importance in many branches of science: mechanical systems and geodesic flows.

Let Q be a d-dimensional smooth compact manifold and take there the local coordinates $q = (q_1, \dots, q_d)$. We can write any $\sigma \in T_q^* Q$ as $\sigma = p \cdot dq$ where $p \in \mathbb{R}^d$ and $dq = (dq_1, \dots, dq_d)$. Therefore, local coordinates on the cotangent bundle $M = T^* Q$ are given by (q, p). Notice that $\omega = dq \wedge dp$ is a symplectic form defined locally on M. For these local coordinates a mechanical system is a Hamiltonian $H \in C^\infty(T^* M)$ given by $H = T + V$, where T is the kinetic energy and $V: Q \to \mathbb{R}$ the potential. The function T is chosen to be homogeneous of degree 2, i.e. $T = \frac{1}{2}\langle p, p \rangle_q$. This is the general setting of most classical mechanics.

The results in the previous sections do not hold if we restrict to mechanical systems, because we would need to perturb in the same class, i.e. on the Riemannian metric $\langle \cdot, \cdot \rangle$ or on the potential V. It is thus an open question whether any sort of generic property would remain true in this context. In particular, we have the following question.

Open problem 7. Can we C^2 approximate any given mechanical system by another mechanical system which has the dichotomy in Theorem 8.6?

A somewhat first step would be to deal with a simpler situation:

Open problem 8. Let Q be a closed surface. Given a C^2 Hamiltonian on T^*Q of the form $H = T$, is there V arbitrarly C^2 small such that $\widetilde{H} = T + V$ has the above mentioned dichotomy?

Geodesic flows on the unit tangent bundle $M = SQ$ are a particular example of Hamiltonian mechanical systems, given by $H = T$. It would be of great interest to answer related questions specifically for those systems.

Acknowledgements MB was partially supported by Fundação para a Ciência e a Tecnologia, SFRH/BPD/20890/2004. JLD was partially supported by Fundação para a Ciência e a Tecnologia through the Program FEDER/POCI 2010.

References

1. Abraham, R., Marsden, J.E.: Foundations of Mechanics, 2nd edn. Benjamin-Cummings (1978)
2. Arbieto, A., Matheus, C.: A pasting lemma and some applications for conservative systems. Ergod. Theory Dyn. Syst. **27**(5), 1399–1417 (2007)
3. Arnaud, M.-C.: The generic symplectic C^1-diffeomorphisms of four-dimensional symplectic manifolds are hyperbolic, partially hyperbolic or have a completely elliptic periodic point. Ergod. Theory Dyn. Syst. **22**(6), 1621–1639 (2002)
4. Arnaud, M.-C.: Le "closing lemma" en topologie C^1. Mém. S.M.F. **74**, 1–120 (1998)
5. Arnaud, M.-C., Bonatti, C., Crovisier, S.: Dynamiques symplectiques génériques. Ergod. Theory Dyn. Syst. **25**, 1401–1436 (2005)
6. Bessa, M.: The Lyapunov exponents of generic zero divergence 3-dimensional vector fields. Erg. Theor. Dyn. Syst. **27**, 1445–1472 (2007)
7. Bessa, M.: Generic incompressible flows are topological mixing. C. R. Acad. Sci. Paris, Ser. I **346**, 1169–1174 (2008)
8. Bessa, M., Lopes Dias, J.: Generic dynamics of 4-dimensional C^2 Hamiltonian systems. Commun. Math. Phys **281**(1), 597–619 (2008)
9. Bessa, M., Lopes Dias, J.: Hamiltonian elliptic dynamics on symplectic 4-manifolds. Proc. Am. Math. Soc. **137**, 585–592 (2009)
10. Bessa, M., Duarte, P.: Abundance of elliptical dynamics on conservative 3-flows. Dyn. Syst. **23**(4), 409–424 (2008)
11. Bessa, M., Rocha, J.: On C^1-robust transitivity of volume-preserving flows. J. Differ. Eqn. **245**(11), 3127–3143 (2008)
12. Bessa, M., Rocha, J.: Anosov versus Heterodimensional cycles: A C^1 dichotomy for conservative maps. http://cmup.fc.up.pt/cmup/bessa/ Preprint (2009)
13. Bessa, M., Ferreira, C., Rocha, J.: On the stability of the set of hyperbolic closed orbits of a Hamiltonian. Math. Proc. Cambridge Phil. Soc. **149**(2), 373–383 (2010)
14. Bochi, J.: Genericity of zero Lyapunov exponents. Erg. Theor. Dyn. Syst. **22**, 1667–1696 (2002)
15. Bochi, J.: C^1-generic symplectic diffeomorphisms: partial hyperbolicity and zero center Lyapunov exponents. J. Inst. Math. Jussieu **9**(1), 49–93 (2010)
16. Bochi, J., Fayad, B.: Dichotomies between uniform hyperbolicity and zero Lyapunov exponents for $SL(2, R)$ cocycles. Bull. Braz. Math. Soc. **37**, 307–349 (2006)

8 Generic Hamiltonian Dynamical Systems: An Overview

17. Bochi, J., Viana, M.: The Lyapunov exponents of generic volume preserving and symplectic maps. Ann. Math. **161**, 1423–1485 (2005)
18. Bochi, J., Viana, M.: Lyapunov exponents: How frequently are dynamical systems hyperbolic? in Advances in Dynamical Systems. Cambridge University Press, Cambridge (2004)
19. Bonatti, C., Crovisier, S.: Récurrence et généricité. Inventiones Math. **158**, 33–104 (2004)
20. Bonatti, C., Díaz, L., Viana, M.: Dynamics beyond uniform hyperbolicity. A global geometric and probabilistic perspective. Encycl. Math. Sc. **102**. Math. Phys. **3**. Springer (2005)
21. Bowen, R.: Equilibrium states and ergodic theory of Anosov diffeomorphisms, Lect. Notes in Math. vol. 470. Springer (1975)
22. Doering, C.: Persistently transitive vector fields on three-dimensional manifolds. Proc. Dyn. Syst. Bifurcation Theory **160**, 59–89 (1987)
23. Ferreira, C.: Genericity of transitive energy levels for Hamiltonians. In Preparation (2010)
24. Horita, V., Tahzibi, A.: Partial hyperbolicity for symplectic diffeomorphisms. Ann. Inst. H. Poincaré Anal. Non Linéaire **23**(5), 641–661 (2006)
25. Hunt, T., MacKay, R.S.: Anosov parameter values for the triple linkage and a physical system with a uniformly chaotic attractor. Nonlinearity **16**, 1499–1510 (2003)
26. Mañé, R. Oseledec's theorem from generic viewpoint. Proceedings of the international Congress of Mathematicians, Warszawa, **2**, 1259–1276, 1983
27. Mañé, R.: The Lyapunov exponents of generic area preserving diffeomorphisms. International Conference on Dynamical Systems, Montevideo, 1995. Res. Notes Math. Ser. **362**, 110–119 (1996)
28. Markus, L., Meyer, K.R.: Generic Hamiltonian Dynamical Systems are neither Integrable nor Ergodic. Memoirs AMS **144** (1974)
29. Mora, L., Romero, N.: Persistence of homoclinic tangencies for area-preserving maps. Ann. Fac. Sc. Toulouse **6**(4), 711–725 (1997)
30. Moser, J.: On the volume elements on a manifold. Trans. Am. Math. Soc. **120**, 286–294 (1965)
31. Newhouse, S.: Quasi-elliptic periodic points in conservative dynamical systems. Am. J. Math. **99**, 1061–1087 (1977)
32. Oseledets, V.I.: A multiplicative ergodic theorem: Lyapunov characteristic numbers for dynamical systems. Trans. Moscow Math. Soc. **19**, 197–231 (1968)
33. Oxtoby, J., Ulam, S.: Measure-preserving homeomorphisms and metrical transitivity. Ann. Math. **42**, 874–920 (1941)
34. Palis, J.: Open questions leading to a global perspective in dynamics. Nonlinearity **21**, 37–43 (2008)
35. Peixoto, M.: On structural stability. Ann. Math. **69**, 199–222 (1959)
36. Peixoto, M.: Structural stability on two-dimensional manifolds. Topology **1**, 101–120 (1962)
37. Pugh, C., Robinson, C.: The C^1 closing lemma, including Hamiltonians. Ergod. Theory Dyn. Syst. **3**, 261–313 (1983)
38. Robinson, C.: Generic properties of conservative systems I and II. Am. J. Math. **92**, 562–603 and 897–906 (1970)
39. Robinson, C.: Lectures on Hamiltonian Systems. Monograf. Mat. IMPA (1971)
40. Saghin, R., Xia, Z.: Partial Hyperbolicity or dense elliptical periodic points for C^1-generic symplectic diffeomorphisms. Trans. A.M.S. **358**(11), 5119–5138 (2006)
41. Vivier, T.: Robustly transitive 3-dimensional regular energy surfaces are Anosov. Institut de Mathématiques de Bourgogne, Dijon, Preprint 412 (2005). http://math.u-bourgogne.fr/topo/prepub/pre05.html
42. Zehnder, E.: Note on smoothing symplectic and volume-preserving diffeomorphisms. In Proc. III Latin Amer. School of Math., Inst. Mat. Pura Aplicada CNPq, Rio de Janeiro, 1976, vol. 597 Lecture Notes in Math., pp. 828–854. Springer, Berlin (1977)

Chapter 9
Microeconomic Model Based on MAS Framework: Modeling an Adaptive Producer

Pavel Brazdil and Frederico Teixeira

Abstract In recent years various methods from the field of artificial intelligence (AI) have been applied to economic problems. The subarea of multiagent systems (MAS) is particularly useful as it enables to simulate individuals or organizations and various interactions among them. In this paper we investigate a scenario with a set of agents, each belonging to a certain sector of activity (e.g. agriculture, clothing, health sector etc.). The agents produce, consume goods or services in their area of activity. Besides, our model includes also the resource of *free time*. The goods and resources are exchanged on a market governed by auction, which determines the prices of all goods. We discuss the problem of developing an *adaptive* producer that exploits *reward-based learning*. This facet enables the agent to exploit previous information gathered and adapt its production to the current conditions. We describe a set of experiments that show how such information can be gathered and explored in decision making. Besides, we describe a scheme that we plan to adopt in a full-fledged experiments in near future.

9.1 Introduction

In recent years various methods from the field of artificial intelligence (AI) [13] have begun to be applied to economic problems [16]. The subareas of AI that have turned out to be useful include multiagent systems, machine learning, planning and optimization among others. The area of multiagent systems (MAS) [15] is useful as

P. Brazdil (✉)
LIAAD-INESC Porto LA, Porto, Portugal
and
FEP, University of Porto, Porto, Portugal
e-mail: pbrazdil@liaad.up.pt

F. Teixeira
FEP, University of Porto, Porto, Portugal
e-mail: jfredericoteixeira@gmail.com

M.M. Peixoto et al. (eds.), *Dynamics, Games and Science I*, Springer Proceedings in Mathematics 1, DOI 10.1007/978-3-642-11456-4_9,
© Springer-Verlag Berlin Heidelberg 2011

this framework enables to simulate individuals or companies and various complex interactions among them. Machine learning (ML) [13, 14] is useful, as the behavior of the agents does not need to be programmed beforehand. The model is constructed by exploiting the observations of the effects of past behavior. The capability to learn permits the agents to optimize certain aspects of their behavior. This is related to the issue of planning and optimization.

The advantage of models is that it enables us to study certain phenomena that are difficult to analyze in a real world due to too many complex interactions. This is particularly evident in the area of economics. Simulation permits us to study the relationship between the state variables characterizing the individual constituents, that is, microeconomic relations. However, the model permits us to observe and analyze certain global trends characterizing a group of agents.

Our aims here are similar to those of Wellman and Hu [6] that provided the initial stimulus for us to develop this work. The issues that these authors addressed and which also concern us are:

1. How can we characterize a group of agents that may change their beliefs?
2. How can an agent change it beliefs by learning?
3. Supposing that each agent is trying to optimize its behavior by searching for optimal actions, can the agent achieve some kind of steady state balance (equilibrium) as a result, in which he may not want to carry out any further changes?
4. Can the whole economic model achieve a steady state balance equilibrium? Or is it more common that the system would be continuously evolving?

These are complex questions and it would be too ambitious to try to provide a general answer to all of them. Our aim here is to provide some answers while focusing on a specific domain – the domain of microeconomics.

Regards issue (4), we share the belief with others [6, 10], that economy is a complex evolving system. Although some equilibria can be attained in a restricted subproblem, in general, the whole system is unlikely to stay in some equilibrium, particularly if the agents themselves are adaptive.

In this paper we explore the notion of equilibrium and in particular conjectural equilibrium, which depends on a set of beliefs held by an agent. This notion was introduced by Hahn [5] in the context of a market model and exploited later by Wellman and Hu [6]. It enables to determine the best action $a_i *$ for each agent a_i, by maximizing a given utility function. As our problem is quite complex, in this paper we do not consider cooperative schemes and the issue of how these could be acquired (i.e. some form of co-learning).

As we will see later, the agent's utility function need not necessarily be the classic utility function that is often exploited in this context. Any more complex utility function different from the classic one, will be referred to here as an extended utility function. Further on we will explain why we need to consider it.

If the agents have an opportunity to act repeatedly in different settings, in game theory this scenario is referred to as a *repeated game*. In such settings the agents can learn from observations of the consequences of past actions. The learning method adopted here falls under the category of *reward-based learning* [4] as it requires a feedback provided by some kind of critic. Here we do not assume an existence of an external critic that attributes rewards, but rather the existence of some function that can characterize states in terms of utilities. The utilities can be compared with the aim to identify the best action.

In this work we prefer to follow Panait et al. [4], and avoid usage of the term reinforcement learning that they attribute to a more specific class of reward-based algorithms (algorithms based on dynamic programming, including Q-learning, Temporal Difference Learning etc.,) which are not exploited here.

The experimental study that we have carried out includes an exchange market, as in [6]. Initial conditions are given and the system stabilizes in an equilibrium state. In addition to the market, we include also production and consumption in our model, as this turns the model more realistic. In general production and consumption destabilize the market equilibrium and force the agents to act. Here we can distinguish between a behavior resulting from a fixed set of beliefs and those that results from changing beliefs acquired by learning (i.e. in our case reward-based learning).

The objective of our research is to reconsider the issues (1), (2) mentioned earlier and provide an answer, backed up by results of an experimental study. More specifically, the interesting questions that arise are:

1. How can we characterize a group of agents active in a microeconomic system, which includes an exchange market, production and consumption?
2. How can we construct an adaptive agent, exploiting a reward-based learning strategy? Which state variables does the agent need to observe? What kind of utility function should the agent use to judge the success of the agents' actions?

Although the work of Wellman and Hu [6] has provided an initial impetus to develop this work, there are many differences that distinguish the two:

- In our model the agents produce, consume and exchange different kinds of goods and services, including agricultural products, clothing, transportation, health, etc. To each of these goods we have attributed initial prices that are related to the current world (as we do not have access to prehistoric data). This has the advantage that we can use common sense to quickly spot errors that may manifest themselves by nonsensical prices.
- One novelty in our system is the introduction of the resource *time*. It is assumed that that each day has a normal duration of 24 h. Part of the day is occupied by work, another part by sleeping (simulating thus what happens in the real world). The remaining part is referred to as *free time*, which is is consumed for leisure activities and is attributed certain utility by the agents. This resource is important for modelling satisfactorily the behavior of a producer/consumer.
- We discuss an extension of the classical utility function, which includes both classical utility and wealth. This extension permits to model satisfactorily the behavior of an adaptive producer/consumer.
- Reward-based learning is used with the objective of developing an adaptive producer agent. This facet enables the agent to exploit previous information gathered in the past in order to adapt its production to the current conditions (e.g. increase production of a certain good by a certain amount).

- We have devised a methodology that can be used to evaluate the performance of our adaptive agent. Basically, the agent is required to act in new settings and we evaluate how good its actions are.

The rest of the paper is organized as follows. In Sect. 9.2 we describe the model of an exchange market and show some typical behavior, including how prices settle to an equilibrium. We discuss also some aggregate measures that can be calculated including for instance utility attributed by an agent to given goods and agent's wealth among others concepts. The model of consumption and production is described in Sect. 9.3. Here we are particularly concerned with the effects of changes in consumption/production on the market, including the prices. Section 9.4 describes the model of an adaptive producer and presents the preliminary results of our experiments.

9.2 Multiagent Model of an Exchange Market

The market is composed by a group of agents that belong to different sectors. Further on, we assume that each agent belongs to a particular sector of activity, such as agriculture, producer of clothing, provider of transportation services, etc. As we will see later, each consumer agent is simultaneously also a producer (in its principal sector of activity). Here we will assume that we have a set of consumer agents identified by a particular sector (the sector for which they are producers) and number within that sector (e.g. AGR1).

Besides agents, the market involves also a set of goods or resources. Our market includes for instance agricultural goods. Regards resources, our model includes for instance transportation resources, money and time. All these can be exchanged on a market. To simplify the following discussion we will use the term *good* to represent both goods and resources. The exchange economy can at time t be described as n-tuple $\langle \mathbf{Q}, \beta, \mathbf{P}, \mathbf{K} \rangle$. Before presenting the details, we just describe the convention used here. Bold capital letters, such as $\mathbf{Q}$, are used to represent vectors (or matrices). Lower case letters, such as $q_{ai,gi}$, for instance, represent the individual items of these vectors (or matrices). The meaning of each item is described below.

- $\mathbf{Q}$ represents a matrix of quantities of goods of agents. The term $q_{ai,gj}$ represents the quantity possessed by agent ai of good gj, where $ai = 1..n$ and $gj = 1..m$.
- Symbol β represents preferences attributed by different agents to different goods. As we will show later, these preference values are used in the calculation of utilities. In general, $\beta_{ai,gj}$ represents the preference of agent ai for good gj, where $ai = 1..n$ and $gj = 1..m$.
- $\mathbf{P}$ represents a price vector of m goods, that it is either given or established by the process that will be described later. So p_{gj} represents the price of good gj, where $gj = 1..m$. We assume that the same price is accepted by all agents in the market (and hence here we do not need $p_{ai,gi}$ that is the price agent ai attributes to good gj).

9 Microeconomic Model Based on MAS Framework: Modeling an Adaptive Producer   143

- Symbol **K** represents so called *background knowledge* of a specific domain. This term is used in machine learning literature and especially in Inductive Logic Programming (ILP) [12]. Here it may be in the form of equations, rules, constraints or other suitable representation. In our case here we use $\mathbf{K_M}$ to represent our underlying assumptions concerning the market. For instance, here it includes a utility function U that takes $q_{ai,gj}$ as an argument and returns a numeric utility value, among other concepts of this kind that we will be described later.

It is assumed that the price vector **P** is available to everyone (i.e. forms part of common blackboard). Regards the rest, it is assumed that the simulated world is partially opaque. So, for instance the values $q_{ai,gj}$ and $\beta_{ai,gj}$ are known only by agent ai (but not some other agent ak etc.).

The initial situation in an exchange market can be represented by $\langle \mathbf{Q^i}, \beta, \mathbf{P^i}, \mathbf{K_M} \rangle$. It may be in equilibrium or not. If the market is in equilibrium, then the agents will not want to exchange any goods. This situation can be described as executing a *null action* in the market. If the system is not in equilibrium, the market will react to achieve equilibrium. This can be represented by $\langle \mathbf{Q^f}, \beta, \mathbf{P^f}, \mathbf{K_M} \rangle$. So the process of transforming the former into the latter can be modelled by a procedure which is called here *exchange.market*$(\mathbf{Q^i}, \beta, \mathbf{P^i}, \mathbf{K_M}, \mathbf{Q^f}, \mathbf{P^f})$, where the first four arguments can be regarded as inputs and the last two as outputs. In this context $\mathbf{Q^i}$ is often referred to as *endowment*, $\mathbf{P^i}$ initial price, $\mathbf{P^f}$ the price determined by the market and $\mathbf{Q^f}$ the *demand*, representing the quantities that the agents would ideally like to have considering the final price $\mathbf{P^f}$. The difference between $\mathbf{Q^f}$ and $\mathbf{Q^i}$ is normally referred to as *excess demand*. The value represents the quantity that the agent wishes to exchange. If it is positive it represents the quantities to be acquired, while if it is negative it represents the quantities to be offered for exchange. It is important to note that the demand depends on the current price, that is, for each price P there will be a specific demand determined by parameters β and $\mathbf{K_M}$.

The excess demand can be seen as a parameter of action *exchange*$(\Delta q^f_{ai,gj})$. If this action is executed by each agent for each good, the equilibrium is reached. One condition for the equilibrium to exist is that for all goods the excess demand of each of the agents is 0, i.e.

$$\forall gj, \forall ai, \Delta q^f_{ai,gj} = 0 \tag{9.1}$$

The procedure that obtains the equilibrium can be implemented in various ways. In the following section we review a description of a method which is based on [6].

9.2.1 Iterative Method for Reaching Equilibrium

Our market model includes, besides the agents, also an auctioneer. The iterative process of obtaining the equilibrium is a WALRAS algorithm [7, 8] which is a variant of *tatonnement* [9]. This involves the following steps:

Continue iterative process until the equilibrium condition (1) has been achieved:

1. Calculation of excess demand: At each iteration k, each agent ai calculates the excess demand $\Delta q^f_{ai,gj}$ for each good gj and communicates these values to the auctioneer.
2. Summing up excess demands: The auctioneer, upon receiving the information about the excess demand of goods from each agent on the market, sums up the total excess demand for each good.
3. Adjustment of prices: For each good, the auctioneer adjusts the prices in order to approximate them to equilibrium prices. That is, if there is an excess (shortage) of supply of some good, the auctioneer decreases (increases) the price, following the basic rules of economics (see e.g. [11]). The new adjusted prices are communicated to each agent in the market.

As we see the algorithm is an iterative process. At each step, the auctioneer adjusts the price partially, but not completely, until the process terminates. More details concerning each step are given in the following sections.

9.2.1.1 Calculation of Excess Demands of Agents

Let us see how each agent ai calculates the excess demand $\Delta q^k_{ai,gj}$ (the superscript k identifies the iteration in between i and f). This step requires that the agent ai calculates the utility value U_{ai} which is the sum of all contributions $U_{ai,gj}$, each representing the utility attributed by the agent to certain quantity of good gj. In this section we adopt the usual scheme which uses the appropriate elements of $\mathbf{Q}$ and β to calculate this. More precisely:

$$U_{ai} = \sum_{gj=1}^{m} u_{ai,gj} = \sum_{gj=1}^{m} \beta_{ai,gj} \times ln(q_{ai,gj}) \tag{9.2}$$

We assume that the method of calculating the utility forms part of the existing background knowledge $\mathbf{K_M}$.

We note that in our model different agents may attribute different utilities to different types of goods (unlike in [6]). So, in our set-up the preferences $\beta_{ai,gj}$ and $\beta_{ai,gk}$ will normally be different. This is useful, as it is more representative of the real world where these values are normally affected by prices. In Sect. 9.2.2 we discuss how these values can be determined. Regards the agents' preferences for the same good, we can provide personalized values, that is, allow that $\beta_{ai,gj} \neq \beta_{ak,gj}$. We have examined such situations and the experimental results are reported later (in Sect. 9.2.3).

The concept of *wealth* is an important one in the exchange of goods. The agent's wealth represents his *budget restriction*, which needs to be taken into account when exchanging goods. Each agent cannot "spend" more than the amount allowed by his budget restriction at each moment of time. This prevents the agents from having negative wealth. This is obviously an assumption that could be relaxed.

9 Microeconomic Model Based on MAS Framework: Modeling an Adaptive Producer   145

The agent's wealth can be calculated from the appropriate elements of $\mathbf{Q^i}$ and $\mathbf{P^i}$. Here we use $\mathbf{P^i}$, as it is the price at the beginning of the process. So, wealth w_{ai}^k representing the budget restriction of agent ai can be calculated as:

$$w_{ai}^k = \sum_{gj=1}^{m} = q_{ai,gj}^i \times p_{gj}^k \qquad (9.3)$$

Each agent faces the question of how to maximize his utility not exceeding his initial budget restriction. So, if one agent has an excess of some good that does not contribute much to its utility, the agent will try to exchange it for another good that could increase its total utility. The agent's aim is to maximize its utility subject to budget constraints. This can be represented as:

$$\arg\max_{q_{ai,gj}\, gj=1..m} u_{ai} = \sum_{gj=1}^{m} \beta_{ai,gj} \times ln(q_{ai,gj}^i) \text{ such that } w_{ai}^k \leq w_{ai}^i \qquad (9.4)$$

For this type of maximization problem an analytical solution exist, which is based on the method of *Lagrange Multipliers* [11]. The ideal quantities of agent ai of good gj representing the agent's demand at iteration k can be calculated as follows:

$$q_{ai,gj}^k = \frac{w_{ai}^k \times \beta_{ai,gj}}{p_{gj}^k \times \sum_{gj=1}^{n} \beta_{ai,gj}} \qquad (9.5)$$

Then, as has been mentioned earlier, excess demand is $\Delta q_{ai,gj}^k = q_{ai,gj}^k - q_{ai,gj}^i$

9.2.1.2 Summing Up the Excess Demand of Goods

As we have mentioned earlier, the auctioneer, upon receiving the information about the excess demand of goods from each agent on the market, sums up the total excess demands for each good. This can be represented by the following equation:

$$\Delta q_{gj}^k = \sum_{ai=1}^{n} \Delta q_{ai,gj}^k \qquad (9.6)$$

9.2.1.3 Adjustment of Prices

The auctioneer adjusts the prices in order to approximate them to equilibrium prices. That is, if there is an excess (shortage) of supply of some good, the auctioneer decreases (increases) the price, following the basic rules of economics (see e.g. [11]). Here we use the following method for the adjustment of price of good gj:

$$p_{gj}^{k+1} = p_{gj}^k \times \left[1 + \frac{\Delta q_{gj}^k}{2q_{gj}}\right], \text{ where } q_{gj} = \sum_{ai=1}^{n} q_{ai,gj}^k = \sum_{ai=1}^{n} q_{ai,gj}^i \qquad (9.7)$$

The exact method of adjustment was not given in [6], perhaps as it was regarded as a simple issue. This turned out not to be as simple as it seems. We have experimented with various alternative ways of carrying out the adjustments. Some of them lead to rather slow changes, requiring thus rather too many cycles before reaching the equilibrium. Others suffered from oscillations around the correct equilibrium value. Method represented by (9.7) has demonstrated advantages in comparison to the others, as it requires relatively few cycles to reach the equilibrium.

This process of price adjustment is repeated for all goods and the new adjusted prices are communicated to each agent in the market.

9.2.2 Determining the Preference Values of Agents

A question that is addressed in this section is how to set the initial values for an exchange market $\langle \mathbf{Q^i}, \beta^i, \mathbf{P^i}, \mathbf{K_M} \rangle$. that would enable us to conduct the experiments. The basic idea that is explored here is the following. First, we set some appropriate values for $\mathbf{Q^0}$ and $\mathbf{P^0}$ using our knowledge of the external world. Then we assume that the prices and quantities will not change and calculate the value of β^i as follows:

$$\beta_{ai,gj}^i = \frac{w_{ai,gj}^0}{w_{ai}^0}, \text{ where } w_{ai,gj}^0 = q_{ai,gj}^0 \times p_{ai,gj}^0 \text{ and } w_{ai}^0 = \sum_{gj=1}^{m} q_{ai,gj}^0 \times p_{qi,gj}^0$$
$$(9.8)$$

So the preference $\beta_{ai,gj}^i$ represents the proportion of agent's wealth considering gi in proportion to the total wealth of his basket of goods. As we assume that the value β_{ai}^i remains fixed for any agent afterwards, we will just use β_{ai} in the following. We note that

$$\beta_{ai} = \sum_{gj=1}^{m} \beta_{ai,gj} = 1 \qquad (9.9)$$

9.2.3 Some Experimental Results with a Simulated Market

We have carried out various experiments. In one series of experiments we have used the following types of goods or services belonging to the following sectors of activity: agriculture, clothing, transportation and health services. Besides, we have used also the following resources: *money* and *free time*. Money was traded as any other good. The resource *free time* was also traded.

The methodology adopted was to use the values $\langle \mathbf{Q^i}, \beta, \mathbf{P^i}, \mathbf{K_M} \rangle$, then alter manually some of them and observe the results. For instance, we have altered some of

9 Microeconomic Model Based on MAS Framework: Modeling an Adaptive Producer 147

the values in $\mathbf{P^i}$ and this way obtained a different initial price vector $\mathbf{P^j}$. In all these cases the procedure found easily the equilibrium. The final price $\mathbf{P^f}$ would be equal to $\mathbf{P^i}$. This is of course no surprise, as we expected the program to behave this way.

We have observed that some agent would offer the resource *free time* on the market for exchange, enabling him to acquire certain goods. The interpretation of this is quite interesting, as it models an offer of work in exchange of goods. In this study we have not modeled more lasting relationships which are normally established in real life. This would involve, for instance, *commitments* among two or more agents to carry out trading in preference to others (adopting the relationship suppliers/consumers), referred to often as *emergent properties* [4] or emergence of *social structures* [18]. We plan to carry out such studies in future.

In another series of experiments we have altered one of the values in the preferences for one of the agents. So, for instance, we have increased the value of for a particular agent ai and some particular good gj, while maintaining condition (9.8), which requires that we decrease some preference values of the other goods. We have observed that this situation leads to an increase of the corresponding price, due to an increased demand of that good.

An interesting set of experiments involves altering the preference for *free time*. If it is increased, the propensity to work of those agents decreases. As our objective here is to study the dynamics of consumption and production, we do not report any more details on this issue here.

9.3 Modeling Consumption and Production and Its Effects on the Market

As we have mentioned earlier, our goal is to model consumption and production and study their effects on the market. We are also interested in the issue of how the behavior of the market can affect the decisions regards consumption and production. To be able to do this we need to introduce several new concepts. Here, in general each concept needs to be indexed with respect to time t. Here the basic time step will be one week, so t will refer to a particular week.

Our microeconomic model is represented by $\langle \mathbf{Q^t}, \beta, \mathbf{P^t}, \mathbf{K_M}, \mathbf{Qc^t}, \mathbf{Qp^t}, \mathbf{K_{ME}} \rangle$, where the first four concepts are similar to the ones discussed earlier in Sect. 9.2.

We note that $\mathbf{P^t}$ represents the *price vector* of m goods at time point t (i.e. week t). However, we need to distinguish between the initial prices of goods entering the market at time point t and the prices determined by the market. For that reason we use another superscript and so $\mathbf{P^{t,i}}$ represents the former and $\mathbf{P^{t,f}}$ the latter. So, for instance, $p_{gj}^{t,k}$ represents the price of good gj at time point t and market iteration k. Some values are dependent on others. In particular, $p_{gj}^{t+1,i} = p_{gj}^{t,f}$, that is, the initial price at the next time point $t+1$ is equal to the final market price at previous time point.

Symbol $\mathbf{Qc^t}$ represents a matrix of consumptions in the time interval beginning at t (i.e. interval between t and $t+1$). So, for instance, $qc_{ai,gj}^t$ represents the consumption of agent ai of good gj during that time interval. Symbol $\mathbf{Qp^t}$ represents a matrix of goods produced in the time interval between t and $t + 1$. So for instance $qp^t_{ai,gj}$ represents the production of agent ai of good gj during that time interval.

Production and consumption affect the quantities that each agent possesses. For all agents and goods the following relationship holds:

$$q^{t+1}_{ai,gj} = q^t_{ai,gj} + qp^t_{ai,gj} - qc^t_{ai,gj} \tag{9.10}$$

For a producer of good gj, the quantity $qp^{t+1}_{ai,gj}$ is normally larger than $qp^t_{ai,gj}$ to enable the agent ai to offer the surplus on the market.

The resources *money* and *free time* require a special consideration. As money is not produced nor consumed, the quantities $qp^t_{ai,m}$ and $qc^t_{ai,m}$ are equal to 0. Regards *free time*, we assume that each agent is attributed a certain amount of free time per week (e.g. certain number of hours per day times the number of days) which is consumed partly in production and partly in leisure activities. The amount of free time contributes to the overall agent's utility.

Symbol $\mathbf{K_{ME}}$ represents the background knowledge representing our underlying assumptions regards the relationship between production and consumption. More details concerning this are given in the next section.

9.3.1 Modeling the Relationship Between Production and Consumption

Without lack of generality let us focus on a particular agent ai involved in the production of good gj. We assume that the consumption at time interval t consists of two parts: the first part is fixed and does not depend on production. As in real life, the person that does not work still needs certain resources to survive. The second part varies with production. The larger the production, the larger is the consumption of certain resources. This can be represented by:

$$qc^t_{ai} = qc^{t,fix}_{ai} + qc^{t,var}_{ai} \tag{9.11}$$

The first term is a vector of individual fixed consumptions of individual goods. The second term is a function of consumption of resources. Here we have adopted a simple model that assumes a linear relationship between the two. That is,

$$qc^{t,var}_{ai} = \sum_{gj=1}^{m} k_{ai,gj} \times qp^t_{ai,gj} \tag{9.12}$$

where $k_{ai,gj}$ is an appropriate constant.

9.3.2 Some Aggregate Measures Characterizing Production and Consumption

In this section we discuss various aggregate measures that can be used to characterize the microeconomy involving production and consumption. The measures can be divided into two groups. The first group involves measures that characterize the production/consumption of a particular good in a particular sector (e.g. agriculture). The second group involves measures that characterize a particular agent, or a group of agents pertaining to a particular sector (e.g. agriculture). Both types of measures are discussed in more detail in the following.

9.3.2.1 Measures Characterizing the Production/Consumption of Particular Good

Goods surplus (gs^t_{gj}): This measure characterizes the relationship between production and consumption of particular good (e.g. agricultural goods). This measure is defined as the ratio of the quantity of goods produced to the quantities consumed:

$$
gs^t_{gj} = \frac{\sum_{ai=1}^{n} qp^t_{ai,gj}}{\sum_{ai=1}^{n} qc^t_{ai,gj}}
\tag{9.13}
$$

We distinguish the following situations. Positive (negative) surplus of good gi occurs if more (less) goods are produced than consumed, that is when gs^t_{gj} is larger (smaller) than 1.

9.3.2.2 Measures Characterizing a Particular Agent (or Group of Agents)

Production share (ps^t_{gj}): This measure is defined as a proportion of goods gj produced by agent ai in a particular sector of activity in relation to total production in that sector. This term has a similar meaning to the *market share* used in economics. We prefer to use the term *production share* as it has a more precise meaning. Not all items produced need to be exchanged on the market.

Relative production share (psr^t_{gj}): It is useful to compare the agent's share of production to the mean value in a particular sector. It is useful to consider whether this value is greater (smaller) than one.

Consumption share (cs^t_{gj}): This measure is defined as a proportion of goods gj consumed by agent ai in a particular sector of activity in relation to total consumption in that sector.

Relative consumption share (csr^t_{gj}): This term describes the agent's share of consumption to the mean value in a particular sector. It is useful to consider if this value is greater (smaller) than one.

Utility (u_{ai}^t): Earlier we have mentioned the concept of utility (see (9.2)) to characterize agent ai. Traditionally this measure is used to characterize the consumer behavior. However, we ote that our consumer agents are also producers (and vice versa).

Relative utility share $(us\,r_{ai}^t)$: It is useful to compare the agent's utility to the mean utility value overall or in a particular sector. If this value is greater (smaller) than one, the agent is in a better (worse) position than the others.

Expected utility change (ϵu_{ai}^t): This measure is defined as ratio of agent's utility generated by production and the utility that is consumed at that time point.

Utility change (Δu_{ai}^t): This measure is defined as a ratio of agent's utilities in two subsequent time intervals (i.e. $t-1$ and t). Increasing (decreasing) utility is characterized by value greater (smaller) than 1.

Wealth (w_{ai}^t): In literature on microeconomics (e.g. [11]) it is often argued that the behavior of producers should be governed by the goal to acquire more wealth and therefore we have adopted this measure here too. Wealth of agent ai at time point t is defined as follows:

$$w_{ai}^t = \sum_{gj=1}^{m} p_{gj}^t \times q_{ai,gj}^t \qquad (9.14)$$

We note that this measure depends on current prices determined by the market.

Relative wealth share $(wr\,s_{ai}^t)$: It is useful to compare the agent's wealth to the mean wealth in a particular sector. If this value is greater (smaller) than one, the agent is richer (poorer) than the others.

Expected wealth change (ϵw_{ai}^t): This measure is defined as the ratio agent's wealth generated by production to the wealth that is consumed at that time point. If it is greater (smaller) than 1, the expectation is that wealth will be generated (spent). This measure describes what some would call *expected productivity*. The expected wealth change may be different to the observed wealth change after goods have been exchanged at the market (see below).

Wealth change (Δw_{ai}^t): This measure is defined as the ratio of agent's wealth in two subsequent time intervals (i.e. $t-1$ and t). This measure will also be referred to as *productivity*.

9.3.3 Detailed-Level States and Generalized States

It is useful to distinguish between the detailed-level states and generalized (hence more abstract) states. The first group involves states described in terms of the basic entities discussed in Sect. 9.3. The second group includes the derived measures discussed in the previous section. Typically one particular generalized state corresponds to many different detailed-level states. For instance, surplus of good gi

may represent many different situations with different quantities of goods that are produced and consumed.

The transformation from detailed-level information to generalized levels is well known in data mining [17]. The data store is referred to as *OLAP data cube* and the transformation from more detail to more general as *drill-up* or *roll-up* operations [3]. Generalized states are useful, as they simplify our reasoning. We have much fewer states to worry about.

In the next section we describe some experiments that were defined in terms of the generalized states (e.g. we consider goods surplus).

9.3.4 Some Experiments Carried Out with Different Production and Consumption Conditions

We have carried out a number of experiments using the implemented system on the lines described above. In one series of experiments we have varied the ratio between the amount of goods produced and consumed. In other words, we have varied conditions leading to production surplus.

Experiments with no surplus, i.e. $gs_{gj=1}$ for all goods resulted in no surprise, as all parameters have maintained their values from one week to another.

Experiments with positive surplus, that is situations where $gs_{gj>1}$ lead to somewhat surprising consequences. In this experiment we have used only 4 agents, one per sector of activity. The sectors used were agriculture, transportation, clothing and health. This scenario is characterized by the situation shown in Fig. 9.1. Despite the fact that prices were decreasing (see Fig. 9.2), wealth maintained its value ($\Delta w_{ai} = 0$) (see Fig. 9.2). This is due to the fact that increase (decrease) in quantities of products is compensated by decreasing (increasing) prices. We note that utility change is positive for each agent, as expected.

This finding has rather dramatic consequences. In microeconomics it is argued that economic agents should base their decisions concerning production on wealth. But the results of our simulation have shown that it may not be an ideal measure of success in some circumstances. This finding was reported in [1, 2].

This problem motivated us to design and adaptive producer which would be capable of learning to adapt its production to particular circumstances. Our lateral aim was to consider various measures to determine which one provides the agent with the best basis for his decisions. This work is described in the next section.

9.4 Modeling an Adaptive Producer

The issue of how to model an adaptive producer relevant, as it is necessary to formulate strategies regarding the best action in a particular situation. This in turn is of relevance to economists and managers in real life.

Fig. 9.1 Utilities of agents

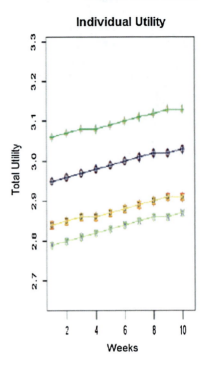

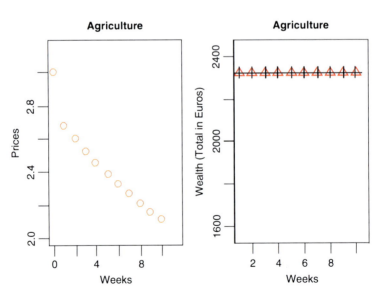

Fig. 9.2 Prices and wealth in agricultural sector in situation of goods surplus

9 Microeconomic Model Based on MAS Framework: Modeling an Adaptive Producer   153

The basic idea that has been explored here involves what in game theory can be referred to as *repeated game* and *fictitious play* [19]. It includes the following main steps:

1. Conceive (generate) different situations of our microeconomic model (e.g. endow the agents with different quantities etc.) and separating some for training and leaving the others for testing.
2. Present these situation to an agent to enable him to characterize these situations using certain aggregate measures.
3. The agent then carries out different actions with the objective of observing the consequences.
4. The consequences of each action are evaluated using certain performance measures and the best action for each situation is identified.
5. The cases identified in the previous step are used to generate a model by employing Machine Learning techniques (e.g. k-nearest-neighbor algorithm, $k - NN$). The model can be used to generate predictions, determining which action is to be executed in which state.
6. The performance of model is evaluated on a given test data. Our aim is to determine whether the agent is capable of learning to act correctly in new situations.
7. Finally we investigate the effect of various parameters and determine how these affect performance.

Each step is described in more detail below.

Step 1: requires that we select one agent for training and conceive different situations of our microeconomic model. In the first set of experiments, the producer of agricultural goods was selected for this aim. We have endowed all agents with certain values of $\langle \mathbf{Q^t}, \beta, \mathbf{P^t}, \mathbf{Qc^t}, \mathbf{Qp^t} \rangle$, and then modified them in various ways. This way we have generated many different situations. Four of them are shown in Table 9.1.

A more general approach involves generating different situations using an automatic process. One possibility of doing this is generating some initial situation and then modify it in various ways. The modifications can performed using a stochastic process, while taking care that some basic constraints do not get violated (e.g. the sum of production shares should always be equal to 1). So, for instance, we can generate situations corresponding to a particular value of goods surplus, relative production/consumption share, productivity, agent's preference for certain goods or its preference for the resource *free time* affecting his willingness to work.

Some of the states generated are used for training the adaptive agent (steps 2 till 5). The remaining states are used for evaluation (step 6).

Step 2: involves presenting each state to an agent to enable him to characterize the situations using various aggregate measures discussed in Sect. 3.2.2. The measures can be divided into two subgroups. The first group includes measures that characterize the conditions of the agent *before* engaging in fictitious play. Here this group includes *goods surplus*, gs_{gj}, characterizing the situation regards a particular good. For instance we can have $gs_{gj} > 1$. The second subgroup includes measures

Table 9.1 Example showing characterization four particular states of the simulated micro-economy, the agents actions and their effects

Situation	Goods surplus		Characterization of the agent				Change of production	Effects of action		Action evaluation using	
	gs^1_{agr}	psr^1_{agr1}	u^1_{agr1}	w^1_{agr1}	$\beta^1_{agr1.agr}$	$\beta^1_{agr1.time}$	Change $k^t_{ai.gj}$	$\Delta u^{1.3}_{agr1}$	$\Delta w^{1.3}_{agr1}$	$\Delta u^{1.3}_{agr1}$	$\Delta w^{1.3}_{agr1}$
1	>1	1	3.12	2324	0.32	0.25	+30%	−0.32	0.00	−	+
1	>1	1	3.19	2324	0.32	0.25	**0%**	1.88	0.00	+	+
1	>1	1	3.16	2324	0.32	0.25	−30%	0.95	0.00	−	+
2	1	1	3.09	2324	0.32	0.25	**+30%**	0.97	0.00	+	+
2	1	1	3.06	2324	0.32	0.25	0%	0.00	0.00	−	+
2	1	1	3.02	2324	0.32	0.25	−30%	−1.32	0.00	−	+
3	<1	2	2.95	2324	0.32	0.25	+30%	−2.03	0.00	−	+
3	<1	2	2.99	2324	0.32	0.25	0%	−0.67	0.00	−	+
3	<1	**2**	3.03	2324	0.32	0.25	**−30%**	0.66	0.00	+	+
4	<1	1	2.95	2324	0.32	0.25	+30%	−2.03	0.00	−	+
4	<1	1	2.99	2324	0.32	0.25	0%	−0.67	0.00	−	+
4	<1	1	2.03	2324	0.32	0.25	−30%	0.33	0.00	+	+

9 Microeconomic Model Based on MAS Framework: Modeling an Adaptive Producer   155

that characterize the agent. This group includes, for instance, *utility change*, Δu_{a1} etc. (see Table 9.1).

Step 3: includes executing different actions. Here it involves increasing (or decreasing) production by $X\%$, or else maintaining it as it was. After the action has been chosen, it is executed for a period of d weeks. In the initial experiments the parameter for increasing (or decreasing) production was set to 30% and the period was set to $d = 3$ (see Table 9.1).

Step 4: involves observing the effects of the execution of actions by the agent. We need some measure(s) enabling the agent to estimate its degree of success. Here we use *utility change*, Δu_{ai}, and/or *wealth change*, Δw_{ai}, in a given time interval (the last two columns in Table 9.1). As we have defined two measures, we have a choice as to which measure to use. In each case the aim was to identify the action that lead to the highest value of this parameter. The best action in each situation is stored for further use by the agent.

Let us consider, for instance, situation 1 in Table 9.1. Let us assume that the effects of the agent's actions measured in terms of utility change. The values indicate that the best action is to maintain the production at the same level. In situation 4, the best action is to decrease the production by 30%.

If we were to use wealth change as a criterion of success, we note that it is difficult to make a decision, as the three possibilities lead to the same result. They are all represented by "+" in the last column in Table 9.1. There is no observable difference between increasing production, maintaining the level or decreasing it.

We are planning to extend the utility measure to incorporate also wealth and this way combine utility and wealth measures and then analyze the advantages from the point of view of the agent,

Step 5: In this step the best actions identified in the previous step are used to generate a model. Here we plan to use two different Machine Learning algorithms. The first one is k-nearest-neighbor algorithm, *k-NN*, which belongs to the family of lazy learning methods [14,17]. Besides this, we plan to use decision trees or rules [14,17] to verify whether the system is capable of generating generalized knowledge.

Step 6: The model generated in the previous step needs to be evaluated. This is done using the test sample of situations obtained in step (1). The aim is to determine in how many situations the agent determines the correct (incorrect) action.

Step 7: We plan to investigate the effect of various parameters and determine how these affect performance. In particular, our aim is to determine how many training situations are needed for the system to achieve a reasonably good performance. This is done by varying the number of examples (situations encountered) in the training set and evaluating the agent's performance.

9.5 Conclusions

In this paper we have investigated a scenario with a set of agents, each belonging to a certain sector of activity (e.g. agriculture, clothing, health sector etc.). The agents produce, consume goods or services in their area of activity. Besides, our model includes also the resource of *free time*. the goods and resources are exchanged on a market, governed by an auction, which determines the prices of all goods. We have described an iterative process that enables the market to reach equilibrium.

We have discussed the problem of developing an *adaptive* producer that exploits *reward-based learning*. This facet enables the agent to exploit previous information gathered and adapt its production to the current conditions.

This aim forced us to develop a set of measures that enable to characterize a particular agent or a group of agents. Some of the measures provide the agent with an estimate of its degree of success. We have drawn attention to the fact that *wealth change* does not, on its own, provide a satisfactory solution in all situations. We have suggested that this measure be complemented by *utility change*.

We have described a set of initial experiments that show how such information can be gathered and exploited by the agent in decision making. Besides, we have also outlined a scheme that we plan to adopt in a full-fledged experiments with an adaptive agent in near future.

Acknowledgements The authors wish to acknowledge the support under grant BII-2009 awarded to Frederico Teixeira and also the pluri-annual support provided by FCT to LIAAD-INESC Porto L.A. Finally, the authors wish to thank colleagues – Pedro Campos, Rui Leite and Rodolfo Matos – for their useful comments and help in preparing this text.

References

1. Teixeira, F., Brazdil, P.: What can a simulation study tell us about the relevance of utility and wealth indicators for decision making concerning production and market behaviour. Poster presented at IJUP Porto and Rutgers University, New Jersey. See http://www.liaad.up.pt/pub/. Institution = LIAAD - Inesc Porto L.A. (2008)
2. Teixeira, F., Araujo, A., Moreira, C.: Microeconomic Model based on MAS framework: Modeling an Adaptive Producer, Poster presented at IJUP Porto. See http://www.liaad.up.pt/pub/. LIAAD - Inesc Porto L.A. (2010)
3. Gray, J., Bosworth, A., Layman, A., Priahesh, H.: Data Cube: A Relational Aggregation Operator Generalizing Group-By, Cross-Tab and Sub-totals. Proc. 12th Int. Conf. Data Eng. IEEE. pp. 152–159 (1995)
4. Panait, L., Luke, S.: Cooperative Multi-Agent Learning: The State of the art. Auton. Agent. Multi Agent Syst. **11**, 387–434 (2005)
5. Hahn, F.H.: Exercises in conjectural equilibrium analysis. Scand. J. Econ. **79**, 210–226 (1977)
6. Wellman, M.P., Hu, J.: Conjectural equilibrium in multiagent learning. Mach. Learn. **33**, 1–23 (1998)
7. Cheng, J.Q., Wellman, M.P.: The WALRAS algorithm: A convergent distributed implementation of general equilibrium outcomes. Comput. Econ. **1**, 1–24 (1998)
8. Wellman, M.P.: A market-oriented programming environment and its application to distributed multicommodity flow problem. J. Artif. Intell. Res. **1**, 1–23 (1993)
9. Negishi, T.: The stability of competitive economy. A survey article. Econometrica. **30**, 635–669 (1962)
10. Blume, L., Durauf, S.: The Economy as an Evolving Complex System, vol. 3. Oxford University Press (2005)
11. Besanko, David, R.B.: Microeconomics. Wiley (2005)
12. Lavrac, N., Dzeroski, S., Horwood, E.: Inductive Logic Programming: Techniques and Applications (1994)

13. Russell, Stuart, J., Norwig, P.: Artificial Intelligence: A Modern Approach. Pearson Education (2003)
14. Mitchell, T.: Machine Learning. McGraw-Hill (1997)
15. Jaque Ferber. Multi-agent Systems: An Introduction to Artificial Intelligence. Addison Wesley (1999)
16. Tesfatsion, L., Judd, K.: Handbook of Computational Economics: Agent-based Computational Economics, vol. 2. North-Holland (2006)
17. Han, J., Kamber, M.: Data Mining: Concepts and Techniques. Morgan Kaufmann (2006)
18. Peyton Young, H.: Individual Strategy and Social Structure: An Evolutionary Theory of Institutions. Princeton University Press (2001)
19. Shoham, Y., Powers, R., Grenager, T.: If multi-agent learning is the answer, what is the question? Artif. Intell. **171**, 365–377 (2007)

Chapter 10
A Tourist's Choice Model

J. Brida, M.J. Defesa, M. Faias, and Alberto A. Pinto

Abstract We present a tourism model where the choice of a resort by a tourist depends not only on the product offered in the resort, but also on the characteristics of the other tourists staying in the resort. In order to explore the effect of the types of the tourists in the allocation of tourists across resorts, we introduce a game theoretical model and describe the relevant Nash equilibria.

10.1 Introduction

Activity in the tourism industry and related areas depends on tourists' preferences for tourism goods and services, or in economic terms, the utility function that represents tourists' tastes. Typically the basic variables incorporated in the utility function correspond to prices and the tourists' specific preferences for these goods and services. Nevertheless, when one makes the decision to travel, there are other fundamental variables to take into account. One such variable is the "characteris-

---

J. Brida (✉)
Free University of Bolzano, Bolzano, Italy
e-mail: JuanGabriel.Brida@unibz.it

M.J. Defesa
Universidad de Alcalá, Alcalá de Henares, Spain
e-mail: mjesus.such@fct.uah.es

M. Faias
Universidade Nova de Lisboa, Lisbon, Portugal
e-mail: mcm@fct.unl.pt

A.A. Pinto
LIAAD-INESC Porto LA e Departamento de Matemática, Faculdade de Ciências, Universidade do Porto, Rua do Campo Alegre, 687, 4169-007, Porto, Portugal
and
Centro de Matemática e Departamento de Matemática e Aplicações, Escola de Ciências, Universidade do Minho, Campus de Gualtar, 4710-057 Braga, Portugal
e-mail: aapinto@fc.up.pt

M.M. Peixoto et al. (eds.), *Dynamics, Games and Science I*, Springer Proceedings in Mathematics 1, DOI 10.1007/978-3-642-11456-4_10,
© Springer-Verlag Berlin Heidelberg 2011

tics", or "type" of other tourists staying at the same resort. For example, young people like to stay together with other young people because of the likelihood of sharing common interests, or a family traveling together will likely prefer resorts where they can find other families so they can take advantage of the enhanced opportunity for their children to socialize with other children through planned activities. There are many other examples where factors such as social status, ethnicity, relationship status and other characteristics play a role in the decision making process. It follows then, in economic terms, that a tourist's selection of destination based on the desire to be with similar types is a variable to include in the utility function. This is valuable information for the tourism industry because it allows for the industry to better target the goods and services of travelers. However, despite the value of this information, there is a dearth of comprehensive studies that measure the distribution of tourist types and how and why these types vary from one destination to another.

Note that the distribution of the different types of tourists reaching a destination affects both the demand and supply side. From the demand perspective, as we have described above, the choice of a particular destination will depend greatly on who the agents believe will be sharing the resort with them. On the supply basis, a destination is characterized by the most frequent type of tourist because this will establish the reputation of the resort. While tourists could evolve from one type to another over time, likewise a resort destination could change it's profile to attract a particular type of clientele. There are several examples of this evolving behavior. Some recent references from the life cycle of a destination perspective are Claver-Cortés et al. [2] for Benidorm (Spain), Liu et al. [7] for Costa Rica. These examples demonstrate a connection between the life cycle of a destination and the changing profile of the traveler from Plog's [8] categorization of travelers, among others.

Our aim is to obtain insights about how the characteristics of tourists play a determined role in allocating are determinant to allocate tourists across resorts. Thus, we consider the model as in [1] where the tourists have a taste type or utility function, the utility function measures the degree of satisfaction that a tourist gets from vacationing at a resort, and we assume that the utility function of a tourist depends not only on the product that is offered in the resort but depends also on certain characteristics of the other tourists staying at the same destination. We refer to these characteristics as crowding types, with the crowding type being a set of observable characteristics of a tourist that affect the welfare of the other tourists, for instance, the example of a family that would prefer to joint a resort with other families in order that their children could play together. In our framework each tourist is characterized by a crowding type which is exogenously assigned.

The concept of crowding type was introduced and explored by Conley and Wooders [3–5], in a cooperative framework. More recently, in Faias and Wooders [6] the crowding type characteristics were exploited in the context of strategic club formation. The Faias and Wooders [6] work provided important motivation for this work due to the similarities in the decision process in choosing a club and choosing a vacation destination. Basically, a resort is comparable to a club where people derive utility not only from the resort or club itself, but also from the interaction with others similar to themselves, who also chose the same resort.

10   A Tourist's Choice Model

In our model the tourists choose resorts, but since the utility they derive depends on the crowding profile of the resort, the utility depends on the choices of the other tourists, therefore we consider a Nash game to model our framework. In the game each tourist chooses a resort within a set of available resorts which constitute the strategy set of each tourist and the payoff of each tourist is given by the utility function of the tourist. The equilibrium in our tourism choice model is an allocation of tourists to the available resorts such that no single tourist has incentive to move to another resort.

The paper is organized as follows, first we state the model by describing the Nash game and then we establish the main theorem which asserts the existence of a Nash equilibrium for the game. Next, we discuss characteristics of the equilibrium by providing sufficient conditions that guarantee the prevalence of a certain equilibrium. These sufficient conditions mainly relate the number of tourists of each crowding type with the relative evaluation that tourists assign to the resort product and to the crowding profile of the resort.

The first result states that if for each tourist the company of the other tourists, that is, their crowding profile, is less important than the resort product then every tourist chooses the resort with the product they prefer. However, when there exists at the same time a group of tourists that values the resort product more and a second group that values the crowding type more our model shows, under certain parameters, that the tourists who value the crowding profile over the resort product end up following the tourists who are choosing their location based on resort product. The final outcome is that a resort could lose all demand for business. We posit that in order to avoid this outcome the resort owner should invest in enhancing the resort product to the extent necessary to cause the tourist previously making a decision based on crowding type to now base his decision on resort product. Resort owners would then avoid losing tourists to alternate locations because they choose these resorts to benefit from the resorts crowding profile. In this way the resort owners or suppliers could avoid losing business caused by the herding effect.

10.2   The Model

We consider an economy with I tourists indexed by $i \in \{1, \ldots, I\} = \mathscr{I}$. The focus of the model is the tourism choice therefore we consider a model with a finite set of tourism resorts where the tourists behave strategically by choosing the resort that gives them the best payoff.

In order to emphasize the role of the key variable that we are adding in this paper, the crowding type, we consider that the tourism supply side consists of four tourism resorts. Each tourism resort is characterized by two features, the location and the offered product. Specifically, we consider a location of either the beach or the mountains, and a hotel with either a disco or a golf course. Thus, the tourists have four tourism resorts available. We denote by BD the tourism resort which is located at the beach and offers hotel and disco, by BG the tourism resort that is located at the beach and offers hotel and golf, by MD the tourism resort that is located at the mountain and offers hotel and disco and by MG the tourism resort that is located at the mountain and offers hotel and golf. Let $\mathcal{R} = \{BD, BG, MD, MG\}$ denotes the set of tourism resorts.

Each agent or tourist chooses a tourism resort. The payoff or welfare that an agent attains when he utilizes the tourism resort depends not only on the physical characteristics of the resort, namely, location and offered product but depends also on some observable characteristics of the other agents that are utilizing the same resort, the crowding types. In the context of our model we consider two illustrative crowding types based on the age of the tourists, namely, let $\mathcal{C} = \{c_y, c_o\}$ be the set of crowding types. If a tourist is characterized by the crowding type c_y, that means that the tourist is a young adult, if the tourist is characterized by the crowding type c_o, that means that the tourist is an older adult. In fact, these are two ordinary observable characteristics of the tourists that could influence the welfare of the other tourists that choose the same resort. Actually, the relevant variable is the number of agents of each crowding type, thus, given a resort $R \in \mathcal{R}$, $m_R = (m_{Rc_y}, m_{Rc_o})$ denotes the crowding profile of the tourists that are in the tourism resort R, m_{Rc_y} is the number of tourists with crowding type c_y in the resort R and m_{Rc_o} is number of tourists with crowding type c_o in the resort R.

The tourists have preferences concerning the tourism resorts and the related crowding profiles and these preferences are described by the taste type. We consider four taste types in the model, $\mathcal{T} = \{t_{BDy}, t_{BGo}, t_{MDy}, t_{MGo}\}$. Each taste type is represented by a utility (payoff) function which assigns the degree of satisfaction attained by a tourist when he joins a tourism resort. This utility or payoff function is determined by two variables, the product and the location of the resort, which is represented by R, and the corresponding crowding profile of the resort, m_R.

Let us now describe the payoff of the four taste types considered in the model.

Tourists with taste type t_{BDy} prefer a resort at the beach with a disco and are indifferent with regard to the other. Furthermore, they prefer the company of young people as opposed to older adults. The payoff function is

$$u_{t_{BDy}}(R, m_R) = f_{BD}(R) + m_{Rc_y} - m_{Rc_o} + C, \quad f_{BD}(R) = \begin{cases} V_{BD} & \text{if } R = BD \\ 0 & \text{if } R \neq BD \end{cases}$$

Tourists with taste type t_{BGo} prefer a resort on the beach with golf and are indifferent with regard to the other. Furthermore, they prefer the company of older people as opposed to yound adults. The payoff function is

$$u_{t_{BGo}}(R, m_R) = f_{BG}(R) - m_{Rc_y} + m_{Rc_o} + C, \quad f_{BG}(R) = \begin{cases} V_{BG} & \text{if } R = BG \\ 0 & \text{if } R \neq BG \end{cases}$$

Tourists with taste type t_{MDy} prefer a resort in the mountains with a disco and are indifferent with regard to the other. Furthermore, they prefer the company of young adults as opposed to older people. The payoff function is

10 A Tourist's Choice Model

$$u_{t_{MDy}}(R, m_R) = f_{MD}(R) + m_{Rc_y} - m_{Rc_o} + C, \quad f_{MD}(R) = \begin{cases} V_{MD} & \text{if } R = MD \\ 0 & \text{if } R \neq MD \end{cases}$$

Tourists with taste type t_{MGo} prefer a resort in the mountains with golf and are indifferent with regard to the other. Furthermore, they prefer the company of older people as opposed to young adults. The payoff function is

$$u_{t_{MGo}}(R, m_R) = f_{MG}(R) + m_{Rc_y} - m_{Rc_o} + C, \quad f_{MG}(R) = \begin{cases} V_{MG} & \text{if } R = MG \\ 0 & \text{if } R \neq MG \end{cases}$$

The constants $V_{BD}, V_{BG}, V_{MD}, V_{MG}$ in each payoff function are positive constants.

In this tourism model each tourist is characterized by two types, the crowding type and the taste type. Let $n(c, t)$ denote the total number of tourist in the economy with crowding type c and taste type t.

The fundamentals of this tourism model, which are the available resorts and the population of tourists, are described by

$$\mathscr{E} = \{\mathscr{R}, \big(n(c, t)\big)_{c \in \mathscr{C}, t \in \mathscr{T}}\}.$$

More precisely,

$$\mathscr{E} = \{\mathscr{R}, \big(n(c_y, t_{BDy}); n(c_y, t_{BGo}); n(c_y, t_{MDy}), n(c_y, t_{MGo}); n(c_o, t_{BDy}); $$
$$n(c_o, t_{BGo}); n(c_o, t_{MDy}); n(c_o, t_{MGo})\big)\}.$$

In order to describe the behavior of tourists we introduce some more notation. Let $\tau : \mathscr{I} \to \mathscr{T}$ be a function that assigns a taste type to each tourist $i \in \mathscr{I}$, that is, $\tau(i) = t$ for some $t \in \mathscr{T}$.

We model the behavior of tourists as a strategic game. The strategy set of each tourist-player is the set of available resorts, $\mathscr{R}$. Each player i chooses a tourism resort, that is, chooses a strategy R_i, with $R_i \in \mathscr{R}$. These choices give rise to a strategy profile, that is, a vector with the strategy of every agent, $(R_1, \ldots, R_i, \ldots, R_I) \in \mathscr{R}^I$. A strategy profile $(R_1, \ldots, R_i, \ldots, R_I)$ defines an allocation of tourists across the resorts and defines also the crowding profile for each resort, that is, the number of members of each crowding type in each resort, m_R. Therefore given a strategy profile $(R_1, \ldots, R_i, \ldots, R_I)$ the payoff of a tourist i is his utility evaluated at (R_i, m_{R_i}), that is, $\Pi^i(R_1, \ldots, R_i, \ldots, R_I) = u_{\tau(i)}(R_i, m_{R_i})$. Observe that m_{R_i} is the number of tourists of each crowding type that have chosen the same resort chosen by tourist i. Thus, the behavior of the tourists in this tourism model $\mathscr{E}$ is described by the game $\mathscr{G} \equiv \{(\mathscr{R}, \Pi^i); i \in \mathscr{I}\}$.

10.3 Equilibrium: Definition and Existence

The goal in this paper is to find an allocation of tourists through the tourism resorts such that given the choices of resort of every tourist, no tourist has incentive to move to another resort. Therefore the suitable equilibrium concept for the game $\mathcal{G}$ is the Nash equilibrium as follows.

Definition. A profile $(R_1^*, \ldots, R_i^*, \ldots, R_I^*) \in \mathcal{R}^I$ is a pure strategy Nash Equilibrium for the game $\mathcal{G} = \{(\mathcal{R}, \Pi^i); i \in \mathcal{I}\}$ if,

$$\Pi^i(R_i^*, R_{-i}^*) = \max_{R_i \in \mathcal{R}} \Pi^i(R_i, R_{-i}^*)$$

for all $i = 1, \ldots, I$.

Theorem. *There exists an equilibrium in mixed strategies for the Nash game $\mathcal{G} = \{(\mathcal{R}, \Pi^i); i \in \mathcal{I}\}$ associated to the tourism model $\mathcal{E}$.*

Proof. For every player i the strategy set $\mathcal{R}$ is finite therefore there exists an equilibrium in mixed strategies. $\qquad\square$

In the next section we present some special cases of our model, specifically, we consider sufficient conditions that guarantee the existence of an equilibrium in pure strategies. These conditions take into consideration the number of agents of each taste and crowding type and the parameters that define the taste types. Moreover, we discuss the characteristics of these pure equilibria.

10.4 Equilibrium Characterization

We assume in the following results that there are no tourists which are of one crowding type and prefer to join resorts with tourists of the other crowding type, that is, we suppose that:

$$n(c_y, t_{BGo}) = 0, \quad n(c_y, t_{MGo}) = 0, \quad n(c_0, t_{BDy}) = 0, \quad n(c_0, t_{MDy}) = 0.$$

Proposition 10.1. *For an economy under the assumptions*

$$(1.a) \ V_{BD} > n(c_y, t_{BDy}) + n(c_y, t_{MDy})$$
$$(1.b) \ V_{MD} > n(c_y, t_{BDy}) + n(c_y, t_{MDy})$$
$$(1.c) \ V_{BG} > n(c_0, t_{vBGo}) + n(c_0, t_{MGo})$$
$$(1.d) \ V_{MG} > n(c_0, t_{BGo}) + n(c_0, t_{MGo})$$

the following distribution of tourists is a Nash equilibrium:

10 A Tourist's Choice Model

	Resorts			
	BD	BG	MD	MG
Tourists	$n(c_y, t_{BDy})$	$n(c_o, t_{BGo})$	$n(c_y, t_{MDy})$	$n(c_o, t_{MGo})$

In this equilibrium all the tourists are separated across the tourism resorts as follows: at the resort BD we have all the young tourists of taste type t_{BDy}, at the resort BG we have all the old tourists of taste type t_{BGo}, at the resort MD we have all the young tourists with taste type t_{MDy} and finally at the resort MG we have all the old tourists with taste type t_{MGo}.

The results in the table are clear and speak for themselves, therefore we will not explain them in detail. Instead, however, we will discuss the intuitive nature of the results.

The intuition of the equilibrium in Proposition 10.1 is the following. Under the assumptions (1.a)–(1.d), for every tourist the corresponding parameter that measures the level of utility relative to the product of the resort is high enough to imply that every tourist values the product of the resort more than the crowding profile. It follows then that in equilibrium every tourist chooses the resort based on their preference for product value.

Proposition 10.2. *For an economy under the assumptions*

$$(6.a) \quad V_{BD} > n(c_y, t_{BDy}) + n(c_y, t_{MDy})$$
$$(6.b) \quad V_{MD} + n(c_y, t_{MDy}) < n(c_y, t_{BDy})$$
$$(6.c) \quad V_{BG} > n(c_o, t_{BGo}) + n(c_o, t_{MGo})$$
$$(6.d) \quad V_{MG} + n(c_o, t_{MGo}) < n(c_o, t_{BGo})$$

the following distribution of tourists is a Nash equilibrium:

	Resorts			
	BD	BG	MD	MG
Tourists	$n(c_y, t_{BDy}); n(c_y, t_{MDy})$	$n(c_o, t_{BGo}); n(c_o, t_{MGo})$		

The assumptions in Proposition 10.2 specify that there is no demand for the mountain destination and the tourists who prefer the beach have strong preference for the resort product and hence choose the beach destination. The tourists who prefer the mountain over the beach are a smaller group with a weaker preference for the tourism product compared to the crowding type component of their payoff function. However this payoff function prevails over the resort product component and as a result tourists who prefer the mountain also end up also choosing the beach destination. The next Proposition demonstrates an analogous result but in this situation there is no demand for the beach destination.

Proposition 10.3. *For an economy under the assumptions*

$$(7.a) \quad V_{BD} + n(c_y, t_{BDy}) < n(c_y, t_{MDy})$$
$$(7.b) \quad V_{MD} > n(c_y, t_{BDy}) + n(c_y, t_{MDy})$$
$$(7.c) \quad V_{BG} + n(c_o, t_{BGo}) < n(c_o, t_{MGo})$$
$$(7.d) \quad V_{MG} > n(c_o, t_{BGo}) + n(c_o, t_{MGo})$$

the following distribution of tourists is a Nash equilibrium:

			Resorts	
	BD	BG	MD	MG
Tourists			$n(c_y, t_{BDy}); n(c_y, t_{MDy})$	$n(c_o, t_{BGo}); n(c_o, t_{MGo})$

The three Propositions above show to what extent the equilibrium depends on the parameters that define the model, namely the parameters that define the evaluation of the product of the resort by the tourists and the number of tourists of each crowding type and taste type.

The proof of the propositions above, that is, the proof that the exhibited equilibra are in fact Nash equilibria is straightforward. Indeed, for every equilibrium when we check if each tourist would became better if he moves to another resort, given the assumptions, the conclusion is always that no tourist would move.

10.5 Conclusion

We conclude that the crowding type variable, which represents the characteristics of tourists that affect the welfare of the other tourists, has a significant effect on the allocation of tourists across resorts. Indeed, the demand for resorts is a result of the tourists' valuation of the crowding profile of the other tourists relative to the resort product. We observe that changes in the number of tourists of each crowding type and changes in the parameters that define the payoff of tourists could in fact change the Nash equilibrium.

Based on the results of our model we suggest when designing a resort the industry should take into account not only the taste type of the tourists but also the crowding type of the tourists that they would like to attract. The equilibria described in this paper suggest how the resort industry might want to focus their investments. For example, the suppliers of the tourism resort MD in the case of Proposition 2, would be better off investing in his product in order to increase the parameter VMD of the tourists who prefer the resorts MD. Otherwise they will lose clientele because these tourists place more value on the crowding profile of the resort than they do on the product of the resort.

Acknowledgements We thank LIAAD-INESC Porto LA, Calouste Gulbenkian Foundation, PRODYN-ESF, FEDER, POFC, POCTI and POSI by FCT and Ministério da Ciência e da Tecnologia, and the FCT Pluriannual Funding Program of the LIAAD-INESC Porto LA and of the Research Centre of Mathematics of University of Minho, for their financial support.

References

1. Brida, J., Defesa, M., Faias, M., Pinto, A.A.: Strategic choice in tourism with differentiated crowding types. Econ. Bull. **30**(2), 1509–1515 (2010)
2. Claver-Cortés, E., Molina-Azorn, J.F., Pereira-Moliner, J.: Competitiveness in mass tourism. Ann. Tourism Res. **34**, 727–745 (2007)
3. Conley, J., Wooders, M.: Taste-homogeneity of optimal jurisdictions in a Tiebout economy with crowding types and endogenous educational investment choices. Ricerche Econ. **50**, 367–387 (1996)
4. Conley, J., Wooders, M.: Equivalence of the core and competitive equilibrium in a Tiebout economy with crowding types. J. Urban. Econ. **41**, 421–440 (1997)
5. Conley, J., Wooders, M.: Tiebout economies with differential inherent types and endogenously chosen crowding characteristics. J. Econ. Theory **98**, 261–294 (2001)
6. Faias, M., Wooders, M.: A strategic model of club formation; existence and characterization of equilibrium (2006). dmat.fct.unl.pt/fct/listarPrePubs.doano=20
7. Liu, Z., Siguaw, J.A, Enz, C.A.: Using tourist travel habits and preferences to assess strategic destination positioning: the case of Costa Rica. Cornell Hospitality Q. **49**, 3 (2008)
8. Plog, C.S.: Why destination areas rise and fall popularity: An update of a Cornell Quarterly classic. Cornell Hospitality Restaur. Adm. Q. **42**(3), 13–24 (2001)

Chapter 11
Computability and Dynamical Systems

J. Buescu, D.S. Graça, and N. Zhong

Abstract In this paper we explore results that establish a link between dynamical systems and computability theory (not numerical analysis). In the last few decades, computers have increasingly been used as simulation tools for gaining insight into dynamical behavior. However, due to the presence of errors inherent in such numerical simulations, with few exceptions, computers have not been used for the nobler task of proving mathematical results. Nevertheless, there have been some recent developments in the latter direction. Here we introduce some of the ideas and techniques used so far, and suggest some lines of research for further work on this fascinating topic.

11.1 Introduction: From Numerics to Dynamics to Computation

In the last century significant developments have been made in the fields of dynamical systems and the theory of computation. Actually, the latter only appeared in the 1930s with the groundbreaking work of Turing, Church and others. These two areas

---

J. Buescu (✉)
DM/FCUL, University of Lisbon, Lisbon, Portugal

and

CMAF, Lisbon, Portugal
e-mail: jbuescu@ptmat.fc.ul.pt

D.S. Graça
DM/Faculdade de Ciências e Tecnologia, Universidade do Algarve, Faro, Portugal
and
SQIG/Instituto de Telecomunicações, Lisbon, Portugal
e-mail: dgraca@ualg.pt

N. Zhong
DMS, University of Cincinnati, Cincinnati, OH 45221-0025, USA
e-mail: ning.zhong@uc.edu

M.M. Peixoto et al. (eds.), *Dynamics, Games and Science I*, Springer Proceedings
in Mathematics 1, DOI 10.1007/978-3-642-11456-4_11,
© Springer-Verlag Berlin Heidelberg 2011

have mostly evolved separately, with very sporadic interactions throughout most of the twentieth century. However, with the advent of fast digital computers and their extensive use as simulation tools, this gap has been narrowing, and some work has been done to establish bridges across it. This paper focuses on this research.

Dynamical systems theory is of interest to computer scientists for a number of reasons. We could point out that computers are used to control continuous processes in everyday life or that silicon is reaching its limits, and new paradigms of computation are now sought (e.g. quantum computation [19]), many of them involving dynamical systems.

However, in this paper we are interested in presenting what the theory of computation has to offer to the dynamical systems community.

The modern theory of dynamical systems began with Poincaré in the late nineteenth century, reached a high level of development in the Russian school by the middle of the twentieth century, and was further developed by western mathematicians and scientists beginning in the 1960s. This development entailed the convergence of two very strong but quite distinct currents: a modeling (numerical) approach and an analytical approach.

On the modeling side, the increasing availability of computational power allowed the numerical study of mathematical models for systems of definite interest in problems of physics, engineering or mathematical sciences in general, showing that these low-dimensional deterministic systems apparently exhibited, in a persistent fashion, a strong form of chaotic behavior. The first and foremost example is of course that of the Lorenz attractor [36], whose display of sensitive dependence on initial conditions led Lorenz himself to coin the term "butterfly effect" to describe this form of chaos. It is far from the only one; soon other model systems were shown to exhibit the same kind of deterministic, low-dimensional chaotic behavior characterized by sensitive dependence on initial conditions. Thus, for instance, the Duffing equation [20], the (nonautonomous) van der Pol system [42] or the Rössler system [47] which arise as (differential) equations of motion for specific physical systems and also discrete time diffeomorphisms or maps, like the Hénon map or the logistic equation, which may be seen as arising directly or indirectly from a Poincaré section of the flow of a differential equation.

On the analytical side, hyperbolic dynamical systems theory began in the Russian school (especially in Anosov's work) and was further developed from the 1960s onward by the Smale school, with the purpose of giving a solid mathematical foundation to the fact that deterministic low-dimensional systems may exhibit persistent chaotic behavior, as evidenced by the wealth of specific examples referred to above. Thus arose the motivation for the main theoretical thrusts in what is nowadays called uniformly hyperbolic dynamical systems theory, leading from Anosov diffeomorphisms to the general theory of hyperbolic systems, whose invariant sets have the structure of a uniform invariant splitting into stable and unstable directions (see Smale [51]). This theory is extremely rich and allowed for the construction and study of very specific instances: the Arnold cat map, the Smale horseshoe and the corresponding symbolic dynamics derived from the associated Markov partitions.

Hyperbolic systems were conceived as an attempt to construct a rigorous theory describing persistent chaotic behavior. There were good grounds to believe that hyperbolic systems coupled with the dynamical equivalence relation of topological conjugacy (corresponding to structural stability) were the appropriate setting for a rigorous theory of chaotic phenomena, since this was the adequate generalization of what was known for two-dimensional systems, namely Peixoto's theorem [41], which states that all planar vector fields have structurally stable perturbations, and one can thus disregard systems which are structurally unstable.

However, further research progressively revealed a vast gap between chaotic behavior as computationally observed in "strange attractors" and the dynamics of hyperbolic systems. Smale himself [50] delivered the first blow when he showed that, in dimension 3 or higher, structurally stable systems are not dense. Thus, even though achieving a complete characterization of hyperbolic systems and their properties was a major accomplishment in dynamical systems, hyperbolicity is too strong a property to characterize a generic set of differential equations or diffeomorphisms.

In particular, the strange attractors arising from the Lorenz system, the Hénon map, the Duffing equation and other computationally well-studied systems, although persistently chaotic, are not hyperbolic and thus fall outside the scope of hyperbolic theory. Indeed, it could have been the case that the Lorenz attractor, in spite of all the numerical studies, did not exist as a (persistent, structurally unstable, chaotic) strange attractor; hyperbolic dynamics simply does not provide an answer. The existence of the Lorenz attractor was, in fact, listed by Steven Smale as one of several challenging problems for the twenty-first century [52].

The way to bridge this gap, within the purely analytical approach, is to extend hyperbolic theory to more general systems. One way to achieve this goal is to allow for *partially hyperbolic* systems, where we require that the flow or map admits an invariant splitting but, instead of requiring uniform rates of expansion and contraction, we allow some directions to have mixed expansive, contractive or neutral behavior in different parts of the system. This approach originated in the works of Pugh–Shub and Mañé in the 1970s.

Yet another way to extend the theory is to use concepts from ergodic theory, where we drop the uniform hyperbolicity requirement and replace it by asymptotic expansion/contraction rates in directions which may depend measurably on the initial point. Such systems are referred to as *non-uniformly hyperbolic*, and the focus of the theory is to construct physical (SRB) invariant measures and more generally equilibrium states, and to study their ergodic properties. The equivalence relation corresponding to structural stability is known as *stochastic stability*.

From the computational point of view much work has also been done in order to bridge this gap. In this approach we need to construct rigorous theoretical methods which allow us to transcend conjectures suggested by more or less precise numerical experiments and prove mathematical results in the most rigorous sense of the term. Paradigmatic in this approach are breakthroughs such as Lanford's computer-assisted proof of the Feigenbaum conjectures [34] and, more recently, W. Tucker's proof that the Lorenz attractor exists [56].

In both cases rigorous computational methods went for beyond educated numerical experiments; they provided deep theoretical insights into the mathematical structure underlying the corresponding dynamical phenomena. In the first case, it substantiated the renormalization interpretation of universality in C^1-unimodal maps: it proved that there is a fixed point of a renormalization operator in a suitable map space with a one-dimensional unstable manifold, the corresponding eigenvalue being the Feigenbaum constant. In the second case, Tucker's work finally provided a proof of the long-standing conjecture that the dynamics of the ordinary differential equations of Lorenz is that of the geometric Lorenz attractor of Williams, Guckenheimer, and Yorke or, in short, that the Lorenz system does indeed contain a persistent strange attractor.

To develop such an approach one must in general leave the realm of plain numerical simulation and look for general statements on computability (in the sense of the theory of computation) of the objects and concepts of dynamical systems theory. Although this is a fairly recent field of research, some promising results have already been achieved. The purpose of this paper is to give an overview of the methods used and results obtained, as well as to point out directions for possible future research.

11.2 Computable Analysis

In the study of differential equations and dynamical systems, scientific computation is playing an ever larger role because most equations cannot be solved explicitly but only approximately by numerical methods. Thus it becomes of central importance to know whether or not the problem being solved is computable. In particular, if a solution is non-computable, then no numerical algorithm computing the solution can always provide approximations with arbitrarily desired precision.

Computability over discrete spaces has been well studied since the 1930s. Although there are several markedly different models which formalize the notion of computability, such as Turing machines, lambda calculus, recursive functions, etc., they all generate the same class of computable functions. This formal notion of computability and the Turing machine model have been accepted by the scientific community as the standard model of computation. Indeed, as the Church-Turing thesis asserts, any intuitively and reasonably computable function is computable by a Turing machine. We refer the reader to [49] for more details on basic results about the theory of computation.

The Turing machine, however, cannot be directly applied to compute real functions because it can only have as input and output a "finite number of bits". To circumvent this, several extensions of the Turing machine model have been proposed. One such extension is the BSS model [4,5]. In the BSS model, a real number can be directly stored on a single cell, so that exact computations over real numbers can be carried out in finite time using infinite-precision arithmetic. Even though this model is algebraically elegant, it has certain weaknesses as a model for scientific computation. For example, the non-computability results obtained in this model do not correspond to computing practice in the real number setting (see [9] for more details), which is undesirable, since identifying non-computable parameters, functions and sets is one of the main objectives in the computability study of continuous structures [10, 44, 59].

11 Computability and Dynamical Systems

Another extension is the Type-2 Turing machine or oracle Turing machine model, which has been developed since the 1950s by many authors. For recent developments and more details about this model, the reader is referred to [31,43,58]. In this model, computations for functions $f : \mathbb{N}^{\mathbb{N}} \to \mathbb{N}^{\mathbb{N}}$ between Baire spaces are explicitly defined via Type-2 machines. Roughly speaking, this means that for any input sequence a in $\mathbb{N}$ on a read-only input tape, the machine computes (in the discrete sense) and writes the sequence $f(a)$ on a one-way output tape. The idea is that the machine keeps reading digits from the input and doing computations (as any computer does) to get partial results written on the output tape. Since the input tape has an infinite number of digits, the computation may require an infinite number of steps to describe the exact output. Because it is desirable to get useful results in finite time, one requires the output tape to be one-way, i.e., the machine cannot change what it has already written on the tape, thus ensuring that one has partially correct results at any given moment (the longer one waits, the more accurate the results are). Computations of real functions $f : A \to B$, $A, B \subseteq \mathbb{R}$, can then be performed by encoding real numbers by sequences of rational numbers and employing a Type-2 machine to compute rational approximations of $f(x)$ with arbitrary precision from a suitable rational approximation of x. The Type-2 Turing machine is used in computable analysis as the model of computation. In this note, we use the computable analysis approach.

In the following, we present the precise definitions for encoding real numbers as well as computable real numbers and computable functions.

Definition 11.1. 1. A sequence $\{r_n\}$ of rational numbers is called a ρ-name of a real number x if there are three functions a, b and c from $\mathbb{N}$ to $\mathbb{N}$ such that for all $n \in \mathbb{N}$, $r_n = (-1)^{a(n)} \frac{b(n)}{c(n)+1}$ and

$$|r_n - x| \le \frac{1}{2^n}. \tag{11.1}$$

2. A double sequence $\{r_{n,k}\}_{n,k\in\mathbb{N}}$ of rational numbers is called a ρ-name for a sequence $\{x_n\}_{n\in\mathbb{N}}$ of real numbers if there are three functions a, b, c from $\mathbb{N}^2$ to $\mathbb{N}$ such that, for all $k, n \in \mathbb{N}$, $r_{n,k} = (-1)^{a(k,n)} \frac{b(k,n)}{c(k,n)+1}$ and

$$\left|r_{n,k} - x_n\right| \le \frac{1}{2^k}.$$

3. A real number x (a sequence $\{x_n\}_{n\in\mathbb{N}}$ of real numbers) is called computable if it has a computable ρ-name, i.e. there is a Type-2 machine that generates the ρ-name without input.

The notion of ρ-name extends in an obvious way to l-vectors. Thus a sequence $\{(r_{1n}, r_{2n}, \ldots, r_{ln})\}_{n\in\mathbb{N}}$ of rational vectors is called a ρ-name of $(x_1, x_2, \ldots, x_l) \in \mathbb{R}^l$ if $\{r_{jn}\}_{n\in\mathbb{N}}$ is a ρ-name of x_j, $1 \le j \le l$. It is easy to see from the definition that a ρ-name of a real number x is simply a code of x by rational numbers.

Next we present a notion of computability for open and closed subsets of $\mathbb{R}^l$ (cf. [58], Definition 5.1.15). We implicitly use ρ-names. For instance, to obtain names of open subsets of $\mathbb{R}^l$, we note that the set of rational balls $B(a,r) = \{x \in \mathbb{R}^l : |x - a| < r\}$, where $a \in \mathbb{Q}^l$ and $r \in \mathbb{Q}$, is a subbase for the standard topology over $\mathbb{R}^l$. Depending on the ρ-names used, we obtain different notions of computability. We omit further details for lack of space.

Definition 11.2. 1. An open set $E \subseteq \mathbb{R}^l$ is called recursively enumerable (r.e. for short) open if there are computable sequences $\{a_n\}$ and $\{r_n\}$, $a_n \in E$ and $r_n \in \mathbb{Q}$, such that
$$E = \cup_{n=0}^{\infty} B(a_n, r_n).$$

Without loss of generality one can also assume that for any $n \in \mathbb{N}$, the closure of $B(a_n, r_n)$, denoted as $\overline{B(a_n, r_n)}$, is contained in E.

2. A closed subset $K \subseteq \mathbb{R}^l$ is called r.e. closed if there exist computable sequences $\{b_n\}$ and $\{s_n\}$, $b_n \in \mathbb{Q}^l$ and $s_n \in \mathbb{Q}$, such that $\{B(b_n, s_n)\}_{n \in \mathbb{N}}$ lists all rational open balls intersecting K.
3. An open set $E \subseteq \mathbb{R}^l$ is called computable (or recursive) if E is r.e. open and its complement E^c is r.e. closed. Similarly, a closed set $K \subseteq \mathbb{R}^l$ is called computable (or recursive) if K is r.e. closed and its complement K^c is r.e. open.

Roughly speaking, an open subset U of $\mathbb{R}^2$ is r.e. if there is a computer program that sketches the image of U by plotting rational open balls on a screen, which will eventually fill up U (but may take infinite time to do so). We may not know how well these balls are filling up U in any finite time if U is merely r.e. On the other hand, if U is recursive, then there is a program that plots the balls filling U up to precision 2^{-k} (in terms of Hausdorff distance) on input k [58].

Definition 11.3. Let A, B be sets, where ρ-names can be defined for elements of A and B. A function $f : A \to B$ is computable if there is a Type-2 machine such that on any ρ-name of $x \in A$, the machine computes as output a ρ-name of $f(x) \in B$.

When dealing with open sets in $\mathbb{R}^l$, we identify a special case of computability, which we call semi-computability. Let $\mathcal{O}(\mathbb{R}^l) = \{O | O \subseteq \mathbb{R}^l$ is open in the standard topology$\}$.

Definition 11.4. A function $f : A \to \mathcal{O}(\mathbb{R}^l)$ is called semi-computable if there is a Type-2 machine such that on any ρ-name of $x \in A$, the machine computes as output two sequences $\{a_n\}$ and $\{r_n\}$, $a_n \in \mathbb{R}^l$ and $r_n \in \mathbb{Q}$ such that

$$f(x) = \cup_{n=0}^{\infty} B(a_n, r_n).$$

Without loss of generality one can also assume that for any $n \in \mathbb{N}$, the closure of $B(a_n, r_n)$ is contained in $f(x)$.

We call this function semi-computable because we can tell in a finite time if a point belongs to $f(x)$, but we have to wait an infinite time to know that it does not belong to $f(x)$.

11 Computability and Dynamical Systems

11.3 Description of Results

In this section we describe some recent results concerning computability of continuous dynamical systems. We consider two types of results: (a) computability of important parameters and sets appearing in dynamical systems, and (2) using dynamical systems as computing models.

In the line of (a), our first result concerns a very basic question – the computability of a single trajectory of a dynamical system defined by a (vector) ODE

$$y' = f(y). \tag{11.2}$$

This may seem like a trivial question – just use a standard numerical algorithm. However, these methods usually require a Lipschitz constant to ensure uniqueness of solutions, which is essential for computation. Since the behavior of a trajectory over time is in general unknown beforehand, one may not have a knowledge of a Lipschitz constant that can be used to compute the entire trajectory (actually it often happens that no such "global" Lipschitz constant exists).

This problem is studied by several authors. In [22], we show that if f is computable and effectively locally Lipschitz (meaning that we can locally compute Lipschitz constants), then we can compute the entire trajectory. This result is extended in [17]. There it is shown that if the solution is unique, then the solution must be computable over its lifespan (the maximal interval on which the solution exists), under the classical conditions ensuring existence of a solution to (11.2) for a given initial point. The idea is to generate all possible "tubes" which cover the solution, and then check if this cover is valid within the desired accuracy. The proof is constructive, although terribly inefficient in practice. Nevertheless, it solves the problem of computing a given trajectory for (11.2).

The result above is not surprising, since the Picard iteration scheme used in the classical existence proof is constructive. However, the issue remains as to whether or not one can compute the lifespan. In [22] we provide a negative answer, showing that even if f is analytic and computable, the lifespan is in general non-computable (i.e. not recursive). However, if f is computable, the lifespan is r.e. The non-computability of the lifespan suggests limitations concerning numerical methods for solving ODE problems, because numerical methods often assume the existence of some time interval where the solution is defined, and this assumption is crucial in error analysis. In the case where the lifespan is non-computable, one may have to settle for a numerical algorithm computing only a local solution.

We have also shown in [22] that the problem of determining whether or not the lifespan is bounded cannot be decided by a Turing machine, even if f is computable and analytic. This result is extended in [24] to the case where f is computable and polynomial. The result is further refined in [46], where it is shown that the set of all initial data generating solutions with lifespans longer than k, $k \in \mathbb{N}$, is in general not computable. The set is however r.e. if f is computable.

Next we describe some results related to the dynamics of a given system. In [61], it is shown that the domain of attraction of a computable and asymptotically stable hyperbolic equilibrium point of the nonlinear system (11.2) is in general not recursive, though it is r.e. This tells us that the domain of attraction can be approximated from the inside on the one hand, but on the other hand there is no algorithm determining how far such an approximation is from filling up this domain. When restricted to planar systems, more can be said. For example, in [25], we show that the operator $\mathscr{F}$ is strictly semi-computable if we consider only structurally stable systems; on the other hand, $\mathscr{F}$ fails to be semi-computable if all C^1 systems are permitted, where $\mathscr{F}$ is the operator that takes two inputs, the description of the flow and a cover of an attractor, and outputs the domain of attraction for the given attractor. In [25] we also demonstrate how to decide whether or not there are limit cycles, and furthermore how to compute hyperbolic ones when given a compact set without an equilibrium point (equilibrium points are computable from f). As a consequence, all kinds of hyperbolic attractors in the plane can be computed, though their domains of attraction cannot.

We now turn to the issue (b) of using dynamical systems as computing models. We have shown in [23] that the evolution of a given Turing machine can be embedded in the dynamics defined by polynomial differential equations, with some degree of robustness to perturbations. In other words, polynomial differential equations can simulate Turing machines. In [7] the following variation of the above result is given: for any given compact set $[a, b] \subseteq \mathbb{R}$, a function $f : [a, b] \to \mathbb{R}$ is computable if and only if it is computable by the "limit dynamics" of polynomial differential equations, i.e., there is a (vector) polynomial p such that given an initial point $x \in [a, b]$, the solution to the initial-value problem $y' = p(t, y)$, $y(0) = (x, y_{2,0}, \ldots, y_{n,0})$, with $y_{2,0}, \ldots, y_{n,0} \in \mathbb{R}$ independent of x, is composed of two components, which we suppose without loss of generality to be y_1, y_2, satisfying

$$|y_1(t) - f(x)| \le y_2(t)$$

and $y_2(t) \to 0$ as $t \to \infty$ (i.e., y_1 converges towards $f(x)$ with error bounded by y_2).

There are interesting results by other authors, usually more related to control theory. Control theory is an interdisciplinary branch of engineering and mathematics that studies how to manipulate the parameters affecting the behavior of a system to produce the desired or optimal outcome. Some good introductions to control theory for mathematicians can be found in [53, 60].

Numerous interesting techniques and results have been obtained over the years by the control theory community. However they have not found their way into the dynamical systems community. In our opinion, this has various causes, ranging from lack of interaction between the two communities and, to some degree, because control theory is more application-oriented. For instance, many results focus on *hybrid systems* (see e.g. [13]), defined as differential equations with discontinuous right hand sides or having (some) discrete variables.

A topic of interest for control theory is *stability* [60]. Usually this notion is related to Lyapunov stability. In [3], the authors consider a particular class of discrete-time dynamical systems, defined by continuous piecewise affine functions. They show

11 Computability and Dynamical Systems

that the stability problem (in their version, "Is the system globally asymptotically stable?") is non-computable, which establishes fundamental limitations for the computation of this kind of problem. Other notions of stability can be considered, e.g. shadowing or robustness, as done in [28]. The author focuses on the *reachability problem*: given some initial point in the state space, does the flow reach some region or point A? In [28] it is shown that the shadowing property is not enough to decide reachability by a computer, while robustness is sufficient.

The previous results rely on the paper [12]. There Collins shows that, in general, the reachable set of some initial region can be semi-computed (technically, lower-computed), but can only be computed under some special conditions. Another interesting algorithm to study the reachability problem is given in [15].

The reachability problem has been one of the most studied problems in the literature, and is interesting for dynamical systems since it has obvious resemblance to the problem of computing the domain of attraction of a given attractor. As a matter of fact, our results about computability of domains of attraction presented in [25] are based on some of these techniques and provide a good example of how control theory may be of use in dynamical systems.

Most results about the reachability problem give rise to undecidability (i.e., cannot be solved by an algorithm) as it is usually easy to encode the evolution of a given Turing machine in the dynamics of the system, e.g. [1, 6, 8, 23, 33, 40] and to show that the reachability problem is equivalent to the Halting Problem, the foremost undecidable problem in the theory of computation, cf. [2].

Despite this undecidability, these results use creative ways to analyze the dynamics of the system. Moreover, they depend critically on the use of exact computations. If some robustness to errors is allowed (in a weaker form than that required by structural stability), then usually the reachability problem is decidable as was mentioned in [28], but previously seen in other classes [23, 38, 39]. This fact was used in [25]. The idea is to cover some region with a grid of points (more precisely, small squares) and follow the individual evolution of each point to get an estimate of the domain of attraction. By using a larger grid (in absolute size) and a thinner mesh size, in the limit one can show rigorously that we compute the domain of attraction, even though exact computation takes "infinite time." Nevertheless, at each point of the computation, we have an estimate of this domain, with the error converging to 0 with time.

Another important area of study in control theory is *controllability* [60]. In controllability the aim is to investigate the possibility of forcing the system into a particular state by using an appropriate control signal. This topic has been partially studied by some members of the dynamical systems community, in control of chaos, which is based on the fact that any chaotic attractor contains an infinite number of unstable periodic orbits, and that one can use small perturbations to stabilize the trajectory into one of these periodic orbits [48].

The literature about computability and controllability focuses essentially on the computation of classes of "controllers" which allow the control of specific classes of systems: hybrid systems [16,37,57] and discrete-time semicontinuous systems [14].

Concerning complex dynamical systems, there is an exciting result by Braverman and Yampolsky [10]. They show that there is no algorithm which computes the Julia set J_c of the quadratic polynomial $f_c(z) = z^2 + c$ from the parameter c, using elaborate arguments involving Julia sets with Siegel disks. This shows that there are limitations when doing accurate computations of those pretty images of Julia sets usually presented to the public.

Other results of interest are those using shifts. In [40], Moore uses generalized shifts to show that basins of attraction, chaotic behavior or even periodicity are non-computable. This kind of result brings to the field of computability questions traditionally related to dynamical systems [29]. In particular, these include deriving necessary conditions for universality [18], computability of entropy [26, 27, 32, 54, 55], and understanding the "edge of chaos [35]."

Also along this line, some work has been done concerning computability of dynamical systems seen from a statistical perspective [21, 30]. We believe this is an interesting and promising topic of research.

11.4 Further Work

The computability theory of continuous dynamical systems is still in an early stage of development, despite notable progress in recent years. Many important fundamental problems have not yet been studied. In general, the problems fall into two categories – computability and computational complexity.

As for computability, one topic of broad scope is to detect non-computable parameters and invariant sets of classical importance and ask further for the fine structure via the theory of degree of unsolvability. Examples are attractors/repellors and their basins in natural families of dynamical systems such as the Hénon attractor, the Rössler attractor, and the Lorenz attractor. Another interesting problem is to identify the analytic/geometric properties that are critical to ensure computability of an object under consideration. For example, in [61] we showed that there exists a C^∞ and polynomial-time computable function f defined on $\mathbb{R}^2$ such that the origin $(0, 0)$ is the only sink of $dx/dt = f(x(t))$, and the domain of attraction of $(0, 0)$ is not computable. However, the issue remains as to whether or not the domain of attraction of a computable polynomial system in the plane is computable.

It could also be interesting to investigate the computability of the dynamical systems used to model the motion of charged particles in modern particle accelerators. These devices (the LHC at CERN, the Tevatron at Fermilab, and many others) are among the most complex machines ever constructed, and numerous numerical codes are used in their design and operation; these numerical algorithms are correspondingly complex. Yet, the computability theory is still lacking.

When it comes to computational complexity, so far as we know, the only major problems which have been investigated are local solutions of the initial value problems for certain ordinary differential equations [31] and Julia sets [10, 45]. There are many processes and sets arising from dynamical systems which have been

11 Computability and Dynamical Systems

proved to be computable but yet their computational complexity remains unknown. One such example is the Smale horseshoe. It can be shown that the horseshoes are computable, uniformly from the horseshoe maps [11]. Nevertheless, the difficulty of the computation is not yet known.

Acknowledgements J. Buescu was partially supported by *Fundação para a Ciência e a Tecnologia*, Financiamento Base 2009 – ISFL/1/209. D. Graça was partially supported by *Fundação para a Ciência e a Tecnologia* and EU FEDER POCTI/POCI via SQIG – Instituto de Telecomunicações. DG was also attributed a Taft Research Collaboration grant which made possible a research visit to U. Cincinnati. N. Zhong was partially supported by the 2009 Taft Summer Research Fellowship.

References

1. Asarin, E., Maler, O.: Achilles and the tortoise climbing up the arithmetical hierarchy. J. Comput. System Sci. **57**(3), 389–398 (1998)
2. Asarin, E., Maler, O., Pnueli, A.: Reachability analysis of dynamical systems having piecewise-constant derivatives. Theoret. Comput. Sci. **138**, 35–65 (1995)
3. Blondel, V.D., Bournez, O., Koiran, P., Tsitsiklis, J.N.: The stability of saturated linear dynamical systems is undecidable. J. Comput. System Sci. **62**, 442–462 (2001)
4. Blum, L., Cucker, F., Shub, M., Smale, S.: Complexity and Real Computation. Springer (1998)
5. Blum, L., Shub, M., Smale, S.: On a theory of computation and complexity over the real numbers: NP-completeness, recursive functions and universal machines. Bull. Am. Math. Soc. **21**(1), 1–46 (1989)
6. Bournez, O.: Achilles and the Tortoise climbing up the hyper-arithmetical hierarchy. Theoret. Comput. Sci. **210**(1), 21–71 (1999)
7. Bournez, O., Campagnolo, M.L., Graça, D.S., Hainry, E.: Polynomial differential equations compute all real computable functions on computable compact intervals. J. Complexity **23**(3), 317–335 (2007)
8. Branicky, M.S.: Universal computation and other capabilities of hybrid and continuous dynamical systems. Theoret. Comput. Sci. **138**(1), 67–100 (1995)
9. Braverman, M., Cook, S.: Computing over the reals: foundations for scientific computing. Notices Amer. Math. Soc. **53**(3), 318–329 (2006)
10. Braverman, M., Yampolsky, M.: Non-computable Julia sets. J. Am. Math. Soc. **19**(3), 551– 0578 (2006)
11. Buescu, J., Graça, D., Zhong, N.: Computability, Noncomputability, and Hyperbolic Systems. preprint (2009)
12. Collins, P.: Continuity and computability of reachable sets. Theor. Comput. Sci. **341**, 162–195 (2005)
13. Collins, P.: Chaotic dynamics in hybrid systems. Nonlinear Dyn. Syst. Theory **8**(2), 169–194 (2008)
14. Collins, P.: Computability of controllers for discrete-time semicontinuous systems. In: Proc. 18th International Symposium on the Mathematical Theory of Networks and Systems (2008)
15. Collins, P.: The reach-and-evolve algorithm for reachability analysis of nonlinear dynamical systems. Electron. Notes Theor. Comput. Sci. **223**, 87–102 (2008)
16. Collins, P.: Controllability and falsification of hybrid systems. In Proc. European Control Conference (2009)
17. Collins, P., Graça, D.S.: Effective computability of solutions of differential inclusions the ten thousand monkeys approach. J. Universal Comput. Sci. **15**(6), 1162–1185 (2009)
18. Delvenne, J.C., Kurka, P., Blondel, V.: Decidability and universality in symbolic dynamical systems. Fund. Inform. **74**(4), 463–490 (2006)

19. Deutsch, D.: Quantum theory, the Church-Turing principle and the universal quantum computer. Proc. R. Soc. Lond. Ser. A A400, 97–117 (1985)
20. Duffing, G.: Erzwungene Schwingungen bei Veranderlicher Eigenfrequenz. Vieweg Braunschweig (1918)
21. Galatolo, S., Hoyrup, M., Rojas, C.: Effective symbolic dynamics, random points, statistical behavior, complexity and entropy. Inform. Comput. (to appear)
22. Graça, D., Zhong, N., Buescu, J.: Computability, noncomputability and undecidability of maximal intervals of IVPs. Trans. Am. Math. Soc. **361**(6), 2913–2927 (2009)
23. Graça, D.S., Campagnolo, M.L., Buescu, J.: Computability with polynomial differential equations. Adv. Appl. Math. **40**(3), 330–349 (2008)
24. Graça, D.S., Campagnolo, M.L., Buescu, J.: Computational bounds on polynomial differential equations. Appl. Math. Comput. **215**(4), 1375–1385 (2009)
25. Graça, D.S., Zhong, N.: Computing domains of attraction for planar dynamics. In: Calude, C.S., Costa, J.F., Dershowitz, N., Freire, E., Rozenberg, G. (eds.) 8th International Conference on Unconventional Computation (UC 2009), LNCS 5715, pp. 179–190. Springer (2009)
26. Hertling, P., Spandl, C.: Computability theoretic properties of the entropy of gap shifts. Fundam. Inf. **83**, 141–157 (2008)
27. Hertling, P., Spandl, C.: Shifts with decidable language and noncombustible entropy. Discrete Math. Theor. Comput. Sci. **10**, 75–94 (2008)
28. Hoyrup, M.: Dynamical systems: stability and simulability. Math. Structures Comput. Sci. **17**, 247–259 (2007)
29. Hoyrupa, M., Kolaka, A., Longo, G.: Computability and the morphological complexity of some dynamics on continuous domains. Theor. Comput. Sci. **398**, 170–182 (2008)
30. Hoyrupa, M., Rojas, C.: Computability of probability measures and martin-lf randomness over metric spaces. Inform. Comput. **207**, 830–847 (2009)
31. Ko, K.I.: Computational Complexity of Real Functions. Birkhauser (1991)
32. Koiran, P.: The topological entropy of iterated piecewise affine maps is uncomputable. Discrete Math. Theor. Comput. Sci. **4**(2), 351–356 (2001)
33. Koiran, P., Moore, C.: Closed-form analytic maps in one and two dimensions can simulate universal Turing machines. Theor. Comput. Sci. **210**(1), 217–223 (1999)
34. Lanford, O.E.: A computer-assisted proof of the feigenbaum conjectures. Bull. AMS **6**, 427–434 (1982)
35. Legenstein, R., Maass, W.: What makes a computational system dynamically powerful? In: Haykin, S., Principe, J.C., Sejnowski, T., Mcwhirter, J. (eds.) New Directions in Statistical Signal Processing: From Systems to Brain, pp. 127–154. MIT (2007)
36. Lorenz, E.N.: Deterministic non-periodic flow. J. Atmos. Sci. **20**, 130–141 (1963)
37. Lygeros, J., Tomlin, C., Sastry, S.: Controllers for reachability specifications for hybrid systems. Automatica **35**, 349–370 (1999)
38. Maass, W., Orponen, P.: On the effect of analog noise in discrete-time analog computations. Neural Comput. **10**(5), 1071–1095 (1998)
39. Maass, W., Sontag, E.: Analog neural nets with gaussian or other common noise distributions cannot recognize arbitrary regular languages. Neural Comp. **11**, 771–782 (1999)
40. Moore, C.: Unpredictability and undecidability in dynamical systems. Phys. Rev. Lett. **64**(20), 2354–2357 (1990)
41. Peixoto, M.: Structural stability on two-dimensional manifolds. Topology **1**, 101–121 (1962)
42. van der Pol, B.: A theory of the amplitude of free and forced triode vibrations. Radio Review **1**, 701–710, 754–762 (1920)
43. Pour-El, M.B., Richards, J.I.: Computability in Analysis and Physics. Springer (1989)
44. Pour-El, M.B., Zhong, N.: The wave equation with computable initial data whose unique solution is nowhere computable. Math. Logic Q. **43**, 499–509 (1997)
45. Rettinger, R., Weihrauch, K.: The computational complexity of some julia sets. In Proc. 35th Annual ACM Symposium on Theory of Computing, pp. 177–185. ACM (2003)
46. Rettinger, R.,Weihrauch, K., Zhong, N.: Topological complexity of blowup problems. J. Univers. Comput. Sci. **15**(6), 1301–1316 (2009)

11 Computability and Dynamical Systems

47. Rössler, O.E.: An equation for continuous chaos. Phys. Lett. A **57**(5), 397–398 (1976)
48. Shinbrot, T., Grebogi, C., Yorke, J.A., Ott, E.: Using small perturbations to control chaos. Nature **363**, 411–417 (1993)
49. Sipser, M.: Introduction to the Theory of Computation, 2nd edn. Course Technology (2005)
50. Smale, S.: Structurally stable systems are not dense. Am. J. Math. **88**, 491–496 (1966)
51. Smale, S.: Differentiable dynamical systems. Bull. Am. Math. Soc. **73**, 747–817 (1967)
52. Smale, S.: Mathematical problems for the next century. Math. Intell. **20**, 7–15 (1998)
53. Sontag, E.D.: Mathematical Control Theory, 2nd edn. Springer (1998)
54. Spandl, C.: Computing the topological entropy of shifts. Math. Log. Quart. **53**(4-5), 493–510 (2007)
55. Spandl, C.: Computability of topological pressure for shifts of finite type with applications in statistical physics. Electr. Notes Theor. Comput. Sci. **202**, 385–401 (2008)
56. Tucker, W.: A rigorous ode solver and smales 14th problem. Found. Comput. Math. **2**(1), 53–117 (2002)
57. Vidal, R., Schaffert, S., Shakernia, O., Lygeros, J., Sastry, S.: Decidable and semi-decidable controller synthesis for classes of discrete time hybrid systems. In Proc. 40th IEEE Conference on Decision and Control, pp. 1243–1248 (2001)
58. Weihrauch, K.: Computable Analysis: An Introduction. Springer (2000)
59. Weihrauch, K., Zhong, N.: Is wave propagation computable or can wave computers beat the Turing machine? Proc. Lond. Math. Soc. **85**(3), 312–332 (2002)
60. Zabczyk, J.: Mathematical Control Theory: An Introduction. Birkhauser (1992)
61. Zhong, N.: Computational unsolvability of domain of attractions of nonlinear systems. Proc. Am. Math. Soc. **137**, 2773–2783 (2009)

Chapter 12
Dynamics and Biological Thresholds

**N.J. Burroughs, M. Ferreira, J. Martins, B.M.P.M. Oliveira,
Alberto A. Pinto, and N. Stollenwerk**

Abstract Our main interest is to study the relevant biological thresholds that appear in epidemic and immunological dynamical models. We compute the thresholds of the SIRI epidemic models that determine the appearance of an epidemic disease. We compute the thresholds of a Tregs immunological model that determine the appearance of an immune response.

M. Ferreira (✉)
Escola Superior de Estudos Industriais e de Gestão do Instituto Politécnico do Porto (IPP),
LIAAD-INESC Porto LA, Escola de Ciências, Universidade do Minho, Braga, Portugal
e-mail: migferreira2@gmail.com

N.J. Burroughs
University of Warwick, Coventry CV4 7AL, UK
e-mail: njb@maths.warwick.ac.uk

J. Martins
LIAAD-INESC Porto LA, Porto, Portugal
and
Department of Mathematics, School of Technology and Management, Polytechnic Institute of
Leiria, Campus 2, Morro do Lena – Alto do Vieiro, 2411-901 Leiria, Portugal
e-mail: jmmartins@estg.ipleiria.pt

B.M.P.M. Oliveira
Faculdade de Ciências da Nutrição e Alimentação da, Universidade do Porto, LIAAD-INESC Porto
LA, Porto, Portugal
e-mail: bmpmo@fcna.up.pt

A.A. Pinto
LIAAD-INESC Porto LA e Departamento de Matemática, Faculdade de Ciências, Universidade
do Porto, Rua do Campo Alegre, 687, 4169-007, Portugal
and
Centro de Matemática e Departamento de Matemática e Aplicações, Escola de Ciências, Universidade do Minho, Campus de Gualtar, 4710-057 Braga, Portugal
e-mail: aapinto@fc.up.pt

N. Stollenwerk
Centro de Matemática e Aplicações Fundamentais, Universidade de Lisboa, Lisbon, Portugal
e-mail: nico@ptmat.fc.ul.pt

M.M. Peixoto et al. (eds.), *Dynamics, Games and Science I*, Springer Proceedings
in Mathematics 1, DOI 10.1007/978-3-642-11456-4_12,
© Springer-Verlag Berlin Heidelberg 2011

12.1 Introduction

The reinfection process in epidemiology and the SIRI model, specifically, have evoked recent interest as a first description of multi-strain epidemics where after an initial infection immunity against one strain only gives partial immunity against a genetically close mutant strain. Models for partial immunization are also of great interest, however for different reasons. Transitions between no-growth, compact growth and annular growth have been observed for these models [8, 10]. In the SIRI model, using pair approximation, we compute analytically the phase transition lines (thresholds) between no-growth and compact growth, between annular growth and compact growth and between no-growth and annular growth [34]. In the other extreme of disease modeling, at the cellular level, we study a model of immune response by T cells with regulatory T cells (Tregs). During a pathogen invasion, T cells specific to the antigen proliferate and promote the removal of the pathogen. However, the immune system can also target self antigens (autoimmunity) and cause tissue damage. Tregs limit autoimmunity with a delicate balance between immune activation and immune response suppression being achieved. We compute the thresholds of a Tregs immunological model, that determine the appearance of an immune response and we present how such a balance between immune activation and immune response suppression is established and controlled [3–7, 26].

12.2 Thresholds for Epidemiological Models

One of the simplest and best studied epidemiological models is the stochastic SIS (susceptible, infected and susceptible again) model. Many authors worked on the SIS model considering only the dynamical evolution of the mean value and the variance of the infected individuals, however the characterization of the SIS model go beyond the consideration of these two moments. In [19, 27], the dynamic equations for all the moments are recursively derived and the stable equilibria manifold computed in the moment closure approximation. Surprisingly, the steady states in the moment closure can be used to obtain good approximations of the quasi-stationary states of the SIS model [13, 22, 23]. The stochastic SIS model can be related with the contact process using creation and annihilation operators [17]. This relation leads to a better characterization of critical thresholds for this and other more complex epidemiological models like reinfection models [33]. One epidemiological model that describes reinfection and partial immunization is the SIRI (susceptible, infected, recovered and again infected) model. The characterization of the thresholds for the spatial stochastic SIRI model can be done in the mean field approximation or in higher order approximations like the pair approximation. In the mean field approximation a first threshold between the disease free state and a non-trivial state with strictly positive endemic equilibrium appear and a second threshold characterized by the ratio between first and secondary infection rate, called the reinfection threshold, also appear [35]. In the pair approximation, we compute analytically

12 Dynamics and Biological Thresholds

the phase transition lines (thresholds) between no-growth and compact growth, between annular growth and compact growth and between no-growth and annular growth [34]. This last phase transition line could only be calculated via a scaling argument [18]. The characterization of the SIRI thresholds in pair approximation improves the mean field results, in which the SIS and the SIR limiting cases have the same critical values for the transition from no-growth to a nontrivial stationary state, and gives a phase transition diagram in better agreement with the phase transition diagrams for similar time discrete models [8] described using stochastic simulations. Further details for both SIS and SIRI thresholds characterization can be found in [25].

For the spatial stochastic SIRI model we consider the following transitions between host classes for N individuals being either susceptible S, infected I by a disease or recovered R

$$S + I \xrightarrow{\beta} I + I$$
$$I \xrightarrow{\gamma} R$$
$$R + I \xrightarrow{\tilde{\beta}} I + I$$
$$R \xrightarrow{\alpha} S$$

resulting in the master equation [1, 40] for variables S_i, I_i and $R_i \in \{0, 1\}$, $i = 1, 2, \ldots, N$, for N individuals eventually on a regular grid, with constraint $S_i + I_i + R_i = 1$ [34]. The first infection $S + I \xrightarrow{\beta} I + I$ occurs with infection rate β, whereas after recovery with rate γ the respective host becomes resistant up to a possible reinfection $R + I \xrightarrow{\tilde{\beta}} I + I$ with reinfection rate $\tilde{\beta}$. Hence, the recovered are only partially immunized. For further analysis of possible stationary states we include a transition from recovered to susceptibles α, which might be simply due to demographic effects (or very slow waning immunity for some diseases). The expectation value for the total number of infected hosts at a given time will be denoted by $\langle I \rangle$. Computing the dynamic evolution of the first moments using the SIRI master equation we obtain an ODE system [18] which includes the pairs like $\langle SI \rangle$, $\langle RI \rangle$, etc.. Hence, either we have to continue to calculate equations for the triples, which will involve even higher clusters, or we can approximate the higher moments by lower ones. The simplest scheme is the mean field approximation where the pairs, like $\langle SI \rangle$, are approximated using the first moments $\langle SI \rangle = Q/N \langle S \rangle \langle I \rangle$, where Q denotes the number of neighbours of each individual, assumed to be constant, and N the population size. In the pair approximation, we go one step further by approximating the triples into pairs [12, 28]. The pair approximation arises as

$$\widetilde{\langle SIR \rangle} \approx \frac{Q - 1}{Q} \cdot \frac{\langle SI \rangle \cdot \langle IR \rangle}{\langle I \rangle} \tag{12.1}$$

from the Bayesian formula for conditional probabilities applied to the local expectation values and a spatial homogeneity argument [18]. The full ODE system in the pair approximation is given by

$$\frac{d}{dt}\langle I\rangle = \beta\langle SI\rangle - \gamma\langle I\rangle + \tilde{\beta}\langle RI\rangle$$

$$\frac{d}{dt}\langle R\rangle = \gamma\langle I\rangle - \alpha\langle R\rangle - \tilde{\beta}\langle RI\rangle$$

$$\frac{d}{dt}\langle SI\rangle = \alpha\langle RI\rangle - (\gamma + \beta)\langle SI\rangle + \beta(Q-1)\langle SI\rangle$$

$$-\beta\frac{Q-1}{Q}\frac{(2\langle SI\rangle + \langle SR\rangle)\cdot\langle SI\rangle}{N-\langle I\rangle - \langle R\rangle} + \tilde{\beta}\frac{Q-1}{Q}\frac{\langle SR\rangle\langle RI\rangle}{\langle R\rangle} \quad (12.2)$$

$$\frac{d}{dt}\langle RI\rangle = \gamma(Q\langle I\rangle - \langle SI\rangle) - (\alpha + 2\gamma + \tilde{\beta})\langle RI\rangle + \beta\frac{Q-1}{Q}\frac{\langle SR\rangle}{N-\langle I\rangle - \langle R\rangle}\langle SI\rangle$$

$$+\tilde{\beta}\frac{Q-1}{Q}\frac{(Q\langle R\rangle - \langle SR\rangle - 2\langle RI\rangle)\cdot\langle RI\rangle}{\langle R\rangle}$$

$$\frac{d}{dt}\langle SR\rangle = \gamma\langle SI\rangle + \alpha(Q\langle R\rangle - 2\langle SR\rangle - \langle RI\rangle)$$

$$-\beta\frac{Q-1}{Q}\frac{\langle SR\rangle}{N-\langle I\rangle - \langle R\rangle}\langle SI\rangle - \tilde{\beta}\frac{Q-1}{Q}\frac{\langle RI\rangle\langle SR\rangle}{\langle R\rangle}.$$

The expressions for all the the moments in stationarity cannot be explicitly computed. In [34], using the information that when $\langle I\rangle^*$ tends to zero the quotient

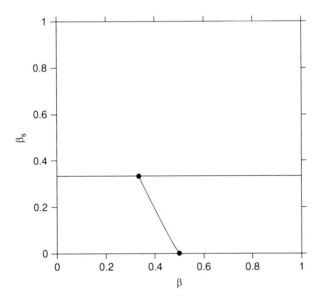

Fig. 12.1 The phase transition line between no-growth and ring-growth determined from the analytic solution for the $\alpha = 0$ case which is explicitly given in (12.3). In addition, we also present the phase transition points for the SIS and SIR limiting cases of the SIRI model

$\langle I \rangle^*/\langle R \rangle^*$ stays finite, the critical curve $\beta(\tilde{\beta})$, in the limit of $\alpha \to 0$, is given by

$$\beta(\tilde{\beta}) = \frac{\gamma^2 Q - \gamma \tilde{\beta}(Q-1)}{\gamma Q(Q-2) + \tilde{\beta}(Q-1)}. \quad (12.3)$$

This threshold is shown graphically in Fig. 12.1, for $\gamma = 1$ and $Q = 4$ appropriated for two dimensional square lattices, as the oblique curve linking the critical points of the SIS limiting case (the case of $\tilde{\beta} = \beta$ and SIR limiting case (where $\tilde{\beta} = 0$).

12.3 Thresholds for Immunological Models

We modeled a population of Tregs (denoted R, R^*) and conventional T cells (T, T^*) with processes shown schematically in Fig. 12.2. Both populations require antigenic stimulation for activation. Levels of antigenic stimulation are denoted a and b for Tregs and conventional T cells, respectively. On activation conventional T cells both secrete IL2 and acquire proliferative capacity in the presence of IL2 while Tregs proliferate less efficiently than normal T cells in the presence of IL2, and they do not secrete IL2. In the model we assumed that T cells activated by exposure to their specific antigen have a cytokine secreting state (a normal activated state) and a non-secreting state to which they revert at a constant rate k; thus in absence of antigen growth halts. Activated Tregs also induce a transition to the (inhibited) nonsecreting state [38], thereby inhibiting T cell growth. This transition rate is assumed proportional to the Treg population density. We assume that T cells regain secretion status with antigen re-exposure. Thus in the presence of costimulation and Tregs, the T cell population is a mixture of partially inhibited, and normal T cells. In [4], we used a generic mechanism that utilises a cytokine (denoted J). Tregs compete for this

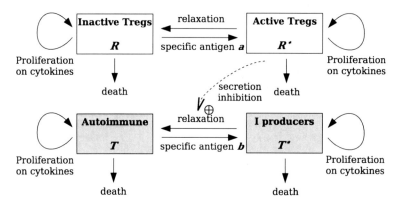

Fig. 12.2 Model schematic showing growth, death and phenotype transitions of the Treg populations R, R^*, and autoimmune T cell T, T^* populations. Cytokine dynamics are not shown: IL2 is secreted by activated T cells T^* and adsorbed by all the T cell populations equally. Reproduced from [3]

cytokine by adsorption and thus population homeostasis is achieved. Alternatively, we can consider an inflow R_{input} of Tregs (as in [5]) instead of the J cytokine (both mechanisms yield similar results). We used a (quadratic) growth limitation mechanism related to Fas-FasL death, that is assumed to act on all T cells equally. Finally, we included an influx of T cells into the tissue (T_{input} in cells per milliliter per day). See [2, 9, 11, 20, 24, 30–32, 36–39] for further biological details. A set of ordinary differential equations is employed to study the dynamics, with a compartment for each T cell population (inactive Tregs R, active Tregs R^*, non secreting T cells T, secreting activated T cells T^*), interleukine 2 density I and the homeostatic Treg cytokine J,

$$\frac{dR}{dt} = (\epsilon\rho(I+J)-\beta(R+R^*+T+T^*)-d_R)R + \hat{k}(R^*-aR) + R_{input},$$

$$\frac{dR^*}{dt} = (\epsilon\rho(I+J)-\beta(R+R^*+T+T^*)-d_{R*})R^* - \hat{k}(R^*-aR),$$

$$\frac{dJ}{dt} = \hat{\sigma}(S-(\hat{\alpha}(R+R^*)+\hat{\delta})J),$$

$$\frac{dT}{dt} = (\rho I - \beta(R+R^*+T+T^*)-d_T)T + k(T^*-bT+\gamma R^*T^*) + T_{input},$$

$$\frac{dT^*}{dt} = (\rho I - \beta(R+R^*+T+T^*)-d_{T*})T^* - k(T^*-bT+\gamma R^*T^*),$$

$$\frac{dI}{dt} = \sigma(T^* - (\alpha(R+R^*+T+T^*)+\delta)I). \tag{12.4}$$

The important aspects of this model are a mechanism to sustain a population of Tregs, secretion inhibition of T cells with a rate that correlates with Treg population size, and growth and competition for IL2 with a higher growth rate of T cells relative to Tregs. Parameters are defined in [4]. Tregs function to limit autoimmune responses determining a delicate balance between appropriate immune activation and immune response suppression. How such a balance is established

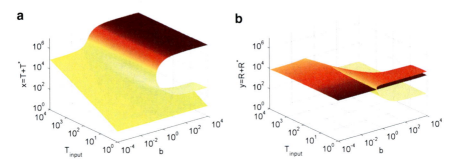

Fig. 12.3 The hysteresis of the equilibria manifold for Thymic inputs $T_{input} \in [1, 10000]$, with the other parameters at their default values. The hysteresis unfolds for high values of the parameter T_{input}. (**a**): Low values of $y = R + R^*$ are reddish and higher values are yellowish. (**b**): Low values of $x = T + T^*$ are reddish and higher values are yellowish

12 Dynamics and Biological Thresholds

and controlled is the central focus of the papers [4–7]. The motivation is the observation that T cell proliferation through cytokines has such a control structure (see Fig. 12.3). The immune response-suppression axis is then a balance between the local numbers of activated T cells (eg from a pathogen encounter) and activated Tregs. This balance can be altered by natural slow changes (as in puberty) or by fast changes (as in infection) [4, 16]. Using the differential equations in (4.4), Burroughs et al. [3] study the bystander proliferation and Burroughs et al. [4] study the cross reactivity.

Acknowledgements Previous versions of this work were presented in the International Congresses of Mathematicians ICM 2006 and 2010, European Conference on Mathematical and Theoretical Biology ECMTB 2008 and ICDEA 2010. This work was highlighted in the article [15] after being presented in ICM 2010. This work was presented in the Encontro Ciência 2008, organized by Alexande Quintanilha, João Sentieiro, Luís Magalhães, Joaquim Cabral e Alberto Pinto. We thank LIAAD-INESC Porto LA, Calouste Gulbenkian Foundation, PRODYN-ESF, FEDER, POFC, POCTI and POCI by FCT and Ministério da Ciência, Tecnologia e Ensino Superior and the FCT Pluriannual Funding Program of the LIAAD-INESC Porto LA and of the Research Centre of Mathematics of the University of Minho and Centro de Matemática e Aplicações Fundamentais da Universidade de Lisboa for their financial support. José Martins also acknowledges the financial support from the FCT grant with reference SFRW/BD/37433/2007. Miguel Ferreira also acknowledges the financial support from the FCT grant with reference SFRH/BD/27706/2006.

References

1. Bailey, N.T.J.: The Elements of Stochastic Processes with Applications to the Natural Sciences. Wiley, New York (1964)
2. Boer, R.J., Hogeweg, P.: Immunological discrimination between self and non-self by precursor depletion and memory accumulation. J. Theor. Biol. **124**, 343 (1987)
3. Burroughs, N.J., Oliveira, B.M.P.M., Pinto, A.A.: Autoimmunity arising from bystander proliferation of T cells in an immune response model. Math. Comput. Model. **53**(7-8), 1389–1393 (2011)
4. Burroughs, N.J., Oliveira, B.M.P.M., Pinto, A.A.: Regulatory T cell adjustment of quorum growth thresholds and the control of local immune responses. J. Theor. Biol. **241**, 134–141 (2006)
5. Burroughs, N.J., Oliveira, B.M.P.M., Pinto, A.A., Ferreira, M.: A transcritical bifurcation in an immune response model. J. Differ. Eqn. Appl. **17**(6), 85–91 (2011)
6. Burroughs, N.J., Oliveira, B.M.P.M., Pinto, A.A., Ferreira, M.: Immune response Dynamics. Math. Comput. Model. **53**(7–8), 1410–1419 (2011)
7. Burroughs, N.J., Oliveira, B.M.P.M., Pinto, A.A., Sequeira, H.J.T.: Sensibility of the quorum growth thresholds controlling local immune responses. Math. Comput. Model. **47**, 714–725 (2008)
8. Dammer, S.M., Hinrichsen, H.: Spreading with immunization in high dimensions. J. Stat. Mech. (2004)
9. Furtado, Glaucia C., Curotto de Lafaille, Maria A., Kutchukhidze, Nino L., Juan, J.: IInterleukin 2 signaling is required for $CD4^+$ regulatory T cell function. J. Exp. Med. **6**, 851–857 (2002)
10. Grassberger, P., Chaté, H., Rousseau, G.: Spreading in media with long-time memory. Phys. Rev. E **55**, 2488–2495 (1997)

11. Hsieh, C.-S., Liang, Y., Tyznik, A.J., Self, S.G., Liggitt, D., Rudensky, A.Y.: Recognition of the peripheral self by naturally arising $CD25^+$ $CD4^+$ T cell receptors. Immunity **21**, 267–277 (2004)
12. Joo, J., Lebowitz, J.L.: Pair approximation of the stochastic susceptible-infected-recovered-susceptible epidemic model on the hypercubic lattice. Phys. Rev. E **70**, 036114 (2004)
13. Kryscio, R., Lefevre, C.: On the extinction of the S-I-S stochastic logistic epidemic. J. Appl. Prob. **26**, 685–694 (1989)
14. Leon, K., Faro, J., Lage, A., Carneiro, J.: Inverse correlation between the incidences of autoimmune disease and infection predicted by a model of T cell mediated tolerance. J. Autoimmun. 31–42 (2004)
15. Leon, K., Lage, A., Carneiro, J.: Tolerance and immunity in a mathematical model of T-cell mediated suppression. J. Theor. Biol. **225**, 107–126 (2003)
16. Leon, K., Perez, R., Lage, A., Carneiro, J.: Modelling T-cell-mediated suppression dependent on interactions in multicellular conjugates. J. Theor. Biol. **207**, 231–254 (2000)
17. Martins, J., Aguiar, M., Pinto, A.A., Stollenwerk, N.: On the series expansion of the spatial SIS evolution operator. J. Differ. Eqn. Appl. **17**(6), 93–104 (2011)
18. Martins, J., Pinto, A.A., Stollenwerk, N.: A scaling analysis in the SIRI epidemiological model. J. Biol. Dyn. **3**(5), 479–496 (2009)
19. Martins, J., Pinto, A.A., Stollenwerk, N.: Stationarity in moment closure and quasi-stationarity of the SIS model. (submitted)
20. Maurus de la Rosa, S., Rutz, S., Dorninger, H., Scheffold, A.: Interleukin-2 is essential for $CD4^+CD25^+$ regulatory T cells function. Eur. J. Immunol. **34**, 2480–2488 (2004)
21. Mudur, G.S.: Maths for movies, medicine & markets. The Telegraph Calcutta, India, 20/09/2010
22. Nåsell, I.: On the quasi-stationary distribution of the stochastic logistic epidemic. Math. Biosci. **156**, 21–40 (1999)
23. Nåsell, I.: The quasi-stationary distribution of the closed endemic SIS model. Adv. Appl. Prob. **28**, 895–932 (1996)
24. Nagata, S.: Fas ligand-induced apoptosis. Annu. Rev. Genet. 29–55 (1999)
25. Pinto, A.A., Aguiar, M., Martins, J., Stollenwerk, N.: Dynamics of Epidemiological Models. Acta Biotheor. **58**(4), 381–389 (2010)
26. Pinto, A.A., Burroughs, N.J., Ferreira, M., Oliveira, B.M.P.M.: Dynamics of immunological models. Acta Biotheor. **58**(4), 391–404 (2010).
27. Pinto, A.A., Martins, J., Stollenwerk, N.: The higher moments dynamic on SIS model. Numerical Analysis and Applied Mathematics. In: Simos, T.E., et al. (eds.) AIP Conference Proceedings, vol. 1168, pp. 1527–1530 (2009)
28. Rand, D.A.: Correlation equations and pair approximations for spatial ecologies. In: Jacqueline McGlade (ed.) Advanced Ecological Theory, pp. 100–142. Blackwell Science, Oxford, London, Edinburgh, Paris (1999)
29. Rogers, Paul R., Dubey, Caroline, Swain and Susan L.: Qualitative changes accompany memory t cell generation: faster, more effective responses at lower doses of antigen. J. Immunol. **164**, 2338–2346 (2000)
30. Sakaguchi, S.: Naturally arising $CD4^+$ regulatory T cells for immunological self-tolerance and negative control of immune responses. Annu. Rev. Immunol. **22**, 531–562 (2004)
31. Shevach, E.M., McHugh, R.S., Piccirillo, C.A., Thornton, A.M.: Control of T-cell activation by CD4(+) CD25(+) suppressor T cells. Immunol. Rev. **182**, 58–67 (2001)
32. Schluns, K.S., Kieper, W.C., Jameson, S.C., Lefrancois, L.: Interleukin-7 mediates the homeostasis of naive and memory CD8 T cells. Nat. Immunol. **1**, 426–432 (2000)
33. Stollenwerk, N., Aguiar, M.: The SIRI stochastic model with creation and annihilation operators. Quant. Methods **1**, 1–10 (2008)
34. Stollenwerk, N., Martins, J., Pinto, A.A.: The phase transition lines in pair approximation for the basic reinfection model SIRI. Phys. Lett. A **371**, 379–388 (2007)
35. Stollenwerk, N., van Noort, S., Martins, J., Aguiar, M., Hilker, F., Pinto, A.A., Gomes, G.: A spatially stochastic epidemic model with partial immunization shows in mean field approximation the reinfection threshold. J. Biol. Dyn. 1–17 (2010)

36. Tanchot, C., Vasseur, F., Pontoux, C., Garcia, C., Sarukhan, A.: Immune regulation by self-reactive T cells is antigen specific. J. Immunol. **172**, 4285–4291 (2004)
37. Thornton, A.M., Donovan, E.E., Piccirillo, C.A., Shevach, E.M.: Cutting edge: IL-2 is critically required for the in vitro activation of CD4(+)CD25(+) T cell suppressor function. J. Immunol. **172**, 6519–6523 (2004)
38. Thornton, A.M., Shevach, E.M.: $CD4^+CD25^+$ immunoregulatory T cells suppress polyclonal T cell activation *in vitro* by inhibiting interleukine 2 production. J. Exp. Med. **188**, 287–296 (1998)
39. Thornton, A.M., Shevach, E.M.: Suppressor effector function of $CD4^+CD25^+$ immunoregulatory T cells is antigen nonspecific. J. Immunol. **164**, 183–190 (2000)
40. van Kampen, N.G.: Stochastic processes in physics and chemistry. North-Holland, Amsterdam (1981)

Chapter 13
Global Convergence in Difference Equations

Elias Camouzis and Gerasimos Ladas

Abstract In this chapter we present some global convergence theorems for difference equations and systems. We also present specific examples of difference equations and systems to illustrate how most of these theorems apply.

13.1 Introduction

In this chapter we present some global convergence theorems for difference equations and systems. We also present specific examples of difference equations and systems to illustrate how most of these theorems apply.

For historical reasons we mention that about 30 years ago, there were almost no global convergence results known in the area of difference equations. Also, 15 years ago, there were only very few such results known. Indeed, R. M. May in [38], p. 839 states: "The response to large amplitude disturbances requires a nonlinear, or global analysis, for which no general techniques are available." Also, Y. Kuang and J. M. Cushing in [15], p. 32 state: "We would like to mention that global stability results for general difference-delay equations are rare."

We believe that the results, which we present in this chapter will have a wide spectrum of applicability in several areas that use difference equations. For the most part the examples which we present here to illustrate the global convergence theorems, are taken from the area of rational difference equations which has motivated these general results during the last 20 years. See [9, 21, 30, 32]. Finally, we believe that further research in rational difference equations will be a great source

---

E. Camouzis (✉)
Department of Mathematics and Natural Sciences, The American College of Greece, Gravias 6 Str.
15342 Aghia Paraskevi, Athens, Greece
e-mail: camouzis@acgmail.gr

G. Ladas
Department of Mathematics, University of Rhode Island,
Kingston, RI 02881-0816, USA
e-mail: gladas@math.uri.edu

M.M. Peixoto et al. (eds.), *Dynamics, Games and Science I*, Springer Proceedings
in Mathematics 1, DOI 10.1007/978-3-642-11456-4_13,
© Springer-Verlag Berlin Heidelberg 2011

194   E. Camouzis and G. Ladas

of motivation for obtaining additional global convergence results. This is because the methodology needed to understand the global character of rational difference equations will also be useful in the analysis of mathematical models that involve difference equations.

In Sect. 13.2 we present the definitions of stability.

In Sect. 13.3 we state the linearized stability result and we state necessary and sufficient conditions for polynomials of degree two, three, and four to have all their roots inside the unit disk.

In Sect. 13.4 we present a comparison result and in Sect. 5 we present the full limiting sequences result.

The most important section in this chapter is Sect. 6. Here we present 24 global convergence theorems and we give several examples to illustrate how most of these theorems apply.

13.2 Definitions of Stability

Let $\bar{x}$ be an equilibrium point of the difference equation

$$x_{n+1} = F(x_n, x_{n-1}, \ldots, x_{n-k}), \quad n = 0, 1, \ldots \tag{13.1}$$

that is

$$F(\bar{x}, \ldots, \bar{x}) = \bar{x}$$

where F is a function that maps some set J^{k+1} into J. The set J is usually an interval of real numbers, or a union of intervals, or a discrete set, such as the set of *integers* $\mathbf{Z} = \{\ldots, -1, 0, 1, \ldots\}$.

Definition 13.1.   1. An equilibrium point $\bar{x}$ of (13.1) is called *locally stable* if, for every $\varepsilon > 0$, there exists $\delta > 0$ such that if $\{x_n\}_{n=-k}^{\infty}$ is a solution of (13.1) with

$$|x_{-k} - \bar{x}| + |x_{1-k} - \bar{x}| + \cdots + |x_0 - \bar{x}| < \delta,$$

then

$$|x_n - \bar{x}| < \varepsilon, \quad \text{for all} \quad n \geq 0.$$

2. An equilibrium point $\bar{x}$ of (13.1) is called *locally asymptotically stable* if, $\bar{x}$ is locally stable, and if in addition there exists $\gamma > 0$ such that if $\{x_n\}_{n=-k}^{\infty}$ is a solution of (13.1) with

$$|x_{-k} - \bar{x}| + |x_{-k+1} - \bar{x}| + \cdots + |x_0 - \bar{x}| < \gamma,$$

then

$$\lim_{n \to \infty} x_n = \bar{x}.$$

13 Global Convergence in Difference Equations 195

3. An equilibrium point $\bar{x}$ of (13.1) is called a *global attractor* if, for every solution $\{x_n\}_{n=-k}^{\infty}$ of (13.1), we have

$$\lim_{n \to \infty} x_n = \bar{x}.$$

4. An equilibrium point $\bar{x}$ of (13.1) is called *globally asymptotically stable* if $\bar{x}$ is locally stable, and $\bar{x}$ is also a global attractor of (13.1).
5. An equilibrium point $\bar{x}$ of (13.1) is called *unstable* if $\bar{x}$ is not locally stable.

13.3 Linearized Stability Analysis

Assume that the function F is continuously differentiable in an open neighborhood of an equilibrium point $\bar{x}$. Let

$$q_i = \frac{\partial F}{\partial u_i}(\bar{x}, \bar{x}, \dots, \bar{x}), \quad \text{for } i = 0, 1, \dots, k$$

denote the partial derivative of $F(u_0, u_1, \dots, u_k)$ with respect to u_i evaluated at the equilibrium point $\bar{x}$ of (13.1). Then the equation

$$y_{n+1} = q_0 y_n + q_1 y_{n-1} + \cdots + q_k y_{n-k}, \quad n = 0, 1, \dots \tag{13.2}$$

is called the *linearized equation of (13.1) about the equilibrium point* $\bar{x}$, and the equation

$$\lambda^{k+1} - q_0 \lambda^k - \cdots - q_{k-1}\lambda - q_k = 0 \tag{13.3}$$

is called the *characteristic equation of (13.2) about* $\bar{x}$.

The following result, known as the *Linearized Stability Theorem*, is useful in determining the local stability character of the equilibrium point $\bar{x}$ of (13.1). See [3, 4, 18, 24, 37, 43, 44].

Theorem 13.1. *(The Linearized Stability Theorem)*
Assume that the function F is a continuously differentiable function defined on some open neighborhood of an equilibrium point $\bar{x}$. Then the following statements are true:

1. *If all the roots of (13.3) have absolute value less than one, then the equilibrium point $\bar{x}$ of (13.1) is locally asymptotically stable.*
2. *If at least one root of (13.3) has absolute value greater than one, then the equilibrium point $\bar{x}$ of (13.1) is unstable.*

The equilibrium point $\bar{x}$ of (13.1) is called *hyperbolic* if no root of (13.3) has absolute value equal to one. If there exists a root of (13.3) with absolute value equal to one, then the equilibrium $\bar{x}$ is called *nonhyperbolic*.

An equilibrium point $\bar{x}$ of (13.1) is called a *saddle point* if it is hyperbolic and if there exists a root of (13.3) with absolute value less than one and another root of (13.3) with absolute value greater than one.

An equilibrium point $\bar{x}$ of (13.1) is called a *repeller* if all roots of (13.3) have absolute value greater than one.

A solution $\{x_n\}_{n=-k}^{\infty}$ of (13.1) is called *periodic with period p* if there exists an integer $p \geq 1$ such that

$$x_{n+p} = x_n, \quad \text{for all} \quad n \geq -k. \tag{13.4}$$

A solution is called periodic with *prime period p* if p is the smallest positive integer for which (13.4) holds.

The next three theorems present necessary and sufficient conditions for all the roots of a real polynomial of degree two, three, or four, respectively, to have modulus less than one.

Theorem 13.2. *Assume that a_1 and a_0 are real numbers. Then a necessary and sufficient condition for all roots of the equation*

$$\lambda^2 + a_1 \lambda + a_0 = 0$$

to lie inside the unit disk is

$$|a_1| < 1 + a_0 < 2.$$

Theorem 13.3. *Assume that a_2, a_1, and a_0 are real numbers. Then a necessary and sufficient condition for all roots of the equation*

$$\lambda^3 + a_2 \lambda^2 + a_1 \lambda + a_0 = 0$$

to lie inside the unit disk is

$$|a_2 + a_0| < 1 + a_1, \quad |a_2 - 3a_0| < 3 - a_1, \quad \text{and} \quad a_0^2 + a_1 - a_0 a_2 < 1.$$

Theorem 13.4. *Assume that a_3, a_2, a_1, and a_0 are real numbers. Then a necessary and sufficient condition for all roots of the equation*

$$\lambda^4 + a_3 \lambda^3 + a_2 \lambda^2 + a_1 \lambda + a_0 = 0$$

to lie inside the unit disk is

$$|a_1 + a_3| < 1 + a_0 + a_2, \quad |a_1 - a_3| < 2(1 - a_0), \quad a_2 - 3a_0 < 3,$$

and

$$a_0 + a_2 + a_0^2 + a_1^2 + a_0^2 a_2 + a_0 a_3^2 < 1 + 2a_0 a_2 + a_1 a_3 + a_0 a_1 a_3 + a_0^3.$$

13 Global Convergence in Difference Equations 197

The following theorem states a sufficient condition for all roots of an equation of any order to lie inside the unit disk. See [11] or ([30], p. 12).

Theorem 13.5. *(C. W. Clark, [11]) Assume that $q_0, q_1, \ldots, q_k$ are real numbers such that*

$$|q_0| + |q_1| + \cdots + |q_k| < 1.$$

Then all roots of (13.3) lie inside the unit disk.

13.4 A Comparison Result

The following comparison result, see ([9], p. 7), is a useful tool in establishing bounds for solutions of nonlinear equations in terms of the solutions of equations with known behavior, for example, linear or Riccati.

Theorem 13.6. *Let I be an interval of real numbers, let k be a positive integer, and let*

$$F : J^{k+1} \to J$$

be a function which is increasing in all of its arguments. Assume that $\{x_n\}_{n=-k}^{\infty}$, $\{y_n\}_{n=-k}^{\infty}$, and $\{z_n\}_{n=-k}^{\infty}$ are sequences of real numbers such that

$$
\begin{cases}
x_{n+1} \le F(x_n, \ldots, x_{n-k}), & n = 0, 1, \ldots \\
y_{n+1} = F(y_n, \ldots, y_{n-k}), & n = 0, 1, \ldots \\
z_{n+1} \ge F(z_n, \ldots, z_{n-k}), & n = 0, 1, \ldots
\end{cases}
$$

and

$$x_n \le y_n \le z_n, \quad \text{for all} \quad -k \le n \le 0.$$

Then

$$x_n \le y_n \le z_n, \quad \text{for all} \quad n > 0. \tag{13.5}$$

13.5 Full Limiting Sequences

The following result about full limiting sequences it is sometimes useful in establishing that all solutions of a given difference equation converge to the equilibrium of the equation. See [19, 28, 42].

Theorem 13.7. *Consider the difference equation*

$$x_{n+1} = F(x_n, x_{n-1}, \ldots, x_{n-k}), \quad n = 0, 1, \ldots \tag{13.6}$$

where $F \in C(J^{k+1}, J)$ for some interval J of real numbers and some non-negative integer k. Let $\{x_n\}_{n=-k}^{\infty}$ be a solution of (13.6). Set $I = \liminf_{n \to \infty} x_n$ and

$S = \limsup_{n \to \infty} x_n$, and suppose that $I, S \in J$. Let $\mathscr{L}_0$ be a limit point of the solution $\{x_n\}_{n=-k}^\infty$. Then the following statements are true:

1. There exists a solution $\{L_n\}_{n=-\infty}^\infty$ of (13.6), called a full limiting sequence of $\{x_n\}_{n=-k}^\infty$, such that $L_0 = \mathscr{L}_0$, and such that for every $N \in \{\ldots, -1, 0, 1, \ldots\}$, L_N is a limit point of $\{x_n\}_{n=-k}^\infty$. In particular,

$$I \le L_N \le S, \quad \text{for all} \quad N \in \{\ldots, -1, 0, 1, \ldots\}.$$

2. For every $i_0 \in \{\ldots, -1, 0, 1, \ldots\}$, there exists a subsequence $\{x_{r_i}\}_{i=0}^\infty$ of $\{x_n\}_{n=-k}^\infty$ such that

$$L_N = \lim_{i \to \infty} x_{r_i + N}, \quad \text{for every} \quad N \ge i_0.$$

13.6 Convergence Theorems

The theorems in this section, for the most part, have been motivated by research in rational difference equations.

The following global attractivity result from [32]–[34], called the *m and M Theorem*, is very useful in establishing convergence results in many situations.

Theorem 13.8. *(Kulenovic, Ladas, and Sizer [32] and [34]) Let $[a, b]$ be a closed and bounded interval of real numbers and let*

$$F \in C([a, b]^{k+1}, [a, b])$$

satisfy the following conditions:

1. *The function $F(z_1, \ldots, z_{k+1})$ is monotonic in each of its arguments.*
2. *For each $m, M \in [a, b]$ and for each $i \in \{1, \ldots, k+1\}$, we define*

$$M_i(m, M) = \begin{cases} M, & \text{if } F \text{ is increasing in } z_i \\ m, & \text{if } F \text{ is decreasing in } z_i \end{cases}$$

and

$$m_i(m, M) = M_i(M, m).$$

Assume that if (m, M) is a solution of the system:

$$\left. \begin{array}{l} M = F(M_1(m, M), \ldots, M_{k+1}(m, M)) \\ m = F(m_1(m, M), \ldots, m_{k+1}(m, M)) \end{array} \right\},$$

then $M = m$.

13 Global Convergence in Difference Equations 199

Then there exists exactly one equilibrium $\bar{x} \in [a, b]$ of the equation

$$x_{n+1} = F(x_n, x_{n-1}, \ldots, x_{n-k}), \quad n = 0, 1, \ldots \tag{13.7}$$

and every solution of (13.7) with initial conditions in $[a, b]$ converges to $\bar{x}$.

The example which motivated Theorem 13.8 is the following:

Example 13.1. Consider the rational difference equation

$$x_{n+1} = \frac{\beta x_n + x_{n-1}}{B x_n + x_{n-1}}, \quad n = 0, 1, \ldots \tag{13.8}$$

with positive parameters β, B and with arbitrary positive initial conditions x_{-1}, x_0. We also assume that $\beta \neq B$ because otherwise the equation is trivial.

Then the following statements are true:

1. The equilibrium $\bar{x}$ of (13.8) is globally asymptotically stable when

$$\beta > B \tag{13.9}$$

or

$$\beta < B \quad \text{and} \quad B < 3\beta + \beta B + 1. \tag{13.10}$$

2. Every solution of (13.8) converges to the equilibrium $\bar{x}$ of (13.8) when

$$\beta < B \quad \text{and} \quad B = 3\beta + \beta B + 1. \tag{13.11}$$

3. Every solution of (13.8) converges to a (not necessarily prime) period-two solution when

$$\beta < B \quad \text{and} \quad B > 3\beta + \beta B + 1. \tag{13.12}$$

Proof. For the details of the proof see Theorem 5.86.1, p. 252 in [9]. $\qquad \square$

The next result applies to difference equations that satisfy the so-called, *negative feedback property.*

Theorem 13.9. *(Hautus and Bolis [22]) Let J be an open interval of real numbers, let $F \in C(J^{k+1}, J)$, and let $\bar{x} \in J$ be an equilibrium point of (13.7). Assume that F satisfies the following two conditions:*

1. *F is increasing in each of its arguments.*
2. *F satisfies the negative feedback property:*

$$(u - \bar{x})[F(u, u, \ldots, u) - u] < 0, \quad \text{for all} \quad u \in I - \{\bar{x}\}.$$

Then the equilibrium point $\bar{x}$ is a global attractor of all solutions of (13.1).

Theorem 13.9 was used by Kuang and Cushing in [31] to establish the global asymptotic stability of the positive equilibrium of the flour beetle model

$$\left.\begin{aligned} L_{n+1} &= bA_n e^{-c_{ea}A_n - c_{el}L_n} \\ B_{n+1} &= (1 - \mu_l)L_n \\ A_{n+1} &= B_n e^{-c_{pa}A_n} + (1 - \mu_a)A_n \end{aligned}\right\}, \quad n = 0, 1, \dots \tag{13.13}$$

when

$$c_{el} = 0.$$

In this case the system reduces to the third-order difference equation

$$A_{n+1} = (1 - \mu_l)A_n + b(1 - \mu_a)A_{n-2}e^{-c_{ea}A_{n-2} - c_{pa}A_n}, \quad n = 0, 1, \dots .$$

The next global convergence result was motivated by Pielou's equation

$$x_{n+1} = \frac{\beta x_n}{1 + x_{n-1}}, \quad n = 0, 1, \dots .$$

For this equation it was established in [36] that every nonnegative solution converges to a finite limit. More precisely, when

$$\beta \leq 1$$

every nonnegative solution converges to the zero equilibrium and when

$$\beta > 1$$

every positive solution converges to the positive equilibrium $\bar{x} = \beta - 1$.

Theorem 13.10. *(Kocic and Ladas [30]) Assume that the following conditions are satisfied:*

1. *$f \in C[(0, \infty) \times (0, \infty), (0, \infty)]$.*
2. *$f(x, y)$ is decreasing in x and strictly decreasing in y.*
3. *$xf(x, x)$ is strictly increasing in x.*
4. *The equation*
$$x_{n+1} = x_n f(x_n, x_{n-1}), \quad n = 0, 1, \dots \tag{13.14}$$
   has a unique positive equilibrium $\bar{x}$.

Then $\bar{x}$ is a global attractor of all positive solutions of (13.14).

The following theorem was motivated by a population model with two age classes. See [23].

13 Global Convergence in Difference Equations

Theorem 13.11. *(Franke, Hoag, and Ladas [23]) Assume that the following conditions are satisfied:*

1. *$f \in C[[0, \infty) \times [0, \infty), (0, \infty)]$.*
2. *$f(x, y)$ is decreasing in each argument.*
3. *$xf(x, y)$ is increasing in x.*
4. *$f(x, y) < f(y, x) \Leftrightarrow x > y$.*
5. *The equation*

$$x_{n+1} = x_{n-1} f(x_{n-1}, x_n), \quad n = 0, 1, \dots \tag{13.15}$$

has a unique positive equilibrium $\bar{x}$.

Then $\bar{x}$ is a global attractor of all positive solutions of (13.15).

The model that motivated Theorem 13.11 is a discrete model with two age classes, adults A_n and juveniles J_n, which interact as follows:

$$\left. \begin{array}{l} A_{n+1} = J_n \\ J_{n+1} = A_n e^{r-(A_n + \alpha J_n)} \end{array} \right\}, \quad n = 0, 1, \dots \tag{13.16}$$

where the two parameters r and α are such that

$$r, \alpha \in (0, 1].$$

Clearly, the variable J_n satisfies the second-order difference equation

$$J_{n+1} = J_{n-1} e^{r-(J_{n-1} + \alpha J_n)}, \quad n = 0, 1, \dots \tag{13.17}$$

for which Theorem 13.11 applies when

$$\alpha < 1$$

and proves that the positive equilibrium

$$\bar{J} = \frac{r}{1 + \alpha}$$

is a global attractor of all positive solutions of (13.17).

The following result of Camouzis and Ladas was motivated by several period-two convergence results in rational difference equations. See Chaps. 4 and 5 in [9]. Thanks to this result, several open problems and conjectures in the literature have now been resolved and the character of solutions of many rational equations has now been established. See Theorems 4.2.2, 4.3.1, 5.74.2, 5.86.1, 5.109.1, and 5.145.2 in [9].

Theorem 13.12. *(Camouzis and Ladas, [2, p. 11] or [8]) Let J be a set of real numbers and let*

$$F : J \times J \to J$$

be a function $F(u, v)$, which decreases in u and increases in v. Then for every solution $\{x_n\}_{n=-1}^{\infty}$ of the equation

$$x_{n+1} = F(x_n, x_{n-1}), \quad n = 0, 1, \ldots \tag{13.18}$$

the subsequences $\{x_{2n}\}_{n=0}^{\infty}$ and $\{x_{2n+1}\}_{n=-1}^{\infty}$ of even and odd terms are eventually monotonic.

Proof. Observe that if for some N the solution $\{x_n\}_{n=-1}^{\infty}$ is such that

$$x_{N+1} \geq x_{N-1} \quad \text{and} \quad x_{N+2} \leq x_N$$

or

$$x_{N+1} \leq x_{N-1} \quad \text{and} \quad x_{N+2} \geq x_N$$

then either $\{x_{N+2n}\}_{n=0}^{\infty}$ is decreasing and $\{x_{N+2n+1}\}_{n=0}^{\infty}$ is increasing or vice versa.

Otherwise, for all $n \geq 0$

$$x_{2n+1} > x_{2n-1} \quad \text{and} \quad x_{2n+2} > x_{2n}$$

or

$$x_{2n+1} < x_{2n-1} \quad \text{and} \quad x_{2n+2} < x_{2n}$$

which implies that the subsequences $\{x_{2n}\}_{n=0}^{\infty}$ and $\{x_{2n-1}\}_{n=0}^{\infty}$ are either both increasing or both decreasing. $\square$

The proof of Statement 3 of Example 13.1 is a direct consequence of the above theorem. Furthermore, the following example provides another illustration of Theorem 13.12.

Example 13.2. Assume that

$$\gamma > \beta + A.$$

Then every positive and bounded solution of the second-order rational difference equation

$$x_{n+1} = \frac{\alpha + \beta x_n + \gamma x_{n-1}}{A + x_n}, \quad n = 0, 1, \ldots, \tag{13.19}$$

converges to the positive equilibrium

$$\bar{x} = \frac{\beta + \gamma - A + \sqrt{(\beta + \gamma - A)^2 + 4\alpha}}{2}.$$

Proof. For the proof see Theorem 4.2.2 in [9]. $\square$

13 Global Convergence in Difference Equations

The next three theorems were motivated by our investigation of the following rational system in the plane:

$$\left.\begin{array}{c} x_{n+1} = \frac{\alpha_1+\gamma_1 y_n}{y_n} \\[2mm] y_{n+1} = \frac{\alpha_2+\beta_2 x_n+\gamma_2 y_n}{A_2+B_2 x_n+C_2 y_n} \end{array}\right\}, \quad n = 0, 1, \ldots . \tag{13.20}$$

The variable y_n satisfies the second-order rational difference equation

$$y_{n+1} = \frac{\alpha + \beta y_n x_{n-1} + \gamma y_{n-1}}{A + B y_n y_{n-1} + C y_{n-1}}, \quad n = 0, 1, \ldots \tag{13.21}$$

which contains 49 special cases of equations each with positive parameters. These special cases were investigated in [1]–[2]. Here we present, as an example, the difference equation

$$x_{n+1} = \frac{\beta x_n x_{n-1}}{1 + x_n x_{n-1}}, \quad n = 0, 1, \ldots . \tag{13.22}$$

By employing the next three theorems we established in [1] that every nonnegative solution of (13.22) has a finite limit.

Theorem 13.13. *(Amleh, Camouzis, and Ladas [1]) Let J be a set of real numbers and let*

$$F : J \times J \rightarrow J$$

be a function $F(u, v)$ which increases in both variables. Then for every solution $\{x_n\}_{n=-1}^{\infty}$ of (13.18) the subsequences $\{x_{2n}\}$ and $\{x_{2n+1}\}$ of even and odd terms are eventually monotonic.

Theorem 13.14. *(Amleh, Camouzis, and Ladas [1]) Assume that the function $f \in C([a, \infty)^2, [a, \infty))$ increases in both variables and that the difference equation*

$$x_{n+1} = f(x_n, x_{n-1}), \quad n = 0, 1, \ldots \tag{13.23}$$

has no equilibrium point in (a, ∞). Let $\{x_n\}_{n=-1}^{\infty}$ be a solution of (13.23) with $x_{-1}, x_0 \in (a, \infty)$. Then

$$\lim_{n \to \infty} x_n = \begin{cases} \infty \text{ if } f(x, x) > x, & \text{for all } x > a. \\ a \text{ if } f(x, x) < x, & \text{for all } x > a. \end{cases}$$

Theorem 13.15. *(Amleh, Camouzis, and Ladas [1]) Assume that $f \in C([0, \infty)^2, [0, \infty))$ increases in both variables and that the difference equation*

$$x_{n+1} = f(x_n, x_{n-1}), \quad n = 0, 1, \ldots \tag{13.24}$$

has two consecutive equilibrium points $\bar{x}_1$ and $\bar{x}_2$, with $\bar{x}_1 < \bar{x}_2$. Also assume that either

$$f(x,x) > x, \quad \text{for} \quad \bar{x}_1 < x < \bar{x}_2 \tag{13.25}$$

or

$$f(x,x) < x, \quad \text{for} \quad \bar{x}_1 < x < \bar{x}_2. \tag{13.26}$$

Then every solution $\{x_n\}_{n=-1}^{\infty}$ of (13.24) with initial conditions

$$x_{-1}, x_0 \in (\bar{x}_1, \bar{x}_2)$$

converges to one of the two equilibrium points, and more precisely the following is true:

$$\lim_{n \to \infty} x_n = \begin{cases} \bar{x}_1, & \text{if (13.26) holds,} \\ \bar{x}_2, & \text{if (13.25) holds.} \end{cases}$$

The above two results extend and generalize to difference equations of higher order.

The next theorem unifies Theorems 13.12 and 13.13 and extends the results to non-autonomous difference equations.

Theorem 13.16. *(Camouzis, [5])*
Let J be a set of real numbers and let

$$f_n : J \times J \to J$$

be a family of functions $f_n(z_1, z_2)$ which increase in z_2 and are monotonic in z_1 throughout J. Also assume that for all $n \in \{0, 1, \ldots\}$ and for some $m \in \{1, 2, \ldots\}$

$$f_{n+m}(z_1, z_2) = f_n(z_1, z_2).$$

Then for every solution $\{x_n\}_{n=-1}^{\infty}$ of the difference equation

$$x_{n+1} = f_n(x_n, x_{n-1}), \quad n = 0, 1, \ldots$$

the 2m-sequences $\{x_{2mn+t}\}_{t=0}^{2m-1}$ are eventually monotonic.

Note that when m is even, m-sequences $\{x_{mn+t}\}_{t=0}^{m-1}$ are eventually monotonic, and when m is odd, $2m$-sequences $\{x_{2mn+t}\}_{t=0}^{2m-1}$ are eventually monotonic.

The following example provides an illustration of Theorem 13.16.

Example 13.3. (See [5]) Assume that the sequences $\{A_n\}_{n=0}^{\infty}$, $\{B_n\}_{n=0}^{\infty}$, and $\{C_n\}_{n=0}^{\infty}$ positive $2m$-periodic sequences. Then every positive solution of the second-order rational difference equation

$$x_{n+1} = \frac{x_{n-1}}{A_n + B_n x_n + C_n x_{n-1}}, \quad n = 0, 1, \ldots, \tag{13.27}$$

converges to a periodic solution with period-$2m$.

13 Global Convergence in Difference Equations 205

The following five results were motivated by our recent investigations on rational difference equations. See [9].

In order to simplify and unify several convergence results for the difference equation

$$x_n = f(x_{n-i_1}, \ldots, x_{n-i_k}), \quad n = 1, 2, \ldots, \tag{13.28}$$

where $k \geq 2$ and the function $f(z_1, \ldots, z_k)$ is monotonic in each of its arguments, we introduce some notation and state several hypotheses.

For every pair of numbers (m, M) and for each $j \in \{1, \ldots, k\}$, we define

$$M_j = M_j(m, M) = \begin{cases} M, & \text{if } f \text{ is increasing in } z_j \\ m, & \text{if } f \text{ is decreasing in } z_j \end{cases}$$

and

$$m_j = m_j(m, M) = M_j(M, m).$$

(H_1) : $f \in C([0, \infty)^k, [0, \infty))$ and $f(z_1, \ldots, z_k)$ is monotonic in each of its arguments.

(H_1^*) : $\in C((0, \infty)^k, (0, \infty))$ and $f(z_1, \ldots, z_k)$ is monotonic in each of its arguments.

(H_1') : $f \in C([0, \infty)^k, [0, \infty))$ and $f(z_1, \ldots, z_k)$ is strictly monotonic in each of its arguments.

(H_1'') : $f \in C((0, \infty)^k, (0, \infty))$ and $f(z_1, \ldots, z_k)$ is strictly monotonic in each of its arguments.

(H_2) : For each $m \in [0, \infty)$ and $M > m$, we assume that

$$f(M_1, \ldots, M_k) \geq M \tag{13.29}$$

implies

$$f(m_1, \ldots, m_k) > m. \tag{13.30}$$

(H_2') : For each $m \in (0, \infty)$ and $M > m$, we assume that

$$f(M_1, \ldots, M_k) \geq M \tag{13.31}$$

implies

$$f(m_1, \ldots, m_k) > m. \tag{13.32}$$

(H_3) : For each $m \in [0, \infty)$ and $M > m$, we assume that either

$$(f(M_1, \ldots, M_k) - M)(f(m_1, \ldots, m_k) - m) > 0 \tag{13.33}$$

or

$$f(M_1, \ldots, M_k) - M = f(m_1, \ldots, m_k) - m = 0. \tag{13.34}$$

(H_3') : For each $m \in (0, \infty)$ and $M > m$, we assume that
either

$$(f(M_1, \ldots, M_k) - M)(f(m_1, \ldots, m_k) - m) > 0 \qquad (13.35)$$

or

$$f(M_1, \ldots, M_k) - M = f(m_1, \ldots, m_k) - m = 0. \qquad (13.36)$$

We also define the following sets:

$$S = \{i_s \in \{i_1, \ldots, i_k\} : f \text{ strictly increases in } x_{n-i_s}\} = \{i_{s_1}, \ldots, i_{s_r}\}$$

and

$$J = \{i_j \in \{i_1, \ldots, i_k\} : f \text{ strictly decreases in } x_{n-i_j}\} = \{i_{j_1}, \ldots, i_{j_t}\}.$$

Clearly when H_1' or H_1'' holds,

$$S \bigcup J = \{i_1, \ldots, i_k\}.$$

(H_4) : The set S consists of even indices only and the set J consists of odd indices only.

(H_5) : Either the set S contains at least one odd index, or the set J contains at least one even index.

(H_6) : The greatest common divisor of the indices in the union of the sets S and J is equal to 1.

The next four theorems have been used to establish global attractivity and period-two convergence results in many special cases of rational equations. See [9].

Theorem 13.17. *(Camouzis and Ladas [9]) The following statements are true:*

1. *Assume that (H_1) and (H_2) hold for the function $f(z_1, \ldots, z_k)$ of (13.28). Then every solution of (13.28) which is bounded from above converges to a finite limit.*
2. *Assume that (H_1^*) and (H_2') hold for the function $f(z_1, \ldots, z_k)$ of (13.28). Then every solution of (13.28) which is bounded from above and from below by positive constants converges to a finite limit.*

The following rational difference equation

$$x_{n+1} = \frac{x_n}{A + B x_n + C x_{n-1}}, \qquad n = 0, 1, \ldots \qquad (13.37)$$

was investigated in [30], where it was shown that when

$$A \in (0, 1)$$

13 Global Convergence in Difference Equations 207

every positive solution converges to the positive equilibrium. We present here a simpler proof, which is based on Theorem 13.17. Note that the change of variables

$$y_n = \frac{1}{x_n},$$

transforms (13.37) to

$$y_{n+1} = B + A y_n + \frac{C y_n}{y_{n-1}}, \quad n = 0, 1, \ldots . \tag{13.38}$$

Also,

$$\frac{y_{n+1}}{y_n} = \frac{B}{y_n} + A + \frac{C}{y_{n-1}}, \quad n = 0, 1, \ldots$$

and so (13.38) becomes

$$y_{n+1} = B + CA + A y_n + \frac{CB}{y_{n-1}} + \frac{C^2}{y_{n-2}}, \quad n \geq 0. \tag{13.39}$$

By employing Theorem 13.17 it follows that every positive solution of (13.39) converges to the positive equilibrium. Therefore, every positive solution of (13.37) also converges to its positive equilibrium.

Theorem 13.18. *(Camouzis and Ladas [9]) Assume that for any of the following three equations of order three:*

$$x_{n+1} = f(x_n, x_{n-1}, x_{n-2}), \quad n = 0, 1, \ldots \tag{13.40}$$

$$x_{n+1} = f(x_n, x_{n-2}), \quad n = 0, 1, \ldots \tag{13.41}$$

or

$$x_{n+1} = f(x_{n-1}, x_{n-2}), \quad n = 0, 1, \ldots \tag{13.42}$$

the hypotheses (H_1'') and (H_3') are satisfied for the arguments shown in the equation and, furthermore, assume that the function f is:

strictly increasing in x_n or x_{n-2}, or strictly decreasing in x_{n-1}.

Then every solution of this equation bounded from below and from above by positive constants converges to a finite limit.

As an illustration of Theorem 13.18 we present the rational difference equation

$$x_{n+1} = \frac{\beta x_n + \delta x_{n-2}}{1 + x_n}, \quad n = 0, 1, \ldots \tag{13.43}$$

with positive parameters and

$$\delta = \beta + 1.$$

Then, by employing Theorem 13.18 it follows that every positive solution of (13.43) converges to a finite limit. For the details of the proof see [9], p. 257.

Theorem 13.19. *(Camouzis and Ladas [9]) Assume that for any of the three equations (13.40), (13.41), or (13.42) the hypotheses (H_1'') and (H_3') are satisfied for the arguments shown in the equation, and, furthermore, assume that the function f is:*

$$\text{strictly decreasing in } x_n, \text{ for } (13.40) \text{ and } (13.41),$$

and

$$\text{strictly increasing in } x_{n-1} \text{ for } (13.40) \text{ and } (13.42),$$

and

$$\text{strictly decreasing in } x_{n-2}.$$

Then every solution of this equation bounded from above and from below by positive constants converges to a (not necessarily prime) period-two solution.

It is interesting to mention that Theorem 13.19 applies to Pielou's equation in the following iterated form

$$x_{n+1} = \frac{\beta x_{n-1}}{1 + x_{n-1}} \cdot \frac{\beta}{1 + x_{n-2}}, \quad n = 0, 1, \dots \tag{13.44}$$

and simplifies substantially the proof given in [36].

Theorem 13.20. *(Camouzis and Ladas [9]) Assume that for any of the three equations (13.40), (13.41), or (13.42) the Hypotheses (H_1') and (H_3) are satisfied for the arguments shown in the equation, and, furthermore, assume that the function f is:*

$$\text{strictly decreasing in } x_n, \text{ for } (13.40) \text{ and } (13.41),$$

and

$$\text{strictly increasing in } x_{n-1} \text{ for } (13.40) \text{ and } (13.42),$$

and

$$\text{strictly decreasing in } x_{n-2}.$$

Then every solution of this equation bounded from above converges to a (not necessarily prime) period-two solution.

An example where Theorem 13.20 applies is the rational difference equation

$$x_{n+1} = \frac{\alpha + \gamma x_{n-1}}{A + B x_n + x_{n-2}}, \quad n = 0, 1, \dots \tag{13.45}$$

with positive parameters and

$$\gamma = A.$$

13 Global Convergence in Difference Equations   209

For this equation Theorem 13.20 implies that every positive solution converges to a (not necessarily prime) period-two solution. See [9], p. 123.

The following general theorem extends and unifies Theorems 13.18–13.20.

Theorem 13.21. *(Camouzis and Ladas [9].) The following statements are true:*

1. *Assume that (H_1'), (H_3), (H_4), and (H_6) hold. Then every bounded solution of (13.28) converges to a (not necessarily prime) period-two solution.*
2. *Assume that (H_1''), (H_3'), (H_4) and (H_6) hold. Then every solution of (13.28) bounded from above and from below by positive constants converges to a (not necessarily prime) period-two solution.*
3. *Assume that (H_1'), (H_3), (H_5), and (H_6) hold. Then every bounded solution of (13.28) converges to a finite limit.*
4. *Assume that (H_1''), (H_3'), (H_5), and (H_6) hold. Then every solution of (13.28) bounded from above and from below by positive constants converges to a finite limit.*

The following result has many interesting applications.

Theorem 13.22. *(El-Metwalli, Grove, Ladas, and Voulov. See [19] and [20]). Let J be an interval of real numbers and let $F \in C(J^{k+1}, J)$. Assume that the following three conditions are satisfied:*

1. *F is increasing in each of its arguments.*
2. *$F(z_1, \ldots, z_{k+1})$ is strictly increasing in each of the arguments $z_{i_1}, z_{i_2}, \ldots, z_{i_l}$, where $1 \le i_1 < i_2 < \ldots < i_l \le k + 1$, and the arguments $i_1, i_2, \ldots, i_l$ are relatively prime.*
3. *Every point c in I is an equilibrium point of*

$$x_{n+1} = F(x_n, x_{n-1}, \ldots, x_{n-k}), \qquad n = 0, 1, \ldots . \tag{13.46}$$

Then every solution of (13.46) has a finite limit.

Example 13.4. Consider the rational difference equation

$$x_n = \frac{1}{\sum_{i=1}^{k} \beta_i x_{n-l_i}}, \qquad n = 0, 1, \ldots, \tag{13.47}$$

where

$$\beta_i > 0, \quad \text{for} \quad i = 1, \ldots, k$$

and

$$l_i \in \{1, 2, \ldots\}.$$

Assume that d_1 and d_2 are the greatest common divisors of the two sets of positive integers:

$$\{l_1, l_2, \ldots, l_k\}$$

and

$$\{l_i + l_j : i, j \in \{1, \ldots, k\}\},$$

respectively. Then the following statements are true:

1. The equilibrium $\bar{x}$ of (13.47) is globally asymptotically stable if and only if

$$d_1 = d_2.$$

2. When

$$d_1 \neq d_2,$$

every solution of (13.47) converges to a (not necessarily prime) period-$(2 \cdot d_1)$ solution.

Proof. The proof is a consequence of Theorem 13.22. For the details see [19] or [21] and [20]. $\square$

We will now state some results about competitive and cooperative systems in the plane. See [10] and [7].

Let I and J be intervals of real numbers and let

$$f : I \times J \to I \quad \text{and} \quad g : I \times J \to J.$$

We say that the system

$$\left.\begin{array}{l} x_{n+1} = f(x_n, y_n) \\[2mm] y_{n+1} = g(x_n, y_n) \end{array}\right\}, \quad n = 0, 1, \ldots \tag{13.48}$$

is *competitive*, when the function $f(x, y)$ is increasing in x and decreasing in y and the function $g(x, y)$ is decreasing in x and increasing in y.

On the other hand, when the function $f(x, y)$ is increasing in x and increasing in y and the function $g(x, y)$ is increasing in x and increasing in y the System (13.48) is called *cooperative*.

The following two results for competitive systems in the plane were first established by DeMottoni and Schiaffino. See [39]. For a more general version of these two theorems see [48]. For their proofs see [7].

Theorem 13.23. *Assume that the System (13.48) is competitive and that for every solution (x_n, y_n) of the System (13.48) the following two statements are satisfied:*

(i) If for some $N \geq 0$,

$$x_{N+1} < x_N \quad \text{and} \quad y_{N+1} < y_N, \tag{13.49}$$

then

$$x_N < x_{N-1} \quad \text{and} \quad y_N < y_{N-1}. \tag{13.50}$$

13 Global Convergence in Difference Equations   211

(ii) *If for some $N \geq 0$,*

$$x_{N+1} > x_N \quad and \quad y_{N+1} > y_N,$$
(13.51)

then

$$x_N > x_{N-1} \quad and \quad y_N > y_{N-1}.$$
(13.52)

Then both sequences $\{x_n\}$ and $\{y_n\}$ are eventually monotonic.

Theorem 13.24. *Assume that the System (13.48) is competitive and that for every solution (x_n, y_n) of the System 13.48 the following two statements are satisfied:*

(i) *If for some $N \geq 0$,*

$$x_{N+2} < x_N \quad and \quad y_{N+2} < y_N,$$
(13.53)

then

$$x_{N+1} > x_{N-1} \quad and \quad y_{N+1} > y_{N-1}.$$
(13.54)

(ii) *If for some $N \geq 0$,*

$$x_{N+2} > x_N \quad and \quad y_{N+2} > y_N,$$
(13.55)

then

$$x_{N+1} < x_{N-1} \quad and \quad y_{N+1} < y_{N-1}.$$
(13.56)

Then the four subsequences $\{x_{2n}\}$, $\{x_{2n+1}\}$, $\{y_{2n}\}$, and $\{y_{2n+1}\}$ are eventually monotonic.

The next two theorems are extensions of Theorems 13.23 and 13.24 to non-autonomous systems.

Theorem 13.25. *Consider the non-autonomous System*

$$\left.\begin{array}{l} x_{n+1} = f_n(x_n^{\uparrow}, y_n^{\downarrow}) \\ \\ y_{n+1} = g_n(x_n^{\downarrow}, y_n^{\uparrow}) \end{array}\right\}, \quad n = 0, 1, \ldots$$
(13.57)

where

$$f_n : I \times J \rightarrow I \quad and \quad g_n : I \times J \rightarrow J, \quad n = 0, 1, \ldots .$$

Also assume that for all $n \in \{0, 1, \ldots\}$ and for some $m \in \{1, 2, \ldots\}$

$$f_{n+m}(z_1, z_2) = f_n(z_1, z_2) \quad and \quad g_{n+m}(z_1, z_2) = g_n(z_1, z_2)$$
(13.58)

and that for every solution (x_n, y_n) of (13.57), the following two statements are satisfied:

(i) *If for some $N \geq 0$,*

$$x_{N+m} < x_N \quad and \quad y_{N+m} < y_N,$$
(13.59)

then

$$x_{N+m-1} < x_{N-1} \quad \text{and} \quad y_{N+m} < y_{N-1}. \tag{13.60}$$

(ii) If for some $N \geq 0$,

$$x_{N+m} > x_N \quad \text{and} \quad y_{N+m} > y_N, \tag{13.61}$$

then

$$x_{N+m-1} > x_{N-1} \quad \text{and} \quad y_{N+m-1} > y_{N-1}. \tag{13.62}$$

Then the $2m$-sequences $\{x_{mn+t}\}_{t=0}^{m-1}$ $\{y_{mn+t}\}_{t=0}^{m-1}$ are eventually monotonic.

Theorem 13.26. *Assume that (13.58) holds and that for every solution (x_n, y_n) of (13.57), the following two statements are satisfied:*

(i) If for some $N \geq 0$,

$$x_{N+m} < x_N \quad \text{and} \quad y_{N+m} < y_N, \tag{13.63}$$

then

$$x_{N+m-1} > x_{N-1} \quad \text{and} \quad y_{N+m} > y_{N-1}. \tag{13.64}$$

(ii) If for some $N \geq 0$,

$$x_{N+m} > x_N \quad \text{and} \quad y_{N+m} > y_N, \tag{13.65}$$

then

$$x_{N+m-1} < x_{N-1} \quad \text{and} \quad y_{N+m-1} < y_{N-1}. \tag{13.66}$$

Then the $4m$-sequences $\{x_{2mn+t}\}_{t=0}^{2m-1}$ and $\{y_{2mn+t}\}_{t=0}^{2m-1}$ are eventually monotonic.

The following two theorems are about cooperative systems in the plane.

Theorem 13.27. *Assume that the System (13.48) is cooperative and that for every solution (x_n, y_n) of the System (13.48) the following two statements are satisfied:*

(i) If for some $N \geq 0$,

$$x_{N+1} < x_N \quad \text{and} \quad y_{N+1} > y_N, \tag{13.67}$$

then

$$x_N < x_{N-1} \quad \text{and} \quad y_N > y_{N-1}. \tag{13.68}$$

(ii) If for some $N \geq 0$,

$$x_{N+1} > x_N \quad \text{and} \quad y_{N+1} < y_N, \tag{13.69}$$

then

$$x_N > x_{N-1} \quad \text{and} \quad y_N < y_{N-1}. \tag{13.70}$$

Then both sequences $\{x_n\}$ and $\{y_n\}$ are eventually monotonic.

13 Global Convergence in Difference Equations 213

Theorem 13.28. *Assume that the System (13.48) is cooperative and that for every solution (x_n, y_n) of the System (13.48) the following two statements are satisfied:*

(i) If for some $N \geq 0$

$$x_{N+2} < x_N \quad and \quad y_{N+2} > y_N, \tag{13.71}$$

then

$$x_{N+1} > x_{N-1} \quad and \quad y_{N+1} < y_{N-1}. \tag{13.72}$$

(ii) If for some $N \geq 0$,

$$x_{N+2} > x_N \quad and \quad y_{N+2} > y_N, \tag{13.73}$$

then

$$x_{N+1} < x_{N-1} \quad and \quad y_{N+1} > y_{N-1}. \tag{13.74}$$

Then the four subsequences $\{x_{2n}\}$, $\{x_{2n+1}\}$, $\{y_{2n}\}$, and $\{y_{2n+1}\}$ are eventually monotonic.

The following two theorems are extensions of Theorems 13.27 and 13.28 to non-autonomous systems.

Theorem 13.29. *Consider the non-autonomous System*

$$\left. \begin{array}{l} x_{n+1} = f_n(x_n^{\uparrow}, y_n^{\uparrow}) \\ y_{n+1} = g_n(x_n^{\uparrow}, y_n^{\uparrow}) \end{array} \right\}, \quad n = 0, 1, \ldots \tag{13.75}$$

where

$$f_n : I \times J \to I \quad and \quad g_n : I \times J \to J, \quad n = 0, 1, \ldots .$$

Also assume that for all $n \in \{0, 1, \ldots\}$ and for some $m \in \{1, 2, \ldots\}$

$$f_{n+m}(z_1, z_2) = f_n(z_1, z_2) \quad and \quad g_{n+m}(z_1, z_2) = g_n(z_1, z_2)$$

and that for every solution (x_n, y_n) of (13.75), the following two statements are satisfied:

(i) If for some $N \geq 0$,

$$x_{N+m} < x_N \quad and \quad y_{N+m} > y_N, \tag{13.76}$$

then

$$x_{N+m-1} < x_{N-1} \quad and \quad y_{N+m} > y_{N-1}. \tag{13.77}$$

(ii) If for some $N \geq 0$,

$$x_{N+m} > x_N \quad and \quad y_{N+m} < y_N, \tag{13.78}$$

then

$$x_{N+m-1} > x_{N-1} \quad and \quad y_{N+m-1} < y_{N-1}. \tag{13.79}$$

Then the $2m$-sequences $\{x_{mn+t}\}_{t=0}^{m-1}$ and $\{y_{mn+t}\}_{t=0}^{m-1}$ are eventually monotonic.

Theorem 13.30. *Consider the non-autonomous System*

$$\left. \begin{array}{l} x_{n+1} = f_n(x_n^{\uparrow}, y_n^{\uparrow}) \\[2mm] y_{n+1} = g_n(x_n^{\uparrow}, y_n^{\uparrow}) \end{array} \right\}, \quad n = 0, 1, \ldots \tag{13.80}$$

where

$$f_n : I \times J \to I \quad and \quad g_n : I \times J \to J, \quad n = 0, 1, \ldots .$$

Also assume that for all $n \in \{0, 1, \ldots\}$ and for some $m \in \{1, 2, \ldots\}$

$$f_{n+m}(z_1, z_2) = f_n(z_1, z_2) \quad and \quad g_{n+m}(z_1, z_2) = g_n(z_1, z_2)$$

and that for every solution (x_n, y_n) of (13.75), the following two statements are satisfied:

(i) If for some $N \geq 0$,

$$x_{N+m} < x_N \quad and \quad y_{N+m} > y_N, \tag{13.81}$$

then

$$x_{N+m-1} > x_{N-1} \quad and \quad y_{N+m} < y_{N-1}. \tag{13.82}$$

(ii) If for some $N \geq 0$,

$$x_{N+m} > x_N \quad and \quad y_{N+m} < y_N, \tag{13.83}$$

then

$$x_{N+m-1} < x_{N-1} \quad and \quad y_{N+m-1} > y_{N-1}. \tag{13.84}$$

Then the $4m$-sequences $\{x_{2mn+t}\}_{t=0}^{2m-1}$ and $\{y_{2mn+t}\}_{t=0}^{2m-1}$ are eventually monotonic.

Finally, we present a quite general extension of the m and M Theorem to systems of difference equations. See ([21], p. 19–21) and [33].

Theorem 13.31. *Assume that for each $i \in \{1, \ldots, k\}$, $[a_i, b_i]$ is a closed and bounded interval of real numbers, and the function*

$$F_i : C([a_1, b_1] \times \ldots \times [a_k, d_k], [a_i, b_i])$$

13 Global Convergence in Difference Equations   215

satisfies the following conditions:

1. $F_i(z_1, \ldots, z_k)$ *is monotonic in each of its arguments.*
2. *For each* $i, j \in \{1, \ldots, k\}$ *and for each* $m_i, M_i \in [a_i, b_i]$, *we define*

$$M_{i,j}(m_i, M_i) = \begin{cases} M_i, & \text{if } F_j \text{ is increasing in } z_i \\ m_i, & \text{if } F_j \text{ is decreasing in } z_i \end{cases}$$

and

$$m_{i,j}(m_i, M_i) = M_{i,j}(M_i, m_i)$$

and we assume that if $m_1, \ldots, m_k$ *and* $M_1, \ldots, M_k$ *is a solution of the system of* $2k$ *equations:*

$$\left. \begin{array}{l} M_i = F_i(M_{1,i}(m_1, M_1), \ldots, M_{k,i}(m_k, M_k)) \\ m_i = F_i(m_{1,i}(m_1, M_1), \ldots, m_{k,i}(m_k, M_k)) \end{array} \right\}, \ i \in \{1, \ldots, k\}$$

then

$$M_i = m_i, \quad \text{for all} \quad i \in \{1, \ldots, k\}.$$

Then the system of k *difference equations*

$$\left. \begin{array}{l} x_{n+1}^1 = F_1(x_n^1, \ldots, x_n^k) \\[2mm] x_{n+1}^2 = F_2(x_n^1, \ldots, x_n^k) \\[1mm] \vdots \\[1mm] x_{n+1}^k = F_k(x_n^1, \ldots, x_n^k) \end{array} \right\}, \ n = 0, 1, \ldots \tag{13.85}$$

with initial condition $(x_0^1, \ldots, x_0^k) \in [a_1, b_1] \times \ldots \times [a_k, b_k]$, *has exactly one equilibrium point* $(\bar{x}^1, \ldots, \bar{x}^k)$ *and every solution of System (13.85) converges to* $(\bar{x}^1, \ldots, \bar{x}^k)$.

We now present a few examples to illustrate how the above convergence theorems apply to systems of difference equations.

Example 13.5. The following system of rational difference equations

$$\left. \begin{array}{l} x_{n+1} = \dfrac{\beta_1 x_n}{A_1 + B_1 x_n + C_1 y_n} \\[4mm] y_{n+1} = \dfrac{\gamma_2 y_n}{A_2 + B_2 x_n + C_2 y_n} \end{array} \right\}, \quad n = 0, 1, \ldots \tag{13.86}$$

with positive parameters, was investigated in [15]. By using Theorem 13.23, it follows that every positive solution of System (13.86) converges to a finite limit.

216   E. Camouzis and G. Ladas

Example 13.6. The corresponding periodically-forced rational system,

$$
\left.
\begin{aligned}
x_{n+1} &= \frac{\beta_{1,n} x_n}{A_{1,n} + B_{1,n} x_n + C_{1,n} y_n} \\[2ex]
y_{n+1} &= \frac{\gamma_{2,n} y_n}{A_{2,n} + B_{2,n} x_n + C_{2,n} y_n}
\end{aligned}
\right\}, \quad n = 0, 1, \ldots \tag{13.87}
$$

where the parameters are positive and m-periodic sequences, was investigated in
[10]. By employing Theorem 13.25 it was shown that every positive solution
converges to a (not necessarily prime) period-m solution.

Example 13.7. Consider the system of rational difference equations

$$
\left.
\begin{aligned}
x_{n+1} &= \frac{\beta_1 x_n + \gamma_1 y_n}{1 + y_n} \\[2ex]
y_{n+1} &= \frac{\beta_2 x_n + \gamma_2 y_n}{1 + x_n}
\end{aligned}
\right\}, \quad n = 0, 1, \ldots \tag{13.88}
$$

with positive parameters. By employing Theorems 13.23, 13.27,and 13.28 the
following statements were established in [10].

1. When
$$
\beta_1, \gamma_2 \in (1, \infty),
$$

   System (13.88) is competitive. It also possesses unbounded solutions in some
   range of its parameters. However, every bounded component of a solution
   converges to a finite limit.
2. When
$$
\beta_1, \gamma_2 \in (0, 1),
$$

   System (13.88) is cooperative. Furthermore, every solution converges to a finite
   limit.
3. When
$$
\beta_1 > 1 \quad \text{and} \quad \gamma_2 < 1,
$$

   the component $\{y_n\}$ of every solution $\{x_n, y_n\}$ is bounded. Also, for some initial
   conditions, the component $\{x_n\}$ is unbounded.
4. When
$$
\beta_1 < 1 \quad \text{and} \quad \gamma_2 > 1,
$$

   the component $\{x_n\}$ of every solution $\{x_n, y_n\}$ is bounded. Also, for some initial
   conditions, the component $\{y_n\}$ is unbounded.
5. When
$$
\beta_1 = 1 \quad \text{and} \quad \gamma_2 < \gamma_1 + 1,
$$

   the solution $\{x_n, y_n\}$ of System (13.88), converges as follows:

$$
\lim_{n \to \infty} x_n = \gamma_1 \quad \text{and} \quad \lim_{n \to \infty} y_n = \frac{\beta_2 \gamma_1}{1 + \gamma_1 - \gamma_2}.
$$

13 Global Convergence in Difference Equations

6. When

$$\beta_1 = 1 \quad \text{and} \quad \gamma_2 \geq \gamma_1 + 1,$$

the solution $\{x_n, y_n\}$ of System (13.88), converges as follows:

$$\lim_{n \to \infty} x_n = \gamma_1 \quad \text{and} \quad \lim_{n \to \infty} y_n = \infty.$$

7. When

$$\gamma_2 = 1 \quad \text{and} \quad \beta_1 < \beta_2 + 1,$$

the solution $\{x_n, y_n\}$ of System (13.88), converges as follows:

$$\lim_{n \to \infty} x_n = \frac{\beta_2 \gamma_1}{1 + \beta_2 - \beta_1} \quad \text{and} \quad \lim_{n \to \infty} y_n = \beta_2.$$

8. When

$$\gamma_2 = 1 \quad \text{and} \quad \beta_1 \geq \beta_2 + 1,$$

the solution $\{x_n, y_n\}$ of System (13.88), converges as follows:

$$\lim_{n \to \infty} x_n = \infty \quad \text{and} \quad \lim_{n \to \infty} y_n = \beta_2.$$

References

1. Amleh, A.M., Camouzis, E., Ladas, G.: On the dynamics of a rational difference equation, Part 1. Int. J. Differ. Equ. **3**, 1–35 (2008)
2. Amleh, A.M., Camouzis, E., Ladas, G.: On the dynamics of a rational difference equation, Part 2. Int. J. Differ. Equ. (2009)
3. Agarwal, R.: Difference Equations and Inequalities, Theory, Methods and Applications. Marcel Dekker, New York (1992)
4. Alligood, K.T., Sauer, T., Yorke, J.A.: CHAOS An Introduction to Dynamical Systems. Springer, New York, Berlin, Heidelberg, Tokyo (1997)
5. Camouzis, E.: Global convergence in a periodically forced rational equation. J. Differ. Equ. Appl. **14**, 1011–1033 (2008)
6. Camouzis, E.: Boundedness of solutions of a rational system of difference equations. Proceedings of the 14th International Conference on Difference Equations and Applications held in Instanbul, Turkey, July 21–25, 2008. Ugur-Bahcesehir University Publishing Company, Istanbul, Turkey Difference Equations and Applications, ISBN 978-975-6437-80-3, 157–164 (2009)
7. Camouzis, E., Kulenović, M.R.S., Ladas, G., Merino, O.: Rational Systems in the Plane. J. Differ. Equ. Appl. **15**, 303–323 (2009)
8. Camouzis, E., Ladas, G.: When does local stability imply global attractivity in rational equations? J. Differ. Equ. Appl. **12**, 863–885 (2006)
9. Camouzis, E., Ladas, G.: Dynamics of Third-Order Rational Difference Equations; With Open Problems and Conjectures. Chapman & Hall/CRC (2007)
10. Camouzis, E., Ladas, G.: Global results on rational systems in the plane, I. J. Differ. Equ. Appl. (2009)
11. Clark, C.W.: A delayed recruitment model of population dynamics with an application to baleen whale populations. J. Math. Biol. **3**, 381–391 (1976)

12. Clark, D., Kulenović, M.R.S.: On a Coupled System of Rational Difference Equations. Comput. Math. Appl. **43**, 849–867 (2002)
13. Clark, D., Kulenović, M.R.S., Selgrade, J.F.: Global asymptotic behavior of a two dimensional difference equation modelling competition. Nonlinear Anal., TMA **52**, 1765–1776 (2003)
14. Clark, C.A., Kulenović, M.R.S., Selgrade, J.F.: On a system of rational difference equations. J. Differ. Equ. Appl. **11**, 565–580 (2005)
15. Cushing, J.M., Henson, S.M., Blackburn, C.C.: Multiple mixed-type attractors in competition models. J. Biol. Dyn. **1**(4), 347–362 (2007)
16. Cushing, J.M., Levarge, S., Chitnis, N., Henson, S.M.: Some discrete competition models and the competitive exclusion principle. J. Differ. Equ. Appl. **10**, 1139–1152 (2004)
17. Dancer, E., Hess, P.: Stability of fixed points for order preserving discrete-time dynamical systems. J. Reine Angew Math. **419**, 125–139 (1991)
18. Elaydi, S.: An Introduction to Difference Equations, 2nd edn. Springer, New York (1999)
19. El-Metwally, H.A., Grove, E.A., Ladas, G.: A global convergence result with applications to periodic solutions. J. Math. Anal. Appl. **245**, 161–170 (2000)
20. El-Metwally, H.A., Grove, E.A., Ladas, G., Voulov, H.D.: On the global attractivity and the periodic character of some difference equations. J. Differ. Equ. Appl. **7**, 837–850 (2001)
21. Grove, E.A., Ladas, G.: Periodicities in Nonlinear Difference Equations. Chapman & Hall/CRC (2005)
22. Hautus, M.L.J., Bolis, T.S.: Solution to problem E2721. Am. Math. Monthly **86**, 865–866 (1979)
23. Franke, J.E., Hoag, J.T., Ladas, G.: Global attractivity and convergence to a two-cycle in a difference equation. J. Differ. Equ. Appl. **5**, 203–210 (1999)
24. Hale, J., Kocak, H.: Dynamics and Bifurcations. Springer, New York, Berlin, Heidelberg, Toyko (1991)
25. Hirsch, M., Smith, H.L.: Monotone Dynamical Systems. In: Canada, A., Drabek, P., Fonda, A. (eds.) Handbook of Differential Equations, vol. 2. Elsevier (2005)
26. Hirsch, M., Smith, H.L.: Monotone Maps: A Review, J. Differ. Equ. Appl. **11**, 379–398 (2005)
27. Hsu, S.B., Smith, H.L., Waltman, P.: Competitive exclusion and coexistence for competitive systems on ordered Banach spaces. Trans. Am. Math. Soc. **348**, 4083–4094 (1996)
28. Karakostas, G.: Convergence of a difference equation via the full limiting sequences method. Differ. Equ. Dyn. Syst. **1**, 289–294 (1993)
29. Kelley, W.G., Peterson, A.C.: Difference Equations. Academic, New York (1991)
30. Kocic, V.L., Ladas, G.: Global Behavior of Nonlinear Difference Equations of Higher Order with Applications. Kluwer, Dordrecht (1993)
31. Kuang, Y., Cushing, J.M.: Global stability in a nonlinear difference-delay equation model of flour beetle population growth. J. Differ. Equ. Appl. **2**, 31–37 (1996)
32. Kulenović, M.R.S., Ladas, G.: Dynamics of Second Order Rational Difference Equations; with Open Problems and Conjectures. Chapman & Hall/CRC (2001)
33. Kulenović, M.R.S., Ladas, G., Overdeep, C.B.: On the dynamics of $x_{n+1} = p_n + \dfrac{x_{n-1}}{x_n}$ with a period-two coefficient. J. Differ. Equ. Appl. **10**, 905–914 (2004)
34. Kulenović, M.R.S., Ladas, G., Sizer, W.S.: On the recursive sequence $x_{n+1} = \dfrac{\alpha x_n + \beta x_{n-1}}{\gamma x_n + C x_{n-1}}$. Math. Sci. Res. Hot-Line **2**(5), 1–16 (1998)
35. Kulenović, M.R.S., Merino, O.: Discrete Dynamical Systems and Difference Equations with Mathematica. Chapman and Hall/CRC, Boca Raton, London (2002)
36. Kuruklis, S.A., Ladas, G.: Oscillation and global attractivity in a discrete delay logistic model. Quart. Appl. Math. **L**, 227–233 (1992)
37. Martelli, M.: Introduction to Discrete Dynamical Systems and Chaos. Wiley, New York (1999)
38. May, R.M.: Host-Parasitoid system in patchy environments. A Phenomenological model. J. Animal Ecol. **47**, 833–843 (1978)
39. deMottoni, P., Schiaffino, A.: Competition systems with periodic coefficients: a geometric approach. J. Math. Biol. **11**, 319–335 (1982)
40. Pielou, E.C.: An Introduction to Mathematical Ecology. Wiley-Interscience, New York (1969)

13 Global Convergence in Difference Equations

41. Pielou, E.C.: Population and Community Ecology. Gordon & Breach, New York (1974)
42. Philos, Ch.G., Purnaras, I.K., Sficas, Y.G.: Global attractivity in a nonlinear difference equation. Appl. Math. Comput. **62**, 249–258 (1994)
43. Robinson, C.: Dynamical Systems, Stability, Symbolic Dynamics, and Chaos. CRC, Boca Raton, FL (1995)
44. Sedaghat, H.: Nonlinear Difference Equations, Theory and Applications to Social Science Models. Kluwer, Dordrecht (2003)
45. Smale, S.: On the differential equations of species in competition. J. Math. Biol. **3**, 5–7 (1976)
46. Smith, H.L.: Invariant curves for mappings. SIAM J. Math. Anal. **17**, 1053–1067 (1986)
47. Smith, H.L.: Periodic competitive differential equations and the discrete dynamics of competitive maps. J. Differ. Equ. **64**, 165–194 (1986)
48. Smith, H.L.: Planar competitive and cooperative difference equations. J. Differ. Equ. Appl. **3**, 335–357 (1998)

Chapter 14
Networks Synchronizability, Local Dynamics and Some Graph Invariants

Acilina Caneco, Sara Fernandes, Clara Grácio, and J. Leonel Rocha

Abstract The synchronization of a network depends on a number of factors, including the strength of the coupling, the connection topology and the dynamical behaviour of the individual units. In the first part of this work, we fix the network topology and obtain the synchronization interval in terms of the Lyapounov exponents for piecewise linear expanding maps in the nodes. If these piecewise linear maps have the same slope $\pm s$ everywhere, we get a relation between synchronizability and the topological entropy. In the second part of this paper we fix the dynamics in the individual nodes and address our work to the study of the effect of clustering and conductance in the amplitude of the synchronization interval.

14.1 Introduction

A network with a complex topology is mathematically described by a graph [3]. Classical random graphs were studied by Paul Erdős and Alfréd Rényi in the late 1950s. Examples of such networks include communication and transportation networks, neural and social interaction networks [1, 13, 20]. Although features of these networks have been studied in the past, it was only recently that massive amount of

---

A. Caneco (✉)
Mathematics Unit, Instituto Superior de Engenharia de Lisboa, Lisbon, Portugal
and
CIMA-UE, Universidade de Évora, Évora, Portugal
e-mail: acilina@deetc.isel.ipl.pt

S. Fernandes and C. Grácio
Department of Mathematics, Universidade de Évora, Évora, Portugal
and
CIMA-UE, Universidade de Évora, Évora, Portugal
e-mail: saf@uevora.pt, mgracio@uevora.pt

J.L. Rocha
Mathematics Unit, Instituto Superior de Engenharia de Lisboa, Lisbon, Portugal
e-mail: jrocha@deq.isel.ipl.pt

M.M. Peixoto et al. (eds.), *Dynamics, Games and Science I*, Springer Proceedings in Mathematics 1, DOI 10.1007/978-3-642-11456-4_14,
© Springer-Verlag Berlin Heidelberg 2011

data are available and computer processing is possible to more easily analyze the behaviour of these networks and verify the applicability of the proposed models. In 1998 Watts and Strogatz [20] proposed the new small-world model to describe many of real networks around us and in 1999 Barabási and Albert [1] proposed the new scale-free model based on preferential attachment. These models reflect the natural and man-made networks more accurately than the classical random graph model. This preferential attachment characteristic leads to the formation of clusters, the nodes with more links have a greater probability of getting new ones.

One of the most important subjects under investigation is the network synchronizability [5, 10, 13, 20]. Pecora and Carroll [16] derived the master stability method. Li and Chen [13] derived synchronization and desynchronization values for the coupling parameter in terms of the network topology and the maximum Lyapunov exponent of the individual chaotic nodes.

In this work we address the study of network synchronizability in two approaches. One is fixing the connection topology and vary the local dynamics in the nodes and the other is consider the local dynamic fixed and vary the structure of the connections. To the first approach, we study, in Sect. 14.2, the synchronization interval considering fixed the network connection topology, for different kinds of local dynamics. Supposing in the nodes, identical piecewise linear expanding maps, with different slopes in each subinterval, we obtained, in Sect. 14.2.1, the synchronization interval in terms of the Lyapunov exponents of these maps. As a particular case, we derive, in Sect. 14.2.2, the synchronization interval in terms of the topological entropy for piecewise linear maps with slope $\pm s$ everywhere and we proved that the synchronizability decreases if the local topological entropy increases. Considering identical chaotic symmetric bimodal maps in the nodes of the network, we express, in Sect. 14.2.3, the synchronization interval in terms of one single critical point of the map. In the second part of this work, we study the network synchronization as a function of the connection topology, fixing the local dynamics. We try to understand the relation of some graph invariants with the spectrum of the Laplacian matrix [2, 3]. We can find a great number of formulas relating some graph invariants with the eigenvalues characterizing the synchronization interval, λ_2 and λ_N, but none, as far as we know, for a relation between these eigenvalues and the clustering formation, neither for a relation between the conductance and the clustering. So, in Sect. 14.3, we perform experimental evaluations, that deepens the understanding of the effect of these quantities on the network synchronizability.

14.2 Network Synchronizability and Local Dynamics

Mathematically, networks are described by graphs and the theory of dynamical networks is a combination of graph theory and nonlinear dynamics. From the point of view of dynamical systems, we have a global dynamical system emerging from the interactions between the local dynamics of the individual elements and graph theory then analyzes the coupling structure.

14 Networks Synchronizability, Local Dynamics and Some Graph Invariants   223

A graph is a set $G = (V(G), E(G))$ where $V(G)$ is a nonempty set of N vertices or nodes and $E(G)$ is the set of m edges or links e_{ij} that connect two vertices v_i and v_j [3]. If the graph is weighted, for each pair of vertices (v_i, v_j) we set a non negative weight a_{ij} such that $a_{ij} = 0$ if the vertices v_i and v_j are not connected. If the graph is not weighted, $a_{ij} = 1$ if v_i and v_j are connected and $a_{ij} = 0$ if the vertices v_i and v_j are not connected. If the graph is not directed, which is the case that we will study, $a_{ij} = a_{ji}$. The matrix $A = A(G) = [a_{ij}]$, where v_i, $v_j \in V(G)$, is called the adjacency matrix. The degree of a node v_i is the number of edges incident on it and is represented by k_i, that is, $k_i = \sum_{i=1}^{i=N} a_{ij}$. The degree distribution is the probability $P(k)$ that a randomly selected node has exactly k edges.

Consider the diagonal matrix $D = D(G) = [d_{ij}]$, where $d_{ii} = k_i$. We call Laplacian matrix to $L = D - A$. The matrix L acts in $\ell^2(V)$ and sometimes is called Kirchhoff matrix of the graph, due to its role in the Kirchhoff Matrix-Tree Theorem. The eigenvalues of L are all real and non negatives and are contained in the interval $[0, \min\{N, 2\Delta\}]$, where Δ is the maximum degree of the vertices. The spectrum of L may be ordered, $\lambda_1 = 0 \leq \lambda_2 \leq \cdots \leq \lambda_N$. The second eigenvalue λ_2 is know as the algebraic connectivity or Fiedler value and plays a special role in the graph theory. As much larger λ_2 is, more difficult is to separate the graph in disconnected parts. The graph is connected if and only if $\lambda_2 \neq 0$. In fact, the multiplicity of the null eigenvalue λ_1 is equal to the number of connected components of the graph. As we will see later, as bigger is λ_2, more easily the network synchronizes.

Consider a network of N identical chaotic dynamical oscillators, described by a connected, unoriented graph, with no loops and no multiple edges. In each node the dynamics of the oscillators is defined by $\dot{x}_i = f(x_i)$, with $f : \mathbb{R}^n \to \mathbb{R}^n$ and $x_i \in \mathbb{R}^n$ the state variables of the node i.

The state equations of this network are

$$\dot{x}_i = f(x_i) + c \sum_{\substack{j=1 \\ j \neq i}}^{N} a_{ij} \Gamma (x_j - x_i), \qquad \text{with } i = 1, 2, \ldots, N \qquad (14.1)$$

where $c > 0$ is the coupling parameter, $A = [a_{ij}]$ is the adjacency matrix and $\Gamma = diag(1, 1, \ldots 1)$. Equation (14.1) can be rewritten as

$$\dot{x}_i = f(x_i) + c \sum_{j=1}^{N} l_{ij} x_j, \qquad \text{with } i = 1, 2, \ldots, N. \qquad (14.2)$$

where $L = (l_{ij}) = D - A$ is the Laplacian matrix or coupling configuration of the network. The network (14.2) achieves asymptotical synchronization if

$x_1(t) = x_2(t) = \ldots = x_N(t) \underset{t\to\infty}{\to} e(t)$, where $e(t)$ is a solution of an isolate node (equilibrium point, periodic orbit or chaotic attractor), satisfying $\dot{e}(t) = f(e(t))$.

14.2.1 Synchronization Interval for Piecewise Linear Maps with Different Slopes

Consider the network (14.2) with identical chaotic nodes. In this work we will consider the network in the discretized form

$$x_i(k+1) = f(x_i(k)) + c \sum_{j=1}^{N} l_{ij} f(x_j(k)), \quad \text{with } i = 1, 2, \ldots, N. \quad (14.3)$$

Let $0 = \lambda_1 < \lambda_2 \le \ldots \le \lambda_N$ be the eigenvalues of the coupling matrix L and let $h_{\max}$ be the Lyapunov exponent of each individual n-dimensional node. If $c > \frac{h_{\max}}{|\lambda_2|}$, then the synchronized states are exponentially stable [13]. We may fix f, the local dynamic in each node and vary the connection topology, L, or fix L and vary f.

In a previous work [4] we have considered, in each node of the network, piecewise linear maps with slope $\pm s$, motivated by the fact that every m-modal map $f: I = [a,b] \subset \mathbb{R} \to I$, with growth rate s and positive topological entropy $h_{top}(f)$ ($\log s = h_{top}(f)$) is, by Theorem 7.4 from Milnor and Thurston [15] and Parry, topologically semi-conjugated to a $p + 1$ piecewise linear map T, with $p \le m$, defined on the interval $J = [0, 1]$, with slope $\pm s$ everywhere and $h_{top}(T) = h_{top}(f) = \log s$. As a generalization, we will consider now a network having in each node a piecewise linear map with different slopes in each subinterval [19].

Let $I \subset \mathbb{R}$ be a compact interval and $f : I \to I$, $f = (f_1, \ldots, f_n)$, a piecewise linear expanding map. The set of n laps of f defines a partition $\mathscr{P}_I = \{I_1, \ldots, I_n\}$ of the interval I. Let a_i, with $i = 1, \ldots, n + 1$, be the discontinuity points and the turning points of the map f. Considering the orbits of these points, we define a Markov partition $\mathscr{P}'_I$ of I. The orbit of each point a_i is defined by

$$o(a_i) = \left\{ x_k^{(i)} : x_k^{(i)} = f^k(a_i), k \in \mathbb{N}_0 \right\}.$$

To simplify the presentation, we consider the points a_1 and a_{n+1} as fixed points of the map f. Let $\{b_1, \ldots, b_{m+1}\} = \{o(a_i) : i = 1, \ldots, n + 1\}$ be the set of the points correspondent to the orbits of the discontinuity points and the turning points, ordered on the interval I. This set allows us to define a subpartition $\mathscr{P}'_I$ of $\mathscr{P}_I$. The subpartition $\mathscr{P}'_I = \{J_1, \ldots, J_m\}$ with $m \ge n$ determines a Markov partition of the interval I. Note that f determines $\mathscr{P}'_I$ uniquely, but the converse is not true. The piecewise linear expanding map f induces a subshift of finite type whose $m \times m$ transition matrix $A = [a_{ij}]$ is defined by

$$a_{ij} = \begin{cases} 1 \text{ if } f\left(int\, J_j\right) \supseteq int\, J_i \\ 0 \text{ otherwise.} \end{cases}$$

We denote this subshift by (Σ_A, σ), where σ is the shift map on $\Sigma_m^{\mathbb{N}}$ defined by $\sigma(x_1 x_2 ...) = x_2 x_3 ...$, where $\Sigma_m = \{1, ..., m\}$ correspondent to the m states of the subshift. The topological entropy of (Σ_A, σ) is $\log \lambda_A$, where λ_A is the spectral radius of the transition matrix A. In [18], using the signal of f', is defined a weighted matrix which describes the transitions between the points $b_1, ..., b_{m+1}$. The relation between the transition matrix and the weighted matrix is established in [18]. This result allows us to compute the topological entropy of a subshift of finite type by a different method. See [18] and [19], for the relation between the kneading data associated to f and the topological entropy. To the subshift (Σ_A, σ) and the Markov partition $\mathcal{P}'_1$, we associated a Lipschitz function $\phi : I \to \mathbb{R}$, [19], defined by

$$\phi = \{\phi_i : J_i \to \mathbb{R}, \ 1 \le i \le m\}$$

where

$$\phi_i(x) = -\beta\, \varphi_i(x) \text{ and } \varphi_i(x) = \log\left|f'_i(x)\right|, \text{ with } \beta \in \mathbb{R}.$$

This function is a weight for the dynamical system associated to (Σ_A, σ) depending on the parameter β. Let $\mathcal{L}^1(I)$ be the set of all Lebesgue integrable functions on I. The transfer operator $L_\phi : \mathcal{L}^1(I) \to \mathcal{L}^1(I)$, associated with f and $\mathcal{P}'_1$,

$$\left(L_{\phi_j} g\right)(x) = \sum_{j=1}^{m} \exp \phi_j \left(f_j^{-1}(x)\right) g\left(f_j^{-1}(x)\right) \chi_{f\left(int\, I_j\right)}$$

where χ_{I_j} is the characteristic function of I_j. In this section we consider a class of one-dimensional transformations that are piecewise linear Markov transformations. Consequently, the transfer operator has the following matrix representation. Let $\mathcal{C}$ be the class of all functions that are piecewise constant on the partition $\mathcal{P}'_1$. The transfer operator has the following matrix representation

$$L_\phi\, g = Q_\beta\, \pi_g$$

with $g \in \mathcal{C}$, where $\mathcal{C}$ is the class of all functions that are piecewise constants, on the partition $\mathcal{P}'_1$ and $\pi_g = (\pi_1, ..., \pi_m)^T$. If D_β is the diagonal matrix defined by

$$D_\beta = (\exp \phi_1, ..., \exp \phi_m)$$

and A is the transition matrix, then the matrix Q_β is the $m \times m$ weighted transition matrix defined by

$$Q_\beta = A\, D_\beta = [q_{ij}] \text{ where } q_{ij} = \frac{a_{ij}}{\left|f'_j\right|^\beta}.$$

By the Ruelle–Perron–Frobenius Theorem there exist $\lambda_\beta > 0$ and $v_\beta \in \mathscr{C}$, with $v_\beta (J_i) > 0$ for all $1 \le i \le m$, such that v_β is the eigenvector of Q_β with largest eigenvalue λ_β, i.e., $Q_\beta v_\beta = \lambda_\beta v_\beta$. This eigenvector of Q_β is used to construct a transition probability matrix, as follows.

Let $\overline{\mu}$ be a measure with support in $\mathscr{P}'_I$, then we denote the adjoint operator of L_ϕ by L^*_ϕ, which is defined by a bounded linear map on measures, i.e.,

$$\left(L^*_\phi \overline{\mu} \right) (g) = \overline{\mu} \left(L_\phi g \right).$$

Note that the adjoint operator L^*_ϕ is represented by the matrix Q^T_β. The eigenvalues of the matrices Q_β and Q^T_β are the same. For the m-dimensional vector space $\mathscr{P}'_I$, we consider two bases

$$\mathscr{B} = \{e_1, \dots, e_m\} \text{ and } \mathscr{B}' = \{e'_1, \dots, e'_m\}.$$

The set of vectors in $\mathscr{B}$ are defined by the column vector $e_j = (0, \dots, 0, 1, 0, \dots, 0)^T$ where 1 is in the jth-position. These vectors correspond to the intervals of the Markov partition. On the other hand, the set of vectors in $\mathscr{B}'$ are defined by $e'_j = (0, \dots, 0, v_j, 0, \dots, 0)^T$, which correspond to the coordinates of the vector v_β. If M_β is the matrix which describes the change from the basis $\mathscr{B}'$ to the basis $\mathscr{B}$, then we define a new matrix, the $m \times m$ matrix

$$R_\beta = M^{-1}_\beta Q_\beta M_\beta = [r_{ij}] \text{ where } r_{ij} = q_{ij} \frac{v_j}{v_i} \text{ with } r_{ij} \ge 0.$$

The matrix R_β is the matrix representation of L_ϕ, with respect to the basis $\mathscr{B}'$. As the matrices Q_β and R_β are similar, the largest eigenvalue λ_β of these matrices is the same. Define a $m \times m$ stochastic matrix $S_\beta = [s_{ij}]$ where

$$s_{ij} = \frac{r_{ij}}{\lambda_\beta} \text{ with } s_{ij} \ge 0 \text{ and } \sum_{j=1}^m s_{ij} = 1.$$

The transpose matrix S^T_β corresponds to a modified or normalized transfer operator, with respect to the basis $\mathscr{B}'$. Let $u'_\beta = (u'_1, \dots, u'_m)$ be the left eigenvector and $v'_\beta = (v'_1, \dots, v'_m)$ be the strictly positive right eigenvector of the matrix R_β. The probability vector $p_\beta = (p_1, \dots, p_m)$ is defined by

$$p_i = \frac{u'_i v'_i}{\sum\limits_{i=1}^m u'_i v'_i}, \text{ such that } \sum_{i=1}^m p_i s_{ij} = p_j \text{ and } \sum_{i=1}^m p_i = 1.$$

This vector defines the unique f-invariant equilibrium state for

14 Networks Synchronizability, Local Dynamics and Some Graph Invariants

$$\phi = -\beta \log |f'(x)|.$$

Note that, if we consider $\mu^* = (u_1 v_1, \ldots, u_m v_m)$, up to a multiplicative constant, then $\mu^* = p_\beta$, see [19] and references therein.

The stochastic matrix S_β and the probability vector p_β allow us to define an invariant probability measure μ_β on the repeller, depending on the parameter β. Let Σ_A and Σ_m be as above. Define μ_β on the semi-algebra of measurable intervals by

$$\mu_\beta \left(\{(x_i)_{i \in \mathbb{N}} \in \Sigma_A : x_q = a_1, \ldots, x_{q+k-1} = a_k, \text{ with } a_k \in \Sigma_m \text{ and } k \in \mathbb{N}\}\right)$$

$$= p_{a_1} S_{a_1 a_2} S_{a_2 a_3} \cdots S_{a_{k-1} a_k}.$$

We call this measure the weighted Markov measure, associated to the weighted one-sided (p_β, S_β)-Markov shift, supported by the repeller. This invariant measure gives nonvanishing probabilities only for the trajectories staying in the repeller.

Lemma 14.1. *The weighted one-sided (p_β, S_β)-Markov shift has Lyapunov exponent $\chi_{\mu_\beta}(f)$ with respect to the measure μ_β, given by*

$$\chi_{\mu_\beta}(f) = \sum_{i=1}^{m} p_i \log \left(|f_i'|\right) \tag{14.4}$$

where the derivative f_i' is evaluated on the interval J_i of the partition $\mathscr{P}_I'$.

See [19] for the proof. Attending that, there exist a unique invariant probability measure μ_1, $(\beta = 1)$ for the map f, generated by the absolutely continuous conditionally invariant measure $\overline{\mu}$ (see Proposition 2 of [19]), we may express the network synchronizability interval in terms of the Lyapunov exponent $\chi_{\mu_1}(f)$.

Theorem 14.1. *Consider the network (14.3), having a connection topology given by some coupling matrix L with eigenvalues $0 = \lambda_1 < \lambda_2 \leq \ldots \leq \lambda_N$ and in each node identical piecewise linear expanding maps f with Lyapunov exponent $\chi_{\mu_1}(f)$ given by (14.4). Then, the network synchronizes if the coupling parameter c verifies*

$$\frac{1 - e^{-\chi_{\mu_1}(f)}}{\lambda_2} < c < \frac{1 + e^{-\chi_{\mu_1}(f)}}{\lambda_N}.$$

Proof. By Lemma 14.1, we have for piecewise linear expanding maps that $h_{\max}(f) = \chi_{\mu_1}(f)$. So, the desired result follows from [13]. $\square$

As an immediate consequence, we have:

Corollary 14.1. *Consider the network (14.3), having a connection topology given by some coupling matrix L with eigenvalues $0 = \lambda_1 < \lambda_2 \leq \ldots \leq \lambda_N$ and in each node identical piecewise linear expanding maps f with Lyapunov exponent $\chi_{\mu_1}(f)$ given by (14.4). Then:*

1. *If the local dynamic is fixed, then, the amplitude of the synchronization interval increases as $\lambda_N - \lambda_2$ decreases, that is, if the network topology is closer to a fully connected graph.*
2. *If $\lambda_N = \lambda_2$, then there exist some non empty synchronization interval, for all $\chi_{\mu_1}(f)$.*
3. *If the connection topology, given by some coupling matrix L, is fixed then, the amplitude of the synchronization interval decreases, as the Lyapunov exponent $\chi_{\mu_1}(f)$ in each node, grows.*

Proof. 1. If the local dynamic is fixed, then $\chi_{\mu_1}(f)$ is constant. The amplitude of the synchronization interval $\dfrac{1 + e^{-\chi_{\mu_1}(f)}}{\lambda_N} - \dfrac{1 - e^{-\chi_{\mu_1}(f)}}{\lambda_2}$ increases if $\lambda_N - \lambda_2$ decreases.

2. If $\lambda_N = \lambda_2$, then the superior bound of the synchronization interval is always larger than the inferior bound, so the synchronization interval is non empty.
3. If the connection topology is fixed, then λ_2 and λ_N are fixed, so the synchronization interval depends only on the local dynamics expressed by $\chi_{\mu_1}(f)$. If the Lyapunov exponent in each node grows, then, $1 + e^{-\chi_{\mu_1}(f)}$ decreases and $1 - e^{-\chi_{\mu_1}(f)}$ increases, so the synchronization interval is smaller. $\quad\square$

14.2.2 Synchronization Interval in Terms of the Topological Entropy

In this section, we consider a particular case, the network described by (14.3), where the function f, representing the local dynamics, is a piecewise linear expanding map with slope $\pm s$ everywhere. Then, the Lyapunov exponent of f is given by

$$h_{\max} = \chi_{\mu_1}(f) = \sum_{i=1}^{m} p_i \log\left(|f_i'|\right) = \sum_{i=1}^{m} p_i \log s = \log s \sum_{i=1}^{m} p_i = \log s.$$
(14.5)

Now we are in position to state that the synchronization interval of these networks may be expressed in terms of the topological entropy $h_{top}(f)$ of the piecewise linear maps, representing the local dynamics in each node.

Corollary 14.2. *Consider the network (14.3), having a connection topology given by some coupling matrix L with eigenvalues $0 = \lambda_1 < \lambda_2 \leq ... \leq \lambda_N$ and in each node identical chaotic piecewise linear map with slope $\pm s$ everywhere. Then, the network synchronizes if the coupling parameter c verifies*

$$\frac{1 - e^{-h_{top}(f)}}{\lambda_2} < c < \frac{1 + e^{-h_{top}(f)}}{\lambda_N}.$$
(14.6)

14 Networks Synchronizability, Local Dynamics and Some Graph Invariants 229

Proof. Attending to (14.5) we have $h_{max} = \log s$ for piecewise linear maps with slope $\pm s$ everywhere. In this case, we have by [15] $\log s = h_{top}(f)$, so the desired result follows from Theorem 14.1. $\qquad\square$

Remark 14.1. As a consequence of Corollaries 14.1 and 14.2, the synchronization interval amplitude decreases as the topological entropy in each node grows, if the connection topology of the network is fixed.

14.2.3 Monotonicity of the Synchronization Interval Amplitude for Symmetric Bimodal Maps

Symbolic dynamics is an important tool to study piecewise monotone interval maps and can be applied to study some characteristics of graphs, see [11] and [12]. Consider a compact interval $I \subset \mathbb{R}$ and a m-modal map $f : I \rightarrow I$, i.e., the map f is piecewise monotone, with m critical points and $m + 1$ subintervals of monotonicity. Suppose $I = [c_0, c_{m+1}]$ can be divided by a partition of points $\mathscr{P} = \{c_0, c_1, \ldots, c_{m+1}\}$ in a finite number of subintervals

$$I_1 = [c_0, c_1], \quad I_2 = [c_1, c_2], \quad \ldots, \quad I_{m+1} = [c_m, c_{m+1}],$$

in such a way that the restriction of f to each interval I_j is strictly monotone, either increasing or decreasing. Assuming that each interval I_j is the maximal interval where the function is strictly monotone, these intervals I_j are called laps of f and the number of distinct laps is called the lap number ℓ, of f. In the interior of the interval I the points $c_1, c_2, \ldots, c_m$, are local minimum or local maximum of f and are called turning or critical points of the function. The limit of the n-root of the lap number of f^n (where f^n denotes the composition of f with itself n times) is called the growth number of f, and its logarithm is the topological entropy

$$s = \lim_{n \to \infty} \sqrt[n]{\ell(f^n)} \quad \text{and} \quad h_{top}(f) = \log s.$$

The kneading matrix associated to a bimodal maps $f_{a,b}$ is (see [11]),

$$N(t) = \begin{pmatrix} N_{11}(t) & N_{12}(t) & N_{13}(t) \\ N_{21}(t) & N_{22}(t) & N_{23}(t) \end{pmatrix}.$$

From $N(t)$ we compute the determinants $D_j(t) = det\,\widehat{N}(t)$, where $\widehat{N}(t)$ is obtained from $N(t)$ removing the j column ($j = 1, 2, 3$), and if the map is decreasing in the first lap, the kneading determinant $D(t)$ verifies the relationship

$$D(t) = \frac{D_1(t)}{1+t} = -\frac{D_2(t)}{1-t} = \frac{D_3(t)}{1+t}. \tag{14.7}$$

The topological entropy of the map $f_{a,b}$ is

$$h_{top}(f) = \log s, \text{ with } s = \frac{1}{t^*} \text{ and } t^* = \min\{t \in [0,1] : D(t) = 0\}.$$

For a bimodal map, the symbolic sequences corresponding to periodic orbits of the critical points c_1, with period p, and c_2 with period k, may be written as

$$\left((AS_1 S_2 ... S_{p-1})^\infty, (BQ_1 ... Q_{k-1})^\infty\right).$$

In that case we have two periodic orbits, but in other cases of bimodal maps we have a single periodic orbit, of period $p + k$, that passes through both critical points (bistable case), for which we write only $(AP_1 ... P_{p-1} BQ_1 ... Q_{k-1})^\infty$.

Now, we will study this two cases, with the additional condition that $p = k$ and the symbols Q_j are the symmetric of the symbols P_j, in the sense of the following definition, [6].

Definition 14.1. Let $f_{a,b}$ be a symmetric bimodal map for which the periodic kneading sequence, with period $q = 2p$, is

$$S = \left((AS_1 S_2 ... S_{p-1})^\infty, (B\hat{S}_1 \hat{S}_2 ... \hat{S}_{p-1}^\infty)^\infty\right) \tag{14.8}$$

or

$$S = (AS_1 S_2 ... S_{p-1} B\hat{S}_1 \hat{S}_2 ... \hat{S}_{p-1})^\infty \tag{14.9}$$

with $S_1, S_2, ..., S_{p-1} \in \{L, M, R\}$, such that

$$\begin{cases} \hat{S}_j = R, \text{ if } S_j = L \\ \hat{S}_j = L \text{ if } S_j = R \quad \text{ and } \quad A \leftrightarrow B. \\ \hat{S}_j = M \text{ if } S_j = M \end{cases} \tag{14.10}$$

The bimodal map $f_{a,b}$ is called a symmetric bimodal map and (14.10) is called a mirror transformation for this map.

Lemma 14.2. Let $S = \left((AS_1 S_2 ... S_{p-1})^\infty, (B\hat{S}_1 \hat{S}_2 ... \hat{S}_{p-1}^\infty)^\infty\right)$ be a symmetric bimodal kneading sequence satisfying the mirror transformation (14.10). Then, we have

$$D_1(t) = D_3(t) = N_{12}(t)(N_{13}(t) - N_{11}(t)) \text{ and } D_2(t) = N_{13}^2(t) - N_{11}^2(t).$$

For the particular case of bistable symmetric bimodal maps we get a similar result.

Lemma 14.3. Let $S = (AS_1 S_2 ... S_{p-1} B\hat{S}_1 \hat{S}_2 ... \hat{S}_{p-1})^\infty$ be a symmetric bimodal kneading sequence with period $q = 2p$, satisfying the mirror transformation (14.10). Then, we have

$$D_1(t) = D_3(t) = N_{12}(t)(N_{13}(t) - N_{11}(t)) \text{ and } D_2(t) = N_{13}^2(t) - N_{11}^2(t).$$

14 Networks Synchronizability, Local Dynamics and Some Graph Invariants   231

With the above lemmas, we are in position to state the following result, concerning the topological entropy of symmetric bimodal maps.

Theorem 14.2. *Let $f_{a,b}$ be a symmetric bimodal map of type (14.8) or (14.9) or in the sense of Definition 14.1 and $D(t)$ the kneading determinant (14.7). Let $s = \frac{1}{t^*}$ be the growth number of $f_{a,b}$, where*

$$t^* = \min\{t \in [0, 1] : N_{13}(t) - N_{11}(t) = 0\}. \tag{14.11}$$

Then, the topological entropy of the map $f_{a,b}$ is $\log s$.

The above results establish that the dynamics of a symmetric bimodal map is determined by only one of its critical point, so it behaves like a unimodal map. See the proofs of the above results in [6]. Consider the network described by (14.3), where the function f, representing the local dynamics, is a piecewise linear map with slope $\pm s$ everywhere. From [15] we know that the topological entropy $h_{top} = \log s$, and attending to Corollary 14.2, the synchronization interval may be expressed in terms of the topological entropy. From Corollary 14.2 we know that the network (14.3) synchronizes, if the coupling parameter c verifies (14.6). As particular cases, one can consider unimodal or bimodal maps in each node of the network. As a consequence of Corollary 14.2 and Theorem 14.2, we may state the following result.

Corollary 14.3. *Consider the network (14.3), having a connection topology given by some coupling matrix L with eigenvalues $0 = \lambda_1 < \lambda_2 \leq ... \leq \lambda_N$ and in each node identical chaotic symmetric bimodal piecewise linear maps $f_{a,b}$, with slope $\pm s$ everywhere. Then, the amplitude of the synchronization interval increases, as t^* (14.11) of the map $f_{a,b}$ grows.*

We have established the synchronization interval in terms of the topological entropy of piecewise linear maps with slope $\pm s$ everywhere. As we know that, every m-modal map with positive topological entropy $h_{top}(f)$ is semiconjugated to a $p + 1$ piecewise linear map T, $(p \leq m)$, with slope $\pm s$ everywhere and $h_{top}(T) = h_{top}(f) = \log s$, we wonder if the above results are valid when in the nodes there are identical m-modal maps. The maps $f : I \to I$ and $T : J \to J$ are semiconjugated if there exist a function h continuous, monotone and onto, $h : I \to J$, such that $T \circ h = h \circ f$. If, in addition, h is a homeomorphism, then f and T are said topologically conjugated. We proved in [7] that in the case of topological conjugacy the synchronization of the two piecewise linear maps T implies the synchronization of the two conjugated m-modal maps f. Furthermore, by a result of [17], if f is topologically transitive, then the mentioned semiconjugacy is in fact a conjugacy. It remains an open problem to find weaker conditions for the relation between the synchronization of piecewise linear maps and the synchronization of the respectively semiconjugated piecewise monotone maps.

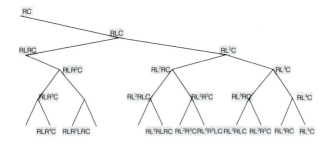

Fig. 14.1 Tree of unimodal admissible trajectories

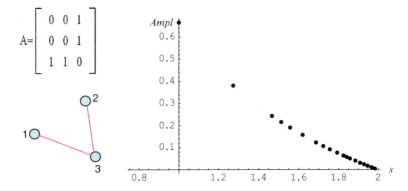

Fig. 14.2 The amplitude of the synchronization interval decreases as the entropy grows along a line in the tree of admissible unimodal maps

14.2.4 Numerical Simulations

Consider a network with a fixed topology and suppose that the nodes are identical chaotic maps. In [12] and [9] we may see a tree (see Fig. 14.1) of admissible kneading sequences associated with unimodal functions f, ordered by its topological entropy. We have computed the synchronization interval along lines where the topological entropy grows in this tree of admissible trajectories for the unimodal piecewise linear map. Our approach consists in determining for each one of the local discrete dynamical system for what values of the coupling parameter, c, there is synchronization, for different coupling scenarios.

Following any horizontal line of this tree, as we go from left to right, the amplitude of the interval decreases as the entropy grows. In Fig. 14.2 we display an example: fixing the graph topology with three nodes and two edges, expressed by the adjacency matrix A and varying the local dynamics on the nodes, following these lines on that tree of admissible trajectories associated with unimodal functions until period 6, we display the variation of the synchronization interval amplitude with the grows number. We confirm, as established in 3 of Corollary 14.1, that the

14 Networks Synchronizability, Local Dynamics and Some Graph Invariants 233

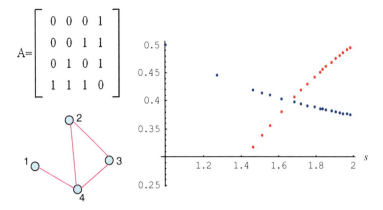

Fig. 14.3 This is a non fully-connected graph with 4 nodes and the synchronization interval does not exist for all s

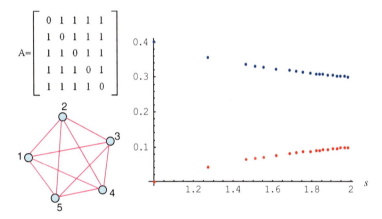

Fig. 14.4 This is a fully-connected graph with 5 nodes and the synchronization interval exist for all s

synchronization interval amplitude decreases when the topological entropy increases. Fixing some other topologies for the graph, described by the adjacency matrix A and varying the local map in the nodes, along the referred lines in the above tree, we display in Figs. 14.3 to 14.5 the amplitude of the synchronization interval. The blue dots represents the upper limit and the red dots represents the lower limit of the synchronization interval. So, the amplitude of the synchronization interval for each value of s is the vertical distance between each pair of one red point and one blue point. When the red dots are above the blue ones, there exist not any synchronization interval. Note that in Fig. 14.4 the graph is fully connected, so $\lambda_2 = \lambda_N$ [10] and the synchronization interval is non empty for all s, which confirms 2 in Corollary 14.1.

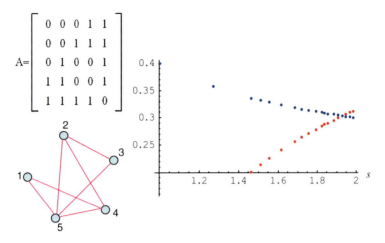

Fig. 14.5 This is a non fully-connected graph with 5 nodes and the synchronization interval does not exist for all s

14.3 Graph Invariants and Synchronization

Consider the graph $G = (V(G), E(G))$ associated to a network, as described above. Among all graph invariants we will pay special attention to the conductance and the quality of the clustering, [14].

There are several definitions of conductance of a graph, [8]. We will use the following

$$\Phi(G) = \min_{U \subset V} \frac{|E(U, V - U)|}{\min\{|E_1(U)|, |E_1(V - U)|\}},$$

where $|E(U, V - U)|$ is the number of edges from U to $V - U$ and $|E_1(U)|$ means the sum of degrees of vertices in U, [2]. A clustering of the graph G is a partition of the vertex set $C = \{C_1, C_2, \ldots, C_k\}$ and $C_i \subset V(G)$ are called clusters. We identify a cluster C_i with the induced subgraph of G, that is, the graph $G(C_i) = \{C_i, E(C_i)\}$. The set of intracluster edges is defined by $Intra(C) = \bigcup_{i=1}^{i=k} E(C_i)$ and the set of intercluster edges by $Inter(C) = \overline{Intra(C)} = E(G) \backslash Inter(C)$. The performance of a clustering should measure the quality of each cluster as well as the cost of the clustering. In [14] this bicriteria is based in a two-parameter definition of a (α, ε)-clustering, where α should measure the quality of the clusters and ε the cost of such partition, that is, the ratio of the intercluster edges to the total of edges in the graph.

Definition 14.2. We call a partition $C = \{C_1, C_2, \ldots, C_k\}$ of V an (α, ε)-clustering if:

1. The conductance of each cluster is at least α

$$\Phi(G(C_i)) \geq \alpha, \text{ for all } i = 1, \ldots, k;$$

14 Networks Synchronizability, Local Dynamics and Some Graph Invariants

2. The fraction of intercluster edges to the total of edges is at most ε

$$\frac{Inter(C)}{|E(C)|} \leq \varepsilon.$$

According to this definition the clustering is good if it maximizes α and minimizes ε. We introduce then a coefficient that accomplish both optimization problems.

Definition 14.3. For an (α, ε)-clustering C, define the performance of C by the ratio

$$R = \frac{\varepsilon}{\alpha}.$$

That means that a clustering is better if it has smaller R.

Our study conduces to the following conclusions:

1. A bad clustering implies a larger synchronization interval.
2. The conductance of the underlying graph is a good parameter to characterize the clustering and the synchronization.

We can see that the curve of R follows the curve of the eigenratio which, in turn, follows the curve of the conductance. Thus the three quantities characterize both the synchronizability and the quality of the clustering. These quantities vary in the opposite direction: better clustering implies poorer synchonization. For the network (14.3) the synchronization interval is (14.6). Fixing the dynamics f in the nodes, the synchronization interval will be as larger as much the eigenratio $r = \lambda_2/\lambda_N$ is bigger. Our conclusions are based in the observance of similar behavior of all three parameters: the conductance Φ, the eigenratio r and the performance of the (α, ε)-clustering R.

We perform an experimental evaluation to observe, for several clustering formation, the effect of the graph conductance and the performance of the (α, ε)-clustering on the synchronizability.

First we consider a complete graph with fifteen nodes and we delete edges, simulating the formation of a certain clustering, until obtain a disconnected graph. In Fig. 14.6 three steps of the process conducing to the partition

$$C = \{\{1, \ldots, 5\}, \{6, \ldots, 9\}, \{10, \ldots, 15\}\}.$$

In Fig. 14.7 are the results for the three parameters.

Next we have considered the formation of a clustering with four clusters $C = \{\{1, 2, 3\}, \{4, 5, 6, 7\}, \{8, 9, 10\}, \{11, 12, 13, 14, 15\}\}$ as can be seen in Fig. 14.8. In Fig. 14.9 are the results for the three parameters in the process of

$$C = \{\{1, 2, 3\}, \{4, 5, 6, 7\}, \{8, 9, 10\}, \{11, 12, 13, 14, 15\}\}$$

clustering formation.

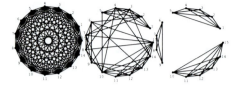

Fig. 14.6 Process of $C = \{\{1, \ldots, 5\}, \{6, \ldots, 9\}, \{10, \ldots, 15\}\}$ clustering formation

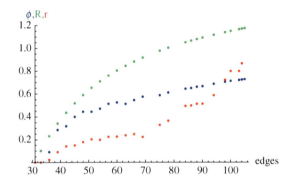

Fig. 14.7 Behavior of the quantities Φ, R and r for the process of $C = \{\{1, \ldots, 5\}, \{6, \ldots, 9\}, \{10, \ldots, 15\}\}$ clustering formation

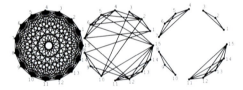

Fig. 14.8 Process of $C = \{\{1, 2, 3\}, \{4, 5, 6, 7\}, \{8, 9, 10\}, \{11, 12, 13, 14, 15\}\}$ clustering formation

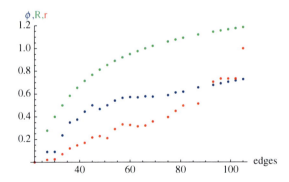

Fig. 14.9 Behavior of the quantities Φ, R and r for the process of $C = \{\{1, 2, 3\}, \{4, 5, 6, 7\}, \{8, 9, 10\}, \{11, 12, 13, 14, 15\}\}$ clustering formation

Fig. 14.10 $C = \{\{1,2,3\},\{4,5,6\},\{7,8,9\},\{10,11,12\},\{13,14,15\}\}$ clustering formation

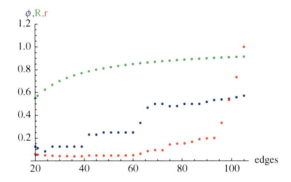

Fig. 14.11 Behavior of the quantities Φ, R and r for the process of $C = \{\{1,2,3\},\{4,5,6\},\{7,8,9\},\{10,11,12\},\{13,14,15\}\}$ clustering formation

Finally we have considered a clustering with 5 clusters $C = \{\{1,2,3\},\{4,5,6\},\{7,8,9\},\{10,11,12\},\{13,14,15\}\}$. See Fig. 14.10.

And we can observe in Fig. 14.11 similar results for the process of $C = \{\{1,2,3\},\{4,5,6\},\{7,8,9\},\{10,11,12\},\{13,14,15\}\}$ clustering formation.

We can observe that, as the cluster formation is more evident (as the ratio R increases), worse is the synchronizability. The same can be observed for the conductance.

Acknowledgements We would like to thank FCT (Portugal), Instituto Superior de Engenharia de Lisboa and CIMA-UE for having in part supported this work.

References

1. Barabási, A.L., Albert, R.: Emergence of scaling in random networks. Science **286**, 509–512 (1999)
2. Bezrukov, S.L.: Edge isoperimetric problems on graphs. Theor. Comput. Sci. **307**(3), 473–492 (2003)
3. Bollobás, B.: Random Graphs. New York (1985)
4. Caneco, A., Fernandes, S., Grácio, C., Rocha, J.L.: Symbolic dynamics and networks. Proc. Int. Workshop Nonlinear Maps Appl. (NOMA'07) 42–45 (2007)
5. Caneco, A., Grácio, C., Rocha, J.L.: Symbolic dynamics and chaotic synchronization in coupled Duffing oscillators. J. Nonlinear Math. Phys. **15**(3), 102–111 (2008)

6. Caneco, A., Grácio, C., Rocha, J.L.: Kneading theory analysis of the duffing equation. Chaos Solitons Fractals **42**(3), 1529–1538 (2009)
7. Caneco, A., Grácio, C., Rocha, J.L.: Topological entropy for the synchronization of piecewise linear and monotone maps. Coupled Duffing oscillators. Int. J. Bifurcat. Chaos **19**(11), 3855–3868 (2009)
8. Fernandes, S., Ramos, J.S.: Second eigenvalue of transition matrix associated to iterated maps. Chaos Solitons Fractals **31**(2), 316–326 (2007)
9. Fernandes, S., Jayachandran, S.: Conductance in discrete dynamical systems, discrete dynamics and difference equations. Proceedings of the Twelfth International Conference on Difference Equations and Applications. World Scientific (2010).
10. Feng, J., Jost, J., Qian, M.: Networks: From Biology to Theory. Springer (2007)
11. Lampreia, J.P., Ramos, J.S.: Symbolic dynamics for bimodal maps. Portugaliae Math. **54**(1), 1–18 (1997)
12. Lampreia, J.P., Severino, R., Ramos, J.S.: Irreducible complexity of iterated symmetric bimodal maps. Discrete Dyn. Nature Soc. **1**, 69–85 (2005)
13. Li, X., Chen, G.: Synchronization and desynchronization of complex dynamical networks: An engineering viewpoint. IEEE Trans. Circ. Syst. I **50**, 1381–1390 (2003)
14. Kannan, R., Vempala, S., Vetta, A.: On Clusterings: Good, bad and spectral. J. ACM **51**(3), 497–515 (2004)
15. Milnor, J., Thurston, W.: On iterated maps of the interval. Dynamical systems (College Park, MD, 1986–1987), 465–563. Lecture Notes in Math, vol. 1342. Springer, Berlin (1988)
16. Pecora, L., Carroll, T.: Master stability functions for synchronized coupled systems. Phys. Rev. Lett. **80**(10), 2109–2112 (1998)
17. Preston, C.: What you need to knead. Adv. Math. **78**, 192–252 (1989)
18. Rocha, J.L., Ramos, J.S.: IFS with overlaps and Hausdorff dimension. Iteration Theory (ECIT/02), Grazer Math. Ber. **346**, 355–376 (2004)
19. Rocha, J.L., Ramos, J.S.: Computing conditionally invariant measures and escape rates. Neural Parallel Sci. Comput. **14**(1), 97–114 (2006)
20. Watts, D.J., Strogatz, S.H.: Collective dynamics of small-world networks. Nature **393**, 440–442 (1998)

Chapter 15
Continuous Models for Genetic Evolution in Large Populations

Fabio A.C.C. Chalub and Max O. Souza

Abstract We consider a recently proposed generalisation of the Kimura equation, a Fokker–Planck type equation describing the evolution of $p(x,t)$, the probability of finding a fraction x of mutants at time t in a population evolving according to standard models in evolutionary biology. We present a detailed description of the solution, and we show that it naturally divides in two different time scales: the first determined by the drift (the natural selection), the second by the diffusion (the genetic drift).

15.1 Introduction

Fokker–Planck equations are ubiquitous in many fields of applied sciences. In this work, we will review a generalisation (introduced in [1,3]) of the celebrated Kimura equation [5] in evolutionary biology, and its underlying mathematical analysis. The results will be presented without proofs. The interested reader is redirected to the original references [2–4].

Let $x \in [0, 1]$ be a fraction of a given gene in a population divided in two types: the mutant and the wild type. We consider the following drift-diffusion (Fokker–Planck) equation.

$$\partial_t p = \frac{\epsilon}{2}\partial_x^2 \left(x(1 - x)p\right) - \partial_x \left(x(1 - x) \left(\psi^{(1)}(x) - \psi^{(2)}(x)\right) p\right), \qquad (15.1)$$

F.A.C.C. Chalub (✉)
Departamento de Matemática and Centro de Matemática e Aplicações, Universidade Nova de Lisboa, Quinta da Torre, 2829-516, Caparica, Portugal
e-mail: chalub@fct.unl.pt

M.O. Souza
Departamento de Matemática Aplicada, Universidade Federal Fluminense, R. Mário Santos Braga, s/n, 22240-920, Niterói, RJ, Brazil
e-mail: msouza@mat.uff.br

M.M. Peixoto et al. (eds.), *Dynamics, Games and Science I*, Springer Proceedings in Mathematics 1, DOI 10.1007/978-3-642-11456-4_15,
© Springer-Verlag Berlin Heidelberg 2011

where $\epsilon > 0$ and $\psi^{(1,2)} : [0, 1] \to \mathbb{R}$ are smooth functions, called in the biological literature the *fitness*, for the mutant and the wild-type, respectively.

Equation (15.1) is supplemented by the conservation laws

$$\partial_t \int_0^1 p(x)\,dx = 0\,, \qquad \partial_t \int_0^1 \phi(x)p(x)\,dx = 0\,, \tag{15.2}$$

where ϕ is the unique solution of

$$\phi'' + \psi\phi' = 0, \qquad \phi(0) = 0, \quad \phi(1) = 1,$$

where $\psi(x) = \psi^{(1)}(x) - \psi^{(2)}(x)$ is the relative fitness of the two species.

The main theorem is given by

Theorem 15.1. *For a given $p^0 \in \mathcal{BM}^+([0, 1])$, there exists a unique solution p to (15.1), with $p \in L^\infty\left([0, \infty); \mathcal{BM}^+([0, 1])\right)$ and such that p satisfies the conservations laws (15.2). The solution can be written as*

$$p(t, x) = q(t, x) + a(t)\delta_0 + b(t)\delta_1,$$

where δ_y denotes the singular measure supported at y, and $q \in C^\infty\left(\mathbb{R}^+; C^\infty((0, 1))\right)$ is a classical solution to (15.1). We also have that $a(t)$ and $b(t)$, belong to $C([0, \infty)) \cap C^\infty(\mathbb{R}^+)$. In particular, we have that

$$p \in C^\infty(\mathbb{R}^+; \mathcal{BM}^+([0, 1])) \cap C^\infty(\mathbb{R}^+; C^\infty((0, 1)))\,.$$

For large time, we have that $\lim_{t\to\infty} q(t, x) = 0$, uniformly, and that $a(t)$ and $b(t)$ are monotonically increasing functions such that:

$$a^\infty := \lim_{t\to\infty} a(t) = \int_0^1 (1 - \phi(x))p^0(x)\,dx \quad \text{and}$$

$$b^\infty := \lim_{t\to\infty} b(t) = \int_0^1 \phi(x)p^0(x)\,dx,$$

Moreover, we have that

$$\lim_{t\to\infty} p(t, \cdot) = a^\infty\delta_0 + b^\infty\delta_1,$$

with respect to the Radon metric. Finally, the convergence rate is exponential.

15.2 Early and Intermediate States

The Early and Intermediate states can be well identified, when $\epsilon \ll 1$.

When $\epsilon = 0$. The solution can be obtained by the characteristic method, and is given by

15 Continuous Models for Genetic Evolution in Large Populations 241

$$p_0(t, x) = p^0(\Phi_{-t}(x)) \left| \frac{\psi(\Phi_{-t}(x))}{\psi(x)} \right| \frac{\Phi_{-t}(x)(1 - \Phi_{-t}(x))}{x(1 - x)}, \tag{15.3}$$

where Φ_t is the flow map of the Replicator differential equation

$$\dot{X} = X(1 - X)\psi(X). \tag{15.4}$$

Now, let p_ϵ denote the solution to (15.1) for $\epsilon > 0$. Then we have

Theorem 15.2. *Let $p_0(t, x)$ and $p_\epsilon(t, x)$ be solutions to (15.1) with $\epsilon = 0$, and $\epsilon > 0$, respectively. Then there exist positive constants C_1 and C_2 such that for $0 \le t \le C_1 \epsilon$, we have*

$$\|p_0(t, \cdot) - p_\epsilon(t, \cdot)\|_\infty \le C_2 \epsilon.$$

Thus, in the very beginning the dynamics is essentially given by the Replicator, (15.4).

On the other hand, in certain cases there will be an intermediate dynamic behaviour as well. Let us suppose that ψ has a single zero, x_0, in $(0, 1)$, and that $\psi'(x_0) < 0$, so that the corresponding equilibrium is stable for the Replicator equation (15.4). Let λ_0 be as defined in Sect. 15.3, and let φ_0 be its associated eigenfunction. Furthermore, let $q(0)$ be the Fourier coefficient of the zeroth order term of q^0. We then have the following:

Theorem 15.3. *Assume that $\psi(x)$ as above. Then, there exist positive constants C_3, C_4 and C_5 such that, for $C_3 < t < C_4 \epsilon^{-1}$, we have*

$$\|q_\epsilon(t, x) - q(0)\varphi_0 e^{-\lambda_0 t}\|_\infty < C_5 \epsilon$$

15.3 Final States

In order to compute the final state, and the rate of convergence to it, given by solutions of (15.1) and (15.2), we introduce the following functional space:

$$\mathscr{D}_s = \left\{ \phi \in L^2([0, 1], \theta dx) \middle| \sum_{j=0}^{\infty} \widehat{\phi(j)} \lambda_j^{s/2} \varphi_j \in L_2([0, 1], \theta dx) \right\}, \quad \widehat{\phi(j)} = (\phi, \varphi_j),$$

with norm given by

$$\|\phi\|_s^2 = \sum_{j=0}^{\infty} \widehat{\phi(j)}^2 \lambda_j^s.$$

Representing the fact that, given a mutant gene, it will eventually be fixed or lost, we conclude that the final state should be a linear combination of Dirac-deltas supported on the boundaries of the domain. More precisely, we have that

Theorem 15.4. *Let ρ denote the Radon metric, and let*

$$p^\infty = a^\infty \delta_0 + b^\infty \delta_1.$$

Let p be the solution to (15.1) obeying conservation laws (15.2) with an initial condition with $q^0 \in \mathcal{BM}^+([0,1])i$. Let λ_0 be the smallest eigenvalue of

$$\begin{cases} -\varphi'' + \frac{1}{4}\left[2\psi' + \psi^2\right]\varphi = \lambda\theta(x)\varphi, \\ \varphi(0) = \varphi(1) = 0, \quad \theta(x) = \frac{1}{x(1-x)}. \end{cases} \tag{15.5}$$

Then, $\lambda_0 > 0$ and there exists a positive constant C, such that

$$\lim_{t\to\infty} e^{\lambda_0 t}\rho(p, p_\infty) \le C. \tag{15.6}$$

In addition, if we assume that

$$w^0 = x(1-x)e^{-\frac{1}{2}\int_0^x \psi(s)\,ds}q^0 \in \mathcal{BM}^+((0,1)) \cap \mathcal{D}_s, \quad s > 0,$$

then there is a constant $C_{0,s}$ such that

$$\rho(p, p_\infty) \le 2C_{0,s}\|w^0\|_s e^{-\lambda_0 t}. \tag{15.7}$$

In particular, (15.6) implies convergence in the Wasserstein metric.

References

1. Chalub, F.A.C.C., Souza, M.O.: The continuous limit of the Moran process and the diffusion of mutant genes in infinite populations. Arxiv preprint math/0602530 (2006)
2. Chalub, F.A.C.C., Souza, M.O.: A non-standard evolution problem arising in population genetics. Commun. Math. Sciences **7(2)**, 489–502 (2009)
3. Chalub, F.A.C.C., Souza, M.O.: From discrete to continuous evolution models: a unifying approach to drift-diffusion and replicator dynamics. Theor. Pop. Biol. **76(4)**, 268–277 (2009)
4. Chalub, F.A.C.C., Souza, M.O.: Multiscaling modelling in evolutionary dynamics. In preparation (2011)
5. Kimura, M.: On the probability of fixation of mutant genes in a population. Genetics **47**, 713–719 (1962)

Chapter 16
Forecasting of Yield Curves Using Local State Space Reconstruction

Eurico O. Covas and Filipe C. Mena

Abstract We examine models of yield curves through chaotic dynamical systems whose dynamics can be unfolded using non-linear embeddings in higher dimensions. We refine recent techniques used in the state space reconstruction of spatially extended time series in order to forecast the dynamics of yield curves. We use daily LIBOR GBP data (January 2007–June 2008) in order to perform forecasts over a one-month horizon. Our method outperforms random walk and other benchmark models on the basis of mean square forecast error criteria.

16.1 Introduction

Yield curve modelling and forecasting has an important role to play in the pricing and risk management of financial instruments. Although a number of past works have addressed the problem of modelling of yield curves, little attention has been paid to the actual forecast of yield curves as a function of both time and maturity.

Diebold and Li [5] consider US government bond data of Fama–Bliss at 17 values of maturity. They use a dynamical Nelson–Siegel model and report better one-year ahead forecasts than previous approaches including linear, random walk and auto regression (AR) models. The model of [5] has also been tested and used with other data (see e.g. [2]).

Stochastic models had been reported to perform well particularly at small forecast horizons, with the random walk models being famously hard to beat (see e.g. [2, 3, 5]). In particular [2] models the Nelson–Siegel parameters using martingales

---

E.O. Covas (✉)
HSBC Bank Plc, Hedging Software Development, 8 Canada Square, Canary Wharf, London, E14 5HQ, UK
e-mail: eurico.covas@hsbcgroup.com

F.C. Mena
Departamento de Matemática, Universidade do Minho, Campus de Gualtar, 4710 Braga, Portugal
e-mail: fmena@math.uminho.pt

M.M. Peixoto et al. (eds.), *Dynamics, Games and Science I*, Springer Proceedings in Mathematics 1, DOI 10.1007/978-3-642-11456-4_16,
© Springer-Verlag Berlin Heidelberg 2011

and finds, in general, lower forecasting errors than the ones using AR(1) estimates for the same parameters.

Here we propose an approach which is radically different from the previous models in this context. Our approach is based on the embedding reconstruction of local states from chaotic dynamical systems theory. The reconstruction preserves the dynamics under smooth coordinate transformations and the theorems Whitney [14], Takens–Mañé [9, 13] and Sauer et al. [12] guarantee the existence of the embedding. However, the theorems indicate an embedding dimension which is sufficient (but not necessary) and is often too high for computational purposes.

In order to find more appropriate dimensions for computations we use a method that results from a refinement of the method described by Parlitz and Merkwirth [11] for the reconstruction of spatiotemporal time series (STTS).

Here we shall take data using the LIBOR official fixing for LIBOR GBP as given by the British Banking Association (BBA). We use daily training sets (with more than 40 maturities) and compare our results with Diebold–Li, random walk, linear (see [5] for a summary of those), spline, AR and Hull–White type models.

16.2 The Parlitz–Merkwirth Method

We shall now describe the method of Parlitz–Merkwirth [11] to reconstruct local state data and the modifications we introduce for our problem.

Let $n = 1, \ldots, N$ and $m = 1, \ldots, M$. Consider a spatially extended time series s which can be represented by a $N \times M$ matrix with components $s_m^n \in \mathbb{R}$. To these components we will call *states* of the STTS.

Consider a number $2I \in \mathbb{N}$ of spatial neighbours of a given s_m^n and a number $J \in \mathbb{N}$ of temporal past neighbours to s_m^n.

For each s_m^n, we define the *super-state* vector $\mathbf{x}(s_m^n)$ with components given by s_m^n, its (nearest) $2I$ spatial neighbours and its J past temporal neighbours, and with K and L being the temporal and spatial lags, correspondingly to $\mathbf{x}(s_n^m)$ to be equal

$$\left\{ s_{m-IL}^n, \ldots, s_m^n, \ldots, s_{m+IL}^n, s_{m-IL}^{n-K}, \ldots, s_m^{n-K}, \ldots, s_{m+IL}^{n-K}, \ldots s_{m-IL}^{n-JK}, \ldots, \right.$$
$$\left. s_m^{n-JK}, \ldots, s_{m+IL}^{n-JK} \right\} \tag{16.1}$$

So the dimension of each $\mathbf{x}(s_n^m)$ is

$$d = (2I + 1)(J + 1)$$

Parlitz–Merkwirth use only rectangular regions for the spatiotemporal neighbours of the centre element s_j^i in order to reconstruct $\mathbf{x}_j^i$. We shall follow one of their suggestions to improve the method by using triangular regions (designated by *light-cones*) and extend this suggestion to semi-triangular regions, i.e. left light cones and right light cones according to their position relative to the central element s_j^i.

16 Forecasting of Yield Curves Using Local State Space Reconstruction   245

Now, for each pair (n, m), there is a one-to-one invertible map f^{-1}

$$f^{-1} : \mathbb{R} \to \mathbb{R}^d$$
$$s_m^n \to \mathbf{x}_m^n \equiv \mathbf{x}(s_m^n)$$

We wish now to approximate $f : \mathbb{R} \to \mathbb{R}^d$.

Take N_{train} time consecutive states s_m^n of s. With those, we form a training set A of super-states $\mathbf{x}_m^n$. We shall reconstruct a given super-state $\mathbf{x}_m^n \in A$ by using its closest past neighbours on A, separated in time by $\tau \in \mathbb{N}$.

We will then approximate f by some unknown function $F : \mathbb{R}^d \to \mathbb{R}$ such that

$$F(x_m^n) = s_m^{n+\tau}$$

There are several ways to do this. Parlitz–Merkwirth proposal is the following: Take a $\mathbf{x}_m^n$. Find the nearest neighbour to $\mathbf{x}_m^n$ on A, say $\mathbf{x}_j^i$, in the euclidean norm. Now, $s_j^{i+\tau}$, which is known a priori, will be an approximation $p_m^{n+\tau}$ for $s_m^{n+\tau}$ i.e.

$$F(x_j^i) \equiv s_j^{i+\tau} \approx s_m^{n+\tau}$$

where x_j^i is the nearest neighbour of x_m^n.

We shall introduce another modification here with respect to the original method by considering the nth nearest neighbouring super state to $\mathbf{x}_m^n$ and then averaging the n values of s obtained in this way in order to get $s_m^{n+\tau}$. This neighbourhood averaging shall also carry some weights according to the Euclidean distance to the central super state x_m^n.

Finally, we shall introduce a smoothing method after we get the data from the above (modified) Parlitz–Merkwirth procedure. For the smoothing we shall use polynomial least squares method (the polynomial order will depend on the case) and the Diebold–Li smoothing which is adapted to yield curve profiles. The embedding theorems do not state how to choose the space and time delays of the embedding. This can be done using the notion of *average mutual information* which has been used widely in the past (see e.g. [7]). This will give us an estimate for the values of the spatial and temporal delays K and L which can then be used to determining I and J (by minimizing the error of the forecasting) and therefore the embedding dimension. Mutual information estimates how measurements of s_j^i at time i are connected to measurements of s_j^{i+L} at time $i + L$.

In order to determine the embedding parameters I and J we shall use the *method of false neighbour detection* proposed by [8] and described in detail in [1].

16.3 Yield Curves and Forecasting Methods

Let $P(t, \tau)$ denote the price of a discount bond of period τ and $y(t, \tau)$ its corresponding yield to maturity. Then the *discount curve* is given by

$$P(t, \tau) = e^{-(\tau - t)\, y(t, \tau)}$$

The relationship between the yield and the forward rates $f(t, \tau)$ is

$$f(t, \tau) = -\frac{P'(t, \tau)}{P(t, \tau)}$$

so yields and forward rates are related by

$$y(t, \tau) = \frac{1}{\tau} \int_t^\tau f(t, u) du$$

which is usually called the *yield curve*. So, one can estimate a smooth discount curve and then use the above formulae to construct the yield curve. We shall use this procedure to get our *daily datasets*.

We shall now briefly describe several methods used in the past to forecast yield curves, namely linear, random walk, Diebold–Li and Hull–White type models.

16.3.1 Linear Models

We shall use two linear models: AR regressions and slope regressions.

On a slope regression the forecasted yield curve results from regressing changes on the curve slope

$$y(t + h/t, \tau) - y(t, \tau) = a_0 + a_1(y(t, \tau) - y(t, \tau_0))$$

for some constants a_0, a_1. This seems a quite naive approach but, surprisingly, gives good results for short-term predictions.

We consider AR type regression on yield levels or on the yield changes (see e.g. [5]). Although the results from these models have been reported to be worse than other models below (see e.g. [5]) we nevertheless test them with our data.

16.3.2 The Diebold–Li Model

Diebold and Li [5] use the well-known Nelson–Siegel [10] model

$$y(t + \frac{h}{t}, \tau) = \beta_1(t) + \beta_2(t) \left(\frac{1 - e^{-\lambda(t)\tau}}{\lambda(t)\tau} \right) + \beta_3(t) \left(\frac{1 - e^{-\lambda(t)\tau}}{\lambda(t)\tau} - e^{-\lambda(t)\tau} \right)$$

but take its β_i parameters as AR(1) processes

$$\beta_1(t + \frac{h}{t}) = c_i + \gamma_i \beta_i(t), \quad i = 1, 2, 3$$

where the coefficients c_i and γ_i are obtained from regressing $\beta_h^{t_f}(t)$ on $\beta_1^{t_f - h}(t)$ for each time step h.

16.3.3 A Hull–White Type Model

We shall use a G2 + + model whose detailed description together with the analogy to the Hull–White model can be found in Brigo and Mercurio [4].

The Hull–White approach is a single-factor, no-arbitrage yield curve model in which the short-term interest rate is the random factor or state variable. The model assumes that the probability distribution of short-term interest rate returns is normally distributed and subject to reversion to the long term mean.

Let $\varphi : [0, T^*] \to \mathbb{R}$ denote a deterministic function and r a short-rate stochastic process (under the risk-adjusted measure Q) given by

$$r(t) = x(t) + \varphi(t), \qquad r(0) = r_0, \ t > 0$$

where the process x satisfies

$$x(t) = -ax(t)dt + \sigma dW(t), \quad x(0) = 0 \tag{16.2}$$

and W is a Brownian motion and a, σ positive constants, defined as the mean reversion and the volatility of the stochastic process. The assumption is that the level of interest rates reverts to a long term level. Mean reversion is referred as the rate that the interest rates revert to its asymptotic level. Volatility refers to the standard deviation of the continuously compounded returns of interest rates within a specific time horizon, typically over one calendar year.

Being t real time and τ maturity time the price of a zero-coupon bond with unit face value is

$$P(t, \tau) = \frac{P^M(0, \tau)}{P^M(0, t)} e^{A(t, \tau)}$$

with $P^M(0, \tau)$ being the term structure of discount factors currently observed and

$$A(t, \tau) = \frac{1}{2}(V(t, \tau) - V(0, \tau) + V(0, t)) - \frac{1 - e^{-a(\tau-t)}}{a} x(t)$$

$$V(t, \tau) = \frac{\sigma^2}{a^2}\left(\tau - t + \frac{2}{a}e^{-a(\tau-t)} - \frac{1}{2a}e^{-2a(\tau-t)} - \frac{3}{2a}\right)$$

16.3.4 Naive Models

By naive we do not mean that they perform worse than other models. They are simplistic in their mathematical formulation but can be quite efficient particularly in short-term unstable periods. These include spline methods and random walk models. The random walk model here will be simply the statement of no change yield forecast

$$y(t + h/t, \tau) = y(t, \tau)$$

These models have been one of the most important competitors for forecasting yield curves.

16.4 Yield Curves Forecasting

This section presents the main results. We perform 30-day forecasts by 1 day concatenated steps. We use data from LIBOR GBP currency data from January 2007 to June 2008, as shown in Fig. 16.1. We compare our results with the benchmark models described in Sect. 16.3.

The random walk models have been hard to beat on the one-year forecast horizon [5] as well as on the one-month ahead forecast (see e.g. [3]). So, following previous authors we will show our error results normalized by the corresponding random walk error. The error measure we shall use is the mean square forecast error.

In order to test the scheme, we shall take a calibration data set and then forecast out-of-sample data which is however known. In this way, we can calculate the error of the prediction at each step by comparing with the true values, i.e. we shall compare the values of the approximation $p_j^{i+\tau}$ with the exact values $s_j^{i+\tau}$, $i \leq n$.

We calibrate our I, J, K, L parameters in (16.1) using mutual information minima for the K and L lags and the false neighbours method for finding the optimal embedding space and time dimensions I and J (see [11] and references therein). Using this calibration approach we get an optimal embedding for a time lag of $L = 7$ business days, a spatial lag of $K = 9$ months in maturity space, $I = 1$, i.e. one spatial neighbour on each side of s_m^n and a total of $J = 3$ time neighbours

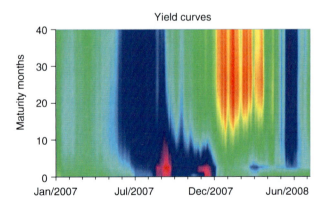

Fig. 16.1 Yield curves from Jan 2007 to June 2008 for LIBOR GBP. We represent here zero coupon rates calculated from the original market data consisting of cash rates, futures, future convexity, swap rates, and turn of year rates. The conversion is done via forward linear interpolation using a conjugate gradient method. The brightest yellow–red colour represent high rates and the blue–green colours represent lower rates. The maximum rate represented here is 6.84% and the lowest is 4.66%. Notice the weak spatio–temporal interaction, where one has migrations across space and time

16 Forecasting of Yield Curves Using Local State Space Reconstruction

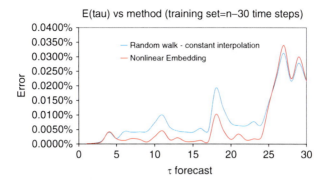

Fig. 16.2 Error of forecast comparison for 30 business days (UK business days) between random walk (i.e. constant forecast) method and our non-linear embedding method. The training set is the one from Fig. 16.1 minus τ time steps. The graph shows that the non-linear embedding method is superior at most forecast horizons, although always closely followed by the random walk method. Notice that at large time horizons, forecasts become meaningless, since the errors grows close to the variance of the full set values

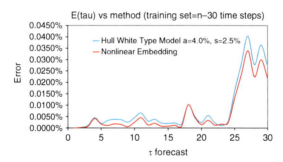

Fig. 16.3 Error of forecast comparison for 30 business days (UK business days) between Brigo–Mercurio formula and our non-linear embedding method. The training set is the one from Fig. 16.1 minus τ time steps. The graph shows that the non-linear embedding method is slightly superior at most forecast horizons, although always very closely followed by the random walk method. Notice that at large time horizons, forecasts become meaningless, since the errors grow close to the variance of the full set values

before s_m^n. Notice that these 4 parameters are not arbitrary or free parameters since the mutual information minimal approach and the false neighbours method should give the optimal embedding.

Our results (see Figs. 16.2 and 16.3) show that the spatiotemporal forecasting is better than both the random walk and Hull–White models.

Notice that for the Hull White model we have (see Fig. 16.4) also explored changing the free parameters mean reversion a and volatility σ, usually calibrated against swaption volatility surfaces. The calibration of the mean reversion and volatility can be arbitrary in the sense it can depend on fine tuning numerical parameters. Furthermore there are many competing methods for calibration, including the possibility of using past data as well as forward data, e.g. calibrating to historical

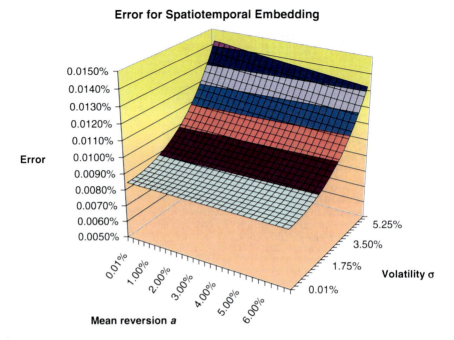

Fig. 16.4 Average error of forecast for 30 business days (UK business days) for the Brigo–Mercurio formula as a function of the short rate's mean reversion a and volatility σ. The training set is the one from Fig. 16.1 minus τ time steps. The plot indicates that the results obtained form the Brigo–Mercurio formula are never better than our non-linear embedding method for this particular training set at all reasonable possible combinations of the unobservable parameters of the model. These unobservable parameters are usually calibrated to swaption volatility matrices, representing the expected future changes of yield curves

time series. To demonstrate that the results are independent of calibration we calculated the forecasting error for the non-linear spatiotemporal method against the Hull–White model for a large range of mean reversions and volatilities. The results in Fig. 16.4 show that the non-linear spatiotemporal forecast is always better independently of the mean reversion and volatility used.

16.5 Conclusion and Future Research

We have implemented a detailed forecasting of yield curve using a novel method based on the work of Parlitz and Merkwith [11] for the reconstruction of spatiotemporal series. While temporal phase-space embeddings have been used extensively for one dimensional time series, the extension of the method to spatiotemporal signals has only recently been attempted. We have applied this deterministic method to forecasting yield curves and compared our results with stochastic methods which

16 Forecasting of Yield Curves Using Local State Space Reconstruction

are usually employed in the literature. A comparison with the random walk (representing the Martingales hypothesis) and Hull–White type models seems to show that the spatiotemporal reconstruction has some potential for applications.

We intend to improve the method using refinements of the embedding set and extend our work to larger spatiotemporal series as well as to other sets of currencies.

Acknowledgements We thank HSBC Bank Plc. for hospitality and for providing the data, Faisal Yousaf at HSBC for interesting discussions and for reading the draft paper and the hive team for providing the analytical software packages for parts of this work. FCM thanks CMAT (University of Minho) and HSBC for financial support. We also thank Christian Merkwirth for useful insights into the Parlitz–Merkwirth method and the numerical approach.

References

1. Abarbanel, H.D.I: Analysis of Observed Chaotic Data. Springer, Berlin (1996)
2. Bauer, M.D.: Forecasting the term structure using Nelson–Siegel factors and combined forecasts, Second Year Applied Project, Econ 220F, UCSD, University of California (2007)
3. Bowsher, C.G., Meeks, R.: The Dynamics of Economic Functions: Modelling and Forecasting the Yield Curve. J. Am. Stat. Assoc. **103**, 1419–1437 (2008)
4. Brigo, D., Mercurio, F.: Interest Rate Models – Theory and Practice, 2nd edn. Springer, Berlin (2006)
5. Diebold, F.X., Li, C.: Forecasting the term structure of government bond yields. J. Econom. **130**, 337–364 (2006)
6. Fraser, A.M., Swinney, H.L.: Phys. Rev. A **33**, 1134 (1980)
7. Kantz, H., Schreiber, T.: Nonlinear Time Series Analysis. Cambridge University Press, Cambridge (1997)
8. Kennel, M.R., Brown, R., Abarbanel, H.D.I.: Determining embedding dimension for phase-space reconstruction using a geometrical construction. Phys. Rev A **45**, 3403–3411 (1992)
9. Mañé, R.: On the dimension of the compact invariant sets of certain nonlinear maps, in *Dynamical Systems and Turbulence*, Lecture Notes in Mathematics, vol. 898, pp. 230–242. Springer, Berlin (1981)
10. Nelson, C., Siegel, A.: Parsimonious Modeling of Yield curves. J. Bus. **60**, 473–489 (1987)
11. Parlitz, U., Merkwirth, C.: Prediction of spatiotemporal time series based on reconstructed local states. Phys. Rev. Lett. **84**, 1890–1893 (2000)
12. Sauer, T., Yorke, J.A., Casdagli, M.: Embedology. J. Stat. Phys. **65**, 579–616 (1991)
13. Takens, F.: Detecting strange attractors in turbulence, in *Dynamical Systems and Turbulence*, Lecture Notes in Mathematics, vol. 898, 366–381. Springer, Berlin (1981)
14. Whitney, H.: Differentiable manifolds. Ann. Math. **37**, 645–680 (1936)

Chapter 17
KAM Theory as a Limit of Renormalization

João Lopes Dias

Abstract This is a brief survey of recent results on the KAM stability of quasiperiodic dynamics using renormalization of vector fields.

17.1 Introduction

For 30 years renormalization ideas have been used in the theory of dynamical systems. After the pioneering work of Feigenbaum [6] in the late 1970s, there has been a number of different applications of renormalization techniques. Its core concept is rescaling. That is, rescaling of space by zooming in a region in phase space; rescaling of time by considering a different time frame, as it takes longer to return to the region. Complicated dynamical behaviour can then turn out to be simpler in the new renormalized system. If by iterating the rescaling one gets convergence, it is a clear hint that the system looks the same in smaller scales. Moreover, if this self similarity is in some sense trivial, one can then hope to prove conjugacy between the systems.

The connection between KAM and renormalization theories has been realized for quite some time. Renormalization approach to KAM has several important advantages. First of all, it provides a unified setting which allows to deal with both the cases of smooth KAM-type invariant tori and non-smooth critical tori. Secondly, the proofs based on renormalizations are conceptually very simple and give a different perspective on the problem of small divisors. For the continuous-time situation, several KAM results for small-divisor problems in quasiperiodic motion have been obtained by studying the stability of trivial fixed sets of renormalization operators (cf. e.g. [7, 16, 19, 22, 23]). There was however a relevant restriction when dealing with multiple frequencies. Because renormalization methods rely fundamentally on the continued fractions expansion of the frequency vector, the lack

---

J.L. Dias
Departamento de Matemática and Cemapre, ISEG, Universidade Técnica de Lisboa, Rua do Quelhas 6, 1200-781 Lisbon, Portugal
e-mail: jldias@iseg.utl.pt

of a multidimensional version of continued fractions was the reason for failing to replicate KAM in its full generality. This limitation was recently overcome in [12] by adapting Lagarias' algorithm [21] and deriving estimates for multidimensional continued fractions (MCF) expansions of diophantine vectors.

In the case of Hamiltonian systems with two degrees of freedom MacKay proposed in the early 1980s a renormalization scheme for the construction of KAM invariant tori [27] (see also [29–31]). The scheme was realized for the construction of invariant curves for two-dimensional conservative maps of the cylinder. An important feature of MacKay's approach is the analysis of both smooth KAM invariant curves and so-called critical curves corresponding to critical values of a parameter above which invariant curves no longer exist. From the point of view of renormalization theory the KAM curves correspond to a trivial linear fixed point for the renormalization transformations, while critical curves give rise to very complicated fixed points with nontrivial critical behavior. MacKay's renormalization scheme was carried out only for a small class of Diophantine rotation numbers with periodic continued fraction expansion (such as the golden mean). Khanin and Sinai studied a different renormalization scheme for general Diophantine rotation numbers [14]. Both of the above early approaches were based on renormalization for maps or their generating functions. Essentially, the renormalization transformations are defined in the space of pairs of mappings which, being iterates of the same map, commute with each other. These commutativity conditions cause difficult technical problems, and led MacKay [28] to propose the development of alternative renormalization schemes acting directly on vector fields. The same idea was realized by Koch [16] who proves a KAM type result for analytic perturbations of linear Hamiltonians $H^0(x, y) = \omega \cdot y$, for frequencies ω which are eigenvectors of hyperbolic matrices in $SL(2, \mathbb{Z})$ with only one unstable direction. Notice that the set of such frequencies has zero Lebesgue measure and in the case $d = 2$ corresponds to vectors with a quadratic irrational slope. Further improvements and applications of Koch's techniques appeared in [1, 7, 17, 22, 23], emphasizing the connection between KAM and renormalization theories.

Other renormalization ideas have appeared in the context of the stability of invariant tori for nearly integrable Hamiltonian systems inspired by quantum field theory and an analogy with KAM theory (see e.g. [2, 8, 9] where it is used a graph representation of the invariant tori in terms of Feynman diagrams).

In Sect. 17.2 we describe a multidimensional continued fractions scheme, which gives estimates to be used in the renormalization. In the remaining sections we include examples of systems and several KAM-type results obtained by renormalization. In particular, in Sect. 17.3 we give a sketch of the proof of almost reducibility for analytic linear skew-product flows (cf. [5]). In Sect. 17.4 we study local conjugacy classes for toroidal flows. Finally, in Sect. 17.5 we present the main ideas for the renormalization proof of the "classical" KAM theorem in the context of Hamiltonian dynamics.

Throughout this text we denote by $Homeo(M)$ and $Diff^r(M)$, $r \in \mathbb{N} \cup \{\infty, \omega\}$, the set of homeomorphisms and C^r-diffeomorphisms on M. Moreover, we add a subscript 0 to distinguish the case of homotopic to the identity maps. Finally,

17 KAM Theory as a Limit of Renormalization — 255

$Vect^r(M)$ stands for the set of C^r-vector fields on M. Recall that the transformation of an arbitrary vector field X on a manifold M by $\psi \in Diff(M)$ is given by

$$\psi^* X = D\psi \circ \psi^{-1} \cdot X \circ \psi^{-1}. \tag{17.1}$$

17.2 Multidimensional Continued Fractions

An essential ingredient of the renormalization scheme is a continued fractions decomposition of vectors, relating the number-theoretical properties of the frequencies and the conjugacy smoothness.

In this section we present the multidimensional continued fractions algorithm introduced in [12] following ideas of Dani [4], Lagarias [21] and Kleinbock–Margulis [15]. In addition, we define the class of diophantine vectors from the properties of the continued fractions expansion.

17.2.1 Flow on Homogeneous Space

Denote by $G = SL(d, \mathbb{R})$, $\Gamma = SL(d, \mathbb{Z})$ and take a fundamental domain $\mathscr{F} \subset G$ of the homogeneous space $\Gamma \backslash G$ (the space of d-dimensional non-degenerate unimodular lattices). On $\mathscr{F}$ consider the flow:

$$\Phi^t : \mathscr{F} \to \mathscr{F}, \quad M \mapsto P(t)ME^t, \tag{17.2}$$

where

$$E^t = diag(e^{-t}, \ldots, e^{-t}, e^{(d-1)t}) \in G$$

and $P(t)$ is the unique family in Γ that keeps $\Phi^t M$ in $\mathscr{F}$ for every $t \geq 0$.

Let $\omega = (\alpha, 1) \in \mathbb{R}^d$. We are interested in the orbit under Φ^t of the matrix

$$M_\omega = \begin{pmatrix} I & \alpha \\ 0 & 1 \end{pmatrix}. \tag{17.3}$$

17.2.2 Growth of the Flow

Let the function $\delta : \Gamma \backslash G \to \mathbb{R}^+$ measuring the shortest vector in the lattice M be

$$\delta(M) = \inf_{k \in \mathbb{Z}^d \backslash \{0\}} \| {}^T k M \|, \tag{17.4}$$

where $\| \cdot \|$ stands for the ℓ_1-norm (in the following we will make use of the corresponding matrix norm taken as the usual operator norm). Notice that $\delta(\Phi^t M_\omega) = \delta(M_\omega E^t)$.

Proposition 17.1 ([12]). *There exist* $C_1, C_2 > 0$ *such that for all* $t \geq 0$

$$\|\Phi^t M_\omega\| \leq \frac{C_1}{\delta(\Phi^t M_\omega)^{d-1}} \quad and \quad \|(\Phi^t M_\omega)^{-1}\| \leq \frac{C_2}{\delta(\Phi^t M_\omega)}. \tag{17.5}$$

17.2.3 Stopping Times

Consider a sequence of times, called *stopping times*,

$$t_0 = 0 < t_1 < t_2 < \cdots \to +\infty \tag{17.6}$$

such that the matrices $P(t)$ in (17.2) satisfy

$$P_n := P(t_n) \neq P(t_{n-1}), \tag{17.7}$$

with $n \in \mathbb{N}$. We also set $P_0 = P(t_0) = I$. The sequence of matrices $P_n \in SL(d, \mathbb{Z})$ are the rational approximates of ω, called the *multidimensional continued fractions expansion*. In addition we define the transfer matrices

$$T_n = P_n P_{n-1}^{-1}, \quad n \in \mathbb{N}, \quad and \quad T_0 = I. \tag{17.8}$$

The flow of M_ω taken at the time sequence is thus the sequence of matrices

$$M_n := \Phi^{t_n} M_\omega = P_n M_\omega E^{t_n}. \tag{17.9}$$

Using some properties of the flow, the above can be decomposed (see [12]) into

$$M_n = \begin{pmatrix} I & \alpha_n \\ 0 & 1 \end{pmatrix} \begin{pmatrix} \Delta_n & 0 \\ {}^{\mathsf{T}}\beta_n & \gamma_n \end{pmatrix} \tag{17.10}$$

with γ_n being the d-th component of the vector $e^{(d-1)t_n} P_n \omega$.

Define $\omega_n = (\alpha_n, 1)$, $\omega_0 = \omega$ and, for $n \in \mathbb{N}$,

$$\omega_n = \eta_n T_n \omega_{n-1}, \tag{17.11}$$

where η_n is a normalization factor.

If $d = 2$ there exists a sequence of stopping times (called Hermitte critical times) that gives an accelerated version of the standard continued fractions of a number α [21].

17 KAM Theory as a Limit of Renormalization

17.2.4 Resonance Cone

Given resonance widths σ, i.e. a sequence $\sigma \colon \mathbb{N}_0 \to \mathbb{R}^+$, define the resonant cones to be

$$I_n^+ = \{k \in \mathbb{Z}^d \colon |k \cdot \omega_n| \le \sigma_n \|k\|\}. \tag{17.12}$$

In addition, let

$$A_n = \sup_{k \in I_n^+ \setminus \{0\}} \frac{\| \, {}^\top T_{n+1}^{-1} k \|}{\|k\|}. \tag{17.13}$$

Proposition 17.2 ([26]). *There is $c > 0$ such that for any $n \in \mathbb{N}_0$*

$$A_n \le c \, e^{-\delta t_{n+1}} \frac{\sigma_n e^{d \delta t_{n+1}} + 1}{\delta(M_n)^{d-1} \delta(M_{n+1})}, \tag{17.14}$$

where $\delta t_{n+1} = t_{n+1} - t_n$.

17.2.5 Diophantine Vectors

A vector $\omega \in \mathbb{R}^d$ is Diophantine with exponent $\beta \ge 0$ if there is a constant $C > 0$ such that

$$|\omega \cdot k| > \frac{C}{\|k\|^{d-1+\beta}}.$$

It is a well known fact that the sets $DC(\beta)$ of Diophantine vectors with exponent $\beta > 0$ are of full Lebesgue measure [3]. On the other hand, the set $DC(0)$ has zero Lebesgue measure. A vector is said to be diophantine if it belongs to $DC = \cup_{\beta \ge 0} DC(\beta)$. The next proposition gives us a complete characterization of diophantine vectors in terms of the behaviour of the flow Φ^t of M_ω.

Proposition 17.3 ([26]). *Let $\beta \ge 0$. Then, $\omega \in DC(\beta)$ iff there is $C' > 0$ such that*

$$\delta(\Phi^t M_\omega) > C' e^{-\theta t}, \quad t \ge 0,$$

with $\theta = \beta/(d + \beta)$.

Proposition 17.4 ([12]). *If $\omega \in DC(\beta)$, $\beta \ge 0$, there are constants $c_i > 0$ such that, for any stopping-time sequence $t \colon \mathbb{N}_0 \to \mathbb{R}$,*

$$\|M_n\| \le c_1 \exp[(d - 1)\theta t_n], \tag{17.15}$$
$$\|M_n^{-1}\| \le c_2 \exp(\theta t_n), \tag{17.16}$$
$$\|T_n\| \le c_5 \exp[(1 - \theta)\delta t_n + d \, \theta \, t_n], \tag{17.17}$$
$$\|T_n^{-1}\| \le c_6 \exp[(d - 1)(1 - \theta)\delta t_n + d \, \theta \, t_n], \tag{17.18}$$

where $\delta t_n = t_n - t_{n-1}$ and $\theta = \beta/(d + \beta)$.

Proposition 17.5 ([26]). *If $\omega \in DC(\beta)$, $\beta \geq 0$, then there is $c > 0$ such that for any $n \in \mathbb{N}_0$,*

$$A_n \leq c\, e^{-(1-\theta)\delta t_{n+1}+d\theta t_n} \left(\sigma_n e^{d\delta t_{n+1}} + 1 \right). \tag{17.19}$$

Possible choices are $t_n = c_1(1+\beta)^n$ and $\sigma_n = e^{-c_2(1+\beta)^n}$ with some $c_i > 0$.

Here we have only discussed the case of Diophantine frequency vectors. However, renormalization can be used for a larger class of vectors, cf. e.g. [10, 11, 18, 20, 24–26].

17.3 Almost Reducibility of Linear Skew-Product Flows

In this section we deal with skew-product vector fields, which are linear differential equations of dimension two, with quasiperiodic coefficients. This is a generalization of the classical Floquet theory. Our goal is to present the main ideas behind renormalization for this kind of dynamics. We present a sketch of a proof on almost reducibility of these systems.

17.3.1 Skew-Product Vector Fields

Consider the manifold $M = \mathbb{T}^d \times SL(2, \mathbb{R})$. Let $Vect_{sw}^r(M)$ be the set of C^r-vector fields on M of the form:

$$X(x, y) = (\omega, f(x)\, y), \quad (x, y) \in M, \tag{17.20}$$

where $\omega \in \mathbb{R}^d \setminus \{0\}$ and $f \in C^r(\mathbb{T}^d, SL(2, \mathbb{R}))$. We will use the following notation

$$X = (\omega, f).$$

Each element of $Vect_{sw}^r(M)$ generates a skew-product flow on M, i.e. a flow of the type

$$\phi^t(x, y) = (x + \omega t, \Phi^t(x)\, y),$$

where $\Phi^t \colon \mathbb{T}^d \to SL(2, \mathbb{R})$.

As we want to preserve the space $Vect_{sw}^r(M)$ under coordinate changes, we consider the set $Diff_{sw}^{r+1}(M)$ of

$$\psi(x, y) = (Tx, F(x)\, y), \quad (x, y) \in M, \tag{17.21}$$

where $F \in C^{r+1}(\mathbb{T}^d, SL(2, \mathbb{R}))$ and $T \in SL(d, \mathbb{Z})$ is a linear automorphism of the torus. For simplicity, we write

$$\psi = (T, F).$$

17 KAM Theory as a Limit of Renormalization

A vector field in the new coordinates is then given by the formula

$$\psi^* X(x, y) = (T\omega, L_\omega F(T^{-1}x) \cdot F(T^{-1}x)^{-1} y + Ad_{F(T^{-1}x)} f(T^{-1}x) \cdot y),$$

(17.22)

where

$$L_\omega = \omega \cdot D = \sum_i \omega_i \, \partial/\partial x_i$$

(17.23)

and $Ad_A b = AbA^{-1}$.

17.3.2 Fibered Rotation Number

Consider the natural projection $p \colon \mathbb{R}^2 \setminus \{0\} \to \mathbb{T}^1$ given by the argument of a vector. The fibered rotation number of the flow generated by $X = (\omega, f) \in Vect^0_{sw}(M)$ is defined to be

$$\rho(X) = \lim_{t \to +\infty} p\left(\frac{\int_0^t f \circ \phi^s(x, y) \, v \, ds}{t}\right)$$

for $(x, y) \in M$ and $v \in \mathbb{R}^2 \setminus \{0\}$. This measures the asymptotic frequency of rotation of the fiber flow in $\mathbb{R}^2$. We will be interested in vector fields for which ρ exists at any point and direction v.

17.3.3 Almost Reducibility

In some cases it is possible to find a diffeomorphism that simplifies X, in particular reducing it to a "constant" vector field. More precisely, we have the following definition.

1. $X \in Vect^r_{sw}(M)$ is C^s-*conjugated to* $Y \in Vect^r_{sw}(M)$ if there is $\psi \in Diff^s_{sw}(M)$ such that $\psi^* X = Y$.
2. X is C^s-*reducible* if its C^s-conjugacy class contains a vector field $Z = (\omega, u)$, with $u \in SL(2, \mathbb{R})$.
3. X is C^s-*almost reducible* if the closure of its C^s-conjugacy class contains a vector field $Z = (\omega, u)$, with $u \in SL(2, \mathbb{R})$.

Theorem 17.1. *Let $\omega \in \mathbb{R}^d$ be Diophantine and $C > 0$. There is $\epsilon > 0$ such that if $f \in C^\omega(\mathbb{T}^d, SL(2, \mathbb{R}))$ is ϵ-C^ω-close to constant and $|\rho(\omega, f)| < C$, then (ω, f) is C^ω-almost reducible.*

Notice that ϵ does not depend on the arithmetical properties of the rotation number. In the remaining part of this section we present the main steps towards the proof of the above theorem.

17.3.4 Non-Homotopic to the Identity Diffeomorphism

Given $m \in \mathbb{Z}^d$, we will also be interested in the following transformation of coordinates:

$$\psi_m = (I, R_m)$$

where $R_m \colon \mathbb{T}^d \to SO(2, \mathbb{R})$ is

$$R_m(x) = \begin{bmatrix} \cos(2\pi m \cdot x) & -\sin(2\pi m \cdot x) \\ \sin(2\pi m \cdot x) & \cos(2\pi m \cdot x) \end{bmatrix}.$$

The action on a vector field $X = (\omega, f)$ is given by

$$\psi_m^* X = \left(\omega, 2\pi m \cdot \omega \begin{bmatrix} 0 & -1 \\ 1 & 0 \end{bmatrix} + Ad_{R_m} f \right).$$

In particular, the rotation number is changed as

$$\rho(\psi_m^* X) = \rho(X) - \frac{1}{2} m \cdot \omega.$$

17.3.5 Lifts and Complexification

Let $r > 0$ and consider the domain

$$\mathcal{D}_r = \{ x \in \mathbb{C}^d \colon \| I m x \| < r/2\pi \} \tag{17.24}$$

for the norm $\|z\| = \sum_i |z_i|$ on $\mathbb{C}^d$. Take a real-analytic map

$$F \colon \mathcal{D}_r \to SL(2, \mathbb{C}),$$

$\mathbb{Z}^d$-periodic, on the form of the Fourier series

$$F(x) = \sum_{k \in \mathbb{Z}^d} F_k e^{2\pi i k \cdot x} \tag{17.25}$$

with $F_k \in SL(2, \mathbb{C})$. The Banach spaces $\mathscr{A}_r$ and $\mathscr{A}_r'$ are the subspaces such that the respective norms

$$\| F \|_r = \sum_{k \in \mathbb{Z}^d} \| F_k \| e^{r \|k\|}, \tag{17.26}$$

$$\| F \|_r' = \sum_{k \in \mathbb{Z}^d} (1 + 2\pi \|k\|) \| F_k \| e^{r \|k\|} \tag{17.27}$$

17 KAM Theory as a Limit of Renormalization

are finite. Here and in the following we use the matrix norm $\|A\| = \max_j \sum_i |A_{i,j}|$ for any square matrix A with entries $A_{i,j}$.

Similarly, define the space $\mathfrak{a}_r$ of real-analytic functions $\mathcal{D}_r \to SL(2,\mathbb{C})$, $\mathbb{Z}^d$-periodic and on the form of Fourier series, having the same type of bounded norm as (17.26). We are interested in vector fields that can be written as

$$X(x, y) = (\omega, f(x)\, y), \quad (x, y) \in \mathcal{D}_r \times SL(2, \mathbb{C}). \tag{17.28}$$

The space of such vector fields is denoted by V_r whenever f is in $\mathfrak{a}_r$. The norm on this space is defined to be

$$\|X\|_r = \|\omega\| + \|f\|_r. \tag{17.29}$$

17.3.6 Uniformization

The theorem below states the existence of a nonlinear change of coordinates isotopic to the identity that cancels the

$$I^- = \{k \in \mathbb{Z}^d : |k \cdot \omega| > \sigma \|k\|\}$$

Fourier modes of a sufficiently close to constant $X \in V_r$, with $\sigma > 0$. We are only eliminating the far from resonance modes, this way avoiding the complications usually related to small divisors.

Let $u \in SL(2, \mathbb{C})$ and

$$B_r(u, \varepsilon) = \{f \in \mathfrak{a}_r : \|f - u\|_r < \varepsilon\}$$

where

$$\varepsilon = \frac{C\sigma^2}{\|\omega\| + \|u\|}. \tag{17.30}$$

In order to simplify notations, here and in the following C stands for some positive universal constant, not necessarily the same.

Theorem 17.2. *Let $|\rho| \le \sigma/4$ and $u \in SL(2, \mathbb{R})$ with eigenvalues $\pm i\rho$. There is an analytic map $\mathfrak{U} \colon B_r(u, \varepsilon) \to \mathscr{A}'_r$ such that*

$$I^- \psi^*(X) = 0 \quad \text{where} \quad \psi = (I, \mathfrak{U}(f))$$

and

$$\|\mathfrak{U}(f) - I\|'_r \le \frac{C}{\sigma}\|I^- X\|_r$$
$$\|\psi^* X - \mathbb{E} X\|_r \le C\|(\mathbb{I} - \mathbb{E})X\|_r. \tag{17.31}$$

Moreover, $\mathfrak{U}(f) \colon \mathbb{R}^d \to SL(2, \mathbb{R})$.

17.3.7 Proof of Theorem 17.2

We now prove Theorem 17.2. □

17.3.7.1 Homotopy Method

The coordinate transformation ψ will be determined by some U in

$$\mathscr{B}_\delta = \left\{ U \in \mathbb{I}^- \mathscr{A}_r' \colon \|U - I\|_r' < \delta \right\},$$

for

$$\delta = C\varepsilon/\sigma < 1. \tag{17.32}$$

Define the operator

$$\begin{aligned} \mathscr{F} \colon \mathscr{B}_\delta &\to \mathbb{I}^- \mathscr{A}_r \\ U &\mapsto \mathbb{I}^- (L_\omega U \cdot U^{-1} + Ad_U f). \end{aligned} \tag{17.33}$$

If U is real-analytic, then $\mathscr{F}(U)$ is also real-analytic. The derivative of $\mathscr{F}$ at U is the linear map from $\mathbb{I}^- \mathscr{A}_r'$ to $\mathbb{I}^- \mathscr{A}_r$ given by

$$D\mathscr{F}(U) H = \mathbb{I}^- (L_\omega H - L_\omega U \cdot U^{-1} H - Ad_U f \cdot H + Hf) U^{-1}. \tag{17.34}$$

We want to find a solution of

$$\mathscr{F}(U_t) = (1 - t)\mathscr{F}(U_0), \tag{17.35}$$

with $0 \le t \le 1$ and initial condition $U_0 = I$. Differentiating the above equation with respect to t, we get

$$D\mathscr{F}(U_t)\frac{dU_t}{dt} = -\mathscr{F}(I). \tag{17.36}$$

Proposition 17.6. *There is $\delta > 0$ such that if $U \in \mathscr{B}_\delta$, then $D\mathscr{F}(U)^{-1} \colon \mathbb{I}^- \mathscr{A}_r \to \mathbb{I}^- \mathscr{A}_r'$ is bounded and*

$$\|D\mathscr{F}(U)^{-1}\| < \delta/\varepsilon.$$

From the above proposition (to be proved in Sect. 17.3.7.2) we integrate (17.36) with respect to t, obtaining the integral equation:

$$U_t = I - \int_0^t D\mathscr{F}(U_s)^{-1} \mathscr{F}(I) \, ds. \tag{17.37}$$

17 KAM Theory as a Limit of Renormalization

In order to check that $U_t \in \mathcal{B}_\delta$ for any $0 \le t \le 1$, we estimate its norm:

$$\begin{aligned}
\|U_t - I\|'_r &\le t \sup_{v \in \mathcal{B}_\delta} \|D\mathscr{F}(v)^{-1}\mathscr{F}(I)\|'_r \\
&\le t \sup_{v \in \mathcal{B}_\delta} \|D\mathscr{F}(v)^{-1}\| \, \|\mathbb{I}^- f\|_r < t\delta\|\mathbb{I}^- f\|_r/\varepsilon,
\end{aligned} \tag{17.38}$$

so, $\|U_t - I\|'_r < \delta$. Therefore, the solution of (17.35) exists in $\mathcal{B}_\delta$ and is given by (17.37). Moreover, if X is real-analytic, then U_t takes real values for real arguments.

In view of

$$\mathbb{I}^+(Ad_U f - u) = \mathbb{I}^+\left[(U - I)f(U^{-1} - I) + (U - I)\widetilde{f} + \widetilde{f}(U^{-1} - I) + \widetilde{f}\right], \tag{17.39}$$

where $\widetilde{f} = f - u$, we get

$$\begin{aligned}
\|U_t^* X - \mathbb{E} X\|_r &\le \|\mathbb{I}^+ L_\omega(U - I) \cdot (U^{-1} - I)\|_r + \|\mathbb{I}^+(Ad_U f - u)\|_r + (1 - t)\|\mathbb{I}^- f\|_r \\
&\le 2\|\omega\| \, \|U\|_r \, \|U - I\|_r\|U - I\|'_r + 2\|U\| \, (\|u\| + \|\widetilde{f}\|)\|U - I\|_r^2 \\
&\quad + \|\widetilde{f}\|_r(1 + 2\|U\|_r)\|U - I\|_r + \|\widetilde{f}\|_r + (1 - t)\|\mathbb{I}^- f\|_r \\
&\le (3 - t)\|\widetilde{f}\|_r.
\end{aligned} \tag{17.40}$$

Theorem 17.2 corresponds to the case $t = 1$.

17.3.7.2 Proof of Proposition 17.6

Lemma 17.1. $D\mathscr{F}(I)^{-1} : \mathbb{I}^- \mathscr{A}_r \to \mathbb{I}^- \mathscr{A}'_r$ is bounded and

$$\|D\mathscr{F}(I)^{-1}\| < \frac{5}{\sigma - 10\|(\mathbb{I} - \mathbb{E})f\|_r}. \tag{17.41}$$

Proof. Let $g = (\mathbb{I} - \mathbb{E})f$. From (17.34) one has

$$\begin{aligned}
D\mathscr{F}(I) H &= \mathbb{I}^-(L_\omega + ad_f) H \\
&= \left[\mathbb{I} + \mathbb{I}^- ad_g \left(L_\omega + ad_u\right)^{-1}\right] (L_\omega + ad_u) H,
\end{aligned} \tag{17.42}$$

where $ad_b A = Ab - bA$. Thus, the inverse of this operator, if it exists, is given by

$$D\mathscr{F}(I)^{-1} = (L_\omega + ad_u)^{-1} \left[\mathbb{I} + \mathbb{I}^- ad_g \left(L_\omega + ad_u\right)^{-1}\right]^{-1}. \tag{17.43}$$

By looking at the spectral properties of the operator $(2\pi i k \cdot \omega I + ad_u)$, with the spectrum of ad_u being $\{0, \pm 4\pi i \rho\}$, it is possible to write

$$(L_\omega + ad_u)H(x) = \sum_{k \in I^-} S\Lambda_k S^{-1} H_k e^{2\pi i k \cdot x} \tag{17.44}$$

where

$$\Lambda_k = (2\pi i) diag(k \cdot \omega, k \cdot \omega, k \cdot \omega + 2\rho, k \cdot \omega - 2\rho) \tag{17.45}$$

and

$$S = \begin{bmatrix} 0 & 1 & -1 & -1 \\ 1 & 0 & i & -i \\ -1 & 0 & i & -i \\ 0 & 1 & 1 & 1 \end{bmatrix}. \tag{17.46}$$

So, we have the linear map from $\mathbb{I}^- \mathscr{A}_r$ to $\mathbb{I}^- \mathscr{A}'_r$,

$$(L_\omega + ad_u)^{-1} F(x) = \sum_{k \in I^-} S\Lambda_k^{-1} S^{-1} F_k e^{2\pi i k \cdot x}. \tag{17.47}$$

Now, for $k \in I^-$,

$$\begin{aligned} \|(L_\omega + ad_u)^{-1} F\|'_r &\leq \frac{4}{2\pi} \sum_{k \in I^-} \frac{1 + 2\pi \|k\|}{|k \cdot \omega|} \|F_k\| e^{r\|k\|} \\ &< \frac{5}{\sigma} \|F\|_r. \end{aligned} \tag{17.48}$$

It is possible to bound from above the norm of ad_g by $2\|g\|_r$. Therefore,

$$\|\mathbb{I}^- ad_g (L_\omega + ad_u)^{-1}\| < \frac{10}{\sigma} \|g\|_r < 1,$$

and

$$\left\| \left[\mathbb{I} + \mathbb{I}^- ad_g (L_\omega + ad_u)^{-1}\right]^{-1} \right\| < \frac{1}{1 - \frac{10}{\sigma} \|g\|_r}.$$

The statement of the lemma is now immediate. $\qquad\square$

As r is constant, in the following we drop it from our notations.

Lemma 17.2. *Given $U \in \mathscr{B}_\delta$, the linear operator $D\mathscr{F}(U) - D\mathscr{F}(I)$ mapping $\mathbb{I}^- \mathscr{A}'_r$ into $\mathbb{I}^- \mathscr{A}_r$, is bounded and*

$$\|D\mathscr{F}(U) - D\mathscr{F}(I)\| < 2\|U\| \left[\|\omega\|(1 + 2\|U\|) + 2\|f\|(1 + \|U\| + \|U\|^2) \right] \|U - I\|. \tag{17.49}$$

Proof. In view of (17.34), we have

$$\begin{aligned} [D\mathscr{F}(U) - D\mathscr{F}(I)] H &= \mathbb{I}^- L_\omega H \cdot (U^{-1} - I) - L_\omega U \cdot U^{-1} H U^{-1} \\ &\quad + H f(U^{-1} - I) + f H - A d_U f \cdot H U^{-1}. \end{aligned} \tag{17.50}$$

17 KAM Theory as a Limit of Renormalization

It is possible to estimate the norms of the above terms by

$$
\begin{aligned}
\|L_\omega H \cdot (U^{-1} - I)\| &\le \|\omega\| \, \|U^{-1} - I\| \, \|H\|', \\
\|L_\omega U \cdot U^{-1} H U^{-1}\| &\le \|\omega\| \, \|U^{-1}\|^2 \|U - I\|' \|H\|, \\
\|H f (U^{-1} - I)\| &\le \|f\| \, \|U^{-1} - I\| \, \|H\|, \\
\|f H - A d_U f \cdot H U^{-1}\| &= \|f H (U^{-1} - I) + f(U^{-1} - I) H U^{-1} \\
&\quad + (U^{-1} - I) f U^{-1} H U^{-1}\| \\
&\le \|f\| \, (1 + \|U^{-1}\| + \|U^{-1}\|^2) \|U^{-1} - I\| \, \|H\|.
\end{aligned}
\tag{17.51}
$$

Finally, notice that $\|U^{-1} - I\| \le \|U^{-1}\| \, \|U - I\| \le 2\|U\| \, \|U - I\|$. $\qquad\square$

Proposition 17.6 now follows from $\|U\| < 1 + \delta$ and

$$
\begin{aligned}
\|D\mathscr{F}(U)^{-1}\| &\le \left(\|D\mathscr{F}(I)^{-1}\|^{-1} - \|D\mathscr{F}(U) - D\mathscr{F}(I)\| \right)^{-1} \\
&< \left\{ \sigma/5 - \varepsilon - 2\delta\|U\| \left[\|\omega\|(1 + 2\|U\|) + 2\|f\|(1 + \|U\| + \|U\|^2) \right] \right\}^{-1} \\
&< \{\sigma/5 - \varepsilon - C \, \delta(\|\omega\| + \|f\|)\}^{-1}.
\end{aligned}
\tag{17.52}
$$

Therefore, for δ and ε as in (17.32) and (17.30), respectively,

$$
\|D\mathscr{F}(U)^{-1}\| < \frac{\delta}{\varepsilon}.
\tag{17.53}
$$

17.3.8 Rescaling

The rescaling that we are interested comes from the continued fractions expansion of ω. That is, we want to use skew diffeomorphisms of the type (T_n, I) where T_n are as in Sect. 17.2. Futhermore, we rescale time by η_n.

Applying the rescaling to a vector field X with no I^- Fourier modes has the effect of improving its analyticity radius and thus C^ω-approximating X to a constant by a factor of order e^{-C/A_n}.

17.3.9 One-Step Renormalization Operator

The renormalization step is briefly summarized below.

1. Let $m = \arg\min\{|k \cdot \omega + 2\rho| : k \in I^-\}$. So, $\|m\| \le \frac{C|\rho|}{1 - \sigma}$ and $|\rho'| = |\rho - \frac{1}{2} m \cdot \omega| \le \sigma/4$.
2. Use ψ_m^* to obtain a vector field with rotation number ρ'. The C^ω-distance between the vector field and a constant will be increased by a factor $e^{C\|m\|}$.

266 J.L. Dias

3. Eliminate the modes in I^-.
4. Use the rescaling introduced in Sect. 17.3.8.

After one step, the vector field will get C^ω-closer to constant if the norm improvement by the rescaling overcomes the opposite effect by ψ_m^*. This indeed holds for ω Diophantine, using the bounds obtained at the end of Sect. 17.2.

Notice that $\|m\|$ only depends on $|\rho|$ and σ. On the other hand, σ is chosen at each step according to the arithmetic properties of ω.

By iterating the renormalization step we are able to show convergence to a trivial limit set, namely a set of constant vector fields. That is, the renormalization contracts a small neighbourhood around that set. We remark that the diameter of that neighbourhood does not depend on the arithmetical properties of ρ, but only on $|\rho|$.

17.4 Conjugacy Classes of Torus Translations

Consider the d-torus $\mathbb{T}^d$. We want to study flows on this manifold. Define the rotation vector of a flow ϕ^t at each $x \in \mathbb{T}^d$ to be the asymptotic direction of the corresponding orbit of the lift $\Phi^t(x)$ to the universal cover:

$$rot(\phi)(x) = \lim_{t \to +\infty} \frac{\Phi^t(x) - x}{t}, \tag{17.54}$$

if the limit exists. If the rotation vector exists at x for a flow ϕ^t generated by a vector field X on $\mathbb{T}^d$ (i.e. $\frac{d}{dt}\phi^t = X \circ \phi^t$), it is the time average of the vector field along the orbit:

$$rot(\phi)(x) = \lim_{t \to +\infty} \frac{1}{t} \int_0^t X \circ \phi^s(x) ds. \tag{17.55}$$

When the rotation vector exists for all $x \in \mathbb{T}^d$, the rotation set of ϕ is

$$rot(\phi) = \{rot(\phi)(x) : x \in \mathbb{T}^d\}. \tag{17.56}$$

Lemma 17.3 ([26]). *Let* $h \in Homeo_0(\mathbb{T}^d)$, $\lambda \neq 0$ *and* $T \in GL(d, \mathbb{Z})$. *If* $rot(\phi) \neq \emptyset$, *then*

$$rot(h^{-1} \circ \phi \circ h) = rot(\phi) \quad and \quad rot(T^{-1} \circ \phi^{\lambda \cdot} \circ T) = \lambda T^{-1} rot(\phi). \tag{17.57}$$

Proposition 17.7 ([26]). *Let* ϕ^t *be the flow generated by* $X \in Vect^0(\mathbb{T}^d)$ *and* $\omega \in \mathbb{R}^d$. *If* $rot\phi = \{\omega\}$, *then*

$$\|\mathbb{E}X - \omega\| \leq d \|X - \mathbb{E}X\|_{C^0}, \tag{17.58}$$

where $\mathbb{E}X = \int_{\mathbb{T}^d} X \, dm$ *and m denotes the Lebesgue measure on* $\mathbb{T}^d$.

17 KAM Theory as a Limit of Renormalization 267

We will be interested in vector fields generating flows that possess the same rotation vector for all orbits. Hence, for a vector field X we will write $rot\, X$ to mean the unique rotation vector associated to the flow generated by X.

The C^ω-conjugacy classes of constant Diophantine vector fields can be described, at least locally, by the rotation vector.

Theorem 17.3 ([26]). *Let $\omega \in \mathbb{R}^d$ be Diophantine. If X is a real-analytic vector field on $\mathbb{T}^d$ sufficiently C^ω-close to constant with unique rotation vector ω, then there exists $h \in Diff_0^\omega(\mathbb{T}^d)$ such that $h^*(X) = \omega$. The conjugacy h depends analytically on X.*

A proof of the above theorem is obtained by comparing the renormalization orbits of X and ω. They get close to each other exponentially fast, and from that we are able to construct an analytic conjugacy.

17.5 Invariant Tori in Phase Space

Let $B \subset \mathbb{R}^d$, $d \geq 2$, be an open set containing the origin, and let H^0 be a real-analytic Hamiltonian function

$$H^0(x, y) = \omega \cdot y + \frac{1}{2}\,{}^\mathsf{T}yQy, \qquad (x, y) \in \mathbb{T}^d \times B, \qquad (17.59)$$

with $\omega \in \mathbb{R}^d$ and a real symmetric $d \times d$ matrix Q. H^0 is said to be non-degenerate if $\det Q \neq 0$.

Theorem 17.4 ([13]). *Suppose H^0 is non-degenerate and ω is Diophantine. If H is a real-analytic Hamiltonian on $\mathbb{T}^d \times B$ sufficiently close to H^0, then the Hamiltonian flow of H leaves invariant a Lagrangian d-dim torus where it is analytically conjugated to the linear flow $\phi_t(x) = x + t\omega$ on $\mathbb{T}^d$, $t \geq 0$. The conjugacy depends analytically on H.*

Hamiltonian vector fields involve more complicated analysis than torus flows since there is extra dynamics on the action direction and we need to preserve the symplectic structure. Our goal is to find an analytic embedding $\mathbb{T}^d \to \mathbb{T}^d \times B$ that conjugates the Hamiltonian flow to the linear flow on the torus given by ω.

We do not work directly with vector fields, instead we renormalize Hamiltonian functions

$$H(x, y) = H^0(x, y) + F(x, y), \qquad (x, y) \in \mathbb{T}^d \times B$$

where F is a sufficiently small analytic perturbation. Using a rescaling of time we may assume that $\omega = (\alpha, 1)$. The perturbation F is decomposed in a Taylor-Fourier series

$$F(x, y) = \sum_{k,v} F_{k,v} y_1^{v_1} \dots y_d^{v_d} e^{2\pi i k \cdot x}$$

where the sum is taken over $k \in \mathbb{Z}^d$ and $v_i \in \mathbb{N} \cup \{0\}$. By the analyticity of F, its modes decay exponentially as $\|k\| \to +\infty$ for fixed v.

Renormalization is an iterative scheme that at each step produces a new Hamiltonian. Suppose that after the $(n-1)$-th step the Hamiltonian is of the form

$$H_{n-1}(x, y) = \omega_{n-1} \cdot y + \frac{1}{2} {}^\top y Q_{n-1} y + F_{n-1}(x, y) \qquad (17.60)$$

where Q_{n-1} is a symmetric matrix with non-zero determinant. Moreover, we assume that F_{n-1} only contains Taylor-Fourier modes in I_{n-1}^+, i.e. satisfying

$$|\omega_{n-1} \cdot k| \le \sigma_{n-1} \|k\| \quad \text{or} \quad \|v\| \ge \tau_{n-1} \|k\|$$

for some $\sigma_{n-1}, \tau_{n-1} > 0$. So, the n-th step is defined by the following operations:

1. Apply a linear operator corresponding to an affine symplectic transformation given by
$$(x, y) \mapsto (T_n^{-1} x, {}^\top T_n y + b_n)$$
for some fixed vector b_n.
2. Rescale the action in order to "zoom in" around the invariant torus.
3. Rescale time (energy) to ensure that the frequency vector is of the form $\omega_n = (\alpha_n, 1)$.
4. Eliminate the (irrelevant) constant mode of the Hamiltonian.
5. Eliminate all the modes outside the resonant cone I_n^+ (thus avoiding dealing with small divisors) by a close to the identity symplectomorphism.

The first transformation above has a conjugate action

$$k \mapsto {}^\top T_n^{-1} k.$$

It follows from the hyperbolicity of T_n that this transformation contracts I_{n-1}^+ if σ_{n-1} and τ_{n-1}^{-1} are small enough. This significantly improves the analyticity domain in the x direction which implies the decrease of the estimates for the corresponding modes. As a result, all modes with $k \ne 0$ become smaller.

Besides the (trivial) case $(k, v) = (0, 0)$ which is dealt by operation (4) above, we control the size of the remaining $k = 0$ modes in different ways. The case

$$S := \sum_i v_i = 1$$

(corresponding to the linear term in the action y) is eliminated by a proper choice of the affine parameter b_n depending on Q_{n-1} and the perturbation. That is, b_n is used to eliminate an unstable direction related to frequency vectors. The quadratic term in the action ($S = 2$) is included in the new symmetric matrix Q_n which has again

17 KAM Theory as a Limit of Renormalization

non-zero determinant and becomes smaller due to the action rescaling. Finally, we show that the action rescaling is also responsible for the decrease of the higher terms $S \geq 3$.

The overall consequence of the iterative scheme just described is that it converges to a limit set of Hamiltonians of the type

$$y \mapsto v \cdot y.$$

That is, the "limit" is a degenerate linear function of the action, and from that we show the existence of an ω-invariant torus for the initial Hamiltonian. To prove convergence we need to find proper choices of σ_n and τ_n as well as of stopping times t_n, which turns out to be possible for Diophantine ω. Roughly, too small values of σ_{n-1} and τ_{n-1}^{-1} make harder to eliminate modes as they are "too" resonant. On the other hand, large values imply that T_n does not contract I_{n-1}^+. Similarly, large $t_n - t_{n-1}$ improve the hyperbolicity of the matrices T_n but worsen the estimates on their norms and consequently enlarge the perturbation.

Acknowledgements Most of the work presented in this text was done in collaboration with K. Khanin and J. Marklof. The author was partially supported by Fundação para a Ciência e a Tecnologia through the Program POCI 2010.

References

1. Abad, J.J., Koch, H.: Renormalization and periodic orbits for Hamiltonian flows. Commun. Math. Phys. **212**, 371–394 (2000)
2. Bricmont, J., Gawędzki, K., Kupiainen, A.: KAM theorem and quantum field theory. Comm. Math. Phys. **201**(3), 699–727 (1999)
3. Cassels, J.W.S.: An Introduction to Diophantine Approximation. Cambridge University Press, Cambridge (1957)
4. Dani, S.G.: Divergent trajectories of flows on homogeneous spaces and Diophantine approximation. J. Reine Angew. Math. **359**, 55–89 (1985)
5. Eliasson, L.H.: Linear quasi-periodic systems–reducibility and almost reducibility. In XIVth International Congress on Mathematical Physics, pp. 195–205. World Scientific, Hackensack, NJ (2005)
6. Feigenbaum, M.J.: Quantitative universality for a class of non-linear transformations. J. Stat. Phys. **19**, 25–52 (1978)
7. Gaidashev, D.: Renormalization of isoenergetically degenerate Hamiltonian flows and associated bifurcations of invariant tori. Discrete Contin. Dyn. Syst. **13**(1), 63–102 (2005)
8. Gallavotti, G.: Twistless KAM tori. Comm. Math. Phys. **164**, 145–156 (1994)
9. Gentile, G., Mastropietro, V.: Methods for the analysis of the lindstedt series for KAM tori and renormalizability in classical mechanics. a review with some applications. Rev. Math. Phys. **8**, 393–444 (1996)
10. Gentile, G.: Resummation of perturbation series and reducibility for Bryuno skew-product flows. J. Stat. Phys. **125**(2), 321–361 (2006)
11. Gentile, G.: Degenerate lower-dimensional tori under the Bryuno condition. Ergod. Theory Dyn. Syst. **27**(2), 427–457 (2007)
12. Khanin, K., Lopes Dias, J., Marklof, J.: Multidimensional continued fractions, dynamic renormalization and KAM theory. Comm. Math. Phys. **207**, 197–231 (2007)

13. Khanin, K., Lopes Dias, J., Marklof, J.: Renormalization of multidimensional Hamiltonian flows. Nonlinearity **19**, 2727–2753 (2006)
14. Khanin, K., Sinai, Ya.: The renormalization group method and Kolmogorov-Arnold-Moser theory. In: Sagdeev, R.Z. (ed.) Nonlinear phenomena in plasma physics and hydrodynamics, pp. 93–118. Mir Moscow (1986)
15. Kleinbock, D.Y., Margulis, G.A.: Flows on homogeneous spaces and Diophantine approximation on manifolds. Ann. Math **148**(2), 339–360 (1998)
16. Koch, H.: A renormalization group for Hamiltonians, with applications to KAM tori. Ergon. Theory Dyn. Syst. **19**, 475–521 (1999)
17. Koch, H.: On the renormalization of Hamiltonian flows, and critical invariant tori. Discrete Contin. Dyn. Syst. **8**(3), 633–646 (2002)
18. Koch, H., Kocic, S.: A renormalization group approach to quasiperiodic motion with Brjuno frequencies. Ergod. Theory Dyn. Syst. **30**, 1131–1146 (2010)
19. Kocić, S.: Renormalization of Hamiltonians for Diophantine frequency vectors and KAM tori. Nonlinearity **18**, 2513–2544 (2005)
20. Kocic, S.: Reducibility of skew-product systems with multidimensional Brjuno base flows. Discrete and Contin. Dyn. Syst. A **29**, 261–283 (2011)
21. Lagarias, J.C.: Geodesic multidimensional continued fractions. Proc. Lond. Math. Soc. **69**, 464–488 (1994)
22. Lopes Dias, J.: Renormalization of flows on the multidimensional torus close to a KT frequency vector. Nonlinearity **15**, 647–664 (2002)
23. Lopes Dias, J.: Renormalization scheme for vector fields on $\mathbb{T}^2$ with a diophantine frequency. Nonlinearity **15**, 665–679 (2002)
24. Lopes Dias, J.: Brjuno condition and renormalisation for Poincaré flows. Discrete Contin. Dyn. Syst. **15**, 641–656 (2006)
25. Lopes Dias, J.: A normal form theorem for Brjuno skew-systems through renormalization. J. Differ. Equ. **230**, 1–23 (2006)
26. Lopes Dias, J.: Local conjugacy classes for analytic torus flows. J. Differ. Equ. (2008)
27. MacKay, R.S.: Renormalisation in Area-Preserving Maps. World Scientific Publishing, River Edge, NJ (1993)
28. MacKay, R.S.: Three topics in Hamiltonian dynamics. In: Aizawa, Y., Saito, S., Shiraiwa, K. (eds.) Dynamical Systems and Chaos, vol. 2. World Scientific (1995)
29. Rand, D.A.: Existence, non-existence and universal breakdown of dissipative golden invariant tori I: Golden critical circle maps. Nonlinearity **5**, 639–662 (1992)
30. Rand, D.A.: Existence, non-existence and universal breakdown of dissipative golden invariant tori II: Convergence of renormalization for mappings of the annulus. Nonlinearity **5**, 663–680 (1992)
31. Rand, D.A.: Existence, non-existence and universal breakdown of dissipative golden invariant tori III: Invariant circles for mappings of the annulus. Nonlinearity **5**, 681–706 (1992)

Chapter 18
An Overview of Optimal Life Insurance Purchase, Consumption and Investment Problems

Isabel Duarte, Diogo Pinheiro, Alberto A. Pinto, and Stanley R. Pliska

Abstract We provide an extension to Merton's famous continuous time model of optimal consumption and investment, in the spirit of previous works by Pliska and Ye, to allow for a wage earner to have a random lifetime and to use a portion of the income to purchase life insurance in order to provide for his estate, while investing his savings in a financial market consisting of one risk-free security and an arbitrary number of risky securities whose diffusive terms are driven by a multi-dimensional Brownian motion. The wage earner's problem is to find the optimal consumption, investment, and insurance purchase decisions in order to maximize expected utility of consumption and of the size of the estate in the event of premature death, and of the size of the estate at the time of retirement. Dynamic programming methods are used to obtain explicit solutions for the case of constant relative risk aversion utility functions, and new results are presented together with the corresponding economic interpretations.

D. Pinheiro (✉)
CEMAPRE, ISEG, Technical University of Lisbon, Lisbon, Portugal
e-mail: dpinheiro@iseg.utl.pt

I. Duarte
Department of Mathematics, University of Minho, Braga, Portugal
e-mail: isabelduarte@math.uminho.pt

A.A. Pinto
LIAAD-INESC Porto LA e Departamento de Matemática, Faculdade de Ciências, Universidade do Porto, Rua do Campo Alegre, 687, 4169–007, Portugal
and
Centro de Matemática e Departamento de Matemática e Aplicações, Escola de Ciências, Universidade do Minho, Campus de Gualtar, 4710–057 Braga, Portugal
e-mail: aapinto@fc.up.pt

S.R. Pliska
Department of Finance, University of Illinois at Chicago, Chicago, IL 60607, USA
e-mail: srpliska@uic.edu

M.M. Peixoto et al. (eds.), *Dynamics, Games and Science I*, Springer Proceedings in Mathematics 1, DOI 10.1007/978-3-642-11456-4_18,
© Springer-Verlag Berlin Heidelberg 2011

18.1 Introduction

We consider the problem faced by a wage earner having to make decisions continuously about three strategies: consumption, investment and life insurance purchase during a given interval of time $[0, \min\{T, \tau\}]$, where T is a fixed point in the future that we will consider to be the retirement time of the wage earner and τ is a random variable representing the wage earner's time of death. We assume that the wage earner receives his income at a continuous rate $i(t)$ and that this income is terminated when the wage earner dies or retires, whichever happens first. One of our key assumptions is that the wage earner's lifetime τ is a random variable and, therefore, the wage earner needs to buy life insurance to protect his family for the eventuality of premature death. The life insurance depends on a premium insurance rate $p(t)$ such that if the insured pays $p(t) \cdot \delta t$ and dies during the ensuing short time interval of length δt then the insurance company will pay one dollar to the insured's estate (so this is like term insurance with an infinitesimal term). We also assume that the wage earner wants to maximize the satisfaction obtained from a consumption process with rate $c(t)$. In addition to consumption and purchase of a life insurance policy, we assume that the wage earner invests the full amount of his savings in a financial market consisting of one risk-free security and a fixed number $N \geq 1$ of risky securities with diffusive terms driven by M-dimensional Brownian motion.

The wage earner is then faced with the problem of finding strategies that maximize the utility of (a) his family consumption for all $t \leq \min\{T, \tau\}$; (b) his wealth at retirement date T if he lives that long; and (c) the value of his estate in the event of premature death. Various quantitative models have been proposed to model and analyze this kind of problem, at least problems having at least one of these three objectives. This literature is highlighted by Yarri [10] who considered the problem of optimal financial planning decisions for an individual with an uncertain lifetime, as well as by Merton [4, 5] who emphasized optimal consumption and investment decisions but did not consider life insurance. These two approaches were combined by Richard [9], who considered a life-cycle life insurance and consumption-investment problem in a continuous time model. Later, Pliska and Ye [6, 7] introduced a continuous-time model that combined the more realistic features of all those in the existing literature and extended the model proposed previously by Richard. While Richard assumed that the lifetime of the wage earner is limited by some fixed number, the model introduced by Pliska and Ye had the key feature that the duration of life is a random variable which takes values in the interval $]0, \infty[$ and is independent of the stochastic process defining the underlying financial market. Moreover, Pliska and Ye made the following refinements to the theory: (a) the planning horizon T is now seen as the moment when the wage earner retires, contrary to Richard's interpretation as maximum life size; and (b) the utility of the wage earner's wealth at the planning horizon T is taken into account as well as the utility of lifetime consumption and the utility of the bequest in the event of premature death.

Whereas Pliska and Ye's financial market involved only one security that was risky, in the present work we study the extension where there is an arbitrary (but finite) number of risky securities. The existence of these extra risky securities gives greater freedom for the wage earner to manage the interaction between his life insurance policies and the portfolio containing his savings invested in the financial market. Some examples of these interactions are described below.

Following Pliska and Ye, we use the model of uncertain lifespan found in reliability theory, commonly used for industrial life-testing and actuarial science, to model the uncertain time of death for the wage earner. This enables us to replace Richard's assumption that lifetimes are bounded with the assumption that lifetimes take values in the interval $]0, \infty[$. We then set up the wage earner's objective functional depending on a random horizon $\min\{T, \tau\}$ and transform it to an equivalent problem having a fixed planning horizon, that is, the wage earner who faces unpredictable death acts as if he will live until some time T, but with a subjective rate of time preferences equal to his "force of mortality" for his consumption and terminal wealth. This transformation to a fixed planning horizon enables us to state the dynamic programming principle and derive an associated Hamilton–Jacobi–Bellman (HJB) equation. We use the HJB equation to derive the optimal feedback control, that is, optimal insurance, portfolio and consumption strategies. Furthermore, we obtain explicit solutions for the family of discounted Constant Relative Risk Aversion (CRRA) utilities and examine the economic implications of such solutions.

In the case of discounted CRRA utilities our results generalize those obtained previously by Pliska and Ye. For instance, we obtain: (a) an economically reasonable description for the optimal expenditure for insurance as a decreasing function of the wage earner's overall wealth; and (b) a more controversial conclusion that possibly an optimal solution calls for the wage earner to sell a life insurance policy on his own life toward the end of his career. Nonetheless, the extra risky securities in our model introduce novel features to the wage earner's portfolio and insurance management interaction such as: (a) a young wage earner with small wealth has an optimal portfolio with larger values of volatility and higher expected returns, with the possibility of having short positions in lower yielding securities; and (b) a wage earner who can buy life insurance policies will choose a more conservative portfolio than a wage earner who is without the opportunity to buy life insurance, the distinction being clearer for young wage earners with low wealth. Full details of our analysis will be provided in a forthcoming paper [1].

This paper is organized as follows. In Sect. 18.2 we describe the problem we address in [1]. Namely, we introduce the underlying financial and insurance markets as well as the problem formulation from the point of view of optimal control. In Sect. 18.3 we see how to use the dynamic programming principle to reduce the optimal control of Sect. 18.2 to one with a fixed planning horizon and then derive an associated HJB equation. We devote Sect. 18.4 to the case of discounted CRRA utilities. We conclude in Sect. 18.5.

18.2 Problem Formulation

In this section, we define the setting in which the wage earner has to make his decisions regarding consumption, investment and life insurance purchase. Namely, we introduce the specifications regarding the financial and insurance markets available to the wage earner. We start with the financial market description, followed by the insurance market and conclude with the definition of a wealth process for the wage earner.

18.2.1 The Financial Market Model

We consider a financial market consisting of one risk-free asset and several risky-assets. Their respective prices $(S_0(t))_{0 \leq t \leq T}$ and $(S_n(t))_{0 \leq t \leq T}$ for $n = 1, \ldots, N$ evolve according to the equations:

$$dS_0(t) = r(t)S_0(t)dt , \qquad\qquad\qquad S_0(0) = s_0 ,$$

$$dS_n(t) = \mu_n(t)S_n(t)dt + S_n(t)\sum_{m=1}^{M} \sigma_{nm}(t)dW_m(t) , \quad S_n(0) = s_n > 0 ,$$

where $W(t) = (W_1(t), \ldots, W_M(t))^T$ is a standard M-dimensional Brownian motion on a probability space $(\Omega, \mathscr{F}, P)$, $r(t)$ is the riskless interest rate, $\mu(t) = (\mu_1(t), \ldots, \mu_N(t)) \in \mathbb{R}^N$ is the vector of the risky-assets appreciation rates and $\sigma(t) = (\sigma_{nm}(t))_{1 \leq n \leq N, 1 \leq m \leq M}$ is the matrix of risky-assets volatilities.

We assume that the coefficients $r(t)$, $\mu(t)$ and $\sigma(t)$ are deterministic continuous functions on the interval $[0, T]$. We also assume that the interest rate $r(t)$ is positive for all $t \in [0, T]$ and the matrix $\sigma(t)$ is such that $\sigma\sigma^T$ is nonsingular for Lebesgue almost all $t \in [0, T]$ and satisfies the following integrability condition

$$\sum_{n=1}^{N} \sum_{m=1}^{M} \int_0^T \sigma_{nm}^2(t)dt < \infty.$$

Furthermore, we suppose that there exists an $(\mathscr{F}_t)_{0 \leq t \leq T}$-progressively measurable process $\pi(t) \in \mathbb{R}^M$, called the market price of risk, such that for Lebesgue-almost-every $t \in [0, T]$ the risk premium $\alpha(t) = (\mu_1(t) - r(t), \ldots, \mu_N(t) - r(t)) \in \mathbb{R}^N$ is related to $\pi(t)$ by the equation

$$\alpha(t) = \sigma(t)\pi(t) \qquad \text{a.s.}$$

and the following two conditions hold

18 An Overview of Optimal Life Insurance Purchase, Consumption

$$\int_0^T \|\pi(t)\|^2 < \infty \qquad \text{a.s.}$$

$$E\left[\exp\left(-\int_0^T \pi(s)dW(s) - \frac{1}{2}\int_0^T \|\pi(s)\|^2\,ds\right)\right] = 1.$$

The existence of such process $\pi(t)$ ensures the absence of arbitrage opportunities in the financial market defined above. See [3] for further details on market viability.

Moreover, throughout the paper we will assume that $(\Omega, \mathscr{F}, P)$ is a filtered probability space and that its filtration $\mathbb{F} = \{\mathscr{F}_t, t \in [0, T]\}$ is the P-augmentation of the filtration generated by the Brownian motion $W(t)$, $\sigma\{W(s), s \leq t\}$ for $t \geq 0$. Each sub-σ-algebra $\mathscr{F}_t$ represents the information known by the agents in the financial market at time t.

18.2.2 The Life Insurance Market Model

We assume that the wage earner is alive at time $t = 0$ and that his lifetime is a non-negative random variable τ defined on the probability space $(\Omega, \mathscr{F}, P)$. Furthermore, we assume that the random variable τ is independent of the filtration $\mathbb{F}$ and has a distribution function $F : [0, \infty) \to [0, 1]$ with density $f : [0, \infty) \to \mathbb{R}^+$ so that

$$F(t) = \int_0^t f(s)ds.$$

We define the *survivor function* $\overline{F} : [0, \infty) \to [0, 1]$ as the probability for the wage earner to survive at least until time t, i.e.

$$\overline{F}(t) = P(\tau \geq t) = 1 - F(t).$$

We shall make use of the *hazard function*, the conditional, instantaneous death rate for the wage earner surviving to time t, that is

$$\lambda(t) = \lim_{\delta t \to 0} \frac{P(t \leq \tau < t + \delta t \mid \tau \geq t)}{\delta t} = \frac{f(t)}{\overline{F}(t)}.$$

Throughout the paper, we will suppose that the hazard function $\lambda : [0, \infty) \to \mathbb{R}^+$ is a continuous and deterministic function such that

$$\int_0^\infty \lambda(t)dt = \infty.$$

These two concepts introduced above are standard in the context of reliability theory and actuarial science. In our case, such concepts enable us to consider an optimal control problem with a stochastic planning horizon and restate it as one with a fixed horizon.

Due the uncertainty concerning his lifetime, the wage earner buys life insurance to protect his family for the eventuality of premature death. The life insurance is available continuously and the wage earner buys it by paying a *premium insurance rate* $p(t)$ to the insurance company. The insurance contract is like term insurance, with an infinitesimally small term. If the wage earner dies at time $\tau < T$ while buying insurance at the rate $p(t)$, the insurance company pays an amount $p(\tau)/\eta(\tau)$ to his estate, where $\eta : [0, T] \to \mathbb{R}^+$ is a continuous and deterministic function which we call the *insurance premium-payout ratio* and is regarded as fixed by the insurance company. The contract ends when the wage earner dies or achieves retirement age, whichever happens first. Therefore, the wage earner's total legacy to his estate in the event of a premature death at time $\tau < T$ is given by

$$Z(\tau) = X(\tau) + \frac{p(\tau)}{\eta(\tau)},$$

where $X(t)$ denotes the wage earner's savings at time t.

18.2.3 The Wealth Process

We assume that the wage earner receives an income $i(t)$ at a continuous rate during the period $[0, \min\{T, \tau\}]$, i.e. the income will be terminated either by his death or his retirement, whichever happens first. Furthermore, we assume that $i : [0, T] \to \mathbb{R}^+$ is a deterministic Borel-measurable function satisfying the integrability condition

$$\int_0^T i(t)\mathrm{d}t < \infty.$$

The *consumption process* $(c(t))_{0 \le t \le T}$ is a $(\mathscr{F}_t)_{0 \le t \le T}$-progressively measurable nonnegative process satisfying the following integrability condition for the investment horizon $T > 0$

$$\int_0^T c(t)\mathrm{d}t < \infty \qquad \text{a.s..}$$

We assume also that the premium insurance rate $(p(t))_{0 \le t \le T}$ is a $(\mathscr{F}_t)_{0 \le t \le T}$-predictable process, i.e. $p(t)$ is measurable with respect to the smallest σ-algebra on $\mathbb{R}^+ \times \Omega$ such that all left-continuous and adapted processes are measurable. In a intuitive manner, a predictable process can be described as such that its values are "known" just in advance of time.

For each $n = 0, 1, \ldots, N$ and $t \in [0, T]$, let $\theta_n(t)$ denote the fraction of the wage earner's wealth allocated to the asset S_n at time t. The *portfolio process* is

18 An Overview of Optimal Life Insurance Purchase, Consumption

then given by $\Theta(t) = (\theta_0(t), \theta_1(t), \dots, \theta_N(t)) \in \mathbb{R}^{N+1}$, where

$$\sum_{n=0}^{N} \theta_n(t) = 1, \qquad 0 \le t \le T. \tag{18.1}$$

We assume that the portfolio process is $(\mathscr{F}_t)_{0 \le t \le T}$-progressively measurable and that for the fixed investment horizon $T > 0$ we have that

$$\int_0^T \|\Theta(t)\|^2 \, dt < \infty \qquad \text{a.s.,}$$

where $\|\cdot\|$ denotes the Euclidean norm in $\mathbb{R}^{N+1}$.

The *wealth process* $X(t), t \in [0, \min\{T, \tau\}]$, is then defined by

$$X(t) = x + \int_0^t [i(s) - c(s) - p(s)] \, ds + \sum_{n=0}^{N} \int_0^t \frac{\theta_n(s) X(s)}{S_n(s)} dS_n(s), \tag{18.2}$$

where x is the wage earner's initial wealth. This last equation can be rewritten in the differential form

$$dX(t) = \left(i(t) - c(t) - p(t) + \left(\theta_0(t) r(t) + \sum_{n=1}^{N} \theta_n(t) \mu_n(t) \right) X(t) \right) dt$$

$$+ \sum_{n=1}^{N} \theta_n(t) X(t) \sum_{m=1}^{M} \sigma_{nm}(t) dW_m(t),$$

where $0 \le t \le \min\{\tau, T\}$.

Using the relation (18.1), we can always write $\theta_0(t)$ in terms of $\theta_1(t), \dots, \theta_N(t)$. From now on, we will define the portfolio process in terms of the *reduced portfolio process* $\theta(t) = (\theta_1(t), \theta_2(t), \dots, \theta_N(t)) \in \mathbb{R}^N$.

18.2.4 The Optimal Control Problem

The wage earner is then faced with the problem of finding strategies that maximize the utility of:

(a) His family consumption for all $t \le \min\{T, \tau\}$.
(b) His wealth at retirement date T if he lives that long.
(c) The value of his estate in the event of premature death.

This problem can be formulated by means of optimal control theory: the wage earner goal is to maximize some cost functional subject to (a) the (stochastic) dynamics of the state variable, i.e. the dynamics of the wealth process $X(t)$ given by (18.2); (b) constraints on the control variables, i.e. the consumption process $c(t)$, the premium insurance rate $p(t)$ and the portfolio process $\theta(t)$; and (c) boundary conditions on the state variables.

Let us denote by $\mathscr{A}(x)$ the set of all admissible decision strategies, i.e. all admissible choices for the control variables $v = (c, p, \theta) \in \mathbb{R}^{N+2}$. The dependence of $\mathscr{A}(x)$ on x denotes the restriction imposed on the wealth process by the boundary condition $X(0) = x$.

The wage earner's problem can then be restated as follows: find a strategy $v = (c, p, \theta) \in \mathscr{A}(x)$ which maximizes the expected utility

$$V(x) = \sup_{v \in \mathscr{A}(x)} E_{0,x} \left[\int_0^{T \wedge \tau} U(c(s), s) \, ds + B(Z(\tau), \tau) I_{\{\tau \leq T\}} \right.$$
$$\left. + W(X(T)) I_{\{\tau > T\}} \right], \tag{18.3}$$

where $T \wedge \tau = \min\{T, \tau\}$, I_A denotes the indicator function of event A, $U(c, \cdot)$ is the utility function describing the wage earner's family preferences regarding consumption in the time interval $[0, \min\{T, \tau\}]$, $B(Z, \cdot)$ is the utility function for the size of the wage earners's legacy in case $\tau < T$ and $W(X)$ is the utility function for the terminal wealth at time $t = T$.

We suppose that U and B are strictly concave on their first variable and that W is strictly concave on its sole variable. In Sect. 18.4 we focus our analysis on the case where the wage earner's preferences are described by discounted CRRA utility functions.

18.3 Stochastic Optimal Control

In this section we describe how dynamic programming can be used to restate the stochastic optimal control problem formulated in the preceding section as one with a fixed planning horizon and then derive the associated HJB equation.

18.3.1 Dynamic Programming Principle

Let us denote by $\mathscr{A}(t, x)$ the set of admissible decision strategies $v = (c, p, \theta)$ for the dynamics of the wealth process with boundary condition $X(t) = x$. For any $v \in \mathscr{A}(t, x)$ we define

$$J(t, x; v) = E_{t,x} \left[\int_t^{T \wedge \tau} U(c(s), s) \, ds + B(Z(\tau), \tau) I_{\{\tau \leq T\}} \right.$$
$$\left. + W(X(T)) I_{\{\tau > T\}} \, \middle| \, \tau > t, \mathscr{F}_t \right]$$

18 An Overview of Optimal Life Insurance Purchase, Consumption

and note that the optimal control problem (18.3) can be restated in dynamic programming form as

$$V(t, x) = \sup_{v \in \mathscr{A}(t,x)} J(t, x; v).$$

The following lemma is the key tool to restating the control problem above as an equivalent one with a fixed planning horizon. See [11] for a proof.

Lemma 18.1. *Suppose that the utility function U is either nonnegative or nonpositive. If the random variable τ is independent of the filtration $\mathbb{F}$, then*

$$J(t, x; v) = E_{t,x}\left[\int_t^T \overline{F}(s,t)U(c(s), s) + f(s,t)B(Z(s), s) \, ds\right.$$

$$\left. + \overline{F}(T,t)W(X(T)) \,\Big|\, \mathscr{F}_t\right],$$

where $\overline{F}(s,t)$ is the conditional probability for the wage earner's death to occur at time s conditional upon the wage earner being alive at time $t \leq s$ and $f(s,t)$ is the corresponding conditional probability density.

Using the previous lemma, one can state the following dynamic programming principle, obtaining a recursive relationship for the maximum expected utility as a function of the wage earner's age and his wealth at that time. (See [11]) for a proof.

Lemma 18.2 (Dynamic programming principle). *For $0 \leq t < s < T$, the maximum expected utility $V(t, x)$ satisfies the recursive relation below*

$$V(t, x) = \sup_{v \in \mathscr{A}(t,x)} E\left[\exp\left(-\int_t^s \lambda(v)dv\right) V(s, X(s))\right.$$

$$\left. + \int_t^s \overline{F}(u,t)U(c(u), u) + f(u,t)B(Z(u), u) \, du \,\Big|\, \mathscr{F}_t\right].$$

The transformation to a fixed planning horizon can then be given the following interpretation: a wage earner facing unpredictable death acts as if he will live until time T, but with a subjective rate of time preferences equal to his "force of mortality" for the consumption of his family and his terminal wealth.

18.3.2 Hamilton–Jacobi–Bellman Equation

The dynamic programming principle enables us to state the HJB equation, a second-order partial differential equation whose "solution" is the value function of the optimal control problem under consideration here. The techniques used in the derivation of the HJB equation and the proof of the next theorem follow closely there in [2, 11, 12]. A complete proof will be provided in [1].

Theorem 18.1. *Suppose that the maximum expected utility V is of class C^2. Then V satisfies the Hamilton–Jacobi–Bellman equation*

$$\begin{cases} V_t(t,x) - \lambda(t)V(t,x) + \sup_{v \in \mathscr{A}(t,x)} \mathscr{H}(t,x;v) = 0 \\ V(T,x) = W(x) \end{cases},$$

where the Hamiltonian function $\mathscr{H}$ is given by

$$\mathscr{H}(t,x;v) = \left(i(t) - c - p + \left(r(t) + \sum_{n=1}^{N} \theta_n(\mu_n(t) - r(t)) \right) x \right) V_x(t,x)$$

$$+ \frac{x^2}{2} \sum_{m=1}^{M} \left(\sum_{n=1}^{N} \theta_n \sigma_{nm}(t) \right)^2 V_{xx}(t,x)$$

$$+ \lambda(t) B\left(x + \frac{p}{\eta(t)}, t \right) + U(c,t).$$

Moreover, an admissible strategy $v^* = (c^*; p^*; \theta^*)$ whose corresponding wealth is X^* is optimal if and only if for a.e. $s \in [t,T]$ and P-a.s. we have

$$V_t(s, X^*(s)) - \lambda(s)V(s, X^*(s)) + \mathscr{H}(s, X^*(s); v^*) = 0.$$

The second part of the theorem above provides a means for deriving the optimal insurance, portfolio and consumption strategies. In particular, we obtain the existence of such optimal strategies under rather weak conditions on the utility functions.

Corollary 18.1. *Suppose that the utility functions U and B are strictly concave on the first variable. Then the Hamiltonian function $\mathscr{H}$ has a regular interior maximum $v^* = (c^*, p^*, \theta^*) \in \mathscr{A}(t,x)$.*

18.4 The Family of Discounted CRRA Utilities

In this section we describe the special case where the wage earner has the same discounted CRRA utility functions for the consumption of his family, the size of his legacy and his terminal wealth.

Assume that $\gamma < 1$, $\gamma \neq 0$ and $\rho > 0$ and let

$$U(c,t) = e^{-\rho t}\frac{c^\gamma}{\gamma}, \qquad B(Z,t) = e^{-\rho t}\frac{Z^\gamma}{\gamma}, \qquad W(X) = e^{-\rho T}\frac{X^\gamma}{\gamma}. \tag{18.4}$$

All details regarding the results in this section and its proofs will be given in a forthcoming paper [1].

18 An Overview of Optimal Life Insurance Purchase, Consumption

18.4.1 The Optimal Strategies

Using the optimality criteria provided in Theorem 18.1, we obtain the following optimal strategies for discounted CRRA utility functions.

Proposition 18.1. *Let ξ denote the non-singular square matrix given by $(\sigma\sigma^T)^{-1}$. The optimal strategies in the case of discounted constant relative risk aversion utility functions are given by*

$$c^*(t,x) = \frac{1}{e(t)}(x + b(t))$$

$$p^*(t,x) = \eta(t)\left((D(t) - 1)x + D(t)b(t)\right)$$

$$\theta^*(t,x) = \frac{1}{x(1 - \gamma)}(x + b(t))\xi\alpha(t),$$

where

$$b(t) = \int_t^T i(s)\exp\left(-\int_t^s r(v) + \eta(v)\,dv\right)ds$$

$$e(t) = \exp\left(-\int_t^T H(v)\,dv\right) + \int_t^T \exp\left(-\int_t^s H(v)\,dv\right)K(s)\,ds$$

$$H(t) = \frac{\lambda(t) + \rho}{1 - \gamma} - \gamma\frac{\Sigma(t)}{(1 - \gamma)^2} - \frac{\gamma}{1 - \gamma}(r(t) + \eta(t))$$

$$D(t) = \frac{1}{e(t)}\left(\frac{\lambda(t)}{\eta(t)}\right)^{1/(1-\gamma)}$$

$$K(t) = \frac{(\lambda(t))^{1/(1-\gamma)}}{(\eta(t))^{\gamma/(1-\gamma)}} + 1$$

$$\Sigma(t) = \alpha^T(t)\xi\alpha(t) - \frac{1}{2}\|\sigma^T\xi\alpha(t)\|^2.$$

Note that the quantities $b(t)$ and $x + b(t)$ are of essential relevance for the definition of the optimal strategies in Proposition 18.1. The quantity $b(t)$, that we will refer to as *human capital*, should be seen as representing the fair value at time t of the wage earner's future income from time t to time T, while the quantity $x + b(t)$ should be thought of as the full wealth (present wealth plus future income) of the wage earner at time t. It is then natural that these two quantities play a central role in the choice of optimal strategies, since they determine the present and future wealth available for the wage earner and his family.

From the explicit knowledge of the optimal strategies, several economically relevant conclusions can be obtained. See Fig. 18.1 for a graphical representation of the optimal life-insurance purchase as a function of age and "full wealth" $x+b(t)$ of the wage earner. The next result provides a qualitative characterization of the optimal life insurance purchase strategy.

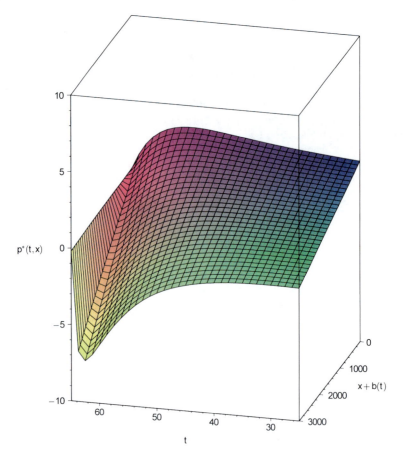

Fig. 18.1 The optimal life-insurance purchase for a wage earner that starts working at age 25 and retires 40 years later. The parameters of the model were taken as $N = M = 2$, $i(t) = 50000\exp(0.03t)$, $r = 0.04$, $\rho = 0.03$, $\gamma = -3$, $\lambda(t) = 0.001 + \exp(-9.5 + 0.1t)$, $\eta(t) = 1.05\lambda(t)$, $\mu_1 = 0.07$, $\mu_2 = 0.11$, $\sigma_{11} = 0.19$, $\sigma_{12} = 0.15$, $\sigma_{21} = 0.17$ and $\sigma_{22} = 0.21$

Corollary 18.2. *Suppose that for all $t \in [0, \min\{T, \tau\}]$ the following two conditions are satisfied:*

(a) $\lambda(t) \leq \eta(t)$.
(b) $H(t) \leq 1$.

Then, the optimal insurance purchase strategy $p^*(t, x)$ is

- *A decreasing function of the total wealth x.*
- *An increasing function of the wage earner's human capital $b(t)$.*
- *Negative for suitable choices of wealth x and "age" t.*

Some comments regarding the assumptions in Corollary 18.2 seem to be necessary. Starting with condition (i), the life insurance company must establish the

18 An Overview of Optimal Life Insurance Purchase, Consumption 283

premium-insurance $\eta(t)$ in such a way that $\lambda(t) \leq \eta(t)$ in order to make a profit (the insurance policy being fair whenever $\lambda(t) \leq \eta(t)$). Regarding condition (ii), we note that the parameters r, ρ, η and λ are usually very small in the real world and, moreover, the relative risk aversion of the wage earner is negative in general. This is consistent with the assumption that $H(t)$ is bounded above by some positive constant.

The extra risky securities in our model introduce novel features to the wage earner's portfolio management, as is exemplified in the following result.

Corollary 18.3. *Let ξ denote the non-singular square matrix given by $(\sigma\sigma^T)^{-1}$ and let $(\xi\alpha(t))_n$ denote the n-th component of the vector $\xi\alpha(t)$. Assume that for every $n \in \{1, \ldots, N\}$ we have $(\xi\alpha(t))_n > 0$ for all $t \in [0, T]$. The optimal portfolio process $\theta^*(t, x) = \left(\theta_1^*, \ldots, \theta_N^*\right)$ is such that for every $n \in \{1, \ldots, N\}$:*

- θ_n^* *is a decreasing function of the total wealth x.*
- θ_n^* *is an increasing function of the wage earner's human capital $b(t)$.*

Furthermore, for every $n, m \in \{1, \ldots, N\}$ the following equalities hold

$$\lim_{x \to 0^+} \theta_n^*(t, x) = +\infty \qquad \lim_{x \to 0^+} \frac{\theta_n^*(t, x)}{\theta_m^*(t, x)} = \frac{(\xi\alpha(t))_n}{(\xi\alpha(t))_m}$$

$$\lim_{x \to \infty} \theta_n^*(t, x) = \frac{(\xi\alpha(t))_n}{1 - \gamma} \qquad \lim_{t \to T} \theta_n^*(t, x) = \frac{(\xi\alpha(T))_n}{1 - \gamma}.$$

The assumption in Corollary 18.3 corresponds to the assumption that, under some correction term determined by the volatility matrix, risky assets have higher expected returns than risk-free assets. One interesting consequence of the previous result is that the optimal strategy for wage earners with small enough wealth is to short the risk-free security and hold an higher amount of risky assets.

Finally, it should be noted that it is also possible to study the qualitative properties of the optimal consumption strategy.

Remark 18.1. Regarding the optimal consumption rate $c^*(t, x)$, we have that this is an increasing function of both the wealth x and the human capital $b(t)$.

18.4.2 The Interaction between Life Insurance Purchase and Portfolio Management

In this section we compare the optimal life-insurance strategies for a wage earner who faces the following two situations:

(a) In the first case, we assume that the wage earner has access to an insurance market as described above and that his goal is to maximize the combined utility of his family consumption for all $t \leq \min\{T, \tau\}$, his wealth at retirement date T if he lives that long, and the value of his estate in the event of premature

284   I. Duarte et al.

death. The optimal strategies for the wage earner in this setting are given in Proposition 18.1.

(b) In the second case, we assume that the wage earner is without the opportunity of buying life insurance. His goal is then to maximize the combined utility of his family consumption for all $t \leq \min\{T, \tau\}$ and his wealth at retirement date T if he lives that long. Similarly to what we have done previously, we translate this situation to the language of stochastic optimal control and derive explicit solutions in the case of discounted CRRA utilities.

We concentrate on the case (b) described above for the moment. Similarly to what was done in case (a), this problem can be formulated by means of optimal control theory. The wage earner's goal is then to maximize a new cost functional subject to

- The dynamics of the state variable, i.e. the dynamics a wealth process $X^0(t)$ given by

$$X^0(t) = x + \int_0^t i(s) - c^0(s) \, ds + \sum_{n=0}^N \int_0^t \frac{\theta_n^0(s) X^0(s)}{S_n(s)} dS_n(s),$$

  where $t \in [0, \min\{T, \tau\}]$ and x is the wage earner's initial wealth.
- Constraints on the remaining control variables, i.e. the consumption process $c^0(t)$ and the reduced portfolio process $\theta^0(t) = \left(\theta_1^0(t), \ldots, \theta_N^0(t)\right) \in \mathbb{R}^N$.
- Boundary conditions on the state variables.

Let us denote by $\mathscr{A}^0(x)$ the set of all admissible decision strategies, i.e. all admissible choices for the control variables $v^0 = (c^0, \theta^0) \in \mathbb{R}^{N+1}$. The dependence of $\mathscr{A}^0(x)$ on x denotes the restriction imposed on the wealth process by the boundary condition $X^0(0) = x$.

The wage earner's problem can then be stated as follows: find a strategy $v^0 = (c^0, \theta^0) \in \mathscr{A}^0(x)$ which maximizes the expected utility

$$V^0(x) = \sup_{v^0 \in \mathscr{A}^0(x)} E_{0,x} \left[\int_0^{T \wedge \tau} U(c^0(s), s) \, ds + W(X^0(T)) I_{\{\tau > T\}} \right], \quad (18.5)$$

where $U(c^0, \cdot)$ is again the utility function describing the wage earner's family preferences regarding consumption in the time interval $[0, \min\{T, \tau\}]$ and $W(X^0)$ is the utility function for the terminal wealth at time $t = T$. As before, we restrict ourselves to the special case where the wage earner has the same discounted CRRA utility functions for the consumption of his family and his terminal wealth given in (18.4).

The same methods that were used in the analysis of case a) enable us to obtain the following result.

Proposition 18.2. *Let ξ denote the non-singular square matrix given by $(\sigma\sigma^T)^{-1}$. The optimal strategies for problem (18.5) in the case where $U(c^0, \cdot)$ and $W(X^0)$ are*

18   An Overview of Optimal Life Insurance Purchase, Consumption   285

the discounted constant relative risk aversion utility functions in (18.4) are given by

$$c^{0*}(t,x) = \frac{1}{e^0(t)}(x + b^0(t))$$

$$\theta^{0*}(t,x) = \frac{1}{x(1-\gamma)}(x + b^0(t))\xi\alpha(t),$$

where

$$b^0(t) = \int_t^T i(s)\exp\left(-\int_t^s r(v)\,dv\right)ds$$

$$e^0(t) = \exp\left(-\int_t^T H^0(v)\,dv\right) + \int_t^T \exp\left(-\int_t^s H^0(v)\,dv\right)ds$$

$$H^0(t) = \frac{\lambda(t)+\rho}{1-\gamma} - \gamma\frac{\Sigma(t)}{(1-\gamma)^2} - \frac{\gamma}{1-\gamma}r(t)$$

and $\Sigma(t)$ is as given in the statement of Proposition 18.1.

Thus, we now have optimal portfolio processes for the two settings (a) and (b) described above. These are given, respectively, in Propositions 18.1 and 18.2.

Theorem 18.2. *Let ξ denote the non-singular square matrix given by $(\sigma\sigma^T)^{-1}$ and $(\xi\alpha(t))_n$ the n-th component of the vector $\xi\alpha(t)$. For each $n \in \{1,\ldots,N\}$, we have that $\theta_n^{0*}(t,x) > \theta_n^*(t,x)$ if and only if $(\xi\alpha(t))_n > 0$.*

The economic implications of the theorem above are made clear in the following result.

Corollary 18.4. *Let ξ denote the non-singular square matrix given by $(\sigma\sigma^T)^{-1}$ and $(\xi\alpha(t))_n$ the n-th component of the vector $\xi\alpha(t)$. Assume that for every $n \in \{1,\ldots,N\}$ we have that $(\xi\alpha(t))_n > 0$. Then, the optimal portfolio of a wage earner with the possibility of buying a life insurance policy is more conservative than the optimal portfolio of the same wage earner if he does not have the opportunity to buy life insurance.*

18.5   Conclusions

We have introduced a model for optimal insurance purchase, consumption and investment for a wage earner with an uncertain lifetime with an underlying financial market consisting of one risk-free security and a fixed number of risky securities with diffusive terms driven by multidimensional Brownian motion. When we restrict ourselves to the case where the wage earner has the same discounted CRRA utility functions for the consumption of his family, the size of his legacy and his terminal wealth, we obtain explicit optimal strategies and describe new properties of these optimal strategies. Namely, we obtain economically relevant conclusions such as: (a) a young wage earner with smaller wealth has an optimal portfolio with larger values of volatility and higher expected returns; and (b) a wage earner who can buy life insurance policies will choose a more conservative portfolio than a similar wage earner who is without the opportunity to buy life insurance.

Acknowledgements We thank the Calouste Gulbenkian Foundation, PRODYN-ESF, FEDER, EDFC, POCTI, and POSI by FCT and Ministério da Ciência, Tecnologia e Ensino Superior, CEMAPRE, Centro de Matemática da Universidade do Minho, Centro de Matemática da Universidade do Porto and LIAAD-INESC Porto LA for their financial support. Part of this research was done during visits by the authors to IMPA (Brazil), The University of Warwick (United Kingdom), Isaac Newton Institute (United Kingdom), IHES (France), CUNY (USA), SUNY (USA) and MSRI (USA), whom we thank for their hospitality. D. Pinheiro's research was supported by FCT - Fundação para a Ciência e Tecnologia program "Ciência 2007". I. Duarte's research was supported by FCT – Fundação para a Ciência e Tecnologia grant with reference SFRH / BD / 33502 / 2008.

References

1. Duarte, I., Pinheiro, D., Pinto, A.A., Pliska, S.R.: Optimal life insurance purchase, consumption and investment on a financial market with multi-dimensional difusive terms. (Submitted, arXiv:q-fin/1102.2263)
2. Fleming, W.H., Soner, H.M.: Controlled Markov Processes and Viscosity Solutions. Springer, New York (2006)
3. Karatzas, I., Shreve, S.: Methods of Mathematical Finance. Springer, New York (1998)
4. Merton, R.C.: Lifetime portfolio selection under uncertainty: the continuous time case. Rev. Econ. Stat. 51247–51257 (1969)
5. Merton, R.C.: Optimum consumption and portfolio rules in a continuous-time model. J. Econ. Theory **3**, 372–413 (1971)
6. Pliska, S.R., Ye, J.: Optimal life insurance purchase and consumption/investment under uncertain lifetime. J. Bank. Finance **31**, 1307–1319 (2007)
7. Pliska, S.R., Ye, J.: Optimal life insurance purchase, consumption and investment. Preprint (2009)
8. Pinto, A.A.: Game Theory and Duopoly Models. Interdisciplinary Applied Mathematics Series. Springer (to appear)
9. Richard, S.F.: Optimal consumption, portfolio and life insurance rules for an uncertain lived individual in a continuous-time model. J. Financ. Econ. **2**, 187–203 (1975)
10. Yaari, M.E.: Uncertain lifetime, life insurance and the theory of the consumer. Rev. Econ. Stud. **32**, 137–150 (1965)
11. Ye, J.: Optimal Life Insurance Purchase, Consumption and Portfolio Under an Uncertain Life. PhD thesis, University of Illinois at Chicago (2006)
12. Yong, J., Zhou, X.Y.: Stochastic Controls: Hamiltonian Systems and HJB Equations. Springer, New York (1999)

Chapter 19
Towards a Theory of Periodic Difference Equations and Its Application to Population Dynamics

Saber N. Elaydi, Rafael Luís, and Henrique Oliveira

Abstract We present a survey of some of the most updated results on the dynamics of periodic and almost periodic difference equations.

19.1 Introduction

In a series of papers, Elaydi and Sacker [13–15, 32] embarked on a systematic study of periodic difference equations or periodic dynamical systems. The authors also wrote a survey [16] which has not been readily available to researches. The main purpose of this survey is to update, extend, and broaden the above-mentioned survey. Since the appearance [16], there have many exciting and new results by many authors as reflected by the extensive list of references.

An emphasis is placed here on bifurcation theory of periodic systems, particularly, those obtained by the authors and their collaborators. In fact, some of the results reported here appear for the first time. A more detailed account of bifurcation theory will appear somewhere else.

Two important omissions should be noted. The first is the extension of Sharkovsky's theorem to periodic difference equations [3]. The second is the study of periodic

---

R. Luís (✉)
Center for Mathematical Analysis, Geometry, and Dynamical Systems, Instituto Superior Tecnico, Technical University of Lisbon, Lisbon, Portugal
e-mail: rafael.luis.madeira@gmail.com

S.N. Elaydi
Department of Mathematics, Trinity University, San Antonio, TX, USA
e-mail: selaydi@trinity.edu

H. Oliveira
Department of Mathematics, Instituto Superior Tecnico, Technical University of Lisbon, Lisbon, Portugal
e-mail: holiv@math.ist.utl.pt

M.M. Peixoto et al. (eds.), *Dynamics, Games and Science I*, Springer Proceedings in Mathematics 1, DOI 10.1007/978-3-642-11456-4_19,
© Springer-Verlag Berlin Heidelberg 2011

systems with the Allee effect [29]. One reason for not including these topics is our self-imposed limitations on the size of the survey. A second reason is the limitation in the expertise of the writers of this survey. We promise the reader to explore these two topics in a forthcoming work.

In Sect. 19.2, we motivate the need for introducing skew-product techniques in the study of nonautonomous difference equations. Section 19.3 develops the basic construction of skew-product dynamical systems.

Subsequently, in Sect. 19.4 our study is focused on periodic difference equations. This section includes two important results in the theory of periodic systems, namely, Lemmas 19.1 and 19.2. In Sect. 19.5, we tackle the question of stability in both the space X and in the skew-product $X \times Y$. The section ends with the fundamental result in Theorem 19.2, which states that in a connected topological space, the period of a globally asymptotically stable periodic orbit must divide the period of the system.

In Sect. 19.6, we extend Singer's theorem to periodic systems. In Sect. 19.7, we develop a bifurcation theory for 2-periodic difference equations. In particular, a unimodal map with the Allee effect is thoroughly analyzed. A bifurcation graph of the parameter space of a 2-periodic system consisting of these maps is developed using the techniques of resultant in Mathematica software.

In Sect. 19.8, we address the question of whether the solutions of bifurcation equations are independent of the phase shifts.

In Sect. 19.9, we present an updated account of results pertaining to attenuance and resonance. The question we tackle here is whether periodic forcing has a deleterious effect on the population (attenuance) or it is advantageous to the population (resonance). In Sect. 19.10, we introduces almost periodicity and contains some of the results obtained in [10]. This is followed by Sect. 19.11 in which the study of stochastic difference equations is conducted.

19.2 Preliminaries

Let X be a topological space and $\mathbb{Z}$ be the set of integers. A discrete dynamical system (X, π) is defined as a map $\pi : X \times \mathbb{Z} \to X$ such that π is continuous and satisfies the following two properties

1. $\pi(x, 0) = x$ for all $x \in X$.
2. $\pi(\pi(x, s), t) = \pi(x, s + t), s, t \in \mathbb{Z}$ and $x \in X$ (the group property).

We say (X, π) is a discrete semidynamical system if $\mathbb{Z}$ is replaced by $\mathbb{Z}^+$, the set of nonnegative integers, and the group property is replaced by the semigroup property.

Notice that (X, π) can be generated by a map f defined as $\pi(x, n) = f^n(x)$, where f^n denotes the nth composition of f. We observe that the crucial property here is the semigroup property.

19 Towards a Theory of Periodic Difference Equations

A difference equation is called autonomous if it is generated by one map such as

$$x_{n+1} = f(x_n), n \in \mathbb{Z}^+. \tag{19.1}$$

Notice that for any $x_0 \in X, x_n = f^n(x_0)$. Hence, the orbit $\mathcal{O}(x_0) = \{x_0, x_1, x_2, \ldots\}$ in (19.1) is the same as the set $\mathcal{O}(x_0) = \{x_0, f(x_0), f^2(x_0), \ldots\}$ under the map f.

A difference equation is called nonautonomous if it is governed by the rule

$$x_{n+1} = F(n, x_n), n \in \mathbb{Z}^+, \tag{19.2}$$

which may be written in the friendlier form

$$x_{n+1} = f_n(x_n), n \in \mathbb{Z}^+, \tag{19.3}$$

where $f_n(x) = F(n, x)$. Here the orbit of a point x_0 is generated by the composition of the sequence of maps $\{f_n\}$. Explicitly,

$$\begin{aligned} \mathcal{O}(x_0) &= \{x_0, f_0(x_0), f_1(f_0(x_0)), f_2(f_1(f_0(x_0))), \ldots\} \\ &= \{x_0, x_1, x_2, \ldots\}. \end{aligned}$$

It should be pointed out here that (19.2) or (19.3) may not generate a discrete semidynamical system as it may not satisfy the semigroup property. The following example illustrates this point.

Example 19.1. Consider the nonautonomous difference equation

$$x_{n+1} = (-1)^n \left(\frac{n+1}{n+2}\right) x_n, x(0) = x_0. \tag{19.4}$$

The solution of (19.4) is

$$x_n = (-1)^{\frac{n(n-1)}{2}} \frac{x_0}{n+1}.$$

Let $\pi(x_0, n) = x_n$. Then

$$\begin{aligned} \pi(\pi(x_0, m), n) &= \pi\left((-1)^{\frac{m(m-1)}{2}} \cdot \frac{x_0}{m+1}, n\right) \\ &= (-1)^{\frac{n(n-1)}{2}} (-1)^{\frac{m(m-1)}{2}} \cdot \frac{x_0}{(n+1)(m+1)} \end{aligned}$$

However,

$$\pi(x_0, m+n) = (-1)^{\frac{(n+m)(n+m-1)}{2}} \frac{x_0}{m+n+1} \neq \pi(\pi(x_0, m), n).$$

19.3 Skew-Product Systems

Consider the nonautonomous difference equation

$$x_{n+1} = F(n, x_n), n \in \mathbb{Z}^+, \tag{19.5}$$

where $F(n, \cdot) \in C(\mathbb{Z}^+ \times X, X) = C$. The space C is equipped with the topology of uniform convergence on compact subsets of $\mathbb{Z}^+ \times X$. Let $F_t(n, \cdot) = F(t + n, \cdot)$ and $\mathscr{A} = \{F_t(n, \cdot) : t \in \mathbb{Z}^+\}$ be the set of translates of F in C. Then $G(n, \cdot) \in \omega(\mathscr{A})$, the omega limit set of $\mathscr{A}$, if for each $n \in \mathbb{Z}^+$,

$$|F_t(n, x) - G(n, x)| \to 0$$

uniformly for x in compact subsets of X, as $t \to \infty$ along some subsequence $\{t_{n_i}\}$. The closure of $\mathscr{A}$ in C is called the hull of $F(n, \cdot)$ and is denoted by $Y = cl(\mathscr{A}) = \mathscr{H}(F)$.

On the space Y, we define a discrete semidynamical system $\sigma : Y \times \mathbb{Z}^+ \to Y$ by $\sigma(H(n, \cdot), t) = H_t(n, \cdot)$; that is σ is the shift map.

For convenience, one may write (19.5) in the form

$$x_{n+1} = f_n(x_n) \tag{19.6}$$

with $f_n(x_n) = F(n, x_n)$.

Define the composition operator Φ as follows

$$\Phi_n^i = f_{i+n-1} \circ \ldots \circ f_{i+1} \circ f_i \equiv \Phi_n(F(i, \cdot)),$$

and the reverse composition operator $\widetilde{\Phi}$ as

$$\widetilde{\Phi}_n^i = f_i \circ f_{i+1} \circ \cdots \circ f_{i+n-1}.$$

When $i = 0$, we write Φ_n^0 as Φ_n and $\widetilde{\Phi}_n^0$ as $\widetilde{\Phi}$.

The skew-product system is now defined as

$$\pi : X \times Y \times \mathbb{Z}^+ \to X \times Y$$

with

$$\pi((x, G), n) = (\Phi_n(G(i, \cdot)), \sigma(G, n)).$$

If $G = f_i$, then $\pi((x, f_i), n) = (\Phi_n^i(x), f_{i+n})$.

The following commuting diagram illustrates the notion of skew-product systems where $\mathscr{P}(a, b) = a$ is the projection map.

19 Towards a Theory of Periodic Difference Equations

$$
\begin{array}{ccc}
X \times Y \times \mathbb{Z}^+ & \xrightarrow{\ \pi\ } & X \times Y \\
\scriptstyle{\mathscr{P} \times id} \downarrow & & \downarrow \scriptstyle{\mathscr{P}} \\
Y \times \mathbb{Z}^+ & \xrightarrow{\ \sigma\ } & Y
\end{array}
$$

For each $G(n, \cdot) \equiv g_n \in Y$, we define the fiber $\mathscr{F}_g$ over G as $\mathscr{F}_g = \mathscr{P}^{-1}(G)$. If $g = f_i$, we write $\mathscr{F}_g$ as $\mathscr{F}_i$.

Theorem 19.1. *[16] π is a discrete semidynamical system.*

Example 19.2 (Example (19.1) revisited). Let us reconsider the nonautonomous difference equation

$$
x_{n+1} = (-1)^n \left(\frac{n+1}{n+2} \right) x_n, \, x(0) = x_0.
$$

Hence, $F(n, x) = (-1)^n \left(\frac{n+1}{n+2} \right) x = f_n(x)$. Its hull is given by $G(n, x) = (-1)^n x$, that is, g_n is a periodic sequence given by $g_0 = g_{2n}, g_1 = g_{2n+1}$, for all $n \in \mathbb{Z}^+$, in which $g_0(x) = x$, and $g_1(x) = -x$.

It is easy to verify that π defined as $\pi((x, f_i), n) = (\Phi_n^i(x), f_{i+n})$ is a semidynamical system.

19.4 Periodicity

In this section our focus will be on p-periodic difference equations of the form

$$
x_{n+1} = f_n(x_n), \tag{19.7}
$$

where $f_{n+p} = f_n$ for all $n \in \mathbb{Z}^+$.

The question that we are going to address is this: What are the permissible periods of the periodic orbits of (19.7)?

We begin by defining an r-periodic cycle (orbit).

Definition 19.1. An ordered set of points $C_r = \{\overline{x}_0, \overline{x}_1, \ldots, \overline{x}_{r-1}\}$ is r-periodic in X if

$$
f_{(i+nr) \bmod p}(\overline{x}_i) = \overline{x}_{(i+1) \bmod r}, n \in \mathbb{Z}^+.
$$

In particular,

$$
f_i(\overline{x}_i) = \overline{x}_{i+1}, 0 \le i \le r - 2,
$$

and

$$
f_t(\overline{x}_{t \bmod r}) = \overline{x}_{(t+1) \bmod r}, r - 1 \le t \le p - 1.
$$

It should be noted that the r-periodic cycle C_r in X generates an s-periodic cycle on the skew-product $X \times Y$ of the form $\widehat{C}_s = \{(\overline{x}_0, f_0), (\overline{x}_1, f_1), \ldots, (\overline{x}_{s \bmod r}, f_{s \bmod p})\}$, where $s = lcm[r, p]$ is the least common multiple of r and p.

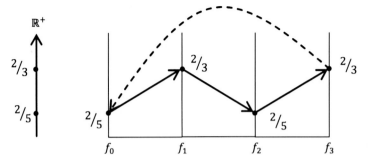

Fig. 19.1 A 2-periodic cycle in a 4-periodic difference equation

The r-periodic orbit C_r is called an *r-geometric cycle*, and the s-periodic orbit $\widehat{C}_r$ is called an *s-complete cycle*.

Example 19.3. Consider the nonautonomous periodic Beverton–Holt equation

$$x_{n+1} = \frac{\mu_n K_n x_n}{K_n + (\mu_n - 1)x_n}, \tag{19.8}$$

with $\mu_n > 1$, $K_n > 0$, $K_{n+p} = K_n$, and $\mu_{n+p} = \mu_n$, for all $n \in \mathbb{Z}^+$.

1. Assume that $\mu_n = \mu > 1$ is constant for all $n \in \mathbb{Z}^+$. Then one may appeal to Corollary 6.5 in [14] to show that (19.8) has no nontrivial periodic cycles of period less than p. In fact, (19.8) has a unique globally asymptotically stable cycle of minimal period p.
2. Assume that μ_n is periodic. Let $\mu_0 = 3$, $\mu_1 = 4$, $\mu_2 = 2$, $\mu_3 = 5$, $K_0 = 1$, $K_1 = \frac{6}{17}$, $K_2 = 2$, and $K_3 = \frac{4}{11}$. This leads to a 4-periodic difference equation. There is, however, a 2-geometric cycle, namely, $C_2 = \{\frac{2}{5}, \frac{2}{3}\}$ (see Fig. 19.1). This 2-periodic cycle in the space X generates the following 4-complete cycle on the skew-product $X \times Y$

$$\widehat{C}_4 = \{(\tfrac{2}{5}, f_0), (\tfrac{2}{3}, f_1), (\tfrac{2}{5}, f_2), (\tfrac{2}{3}, f_3)\},$$

where $f_0(x) = \frac{3x}{1+2x}$, $f_1(x) = \frac{24x}{6+51x}$, $f_2(x) = \frac{4x}{2+x}$, and $f_3(x) = \frac{5x}{1+11x}$.

We are going to provide a deeper analysis of the preceding example. Let $d = \gcd(r, p)$ be the greatest common divisor of r and p, $s = lcm[r, p]$ be the least common multiple of r and p, $m = \frac{p}{d}$, and $\ell = \frac{s}{p}$. The following result is one of two crucial lemmas in this survey.

Lemma 19.1. *[14] Let $C_r = \{\overline{x}_0, \overline{x}_1, \ldots, \overline{x}_{r-1}\}$ be a set of points in a metric space X. Then the following statements are equivalent.*

19 Towards a Theory of Periodic Difference Equations

1. C_r is a periodic cycle of minimal period r.

2. For $0 \le i \le r - 1$, $f_{(i+nd) \bmod p}(\overline{x}_i) = \overline{x}_{(i+1) \bmod r}$.

3. For $0 \le i \le r - 1$, the graphs of the functions

$$f_i, f_{(i+d) \bmod p}, \ldots, f_{(i+(m-1)d) \bmod p}$$

intersect at the ℓ points

$$(\overline{x}_i, \overline{x}_{(i+1) \bmod r}), (\overline{x}_{(i+d) \bmod r}, \overline{x}_{(i+1+d) \bmod r}), \ldots,$$

$$(\overline{x}_{(i+(\ell-1)d) \bmod r}, \overline{x}_{(i+(\ell-1)d+1) \bmod r})$$

Corollary 19.1. *[29] Assume that the one-parameter family $F(\alpha, x)$ is one to one in α. Let $f_n(x_n) = F(\alpha_n, x_n)$. Then if the p-periodic difference equation, with minimal period p,*

$$x_{n+1} = f_n(x_n) \tag{19.9}$$

has a nontrivial periodic cycle of minimal period r, then $r = tp$, $t \in \mathbb{Z}^+$.

Proof. Suppose that (19.9) has a periodic cycle $C_r = \{\overline{x}_0, \overline{x}_1, \ldots, \overline{x}_{r-1}\}$ of period $r < p$, and let $d = \gcd(r, p)$, $s = lcm[r, p]$, $m = \frac{p}{d}$, and $\ell = \frac{s}{p}$. Then by Lemma 19.1, the graphs of the maps $f_0, f_d, \ldots, f_{(m-1)d}$ must intersect at the points $(\overline{x}_0, \overline{x}_1), (\overline{x}_d, \overline{x}_{d+1}), \ldots, (\overline{x}_{(\ell-1)d}, \overline{x}_{(\ell-1)d+1})$.

Since $F(\alpha, x)$ is one to one in α, the maps $f_0, f_d, \ldots, f_{(m-1)d}$ do not intersect, unless they are all equal. Similarly, one may show that $f_i = f_{i+d} = \ldots = f_{i+(m-1)d}$. This shows that (19.9) is of minimal period d, a contradiction. Hence r is equal to p or a multiple of p. $\qquad\square$

Applying Corollary 19.1 to the periodic Beverton–Holt equation with $K_{n+p} = K_n, \mu_n = \mu$, for all $n \in \mathbb{Z}^+$, shows that the only possible period of a nontrivial periodic cycle is p. However, for the case μ_n and K_n are both periodic of common period p, the situation is murky as was demonstrated by Example 19.3, case 2.

For the values $\mu_0 = 3$, $\mu_1 = 4$, $\mu_2 = 2$, $\mu_3 = 5$, $K_0 = 1$, $K_1 = 6/17$, $K_2 = 2$, and $K_3 = 4/11$, we have $f_0(x) = \frac{3x}{1+2x}$, $f_1(x) = \frac{24x}{6+51x}$, $f_2(x) = \frac{4x}{2+x}$, and $f_3(x) = \frac{5x}{1+11x}$. Let $\mathcal{F} = \{f_0, f_1, f_2, f_3\}$. Clearly $x^* = 0$ is a fixed point of the periodic system $\mathcal{F}$. To have a positive fixed point (period 1) or a periodic cycle of period 3, we must have the graphs of f_0, f_1, f_2, f_3 intersect at points $(\overline{x}_0, \overline{x}_1), (\overline{x}_1, \overline{x}_2), \ldots, (\overline{x}_{\ell-1}, \overline{x}_\ell)$, where $\ell = 1$ or $\ell = 3$. Simple computation shows that this is not possible. Moreover, one may show that the graphs of f_0 and f_2 intersect at the points $(2/5, 2/3)$ and the graphs of f_1 and f_3 intersect at the points $(2/3, 2/5)$. Hence $C_2 = \{2/5, 2/3\}$ is a 2-periodic cycle. Moreover, the equation has the 4-periodic cycle

$$\left\{ \frac{238}{361}, \frac{119}{298}, \frac{238}{417}, \frac{238}{607} \right\}.$$

Suppose that the p-periodic difference equation

$$x_{n+1} = f_n(x_n), \ f_{n+p} = f_n, n \in \mathbb{Z}^+ \tag{19.10}$$

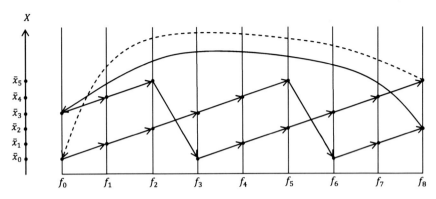

Fig. 19.2 A 6-periodic cycle in a 9-periodic system

has a periodic cycle of minimal period r. Then the associated skew-product system π has a periodic cycle of period $s = lcm[r, p]$ (s-complete cycle). There are p fibers $\mathscr{F}_i = \mathscr{P}^{-1}(f_i)$. Are the s periodic points equally distributed on the fibers? i.e. is the number of periodic points on each fiber equal to $\ell = s/p$?

Before giving the definitive answer to this question, let us examine the diagram present in Fig. 19.2 in which $p = 9$, and $r = 6$.

There are two points $\left(2 = \frac{lcm[6,9]}{9}\right)$ on each fiber. Since $d = \gcd(6, 9) = 3$, the graphs f_0, f_3, and f_6 intersect at the two points $(\overline{x}_0, \overline{x}_1)$, $(\overline{x}_3, \overline{x}_4)$; the graphs f_1, f_4, and f_7 intersect at the two points $(\overline{x}_1, \overline{x}_2)$, $(\overline{x}_4, \overline{x}_5)$; and the graphs f_2, f_5, f_8 intersect at the points $(\overline{x}_2, \overline{x}_3)$, $(\overline{x}_5, \overline{x}_0)$.

Note that the number of periodic points on each fiber is 2, which is $\ell = \frac{lcm[r,p]}{p}$. The following crucial lemma proves this observation.

Lemma 19.2. *[13] Let $s = lcm[r, p]$. Then the orbit of $(\overline{x}_i, f_i)$ in the skew-product system intersect each fiber $\mathscr{F}_j$, $j = 0, 1, \ldots, p-1$, in exactly $\ell = s/p$ points and each of these points is periodic under the skew-product π with period s.*

Proof. Let $C_r = \{\overline{x}_0, \overline{x}_1, \ldots, \overline{x}_{r-1}\}$ be a periodic cycle of minimal period r. Then the orbit of $(\overline{x}_0, f_0)$ in the skew-product has a minimal period $s = lcm[r, p]$. Now $\mathscr{S} = \mathscr{O}((\overline{x}_0, f_0)) = \{\pi((\overline{x}_0, f_0), n) : n \in \mathbb{Z}^+\} \subset X \times Y$ is minimal, invariant under π and has s distinct points.

For each i, $0 \leq i \leq p - 1$, the maps

$$f_i : \mathscr{S} \cap \mathscr{F}_i \to \mathscr{S} \cap \mathscr{F}_{(i+1) \bmod p} \tag{19.11}$$

are surjective. We now show that it is injective.

Let N_i be cardinality of $\mathscr{S} \cap \mathscr{F}_i$. Then N_i is a non-increasing integer valued function and thus stabilizes at some fixed value from which it follows that N_i is constant. Thus each $\mathscr{S} \cap \mathscr{F}_i$ contains the same number of points, namely $\ell = s/p$. □

19 Towards a Theory of Periodic Difference Equations

19.5 Stability

We begin this section by stating the basic definitions of stability.

Definition 19.2. Let $C_r = \{\overline{x}_0, \overline{x}_1, \ldots, \overline{x}_{r-1}\}$ be an r-periodic cycle in the p-periodic equation (19.10) in a metric space (X, ρ) and $s = lcm[r, p]$ be the least common multiple of p and r. Then

1. C_r is stable if given $\epsilon > 0$, there exists $\delta > 0$ such that

$$\rho(z, \overline{x}_{i \ mod \ r}) < \delta \text{ implies } \rho(\Phi_n^i(z), \Phi_n^i(\overline{x}_{i \ mod \ r})) < \epsilon$$

   for all $n \in \mathbb{Z}^+$, and $0 \leq i \leq p - 1$. Otherwise, C_r is said unstable.
2. C_r is attracting if there exists $\eta > 0$ such that

$$\rho(z, \overline{x}_{i \ mod \ r}) < \eta \text{ implies } \lim_{n \to \infty} \Phi_{ns}^i(z) = \overline{x}_{i \ mod \ r}.$$

3. We say that C_r is asymptotically stable if it is both stable and attracting. If in addition, $\eta = \infty$, C_r is said to be globally asymptotically stable.

Lemma 19.3. [29] An r-periodic cycle $C_r = \{\overline{x}_0, \overline{x}_1, \ldots, \overline{x}_{r-1}\}$ in (19.10) is

1. Asymptotically stable if $|\prod_{i=0}^{s} f'_{i \ mod \ p}(\overline{x}_{i \ mod \ r})| < 1$.
2. Unstable if $|\prod_{i=0}^{s} f'_{i \ mod \ p}(\overline{x}_{i \ mod \ r})| > 1$.

where $s = lcm[r, p]$ is the least common multiple of p and r.

Consider the skew-product system π on $X \times Y$ with X a metric space with metric ρ, $Y = \{f_0, f_1, \ldots, f_{p-1}\}$ equipped with the discrete metric $\widetilde{\rho}$, where

$$\widetilde{\rho}(f_i, f_j) = \begin{cases} 0 \text{ if } i = j \\ 1 \text{ if } i \neq j \end{cases}.$$

Define a metric D on $X \times Y$ as

$$D\left((x, f_i), (y, f_j)\right) = \rho(x, y) + \widetilde{\rho}(f_i, f_j).$$

Let $\pi^1(x, f) = \pi((x, f), 1)$, then $\pi^n(x, f) = \pi((x, f), n)$. Thus $\pi^1 : X \times Y \to X \times Y$ is a continuous map which generates an autonomous system on $X \times Y$. Consequently, the stability definitions of fixed points and periodic cycles follow the standard ones that may be found in [9, 11].

Now we give a definition of stability for a complete periodic cycle in the skew-product system.

Definition 19.3. A complete periodic cycle $\widehat{C}_s = \{(\overline{x}_0, f_0), \ldots, (\overline{x}_{s \ mod \ r}, f_{s \ mod \ p})\}$ is

1. Stable if given $\epsilon > 0$, there exits $\delta > 0$, such that

$$D((z, f_i), (\overline{x}_0, f_0)) < \delta \text{ implies } D(\pi^n(z, f_i), \pi^n(\overline{x}_0, f_0)) < \epsilon, \forall n \in \mathbb{Z}^+.$$

Otherwise, $\widehat{C}_s$ is said unstable.

2. Attracting if there exists $\eta > 0$ such that

$$D((z, f_i), (\overline{x}_0, f_0)) < \eta \text{ implies } \lim_{n \to \infty} \pi^{ns}(z, f_i) = (\overline{x}_0, f_0).$$

3. Asymptotically stable if it is both stable and attracting. If in addition, $\eta = \infty$, $\widehat{C}_s$ is said to be globally asymptotically stable.

Since $f_{i+ns} = f_i$ for all n, it follows from the above convergence that $f_i = f_0$. Hence, stability can occur only on each fiber $X \times \{f_i\}, 0 \leq i \leq p - 1$.

It should be noted that one may reformulate Lemma 19.3 in the setting of the skew-product theorem. However, to do so, one needs to develop the notion of derivative in the space $X \times Y$.

Definition 19.4. Let $g = \pi^p : X \times Y \to X \times Y$ defined as $g(x, f_i) = (\Phi_p^i(x), f_i)$. The generalized derivative of g is defined as $g'(x, f_i) = \frac{d}{dx} \left(\Phi_p^i(x) \right) = \left(\Phi_p^i \right)'(x)$.

Lemma 19.4. *A complete periodic cycle* $\widetilde{C}_s = \{(\overline{x}_0, f_0), \dots, (\overline{x}_{s \bmod r}, f_{s \bmod p})\}$ *of the skew-product system π on $X \times Y$ is*

1. *Asymptotically stable if* $|\prod_{i=0}^{s} f'_{i \bmod p}(\overline{x}_{i \bmod r})| < 1$,
2. *Unstable if* $|\prod_{i=0}^{s} f'_{i \bmod p}(\overline{x}_{i \bmod r})| > 1$,

where $s = lcm[r, p]$ is the least common multiple of p and r.

We are now ready to state our main result in this survey.

Theorem 19.2. *[13] Assume that X is a connected metric space and each $f_i \in Y$ is a continuous map on X, with $f_{i+p} = f_i$. Let $C_r = \{\overline{x}_0, \overline{x}_1, \dots, \overline{x}_{r-1}\}$ be a periodic cycle of minimal period r. If C_r is globally asymptotically stable, then r divides p. Moreover, $r = p$ if the sequence $\{f_n\}$ is a one-parameter family of maps $F(\mu_n, x)$ and F is one to one with respect to μ.*

Proof. The skew-product system π on $X \times Y$ has the periodic orbit

$$\{(\overline{x}_0, f_0), (\overline{x}_1, f_1), \dots, (\overline{x}_{s \bmod r}, f_{s \bmod p})\}$$

which is globally asymptotically stable. But as we remarked earlier, globally stability can occur only on fibers. By Lemma 19.2, there are $\ell = s/p$ points on each fiber. If $\ell > 1$, we have a globally asymptotically ℓ-periodic cycle in the connected metric space $X \times \{f_i\}$ under the map π^p. This violates Elaydi–Yakubu Theorem [12]. Hence $\ell = 1$ and consequently $r|p$.

Note that by Lemma 19.1, the graphs of the maps $f_i, f_{i+d}, \dots, f_{i+(m-1)d}$, $0 \leq i \leq p - 1$, must intersect at ℓ points. However, since $\{f_i\}$ is a one parameter

19 Towards a Theory of Periodic Difference Equations

family of maps $F(\mu_n, x)$ where F is one to one with respect to the parameter μ, it follows that $f_i = f_{i+d}, 0 \leq i \leq p - 1$. This implies that d is the period of our system and since p is the minimal period of the system, this implies that $d = p$. Hence $r = p$. $\qquad\square$

19.6 An Extension of Singer's Theorem

One of the well known work done by Singer is present in his famous paper [33] and currently known by Singer's theorem. It is a useful tool in finding an upper bound for the number of stable cycles in autonomous difference equations. In this section we present the natural extension of this theorem to the periodic nonautonomous difference equations.

Recall that the Schwarzian derivative, Sf, of a map f at x is defined as

$$Sf(x) = \frac{f'''(x)}{f'(x)} - \frac{3}{2} \left(\frac{f''(x)}{f'(x)} \right)^2.$$

Let $f : I \to I$ be a C^3 map with a negative Schwarzian derivative for all $x \in I$, defined on the closed interval I. If f has m critical points in I, then f has at most $m + 2$ attracting period cycles of any given period.

Now consider the p-periodic system $\mathscr{F} = \{f_0, f_1, f_2, \ldots, f_{p-1}\}$ of continuous maps defined on a closed interval I.

Assume that there are m_i critical points for the map $f_i, 0 \leq i \leq p - 1$. On the fiber $\mathscr{F}_0 = I \times f_0$, there are m_0 critical points of f_0, at least m_1 critical points consisting of all the pre-images under f_0 of the m_1 critical points of $f_1, \ldots$ and at least m_{p-1} critical points that consist of all the pre images, under Φ_{p-2}, of the m_{p-1} critical points of f_{p-1}. Since each critical point of Φ_p is mapped, under compositions of our maps, to one of the original critical points of one of the maps f_i, it follows that the number of significant critical points is $\sum_{i=0}^{p-1} m_i$.

By Singer's Theorem, there are at most $\left[\sum_{i=0}^{P-1} m_i + 2 \right]$ attracting periodic cycles of any given period. Notice that periodic cycles that appear on fiber $\mathscr{F}_i$ are just phase shifts of periodic cycles that appear on fiber $\mathscr{F}_0$. Hence we conclude that there are at most $\sum_{i=0}^{P-1} m_i + 2$ attracting cycles of any given period (See [2] for details).

So a consequence of this extension, one may show that if the maps are the logistic maps

$$f_i(x) = \mu_i x (1 - x), \mu_i > 0, 0 \leq i \leq p - 1,$$

defined on the interval $[0, 1]$, then the p-periodic system $\{f_0, f_1, \ldots, f_{p-1}\}$ has at most p-attracting cycles of any given period r. Notice that each map f_i has one critical point, $x = 1/2$, and the boundary points 0 and 1 are attracted only to 0.

19.7 Bifurcation

The study of various notions of bifurcation in the setting of discrete nonautonomous systems is still at its infancy stage. The main contribution in this area are the papers by Henson [24], Al-Sharawi and Angelos [2], Oliveira and D'Aniello [30], and recently Luís, Elaydi and Oliveira [29].

The main objective in this section is to give the pertinent definitions, notions, terminology and results done in [29]. Though our focus here will be on 2-periodic systems, the ideas presented can be easily extended to the general periodic case.

Throughout this section we assume that the maps f_0 and f_1 arise from a one-parameter family of maps such that $f_1 = f_{\alpha_1}$ and $f_0 = f_{\alpha_0}$ with $\alpha_0 = q\alpha_1$ for some real number $q > 0$. Thus one may write, without loss of generality, our system as $\mathscr{F} = \{f_0, f_1\}$.

Moreover, we assume that the one-parameter family of maps is one-to-one with respect to the parameter. Let $C_r = \{\overline{x}_0, \overline{x}_1, \dots, \overline{x}_{r-1}\}$ be an r-periodic cycle of $\mathscr{F}$. Then by Corollary 19.1 the latter assumption implies that $r = 2m, m \geq 1$.

With $\Phi_2 = f_1 \circ f_0$, one may write the orbit of $\overline{x}_0$ as (see Fig. 19.3)

$$\mathcal{O}(\overline{x}_0) = \{\overline{x}_0, f_0(\overline{x}_0), \Phi_2(\overline{x}_0), f_0 \circ \Phi_2(\overline{x}_0), \Phi_4(\overline{x}_0), \dots, \Phi_{2(m-1)}(\overline{x}_0),$$
$$\times f_0 \circ \Phi_{2(m-1)}(\overline{x}_0)\} = \{\overline{x}_0, \Phi_1(\overline{x}_0), \Phi_2(\overline{x}_0), \dots, \Phi_{2m-1}(\overline{x}_0)\} \quad (19.12)$$

Equivalently, one may write the sequence of points given in (19.12) as

$$\mathcal{O}(\overline{x}_1) = \{f_1 \circ \widetilde{\Phi}_{2(m-1)}(\overline{x}_1), \overline{x}_1, f_1(\overline{x}_1), \widetilde{\Phi}_2(\overline{x}_1), f_1 \circ \widetilde{\Phi}_2(\overline{x}_1), \dots, \widetilde{\Phi}_{2(m-1)}(\overline{x}_1)\}$$
$$= \{\widetilde{\Phi}_{2m-1}(\overline{x}_1), \overline{x}_1, \widetilde{\Phi}_1(\overline{x}_1), \widetilde{\Phi}_{2m}(\overline{x}_1), \dots, \widetilde{\Phi}_{2m-2}(\overline{x}_1)\} \quad (19.13)$$

where $\widetilde{\Phi} = f_0 \circ f_1$. Hence the order of the composition is irrelevant to the dynamics of the system.

The dynamics of $\mathscr{F}$ depends very much on the parameter as the qualitative structure of the dynamical system changes as the parameter changes. These qualitative changes in the dynamics of the system are called bifurcations and the parameter values at which they occur are called bifurcation points. For autonomous systems or single maps the bifurcation analysis may be found in Elaydi [11].

In a one-dimensional systems generated by a one-parameter family of maps f_α, a bifurcation at a fixed point x^* occurs when $\frac{\partial f}{\partial x}(\alpha^*, x^*) = 1$ or -1 at a bifurcation point α^*. The former case leads to a saddle-node bifurcation, while the latter case leads to a period-doubling bifurcation.

Now we are going to extend this analysis to 2-periodic difference equations or $\mathscr{F} = \{f_0, f_1\}$. To simplify the notation we write $\Phi_2(\alpha, x)$ instead of $\Phi_2(x)$ and $\widetilde{\Phi}_2(\alpha, x)$ instead of $\widetilde{\Phi}_2(x)$. Then $\Phi_{2m}(\overline{x}_{2i}) = \overline{x}_{(2i)modr}$ and $\widetilde{\Phi}_{2m}(\overline{x}_{2i+1}) = \overline{x}_{(2i+1)modr}$, $1 \leq i \leq m$. In general, we have $\Phi_{2nm}(\overline{x}_{2i}) = \overline{x}_{(2i)modr}$ and $\widetilde{\Phi}_{2nm}(\overline{x}_{2i+1}) = \overline{x}_{(2i+1)modr}$, $n \geq 1$.

19 Towards a Theory of Periodic Difference Equations

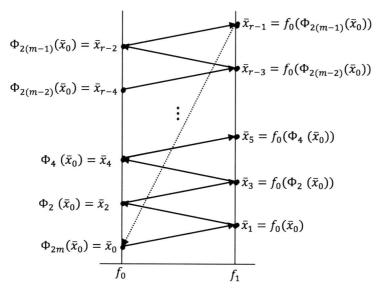

Fig. 19.3 Sequence of the periodic points $\{\bar{x}_0, \bar{x}_1, \ldots, \bar{x}_{r-1}\}$ in the 2-periodic system $\mathscr{F} = \{f_0, f_1\}$ illustrated in the fibers, where $\Phi_2 = f_1 \circ f_0$ and $r = 2m, m \geq 1$

Assuming $\frac{\partial \Phi_{2m}}{\partial x}(\bar{\alpha}, \bar{x}_0) = 1$ at a bifurcation point $\bar{\alpha}$, by the chain rule, we have

$$\frac{\partial \Phi_2}{\partial x}(\bar{\alpha}, \bar{x}_{2m-2}) \frac{\partial \Phi_2}{\partial x}(\bar{\alpha}, \bar{x}_{2m-4}) \ldots \frac{\partial \Phi_2}{\partial x}(\bar{\alpha}, \bar{x}_2) \frac{\partial \Phi_2}{\partial x}(\bar{\alpha}, \bar{x}_0) = 1$$

or

$$f_1'(\bar{x}_{2m-1}) f_0'(\bar{x}_{2m-2}) f_1'(\bar{x}_{2m-3}) f_0'(\bar{x}_{2m-4}) \ldots f_1'(\bar{x}_3) f_0'(\bar{x}_2) f_1'(\bar{x}_1) f_0'(\bar{x}_0) = 1 \tag{19.14}$$

Applying f_0 on both sides of the identity $\Phi_{2m}(\bar{\alpha}, \bar{x}_0) = \bar{x}_0$, yields $\widetilde{\Phi}_{2m}(\bar{\alpha}, \bar{x}_1) = \bar{x}_1$. Differentiating both sides of this equation yields

$$\frac{\partial \widetilde{\Phi}_2}{\partial x}(\bar{\alpha}, \bar{x}_{2m-1}) \frac{\partial \widetilde{\Phi}_2}{\partial x}(\bar{\alpha}, \bar{x}_{2m-3}) \ldots \frac{\partial \widetilde{\Phi}_2}{\partial x}(\bar{\alpha}, \bar{x}_3) \frac{\partial \widetilde{\Phi}_2}{\partial x}(\bar{\alpha}, \bar{x}_1) = 1$$

or equivalently

$$f_0'(\bar{x}_0) f_1'(\bar{x}_{2m-1}) f_0'(\bar{x}_{2m-2}) f_1'(\bar{x}_{2m-3}) \ldots f_0'(\bar{x}_4) f_1'(\bar{x}_3) f_0'(\bar{x}_2) f_1'(\bar{x}_1) = 1. \tag{19.15}$$

Hence (19.14) is equivalent to (19.15). More generally the following relation holds

$$\frac{\partial \Phi_{2m}}{\partial x}(\bar{\alpha}, \bar{x}_{2j}) = \frac{\partial \widetilde{\Phi}_{2m}}{\partial x}(\bar{\alpha}, \bar{x}_{2j-1}), j \in \{1, 2, \ldots, m\}. \tag{19.16}$$

Now we are ready to write the two main results of this section.

Theorem 19.3 (Saddle-node Bifurcation for 2-periodic systems [29]). *Let* $C_r = \{\overline{x}_0, \overline{x}_1, \ldots, \overline{x}_{r-1}\}$ *be a periodic r-cycle of $\mathscr{F}$. Suppose that both $\frac{\partial^2 \Phi_2}{\partial x^2}$ and $\frac{\partial^2 \Phi_2}{\partial^2}$ exist and are continuous in a neighborhood of a periodic orbit such that $\frac{\partial \Phi_{2m}}{\partial x}(\overline{\alpha}, \overline{x}_0) = 1$ for the periodic point $\overline{x}_0$. Assume also that*

$$A = \frac{\partial \Phi_{2m}}{\partial \alpha}(\overline{\alpha}, \overline{x}_0) \neq 0 \text{ and } B = \frac{\partial^2 \Phi_{2m}}{\partial x^2}(\overline{\alpha}, \overline{x}_0) \neq 0.$$

Then there exists an interval J around the periodic orbit and a C^2-map $\alpha = h(x)$, where $h : J \to \mathbb{R}$ such that $h(\overline{x}_0) = \overline{\alpha}$, and $\Phi_{2m}(x, h(x)) = x$. Moreover, if $AB < 0$, the periodic points exists for $\alpha > \overline{\alpha}$, and, if $AB > 0$, the periodic points exists for $\alpha < \overline{\alpha}$.

When $\frac{\partial \Phi_{2m}}{\partial x}(\overline{\alpha}, \overline{x}_0) = 1$ but $\frac{\partial \Phi_{2m}}{\partial \alpha}(\overline{\alpha}, \overline{x}) = 0$, two types of bifurcations appear. The first is called transcritical bifurcation which occurs when $\frac{\partial^2 \Phi_{2m}}{\partial x^2}(\overline{\alpha}, \overline{x}_0) \neq 0$ and the second is called pitchfork bifurcation which appears when $\frac{\partial^2 \Phi_{2m}}{\partial x^2}(\overline{\alpha}, \overline{x}_0) = 0$. For more details about this two types of bifurcation see Table 2.1 in [11, pp. 90], and [30]. In the former work the author presents many cases for autonomous maps while in the latter article the authors study the pitchfork bifurcation for nonautonomous 2-periodic systems in which the maps have negative Schwarzian derivative.

The next result gives the conditions for the period-doubling bifurcation.

Theorem 19.4 (Period-Doubling Bifurcation for 2-periodic systems [29]). *Let* $C_r = \{\overline{x}_0, \overline{x}_1, \ldots, \overline{x}_{r-1}\}$ *be a periodic r-cycle of $\mathscr{F}$. Assume that both $\frac{\partial^2 \Phi_2}{\partial x^2}$ and $\frac{\partial \Phi_2}{\partial \alpha}$ exist and are continuous in a neighborhood of a periodic orbit, $\frac{\partial \Phi_{2m}}{\partial x}(\overline{\alpha}, \overline{x}_0) = -1$ for the periodic point $\overline{x}_0$ and $\frac{\partial^2 \Phi_{4m}}{\partial \alpha \partial x}(\overline{\alpha}, \overline{x}_0) \neq 0$. Then, there exists an interval J around the periodic orbit and a function $h : J \to \mathbb{R}$ such that $\Phi_{2m}(x, h(x)) \neq x$ but $\Phi_{4m}(x, h(x)) = x$.*

Now we are going to apply these two results with an interesting example from [29]. First we need the following definition.

Definition 19.5. A unimodal map is said to have the Allee[1] effect if it has three fixed points $x_1^* = 0$, $x_2^* = A$, and $x_3^* = K$, with $0 < A < K$, in which x_1^* is asymptotically stable, x_2^* is unstable, and x_3^* may be stable or unstable.

Remark 19.1. Note that if $\mathscr{F}$ is a periodic set formed by unimodal Allee maps, neither the zero fixed point nor the threshold point can contribute to bifurcation,

---

[1] The Allee effect is a phenomenon in population dynamics attributed to the American biologist Warder Clayde Allee 1885–1955 [1]. Allee proposed that the per capita birth rate declines at low density or population sizes. In the languages of dynamical systems or difference equations, a map representing the Allee effect must have tree fixed points, an asymptotically stable zero fixed point, a small unstable fixed point, called the threshold point, and a bigger positive fixed point, called the carrying capacity, that is asymptotically stable at least for smaller values of the parameters.

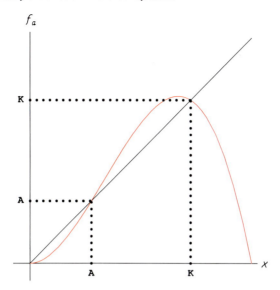

Fig. 19.4 A unimodal Allee map with three fixed points 0, A and k

since the former is always asymptotically stable and the latter is always unstable. Hence bifurcation may only occur at the carrying capacity of $\mathscr{F}$.

Example 19.4. [29] Consider the 2-periodic system $\mathscr{W} = \{f_0, f_1\}$, where

$$f_i(x) = a_i x^2 (1-x), i = 0, 1$$

in which $x \in [0, 1]$ and $a_i > 0$, $i = 0, 1$. For an individual map f_i, if $a_i < 4$ we have a globally asymptotically stable zero fixed point and no other fixed point. At $a_i = 4$ an unstable fixed point is born after which f_i becomes a unimodal map with an Allee effect (see Fig. 19.4). Henceforth, we will assume that $a_0, a_1 > 4$.

Since 0 is the only fixed point under the system $\mathscr{W}$, we focus our attention on 2-periodic cycles $\{\overline{x}_0, \overline{x}_1\}$ with $f_0(\overline{x}_0) = \overline{x}_1$, and $f_1(\overline{x}_1) = \overline{x}_0$.

A Saddle-node bifurcation occurs when $\frac{\partial}{\partial t}(\Phi_2(t))\big|_{t=\overline{x}_0} = \Phi'_2(\overline{x}_0) = 1$, and a period-doubling bifurcation occurs when $\frac{\partial}{\partial t}(\Phi_2(t))\big|_{t=\overline{x}_0} = \Phi'_2(\overline{x}_0) = -1$.

For the saddle-node bifurcation we then solve the equations

$$\begin{cases} \overline{x}_0 = f_1(f_0(\overline{x}_0)) \\ f'_1(f_0(\overline{x}_0)) f'_0(\overline{x}_0) = 1 \end{cases} \tag{19.17}$$

and for the period-doubling bifurcations we solve the equations

$$\begin{cases} \overline{x}_0 = f_1(f_0(\overline{x}_0)) \\ f'_1(f_0(\overline{x}_0)) f'_0(\overline{x}_0) = -1 \end{cases} \tag{19.18}$$

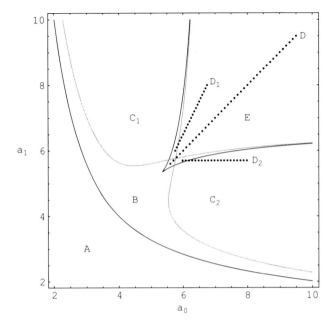

Fig. 19.5 Bifurcations curves for the 2-periodic nonautonomous difference equation with Allee effects $x_{n+1} = a_n x_n^2 (1 - x_n)$, in the (a_0, a_1)-plane, where $a_{n+2} = a_n$ and $x_{n+2} = x_n$

Using the command "resultant"[2] in Mathematica or Maple Software, we eliminate the variable $\bar{x}_0$ in (19.17) and (19.18). Equation (19.17) yields

$$16777216 + 16384 a_0 a_1 - 576000 a_0^2 a_1 + 84375 a_0^3 a_1 - 576000 a_0 a_1^2$$
$$+ 914 a_0^2 a_1^2 - 350 a_0^3 a_1^2 + 84375 a_0 a_1^3 - 350 a_0^2 a_1^3 + 19827 a_0^3 a_1^3$$
$$- 2916 a_0^4 a_1^3 - 2916 a_0^3 a_1^4 + 432 a_0^4 a_1^4 = 0$$

while (19.18) yields

$$100000000 - 120000 a_0 a_1 - 2998800 a_0^2 a_1 + 453789 a_0^3 a_1 - 2998800 a_0 a_1^2$$
$$- 4598 a_0^2 a_1^2 + 2702 a_0^3 a_1^2 + 453789 a_0 a_1^3 + 2702 a_0^2 a_1^3 + 89765 a_0^3 a_1^3$$
$$- 13500 a_0^4 a_1^3 - 13500 a_0^3 a_1^4 + 2000 a_0^4 a_1^4 = 0$$

For each one of these last two equations we invoke the implicit function theorem to plot, in the (a_0, a_1)-plane, the bifurcation curves (see Fig. 19.5). The black curves are the solution of the former equation at which saddle-node bifurcation

---

[2] The command "resultant" is a powerful tool that helps us in finding the implicit solutions for a polynomial equations with low degree. We are not aware of similar techniques that work for nonpolynomial equation such the Ricker map $R_p(x) = x e^{p-x}$, $p, x > 0$.

19 Towards a Theory of Periodic Difference Equations 303

occurs, while the gray curves are the solution of the latter equation at which period-doubling bifurcations occurs. The black cusp is the curve of pitchfork bifurcation. In the regions identified by letters one can conclude the following.

- If $a_0, a_1 \in A$ then the fixed point $x^* = 0$ is globally asymptotically stable.
- If $a_0, a_1 \in B \backslash D$ then there are two 2-periodic cycles, one attracting and one unstable.
- If $a_0, a_1 \in D$ then there are two attracting 2-periodic cycles (from the pitchfork bifurcation) and two unstable 2-periodic cycles.
- If $a_0, a_1 \in (C_1 \cup C_2) \backslash (D_1 \cup D_2)$ then there is an attracting 4-periodic cycle (from the period doubling bifurcation) and two unstable 2-periodic cycles.
- If $a_0, a_1 \in D_1 \cup D_2$ then there are two attracting 4-periodic cycles (from pitchfork bifurcation) and two unstable 2-periodic cycles.
- If $a_0, a_1 \in E$ then there are two attracting 8-periodic cycles (from period doubling bifurcation), two attracting 4-periodic cycles (from pitchfork bifurcation), and four unstable 2-periodic cycles.

It should be noted here that the bifurcation curves for the system $\mathscr{W}$ in Fig. 19.5 are incomplete. If we want to draw more bifurcation curves in the space of the parameters we must do the same for 4-periodic cycles, 8-periodic cycles, and so on. Finding the implicit solutions of these two new equations involve horrendous computations. The command "resultant" does not produce answers after certain values of the degree of the polynomial. So, for the system $\mathscr{W}$, unfortunately we are unable to draw these curves for the 4-periodic cycle. However, it should be noted that AlSharawi and Angelos [2] have used the command "resultant" to investigate the bifurcations of the periodically forced logistic map, and they were able to draw these curves for the 4-periodic cycles of the 2-periodic system. Moreover, these authors drew the bifurcation surfaces for the 3-periodic cycle of the 3-periodic system in the three dimensional space of the parameters.

Finally, we should mention that Grinfeld et al. [20] have used the command "resultant" much earlier to study the bifurcation of 2-periodic logistic systems.

19.8 A Note on Bifurcation Equations

In [5] the authors study the symmetry of degenerate bifurcation equations of periodic orbits in a nonautonomous system with respect to the order of the composition. They proved that the cyclic permutation in the order of the composition do not affect the solutions of the bifurcations in the parameter space.

In order to see this last observation, let $f_0, f_1, \ldots, f_{p-1}$ be a collection of maps

$$
\begin{aligned}
f_j : I_j \times \mathbb{R}^K &\longrightarrow \mathbb{R} \\
(x, \lambda) &\longrightarrow f_j(x, \lambda)
\end{aligned}
$$

where λ is a parameter vector, the fiber $\mathcal{F}_j = I_j \times \{f_j\}$ and $f_j \in \mathcal{C}^m\left(I_j, \mathbb{R}^K\right)$, $j = 0, 1, \ldots, p-1$.

We are concerned with the bifurcations that can occur, in particular with the bifurcations with higher degeneracy conditions on the derivatives of the iteration variable x and not on the degeneracy conditions on the parameters.

These below are the bifurcation equations with the most degenerate conditions that appear with j fixed, $0 \leq j \leq p-1$

$$\Phi_{kp}^j(x) = x, \tag{19.19}$$

$$\frac{d\Phi_{kp}^j}{dx}(x) = 1,$$

$$\frac{d^2\Phi_{kp}^j}{dx^2}(x) = 0,$$

$$\frac{d^3\Phi_{kp}^j}{dx^3}(x) = 0,$$

$$\vdots$$

$$\frac{d^m\Phi_{kp}^j}{dx^m}(x) = 0.$$

These equations have different solution in terms of x, depending on the j we choose.

A natural question arises:

Do the solutions in the parameter space depend on the particular choice of Φ_p^j?

This question was posed in [4, 30] and, was positively solved for $p = 2$, and degeneracy conditions of order $m = 2, 3$, that is, for pitchfork and swallowtail, respectively.

We now present the following lemma that is useful for solving general problems of the symmetry of the bifurcation equations.

Lemma 19.5. *Let $m \geq 1$ and let φ and ψ be real maps satisfying the conditions:*

1. *There exists a such that $\psi(a) = a$ and ψ is a Lipschitz homeomorphism in some open interval $I \ni a$.*
2. *φ is a Lipschitz homeomorphism with Lipschitz constant L in a open neighborhood I_φ of a point a such that $\psi(a) = a$ and ψ is a Lipschitz homeomorphism in some open interval $I \ni a.a$ such that $\varphi(a) = b$. Let φ^{-1} be its inverse in another open neighborhood of b, φ^{-1} is also Lipschitz continuous with constant M.*
3.

$$\lim_{x \to a} \frac{|\psi(x) - x|}{|x - a|^m} = 0.$$

19 Towards a Theory of Periodic Difference Equations

Then the conjugate $\widetilde{\psi}$ of ψ by the homeomorphism φ

$$\widetilde{\psi} = \varphi \psi \varphi^{-1}$$

satisfies

$$\lim_{y \to b} \frac{\left|\widetilde{\psi}(y) - y\right|}{|y - b|^m} = 0. \tag{19.20}$$

Using Lemma 19.5 one can prove the following theorem.

Theorem 19.5. *[5] Let $f_0, f_1 \dots, f_{p-1}$ be p functions with a sufficient number of derivatives satisfying the conditions:*

1. *There exists $\overline{x}_0, \overline{x}_1, \dots, \overline{x}_{p-1}$, fixed points of $\Phi_p, \Phi_p^1, \dots, \Phi_p^{p-1}$, respectively, that is*

$$\Phi_p^j(\overline{x}_j) = \overline{x}_j$$

2. *The first bifurcation condition holds*

$$\left. \frac{d\Phi_p(x)}{dx} \right|_{x=\overline{x}_0} = \prod_{j=0}^{p-1} \frac{df_j}{dx}(\overline{x}_j) = 1.$$

3. *Higher degeneracy conditions hold for Φ_p*

$$m \geq 2 : \left. \frac{d^n \Phi_p}{dx^n}(x) \right|_{x=\overline{x}_0} = 0, \qquad 2 \leq n \leq m.$$

Then the composition operator Φ_p^j, $0 \leq j \leq p-1$ satisfies

$$\left. \frac{d^n \Phi_p^j}{dx^n}(x) \right|_{x=\overline{x}_j} = 0, \text{ for } 2 \leq n \leq m.$$

In the case of periodic systems with period two the result is a particular case of the previous theorem.

Corollary 19.2. *[5] Let f_0 and f_1 be maps with a sufficient number of derivatives satisfying the conditions:*

1. *$(f_1 \circ f_0)(\overline{x}_0) = \overline{x}_0$ and $(f_0 \circ f_1)(\overline{x}_1) = \overline{x}_1$.*
2. *$\left. \frac{d(f_1 \circ f_0)}{dx}(x) \right|_{x=\overline{x}_0} = f_1'(\overline{x}_1) f_0'(\overline{x}_0) = 1$.*
3. *Fixed $m \geq 2$: $\left. \frac{d^n(f_1 \circ f_0)}{dx^n}(x) \right|_{x=\overline{x}_0} = 0$ for $2 \leq n \leq m$.*

Then the reverse composition $f_0 \circ f_1$ satisfies

$$\frac{d^n (f_0 \circ f_1)}{dx^n} (x)\bigg|_{x=\overline{x}_1} = 0, \text{ for } 2 \leq n \leq m.$$

Example 19.5. [5] We will prove directly that the second and third derivatives of alternating maps are both zero, regardless of the order of composition. We do this directly using the higher order chain rule, or Faà di Bruno formula [25] fist proved in [28].

Let f_0 and f_1 be $\mathbf{C}^3$ functions satisfying the conditions:

1. $(f_0 \circ f_0)(\overline{x}_0) = \overline{x}_0$ and $(f_0 \circ f_1)(\overline{x}_1) = \overline{x}_1$ which is $f_0(\overline{x}_0) = \overline{x}_1$ and $f_1(\overline{x}_1) = \overline{x}_0$.
2. $\frac{d(f_1 \circ f_0)}{dx}(x)\big|_{x=\overline{x}_0} = f_1'(\overline{x}_1) f_0'(\overline{x}_0) = 1.$
3. $\frac{d^m (f_1 \circ f_0)}{dx^m}(x)\big|_{x=\overline{x}_0} = 0$ for $m = 2, 3.$

Let us recall the formula of Faà di Bruno for the derivatives of the composition

$$\frac{d^m (f_1 \circ f_0)}{dx^m} (x) = m! \sum_{n=1}^{m} f_1^{(n)} (f_0(x)) \prod_{j=1}^{m} \frac{1}{b_j!} \left(\frac{f_0^{(j)}(x)}{j!} \right)^{b_j}, \qquad (19.21)$$

where the sum is over all different solutions b_j in nonnegative integers of the equation

$$\sum_{j=1}^{m} j b_j = m, \text{ and } n := \sum_{j=1}^{m} b_j.$$

To avoid to overload this example with indexes we use the notation

$$\frac{d^m (f_1 \circ f_0)}{dx^m} (x)\bigg|_{x=\overline{x}_0} = (f_1 f_0)_m, \qquad \frac{d^m (f_0 \circ f_1)}{dx^m} (x)\bigg|_{x=\overline{x}_1} = (f_0 f_1)_m.$$

With this notation the Faà di Bruno Formula computed at the conditions of the problem is

$$(f_1 f_0)_m = m! \sum_{n=1}^{m} f_1^{(n)}(\overline{x}_1) \prod_{j=1}^{m} \frac{1}{b_j!} \left(\frac{f_0^{(j)}(\overline{x}_0)}{j!} \right)^{b_j} \qquad (19.22)$$

and

$$(f_0 f_1)_m = m! \sum_{n=1}^{m} f_0^{(n)}(\overline{x}_0) \prod_{j=1}^{m} \frac{1}{b_j!} \left(\frac{f_1^{(j)}(\overline{x}_1)}{j!} \right)^{b_j} \qquad (19.23)$$

Condition 2 in this notation is now

$$f_0'(\overline{x}_0) f_1'(\overline{x}_1) = 1. \qquad (19.24)$$

19 Towards a Theory of Periodic Difference Equations   307

Let us consider the first cases. Let $m = 2$, we will use the formula (19.21), so we have to solve the equation

$$b_1 + 2b_2 = 2,$$

for all possible values of the vector (b_1, b_2) with nonnegative integers. The only solutions are $b_1 = 0, b_2 = 1$ which gives $n = 1$ and $b_1 = 2, b_2 = 0$ with $n = 2$, so we have

$$(f_1 f_0)_2 = 2! \left(f_1'(\overline{x}_1) \frac{1}{0!} \left(\frac{f_0'(\overline{x}_0)}{1!} \right)^0 \frac{1}{1!} \left(\frac{f_0''(\overline{x}_0)}{2!} \right)^1 \right)$$
$$+ 2! \left(f_1''(\overline{x}_1) \frac{1}{2!} \left(\frac{f_0'(\overline{x}_0)}{1!} \right)^2 \frac{1}{0!} \left(\frac{f_0''(\overline{x}_0)}{2!} \right)^0 \right) = 0$$
$$= f_1'(\overline{x}_1) f_0''(\overline{x}_0) + f_1''(\overline{x}_1)(f_0'(\overline{x}_0))^2$$
$$= f_1'(\overline{x}_1) f_0''(\overline{x}_0) + \frac{f_1''(\overline{x}_1)}{f_1'(\overline{x}_1)} = 0 \tag{19.25}$$

and

$$(fg)_2 = 2! \left(f_0'(\overline{x}_0) \frac{1}{0!} \left(\frac{f_1'(\overline{x}_1)}{1!} \right)^0 \frac{1}{1!} \left(\frac{f_1''(\overline{x}_1)}{2!} \right)^1 \right)$$
$$+ 2! \left(f_0''(\overline{x}_0) \frac{1}{2!} \left(\frac{f_1'(\overline{x}_1)}{1!} \right)^2 \frac{1}{0!} \left(\frac{f_1''(\overline{x}_1)}{2!} \right)^0 \right)$$
$$= f_0'(\overline{x}_0) f_1''(\overline{x}_1) + f_0''(\overline{x}_0)(f_1'(\overline{x}_1))^2$$
$$= \frac{f_1''(\overline{x}_1)}{f_1'(\overline{x}_1)} + f_0''(\overline{x}_0)(f_1'(\overline{x}_1))^2. \tag{19.26}$$

Using Cramer's rule we solve the system with (19.24) and (19.25) for $f_1''(\overline{x}_1)$, we get

$$f_1''(\overline{x}_1) = \frac{1}{(f_0'(\overline{x}_0))^3} \begin{vmatrix} 0 & f_0''(\overline{x}_0) \\ 1 & f_0'(\overline{x}_0) \end{vmatrix} = -\frac{f_0''(\overline{x}_0)}{(f_0'(\overline{x}_0))^3},$$

substituting $f_1'(\overline{x}_1)$ and $f_1''(\overline{x}_1)$ in (19.26) we get

$$(f_0 f_1)_2 = -f_0'(\overline{x}_0) \frac{f_0''(\overline{x}_0)}{(f_0'(\overline{x}_0))^3} + f_0''(\overline{x}_0) \frac{1}{(f_0'(\overline{x}_0))^2} = 0.$$

Now we consider the case $m = 3$

$$(f_1 f_0)_3 = f_1'(\overline{x}_1) f_0'''(\overline{x}_0) + 3 f_1''(\overline{x}_1) f_0'(\overline{x}_0) f_0''(\overline{x}_0) + f_1'''(\overline{x}_1)(f_0'(\overline{x}_0))^3$$

(19.27)

$$= f_1'(\overline{x}_1) f_0'''(\overline{x}_0) + \frac{3 f_1''(\overline{x}_1) f_0''(\overline{x}_0)}{f_1'(\overline{x}_1)} + \frac{f_1'''(\overline{x}_1)}{(f_1'(\overline{x}_1))^3} = 0.$$

We use Cramer's rule to solve the system consisting of (19.24), (19.25) and (19.27) for $f_1'''(\overline{x}_1)$

$$f_1'''(\overline{x}_1) = \frac{1}{(f_0'(\overline{x}_0))^6} \begin{vmatrix} 0 & 3 f_0'(\overline{x}_0) f_0''(\overline{x}_0) & f_0'''(\overline{x}_0) \\ 0 & (f_0'(\overline{x}_0))^2 & f_0''(\overline{x}_0) \\ 1 & 0 & f_0'(\overline{x}_0) \end{vmatrix}$$

$$= -\frac{f_0'''(\overline{x}_0)}{(f_0'(\overline{x}_0))^4} - \frac{3 \left(-\frac{f_0''(\overline{x}_0)}{(f_0'(\overline{x}_0))^3} \right) f_0''(\overline{x}_0)}{(f_0'(\overline{x}_0))^2}$$

$$= \frac{3(f_0''(\overline{x}_0))^2}{(f_0'(\overline{x}_0))^5} - \frac{f_0'''(\overline{x}_0)}{(f_0'(\overline{x}_0))^4}.$$

In the case of the reverse order composition the third derivative (substituting $f_1'(\overline{x}_1)$, $f_1''(\overline{x}_1)$ and $f_1'''(\overline{x}_1)$ by the solutions obtained previously) is given by

$$(f_0 f_1)_3 = f_0'(\overline{x}_0) f_1'''(\overline{x}_1) + 3 f_0''(\overline{x}_0) f_1'(\overline{x}_1) f_1''(\overline{x}_1) + f_0'''(\overline{x}_0)(f_1'(\overline{x}_1))^3$$

$$= f_0'(\overline{x}_0) \left(\frac{3(f_0''(\overline{x}_0))^2}{(f_0'(\overline{x}_0))^5} - \frac{f_0'''(\overline{x}_0)}{(f_0'(\overline{x}_0))^4} \right)$$

$$+ 3 f_0''(\overline{x}_0) \frac{1}{f_0'(\overline{x}_0)} \left(-\frac{f_0''(\overline{x}_0)}{(f_0'(\overline{x}_0))^3} \right) + \frac{f_0'''(\overline{x}_0)}{(f_0'(\overline{x}_0))^3}$$

$$= 0.$$

(19.28)

Finally we end this section presenting the extension of Theorem 19.5 to the periodic case that answers the question posed in the beginning of this section.

Theorem 19.6. *[5] Let $f_0, f_1, \ldots, f_{p-1}$ be maps with a sufficient number of derivatives satisfying the conditions:*

1. There are periodic orbits with period k for the compositions (kp for the iterates)

$$\Phi_{kp}(\overline{x}_0) = \overline{x}_0,$$
$$\Phi_{kp}^1(\overline{x}_1) = \overline{x}_1$$

$$\vdots$$

$$\Phi_{kp}^{p-1}(\overline{x}_{p-1}) = \overline{x}_{p-1}$$

19 Towards a Theory of Periodic Difference Equations

2. *The first bifurcation condition holds*

$$\frac{d\Phi_{kp}}{dx}(x)\Big|_{x=\bar{x}_0} = 1.$$

3. *Higher degeneracy conditions hold.*
   Fixed $m \geq 2$:

$$\frac{d^n\Phi_{kp}}{dx^n}(x)\Big|_{x=\bar{x}_0} = 0, \qquad 2 \leq n \leq m.$$

Then Φ_{kp}^j, with $j = 1,\ldots, p-1$, satisfies

$$\frac{d^n\Phi_{kp}^j}{dx^n}(x)\Big|_{x=\bar{x}_j} = 0, \text{ for } 2 \leq n \leq m.$$

Now we give an example for a 2-periodic system where the maps do not arise from a family of maps, one is unimodal and the other is bimodal.

Example 19.6. [30]
Let us now consider the maps

$$f_0 : [-1,1] \times [1,4] \longrightarrow \qquad \mathbb{R}$$
$$(x,\lambda_0) \qquad \longrightarrow \lambda_0 x^3 + (1-\lambda_0)x$$

and

$$f_1 : [-1,1] \times [0,2] \longrightarrow \qquad \mathbb{R}$$
$$(x,\lambda_0,\lambda_1) \qquad \longrightarrow -\lambda_1 x^2 - 1 + \lambda_1 \ .$$

The composition operator Φ_2 is now defined $\Phi : [-1,1] \times [1,4] \times [0,2] \longrightarrow \mathbb{R}$ such that

$$\Phi_2(x,\lambda_0,\lambda_1) = f_1(f_0(x,\lambda_0),\lambda_1)$$
$$= -\lambda_1(\lambda_0 x^3 + (1-\lambda_0)x)^2 - 1 + \lambda_1$$

We consider the pitchfork bifurcation problem. In this case we have $m = 2$. The bifurcation equations are

$$\Phi_{2k}(x,\lambda_0,\lambda_1) = x, \qquad\qquad (19.29)$$
$$\frac{d\Phi_{2k}}{dx}(x,\lambda_0,\lambda_1) = 1,$$
$$\frac{d^2\Phi_{2k}}{dx^2}(x,\lambda_0,\lambda_1) = 0,$$

where we assume that there are no more degeneracy conditions. This problem has two solutions, respectively

$$\overline{x}_0 \doteq -0.247674$$
$$(\lambda_0, \lambda_1) \doteq (2.85032, 0.90883)$$

and

$$\overline{y}_0 \doteq 0.620345$$
$$(\lambda_0, \lambda_1) \doteq (2.20004, 1.70216).$$

Hence there are two pitchfork bifurcation points.

Example 19.7. [30] We can also study $\widetilde{\Phi}_2 = f_0 \circ f_1$, with the same families of Example 19.6, the composition appearing now in the reverse order. It is possible to show, with much more cumbersome computations if treated directly, that this problem has two pitchfork bifurcation points. As in the previous example, exactly at the same values of the parameters

$$\overline{x}_0 \doteq 0.414971$$
$$(\lambda_0, \lambda_1) \doteq (2.85032, 0.90883)$$

and

$$\overline{y}_1 \doteq -0.219234$$
$$(\lambda_0, \lambda_1, x) \doteq (2.20004, 1.70216).$$

19.9 Attenuance and Resonance

In this section we study the attenuance and the resonance of some periodic models, that is, we compare the average of the carrying capacities with the average of the periodic cycle.

19.9.1 The Beverton–Holt Equation

In [8] Cushing and Henson conjectured that a nonautonomous p-periodic Beverton–Holt equation with periodically varying carrying capacity must be attenuant. This means that if $C_p = \{\overline{x}_0, \overline{x}_1, \ldots, \overline{x}_{p-1}\}$ is its p-periodic cycle, and $K_i, 0 \le i \le p-1$ are the carrying capacities, then

$$\frac{1}{p} \sum_{i=0}^{p-1} \overline{x}_i < \frac{1}{p} \sum_{i=0}^{p-1} K_i. \tag{19.30}$$

19 Towards a Theory of Periodic Difference Equations 311

Since the periodic cycle C_p is globally asymptotically stable on $(0, \infty)$, it follows that for any initial population density x_0, the time average of the population density x_n is eventually less than the average of the carrying capacities, i.e.,

$$\lim_{n \to \infty} \frac{1}{n} \sum_{i=0}^{n-1} x_i < \frac{1}{p} \sum_{i=0}^{p-1} K_i. \tag{19.31}$$

Equation (19.31) gives a justification for the use of the word "attenuance" to describe the phenomenon in which a periodically fluctuation carrying capacity of the Beverton–Holt equation has a deleterious effect on the population. This conjecture was first proved by Elaydi and Sacker in [13, 14] and independently by Kocic [26] and Kon [27]. The following theorem summarizes our findings.

Theorem 19.7. *[14] Consider the p-periodic Beverton–Holt equation*

$$x_{n+1} = \frac{\mu K_n x_n}{K_n + (\mu - 1)x_n}, n \in \mathbb{Z}^+, \tag{19.32}$$

where $\mu > 1$, $K_{n+p} = K_n$, *and* $K_n > 0$. *Then (19.32) has a globally asymptotically stable p-periodic cycle. Moreover, (19.32) is attenuant.*

Kocic [26], however gave the most elegant proof for the presence of attenuance. Utilizing effectively the Jensen's inequality, he was able to give the following more general result.

Theorem 19.8. *[26] Assume that* $\mu > 1$ *and* $\{K_n\}$ *is a bounded sequence of positive numbers*

$$0 < \alpha < K_n < \beta < \infty.$$

Then for every positive solution $\{x_n\}$ *of (19.32) we have*

$$\limsup_{n \to \infty} \frac{1}{n} \sum_{i=0}^{n-1} \overline{x}_i \le \limsup_{n \to \infty} \frac{1}{n} \sum_{i=0}^{n-1} K_i. \tag{19.33}$$

19.9.2 Neither Attenuance nor Resonance

By a simple trick, Sacker [31] showed that neither attenuance nor resonance occurs when periodically forcing the Ricker maps

$$R(x) = xe^{p-x}.$$

So consider the k-periodic system

$$x_{n+1} = x_n e^{p_n - x_n}, p_{n+k} = p_n, n \in \mathbb{Z}^+. \tag{19.34}$$

If $0 < p_n < 2$, (19.34) has a globally asymptotically stable k-periodic cycle [31]. Let $C_k = \{\overline{x}_0, \overline{x}_1, \ldots, \overline{x}_{k-1}\}$ be this unique k-periodic cycle. Then

$$\overline{x}_0 = \overline{x}_k = \overline{x}_{k-1} e^{p_{k-1} - \overline{x}_{k-1}}$$
$$= \overline{x}_{k-2} e^{p_{k-2} - \overline{x}_{k-2}} e^{p_{k-1} - \overline{x}_{k-1}},$$

and by iteration we get

$$\overline{x}_0 = \overline{x}_0 e^{\sum_{i=0}^{k-1} p_i - \sum_{i=0}^{k-1} \overline{x}_i}.$$

Hence

$$\frac{1}{k} \sum_{i=0}^{k-1} p_i = \frac{1}{k} \sum_{i=0}^{k-1} \overline{x}_i,$$

i.e., neither attenuance nor resonance.

19.9.3 An Extension: Monotone Maps

Using an extension to monotone maps, Kon [27] considered a p-periodic difference equation of the form

$$x_{n+1} = g\left(x_n / K_n\right) x_n, n \in \mathbb{Z}^+, \tag{19.35}$$

where $K_{n+p} = K_n$, $K_n > 0$, $x_0 \in [0, \infty)$ and $g : \mathbb{R}^+ \to \mathbb{R}^+$ is a continuous function which satisfies the following properties

- $g(1) = 1$.
- $g(x) > 1$ for all $x \in (0, 1)$.
- $g(x) < 1$ for all $x \in (1, \infty)$.

Theorem 19.9. [27] Let $C_r = \{\overline{x}_0, \overline{x}_1, \ldots, \overline{x}_{r-1}\}$ be a positive r-periodic cycle of (19.35) such that $K_i \neq K_{i+1}$ for some $0 \leq i \leq p - 1$. Assume that $zg(z)$ is strictly concave on an interval (a, b), $0 < a < b$ containing all points $\frac{\overline{x}_i}{K_i} \in (a, b)$, $1 \leq i \leq rp$. Then the cycle C_r is attenuant.

This theorem provides an alternative proof of the attenuance of the periodic Bevertob–Holt equation (19.32).

Consider the equation [27]

$$x_{n+1} = \left(\frac{x_n}{K_n}\right)^{a-1}, 0 < a < 1, \tag{19.36}$$

where $K_{n+p} = K_n$, $n \in \mathbb{Z}^+$, and $K_i \neq K_{i+1}$ for some $i \in \mathbb{Z}^+$. The maps belong to the class $\mathscr{K}$ and satisfy the assumption of the preceding theorem. Consequently, (19.36) has a globally asymptotically stable p-periodic cycle that is attenuant.

19 Towards a Theory of Periodic Difference Equations   313

19.9.4 The Loss of Attenuance: Resonance

Consider the periodic Beverton–Holt equation (19.32) in which both parameters μ_n and K_n are periodic of common period p. This equation may be attenuant or resonant. In fact, when $p = 2$, Elaydi and Sacker [14] showed that

$$\overline{x} = \overline{K} + \sigma \frac{K_0 - K_1}{2} - \Delta \frac{(\mu_0 - 1)(\mu_1 - 1)}{2(\mu_0 \mu_1 - 1)} (K_0 - K_1)^2 \tag{19.37}$$

where

$$\overline{x} = \frac{\overline{x}_0 + \overline{x}_1}{2} \text{ and } \overline{K} = \frac{K_0 + K_1}{2},$$

$$\sigma = \frac{\mu_1 - \mu_0}{\mu_0 \mu_1 - 1}, 0 \le |\sigma| < 1,$$

and

$$\Delta = \frac{\mu_0(\mu_1^2 - 1)K_0 + \mu_1(\mu_0^2 - 1)K_1}{\mu_0(\mu_1 - 1)^2 K_0^2 + (\mu_0 - 1)(\mu_1 - 1)(\mu_0 \mu_1 + 1)K_0 K_1 + \mu_1(\mu_0 - 1)^2 K_1^2} > 0.$$

It follows that attenuance is present if either $(\mu_1 - \mu_0)(K_0 - K_1) < 0$ (out of phase) or the algebraic sum of the last two terms in (19.37) is negative. On the other hand, resonance is present if the algebraic sum of the last two terms in (19.37) is positive.

Notice that if $\mu_0 = \mu_1 = \mu$ with $p = 2$, then we have

$$\frac{1}{p} \sum_{i=0}^{p-1} \overline{x}_i = \frac{1}{p} \sum_{i=0}^{p-1} K_i - \frac{\mu(K_0 + K_1)(K_1 - K_0)^2}{2 \left[\mu K_0^2 + (\mu^2 + 1) K_0 K_1 + \mu K_1^2 \right]},$$

which gives an exact expression for the difference in the averages.

Remark 19.2. Now for $\mu_0 = 4$, $\mu_1 = 2$, $K_0 = 11$, and $K_1 = 7$, we have resonance as $\frac{1}{2} \sum_{i=0}^{1} \overline{x}_i \approx 9.23$ and $\frac{1}{2} \sum_{i=0}^{1} K_i = 9$. On the other hand, one can show that for $\mu_0 = 2$, $\mu_1 = 4$, $K_0 = 11$, and $K_1 = 7$, we have attenuance as may be seen from (19.37).

19.9.5 The Signature Functions of Franke and Yakubu

In [19], the authors gave a criteria to determine attenuance or resonance for the 2-periodic difference equation

$$x_{n+1} = x_n g(K_n, \mu_n, x_n), n \in \mathbb{Z}^+, \tag{19.38}$$

where $K_n = K(1 + \alpha(-1)^n)$, $\mu_n = \mu(1 + \beta(-1)^n)$, and $\alpha, \beta \in (-1, 1)$.

Define the following

$$\omega_1 = \frac{\left(k\frac{\partial^2 g}{\partial x^2} + 2\frac{\partial g}{\partial x}\right)\left(\frac{K^2\frac{\partial g}{\partial K}}{2+K\frac{\partial g}{\partial x}}\right)^2 + \left(2K\frac{\partial g}{\partial K} + 2K^2\frac{\partial^2 g}{\partial x\partial K}\right) + K^3\frac{\partial^2 g}{\partial K^2}}{-2K\frac{\partial g}{\partial x}}, \qquad (19.39)$$

$$\omega_2 = \frac{-\left(\mu\frac{\partial g}{\partial \mu} + K\mu\frac{\partial^2 g}{\partial x\partial \mu}\right)\left(\frac{-K^2\frac{\partial g}{\partial K}}{2+K\frac{\partial g}{\partial x}}\right) + K^2\mu\frac{\partial^2 g}{\partial K\partial \mu}}{K\frac{\partial g}{\partial x}}, \qquad (19.40)$$

and

$$\mathcal{R}_d = sign(\alpha(\omega_1\alpha + \omega_2\beta)). \qquad (19.41)$$

Theorem 19.10. *[19] If for $\alpha = 0, \beta = 0$, K is hyperbolic fixed point of (19.38), then for all sufficiently small $|\alpha|$ and $|\beta|$, (19.38), with $\alpha, \beta \in (-1, 1)$, has an attenuant 2-periodic cycle if $\mathcal{R}_d < 0$ and a resonant 2-periodic cycle if $\mathcal{R}_d > 0$.*

To illustrate the effectiveness of this theorem, let us to consider the logistic equation

$$x_{n+1} = x_n\left[1 + \mu(1 + \beta(-1)^n)\left(1 - \frac{x_n}{K(1 + \alpha(-1)^n)}\right)\right]. \qquad (19.42)$$

For $0 < \mu < 2$ (19.42) has an asymptotically stable 2-periodic cycle. Using formulas (19.39) and (19.40), one obtains

$$\omega_1 = \frac{-8K}{(\mu - 2)^2} \quad \text{and} \quad \omega_2 = \frac{-4K}{\mu - 2}.$$

Assume that $\alpha > 0$ and $0 < \mu < 2$. Using (19.41) yields

$$\mathcal{R}_d = sign\left(\frac{2}{\mu - 2}\alpha + \beta\right) = sign\left(\beta - \frac{2}{2 - \mu}\alpha\right).$$

Hence we have attenuance if $\beta < \frac{2}{2-\mu}\alpha$, i.e., if the relative strength of the fluctuation of the demographic characteristic of the species is weaker than $\frac{2}{2-\mu}$ times the relative strength of the fluctuation of the carrying capacity. On the other hand if $\beta > \frac{2}{2-\mu}\alpha$ we obtain resonance.

Notice that if $\alpha = 0$ (the carrying capacity is fixed), then we have resonance if $\beta > 0$ and we have attenuance if $\beta < 0$. For the case that $\beta = 0$ (the intrinsic growth rate is fixed), we have attenuance.

Finally, we note that Franke and Yakubu extended their study to periodically forced Leslie model with density-dependent fecundity functions [18]. The model is of the form

19 Towards a Theory of Periodic Difference Equations — 315

$$x_{n+1}^1 = \sum_{i=1}^{s} x_n^i g_n^i(x_n^i) = \sum_{i=1}^{s} f_n^i(x_n^i)$$

$$x_{n+1}^2 = \lambda_1 x_n^1$$

$$\vdots$$

$$x_{n+1}^s = \lambda_{s-1} x_n^{s-1},$$

where f_n^i is of the Beverton–Holt type. Results similar to the one-dimensional case where each f_n^i is under compensatory, i.e.,

$$\frac{\partial f_n^i(x_i)}{\partial x_i} > 0, \frac{\partial^2 f_n^i(x_i)}{\partial x_i^2} < 0,$$

and $\lim_{x_i \to \infty} f_n^i(x_i)$ exists for all $n \in \mathbb{Z}^+$.

19.10 Almost Periodic Difference Equations

In this section we extend our study to the almost periodic case. This is particularly important in applications to biology in which habitat's fluctuations are not quite periodic.

But in order to embark on this endeavor, one needs to almost reinvent the wheel. The problem that we encounter here is that the existing literature deals exclusively with almost periodic fluctuations (sequences) on the real line $\mathbb{R}$ (on the integers $\mathbb{Z}$). To have meaningful applications to biology, we need to study almost periodic fluctuations or sequence on $\mathbb{Z}^+$ (the set of nonnegative integers). Such a program has been successfully implemented in [10]. Our main objective here is to report to the reader a brief but through exposition of these results.

We start with the following definitions from [17, 21].

Definition 19.6. An $\mathbb{R}^k$-valued sequence $x = \{x_n\}_{n \in \mathbb{Z}^+}$ is called Bohr almost periodic if for each $\epsilon > 0$, there exists a positive integer $T_0(\epsilon)$ such that among any $T_0(\epsilon)$ consecutive integers, there exists at least one integer τ with the following property:

$$\| x_{n+\tau} - x_n \| < \epsilon, \forall n \in \mathbb{Z}^+.$$

The integer τ is then called an ϵ-period of the sequence $x = \{x_n\}_{n \in \mathbb{Z}^+}$.

Definition 19.7. An $\mathbb{R}^k$-value sequence $x = \{x_n\}_{n \in \mathbb{Z}^+}$ is called Bochner almost periodic if for every sequence $\{h(n)\}_{n \in \mathbb{Z}^+}$ of positive integers there exists a subsequence $\{h_{n_i}\}$ such that $\{x_{n+n_i}\}_{n_i \in \mathbb{Z}^+}$ converges uniformly in $n \in \mathbb{Z}^+$.

In [10] it was shown that the notions of Bohr almost periodicity and Bochner almost periodicity are equivalent.

Now a sequence $f : \mathbb{Z}^+ \times \mathbb{R}^k \to \mathbb{R}^k$ is called almost periodic in $n \in \mathbb{Z}^+$ uniformly in $x \in \mathbb{R}^k$ if for each $\varepsilon > 0$, there exists $T_0(\varepsilon) \in \mathbb{Z}^+$ such that among $T_0(\varepsilon)$ consecutive integers there exists at least one integer s with

$$\| f(n + s, x) - f(n, x) \| < \varepsilon$$

for all $x \in \mathbb{R}^k$, and $s \in \mathbb{Z}^+$.

Now consider the almost periodic difference equations

$$x_{n+1} = A_n x_n \tag{19.43}$$

$$y_{n+1} = A_n y_n + f(n, y_n), \tag{19.44}$$

where A_n is a $k \times k$ almost periodic matrix on $\mathbb{Z}^+$, and $f : \mathbb{Z}^+ \times \mathbb{R}^k \to \mathbb{R}^k$ is almost periodic.

Let $\Phi(n, s) = \prod_{r=s}^{n-1} A_r$ be the state transition matrix of (19.43). Then (19.43) is said to posses a regular exponential dichotomy [23] if there exist a $k \times k$ projection matrix $P_n, n \in \mathbb{Z}^+$, and positive constants M and $\beta \in (0, 1)$ such that the following properties hold:

1. $A_n P_n = P_{n+1} A_n$.
2. $\| X(n, r) P_r x \| \leq M \beta^{n-r} \| x \|, 0 \leq r \leq n, x \in \mathbb{R}^k$.
3. $\| X(r, n) (I - P_n) x \| \leq M \beta^{n-r} \| x \|, 0 \leq r \leq n, x \in \mathbb{R}^k$.
4. The matrix A_n is an isomorphism from $R(I - P_n)$ onto $R(I - P_{n+1})$, where $R(B)$ denotes the range of the matrix B.

We are now in a position to state the main stability result for almost periodic systems.

Theorem 19.11. *Suppose that (19.43) possesses a regular exponential dichotomy with constant M and β and f is a Lipschitz with a constant Lipschitz L. Then (19.44) has a unique globally asymptotically stable almost periodic solution provided*

$$\frac{M \beta L}{1 - \beta} < 1.$$

Proof. Let $AP(\mathbb{Z}^+)$ be the space of almost periodic sequences on $\mathbb{Z}^+$ equipped with the topology of the supremum norm. Define the operator Γ on $AP(\mathbb{Z}^+)$ by letting

$$(\Gamma \varphi)_n = \sum_{r=0}^{n-1} \left(\prod_{s=r}^{n-1} \right) A_s f(r, \varphi_r).$$

Then $\Gamma : AP(\mathbb{Z}^+) \to AP(\mathbb{Z}^+)$ is well defined. Moreover Γ is a contraction. Using the Banach fixed point theorem, we obtain the desired conclusion. $\square$

The preceding result may be applied to many populations models. However, we will restrict our treatment here on the almost periodic Beverton–Holt equation with overlapping generations

19 Towards a Theory of Periodic Difference Equations

$$x_{n+1} = \gamma_n x_n + \frac{(1 - \gamma_n)\mu K_n x_n}{(1 - \gamma_n)K_n + (\mu - 1\gamma_n)x_n} \tag{19.45}$$

with $K_n > 0$ and $\gamma_n \in (0, 1)$ are almost periodic sequences, and $\mu > 1$. As before μ and K denote the intrinsic growth rate and the carrying capacity of the population, respectively, while γ is the survival rate of the population from one generation to the next.

The following result follows from Theorem 19.11

Theorem 19.12. *Equation (19.45) has a unique globally asymptotically stable almost periodic solution provided that*

$$\sup\{\gamma_n : n \in \mathbb{Z}^+\} < \frac{1}{1 + \mu}$$

To this end, we have addressed the question of stability and existence of almost periodic solution of almost periodic difference equation. We now embark on the task of the determination of whether a system is attenuant or resonant.

Let $\{\mu_n\}_{n \in \mathbb{Z}^+}$ be an almost periodic sequence on $\mathbb{Z}^+$. Then we define its mean value as

$$M(\mu_n) = \lim_{n \to \infty} \frac{1}{m} \sum_{r=1}^{m} \mu_{n+r} \tag{19.46}$$

It may be shown that $M(\mu_n)$ exists [10].

Let $\{\overline{x}_n\}$ be the almost periodic solution of a given almost periodic system. Then we say that the system is

1. Attenuant if $M(\overline{x}_n) < M(K_n)$.
2. Resonant if $M(\overline{x}_n) > M(K_n)$.

Theorem 19.13. *[10] Suppose that $\{K_n\}_{n \in \mathbb{Z}^+}$ is almost periodic, $K_n > 0$, $\mu > 1$, and $\gamma_n = \gamma \in (0, 1)$. Then*

1. $\limsup\limits_{n \to \infty} \frac{1}{n} \sum_{m=0}^{n-1} x_m \leq \limsup\limits_{n \to \infty} \frac{1}{n} \sum_{m=0}^{n-1} K_m$ *for any solution x_n of (19.45).*
2. $M(\overline{x}_n) \leq M(K_n)$ *if $\overline{x}_n$ is the unique almost periodic solution of (19.45).*

19.11 Stochastic Difference Equations

In [22] the authors investigated the stochastic Beverton–Holt equation and introduced new notions of attenuance and resonance in the mean.

Following on the same lines [6] the authors investigated the stochastic Beverton–Holt equation with overlapping generations.

318   S.N. Elaydi et al.

In this section, we will consider the latter study and consider the equation

$$x_{n+1} = \gamma_n x_n + \frac{(1 - \gamma_n)\mu K_n x_n}{(1 - \gamma_n)K_n + (\mu - 1 + \gamma_n)x_n}. \tag{19.47}$$

Let $L^1(\Omega, \upsilon)$ be the space of integrable functions on a measurable space $(\Omega, \mathscr{F}, \upsilon)$ equipped with its natural norm given by

$$\| f \|_1 = \int_\Omega f(x)d\upsilon.$$

Let

$$\mathscr{D}(E) := \{f \in L^1(E, \upsilon) : f \geq 0 \text{ and } \int_\Omega f d\upsilon\}$$

be the space of all densities on Ω.

Definition 19.8. Let $\mathscr{D} : L^1(\Omega, \upsilon) \to L^1(\Omega, \upsilon)$ be a Markov operator. Then $\{\mathscr{D}^n\}$ is said to be asymptotically stable if there exists $f^* \in \mathscr{D}$ for which

$$\mathscr{D} f^* = f^*$$

and for all $f \in \mathscr{D}$,

$$\lim_{n \to \infty} \| \mathscr{D}^n f - f^* \|_1 = 0.$$

We assume that both the carrying capacity K_n and the survival rate γ_n are random and for all n, (K_n, γ_n) is chosen independently of (x_0, K_0, γ_0), (x_1, K_1, γ_1), ..., $(x_{n-1}, K_{n-1}, \gamma_{n-1})$ from a distribution with density $\Phi(K, \gamma)$.

The joint density of x_n, K_n, γ_n is $f_n(x)\Phi(K, \gamma)$, where f_n is the density of x_n. Furthermore, we assume that

$$E|K_n| < \infty, E|x_0| < \infty$$

and $K^2\Phi(K, \gamma)$ is bounded above independently of γ and that Φ is supported on the product interval

$$[K_{\min}, \infty) \times [\gamma_{\min}, \infty),$$

for some $K_{\min} > 0$ and $\gamma_{\min} > 0$.

Moreover, we assume there exists an interval $(K_l, K_u) \subset \mathbb{R}^+$ on which Φ is positive everywhere for all γ.

Let h be an arbitrary bounded and measurable function on $\mathbb{R}^+$ and define $b(K_n, \gamma_n, x_n)$ to be equal to the right-hand side of (19.47). The expected value of h at time $n + 1$ is then given by

$$E[h(x_{n+1})] = \int_0^\infty h(x) f_{n+1}(x)dx. \tag{19.48}$$

19 Towards a Theory of Periodic Difference Equations

Furthermore, because of (19.47) and the fact that the joint density of x_n, and γ_n is just $f_n(x)\Phi(K,\gamma)$, we also have

$$
\begin{aligned}
E[h(x_{n+1})] &= E[h(b(K_n,\gamma_n,x_n))] \\
&= \int_0^\infty \int_0^1 \int_0^\infty h(b(K,\gamma,y))f_n(y)\Phi(K,\gamma)dy\,d\gamma\,dy.
\end{aligned}
$$

Let us define $K = K(x,\gamma,y)$ by the equation

$$
x = \frac{(1-\gamma)\mu K y}{(1-\gamma)K + (\mu-1+\gamma)y} + \gamma y. \tag{19.49}
$$

Solving explicitly this equation for K yields

$$
K = \frac{(\mu-1+\gamma)y(x-\gamma y)}{(1-\gamma)[\mu y - (x-\gamma y)]}. \tag{19.50}
$$

By a change of variables, this can be written as

$$
E[h(x_{n+1})] = \iiint_{\{(x,\gamma,y):0<x-\gamma y<\mu y\}} h(x)f_n(y)\Phi(K,\gamma)\frac{dk}{db(K,\gamma,y)}dx\,d\gamma\,dy.
$$

A simple calculation yields

$$
E[h(x_{n+1})] = \mu \int_0^\infty \left\{ \iint_A \frac{1-\gamma}{(\mu-1+\gamma)}\frac{1}{(x-\gamma y)^2} f_n(y)K^2\Phi(K,\gamma)d\gamma\,dy \right\} dx,
$$

where

$$
A = \{(\gamma,y):0<x-\gamma y<\mu y\}. \tag{19.51}
$$

Equating the above equations, and using the fact that h was an arbitrary, bounded, measurable function, we immediately obtain

$$
f_{n+1}(x) = \mu \iint_A \frac{1-\gamma}{(\mu-1+\gamma)}\frac{1}{(x-\gamma y)^2} f_n(y)K^2\Phi(K,\gamma)d\gamma\,dy.
$$

Let $\mathscr{P}: L^1(\mathbb{R}^+) \to L^1(\mathbb{R}^+)$ be defined by

$$
\mathscr{P}f(x) = \mu \iint_A \frac{1-\gamma}{(\mu-1+\gamma)}\frac{1}{(x-\gamma y)^2} f(y)K^2\Phi(K,\gamma)d\gamma\,dy, \tag{19.52}
$$

where $k = K(x,\gamma,y)$ is defined by (19.50) and A in (19.51).

We can now state the main theorem of this section

Theorem 19.14. *[6] The Markov operator $\mathscr{P}: L^1(\mathbb{R}^+) \to L^1(\mathbb{R}^+)$ defined by (19.52) is asymptotically stable.*

For the case when $\gamma_n = \gamma$ is a constant and K_n is a random sequence, the following attenuance result was obtain.

For almost every $w \in \Omega$ and $x \in \mathbb{R}^+$

$$\lim_{n \to \infty} \frac{1}{n} \sum_{i=0}^{n-1} x_i(w, x) < \lim_{n \to \infty} \frac{1}{n} \sum_{i=0}^{n-1} K_i(w),$$

that is we have attenuance in the mean.

It is still an open problem to determine the attenuance or resonance when both γ_n and K_n are random sequences on $\mathbb{Z}^+$.

References

1. Allee, W.C.: The Social Life of Animals. William Heinemann, London (1938)
2. AlSharawi, Z., Angelos, J.: On the periodic logistic equation. Appl. Math. Comput. **180**(1), 342–352 (2006)
3. AlSharawi, Z., Angelos, J., Elaydi, S., Rakesh, L.: An extension of Sharkovsky's theorem to periodic difference equations. J. Math. Anal. Appl. **316**, 128–141 (2006)
4. D'Aniello, E., Oliveira, H.: Swallowtail bifurcation for non-autonomous interval maps, Center for Mathematical Analysis, Geometry, and Dynamical Systems, Instituto Superior Tecnico, Thecnical University of Lisbon, Lisbon, Portugal, Preprint
5. D'Aniello, E., Oliveira, H.: Bifurcation equations and cyclic permutations of periodic non autonomous system, Submitted (2010)
6. Bezandry, H., Diagana, T., Elaydi, S.: On the stochastic Beverton–Holt equation with survival rates. J. Differ. Equ. Appl. **14**(2), 175–190 (2007)
7. Cushing, J.M.: Oscillations in age-structured population models with an Allee effect. J. Comput. Appl. Math. **52**, 71–80 (1994)
8. Cushing, J.M., Henson, S.M.: Aperiodically forced Beverton–Holt equation. J. Differ. Equ. Appl. **8**, 1119–1120 (2002)
9. Devaney, R.L.: An introduction to Chaotic Dynamical Systems, 2nd edn. Perseus Books Publishing, L.L.C. (1989)
10. Diagana, T., Elaydi, S., Yakubu, A.: Populations models in almost periodic environments. J. Differ. Equ. Appl. **13**(4), 289–308 (2007)
11. Elaydi, S.: Discrete Chaos: With Applications in Science and Engineering, 2nd edn. Chapman and Hall/CRC (2008)
12. Elaydi, S., Yakubu, A.: Global stability of cycles: Lotka–Volterra competition model with stocking. J. Differ. Equ. Appl. **8**(6), 537–549 (2002)
13. Elaydi, S., Sacker, R.: Global stability of periodic orbits of nonautonomous difference equations and populations biology. J. Differ. Equ. **208**, 258–273 (2005)
14. Elaydi, S., Sacker, R.: Nonautonomous Beverton–Holt equations and the Cushing–Henson conjectures. J. Differ. Equ. Appl. **11**(4-5), 337–346(10) (2005)
15. Elaydi, S., Sacker, R.: Periodic difference equations, populations biology and the Cushing–Henson conjecture. Math. Biosci. **201**, 195–2007 (2006)
16. Elaydi, S., Sacker, R.: Skew-product dynamical systems: Applications to difference equations. Proceedings of the Second Annual Celebration of Mathematics, United Arab Emirates (2005)
17. Fink, A.M.: Almost periodic differential equations, Lecture Notes in Mathematics, vol. 377. Springer, New York, Berlin (1974)
18. Franke, J., Yakubu, A.: Globally attracting attenuant versus resonant cycles in periodic compensatory Leslie models. Math. Biosci. **204**(1), 1–20 (2006)

19 Towards a Theory of Periodic Difference Equations

19. Franke, J., Yakubu, A.: Using a signature function to determine resonant and attenuant 2-cycles in the Smith–Slatkin population model. J. Differ. Equ. Appl. **13**(4), 289–308 (2007)
20. Grinfeld, M., Knight, P.A., Lamba, H.: On the periodically perturbed logistic equation. J. Phys. A Math. Gen. **29**, 8035–8040 (1996)
21. Halanay, A., Rasvan, V.: Stability and stable oscillations in discrete time systems, Advances in Discrete Mathematics and Applications, vol. 2. Gordon and Breach Science Publication (2000)
22. Haskell, C., Sacker, R.: The Stochastic Beverton–Holt equation and the M. Neubert conjecture. J. Dyn. Differ. Equ. Elsevier, **17**(4), 825–842 (2005)
23. Henry, D.: Geometric theory of semilinear parabolic equations, Lecture Notes in Mathematics. Springer, New York, Berlin (1981)
24. Henson, S.: Multiple attractors an resonance in periodically forced population models. Physica D. **140**, 33–49 (2000)
25. Johnson, W.: The Curious History of Faà di Bruno's Formula. Am. Math. Month. **109**, 217–234 (2002)
26. Kocic, V.: A note on the nonautonomous Beverton–Holt model. J. Differ. Equ. Appl. **11**(4-5), 415–422 (2005)
27. Kon, R.: A note on attenuant cycles of population models with periodic carrying capacity. J. Differ. Equ. Appl. **10**(8), 791–793 (2004)
28. Silvestre-François Lacroix, Traité du Calcul différentiel et du Calcul intégral, 2e édn. Revue et Augmentée, vol. 1. Chez Courcier, Paris (1810); vol. 2, Mme. Courcier, Paris (1814); vol. 3, Mme. Courcier, Paris (1819)
29. Luís, R., Elaydi, S., Oliveira, H.: Nonautonomous periodic systems with Allee effects, J. Differ. Equ. Appl. **16**(10), 1179–1196 (2010)
30. Oliveira, H., D'Aniello, E.: Pitchfork bifurcation for non autonomous interval maps. J. Differ. Equ. Appl. **15**(3), 291–302 (2009)
31. Sacker, R.: A Note on Periodic Ricker Maps. J. Differ. Equ. Appl. **13**(1), 89–92 (2007)
32. Sacker, R., Elaydi, S.: Population models with Allee effects: A new model, J. Biol. Dyn. **4**(4), 397–408 (2010)
33. Singer, D.: Stable orbits and bifurcation of maps of the interval. SIAM J. Appl. Math. **35**, 260–267 (1978)

Chapter 20
Thompson's Group, Teichmüller Spaces, and Dual Riemann Surfaces

Edson de Faria

Abstract In this paper we present a brief survey of the role played by Richard Thompson's group F in the study of the dynamical classification of certain conformal repellers, as first described in de Faria et al. (Contemp. Math., 355:166–1855, 2004). We exhibit a faithful and discrete action of F in the asymptotic Teichmüller spaces of such conformal repellers. An important ingredient to monitor such actions is the complex scaling function of the repeller, defined on its dual Riemann surface. We ask for generalizations to more general repellers, and formulate some open questions.

20.1 Introduction

The so-called *chameleon* groups of Richard Thompson were introduced in 1965, in a set of unpublished notes. These groups have since appeared in many different contexts and guises, ranging from logic to algebraic topology and geometry. An extremely elegant exposition of the basic theory of such groups is given in [2]. In a special conference held at the American Institute of Mathematics of Palo Alto in 2004, some of the main experts on the subject attempted to take stock of everything then known about Thompson's group. The first connection between F, the smallest of Thompson's groups, and the Teichmüller theory of some very simple dynamical systems was presented in that conference, and published in [5]. In that paper, it was established that F acts faithfully and discretely in the asymptotic Teichmüller space of a certain complex dynamical system, a Cantor repeller. In the present paper, we briefly survey the results of [5], and then indicate how they can be adapted to prove that F acts faithfully and discretely also in the asymptotic Teichmüller space of the expanding map $z \mapsto z^2$. The geometric action is through piecewise affine motions of invariant Carleson boxes, as indicated in Sect. 20.5.

E. de Faria
IME-USP, Rua do Matão, 1010 Butantã, 05508-090 – Sao Paulo, SP, Brazil
e-mail: edson@ime.usp.br

M.M. Peixoto et al. (eds.), *Dynamics, Games and Science I*, Springer Proceedings in Mathematics 1, DOI 10.1007/978-3-642-11456-4_20,
© Springer-Verlag Berlin Heidelberg 2011

20.2 Thompson's F Group

There are three Thompson groups, usually denoted F, T and V in the literature, and one has $F \subset T \subset V$. In this paper, we are concerned with F only. All of Thompson's groups have several "incarnations". We shall describe two such for F. Many more equivalent definitions exist. See [2] and the references therein.

20.2.1 First Incarnation

The easiest way to define F is perhaps the least enlightening. Let $\text{PL}^+(I)$ denote the group of orientation-preserving, piecewise linear homeomorphisms of the unit interval $I = [0, 1]$. We define F to be the subgroup of $\text{PL}^+(I)$ consisting of those $\varphi \in \text{PL}^+(I)$ which have finitely many break-points, all at dyadic rationals, and whose slopes on linear pieces are given by powers of 2. Thus, if $\varphi \in F$, and if $J \subset I$ is a subinterval such that $\varphi|_J$ is linear, then

$$\varphi(x) = 2^m x + \frac{p}{2^n}$$

for all $x \in J$, for some $m, p \in \mathbb{Z}$ and some $n \in \mathbb{N}$.

Proposition 20.1. *For each $\varphi \in F$ there exist two finite partitions of I, say $0 = x_0 < x_1 < \cdots < x_n = 1$ and $0 = y_0 < y_1 < \cdots y_n = 1$, with the following properties.*

(a) Both partitions are standard dyadic partitions of I, i.e. each one of the intervals $[x_i, x_{i+1}], [y_i, y_{i+1}]$ is of the form $[a2^{-k}, (a+1)2^{-k}]$ for non-negative integers a, k.

(b) Each restriction $\varphi|_{[x_i, x_{i+1}]}$ is affine and $\varphi([x_i, x_{i+1}]) = [y_i, y_{i+1}]$, for $i = 0, 1 \ldots, n-1$ (in particular, $\varphi(x_i) = y_i$ for all i).

The proof of this basic result can either be worked out as an exercise, or else looked up in [2].

20.2.2 Second Incarnation

Let us consider now a second way to define F, one which makes the combinatorial structure of this group more apparent. Let us first look at finite rooted binary trees in the plane, up to isotopy. We think of each such tree as made up of finitely many *carets* (wedges $\wedge$ made of two line segments). The topmost vertex of the topmost caret is called the *root* of the tree. The free edges are called *leaves*. We think of the leaves as ordered from left to right (relative to the way they are deployed in the plane). Note that if there are n carets in the tree, then there are $n + 1$ leaves, which we label $1, 2, \ldots, n$ from left to right. Let us call $\mathscr{P}$ the set of such trees up to

Fig. 20.1 An example of equivalence

isotopy (the isotopy should preserve the roots and the relative order of leaves). Now let us denote by $\mathscr{F} \subset \mathscr{P} \times \mathscr{P}$ the set of pairs of (isotopy classes of) trees in which both components have exactly the same number of carets (or leaves). We introduce in $\mathscr{F}$ an equivalence relation as follows.

First, define a *birth* to be the operation of adding a caret to a given tree through one of its free vertices (free ends of leaves). Given two pairs of trees in $\mathscr{F}$, say (R_1, D_1) and (R_2, D_2), we declare them to be equivalent, $(R_1, D_1) \sim (R_2, D_2)$, if R_1 can be obtained from R_2 through a sequence of births (or vice-versa) *and* D_1 can be obtained from D_2 through a sequence of births *following the same order* as the first sequence of births (or vice-versa). Here, *order* means order with respect to the numbering of the leafs of a tree. An example of equivalence is shown in Fig. 20.1. This is an equivalence relation, as the reader can easily convince himself. Let F be the quotient $\mathscr{F}/\sim$. The element of F corresponding to a pair (R, D) will be denoted $[R_1, D_1]$.

Let us now define the group operation on this set F of equivalence classes. Given two pairs of trees (R_1, D_1) and (R_2, D_2) representing two elements of F, we look at D_1 and R_2. These in general will be different trees. By a suitable sequence of births on D_1 and a corresponding suitable sequence of births in R_2, we can make both trees *look the same* (see [2] for details). In order words, we get a tree T from D_1 through a finite sequence of births, and the same tree T from R_2 through another finite sequence of births. Let $\widetilde{R}_1$ and $\widetilde{D}_2$ be such that $(R_1, D_1) \sim (\widetilde{R}_1, T)$ and $(T, \widetilde{D}_2) \sim (R_2, D_2)$, and define

$$[R_1, D_1] \circ [R_2, D_2] = [\widetilde{R}_1, \widetilde{D}_2].$$

This operation is compatible with the equivalence relation just introduced, and it is associative. It has an identity element, namely $e = [\wedge, \wedge]$. If $[R, D] \in F$, then its inverse is simply $[D, R]$. Thus, F is a group under this operation. This composition law is illustrated in Fig. 20.2.

We claim it is the same group of the previous definition, up to isomorphism. What is the relationship between both definitions? Given $\varphi : I \to I$, an element in the first incarnation of F, let P_D^φ and P_R^φ be the partitions in the domain and range of φ given by Proposition 20.1. Each such partition being a standard dyadic partition, they both give rise to binary trees D^φ, R^φ in an obvious way (and with the same number of leaves). The map $\varphi \to [D^\varphi, R^\varphi]$ is an isomorphism between both incarnations of F.

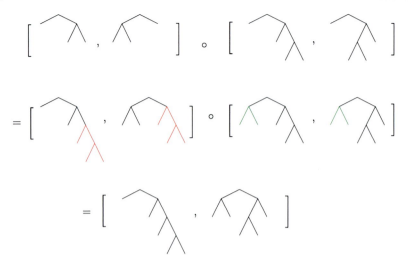

Fig. 20.2 The composition law

Here is a brief description of the inverse isomorphism to this isomorphism. First, given a rooted binary tree T having n carets and leaves labeled $1, 2, \ldots, n+1$, let $v_1, v_2, \ldots, v_{n+1}$ be the free vertices of T, ordered according to the leaves to which they belong. We define the *weight* of v_j to be $\omega(v_j) = 2^{-\ell(v_j)}$, where $\ell(v_j)$ is the integer length (number of edges) of the shortest path joining the root of T to the vertex v_j. Note that $\sum_{j=1}^{n+1} \omega(v_j) = 1$. Now define the partition P_T associated to T as follows: put $x_0 = 0$ and for each $j = 1, 2, \ldots, n+1$ let

$$x_j = \sum_{i=1}^{j} \omega(v_i).$$

The resulting partition P_T is a standard dyadic partition of I. Now, if $[R, D]$ is an element of the second incarnation of F, we consider the associated pair (P_D, P_R) of standard partitions and let $\varphi_{[R,D]} : I \to I$ be the unique orientation-preserving, piecewise affine homeomorphism that sends each atom of P_D to an atom of P_R. This map $[R, D] \to \varphi_{[R,D]}$ is the inverse map of the map we just defined in the previous paragraph.

20.2.3 Generators and Relations

The group F is finitely generated; in fact, it is generated by just two elements, which we call here φ_0 and φ_1.

Let us, more generally, define for each $n \geq 0$ an element $\varphi_n \in F$ as follows. Let R_n be the rooted finite binary tree with $n+2$ carets and free vertices $v_1, \ldots, v_{n+3}$ whose associated weights are given by $\omega(v_j) = 2^{-j}$ for $j = 1, 2, \ldots, n+2$ and

20 Thompson's Group, Teichmüller Spaces, and Dual Riemann Surfaces

$\omega(v_{n+3}) = 2^{-n-2}$. Geometrically, this tree is obtained starting from the top caret and then attaching each successive caret to the right-hand leaf of the previous one. Likewise, let D_n be the tree with $n + 2$ carets and vertices $v_1', \ldots, v_{n+3}'$ whose associated weights are given by $\omega(v_j') = 2^{-j}$ for $j = 1, 2, \ldots, n$, $\omega(v_{n+1}) = \omega(v_{n+2}) = 2^{-n-2}$, and $\omega(v_{n+3}) = 2^{-n-1}$. Geometrically, this tree looks almost like R_n, except that the last caret is attached to the *left-hand* leaf of the last-but-one. We define φ_n to be the element of F determined by this pair of trees, in other words

$$\varphi_n = [R_n, D_n] \in F.$$

Working out from the definition of the composition law, it is not difficult to to prove that, for all $n \geq 0$ and all $0 \leq k \leq n$, we have

$$\varphi_k \, \varphi_{n+1} \, \varphi_k^{-1} = \varphi_{n+2}. \tag{20.1}$$

It can also be shown that these relations form an infinite presentation of F. Note that, once φ_0 and φ_1 are given, (20.1) shows that $\varphi_2, \varphi_3, \ldots$ are inductively defined from those two elements. Thus, F is generated by two elements subject to the infinitely many relations (20.1). Surprisingly, it turns out that F is *finitely presented*. Indeed, two relations involving commutators suffice, and the group F can be described as

$$F = \left\langle \varphi_0, \varphi_1 : [\varphi_0 \varphi_1^{-1}, \varphi_0^{-1} \varphi_1 \varphi_0], [\varphi_0 \varphi_1^{-1}, \varphi_0^{-2} \varphi_1 \varphi_0^2] \right\rangle.$$

For a proof of these facts and more, see [2, Theorem 3.1].

20.2.4 More Properties of F

Let us enumerate a few further remarkable properties enjoyed by the group F. They will not be used here, but are a part of the general culture about Thompson's group. For proofs of the first six properties, the reader should consult [2]. The seventh property is a very nice theorem due to E. Ghys and V. Sergiescu, whose proof can be found in [7].

1. The group F is *almost abelian*, in the sense that $F/[F, F] \cong \mathbb{Z} \oplus \mathbb{Z}$. In fact, there is a homomorphism $\theta : F \to \mathbb{Z} \oplus \mathbb{Z}$ given by

$$\theta(\varphi) = \left(\log_2 \varphi'(0), \log_2 \varphi'(1) \right).$$

   One easily checks that the kernel of this homomorphism is precisely the commutator subgroup $[F, F]$.
2. The commutator subgroup $[F, F]$ is a simple group.
3. Every proper quotient group of F is abelian.
4. The group F has exponential growth.
5. Every non-abelian subgroup of F contains a free abelian group of infinite rank.

328   E. de Faria

6. As it follows easily from 5, F does not contain a copy of the free group on two generators.
7. The group F is isomorphic to a discrete subgroup of C^3 diffeomorphisms of the unit circle (this is proved in [7]).

For more on the *geometry* of Thompson's group, see [9].

20.3 Conformal Repellers, and Dual Riemann Surfaces

The concept of conformal repeller is fairly broad. Let us agree to the following definition.

Definition 20.1. A conformal repeller consists of an open set $U \subset \widehat{\mathbb{C}}$, a compact subset $K \subset U$ and a pseudo-semigroup G of conformal transformations $g : U_g \to \widehat{\mathbb{C}}$, where each $U_g \subset U$ is open, such that

(a) The set K is G-invariant: $GK \subset K$.
(b) For each $g \in G$, the restriction $g|_{K \cap U_g}$ is expanding, i.e. $|g'(x)| \geq \lambda > 1$ for all $x \in K \cap U_g$.

The constant λ is uniform (that is to say, independent of g).

The expression *pseudo-semigroup* above means that, in general, the maps of a conformal repeller are non-invertible, and their compositions are only partially defined (i.e. we compose them wherever we can). This covers several different situations, ranging from Fuchsian groups to expanding analytic circle maps (see [3]). In the present paper, we only care about two particularly simple situations: Cantor repellers and expanding circle maps.

20.3.1 Cantor Repellers

We borrow the following definition from [3] (see also [4]).

Definition 20.2. A Cantor repeller consists of two open sets $U, V \subseteq \mathbb{C}$ and a holomorphic map $f : U \to V$ satisfying the following conditions:

1. The domain U is the union of Jordan domains $U_0, U_1, \ldots, U_{m-1}$ (for some $m \geq 2$) having pairwise disjoint closures.
2. The co-domain V is the union of Jordan domains $V_0, V_1, \ldots, V_{M-1}$ (for some $M \geq 1$) having pairwise disjoint closures.
3. For each $i \in \{0, 1, \ldots, m - 1\}$ there exists $j(i) \in \{0, 1, \ldots, M - 1\}$ such that $f|_{U_i}$ maps U_i conformally onto $V_{j(i)}$.
4. We have $\overline{U} \subset V$.
5. The limit set $\Lambda_f = \cap_{n \geq 0} f^{-n}(V)$ has the locally eventually onto property (meaning that every open set intersecting Λ_f, no matter how small, is eventually mapped by iteration onto an open set containing Λ_f).

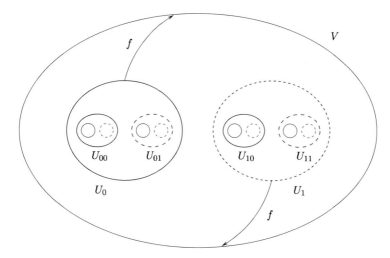

Fig. 20.3 The standard triadic repeller

Note that the invariant set Λ_f is a compact, perfect set without isolated points, i.e. a Cantor set (hence the name).

The Cantor repeller that will matter the most for us is the *standard triadic repeller*, in which $m = 2$ and $M = 1$, and the limit set is the standard triadic Cantor set on the real line. See Fig. 20.3. This terminology may seem a little idiosyncratic, for clearly the Riemann surface of a triadic repeller has the structure of a *binary tree*.

20.3.2 Jordan Repellers

A Jordan repeller is a conformal repeller $f : U \to V$ where $U, V \subset \widehat{\mathbb{C}}$ are *annuli* with $\overline{U} \subset V$, no component of $\widehat{\mathbb{C}} \setminus U$ is contained in V, and the map f is a proper covering map of degree $d \geq 2$ onto V. In this case, the limit set is

$$\Lambda_f = \bigcap_{n \geq 0} f^{-n}(V).$$

Topologically, Λ_f is a Jordan curve, hence the name. But it is in general quite wild, typically a non-rectifiable, nowhere differentiable curve. A simpler situation is the case when we have a symmetry, say when Λ_f is a circle, and U, V and f are all symmetric about this circle. This is the case, for example, of the map $f(z) = z^2$ restricted to the annulus $U = \{z : 2^{-1} < |z| < 2\}$, which is mapped onto the annulus $V = \{z : 4^{-1} < |z| < 4\}$ with degree 2. Here the invariant curve is $\Lambda_f = \mathbb{T}^1$, the unit circle.

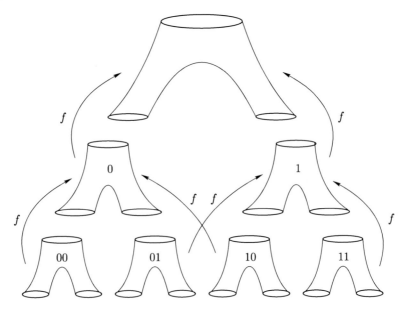

Fig. 20.4 The triadic repeller's chopped up tree

20.3.3 Dual Riemann Surfaces

In [5], we constructed a dynamical action of Thompson's group F in the asymptotic Teichmüller space of the standard triadic Cantor repeller, as outlined in Sect. 20.5. A crucial ingredient was the construction of a dual repeller, defined on a dual Riemann surface. This dual Riemann surface was constructed using the decomposition of the original Riemann surface into dynamically defined "pairs of pants". The idea was to chop the original surface along such pairs of pants, and re-glue them together using the dynamics itself. The waist curves of all pairs of pants are dynamically related: any waist curve an be mapped to any other waist curve by a suitable composition of iterates of f with (suitably chosen) inverse branches. See [5] for the exact construction. Thus, dynamical relationships between two pairs of pants (say in consecutive levels of the hierarchy) are transformed into spatial relationships (say adjacent pairs of pants). This is indicated in Fig. 20.4. This dual operation turns out to be involutive (up to conformal equivalence). In other words, if one performs the dual construction on the dual system, one gets back the original system up to conformal conjugacy.

This works fine for the standard triadic repeller, as shown in detail in [5], but doesn't work for all Cantor repellers, as the example in the following section shows.

20.3.4 The Fibonacci Repeller

Let us consider the situation depicted in Fig. 20.5. We have a Cantor repeller whose domain consists of three disks D_0, D_1, D_2 with pairwise disjoint closures, and they

20 Thompson's Group, Teichmüller Spaces, and Dual Riemann Surfaces

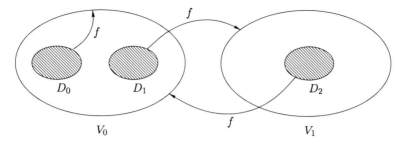

Fig. 20.5 The Fibonacci repeller

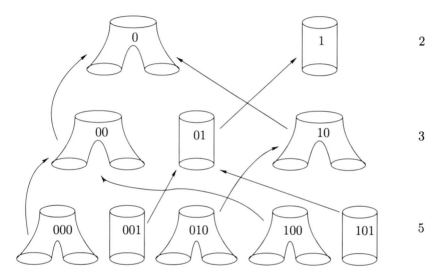

Fig. 20.6 The Fibonacci chopped up tree

map out to two larger disjoint disks V_0 and V_1, with $D_0 \cup D_1 \subset V_0$, $D_2 \subset V_1$, $f(D_0) = V_0 = f(D_2)$ and $f(D_1) = V_1$. This repeller is called the Fibonacci repeller, for reasons that will be clear in a moment. The (disconnected) Riemann surface of this Cantor repeller has the hierarchical structure shown in Fig. 20.6. The dynamics of f yields, just as in the case of the standard triadic repeller, a decomposition of the Riemann surface into connectors, which in this case are of two types: either cylinders or pairs of pants. At each level, the number of connectors making up the hierarchical structure at that level is a Fibonacci number, hence the name.

Now, if one tries to apply to this repeller the dual Riemann surface construction outlined for the standard triadic repeller, one quickly runs into trouble. For instance, the connectors labeled "001" and "101" should both be connected to the connector labeled "01" along their waist curves. Since the connector "01" is a cylinder, we are thus required to glue two curves (the waists of "001" and "101") onto the same boundary curve of "01". The situation gets even worse if we look at lower levels of the tree. The conclusion is that the dual to the Fibonacci Riemann surface is no longer a Riemann surface. It is rather a non-Hausdorff space, a *branched Riemann surface*.

What is the difference between this situation and that of the standard triadic repeller? Why can we form the dual surface in that situation but not in this one? Note that the dynamics in the limit set in the Fibonacci repeller is encoded by a subshift of finite type with matrix

$$\begin{pmatrix} 1 & 1 & 0 \\ 0 & 0 & 1 \\ 1 & 1 & 0 \end{pmatrix},$$

whereas in the standard triadic case the matrix is that of a full 2-shift. Motivated by this, we formulate the following problem.

20.1. Find a necessary and sufficient condition on the subshift of finite type associated to a Cantor repeller for the existence of a dual Riemann surface for the repeller.

20.4 Asymptotic Teichmüller Spaces

We can define various Teichmüller spaces in a given geometric or dynamical situation.

20.4.1 Teichmüller Spaces without Dynamics

We recall that, if X is a Riemann surface with ideal boundary $\partial_\infty X$, its *Teichmüller space* $T(X)$ is the space of equivalence classes of pairs (h, Y), where $h : X \to Y$ is a quasiconformal homeomorphism, the equivalence between two such pairs (h_0, Y_0) and (h_1, Y_1) being that there exists a *conformal* map $c : Y_0 \to Y_1$ such that $c \circ h_0$ is homotopic to h_1 relative to $\partial_\infty X$ (in other words, there exists an isotopy $\phi_t : X \to Y_1$ such that $\phi_0 = c \circ h_0$, $\phi_1 = h_1$ and $\phi_t(p) = h_1(p)$ for all $p \in \partial_\infty X$, for all $0 \le t \le 1$. The *co-asymptotic Teichmüller space* $AT(X)$ is defined in the same way, replacing the word "conformal" by "asymptotically conformal". By an asymptotically conformal map between Riemann surfaces we mean a quasiconformal map which is closer and closer to being conformal outside large compact sets in the domain.

We also define $T_0(X) \subset T(X)$ to be the subspace of *asymptotically conformal classes* (by which we mean classes represented by a pair (h, Y) where $h : X \to Y$ is an asymptotically conformal homeomorphism). The Teichmüller space $T(X)$ has a natural metric on it, the so-called Teichmüller metric, making it into a complete metric space. The space $T_0(X)$ is easily seen to be a closed subspace under this metric. The same holds for $AT(X)$.

We can also incorporate symmetries into these spaces. The most relevant for us is the case of an *anti-conformal involution* $j : X \to X$, i.e. an anti-conformal map j

20 Thompson's Group, Teichmüller Spaces, and Dual Riemann Surfaces

with $j \circ j = \mathrm{id}_X$. One can then define $T(X, j)$, $AT(X, j)$ and $T_0(X, j)$ in the obvious way, making all the appropriate equivalences respect this involution.

The case we are most interested here is when $X = \widehat{\mathbb{C}} \setminus \Lambda$, where Λ is a Cantor set (which will turn out to be the limit set of a Cantor repeller) lying on the real line in the complex plane. Here, of course, there is a natural identification between Λ and $\partial_\infty X$, and the involution j is the standard complex conjugation.

20.4.2 Teichmüller Space with Dynamics

Let us now suppose that we are given a dynamical system in the plane which, for definiteness, we assume to be a Cantor repeller (Ω_f, f). Here $\Omega_f = \cup D_i$ is a union of topological disks with pairwise disjoint closures, and $f_i = f|_{D_i}$ ($i = 1, \ldots, N$, $N \geq 2$) is a quasiconformal homeomorphism onto a topological disk V_{k_i}, and $\cup V_j \supset \overline{\Omega_f}$. If we look at the inverse maps $f_i^{-1} : V_{k_i} \to D_i$ and iterate them, we get an invariant Cantor set

$$\Lambda_f = \bigcap_{n=1}^{\infty} \bigcup_{(i_1, \ldots, i_n)} f_{i_1}^{-1} \circ \cdots \circ f_{i_n}^{-1}(D_{k_{i_n}}),$$

where in the right-hand side the union, for each n, is over all admissible n-tuples $(i_1, \ldots, i_n)$ (i.e. those for which the indicated composition of inverse maps makes sense). This is the limit set of the repeller.

We make two extra hypotheses about this repeller. The first hypothesis is that (Ω_f, f) is *uniformly asymptotically conformal* (abbreviated UAC), meaning that for each $\epsilon > 0$ there exists a neighborhood $U \supset \Lambda_f$ such that, for all $z \in U \setminus \Lambda_f$ and all words w in the alphabet $\{f_1, \ldots, f_N ; f_1^{-1}, \ldots, f_N^{-1}\}$ with $w(z) \in U$, we have $K_w(z) \leq 1 + \epsilon$ (here $K_w(z)$ denotes the quasiconformal distortion of w at z). The second hypothesis is that (Ω_f, f) is symmetric with respect to an anti-conformal involution $j_f : \Omega_f \to \Omega_f$. This front-to-back symmetry of our repeller avoids topological considerations having to do with the braid group, cf. [5].

We can consider the *waist curves* of our repeller: these are the curves obtained as inverse images of the Jordan curves ∂D_i under the various inverse branches of f. These curves deploy themselves according to a hierarchy (following the combinatorics of the repeller).

Now, we wish to define the asymptotic Teichmüller space of a UAC repeller. Let us consider the class of all repellers as above which are quasiconformally conjugate to a given one, say (Ω_f, f). Given two repellers (Ω_{g_1}, g_1) and (Ω_{g_2}, g_2) in this class, we declare them to be equivalent, $(\Omega_{g_1}, g_1) \sim (\Omega_{g_2}, g_2)$, if there exists an asymptotically conformal map $c : \Omega_f \to \Omega_g$ such that

1. $c \circ g_1 = g_2 \circ c$.
2. c preserves the hierarchical structure of waist curves of both repellers.
3. $c \circ j_{g_1} = j_{g_2} \circ c$.

This is an equivalence relation, and the quotient space is called the *asymptotic Teichmüller space* of the given repeller, and it is denoted $AT(\Omega_f, f)$. This Teichmüller space (like all others deserving the name) has a natural metric, the *Teichmüller metric*, defined as follows:

$$d_T((\Omega_{g_1}, g_1), (\Omega_{g_1}, g_1)) = \inf_{\phi} \frac{1}{2} \log H(\phi),$$

where the infimum is taken over all quasiconformal homeomorphisms $\phi : \Omega_{g_1} \to \Omega_{g_2}$ satisfying

1. $\phi \circ g_1 = g_2 \circ \phi$.
2. $\phi \circ j_{g_1} \simeq j_{g_2} \circ \phi$ (rel. Λ_{g_1}).
3. ϕ preserves the hierarchical structure of waist curves.

Here, $H(\phi)$ denotes the *boundary dilatation* of ϕ, namely,

$$H(\phi) = \inf_{E} \sup \{ K_{\phi}(z) : z \in \Omega_{g_1} \setminus E \},$$

the infimum ranging over all compact subsets $E \subset \Omega_{g_1}$.

One has the following standard result.

Theorem 20.1. *The space $AT(\Omega_f, f)$ with the metric defined above is a complete metric space.*

Once again, we refer to [5] for a proof. Similar definitions and results can be stated for other repellers, such as Jordan repellers. The appropriate notion of UAC system, and the corresponding asymptotic Teichmüller space, can be defined as above, *mutatis mutandis*. For more on the Teichmüller theory of UAC systems, see [6].

20.5 Actions of F

In this section we indicate how F acts on certain Teichmüller spaces, with and without dynamics. In both contexts, the Riemann surfaces involved have an underlying binary tree structure.

20.5.1 Geometric Actions

There is a natural action of F on $T_0(\Omega, j)$ when Ω is the binary Riemann surface arising as the complement of the standard triadic Cantor set (viewed as a subset of the plane), and j is the obvious involution (complex conjugation). This action is given geometrically by a *faithful representation* $\pi : F \to MCG(\Omega)$ of F into the mapping class group of Ω. The effect of the generators φ_0 and φ_1 of Thompson's

Fig. 20.7 The basic Thompson moves

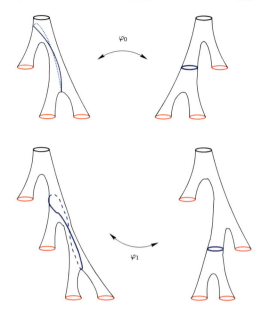

group in this situation can be seen in Fig. 20.7. It is achieved through a suitable *squeezing of geodesics into waist curves*. The quasiconformal distortion of each map in $\pi(F)$ is compactly supported by construction. The details of this representation, and the proof that it is faithful, are given in [5]. For a similar result for Thompson's T group (and much more on the representation theory of T), see [8].

20.5.2 Dynamical Actions

Let us now take the dynamics of repellers into account. We can define interesting dynamical actions of Thompson's group F as follows.

First, let us consider the case of the standard triadic repeller (Ω, f). The action of F on $AT(\Omega, f))$ is defined via the geometric action previously introduced, but on the *dual* system instead of the system itself – i.e. on $T_0(\Omega^*)$ instead of $T_0(\Omega)$. More precisely, we know that there is a faithful representation $\pi : F \to MCG(\Omega^*)$. Hence we define an action

$$\alpha : F \times AT(\Omega, f) \to AT(\Omega, f)$$

in the following way (see [5, p. 182]): if $[g] \in AT(\Omega, f)$, then

$$\alpha(\varphi, [g]) = [(g^* \circ \pi(\varphi^{-1}))^*].$$

In [5], we prove the following result.

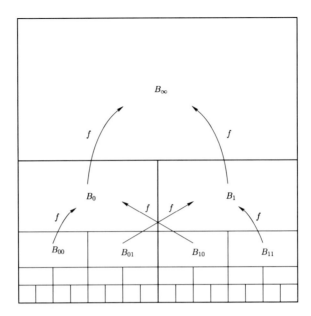

Fig. 20.8 Invariant Carleson boxes

Theorem 20.2. *The action of F defined above is faithful and discrete.*

In order to monitor this group action, we have used the so-called *scaling function* of the repeller. See [5, pp. 182–184] for details. The concept of scaling function for real repellers is due to D. Sullivan [10] (generalizing an idea due to M. Feigenbaum for quadratic maps).

Something completely analogous can be done in the case of the Jordan repeller given by the expanding map $z \mapsto z^2$ on the circle. Here the connectors (making up the Riemann surface Ω of the previous situation) are replaced by the Carleson boxes of Fig. 20.8. The duality construction works in the same way: one glues Carleson boxes along their edges using the dynamics and then applies the measurable Riemann mapping theorem, in order to get the dual system. The only non-trivial thing one must check in the construction is that one can still realize Thompson moves in this setting. That this can be done is shown in Fig. 20.9. There we are looking at three consecutive generations of Carleson boxes. The map realizing the basic move is piecewise affine. The inverted L-shaped region (the union of a Carleson box with one of its children) in the domain is mapped to the L-shaped region on the right. Each of the six triangles labeled 1–6 making up the inverted L-shaped region in the domain is mapped *by an affine map* onto the corresponding triangle making up the L-shaped region in the range. This forces the dashed L-shaped region in the domain to be mapped onto the dashed inverted L-shaped region in the range, also in piecewise affine fashion. This allows each of the remaining Carleson boxes in the domain (not shown) to be mapped exactly onto a corresponding Carleson box in the range (also not shown). Moreover, the only quasiconformal distortion happens

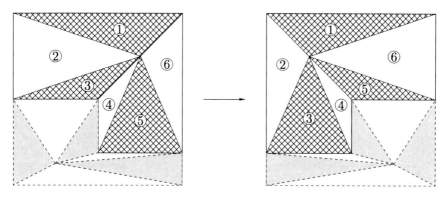

Fig. 20.9 Thompson moves on Carleson boxes

at the 3 levels where the L-shaped regions involved in the process live. At all other levels, Carleson boxes are mapped *conformally* onto other Carleson boxes. Thus, our basic Thompson moves are realized as *asymptotically conformal maps*, as they should.

General Thompson moves arise as compositions of such basic Thompson moves, happening at various levels of the tree structure provided by the invariant system of Carleson boxes. Thus, it is not difficult to guess that a result similar to Theorem 20.2 holds true here. Indeed, we have the following.

Theorem 20.3. *Thompson's group acts faithfully and discretely in the asymptotic Teichmüller space of the expanding map $z \mapsto z^2$ on the unit circle.*

We hope that the brief sketch of the main idea given above, leading to a proof of Theorem 20.3, has been convincing enough. The actual details will appear elsewhere. Note that we are *not* claiming, by any means, that F comprises the entire automorphism group of the asymptotic Teichmüller space in question.

What about other repellers, and other Thompson-like groups, such as general FP^∞ groups in the sense of Brown and Geoghegan in [1]?

20.2. Describe similar actions of such Thompson-like groups on other Cantor and Jordan repellers. Are these actions faithful and discrete?

Acknowledgements The author wishes to thank Alberto Pinto for urging him to write this paper, and Peter Hazard for pointing out several mistakes in an earlier version of the manuscript. This work received the financial support of FAPESP through *Projeto Temático Dinâmica em Baixas Dimensões*.

References

1. Brown, K.S., Geoghegan, R.: An infinite dimensional torsion-free FP^∞ group. Invent. Math. **77**, 367–381 (1984). MR **85m:**20073
2. Cannon, J.W., Floyd, W.J., Parry, W.R.: Notes on Richard Thompson's groups F and T. L'Enseignement Math. **42**, 215–256 (1996) MR **98g:**20058

3. de Faria, E.: An introduction to the thermodynamics of conformal repellers. São Paulo J. Math. Sci. (2010) (to appear)
4. de Faria, E., de Melo, W.: Mathematical Tools for One-dimensional Dynamics. Cambridge Studies in Advanced Mathematics, vol. 115. Cambridge University Press, Cambridge (2008)
5. de Faria, E., Gardiner, F., Harvey, W.: Thompson's group as a Teichmüller mapping class group. Contemp. Math. **355**, 165–185 (2004)
6. Gardiner, F.P., Sullivan, D.P.: Lacunary series as quadratic differentials in conformal dynamics. Contemp. Math. **169**, 307–330 (1994). MR **95h:**30020
7. Ghys, E., Sergiescu, V.: Sur un groupe remarkable de difféomorphismes du cercle. Comm. Math. Helv. **66**, 185–239 (1987)
8. Kapoudjian, C., Sergiescu, V.: An extension of the Bureau representation to a mapping class group associated to Thompson's group T. Prépublication **247**, Laboratoire de Mathematiques Emile Picard, Univ. Paul Sabatier Toulouse III (2002)
9. Martin, X.: Sur la géométrie du groupe de Thompson. C. R. Acad. Sci. Paris **333**, 773–778 (2001). MR **2002j:**20074
10. Sullivan, D.: Differentiable structures on fractal-like sets determined by intrinsic scaling functions on dual Cantor sets. Proc. Symp. Pure Math., AMS **48**, 15–23 (1988). MR **90k:**58141

Chapter 21
Bargaining Skills in an Edgeworthian Economy

M. Ferreira, B. Finkenstädt, B.M.P.M. Oliveira, Alberto A. Pinto, and A.N. Yannacopoulos

Abstract We present a model of an Edgeworthian exchange economy where two goods are traded in a market place. For a specific class of random matching Edgeworthian economies, the expectation of the limiting equilibrium price coincides with that of related Walrasian economies. The novelty of our model is that we assign a bargaining skill factor to each participant which introduces a game, similar to the prisoner's dilemma, into the usual Edgeworth exchange economy. We analyze the effect of the bargaining skill factor on the amount of goods acquired and the overall increase in the utility of the consumer. Finally, we let the bargaining skills of the participants evolve with subsequent trades and study the impact of this change over time.

M. Ferreira (✉)
Escola Superior de Estudos Industriais e de Gestão do Instituto Politécnico do Porto (IPP), LIAAD-INESC Porto LA, Escola de Ciências, Universidade do Minho, Braga, Portugal
e-mail: migferreira2@gmail.com

B. Finkenstädt
Department of Statistics, University of Warwick, Coventry CV4 7AL, UK
e-mail: b.f.finkenstadt@warwick.ac.uk

B.M.P.M. Oliveira
Faculdade de Ciências da Nutrição e Alimentação da, Universidade do Porto, LIAAD-INESC Porto LA, Porto, Portugal
e-mail: bmpmo@fcna.up.pt

Alberto A. Pinto
LIAAD-INESC Porto LA e Departamento de Matemática, Faculdade de Ciências, Universidade do Porto, Rua do Campo Alegre, 687, 4169-007, Portugal
and
Centro de Matemática e Departamento de Matemática e Aplicações, Escola de Ciências, Universidade do Minho, Campus de Gualtar, 4710-057 Braga, Portugal
e-mail: aapinto@fc.up.pt

A.N. Yannacopoulos
Department of Statistics, Athens University of Economics and Business, Patission 76, 10434 Athens, Greece
e-mail: ayannaco@aueb.gr

M.M. Peixoto et al. (eds.), *Dynamics, Games and Science I*, Springer Proceedings in Mathematics 1, DOI 10.1007/978-3-642-11456-4_21,
© Springer-Verlag Berlin Heidelberg 2011

21.1 Introduction

In most economies three basic activities occur: production, exchange, and consumption. In this paper we focus on the case of a pure exchange economy where individuals trade their goods in the market place for mutual advantage. There are two different approaches for modelling the nature of these economic activities: (1) The *Walrasian general equilibrium model* which assumes that consumers are passive price takers. They regard a given set of prices as parameters in determining their optimal net demands and supplies. The equilibrium price is set such that the market clears. Then the consumers change their endowments by the allocations determined by the equilibrium price. A mechanism that leads to the equilibrium price can be achieved, for instance, through an auctioneer who collects all the offers and demands for each good and adjusts the price vector to clear the market; and (2) The *Edgeworthian concept* considers consumers as active market participants trading with each other in an attempt to reach a higher level of utility. According to this model, an equilibrium is achieved when no person participating in the market can become better off without another person becoming worse off. We will look at the models in this perspective. An accepted and effective approach to pursuing this line of research is through the use of dynamic matching games, in which agents meet randomly, and exchange rationally, according to local rules. This general approach started with the seminal work *Mathematical Pshycics* of Francis Ysidro Edgeworth in 1881, [6, 8], and were further advanced by a number of researchers, including Aliprantis et al. [1], Aumann and Shapley [3,4], Binmore and Herrero [5], Gale [11], McLennan and Schonnenschein [13], Mas-Colell [14] and Rubinstein and Wolinsky [16]. The random matching game consists of agents paired at random who exchange goods at the bilateral Walras equilibrium price, which is the price at the core, such that the market locally clears. The choice for this scenario, is inspired by the work of Binmore and Herrero [5]. Under certain symmetry conditions on the initial endowments and the agents preferences, Ferreira et al. [9] show that the expectation of the logarithm of the equilibrium price, obtained as a limit for the repeated game as the number of trades tends to infinity, is equal to the expectation of the logarithm of the Walrasian equilibrium price.

Here, we assign to each participant a bargaining skill factor in the trade deviating from bilateral equilibrium model. The bargaining skill affects the trade, for instance two less skilled participants will split the benefits by choosing the bilateral competitive equilibrium, a more skilled participant and a less skilled participant will split the benefits with an advantage for the more skilled participant, and two more skilled participants are penalized by not trading. Hence, the participants are playing a game in the core alike the prisoner's dilemma, where the bargaining skill factor determines their strategy. The more skilled participants correspond to the non cooperative players and the less skilled ones correspond to the cooperative players in the prisoner's dilemma. We analyze the effect of the bargaining skill factor in the variation between the limit allocation and the initial endowment of each individual. We also examine the impact of the bargaining skill factor on the utility value obtained by the participants. For some parameter values, we find that it is better to be in the minority. When the more skilled participants are in minority the increase in the value of their utility is larger than the increase of the less skilled participants. However, When the less skilled participants are in the minority, the increase in the value of the utility for each less skilled participant is larger than the increase in the value of the utilities of the more skilled participants (the majority).

We study the evolution of the bargaining skill factor, inspired by the works of Durlauf [7]. We allow the bargaining skill of each participant to evolve over time. In each iteration we choose a pair of participants to trade and, after the trade, the bargaining skill can evolve following one of two possible rules: (a) the bargaining skills of the participants decrease if they were able to trade and increases if they were unable to trade; or (b) the bargaining skills of the participants increases if they were able to trade and decreases if they were unable to trade. We observe convergence of the bargaining skills to one of the two extreme limit values in model (a) and convergence of the bargaining skills to the middle point value in model (b).

The paper is organized as follows: in Sect. 21.2, we describe the Edgeworth model. In Sect. 21.3 we present our main result that relates the Walrasian Equilibrium price with the limiting bilateral equilibrium price. In Sect. 21.4, we incorporate the trade deviating from bilateral equilibrium model and we observe the relationship between the increase in the utilities and the bargaining skills. In Sect. 21.5 we present the evolutionary rules for the bargaining skills.

21.2 Edgeworth Model

We look at a pure exchange economy $(\Im, X_i, \succeq_i, w_i)$ where $\Im$ is the population of agents, each of them characterized by a consumption set $X_i \in R_+^2$ and $\succeq_i$ the individual preferences are given accordingly to the Cobb–Douglas utility function. So, an exchange economy in which some given amounts of goods X and Y are distributed among n individuals (individual i owns an initial endowment $\bar{x}_i$, $\bar{y}_i$ of good X and Y respectively) is considered. Note that the initial endowments $(\bar{x}_i, \bar{y}_i) \in int(X_i)$.

The Cobb–Douglas utility function is a model which is so well known in economic theory and thus requires almost no explanation. As stated in the comprehensive review paper of Lloyd [12] the Cobb–Douglas function was proposed long before it was formally tested by Cobb and Douglas, and influenced significantly the work of Mill, Pareto, Wicksell, Von Thünen for various reasons serves as the standard test bed for a great number of studies in mathematical economics. One of these reasons is that it is mathematically simple, yet captures important theoretical issues such as constant marginal rate of substitution. However, this is not the sole reason for its generalized use. Recent results of Voorneveld [18] show that the utility function being of the Cobb–Douglas form is equivalent to the preferences of the agents having the property of strict monotonicity, homotheticity in each coordinate and upper semicontinuity. These properties are rather generic properties for preferences, and are very reasonable assumptions. Furthermore, there is empirical evidence [15], demonstrating that with very large and increasing per capita income the utility function becomes asymptotically indistinguishable from Cobb–Douglas. These very interesting results shed new light on the Cobb–Douglas utility function and provide further justification for its use as a standard model, in addition to its apparent analytical simplicity.

We assume individual i obtains utility from the quantities x_i and y_i according to the Cobb–Douglas utility function

$$U_i(x_i, y_i) = x_i^{\alpha_i} y_i^{1-\alpha_i}, 0 < \alpha_i < 1 \qquad (21.1)$$

which represents strictly convex, continuous and nondecreasing preferences where α_i defines the preferences of the goods X and Y for participant i. The bilateral equilibrium price of a pair of participants (i, j) is the Walrasian equilibrium price of the economy consisting only of this pair of participants (i, j), and so, not taking into account the other participants of the economy. Considering the good X to be the *numeraire*, it is well known that the bilateral equilibrium price p is given by

$$p = \frac{\alpha_i y_i + \alpha_j y_j}{(1 - \alpha_i)x_i + (1 - \alpha_j)x_j} \qquad (21.2)$$

where p is the price of the good Y. The bilateral trade is the well known scenario analyzed in the Edgeworth box diagram (see Fig. 21.1). The horizontal axis represents the amount of good X and the vertical represents the amount of good Y of participant i. The point $(x_i + x_j, y_i + y_j)$ is the vertex opposite to the origin. The horizontal and vertical lines starting at the opposite vertex are the axes representing the amounts of good X and Y, respectively, of participant j. We represent in the Edgeworth box the indifference curves for both participants passing through the

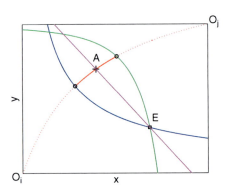

Fig. 21.1 Edgeworth Box with the indifference curves for participant i (*blue convex curve*) and j (*green concave curve*). The red curve is the core and the red dots represent the contract curve. The slope of the pink segment line is the bilateral equilibrium price. The interception point (A) of the core with the pink segment line determines the new allocations and the square (E) marks the initial endowments

21 Bargaining Skills in an Edgeworthian Economy

point corresponding to the initial endowments of both participants. The core is the curve where the indifference curves of both participants are tangent and such that the utilities of both participants are greater or equal to the initial ones. The bilateral price determines a segment of allocations that pass through the point corresponding to the initial endowments. The interception of this segment with the core determines the new allocations of the two participants.

In the random matching Edgeworthian economies, agents start with a set of initial endowments $(x_i(0), y_i(0))$. Then, these agents trade in random pairs, chosen with the same probability, and after the t trade end up with a consumption bundle $(x_i(t), y_i(t))$ which is traded in the bilateral equilibrium price $p(t)$ given by the formula (21.2) with x_i, x_j, y_i, y_j substituted by $x_i(t-1), x_j(t-1), y_i(t-1), y_j(t-1)$. On each trade only two randomly chosen agents i, j exchange goods and the consumption bundles of the other agents remain unchanged. The demand of the agents on the two goods is a stochastic process and that turns the price $p(t)$ into a stochastic process as well.

This raises an interesting question in this context:

Does there exist a limiting price $p_\infty = \lim_{t \to \infty} p(t)$ and if so how would that compare to the Walrasian equilibrium price, where all agents meet simultaneously and trade at that time?

An answer to the first question has been given by a number of authors, see for example [11] and references therein. According to this previous work, p_∞ exists almost surely, and it is a random variable. However, it depends on the actual game of the play, which is dependent on the exact order of the random pairing of the agents.

The aim of the present work is to provide some results on the expectation of this random variable p_∞, and see how it compares to the Walrasian price p_w. In particular, under some rather general symmetry conditions on the initial endowments of the agents and distribution of initial preferences, Ferreira et al. [9] show that the expectation of the logarithm of p_∞ equals the logarithm of the Walrasian price for the same initial endowments of the agents (in Fig. 21.2). This is an interesting result, in the sense that even though the agents meet and trade myopically in random pairs, they somehow "self-organize" and the expected limiting price equals that of a market where a central planner announces prices and all the agents conform to them through utility maximization, as happens in the Walrasian model. When this asymmetry is broken, the discrepancy $E(\log p_\infty) - \log p_w$ deviates from zero (compare Figs. 21.2c with d).

The main reason why organizing behaviour is observed is the symmetry in the endowments and preferences of the agents that, as will become clear from our analysis in the next section, imposes global constraints in the market, in the sense that it enforces each agent to have a *mirror*, or a *dual agent*.

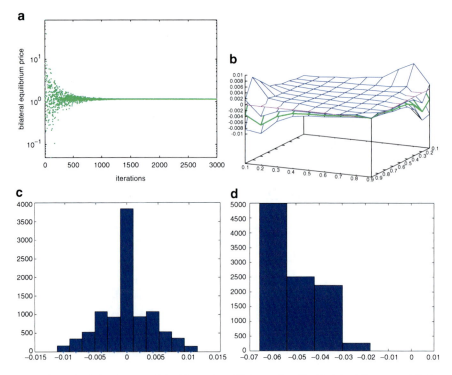

Fig. 21.2 (**a**) Variation in time of the bilateral equilibrium price. The limiting bilateral equilibrium price p_∞ is a value near the Walrasian price of the initial endowments p_w. The green dots represent the bilateral equilibrium price at each iteration. (**b**) Dependency of the $E(\log p_\infty) - \log p_w$ with the preferences of the agents. We consider a market with two types of agents with the preferences α of half agents in the x axis and the preferences of the other half in the y axis. The green surface represents the mean of 100 simulations and the blue surface is the 95% confidence interval. (**c**) Histogram for the $E(\log p_\infty) - \log p_w$ for 10,000 simulations of a market with 4 participants with preferences $\alpha_1 = \alpha_2 = 0.1$ and initial endowments $x_i = y_i = 1/4$, $i = 1, 2, 3, 4$. (**d**) Histogram for the $E(\log p_\infty) - \log p_w$ for 10,000 simulations of a market with 3 participants with preferences $\alpha_1 = \alpha_2 = 0.3$ and $\alpha_3 = 0.9$

21.3 Statistical duality

We introduce the concept of *duality* in the market. We assume that the collection of agents is completely characterized by their preferences α, and their endowments (x, y) in the 2 goods. We can define a probability distribution function on (α, x, y) space, $f(\alpha, x, y)$ which determines the probability that a chosen agent has preferences in $(\alpha, \alpha + d\alpha) \times (x, x + dx) \times (y, y + dy)$. We assume that the probability distribution has compact support, and the support in (x, y) is bounded away from zero.

21 Bargaining Skills in an Edgeworthian Economy

Definition 21.1. We say that a market satisfies the p - *statistical duality* condition if the probability function has the symmetry property

$$f(\alpha, x, y) = f\left(1 - \alpha, \frac{y}{p}, p\,x\right)$$

where $p \in R^+$.

The p-statistical duality property means that each agent with characteristics (α, x, y) has a mirror agent with characteristics $(1 - \alpha, y/p, p\,x)$ with the same probability under f. The class of probability functions $f(\alpha, x, y)$ of the form $f_1(\alpha)$ $f_2(x, y)$ with the property that $f_1(\alpha) = f_1(1 - \alpha)$ and $f_2(x, y) = f_2(y/p, p\,x)$ satisfies the p-statistical duality. A common probability function f_2, satisfying the above condition, is the uniform distribution. Another common example of a probability function satisfying the p-statistical duality is used in Corollary 21.1, below, and determines the most well known matching technology used in random matching games with N agents.

Statistical duality guarantees that the prices observed in the random matching Edgeworthian economy coincide in expectation with those of the Walrasian economy. For each collection of agents, let p_w denote the Walrasian equilibrium price of the initial market.

Theorem 21.1. *Assume a market consisting of a finite number N of agents, such that p-statistical duality holds for the initial endowments, then*

$$E[\ln(p(t))] = \bar{E}[\ln(p_w)] = \ln(p), \ \text{for all} \ t \in \{1, 2, \ldots, +\infty\}.$$

Furthermore,

$$E[\ln(p_\infty)] = \ln(p)$$

where $\bar{E}$ is the expectation over the distribution of agents and E is expectation over the distribution of agents and over all possible runs of the game.

See proof of Theorem 21.1 in [9]. The advantage of using the logarithm of the price is that if we consider the other good to be the enumeraire, the absolute value of the logarithm of the price remains the same and just the sign of the value of the logarithm of the price changes.

An example of an economy with the p-statistical duality property is an economy where with probability 1 we start with a sample of $N = 2\,M$ agents where M agents have characteristics $(a_i, x_i, y_i), i = 1, \ldots, M$, and the remaining M agents have characteristics $(a_{i+M}, x_{i+M}, y_{i+M}) = (1 - a_i, y_i/p, p\,x_i), i = 1, \ldots, M$. In other words, in this economy, each agent has a dual agent, i.e. agent i is dual to agent $i + M$ where $i = 1, \ldots, M$.

Corollary 21.1. *Assume a market consisting of a finite number $N = 2M$ of agents, such that M agents have characteristics $(a_i, x_i, y_i), i = 1, \ldots, M$, and the remaining M agents have characteristics $(a_{i+M}, x_{i+M}, y_{i+M}) = (1 - a_i, y_i/p, p\, x_i), i = 1, \ldots, M$, then*

$$E[\ln(p(t)] = \ln(p_w) = \ln(p), \text{ for all } t \in \{1, 2, \ldots, +\infty\}.$$

Furthermore,

$$E[\ln(p_\infty)] = \ln(p)$$

where E is the expectation over all possible runs of the game.

The theorem can also be shown to hold for a generalized random matching economy in which agents do not only meet in pairs. In this game, we initially pick N agents and then for each trading date we pick randomly $M \leq N$ agents, that decide to trade on the competitive price for the local market consisting only of these M agents. The number M may change with t. Then, under our statistical duality condition, it may be shown that the stated result holds.

Remark 21.1. Theorem 21.1, in its specific form is limited to the particular form of the Cobb–Douglas utility function, which was assumed to model the preferences of the agents in the market game. However, this does not reduce the interest and the importance of this result for the following three reasons.

(a) The Cobb–Douglas utility is a very important choice for the utility function for the reasons stated in Sect. 21.2. Therefore, the complete understanding of the effect of symmetry for the asymptotic results of the market game, for this choice of utility function, offered by Theorem 21.1, is interesting in its own right as it provides a "realistic" scenario under which Edgeworthian and Walrasian considerations asymptotically coincide.

(b) The important assumption for Theorem 21.1 to hold true seems to be the constancy of the marginal rates of substitution, a fact that is guaranteed by the special choice of the Cobb–Douglas utility function. However, one may try and modify Theorem 21.1, for the case where the marginal rates of substitution that are close to having this property, e.g. it is slowly varying. Then, by proper assumptions that allow us to control such deviations, one could obtain approximate symmetry with results similar to Theorem 21.1. However, such considerations are beyond the scope of the present work.

(c) Theorem 21.1 may be modified to hold for more general situations. Such situations will involve the use of different types of utility functions than the Cobb–Douglas. The modifications could be in the type and rate of convergence. For instance the convergence of $\ln(p)$ to the Walrasian value in the mean, is special for the choice of the Cobb–Douglas function, and the choice of the logarithmic function is dictated by the specific dynamics and symmetries of this utility function. The same holds true for the definition of p-statistical duality.

It is conceivable that different choices of the utility function for the agents will require the choice of different symmetry conditions and different functions than the logarithmic for the quantification of the convergence of the Edgeworthian to the Walrasian price. However, what is important is that symmetries in the distribution of preferences and initial endowments will still dictate the asymptotic coincidence of the prices predicted by the Edgeworth and the Walras scenario. The full treatment of this point is beyond the scope of the present paper and will be treated elsewhere.

21.4 Deviating from the Bilateral Equilibrium

The model with *trade deviating from bilateral equilibrium* is similar to the Edgeworth model. The difference is that, in this model, we introduce a new parameter g_i representing the bargaining skill of each participant. If two less skilled $g_i = g_j = 0$ participants meet they will trade in the point of the core determined by their bilateral equilibrium price, as in the Edgeworth model. However, if a more skilled participant $g_i = 1$ meets a less skilled $g_i = 0$ participant, they will trade in a point of the core between the point determined by their bilateral equilibrium price and the interception of the core with the indifference curve of the less skilled participant (see Fig. 21.3), traducing an advantage to the more skilled participant. Finally, if both participants are highly skilled $g_i = g_j = 1$ they are penalized by not being able to trade. This is similar to the "prisoner's dilemma", where two non-cooperative players are penalized, a non-cooperative player has a better payoff than a cooperative player, and two cooperative players have a better payoff than when

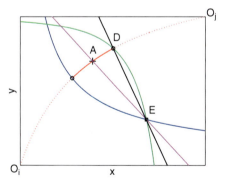

Fig. 21.3 Edgeworth Box with the indifference curves for the more skilled participant i (*blue convex curve*) and less skilled participant j (*green concave curve*). The red curve is the core and the red dots represent the contract curve. The slope of the pink line (that intercepts the core in point A) is the bilateral equilibrium price. The slope of the black line (that intercepts the core in point D) is the price that gives the greatest advantage to the more skilled participant. The final allocation will be a point in the core inside the curve AD and the square marks the initial endowments (E)

they meet a non-cooperative player but still worse than the payoff of the non-cooperative player. We consider that the bargaining skill g is a continuous variable, where higher values mean better bargaining skills. Without loss of generality we can consider that $g_i \geq g_j$. We impose that trade only occurs if $g_i + g_j \leq 1$ and $g_i - g_j \in [0, 1]$. The pair trades to the point in the core determined by the price $\log p_g = \eta \log p + (1 - \eta) \log m_j$ where p is the bilateral equilibrium price, $\eta = g_i - g_j$ and m_j is the maximum price at which the participant j accepts to trade determined by the interception of the core with the indifference curve of participant j (equal to the slope of the line [ED] in Fig. 21.3).

We study the effect of the bargaining skills on the increase of the value of the utility of the participants. Let the variation of the utility function of a participant $u_f - u_0$ be the difference between the limit value of the utility function and the initial value of the utility function. We present, in Fig. 21.4, two cumulative distribution functions of the variation of the utility functions, one corresponding to the less skilled participants ($g_i = 0.25$) and the other corresponding to the more skilled participants ($g_i = 0.75$). This function indicates the proportion of participants who have variations of the utility function less than or equal to its argument. In Fig. 21.4a there are 20% of more skilled participants. We observe that the median of the variation of the utility function is higher for the more skilled participants. On the other hand, in Fig. 21.4b, there are 80% of more skilled participants, and we observe that the median of the variation of the utility function is lower for the more skilled participants. We notice that the strategy followed by the minority is the one that provides a higher median variation in the utility function.

When we compare different values assigned to the bargaining skills of the participants, we observe distinct behaviors. When $g_i = 0$, for the less skilled participants and $g_i = 1$ for the more skilled participants, the trade gives the most advantage possible to the more skilled participants (the final allocation is represented by point D in Fig. 21.3). We see that the more skilled participants have a larger median increase

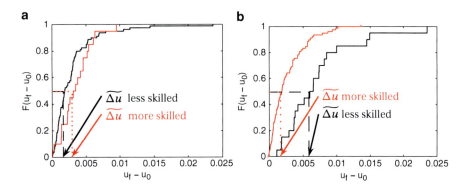

Fig. 21.4 Cumulative distribution function of the variation of the utility (defined as $u_f - u_0$) for the less skilled participants (*black*) and for the more skilled participants (*red*), with $g_i \in \{0.25; 0.75\}$. (**a**) Simulation with 20 more skilled participants and 80 less skilled participants; (**b**) Simulation with 80 more skilled participants and 20 less skilled participants

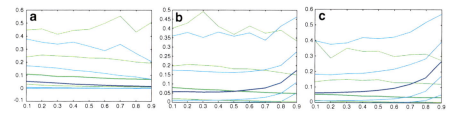

Fig. 21.5 Variation of the utility ($u_f - u_0$) for the less skilled participants (*blue/cyan*) and for the more skilled participants (*green / yellow*). Data from 100 simulations with 100 participants when the fraction of more skilled participants is $0.1, 0.2, \ldots, 0.9$. Each set of participants has lines for the minimum percentile, 5%, median (*thick line*), percentile 95% and maximum. (**a**) Advantage to the more skilled participants when $g_i \in \{0; 1\}$; (**b**) Advantage to the minority when $g_i \in \{0.25; 0.75\}$; (**c**) Advantage to the less skilled participants when $g_i \in \{0.499; 0.501\}$

in the utility (Fig. 21.5a) for all fractions between 0.1 and 0.9 of more skilled participants. In the opposite case, when $g_i = 0.499$, for the less skilled participants and $g_i = 0.501$ for the more skilled participants, the trade gives a very small advantage to the more skilled participants (the final allocation is near point A in Fig. 21.3). In this case, the more skilled participants have a smaller median increase in the utility (Fig. 21.5c) for all fractions between 0.1 and 0.9 of more skilled participants, due to the impact of the penalization of no trade between them. If we consider $g_i = 0.25$, for the less skilled participants and $g_i = 0.75$ for the more skilled participants, the trade gives an intermediate advantage to the more skilled participants (the final allocation is a point in the core roughly midway between A and D in Fig. 21.3). We observe that the group in minority has the advantage. Namely, for fractions of more skilled participants between 0.1 and 0.4, these have a higher median increase in the utility and for fractions of more skilled participants between 0.6 and 0.9, these have a lower median increase in the utility. For fractions of more skilled participants near 0.5, the median increase in the utility of the more skilled participants is similar to the less skilled participants (Fig. 21.5b).

21.5 Evolution of the Bargaining Skills

In this section, at each iteration, a random pair of participants (i, j) is chosen with trade occurring deviating from the bilateral equilibrium price (as in the previous section). We consider that the bargaining skill g is a continuous variable, where higher values mean better bargaining skills. After the trade we allow evolution on the bargaining skills of the participants of the form $g_i(t+1) = g_i(t) + \epsilon_{ij} g_i(t)(1 - g_i(t))$ according to two distinct rules: (a) the bargaining skills of the participants increase if they were not able to trade and decrease if they were able to trade; (b) the bargaining skills of the participants decreases if they were not able to trade and increase otherwise.

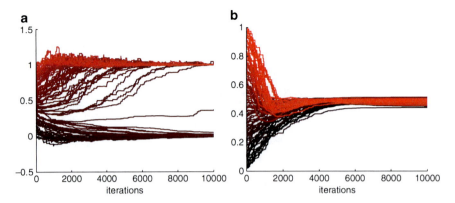

Fig. 21.6 Variation of the bargaining skills with time. (**a**) The bargaining skills decrease when trade is allowed and increase otherwise; (**b**) The bargaining skills increase when trade is allowed and decrease otherwise

In case (a) if the sum of their bargaining skills is above the cut point, the participants are not allowed to trade and both participants' bargaining skills are increased (we choose $\epsilon_{ij} = \epsilon > 0$). Otherwise, if the sum of their bargaining skills is below the cut point, the participants will be allowed to trade with advantage to the more skilled participant. After the trade, the bargaining skills of both participants decrease ($\epsilon_{ij} = -\epsilon$). In this case we observe (see Fig. 21.6a) that the participants' bargaining skills converge to one of the two extreme limit values 0 or 1.

In case (b) if the sum of their bargaining skills is above the cut point, the participants are not allowed to trade and their bargaining skills are decreased. Otherwise, if the sum of their bargaining skills is below the cut point, the participants will be allowed to trade with advantage to the more skilled participant. After the trade, the bargaining skills of both participants increase. In this case (see Fig. 21.6b) we observe that the bargaining skills converge to one limit value 1/2.

21.6 Conclusion

We presented a model of an Edgeworthian exchange economy where two goods are traded in a market place. Under symmetry conditions, prices in a random exchange economy with two goods, where the agents preferences are characterized by the Cobb–Douglas utility function, converge to the Walrasian price. Under proper symmetry conditions, we conjecture that this result holds for the case of n goods.

The novelty of our model is that we associated a bargaining skill factor to each participant which invokes a game similar to the prisoner's dilemma into the usual Edgeworth exchange economy. We analyzed the effect of the bargaining skill factor in the variations of the individual's amount of goods and in the increase of the value of their utilities. For the scenarios presented in this work, it is better to be in the minority. For instance, if there are a greater number of more skilled participants, the increase in the value of their utilities is smaller then the increase in the value of the utilities of the less skilled participants who are in the minority.

The two evolutionary rules presented result in different behaviors. If the bargaining skill decreases when trade is allowed (and increases otherwise), the bargaining skills of the participants converge over time to one of two extreme limit values, however, if the bargaining skill increases when trade is allowed (and decreases otherwise), we observe that the bargaining skill converges over time to the middle value.

Acknowledgements Previous versions of this work were presented in the International Congresses of Mathematicians ICM 2006 and 2010, International Congress for Industrial and Applied Mathematics 2007, the Second Brazilian Workshop of the Game Theory Society in honor of John Nash and EURO 2010, ICDEA 2009, SAMES 2006 and VII JOLATE. We thank LIAAD-INESC Porto LA, Calouste Gulbenkian Foundation, PRODYN-ESF, FEDER, POFC, POCTI and POSI by Fundação para a Ciência e a Tecnologia (FCT) and Ministério da Ciência e da Tecnologia, and the FCT Pluriannual Funding Program of the LIAAD-INESC Porto LA and of the Research Centre of Mathematics of the University of Minho, for their financial support. Bruno Oliveira gratefully acknowledges financial support from PRODEP III by FSE and EU and Centro de Matemática da Universidade do Porto and Miguel Ferreira gratefully acknowledges financial support from FCT given through a PhD scholarship.

References

1. Aliprantis, C.D., Brown, D.J., Burkinshaw, O.: Edgeworth equilibria. Econometrica **55**, 1109–1137 (1987)
2. Aumann, R.: Existence of competitive equilibria in markets with a continuum of traders. Econometrica **34**(1), 1–17 (1966)
3. Aumann, R.: Correlated equilibrium as an expression of Bayesian rationality. Econometrica **55**, 1–18 (1987)
4. Aumann, R., Shapley, L.: Values of non atomic games. Princeton University Press (1974)
5. Binmore, K., Herrero, M.: Matching and bargaining in dynamic markets. Rev. Econ. Stud. **55**, 17–31 (1988)
6. Debreu, G., Scarf, H.: A limit theorem on the core of an economy. Int. Econ. Rev. **4**, 235–246 (1963)
7. Durlauf, S., Young, P.: Social Dynamics: Economic Learning and Social Evolution. MIT, Cambridge and London (2001)
8. Edgeworth, F.Y.: Mathematical Psychics: An Essay on the Application of Mathematics to the Moral Sciences. Kegan Paul, London (1881)
9. Ferreira, M., Finkenstädt, B., Oliveira, B.M.P.M., Pinto, A.A., Yannacopoulos, A.N.: On the convergence to Walrasian prices in random matching Edgeworthian economies (submitted)
10. Finkenstädt, B.: Nonlinear Dynamics in Economics, A theoretical and statistical approach to agricultural markets. Lecture Notes in Economics and Mathematical Systems (1995)
11. Gale, D.: Strategic Foundations of Perfect Competition. Cambridge University Press, Cambridge (2000)
12. Lloyd, P.J.: The origins of the von Thünen–Mill–Pareto–Wicksell–Cobb–Douglas function. Hist. Polit. Econ. **33**, 1–19 (2001)
13. McLennan, A., Schonnenschein, H.: Sequential bargaining as a non cooperative foundation for Walrasian equilibria. Econometrica **59**, 1395–1424 (1991)

14. Mas-Colell, A.: An equivalence theorem for a bargaining set. J. Math. Econ. **18**, 129–139 (1989)
15. Powell, A.A., McLaren, K.R., Pearson, K.R., Rimmer, M.T.: Cobb–Douglas Utility Eventually!, Monash University, Department of Econometrics and Business Statistics, Working Paper (2002)
16. Rubinstein, A., Wolinsky, A.: Decentralized trading, strategic behaviour and the Walrasian outcome. Rev. Econ. Stud. **57**, 63–78 (1990)
17. Shubik, M.: Edgeworth's market games. In: Luce, R.D., Tucker, A.W. (eds.) Contributions to the theory of games IV. Princeton University Press (1959)
18. Voorneveld, M.: From preferences to Cobb–Douglas utility, SSE/EFI Working Paper Series in Economics and Finance, 701 (2008)

Chapter 22
Fractional Analysis of Traffic Dynamics

Lino Figueiredo and J.A. Tenreiro Machado

Abstract This article presents a dynamical analysis of several traffic phenomena, applying a new modelling formalism based on the embedding of statistics and Laplace transform. The new dynamic description integrates the concepts of fractional calculus leading to a more natural treatment of the continuum of the Transfer Function parameters intrinsic in this system. The results using system theory tools point out that it is possible to study traffic systems, taking advantage of the knowledge gathered with automatic control algorithms.

22.1 Dynamical Analysis

In the dynamical analysis of Traffic can be applied the tools of systems theory. In this line of thought, a set of simulation experiments are developed in order to estimate the influence of the vehicle speed $v(t;x)$, the road length l and the number of lanes n_l in the traffic flow $\phi(t;x)$ at time t and road coordinate x. For a road with n_l lanes the Transfer Function (TF) between the flow measured by two sensors is calculated by the expression:

$$G_{r,k}(s;x_j,x_i) = \Phi_r(s;x_j)/\Phi_k(s;x_i) \tag{22.1}$$

where $k, r = 1, 2, \ldots, n_l$ define the lane number and, x_i and x_j represent the road coordinates ($0 \leq x_i \leq x_j \leq l$). The Fourier Transform for each traffic flow is:

$$\Phi_r(s;x_j) = F\{\phi_r(t;x_j)\}$$
$$\Phi_k(s;x_i) = F\{\phi_k(t;x_i)\} \tag{22.2}$$

L. Figueiredo (✉) and J.A.T. Machado
Department of Electrical Engineering, ISEP-Institute of Engineering of Porto, Rua Dr. Antonio Bernardino de Almeida, 4200-072 Porto, Portugal
e-mail: lbf@isep.ipp.pt, jtm@isep.ipp.pt

M.M. Peixoto et al. (eds.), *Dynamics, Games and Science I*, Springer Proceedings in Mathematics 1, DOI 10.1007/978-3-642-11456-4_22,
© Springer-Verlag Berlin Heidelberg 2011

It should be noted that the traffic flow is a time variant system but, in the sequel, it is shown that the Fourier transform can be used to analyse the system dynamics.

The first group of experiments considers a one-lane road (i.e., $k = r = 1$) with length $l = 1,000$ m. Across the road are placed n_s sensors equally spaced. The first sensor is placed at the beginning of the road (*i.e.*, at $x_i = 0$) and the last sensor at the end (i.e., at $x_j = l$). Therefore, we calculate the *TF* between two traffic flows at the beginning and the end of the road such that, $\phi_1(t; 0) \in [0.12, 1]$ vehicles s^{-1} for vehicle speed $v_1(t; 0) \in [30, 70]$ km h^{-1} that is, for $v_1(t; 0) \in [v_{av} - \Delta v, v_{av} + \Delta v]$, where $v_{av} = 50$ km h^{-1} is the average vehicle speed and $\Delta v = 20$ km h^{-1} is the maximum speed variation. These values are generated according to a uniform probability distribution function.

The polar plot result for the *TF* between the traffic flow at the beginning and end of the one-lane road $G_{1,1}(s; 1000, 0) = \Phi_1(s; 1000)/\Phi_x(s; 0)$ is distinct from those usual in systems theory revealing a large variability, as revealed by Fig. 22.1. Moreover, due to the stochastic nature of the phenomena involved different experiments using the same input range parameters result in different *TF*s.

In fact traffic flow is a complex system but it was shown [1] that, by embedding statistics and Fourier transform (leading to the concept of Statistical Transfer Function (*STF*)) [3], we could analyse the system dynamics in the perspective of systems theory. To illustrate the proposed modelling concept (*STF*), the simulation was repeated for a sample of $n = 2,000$ vehicles and it was observed the existence of a convergence of the *STF*, $T_{1,1}(s; 1,000, 0)$, as show in Fig. 22.2, for a one-lane road with length $l = 1,000$ m $\phi_1(t; 0) \in [0.12, 1]$ vehicles s^{-1} and $v_1(t; 0) \in [30, 70]$ km h^{-1}.

The chart has characteristics similar to those of a low-pass filter with time delay, common in systems involving transport phenomena. Nevertheless, in our case we

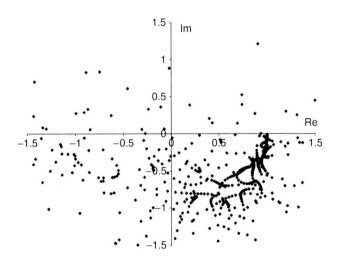

Fig. 22.1 Polar diagram of *TF* for $n = 1$ experiment, with $\phi_1(t; 0) \in [0.12, 1]$ vehicles s^{-1} and $v_1(t; 0) \in [30, 70]$ km h^{-1} ($v_{av} = 50$ km h^{-1}, $\Delta v = 20$ km h^{-1}, $l = 1,000$ m and $n_l = 1$)

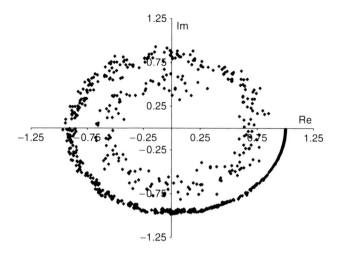

Fig. 22.2 Polar diagram of *STF* for $n = 2{,}000$ experiments, with $\phi_1(t;0) \in [0.12, 1]$ vehicles s^{-1} and $v_1(t;0) \in [30, 70]$ km h^{-1} ($v_{av} = 50$ km h^{-1}, $\Delta v = 20$ km h^{-1}, $l = 1{,}000$ m and $n_l = 2$)

need to include the capability of adjusting the description to the continuous variation of the system working conditions. This requirement precludes the adoption of the usual integer-order low-pass filter and points out the need for the adoption of a fractional-order *TF*. Therefore, in this case we adopt a fractional low pass system [2] with time delay:

$$T_{1,1}(s; 1000, 0) = \frac{k_B \, e^{-\tau s}}{\left(\frac{s}{p} + 1\right)^\alpha} \qquad (22.3)$$

With this description we get not only a superior adjustment of the numerical data, impossible with the discrete steps in the case of integer-order *TF*, but also a mathematical tool more adapted to the dynamical phenomena involved. For fitting expression (22.3) with the numerical data it is adopted a two-step method based on the minimization of the quadratic error. In the first phase (k_B, p, α) are obtained through error amplitude minimization of the Bode diagram. Once established (k_B, p, α), in a second phase, τ is estimated through the error minimization in the Polar diagram. For the numerical parameters of Fig. 22.2 we get $k_B = 1.0$, $\tau = 96.0$ s, $p = 0.07$ and $\alpha = 1.5$. The parameters (τ, p, α) vary with the average speed v_{av} and its range of variation Δv, the road length l and the input vehicle flow ϕ_1. For example, Fig. 22.3 shows (τ, p, α) versus Δv (with $v_{av} = 50$ km h^{-1}).

It is interesting to note that $(\tau, p) \to (\infty, 0)$, when $\Delta v \to v_{av}$, and $(\tau, p) \to (lv_{av}^{-1}, \infty)$, when $\Delta v \to 0$. These results are consistent with our experience that suggests a pure transport delay $T(s) \approx e^{-\tau s}$ ($\tau = lv_{av}^{-1}$), $\Delta v \to 0$ and $T(s) \approx 0$, when $\Delta v \to v_{av}$ (because of the existence of a blocking cars, with zero speed, on the road).

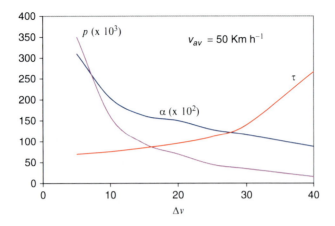

Fig. 22.3 Parameters τ, p and α with $n_l = 1, l = 1{,}000$ m and $\phi_1(t;0) \in [0.12, 1]$ vehicles s^{-1} versus Δv (with $v_{av} = 50$ km h^{-1})

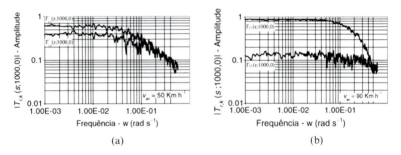

Fig. 22.4 Bode diagram of $T_{r,k}(s; 1{,}000, 0)$ for (**a**) $v_{av} = 50$ km h^{-1} and (**b**) $v_{av} = 90$ km h^{-1}, $n_l = 2, l = 1{,}000$ m, $\phi_k(t;0) \in [0.12, 1]$ vehicles s^{-1}, $\Delta v = 20$ km h^{-1}, $k = 1, 2$

In a second group of experiments are analyzed the characteristics of the *STF* matrix for roads with two lanes considering identical traffic conditions (i.e., $\phi_k(t;0) \in [0.12, 1]$ vehicles s^{-1}, $k = 1, 2$, $l = 1{,}000$, $\Delta v = 20$ km h^{-1}). Figure 22.4a depicts the amplitude Bode diagram of $T_{1,1}(s; 1{,}000, 0)$ and $T_{1,2}(s; 1{,}000, 0)$ for $v_{av} = 50$ km h^{-1} (i.e., $v_k(t;0) \in [30, 70]$ km h^{-1}).

Figure 22.4a shows that $T_{1,1}(s; 1{,}000, 0) \approx T_{2,2}(s; 1{,}000, 0)$ and $T_{1,2}(s; 1{,}000, 0) \approx T_{2,1}(s; 1{,}000, 0)$. This property occurs because SITS uses a lane change logic where, after the overtaking, the vehicle tries to return to the previous lane. Therefore, lanes 1 and 2 have the same characteristics leading to identical *STF*.

The Fig. 22.4b presents the amplitude Bode diagram of $T_{1,1}(s; 1{,}000, 0)$ and $T_{1,2}(s; 1{,}000, 0)$ for $v_{av} = 90$ km h^{-1} (i.e., $v_k(t;0) \in [70, 110]$ km h^{-1}).

Comparing Fig. 22.4a and these results, we conclude that the transfer matrix elements vary significantly with v_{av}. Moreover, the *STF* parameter dependence is similar to the one-lane case represented previously.

22.2 Conclusions

In this paper several experiments were carried out in order to analyse the dynamics of the traffic systems. Bearing these ideas in mind it was adopted a formalism based on the tools of systems theory. Moreover, the new dynamic description integrated the concepts of fractional calculus lead to a more natural treatment of the continuum of the TF parameters intrinsic in this system. The results pointed out that it is possible to study traffic systems.

References

1. Figueiredo, L., Machado, J., Ferreira J.: Dynamical analysis of freeway traffic. IEEE Trans. Intell. Transp. Syst. **5**(4), ISSN: 1524-9050, 259–266 (2004)
2. Machado, J.: A probabilistic interpretation of the fractional-order differentiation. J. Fractional Calc. Appl. Anal. **6**(1), 73–80 (2003)
3. Machado, J., Galhano A.: A statistical perspective to the Fourier Analysis of mechanical manipulators. J. Syst. Anal. Model. Simul. **33**, 373–384 (1998)

Chapter 23
The Set of Planar Orbits of Second Species in the RTBP

Joaquim Font, Ana Nunes, and Carles Simó

Abstract We present a brief summary of the conclusions of our work on the set of orbits of the planar circular restricted three body problem which undergo consecutive close encounters with the small primary, or orbits of second species. In this study, the value of the Jacobi constant is fixed, and we consider consecutive close encounters which occur within a maximal time interval. With these restrictions, the full set of orbits of second species is found numerically from the intersections of the stable and unstable manifolds of the collision singularity on the surface of section that corresponds to passage through the pericentre. A "skeleton" of this set of curves can be computed from the solutions of the two-body problem. The set of intersection points found in this limit corresponds to the S-arcs and T-arcs of Hénon's classification which verify the energy and time constraints, and can be used to construct an alphabet to describe the orbits of second species. We find periodic orbits that combine S-type and T-type quasi-homoclinic arcs and we determine the symbolic dynamics of the full set of orbits of second species.

23.1 A Summary of Results about the Problem

In his study of the three body problem, Poincaré conjectured the existence of periodic orbits that for small values of two of the masses are close to sequences of arcs of Keplerian ellipses, glued together at singularities that correspond to collisions of the two zero mass bodies [1]. In the perturbed problem, these singularities would be replaced by close encounters that would shift the orbital elements of the approximate

---

A. Nunes (✉)
CFTC/Departamento de Física, Faculdade de Ciências, Universidade de Lisboa, Av. Prof. Gama Pinto 2, 1649-003 Lisbon, Portugal
e-mail: anunes@ptmat.fc.ul.pt

J. Font and C. Simó
Departament de Matemàtica Aplicada i Anàlisi, Universitat de Barcelona, Gran Via 587, 08007 Barcelona, Spain
e-mail: quim@maia.ub.es, carles@maia.ub.es

M.M. Peixoto et al. (eds.), *Dynamics, Games and Science I*, Springer Proceedings in Mathematics 1, DOI 10.1007/978-3-642-11456-4_23,
© Springer-Verlag Berlin Heidelberg 2011

359

Keplerian orbits. Poincaré named these periodic orbits "of second species", to distinguish them from the periodic orbits "of first species", which do not involve passages close to singularities.

The challenge of providing rigorous proof of the existence, and a characterization, of the periodic orbits of second species has been taken in the simpler setting of the planar circular restricted three body problem (PCRTBP) in several analytic and numerical studies during the past 40 years, starting with Hénon's paper on the description of the families of limit orbits with consecutive collisions [2, 3]. These limit orbits are formed by arcs, which are pieces of Keplerian ellipses that begin and end in a collision, and an arc is called of type S (resp. T) if it begins and ends at different points (resp. at the same point) on the ellipse. In the rotating frame of reference of the restricted three body problem orbits formed by S arcs are symmetric with respect to the line joining the two primaries, while orbits that contain T arcs are in general not symmetric.

The existence of a large set of periodic and chaotic orbits of second species that are close to sequences of arcs of type T was proved in [4] for the circular problem, and a similar result was later reported in [5], where a numerical study of these in general asymmetric periodic orbits was also presented. The results of [5] were extended in [6] to include the orbits that converge to sequences that combine arcs of type S and arcs of type T, and the symbolic dynamics for the *whole set* of planar orbits of second species, that is, orbits either periodic or chaotic with infinitely many close encounters with the small primary, was obtained: every infinite periodic sequence of S- and T-arcs which does not contain two identical T-arcs in succession is the limit when the mass parameter $\mu \to 0$ of a family of periodic orbits of the planar restricted three body problem. The symbolic dynamics on this large subshift of the full shift implies, in particular, the existence for μ small enough of periodic orbits that combine arcs of types S and T. Here we present a brief summary of the approach followed in [6], and a numerical example of this type of periodic orbits.

Consider the PCRTBP, choosing as usual appropriate mass, length and time units so that the masses of the two primaries are $m_1 = 1 - \mu$ for the large primary, and $m_2 = \mu$ for the small primary M, the angular velocity of their motion around the fixed centre of mass is 1 and their positions in the plane are given in synodic coordinates by $(\mu, 0)$ and $(\mu - 1, 0)$, respectively. We have studied the intersection of the stable and unstable invariant manifolds of collision at M with the passage through the pericentre surface of section, as a means to characterize the set of all the orbits that, for a given value of the Jacobi constant C_J and of the mass parameter μ, undergo consecutive close encounters with the small primary M. This intersection can be obtained numerically using the regularized equations of motion and the symmetry of the system. The first intersection of the stable and unstable manifolds of collision with the passage through the pericentre surface of section is shown in Fig. 23.1, where we have imposed a bound of 8π on the integration time. The fifteen homoclinic points where these curves intersect each other correspond to ejection-collision orbits with exactly one intermediate passage through the pericentre. Eleven out of these fifteen orbits take less than 8π time units from ejection to collision.

Fig. 23.1 The first intersection of the stable (*dashed line*) and unstable (*full line*) manifold of collision with the passage through the pericentre surface of section for $C_J = 2.8$ and $\mu = 10^{-4}$

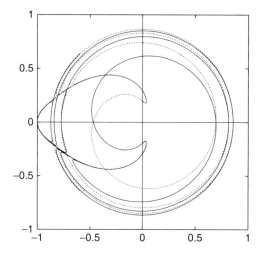

We restrict the study to orbits such that consecutive encounters occur within time intervals smaller than a multiple q of 2π, and such that the number of passages through the pericentre in between close encounters does not exceed p. For $C_J = 2.8$ and within the imposed cut-off of $q = p = 4$, the homoclinic points are organized in 41 families, each of which corresponds to a basic homoclinic orbit, that is, a homoclinic orbit that involves no intermediate close encounters. In the limit when $\mu \to 0$, these basic homoclinic orbits tend to the orbits of the two-body problem associated with the S-arcs and T-arcs of Hénon's classification that are compatible with the constraints on the value of C_J, and with the bounds imposed on time and on the number of passages through the pericentre. For $C_J = 2.8$, and $p = q = 4$ the 41 basic homoclinic orbits of the rotating two-body problem are 18 T-arcs (T, p, q, s), where p and q are the integers that denote the number of full turns of the massless body and of the reference frame, respectively, that take place along the orbit and $s = +1$ (resp. $s = -1$) for the ingoing (resp. outgoing) orbits; and 23 S-arcs (S, p, q, s).

Let the set of initial conditions exiting the circle $\mathscr{C}$ of radius μ^α around the small primary M be parameterized by the angles ϕ and ψ. To each one of the admissible 41 basic homoclinic orbits for $C_J = 2.8$ and $p = q = 4$ corresponds a (X, p, q, s)-*homoclinic strip*, $X \in \{T, S\}$ and $s = \pm 1$ in the (ϕ, ψ) torus of initial conditions of orbits that leave $\mathscr{C}$ and return to $\mathscr{C}$ and are in the meantime close to the basic homoclinic orbit (X, p, q, s) of the unperturbed problem. The return map on the homoclinic strips can be approximated analytically by taking the composition of two different approximate integrable problems in and out of the circle $\mathscr{C}$ and tuning the parameter α to $2/5$. It is a horseshoe map on a subset of the set of all the homoclinic strips. The symbolic dynamics in the small μ limit can be obtained from this approximation of the return map. It differs from the full shift on the alphabet of 41 symbols that corresponds to the admissible two-body problem homoclinic arcs

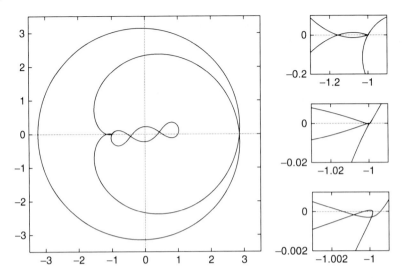

Fig. 23.2 A periodic orbits which combines a T-arc with an S-arc. On the right, and from top to bottom, we display successive magnifications of the passages close to collision

only in that every transition $(T, p, q, s) \to (T, p, q, s)$ is forbidden. In general the following holds:

Theorem 23.1. *For a given upper bound N on p and q, μ can be chosen small enough so that, for C_J outside a finite union of small intervals, the set of all the orbits of the second species is described by a subshift on the alphabet of all the admissible S- and T-arcs of the two body problem, for which the only forbidden transition is the one that concatenates two identical resonant homoclinic arcs.*

For the selected value of $C_J = 2.8$ and for $\mu = 10^{-4}$, the shift that describes the dynamics of the planar orbits of second species built with arcs that verify the constraints imposed on time and number of intermediate pericentre passages is still a large subshift of the full shift, although a few other transitions besides $(T, p, q, s) \to (T, p, q, s)$ have to be excluded. In particular, there are periodic orbits that combine T-arcs and S-arcs, such as the one shown in Fig. 23.2.

Acknowledgements This work was partially supported by grants CRUP-E18-05 and FCT/POCTI/ISFL/ 2/618 (Portugal), for the second author, and MTM2006-05849/Consolider (Spain) and CIRIT 2005 SGR-1028 (Catalonia) for the first and third authors. The use of the cluster facility HIDRA of the Dynamical Systems group at the Universitat de Barcelona is also acknowledged.

References

1. Poincaré, H.: Les Méthodes Nouvelles de la Mécanique Celeste, Tome III. Gauthier-Villars, Paris (1899)
2. Hénon, M.: Sur les orbites interplanétaires que rencontrent deux fois la Terre. Bull. Astron. **3**, 377–393 (1968)

23 The Set of Planar Orbits of Second Species in the RTBP

3. Hénon, M.: Generating Families in the Restricted Three-Body Problem. Springer, Berlin Heidelberg (1997)
4. Bolotin, S.V., Mackay, R.S.: Periodic and chaotic trajectories of the second species for the n-centre problem. Celestial Mech. Dyn. Astron. **77**, 49–75 (2000)
5. Font, J., Nunes, A., Simó, C.: Consecutive quasi-collisions in the planar circular RTBP. Nonlinearity **15**, 115–142 (2002)
6. Font, J., Nunes, A., Simó, C.: A numerical study of the orbits of second species of the planar circular RTBP. Celestial Mech. Dyn. Astron. **103**, 143–162 (2009)

Chapter 24
Statistical Properties of the Maximum for Non-Uniformly Hyperbolic Dynamics

Ana Cristina Moreira Freitas, Jorge Milhazes Freitas, and Mike Todd

Abstract We study the asymptotic distribution of the partial maximum of observable random variables evaluated along the orbits of some particular dynamical systems. Moreover, we show the link between Extreme Value Theory and Hitting Time Statistics for discrete time non-uniformly hyperbolic dynamical systems. This relation allows to study Hitting Time Statistics with tools from Extreme Value Theory, and vice versa.

24.1 Introduction

Consider a discrete time dynamical system $(\mathscr{X}, \mathscr{B}, \mu, f)$, where $\mathscr{X}$ is a d-dimensional Riemannian manifold, $\mathscr{B}$ is the Borel σ-algebra, $f : \mathscr{X} \to \mathscr{X}$ is a measurable map and μ an f-invariant probability measure, absolutely continuous with respect to Lebesgue measure (acip), with density denoted by $\rho = \frac{d\mu}{d\mathrm{Leb}}$.

Given an observable $\varphi : \mathscr{X} \to \mathbb{R} \cup \{\pm\infty\}$ achieving a global maximum at $\zeta \in \mathscr{X}$ (we allow $\varphi(\zeta) = +\infty$), consider the stationary stochastic process $X_0, X_1, \ldots$ defined by

$$X_n = \varphi \circ f^n, \quad \text{for each } n \in \mathbb{N}. \tag{24.1}$$

Here, we are particularly interested on the statistical properties of the partial maximum

$$M_n := \max\{X_0, \ldots, X_{n-1}\},$$

when properly normalised.

A.C.M. Freitas (✉)
Centro de Matemática & Faculdade de Economia da, Universidade do Porto, Rua Dr. Roberto Frias, 4200-464 Porto, Portugal
e-mail: amoreira@fep.up.pt

J.M. Freitas and M. Todd
Centro de Matemática da, Universidade do Porto, Rua do Campo Alegre 687, 4169-007 Porto, Portugal
e-mail: jmfreita@fc.up.pt, mtodd@fc.up.pt

M.M. Peixoto et al. (eds.), *Dynamics, Games and Science I*, Springer Proceedings in Mathematics 1, DOI 10.1007/978-3-642-11456-4_24,
© Springer-Verlag Berlin Heidelberg 2011

The study of distributional properties of the higher order statistics of a sample, like the maximum of the sample, is the purpose of Extreme Value Theory (EVT).

24.2 Extreme Value Laws

We are interested in knowing if there are normalising sequences $\{a_n\}_{n\in\mathbb{N}} \subset \mathbb{R}^+$ and $\{b_n\}_{n\in\mathbb{N}} \subset \mathbb{R}$ such that

$$\mu\left(\{x : a_n(M_n - b_n) \le y\}\right) = \mu\left(\{x : M_n \le u_n\}\right) \to H(y), \tag{24.2}$$

for some non-degenerate distribution function (d.f.) H, as $n \to \infty$. Here $u_n := u_n(y) = y/a_n + b_n$ is such that

$$n\mu(X_0 > u_n) \to \tau, \quad \text{as } n \to \infty, \tag{24.3}$$

for some $\tau = \tau(y) \ge 0$ and in fact $H(y) = H(\tau(y))$. When this happens we say that we have an *Extreme Value Law* (EVL) for M_n. Note that, clearly, we must have $u_n \to \varphi(\zeta)$, as $n \to \infty$.

Classical Extreme Value Theory asserts that there are only three types of non-degenerate asymptotic distributions for the maximum of an independent and identically distributed (i.i.d.) sample under linear normalisation. They will be referred to as *classical* EVLs and we denote them by:

Type 1: $EV_1(y) = e^{-e^{-y}}$, $y \in \mathbb{R}$; this is also known as the Gumbel extreme value distribution (e.v.d.)

Type 2: $EV_2(y) = \begin{cases} e^{-y^{-\alpha}}, & y > 0 \\ 0, & y \le 0 \end{cases}$ ($\alpha > 0$); this family of distribution

functions is known as the *Fréchet* e.v.d.

Type 3: $EV_3(y) = \begin{cases} e^{-(-y)^{\alpha}}, & y \le 0 \\ 1, & y > 0 \end{cases}$ ($\alpha > 0$); this family of distribution

functions is known as the *Weibull* e.v.d.

It is also known that the limiting behaviour for maxima of a stationary process can be reduced, under adequate conditions on the dependence structure, to the Classical Extreme Value Theory for sequences of i.i.d. random variables (r.v.). Hence, to the stationary process $X_0, X_1, \ldots$ defined in (24.1) we associate an i.i.d. sequence $Z_0, Z_1, \ldots$, whose distribution function (d.f.) is the same of X_0 and whose partial maximum we define as

$$\hat{M}_n := \max\{Z_0, \ldots, Z_{n-1}\}. \tag{24.4}$$

Let us focus on the conditions that allow us to relate the asymptotic distribution of M_n with that of $\hat{M}_n$. Following [14] we refer to these conditions as

24 Statistical Properties of the Maximum for Non-Uniformly Hyperbolic Dynamics 367

$D(u_n)$ and $D'(u_n)$, where u_n is a suitable sequence of thresholds converging to $\max_{x \in [-1,1]} X_0(x)$, as n goes to ∞. $D(u_n)$ imposes a certain type of distributional mixing property. Essentially, it says that the dependence between some special type of events fades away as they become more and more apart in the time line. $D'(u_n)$ restricts the appearance of clusters, that is, it makes the occurrence of consecutive 'exceedances' of the level u_n an unlikely event.

Let the d.f. F be given by

$$F(u) = \mu(X_0 \leq u).$$

We say that an *exceedance* of the *level* u_n occurs at time i if $X_i > u_n$. The probability of such an exceedance is $1 - F(u_n)$ and so the mean value of the number of exceedances occurring up to n is $n(1 - F(u_n))$. The sequences of levels u_n we consider are such that

$$n(1 - F(u_n)) \to \tau, \quad \text{as } n \to \infty,$$

for some $\tau \geq 0$, which means that, in a time period of length n, the expected number of exceedances is approximately τ.

The original condition $D(u_n)$ from [14], which we will denote by $D_1(u_n)$, is a type of uniform mixing requirement specially adapted to Extreme Value Theory. In this context, the events of interest are those of the form $\{X_i \leq u\}$ and their intersections. Observe that $\{M_n \leq u\}$ is just $\{X_0 \leq u, \ldots, X_{n-1} \leq u\}$. A natural mixing condition in this context is the following. Let $F_{i_1,\ldots,i_n}(x_1, \ldots, x_n)$ denote the joint d.f. of $X_{i_1}, \ldots, X_{i_n}$, and set $F_{i_1,\ldots,i_n}(u) = F_{i_1,\ldots,i_n}(u, \ldots, u)$.

Condition $D_1(u_n)$. We say that $D_1(u_n)$ holds for the sequence $X_0, X_1, \ldots$ if for any integers $i_1 < \ldots < i_p$ and $j_1 < \ldots < j_k$ for which $j_1 - i_p > m$, and any large $n \in \mathbb{N}$,

$$\left| F_{i_1,\ldots,i_p,j_1,\ldots,j_k}(u_n) - F_{i_1,\ldots,i_p}(u_n) F_{j_1,\ldots,j_k}(u_n) \right| \leq \gamma(n, m),$$

where $\gamma(n, m_n) \to 0$, as $n \to \infty$, for some sequence $m_n = o(n)$.

In [8], we verify that condition $D_1(u_n)$ can be stated in a weaker form and the result still prevails. The advantage of having this weaker requirement is that, in the context of Dynamical Systems, the new condition should follow from decay of correlations. Motivated by the work of [6], we proposed, in [8], the following weaker version of condition $D_1(u_n)$, that we denote by $D_2(u_n)$:

Condition $D_2(u_n)$. We say that $D_2(u_n)$ holds for the sequence $X_0, X_1, X_2, \ldots$ if for any integers ℓ, t and n

$$\mu\left(\{X_0 > u_n\} \cap \{\max\{X_t, \ldots, X_{t+l-1}\} \leq u_n\}\right)$$
$$- \mu(\{X_0 > u_n\})\mu(\{M_\ell \leq u_n\}) \leq \gamma(n, t),$$

where $\gamma(n, t)$ is nonincreasing in t for each n and $n\gamma(n, t_n) \to 0$ as $n \to \infty$ for some sequence $t_n = o(n)$.

368   A.C.M. Freitas et al.

The sequence u_n is such that the average number of exceedances in the time interval $\{0, \ldots, [n/k]\}$ is approximately τ/k, which goes to zero as $k \to \infty$. However, the exceedances may have a tendency to be concentrated in the time period following the first exceedance at time 0. To avoid this we introduce

Condition $D'(u_n)$. We say that $D'(u_n)$ holds for the sequence $X_0, X_1, X_2, \ldots$ if

$$\lim_{k \to \infty} \limsup_{n \to \infty} n \sum_{j=1}^{[n/k]} \mu(\{X_0 > u_n\} \cap \{X_j > u_n\}) = 0.$$

This guarantees that the exceedances should appear scattered through the time period $\{0, \ldots, n-1\}$.

In [8] we have shown that M_n and $\hat{M}_n$, conveniently normalised, have the same asymptotic distribution under $D_2(u_n)$ and $D'(u_n)$.

Theorem 24.1. *[8, Theorem 1] Let $(u_n)_{n \in \mathbb{N}}$ be such that $n\mu(X > u_n) = n(1 - F(u_n)) \to \tau$, as $n \to \infty$, for some $\tau \geq 0$. Assume that conditions $D_2(u_n)$ and $D'(u_n)$ hold. Then*

$$\lim_{n \to \infty} \mu(M_n \leq u_n) = \lim_{n \to \infty} \mu(\hat{M}_n \leq u_n).$$

24.3 Extreme Value Laws for Benedicks-Carleson Quadratic Maps

In [7], we consider the quadratic maps $f_a(x) = 1 - ax^2$ on $I = [-1, 1]$, with $a \in \mathscr{BC}$, where $\mathscr{BC}$ is the Benedicks-Carleson parameter set introduced in [1]. The set $\mathscr{BC}$ has positive Lebesgue measure and is built in such a way that, for every $a \in \mathscr{BC}$, the Collet-Eckmann condition holds: there is exponential growth of the derivative of f_a along the critical orbit, i.e., there is $c > 0$ such that

$$\left| \left(f_a^n \right)' (f_a(0)) \right| \geq e^{cn},$$

for all $n \in \mathbb{N}$. This property guarantees not only the non-existence of an attracting periodic orbit but also the existence of an ergodic f_a-invariant probability measure μ_a that is absolutely continuous with respect to Lebesgue measure on $[-1, 1]$.

In [7], we consider the particular stationary stochastic process $X_0, X_1, \ldots$ defined by

$$X_n = f_a(X_{n-1}) = f_a^n(X_0), \quad \text{for each } n \in \mathbb{N}. \tag{24.5}$$

This way, we obtain a stationary stochastic process $X_0, X_1, \ldots$ with common marginal d.f. given by $G_a(x) = \mu_a\{X_0 \leq x\}$.

The main result of [7] states that the limiting law of M_n is the same as if $X_0, X_1, \ldots$ were independent with the same d.f. G_a. In fact we verify that,

24 Statistical Properties of the Maximum for Non-Uniformly Hyperbolic Dynamics   369

under appropriate normalisation, the asymptotic distribution of M_n is of Type III (Weibull). As usual, we denote by G_a^{-1} the generalised inverse of the d.f. G_a, which is to say that $G_a^{-1}(y) := \inf\{x : G_a(x) \geq y\}$.

Theorem 24.2. *[8, Theorem A] For each $a \in \mathcal{BC}$ and every stationary stochastic process $X_0, X_1, \ldots$ given by (24.5), we have:*

$$\mu_a\{a_n(M_n - 1) \leq x\} \rightarrow H(x) = \begin{cases} e^{-(-x)^{1/2}}, & x \leq 0 \\ 1, & x > 0 \end{cases},$$

where $a_n = \left(1 - G_a^{-1}\left(1 - \frac{1}{n}\right)\right)^{-1}$.

Based on Theorem 24.1, our strategy to prove Theorem 24.2 was the following:

- Compute the limiting distribution of $\hat{M}_n$ defined as in (24.4) and the associated normalising sequences a_n and b_n, that is, the sequences a_n and b_n such that $\mu_a\{a_n(\hat{M}_n - b_n) \leq x\} \rightarrow H(x)$ for some non-degenerate d.f. H.

- Show that conditions $D(u_n)$ and $D'(u_n)$ are valid for the stochastic process $X_0, X_1, \ldots$ defined in (24.5), where $u_n := u_n(x) = x/a_n + b_n$ is such that $n\mu_a(X > u_n) \rightarrow \tau$, as $n \rightarrow \infty$, for some $\tau \geq 0$.

24.4 General Characterisation of the Observables

We next turn again to the general case where $\mathscr{X}$ is a d-dimensional Riemannian manifold, $f : \mathscr{X} \rightarrow \mathscr{X}$ is a measurable map and

$$X_n = \varphi \circ f^n, \quad \text{for each } n \in \mathbb{N},$$

for an observable $\varphi : \mathscr{X} \rightarrow \mathbb{R} \cup \{\pm\infty\}$ achieving a global maximum at $\zeta \in \mathscr{X}$.
We assume that the observable φ is of the form

$$\varphi(x) = g(\text{dist}(x, \zeta)), \tag{24.6}$$

where 'dist' denotes the usual Euclidean distance, the function $g : [0, +\infty) \rightarrow \mathbb{R} \cup \{+\infty\}$ has a global maximum at 0, is a strictly decreasing bijection $g : V \rightarrow W$ in a neighbourhood V of 0 and has one of the following three types of behaviour:

Type 1: there exists some strictly positive function $p : W \rightarrow \mathbb{R}$ such that for all $y \in \mathbb{R}$

$$\lim_{s \rightarrow g_1(0)} \frac{g_1^{-1}(s + yp(s))}{g_1^{-1}(s)} = e^{-y};$$

Type 2: $g_2(0) = +\infty$ and there exists $\beta > 0$ such that for all $y > 0$

$$\lim_{s \to +\infty} \frac{g_2^{-1}(sy)}{g_2^{-1}(s)} = y^{-\beta};$$

Type 3: $g_3(0) = D < +\infty$ and there exists $\gamma > 0$ such that for all $y > 0$

$$\lim_{s \to 0} \frac{g_3^{-1}(D - sy)}{g_3^{-1}(D - s)} = y^\gamma.$$

Examples of each one of the three types are as follows: $g_1(x) = -\log x$, $g_2(x) = x^{-1/\alpha}$ for some $\alpha > 0$ and $g_3(x) = D - x^{1/\alpha}$ for some $D \in \mathbb{R}$ and $\alpha > 0$.

Observe that if at time $j \in \mathbb{N}$ we have an exceedance of the level u (sufficiently large), i.e., $X_j(x) > u$, then we have an entrance of the orbit of x into the ball of radius $g^{-1}(u)$ around ζ, $B_{g^{-1}(u)}(\zeta)$, at time j.

Based on this fact, we recently demonstrated in [9] the link between Extreme Value Theory and Hitting Time Statistics (HTS).

24.5 Hitting Time Statistics

For a set $A \subset \mathcal{X}$ we let $r_A(y)$ denote the *first hitting time to* A of the point y, that is,

$$r_A(y) = \min\{j \in \mathbb{N} : f^j(y) \in A\}.$$

We are interested in the fluctuations of this functions as the set A shrinks. Firstly we consider the Return Time Statistics (RTS) of this system. Let μ_A denote the conditional measure on A, i.e., $\mu_A := \frac{\mu|_A}{\mu(A)}$. By Kac's Lemma, the expected value of r_A with respect to μ is $\int_A r_A \, d\mu_A = 1/\mu(A)$. So in studying the fluctuations of r_A on A, the relevant normalising factor is $1/\mu(A)$.

Given a sequence of sets $\{U_n\}_{n \in \mathbb{N}}$ so that $\mu(U_n) \to 0$, the system has *Return Time Statistics* $G(t)$ for $\{U_n\}_{n \in \mathbb{N}}$ if for all $t \geq 0$ the following limit exists and equals $G(t)$:

$$\lim_{n \to \infty} \mu_{U_n} \left(r_{U_n} \geq \frac{t}{\mu(U_n)} \right). \tag{24.7}$$

We say that $(\mathcal{X}, f, \mu)$ has *Return Time Statistics* $G(t)$ *to balls at* ζ if for any sequence $\{\delta_n\}_{n \in \mathbb{N}} \subset \mathbb{R}^+$ such that $\delta_n \to 0$ as $n \to \infty$ we have RTS $G(t)$ for $U_n = B_{\delta_n}(\zeta)$.

If we study r_A defined on the whole of $\mathcal{X}$, i.e., not simply restricted to A, we are studying the Hitting Time Statistics. Note that we will use the same normalising factor $1/\mu(A)$ in this case.

24 Statistical Properties of the Maximum for Non-Uniformly Hyperbolic Dynamics   371

Analogously to the above, given a sequence of sets $\{U_n\}_{n\in\mathbb{N}}$ so that $\mu(U_n) \to 0$, the system has *Hitting Time Statistics* $G(t)$ for $\{U_n\}_{n\in\mathbb{N}}$ if for all $t \geq 0$ the following limit is defined and equals $G(t)$:

$$\lim_{n\to\infty} \mu\left(r_{U_n} \geq \frac{t}{\mu(U_n)}\right). \tag{24.8}$$

HTS to balls at a point ζ is defined analogously to RTS to balls.

In [10], it was shown that the limit for the HTS defined in (24.8) exists if and only if the limit for the analogous RTS defined in (24.7) exists. Moreover, they show that the HTS distribution exists and is exponential (*i.e.,* $G(t) = \mathrm{e}^{-t}$) if and only if the RTS distribution exists and is exponential.

For many mixing systems it is known that the HTS are exponential around almost every point. For example, this was shown for Axiom A diffeomorphisms in [11], transitive Markov chains in [15] and uniformly expanding maps of the interval in [5].

For non-uniformly hyperbolic systems less is known. A major breakthrough in the study of HTS/RTS for non-uniformly hyperbolic maps was made in [12], where they gave a set of conditions which, when satisfied, imply exponential RTS to cylinders and/or balls. Their principal application was to maps of the interval with an indifferent fixed point.

Another important paper was [2], in which they showed that the RTS for a map are the same as the RTS for the first return map. (The first return map to a set $U \subset \mathscr{X}$ is the map $F = f^{r_U}$.) Since it is often the case that the first return maps for non-uniformly hyperbolic dynamical systems are much better behaved (possibly hyperbolic) than the original system, this provided an extremely useful tool in this theory. For example, they proved that if $f : I \to I$ is a unimodal map for which the critical point is nowhere dense, and for which an acip μ exists, then the relevant first return systems (U, F, μ_U) have a 'Rychlik' property. They then showed that such systems, studied in [16], must have exponential RTS, and hence the original system (I, f, μ) also has exponential RTS (to balls around μ-a.e. point).

The presence of a recurrent critical point means that the first return map itself will not satisfy this Rychlik property. To overcome this problem in [4] special induced maps, (U, F), were used, where for $x \in U$ we have $F(x) = f^{\mathrm{ind}(x)}(x)$ for some inducing time $\mathrm{ind}(x) \in \mathbb{N}$ that is not necessarily the first return time of x to U. The fact that these particular maps can be seen as first return maps in the canonical Markov extension, the 'Hofbauer tower', meant that they were still able to exploit the main result of [2] to get exponential RTS around μ-a.e. point for unimodal maps $f : I \to I$ with an acip μ as long as f satisfies a polynomial growth condition along the critical orbit. In [3] this result was improved to include any multimodal map with an acip, irrespective of the growth along the critical orbits, and of the speed of mixing.

24.6 The Link between Hitting Time Statistics and Extreme Value Theory

In [9] we establish two main results. In the first one, we obtain EVLs from HTS.

Theorem 24.3. *[9, Theorem 1] Let $(\mathcal{X}, \mathcal{B}, \mu, f)$ be a dynamical system where μ is an acip, and consider $\zeta \in \mathcal{X}$ for which Lebesgue's Differentiation Theorem holds.*

- *If we have HTS to balls centred on $\zeta \in \mathcal{X}$, then we have an EVL for M_n which applies to the observables (24.6) achieving a maximum at ζ.*
- *If we have exponential HTS ($G(t) = e^{-t}$) to balls at $\zeta \in \mathcal{X}$, then we have an EVL for M_n which coincides with that of $\hat{M}_n$. In particular, this EVL must be one of the 3 classical types. Moreover, if g is of type g_i, for some $i \in \{1, 2, 3\}$, then we have an EVL for M_n of type EV_i.*

We next define a class of multimodal interval maps $f : I \to I$. We denote the finite set of critical points by Crit. We say that $c \in$ Crit is *non-flat* if there exists a diffeomorphism $\psi_c : \mathbb{R} \to \mathbb{R}$ with $\psi_c(0) = 0$ and $1 < \ell_c < \infty$ such that for x close to c, $f(x) = f(c) \pm |\psi_c(x - c)|^{\ell_c}$. The value of ℓ_c is known as the *critical order* of c. Let

$$NF^k := \left\{ f : I \to I : f \text{ is } C^k, \text{ each } c \in \text{Crit is non-flat and} \right.$$

$$\left. \inf_{f^n(p)=p} |Df^n(p)| > 1 \right\}.$$

The following is a simple corollary of Theorem 24.3 and [3, Theorem 3]. It generalises the result of Collet in [6] from unimodal maps with exponential growth on the critical point to multimodal maps where we only need to know that there is an acip.

Corollary 24.1. *Suppose that $f \in NF^2$ and f has an acip μ. Then (I, f, μ) has an EVL for M_n which coincides with that of $\hat{M}_n$, and this holds for μ-a.e. $\zeta \in \mathcal{X}$ fixed at the choice of the observable in (24.6). Moreover, the EVL is of type EV_i when the observables are of type g_i, for each $i \in \{1, 2, 3\}$.*

In the next result, we show how to get HTS from EVLs.

Theorem 24.4. *[9, Theorem 2] Let $(\mathcal{X}, \mathcal{B}, \mu, f)$ be a dynamical system where μ is an acip and consider $\zeta \in \mathcal{X}$ for which Lebesgue's Differentiation Theorem holds.*

- *If we have an EVL for M_n which applies to the observables (24.6) achieving a maximum at $\zeta \in \mathcal{X}$ then we have HTS to balls at ζ.*
- *If we have an EVL for M_n which coincides with that of $\hat{M}_n$, then we have exponential HTS ($G(t) = e^{-t}$) to balls at ζ.*

The following is immediate by the above and Theorem 24.1.

24 Statistical Properties of the Maximum for Non-Uniformly Hyperbolic Dynamics

Corollary 24.2. *Let* $(\mathscr{X}, \mathscr{B}, \mu, f)$ *be a dynamical system where* μ *is an acip and consider* $\zeta \in \mathscr{X}$ *for which Lebesgue's Differentiation Theorem holds. If* $D_2(u_n)$ *(or* $D_1(u_n)$*) and* $D'(u_n)$ *hold for a stochastic process* $X_0, X_1, \ldots$ *defined by* (24.1) *and* (24.6), *where* u_n *is a sequence of levels satisfying* (24.3), *then we have exponential HTS to balls at* ζ.

The following is an immediate corollary of Theorems 24.4 and 24.2.

Corollary 24.3. *For every Benedicks-Carleson quadratic map* f_a *(with* $a \in \mathscr{BC}$*) we have exponential HTS to balls around the critical point or the critical value.*

The next result is a byproduct of Theorems 24.3, 24.4 and the fact that under $D_1(u_n)$ the only possible limit laws for partial maximums are the classical EV_i for $i \in \{1, 2, 3\}$.

Corollary 24.4. *Let* $(\mathscr{X}, \mathscr{B}, \mu, f)$ *be a dynamical system,* μ *is an acip and consider* $\zeta \in \mathscr{X}$ *for which Lebesgue's Differentiation Theorem holds. If* $D_1(u_n)$ *holds for a stochastic process* $X_0, X_1, \ldots$ *defined by* (24.1) *and* (24.6), *where* u_n *is a sequence of levels satisfying* (24.3), *then the only possible HTS to balls around* ζ *are of exponential type, meaning that, there is* $\theta > 0$ *such that* $G(t) = e^{-\theta t}$.

As we have seen, the relation established in Theorems 24.3 and 24.4 allows to study Hitting Time Statistics with tools from Extreme Value Theory, and vice-versa. In [9], we also give applications of this theory to higher dimensional examples, for which we also obtain classical extreme value laws and exponential hitting time statistics (for balls). In the same work, we extend these ideas to the subsequent returns to asymptotically small sets, linking the Poisson statistics of both processes. More precisely, we show that the point process of hitting times has a Poisson limit if and only if the point process of exceedances has a Poisson limit.

Extreme Value Laws have also recently been proved in [13] for continuous time dynamical systems (flows) with acips, as well as for potentials like those presented here, but with multiple maxima.

Acknowledgements MT is supported by FCT grant SFRH/BPD/26521/2006. All three authors are supported by FCT through CMUP.

References

1. Benedicks, M., Carleson, L.: On iterations of $1 - ax^2$ on $(-1, 1)$. Ann. Math. **122**, 1–25 (1985)
2. Bruin, H., Saussol, B., Troubetzkoy, S., Vaienti, S.: Return time statistics via inducing. Ergod. Theory Dyn. Syst. **23**, 991–1013 (2003)
3. Bruin, H., Todd, M.: Return time statistics of invariant measures for interval maps with positive Lyapunov exponent. Stoch. Dyn. **9**(1), 81–100 (2009)
4. Bruin, S., Vaienti, S.: Return time statistics for unimodal maps. Fund. Math. **176**, 77–94 (2003)
5. Collet, P.: Some ergodic properties of maps of the interval. Dynamical Systems (Temuco, 1991/1992), (Travaux en cours, 52), pp. 55–91. Herman, Paris (1996)

6. Collet, P.: Statistics of closest return for some non-uniformly hyperbolic systems. Ergod. Theory Dyn. Syst. **21**, 401–420 (2001)
7. Freitas, A.C.M., Freitas, J.M.: Extreme values for Benedicks Carleson maps. Ergod. Theory Dyn. Syst. **28**, 1117–1133 (2008)
8. Freitas, A.C.M., Freitas, J.M.: On the link between dependence and independence in Extreme Value Theory for Dynamical Systems. Stat. Probab. Lett. **78**, 1088–1093 (2008)
9. Freitas, A.C.M., Freitas, J.M., Todd, M.: Hitting time statistics and extreme value theory. Probab. Theory Relat. Fields **147**(3), 675–710 (2010)
10. Haydn, N., Lacroix, Y., Vaienti, S.: Hitting and return times in ergodic dynamical systems. Ann. Probab. **33**, 2043–2050 (2005)
11. Hirata, M.: Poisson law for Axiom A diffeomorphisms. Ergod. Theory Dyn. Syst. **13**, 533–556 (1993)
12. Hirata, M., Saussol, B., Vaienti, S.: Statistics of return times: a general framework and new applications. Comm. Math. Phys. 206, 33–55 (1999)
13. Holland, M., Nicol, M., Torok, A.: Extreme value distributions for non-uniformly expanding systems. Trans. Am. Math. Soc. (2010, to appear)
14. Lindgren, G., Leadbetter, M.R., Rootzén, H.: Extremes and related properties of random sequences and processes. Springer Series in Statistics. Springer, New York, Berlin (1983)
15. Pitskel, B.: Poisson limit law for Markov chains. Ergod. Theory Dyn. Syst. **11**, 501–513 (1991)
16. Rychlik, M.: Bounded variation and invariant measures. Studia Math. **76**, 69–80 (1983)

Chapter 25
Adaptive Learning and Central Bank Inattentiveness in Optimal Monetary Policy

Orlando Gomes, Vivaldo M. Mendes, and Diana A. Mendes

Abstract This paper analyzes the dynamic properties of a standard New Keynesian monetary policy model in which private agents expectations are formed under a learning mechanism while the central bank believes they follow the hypothesis of rational expectations. By assuming a gain sequence that is asymptotically constant, explicit local and global stability conditions are derived. The main results are that stability is guaranteed even in cases in which full convergence to the rational expectations equilibrium is not attainable; furthermore, endogenous business cycles are likely to arise.

25.1 Introduction

A large amount of literature on the formation of macroeconomic expectations through learning has been produced over the last few years. The motivation for this literature can be found in the seminal paper by Marcet and Sargent [7], who questioned the plausibility of the notion of rational expectations as developed and applied by Muth [8] and Lucas [6].

Under learning, instead of knowing the true process underlying the evolution of economic aggregates, the agents will choose a rule that is used to predict future

---

O. Gomes (✉)
Lisbon Higher Institute of Accountancy and Administration (ISCAL-IPL), Av. Miguel Bombarda, 20, 1069-035 Lisbon, Portugal and UNIDE/ISCTE
e-mail: omgomes@iscal.ipl.pt

V.M. Mendes
Department of Economics, Instituto Superior de Ciências do Trabalho e da Empresa, Avenida das Forças Armadas, 1649-026 Lisbon, Portugal
e-mail: vivaldo.mendes@iscte.pt

D.A. Mendes
Department of Quantitative Methods, IBS – ISCTE Business School, ISCTE, Avenida das Forças Armadas, 1649-026 Lisbon, Portugal
e-mail: diana.mendes@iscte.pt

M.M. Peixoto et al. (eds.), *Dynamics, Games and Science I*, Springer Proceedings in Mathematics 1, DOI 10.1007/978-3-642-11456-4_25,
© Springer-Verlag Berlin Heidelberg 2011

outcomes based on past information. As new information arrives and becomes available, the learning skills improve and the rule is updated. A continuing process of improved learning will then probably lead to an asymptotic long run fixed point that may coincide with the rational expectations equilibrium (REE). This hypothetical convergence to the REE is one of the most relevant properties of learning schemes, as initially remarked by Marcet and Sargent [7], or by Beeby et al. [2, pp. 5] who pointed out that "the attraction of learning then is that it allows agents to make mistakes in the short-run, but not in the long-run".

There are several ways in which learning can be modeled in the field of macroeconomics. The one that has received more attention in the literature is adaptive learning, and it is in this learning mechanism that we will focus our attention.[1]

Results other than the fixed point associated with full rationality are obtainable in adaptive learning settings. Such results may include, as in this paper, periodic and aperiodic long run cycles. The eventual presence of cycles is dependent on the specific form of the gain sequence measuring the sensitivity of estimates to new data. As new information adds to the existing one, the gain should be decreasing, shrinking asymptotically towards zero. Nevertheless, model misspecification or some kind of imperfect knowledge assumption lead us to accept that the gain sequence may not effectively fall to zero. The idea of constant gain learning – i.e., of persistent learning dynamics – does not seem an unreasonable assumption, and in many settings it can be more appropriate than a simple complete learning scheme with convergence to the REE. Constant gain forms the crucial element upon which the basic results of the paper are derived.[2]

The remainder of the paper is organized as follows. Section 25.2 briefly presents the benchmark optimal monetary policy model. Section 25.3 studies the dynamic behavior of the model under learning, assuming that the monetary authority overlooks such learning process by private agents. Section 25.4 concludes.

25.2 The Optimal Monetary Policy Model

The benchmark model is a fully deterministic version of the New Keynesian monetary policy problem (see Woodford [9]). The state of the economy is given by two dynamic equations. First, an IS equation, which establishes a relation of opposite sign between the output gap, x_t, and the expected real interest rate, $i_t - E_t \pi_{t+1}$. The output gap is defined as the difference in logs between effective output and some measure of potential output; the inflation rate, π_t, is simply the variation rate of the price level, while in the real interest rate expression, i_t represents the nominal

---

[1] An extensive survey on macroeconomic issues where adaptive learning is involved is presented in Evans and Honkapohja [4].

[2] See Cellarier [3], for the development of a model (in the case, a growth model) where cycles arise as a direct consequence of constant gain. For a thorough and rigorous discussion of the implications of adaptive learning over long-run dynamics in general equilibrium settings, see Grandmont [5].

25 Adaptive Learning and Central Bank Inattentiveness in Optimal Monetary Policy   377

interest rate and E_t is the expectations operator. The complete IS relation is given by,

$$x_t = -\varphi(i_t - E_t\pi_{t+1}) + E_t x_{t+1}, \quad x_0 \text{ given.} \tag{25.1}$$

where $\varphi > 0$ is an elasticity parameter.

On the supply side, we assume a New Keynesian Phillips curve, according to which there is a positive relation between the contemporaneous values of inflation and the output gap. The current value of inflation also suffers the influence of the expected value of inflation for the next period. The equation is

$$\pi_t = \lambda x_t + \beta E_t \pi_{t+1}, \quad \pi_0 \text{ given.} \tag{25.2}$$

In (25.2), parameter $\lambda \in (0, 1)$ is a measure of price flexibility. The closer this value is to zero, the stronger is the degree of price stickiness or sluggishness. Constant $\beta \in (0, 1)$ is the intertemporal discount factor.

The monetary authority is supposed to control the value of the nominal interest rate in order to attain some policy goals. We consider that the central bank aims at an inflation rate level π^* and at an output gap x^* (the current practice of monetary authorities points to low but positive inflation and output gap targets). The central bank also attributes different degrees of relevance to the two policy goals. The objective function is

$$V_0 = E_0 \left\{ -\frac{1}{2} \sum_{t=0}^{+\infty} \beta^t \left[(\pi_t - \pi^*)^2 + a(x_t - x^*)^2 \right] \right\} \tag{25.3}$$

where parameter $a > 0$ represents the weight of the output gap objective, relatively to the inflation goal, in the monetary authority objective function.

By maximizing V_0 subject to (25.1) and (25.2), the central bank chooses the optimal path for the nominal interest rate. Computing first-order optimality conditions, one obtains the dynamic relation

$$E_t x_{t+1} = \left(1 + \frac{\lambda^2}{a\beta} \right) x_t - \frac{\lambda}{a\beta} \pi_t + \frac{\lambda}{a} \pi^* \tag{25.4}$$

The dynamics of the monetary policy problem are addressable with the information given by the Phillips curve in (25.2) and by (25.4). Defining the steady state as the point $(\overline{x}, \overline{\pi})$ such that $\overline{x} \equiv x_t = E_t x_{t+1}$ and $\overline{\pi} \equiv \pi_t = E_t \pi_{t+1}$, one encounters the result $(\overline{x}, \overline{\pi}) = \left(\frac{1-\beta}{\lambda} \pi^*; \pi^* \right)$.

The system can be presented in matricial form by

$$\begin{bmatrix} E_t x_{t+1} - \frac{1-\beta}{\lambda} \pi^* \\ E_t \pi_{t+1} - \pi^* \end{bmatrix} = \begin{bmatrix} 1 + \frac{\lambda^2}{a\beta} & -\frac{\lambda}{a\beta} \\ -\frac{\lambda}{\beta} & \frac{1}{\beta} \end{bmatrix} \cdot \begin{bmatrix} x_t - \frac{1-\beta}{\lambda} \pi^* \\ \pi_t - \pi^* \end{bmatrix}. \tag{25.5}$$

Let J be the Jacobian matrix in system (25.5). This possesses two eigenvalues, $0 < \varepsilon_1 < 1$ and $\varepsilon_2 > 1$, such that

$$\varepsilon_1, \varepsilon_2 = \frac{a(1+\beta) + \lambda^2}{2a\beta} \mp \sqrt{\left[\frac{a(1+\beta) + \lambda^2}{2a\beta}\right]^2 - \frac{1}{\beta}} \qquad (25.6)$$

Under the case in which the central bank assumes that private expectations about inflation follow the hypothesis of rational expectations, the system is characterized by a saddle-path stable equilibrium: in the two-dimensional space that defines the system, one direction is stable while the other is unstable. By computing the eigenvectors associated with each of the eigenvalues, the following expressions are derived for the stable and unstable trajectories, respectively,

$$x_t = \frac{\beta(1-\varepsilon_1)}{\lambda}\pi^* - \frac{1-\beta\varepsilon_1}{\lambda}\pi_t \qquad (25.7)$$

$$x_t = -\frac{\beta(\varepsilon_2-1)}{\lambda}\pi^* + \frac{\beta\varepsilon_2-1}{\lambda}\pi_t \qquad (25.8)$$

Replacing the output gap expressions in (25.7) and (25.8) into the Phillips curve (25.2), one finds, respectively,

$$E_t\pi_{t+1} = \varepsilon_1\pi_t + (1-\varepsilon_1)\pi^* \qquad (25.9)$$

$$E_t\pi_{t+1} = \varepsilon_2\pi_t - (\varepsilon_2-1)\pi^* \qquad (25.10)$$

where (25.9) is stable (it corresponds to the inflation dynamics when the stable path is followed) and (25.10) is unstable (it corresponds to the inflation dynamics when the unstable path is followed).

25.3 Inattentive Central Bank

The central bank intertemporal optimization problem may be solved by assuming that private agents have rational expectations, as in the previous section. However, although the central bank may stick to this belief, private agents might act differently and use some kind of learning rule to form expectations concerning inflation. Here, we follow the mechanism of expectations formation used in Adam et al. [1].

Expectations concerning next period inflation are formed using present and past information. We specify expectations under learning as $E_t\pi_{t+1} = b_t\pi_t$, where b_t is an estimator of inflation based on past information. The mechanism of learning obeys to the rule

$$b_t = b_{t-1} + \sigma_t\left(\frac{\pi_{t-1}}{\pi_{t-2}} - b_{t-1}\right), \quad b_0 \text{ given.} \qquad (25.11)$$

25 Adaptive Learning and Central Bank Inattentiveness in Optimal Monetary Policy

Variable $\sigma_t \in [0, 1]$ is attached to the notion of gain sequence. Convergence to the REE implies $\sigma_t \to 0$; if $\sigma_t \to \bar{\sigma} \in (0, 1)$, constant gain holds. The value $\bar{\sigma}$ may be interpreted as a measure of the quality of the learning process; the closer it is to zero, the more efficient is learning (or, in other words, less relevant is the loss of memory).

25.3.1 Local Stability

Under our setting, the monetary authority sets the interest rate in an optimal trajectory, which has two arms, one stable, (25.9), and the other unstable, (25.10). Although these equations arise from the assumption that the central bank follows perfect foresight, the private economy effectively learns over time, and therefore the estimator b_t may be presented as $b_t = \frac{E_t \pi_{t+1}}{\pi_t} = \varepsilon_1 + (1 - \varepsilon_1)\frac{\pi^*}{\pi_t}$ (if the stable trajectory is followed) and $b_t = \frac{E_t \pi_{t+1}}{\pi_t} = \varepsilon_2 - (\varepsilon_2 - 1)\frac{\pi^*}{\pi_t}$ (if the unstable trajectory is followed). Replacing these expressions in (25.11), one arrives to the following system,

$$
\begin{cases}
\pi_{t+1} = \dfrac{(1-\varepsilon_i)\pi^*}{\sigma_{t+1}\left(\frac{\pi_t}{z_t}-\varepsilon_i\right)+(1-\sigma_{t+1})(1-\varepsilon_i)\frac{\pi^*}{\pi_t}} & i = 1, 2 \\
z_{t+1} = \pi_t
\end{cases}
\tag{25.12}
$$

Variable z_t is defined as the inflation rate in period $t - 1$. To synthesize, we must stress that system (25.12) characterizes the admissible inflation rate paths when (a) the central bank adopts an optimal interest rate rule, assuming that private agents are fully rational regarding their expectations; (b) agents predict inflation rates under a learning scheme.

Both cases in (25.12) have a unique steady state point $(\bar{\pi}, \bar{z}) = (\pi^*, \pi^*)$. In the vicinity of this steady state we can study the stability of the system by considering the following linearization:

$$
\begin{bmatrix} \pi_{t+1} - \pi^* \\ z_{t+1} - \pi^* \end{bmatrix} = \begin{bmatrix} (1 - \bar{\sigma}) - \frac{\bar{\sigma}}{1-\varepsilon_i} & \frac{\bar{\sigma}}{1-\varepsilon_i} \\ 1 & 0 \end{bmatrix} \cdot \begin{bmatrix} \pi_t - \pi^* \\ z_t - \pi^* \end{bmatrix}, \ i = 1, 2
\tag{25.13}
$$

A first relevant result is straightforward to obtain from (25.13),

Proposition 25.1. *In the optimal monetary policy problem in which the monetary authority overlooks the evidence that the private economy forms expectations through learning, the following local stability results are obtained:*

Case 1. Under (25.9),

- *If $\bar{\sigma} < \frac{2(1-\varepsilon_1)}{3-\varepsilon_1}$, the system is stable.*
- *If $\bar{\sigma} = \frac{2(1-\varepsilon_1)}{3-\varepsilon_1}$, the system undergoes a flip bifurcation.*
- *If $\bar{\sigma} > \frac{2(1-\varepsilon_1)}{3-\varepsilon_1}$, the system is saddle-path stable.*

Case 2. Under (25.10),

- *If $\bar{\sigma} < \varepsilon_2 - 1$, the system is stable.*
- *If $\bar{\sigma} = \varepsilon_2 - 1$, the system undergoes a Neimark–Sacker bifurcation.*
- *If $\bar{\sigma} > \varepsilon_2 - 1$, the system is unstable.*

Proof. Trace and determinant of the Jacobian matrix in system (25.13) are $Tr(J) = (1 - \bar{\sigma}) - \frac{\bar{\sigma}}{1-\varepsilon_i}$ and $Det(J) = -\frac{\bar{\sigma}}{1-\varepsilon_i}$. Stability conditions of two-dimensional discrete time systems are the following: $1 - Tr(J) + Det(J) > 0$, $1 + Tr(J) + Det(J) > 0$ and $1 - Det(J) > 0$. These expressions correspond, in the present case, respectively to $\bar{\sigma} > 0$, $2 - \bar{\sigma} - 2\frac{\bar{\sigma}}{1-\varepsilon_i} > 0$ and $1 + \frac{\bar{\sigma}}{1-\varepsilon_i} > 0$. For $i = 1$, the first and the third inequalities are satisfied; the second requires $\bar{\sigma} < \frac{2(1-\varepsilon_1)}{3-\varepsilon_1}$, as specified in the proposition. If the opposite condition holds, then the system is saddle-path stable [because condition $1 + Tr(J) + Det(J) > 0$ is violated]. In the point in which $1 + Tr(J) + Det(J) = 0$, the system undergoes a flip bifurcation.

For $i = 2$, the first and the second stability conditions are satisfied, while the third requires $\bar{\sigma} < \varepsilon_2 - 1$. If $\bar{\sigma} > \varepsilon_2 - 1$, then $Det(J) > 1$, and therefore the system falls in the instability region. When $\bar{\sigma} = \varepsilon_2 - 1$ (i.e., $Det(J) = 1$), the eigenvalues of the Jacobian matrix turn into two complex conjugate values with modulus equal to 1, and the system undergoes a Neimark–Sacker bifurcation. □

If we combine the trace and the determinant expressions of the Jacobian matrix in (25.13), the equation $Det(J) = Tr(J) - (1 - \bar{\sigma})$ is obtained. This relation is depicted graphically in Fig. 25.1.

The three lines that form the inverted triangle, in Fig. 25.1, are bifurcation lines. The area inside the triangle corresponds to the region of stability (two eigenvalues inside the unit circle). The bold line relates to the location of system (25.13) in terms of the trace-determinant relation.

Especially relevant is the fact that, in both cases in Proposition 25.1, stability holds for low values of $\bar{\sigma}$ (near zero). This means that the learning process does not

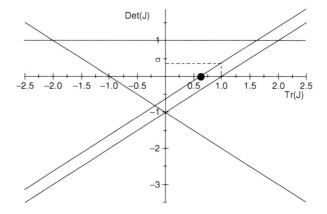

Fig. 25.1 Local inflation dynamics under learning

25 Adaptive Learning and Central Bank Inattentiveness in Optimal Monetary Policy

need to be fully efficient (i.e. to converge to the REE) to lead to the stable outcome of rational expectations. If some memory loss is considered, it does not imply a departure from the benchmark result where convergence to the target value of the inflation rate is achieved. When learning inefficiency passes a given threshold (those referred to in the proposition), then inflation stability is lost, and inflation does no longer converge to its target value. This result seems to be economically relevant: it says that agents do not need to be completely efficient when learning, but they need to be almost efficient in order to be possible to attain the desired policy result.

25.3.2 Global Dynamics

Local dynamics indicate that we are in the presence of points of bifurcation. When considering any of (25.9) and (25.10), the introduction of the learning mechanism induces the presence of a change in the qualitative nature of the dynamics as one varies the long run value of the gain variable. Given the nonlinear nature of the first equation in system (25.12), one might expect such shifts in the topological properties of the model to produce endogenous fluctuations. In this section, we take some reasonable parameter values to explore the global properties of the system. It is found that as one passes from a local area of stability to an area of instability or saddle-path stability, this is translated, in terms of global dynamics, as the transition from a fixed-point steady state into areas of periodic and aperiodic cycles that exist in a given region before instability eventually sets in.

To address global dynamics, we take the steady state gain value $\bar{\sigma}$ as the bifurcation parameter. The other parameters assume reasonable quarterly values: $\pi^* = 0.005$, $a = 0.25$, $\lambda = 0.024$, $\beta = 0.99$. Recall that the eigenvalues ε_1 and ε_2 are the ones in (25.6); hence, for the assumed parameter specifications, one has $\varepsilon_1 = 0.9576$ and $\varepsilon_2 = 1.0549$. In the graphical presentation that follows we assume the initial values $\pi_0 = z_0 = 0.004$ (the initial value of σ_t is irrelevant as long as $\sigma_t \in (0, 1)$).

The graphical analysis consists in presenting bifurcation diagrams for the inflation rate and considering both values of ε_i (Fig. 25.2). We observe that for low values of the long term gain variable, a stable fixed point is obtained. This confirms that if we are near the REE long term outcome, then the system is stable. As we depart from such outcome, a period-two cycle becomes dominant and regions of aperiodic motion will also arise. These, however, are relatively small. Chaotic motion may eventually exist.

The diagrams in Fig. 25.2 can be analyzed together with the local dynamics results in Proposition 1. For $\varepsilon_1 = 0.9576$, the system is stable if $\bar{\sigma}$ is lower than 0.0415. A bifurcation occurs at $\bar{\sigma} = 0.0415$, and this can be confirmed by observing the upper panel of Fig. 25.2. To the right of this point, local dynamics led to a result of saddle-path stability, that we verify to be a region of cyclical motion. In what concerns the second case, $\varepsilon_2 = 1.0549$, the Neimark–Sacker bifurcation occurs when $\bar{\sigma} = 0.0549$. To the left of this point we have stability, and to the right instability prevails (locally) and cycles are evidenced (globally).

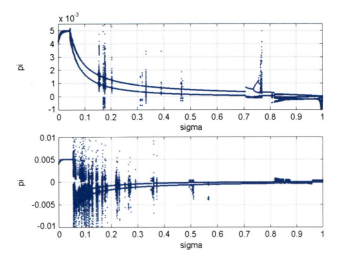

Fig. 25.2 Bifurcation diagram (π_t, σ); $\varepsilon = 0.9576$ (*first panel*); $\varepsilon = 1.0549$ (*second panel*)

25.4 Conclusion

In this paper, we have analyzed an environment where private agents learn but the central bank does not take this into account. The dynamic analysis of the monetary policy model found, locally, bifurcation points separating regions of stability from saddle-path stability/instability. A global approach allowed us to realize that the bifurcations separate regions of fixed point stability from areas where endogenous fluctuations are observed. Dynamics with period 2 cycles is a predominant result but higher periodicity cycles and complete a-periodicity are also revealed. Particularly important is that stability is found solely for low values of the gain variable (i.e., near the REE), meaning that a stable fixed point outcome is directly associated with a high quality learning. The obtained results seem to corroborate the idea, which is pervasive in the literature, that endogenous cycles in adaptive learning settings arise under constant gain.

Acknowledgements The authors would like to thank the organizers and the participants of the Conference "Dynamics & Applications in honour of Mauricio Peixoto and David Rand", University of Minho, Braga, September 2008.

References

1. Adam, K., Marcet, A., Nicolini, J.P.: Stock Market Volatility and Learning. European Central Bank working paper (2008)
2. Beeby, M., Hall, S.G., Henry, B.: "Rational Expectations and Near Rational Alternatives: How Best to Form Expectations. European Central Bank working paper series, 086 (2001)

25 Adaptive Learning and Central Bank Inattentiveness in Optimal Monetary Policy 383

3. Cellarier, L.: Constant gain learning and business cycles. J. Macroecon. **28**, 51–85 (2006)
4. Evans, G.W., Honkapohja, S.: Learning and Expectations in Macroeconomics. Princeton University Press, Princeton, New Jersey (2001)
5. Grandmont, J.-M.: Expectations formation and stability in large socio-economic systems. Econometrica **66**, 741–781 (1998)
6. Lucas, R.E.: Expectations and the neutrality of money. J. Econ. Theory **4**, 103–124 (1972)
7. Marcet, A., Sargent, T.: Convergence of least squares learning mechanisms in self-referential linear stochastic models. J. Econ. Theory **48**, 337–368 (1989)
8. Muth, J.F.: Rational expectations and the theory of price movements. Econometrica **29**, 315–335 (1961)
9. Woodford, M.: Interest and Prices: Foundations of a Theory of Monetary Policy. Princeton University Press, Princeton, New Jersey (2003)

Chapter 26
Discrete Time, Finite State Space Mean Field Games

Diogo A. Gomes, Joana Mohr, and Rafael Rigão Souza

Abstract In this paper we report on some recent results for mean field models in discrete time with a finite number of states. These models arise in situations that involve a very large number of agents moving from state to state according to certain optimality criteria. The mean field approach for optimal control and differential games (continuous state and time) was introduced by Lasry and Lions (C. R. Math. Acad. Sci. Paris, 343(9):619–625, 2006; 343(10):679–684, 2006; Jpn. J. Math., 2(1):229–260, 2007). The discrete time, finite state space setting is motivated both by its independent interest as well as by numerical analysis questions which appear in the discretization of the problems introduced by Lasry and Lions. We address existence, uniqueness and exponential convergence to equilibrium results.

26.1 Introduction

In this paper we report on our recent results on mean field model for discrete time with a finite number of states, dynamic games. The mean field approach for optimal control and differential games (continuous state and time) was introduced by Lasry and Lions [1–3]. In the continuous state and time setting, mean field problems gives rise to Hamilton–Jacobi equations coupled with transport equations. The discrete time, finite state space setting is motivated both by its independent interest as well as by numerical analysis questions which appear in the discretization of the problems introduced by Lasry and Lions. The discretization of these models has been studied by I. Capuzzo–Dolcetta and Y. Achdou.

Let $d > 1$ and $N \geq 1$ be natural numbers, representing, respectively, the number of possible states in which certain agents can be at any given time, and the total

---

D.A. Gomes (✉)
Instituto Superior Técnico, Av. Rovisco Pais, 1049-001 Lisbon, Portugal
e-mail: dgomes@math.ist.utl.pt

J. Mohr and R.R. Souza
Instituto de Matemática, UFRGS, 91509-900, Porto Alegre, Brazil
e-mail: joanamohr@yahoo.com.br, rafars@mat.ufrgs.br

M.M. Peixoto et al. (eds.), *Dynamics, Games and Science I*, Springer Proceedings in Mathematics 1, DOI 10.1007/978-3-642-11456-4_26,
© Springer-Verlag Berlin Heidelberg 2011

duration of the process. Let π^0 and V^N be given d-dimensional vectors. We suppose that π^0 is a probability vector, the initial probability distribution of agents among states, and that V^N, the terminal cost, is an arbitrary vector. A solution to the mean field game is a sequence of pairs of d-dimensional vectors

$$\{(\pi^n, V^n) \; ; \; 0 \le n \le N\},$$

where π^n is the probability distribution of agents among states at time n and V_j^n is the expected minimum total cost for an agent at state j, at time n. These pairs must satisfy certain optimality conditions that we describe in what follows: at every time step, the agents in state i choose a transition probability, P_{ij}, from state i to state j. Given the transition probabilities P_{ij}^n at time $0 \le n < N$, the distribution of agents at time $n+1$ is simply

$$\pi_j^{n+1} = \sum_i \pi_i^n P_{ij}^n \, .$$

Associated to this choice there is a transition cost $c_{ij}(\pi, P)$. In the special case in which c_{ij} only depends on π and on the ith line of P we use the simplified notation $c_{ij}(\pi, P_i.)$. This last case arises when the choices of players in states $j \neq i$ do not influence the transition cost to an agent in state i. Let $e_i(\pi, P, V)$ be the average cost that agents which are in state i incur when matrix P is chosen, given the current distribution π and the cost vector V at the subsequent instant. We assume that

$$e_i(\pi, P, V) = \sum_j c_{ij}(\pi, P) P_{ij} + V_j P_{ij} \, .$$

Define the probability simplex $\mathbb{S} = \{(q_1, \ldots, q_d) \; ; \; q_j \ge 0 \, \forall j, \sum_{j=1}^d q_j = 1\}$. The set of $d \times d$ stochastic matrices is identified with $\mathbb{S}^d$. Given a stochastic matrix $P \in \mathbb{S}^d$ and a probability vector $q \in \mathbb{S}$, we define $\mathscr{P}(P, q, i)$ to be the $d \times d$ stochastic matrix obtained from P by replacing the ith row by q, and leaving all others unchanged.

Definition 26.1. Fix a probability vector $\pi \in \mathbb{S}$ and a cost vector $V \in \mathbb{R}^d$. A stochastic matrix $P \in \mathbb{S}^d$ is a *Nash minimizer* of $e(\pi, \cdot, V)$ if for each $i \in \{1, \ldots, d\}$ and any $q \in \mathbb{S}$

$$e_i(\pi, P, V) \le e_i(\pi, \mathscr{P}(P, q, i), V).$$

Under the uniqueness of a Nash minimizer for e, we can define the (backwards) evolution operator for the value function

$$\mathscr{G}_\pi(V) = e(\pi, \bar{P}, V), \tag{26.1}$$

as well as the (forward) evolution operator for π

$$\mathscr{K}_V(\pi) = \pi \bar{P}. \tag{26.2}$$

26 Discrete Time, Finite State Space Mean Field Games

Since the operator $\mathscr{G}_\pi$ commutes with addition with constants, it can be regarded as a map from $\mathbb{R}^d/\mathbb{R}$ to $\mathbb{R}^d/\mathbb{R}$. Here $\mathbb{R}^d/\mathbb{R}$ is the set of equivalence classes of vectors in $\mathbb{R}^d$ whose components differ by the same constant. In $\mathbb{R}^d/\mathbb{R}$ we define the norm

$$\|\psi\|_\# := \inf_{\lambda \in \mathbb{R}} \|\psi + \lambda\|, \tag{26.3}$$

We will be interested in this paper both in stationary solutions and in the terminal initial value problem, as we define next,

Definition 26.2. A pair of vectors $(\bar\pi, \bar V)$ is a stationary solution to the mean field game if there exists a constant $\bar\lambda$, called critical value, such that

$$\begin{cases} \bar V = \mathscr{G}_{\bar\pi}(\bar V) + \bar\lambda \\ \\ \bar\pi = \mathscr{K}_{\bar V}(\bar\pi). \end{cases} \tag{26.4}$$

Definition 26.3. A sequence of pairs of d-dimensional vectors

$$\{(\pi^n, V^n)\,;\, 0 \le n \le N\}$$

is a *solution of the mean field game* if for every $0 \le n \le N - 1$

$$\begin{cases} V^n = \mathscr{G}_{\pi^n}(V^{n+1}) \\ \\ \pi^{n+1} = \mathscr{K}_{V^{n+1}}(\pi^n). \end{cases} \tag{26.5}$$

26.2 Main Assumptions

In this section, for the convenience of the reader, we list the main assumptions that will be used for the statement of the main results.

Assumption 1. *For each $\pi \in \mathbb{S}$, $V \in \mathbb{R}^d$, $P \in \mathbb{S}^d$, and each index $1 \le i \le d$, the mapping $q \mapsto e_i(\pi, \mathscr{P}(P, q, i), V)$, defined for $q \in \mathbb{S}$, and taking values on $\mathbb{R}$, is convex.*

Assumption 2. *The map $P \mapsto e_i(\pi, P, V)$ is continuous for all i.*

Theorem 26.1. *Suppose that assumptions 1 and 2 hold. Then, for any pair of vectors π and V there exists a Nash minimizer P of $e(\pi, \cdot, V)$.*

Definition 26.4. A function $g : \mathbb{R}^{d \times d} \to \mathbb{R}^{d \times d}$ is diagonally convex if for all $P^1, P^2 \in \mathbb{R}^d$, $P^1 \ne P^2$, we have

$$\sum_{ij}(P_{ij}^1 - P_{ij}^2)(g_{ij}(P^1) - g_{ij}(P^2)) > 0.$$

Assumption 3. *Let*

$$g_{ij}(P) = \frac{\partial e_i(\pi, P, V)}{\partial P_{ij}}.$$

Then g is diagonally convex.

Theorem 26.2. *Suppose assumptions 1–3 hold. Then there exists a unique transition matrix P which is a Nash minimizer of* $e(\pi, \cdot, V)$.

Assumption 4. *For each index* $1 \leq i \leq d$, $e_i : \$ \times \$^d \times \mathbb{R}^d \to \mathbb{R}$ *is a continuous function.*

Denote by $\rho_{i,i'}(P)$ the matrix we obtain from P by replacing its i'th row by its ith row, and leaving all other rows (including the ith) unchanged.

Assumption 5. *There exists* $C > 0$ *such that for all* i *and* i', *and any* $\pi \in \$$, $P \in \$^d$

$$\sum_j \left| c_{ij}(\pi, P) - c_{i'j}(\pi, \rho_{i,i'}(P)) \right| P_{ij} \leq C. \tag{26.6}$$

Assumption 6. *The cost* $c_{ij}(\pi, P_i.)$ *depends on* π *and, for each* i, *only on the* ith *line of P.*

Assumption 7. *There exists a constant* $\gamma > 0$ *such that*

$$\tilde{\pi} \cdot (\mathscr{G}_{\tilde{\pi}}(V) - \mathscr{G}_{\pi}(V)) + \pi \cdot (\mathscr{G}_{\pi}(\tilde{V}) - \mathscr{G}_{\tilde{\pi}}(\tilde{V})) \geq \gamma \|\pi - \tilde{\pi}\|^2,$$

for any $V, \tilde{V} \in \mathbb{R}^d$ *and all* $\pi, \tilde{\pi} \in \$$.

Assumption 8. *For all* $\pi \in \$$ *and all* $V^1, V^2 \in \mathbb{R}^d$ *we have*

$$\pi \cdot (\mathscr{G}_{\pi}(V^2) - \mathscr{G}_{\pi}(V^1)) + \mathscr{K}_{V^1}(\pi)(V^1 - V^2) \leq -\gamma_{\pi} \|V^1 - V^2\|_{\#}^2.$$

Assumption 9. *There exists* $K > 0$ *such that for all* $\pi, \tilde{\pi} \in \$$, *and for any matrix* $P \in \$^d$

$$\left| c_{ij}(\pi, P) - c_{ij}(\tilde{\pi}, P) \right| \leq K. \tag{26.7}$$

Note that the previous assumption holds if c_{ij} is bounded, for instance.

26.3 Main Results

We finally describe our main results for these models. The first two theorems concern the existence and uniqueness of, respectively, stationary solutions and the initial-terminal value problems. The uniqueness proofs follow the monotonicity methods introduced by Lasry and Lions, see [1–3].

26 Discrete Time, Finite State Space Mean Field Games

Theorem 26.3. *Suppose that assumptions 1, 2, 3, 4, and 5 hold. Then there exists a pair of vectors $(\bar{\pi}, \bar{V})$, a constant $\bar{\lambda}$ and a transition matrix $\bar{P}$ such that for all i,*

$$\mathcal{G}_{\bar{\pi}}(\bar{V})_i = \sum_j c_{ij}(\bar{\pi}, \bar{P})\bar{P}_{ij} + \bar{V}_j \bar{P}_{ij} = \bar{V}_i + \bar{\lambda},$$

and $\bar{\pi} = \bar{\pi}\bar{P}$.

Assume further that 6, 7 and 8 hold then there is a unique stationary solution, up to addition of a constant in $\bar{V}$.

Theorem 26.4. *Suppose assumptions 1, 2, 3, and 4 hold. Then for any initial probability vector $\tilde{\pi} \in \mathbb{S}$ and terminal cost $\tilde{V}$ there exists a solution*

$$\{(\pi^n, V^n) ; 0 \leq n \leq N\}$$

to the initial-terminal value problem for the mean field game with $\pi^0 = \tilde{\pi}$ and $V^N = \tilde{V}$.

Assume further that 6 and 7 hold. Then the solution to the initial-terminal value problem is unique.

One of our main contributions is the exponential convergence to equilibrium for the initial-terminal value problem. Our setting is the following: consider a initial-terminal value problem with initial data π^{-N} and terminal data V^N. We will now study conditions under which $\pi^0 \to \bar{\pi}$ and $V^0 \to \bar{V}$ where $(\bar{\pi}, \bar{V})$ are stationary solutions, as $N \to \infty$.

Theorem 26.5. *Suppose assumptions 1,2, 3, and 6, 7, 8 and 9 hold. Fix $\tilde{V}, \tilde{\pi}$. Given $N > 0$, denote by (π_N^0, V_N^0) the solution of the mean field game at time 0 that has initial distribution $\pi^{-N} = \tilde{\pi}$ and terminal cost $V^N = \tilde{V}$.*

Then, as $N \to \infty$

$$V_N^0 \to \bar{V} \ (in \ \mathbb{R}^d/\mathbb{R}), \quad \pi_N^0 \to \bar{\pi}$$

where $\bar{V}$ and $\bar{\pi}$ is the unique stationary solution.

Acknowledgements D.G. was partially supported by the CAMGSD/IST through FCT Program POCTI/FEDER and by grant DENO/FCT-PT (PTDC/EEA-ACR/67020/2006).R.R.S was partially supported by CAPES, PROCAD, Projeto Universal CNPq 471473/2007-3. J.M: is beneficiary of a CNPq PhD scholarship.

References

1. Lasry, J.-M., Lions, P.-L.: Jeux à champ moyen. I. Le cas stationnaire. C. R. Math. Acad. Sci. Paris **343**(9), 619–625 (2006)
2. J.-M. Lasry, Lions, P.-L.: Jeux à champ moyen. II. Horizon fini et contrôle optimal. C. R. Math. Acad. Sci. Paris **343**(10), 679–684 (2006)
3. Lasry, J.-M., Lions, P.-L.: Mean field games. Jpn. J. Math. **2**(1), 229–260 (2007)

Chapter 27
Simple Exclusion Process: From Randomness to Determinism

Patrícia Gonçalves

Abstract In this work I introduce a classical example of an Interacting Particle System: the Simple Exclusion Process. I present the notion of hydrodynamic limit, which is a Law of Large Numbers for the empirical measure and an heuristic argument to derive from the microscopic dynamics between particles a partial differential equation describing the evolution of the density profile. For the Simple Exclusion Process, in the Symmetric case ($p = 1/2$) we will get to the heat equation while in the Asymmetric case ($p \neq 1/2$) to the inviscid Burgers equation. Finally, I introduce the Central Limit Theorem for the empirical measure and the limiting process turns out to be a solution of a stochastic partial differential equation.

27.1 Introduction

In this work I am presenting some well known results and some of the latest developments on a classical interacting particle system: the simple exclusion process (SEP). Interacting particle systems were introduced by Spitzer in the late 1970s and since then, their study has attracted the attention of researchers of several fields of Mathematics. The problems that initially appeared, have arisen from the physicists and the goal was to give precise answers to conjectures and experiments done by the physics community. Now I describe the idea behind the problems that we usually deal with. Suppose that one is interested in analyzing the evolution of some physical system, constituted by a large number of components, for example, a fluid or a gas. Due to the large number of molecules it becomes hard to analyze the microscopic evolution of the system, and as a consequence it is more relevant to analyze the macroscopic evolution of the structure of that system. Following the approach proposed by Boltzmann from the Statistical Mechanics, first one finds the equilibrium states of this physical system and characterizes them through macroscopic quantities, called thermodynamical quantities that one is interested in analyzing

---

P. Gonçalves
Centro de Matemática da, Universidade do Minho, Campus de Gualtar, 4710-057 Braga, Portugal
e-mail: patg@math.uminho.pt

M.M. Peixoto et al. (eds.), *Dynamics, Games and Science I*, Springer Proceedings
in Mathematics 1, DOI 10.1007/978-3-642-11456-4_27,
© Springer-Verlag Berlin Heidelberg 2011

such as the pressure, temperature, density... The natural question that follows is to analyze the behavior of that physical system out of equilibrium. The characterization and study of phenomena out of equilibrium is one of the biggest challenge of the Statistical Physics and despite its long history, nowadays, it still has not been found a satisfactory answer to this kind of problems. From this approach some differential equations arise that provide some information about the macroscopic evolution of the thermodynamical quantities of the system. Usually, and at least heuristically, these equations can be deduced from the scaling limit of a system and this deduction gives validity to this equation. When approaching these problems, due to the huge complexity of its analysis, some simplifications need to be introduced. With that purpose, usually one assumes that the underlying microscopic dynamics, i.e. the dynamics between molecules, is stochastic, in such a way that a probabilistic analysis of the system can be done. Assuming that the particles (or molecules) behave as interacting random walks subjected to random local restrictions, arise the so called Interacting Particle Systems [6]. Nowadays, there exists a well developed theory to deal with this kind of problems, that consists on the microscopic analysis of a particle system – a continuous time Markov process, whose macroscopic evolution of the density profile is governed by one (or system) partial differential equation, denominated by *Hydrodynamic Limit* [5]. This research field, deals and answers to the discretization of several partial differential equations, which have solutions with different qualitative behavior and whose microscopic dynamics has originated the study of different particle systems with hydrodynamic behavior.

Usually, for all the studied systems, the behavior of the associated partial differential equation, gives information about the behavior of the particle system. Nevertheless, there are several hard phenomena, that are very difficult to analyze in the analytical point of view of the solutions of the partial differential equation which can be analyzed through the study of the underlying microscopic system, as for example the partial differential equations that exhibit shocks, as the Burgers equation [2] (see [3] and references therein). The development of this theory has also provided some answers to questions related to the behavior of physical systems out of equilibrium, see [10].

Here is an outline of these notes. On the second section I introduce the simple exclusion process, generator and the invariant measures. On the third section I give the notion of hydrodynamic limit and an heuristic argument to get to the hydrodynamic equation for two cases, symmetric and asymmetric jumps. Then I give the notion of equilibrium and non-equilibrium fluctuations. Finally at section five by superposing both dynamics one obtains the WASEP and the results above are also stated for this process.

27.2 The Simple Exclusion Process

In this section I introduce the one-dimensional Exclusion Process. In this process, particles evolve on $\mathbb{Z}$ according to interacting random walks with an exclusion rule which prevents more than one particle per site. The dynamics can be informally

27 Simple Exclusion Process: From Randomness to Determinism

described as follows. Fix a probability $p(\cdot)$ on $\mathbb{Z}$. Each particle, independently from the others, waits a mean one exponential time, at the end of which being at the site x it jumps to $x + y$ at rate $p(y)$. If the site is occupied the jump is suppressed to respect the exclusion rule. In both cases, the particle waits a new exponential time. The space state of the Markov process η_t is $\{0, 1\}^{\mathbb{Z}}$ and we denote the configurations by the Greek letter η, so that $\eta(x) = 0$ if the site x is vacant and $\eta(x) = 1$ otherwise. The case in which $p(y) = 0 \; \forall |y| > 1$ is referred as the Simple Exclusion process (SEP) and for the Asymmetric Simple Exclusion process (ASEP) the probability $p(\cdot)$ is such that $p(1) = p$, $p(-1) = 1 - p$ with $p \neq 1/2$ while in the Symmetric Simple Exclusion process (SSEP) $p = 1/2$. The case in which $p = 1$ is denoted by TASEP and means totally asymmetric simple exclusion process, since particles can perform jumps only to the right.

The dynamics of the SEP can be translated by means of a generator given on local functions by

$$\mathcal{L} f(\eta) = \sum_{x \in \mathbb{Z}} \sum_{y = x \pm 1} c(x, y, \eta)[f(\eta^{x,y}) - f(\eta)],$$

where $c(x, y, \eta) = p(x, y)\eta(x)(1 - \eta(y))$ and

$$\eta^{x,y}(z) = \begin{cases} \eta(z), & \text{if } z \neq x, y \\ \eta(y), & \text{if } z = x \\ \eta(x), & \text{if } z = y \end{cases} .$$

To keep notation simple we denote by $\mathcal{L}_S$ ($\mathcal{L}_A$) the generator of the SSEP (ASEP).

Before proceeding we give the definition of an equilibrium state of the system. Let η. denote a Markov Process with generator Ω and semigroup $(S(t))_{t \geq 0}$. Let $\mathcal{P}$ denote the set of probability measures on $\{0, 1\}^{\mathbb{Z}}$. A probability measure $\mu \in \mathcal{P}$ is said to be an invariant measure for the Markov process if $\mu S(t) = \mu$ for all $t \geq 0$, which is the same as saying that the distribution of $(\eta_t)_t$ does not depend on the time t. There is a nice criterium to find equilibrium states for a Markov Process and we recall it from [6]:

Proposition 27.1. *Let $\mathcal{I}$ denote the set of probability measures in $\{0, 1\}^{\mathbb{Z}}$ and η. be a Markov Process with generator Ω. Then*

$$\mathcal{I} = \{\mu : \int \mathcal{L} f(\eta)\mu(d\eta) = 0, \forall f \text{ local}\}. \tag{27.1}$$

For $0 \leq \alpha \leq 1$, denote by ν_α the Bernoulli product measure on $\{0, 1\}^{\mathbb{Z}}$ with density α. This means that the random variables $(\eta(x))_{x \in \mathbb{Z}}$ are independent with Bernoulli distribution:

$$\nu_\alpha(\eta(x) = 1) = \alpha \tag{27.2}$$

It is known that ν_α is an invariant measure for the SEP and in fact, that all invariant and translation invariant measures are convex combinations of ν_α if $p(.)$ is such that $p_t(x, y) + p_t(y, x) > 0$, $\forall x, y \in \mathbb{Z}^d$ and $\sum_x p(x, y) = 1$, $\forall y \in \mathbb{Z}^d$.

27.3 Hydrodynamic Limit

27.3.1 *From Microscopic to Macroscopic*

In order to investigate the hydrodynamic limit, we need to settle some notation. We are going to consider the physical system evolving in a continuum space – the *macroscopic space*. The idea is to discretize this set by relating it to another one, but this last being a discrete set – the *microscopic space*. In the discrete space we define a particle system and since we want to study the temporal evolution of the density profile we have two different scales for time as well: a *macroscopic time* denoted by t and a microscopic time denoted by $t\theta(N)$. This function $\theta(N)$ depends on the subjacent microscopic dynamics and as we will see, for the SSEP we need $\theta(N) = N^2$ while for the ASEP $\theta(N) = N$ is enough. In order to simplify the exposition we suppose that we take the macroscopic space to be the one-dimensional torus $\mathbb{T}$. Then, we fix an integer N and split it in small interval of size $\frac{1}{N}$. The relation between this two sets is that if $u \in \mathbb{T}$ it corresponds to $[uN]$ in the microscopic space while if $x \in \mathbb{T}_N$ it corresponds to x/N in the macroscopic space $\mathbb{T}$.

Suppose now that the simple exclusion process is evolving on $\mathbb{T}_N$. For a given configuration η we define the the *empirical measure* π^N as the positive measure on $\mathbb{T}$ which gives to each particle a mass $1/N$, namely

$$\pi^N(\eta, du) = \frac{1}{N} \sum_{x \in \mathbb{T}_N} \eta(x)\delta_{\frac{x}{N}}(du), \tag{27.3}$$

where δ_u denotes the Dirac measure at u. Then we consider the time evolution of this measure defined by

$$\pi_t^N(du) = \frac{1}{N} \sum_{x \in \mathbb{T}_N} \eta_t(x)\delta_{\frac{x}{N}}(du) \tag{27.4}$$

where as usual η_t is the process at time t which is generated by $\mathscr{L}$ when the configuration at time zero is η.

Fix now, an initial profile $\rho_0 : \mathbb{T} \to [0, 1]$ and denote by $(\mu_N)_{N \geq 1}$ a sequence of probability measures on $\{0, 1\}^{\mathbb{T}_N}$. Depending on the model itself the initial profile ρ_0 needs to satisfy certain conditions that we shall impose later.

Assume that at time 0, the system starts from a initial measure μ_N that is associated to the initial profile ρ_0, ie the empirical measure at time 0 satisfies a law of large numbers:

27 Simple Exclusion Process: From Randomness to Determinism

Definition 27.1. A sequence $(\mu^N)_{N \geq 1}$ is associated to ρ_0, if for every continuous function $H : \mathbb{T} \to \mathbb{R}$ and for every $\delta > 0$

$$\lim_{N \to +\infty} \mu_N \left[\eta : \left| \frac{1}{N} \sum_{x \in \mathbb{T}_N} H\left(\frac{x}{N}\right) \eta(x) - \int_{\mathbb{T}^d} H(u) \rho_0(u) du \right| > \delta \right] = 0. \quad (27.5)$$

Note that the first term corresponds to the integral of H with respect to π^N, thus the above definition corresponds to asking that the sequence $\pi^N(\eta, du)$ converges in μ_N-probability to $\rho_0(u)du$.

The goal in hydrodynamic limit consists in showing that, if at time $t = 0$ the empirical measures are associated to some initial profile ρ_0, at the macroscopic time t (i.e. the microscopic time $t\theta(N)$) they are associated to a profile ρ_t which is the solution of the some partial differential equation. In other words the aim is to prove that the random measures $\pi^N_{t\theta(N)}$ converge in probability to the deterministic measure $\rho(t, u)du$, which is absolutely continuous with respect to the Lebesgue measure whose density evolves according to some partial differential equation - called hydrodynamic equation.

For the SSEP it was shown that starting from a sequence of measures $(\mu_N)_N$ associated to a profile $\rho_0(\cdot)$, under the *parabolic time scale* tN^2,

$$\pi^N_{tN^2} \xrightarrow[N \to +\infty]{} \rho(t, u)du \quad (27.6)$$

in $\mu^N S^S_N(t)$-probability, where $\rho(t, u)$ is a weak solution of the parabolic equation

$$\partial_t \rho(t, u) = \frac{1}{2} \Delta \rho(t, u) \quad (27.7)$$

and S^S_N is the semigroup associated to the generator $\mathscr{L}_S$.

For a proof of last result one can see for example Chap. 4 of [5] where the entropy method is applied.

On the other hand, for the ASEP starting from a sequence of measures $(\mu_N)_N$ associated to a profile $\rho_0(.)$ and some additional hypotheses (see [8]) under the *hyperbolic time scale* tN

$$\pi^N_{tN} \xrightarrow[N \to +\infty]{} \rho(t, u)du, \quad (27.8)$$

in $\mu^N S^A_N(t)$-probability, where $\rho(t, u)$ is the entropy solution of the hyperbolic equation

$$\partial_t \rho(t, u) + (p - q)(1 - 2\rho(t, u))\nabla \rho(t, u) = 0 \quad (27.9)$$

known as the inviscid *Burgers equation* and $S^A_N(t)$ is the semigroup associated to the generator $\mathscr{L}_A$.

For a proof of last result we refer the interested reader to [8].

27.3.2 Hydrodynamic Equation

As we have seen above, for the simple exclusion process we obtain the heat equation when considering symmetric jump rates while in the asymmetric jumps one gets to the inviscid Burgers equation. So one can ask, why defining similar microscopic jump rates can we get to completely different macroscopic behaviors. Here I present an heuristic argument relying on the microscopic dynamics to get the hydrodynamic equation for the two different processes, see [5].

In a general setting let η_t denote a Markov process whose generator is denoted by Ω and suppose it is evolving on the microscopic time scale $t\theta(N)$. It is known from the classical theory of Markov processes that, for a test function $H : \mathbb{T} \to \mathbb{R}$

$$M_t^{N,H} = < \pi_t^N, H > - < \pi_0^N, H > - \int_0^t \Omega < \pi_s^N, H > ds \qquad (27.10)$$

is a martingale with respect to the natural filtration $\mathscr{F}_t = \sigma(\eta_s, s \le t)$, whose quadratic variation is given by

$$\int_0^t \Omega(< \pi_s^N, H >)^2 - 2 < \pi_s^N, H >)\Omega < \pi_s^N, H > ds. \qquad (27.11)$$

Here $< \pi_t^N, H >$ denotes the integral of H with respect to π_t^N. Using the explicit definition of the empirical measure, the integral part of the martingale is written as

$$\int_0^t \frac{1}{N} \sum_{x \in \mathbb{T}_N} H\left(\frac{x}{N}\right) \Omega \eta_s(x) ds. \qquad (27.12)$$

Consider now the SEP with generator $\mathscr{L}$. It is easy to see that

$$\mathscr{L}(\eta(x)) = W_{x-1,x}(\eta) - W_{x,x+1}(\eta), \qquad (27.13)$$

where for a site x and a configuration η, $W_{x,x+1}(\eta)$ is the instantaneous current between the sites x and $x + 1$, namely

$$W_{x,x+1}(\eta) = p(x, x + 1)\eta(x)(1 - \eta(x + 1)) - p(x + 1, x)\eta(x + 1)(1 - \eta(x)). \qquad (27.14)$$

Since the generator applied to $\eta_s(x)$ is written as gradient, this allows us to perform a summation by parts in the integral part of the martingale:

$$M_t^{N,H} = < \pi_t^N, H > - < \pi_0^N, H > - \int_0^t \frac{1}{N^2} \sum_{x \in \mathbb{T}_N} \nabla^N H\left(\frac{x}{N}\right) W_{x,x+1}(\eta_s) ds, \qquad (27.15)$$

where $\nabla^N H$ denotes the discrete derivative of H.

27 Simple Exclusion Process: From Randomness to Determinism

Now we restrict ourselves to the SSEP. In this case the instantaneous current between the sites x and $x + 1$, denoted by $W^S_{x,x+1}$ is given by a gradient:

$$W^S_{x,x+1}(\eta) = \frac{1}{2}(\eta(x) - \eta(x+1)).$$

This allows us to perform another summation by parts and write the martingale as

$$M_t^{N,H} = <\pi_t^N, H> - <\pi_0^N, H> - \int_0^t \frac{1}{2N^3} \sum_{x \in \mathbb{T}_N} \Delta_N H\left(\frac{x}{N}\right) \eta_s(x) ds,$$

$$(27.16)$$

where $\Delta_N H$ denotes the discrete laplacian of H.

Since we want to close the integral part of the martingale in terms of the empirical measure we have to rescale time by tN^2. This together with a change of variables gives us that

$$M_{tN^2}^{N,H} = \frac{1}{N} \sum_{x \in \mathbb{T}_N} H\left(\frac{x}{N}\right)\left(\eta_{tN^2}(x) - \eta_0(x)\right)$$

$$- \int_0^t \frac{1}{N} \sum_{x \in \mathbb{T}_N} \eta_{sN^2}(x) \frac{1}{2} \Delta^N H\left(\frac{x}{N}\right) ds. \qquad (27.17)$$

Since this martingale vanishes at time 0 its expectation is equal to zero uniformly in time.

Now we recall the notion of *conservation of local equilibrium* which means, loosely speaking, that for a macroscopic time t, the expectation of η_{tN^2} with respect to the distribution of the system at the microscopic time $t\theta(N)$ is close to the expectation of $\eta(0)$ with respect to $v_{\rho(t,x/N)}$.

Then applying expectation to the equality above, we obtain:

$$\frac{1}{N} \sum_{x \in \mathbb{T}_N} H\left(\frac{x}{N}\right)\left(\rho(t, x/N) - \rho(0, x/N)\right) = \int_0^t \frac{1}{N} \sum_{x \in \mathbb{T}_N} \rho(s, x/N) \frac{1}{2} \Delta^N H\left(\frac{x}{N}\right) ds$$

$$(27.18)$$

Taking the limit as $N \to +\infty$ if follows that $\rho(t, u)$ is a weak solution of the heat equation:

$$\begin{cases} \partial_t \rho(t, u) = \frac{1}{2} \Delta \rho(t, u) \\ \rho(0, \cdot) = \rho_0(\cdot) \end{cases}. \qquad (27.19)$$

On the other hand for the ASEP, the instantaneous current between x and $x + 1$, here denoted by $W^A_{x,x+1}(\eta)$ is given by

$$W^A_{x,x+1}(\eta) = p\eta(x)(1 - \eta(x+1)) - q\eta(x+1)(1 - \eta(x)). \qquad (27.20)$$

398   P. Gonçalves

This allows *just one* summation by parts which together with the re-scaling of time by tN and the convergence to local equilibrium gives us

$$\frac{1}{N}\sum_{x\in\mathbb{T}_N}H\left(\frac{x}{N}\right)\left(\rho(t,x/N)-\rho(0,x/N)\right)$$

$$+\int_0^t\frac{1}{N}\sum_{x\in\mathbb{T}_N}F(\rho(s,x/N))\nabla^N H\left(\frac{x}{N}\right)ds=0$$

where $F(\rho)=(p-q)\rho(1-\rho))$. Taking the limit as $N\to+\infty$ it follows that $\rho(t,u)$ is a weak solution of the inviscid Burgers equation:

$$\begin{cases}\partial_t\rho(t,u)+\nabla F(\rho(t,u))=0\\ \rho(0,\cdot)=\rho_0(\cdot)\end{cases}.\qquad(27.21)$$

27.4 Central Limit Theorem for the Empirical Measure

27.4.1 Equilibrium Case

Fix $\alpha\in(0,1)$ and take the SEP starting from the invariant state ν_α. Let $k\in\mathbb{N}$ and denote by $\mathcal{H}_k$ the Hilbert space induced by $S(\mathbb{R})$ and $<f,g>_k=<f,(x^2-\Delta)^k g>$, where $<\cdot,\cdot>$ denotes the inner product of $L^2(\mathbb{R})$ and by $\mathcal{H}_{-k}$ the dual of $\mathcal{H}_k$, relatively to this inner product. For a Markov process $\eta.$ define the density fluctuation field acting on functions $H\in S(\mathbb{R})$ as

$$Y_t^N(H)=\frac{1}{\sqrt{N}}\sum_{x\in\mathbb{Z}}H\left(\frac{x}{N}\right)(\eta_t(x)-\alpha).\qquad(27.22)$$

Consider the function below

$$D(\mathbb{R}_+,\{0,1\}^{\mathbb{Z}})\longrightarrow D(\mathbb{R}_+,\mathcal{H}_{-k})$$

$$\eta.\longrightarrow Y^N(\eta.)$$

and let $\mathbb{P}_{\nu_\alpha}^N$ the probability measure on $D(\mathbb{R}_+,\{0,1\}^{\mathbb{Z}})$ induced by ν_α and by the Markov process $\eta.$ speeded up by $t\theta(N)$; Q_N be the probability measure on $D(\mathbb{R}_+,\mathcal{H}_{-k})$ induced by the density fluctuation Y^N and ν_α.

Theorem 27.1. *(Ravishankar [7]) Fix an integer $k>3$. Let $\eta.$ be the SSEP evolving on the parabolic time scale tN^2 starting from ν_α and let Q_N be the probability measure on $D(\mathbb{R}_+,\mathcal{H}_{-k})$ induced by the density fluctuation Y^N and ν_α. Let Q be the probability measure on $C(\mathbb{R}^+,\mathcal{H}_{-k})$ corresponding to a stationary mean zero*

27 Simple Exclusion Process: From Randomness to Determinism   399

generalized Ornstein–Uhlenbeck process with characteristics $\mathfrak{A} = 1/2\Delta$ and $\mathfrak{B} = \sqrt{\chi(\alpha)}$. Then $(Q_N)_N$ converges weakly to Q.

Theorem 27.2. *(G. [4]) Fix an integer $k > 2$. Let $\eta.$ be the ASEP evolving on the hyperbolic time scale tN starting from ν_α and let Q_N be the probability measure on $D(\mathbb{R}_+, \mathscr{H}_{-k})$ induced by the density fluctuation Y^N and ν_α. Let Q be the probability measure on $C(\mathbb{R}^+, \mathscr{H}_{-k})$ corresponding to a stationary Gaussian process with mean 0 and covariance given by*

$$E_Q[Y_t(H)Y_s(G)] = \chi(\alpha) \int_\mathbb{R} H(u + v(t - s))G(u)du \qquad (27.23)$$

for every $0 \le s \le t$ and H, G in $\mathscr{H}_k$. Here $\chi(\alpha) = \mathbf{Var}(\nu_\alpha, \eta(0)) = \alpha(1 - \alpha)$ and $v = (p - q)(1 - 2\alpha)$. Then, $(Q_N)_N$ converges weakly to Q.

In order to complete the exposition I just give a short presentation of the proof. The idea is to verify that $(Q_N)_N$ is tight and to characterize the limit field. The proof of tightness is technical and details can be found in [5]. So we proceed by characterizing the limit field.

We start by the symmetric case. Fix $H \in S(\mathbb{R})$ and note that

$$M_t^{N,H} = Y_t^N(H) - Y_0^N(H) - \int_0^t \frac{1}{2\sqrt{N}} \sum_{x \in \mathbb{Z}} \Delta_N H\left(\frac{x}{N}\right)\eta_s(x)ds \qquad (27.24)$$

$$N_t^{N,H} = (M_t^{N,H})^2 - \int_0^t \frac{1}{N} \sum_{x \in \mathbb{Z}} \left(\nabla^N H\left(\frac{x}{N}\right)\right)^2$$
$$\times \left[c^S(x, x + 1, \eta_s) + c^S(x + 1, x, \eta_s)\right]ds, \qquad (27.25)$$

are martingales with respect to the filtration $\mathscr{F}_t = \sigma(\eta_s, s \le t)$.
Here $c^S(x, x + 1, \eta)$ denotes the jump rate from x to $x + 1$ in η. It is easy to show that

$$\lim_{N \to +\infty} \mathbb{E}_{\nu_\alpha}\left[\int_0^t \frac{1}{N} \sum_{x \in \mathbb{Z}} H\left(\frac{x}{N}\right)\left(c^S(x, x + 1, \eta) - \frac{1}{2}\alpha(1 - \alpha)\right)\right]^2 = 0. \quad (27.26)$$

Then, the limit of the martingale $N_t^{N,H}$ denoted by N_t^H equals to

$$(M_t^H)^2 - ||\mathfrak{B}H||_2^2 t, \qquad (27.27)$$

where M_t^H denotes the limit of the martingale $M_t^{N,H}$ and $\mathfrak{B} = \sqrt{\chi(\alpha)}\nabla$, with $\chi(\alpha) = \alpha(1 - \alpha)$. Note that

$$M_t^H = Y_t(H) - Y_0(H) - \int_0^t Y_s(\mathfrak{A}H)ds \qquad (27.28)$$

where $\mathfrak{A} = \frac{1}{2}\Delta$. For each $H \in S(\mathbb{R})$, $B_t^H = ||\mathfrak{B}H||_2^{-1}M_t^H$ is a martingale whose quadratic variation is equal to t which implies that B_t^H is a Brownian motion. Then

$$Y_t(H) = Y_0(H) - \int_0^t Y_s(\mathfrak{A}H)ds + ||\mathfrak{B}H||_2 B_t^H, \qquad (27.29)$$

which means that Y_t satisfies:

$$dY_t = \frac{1}{2}\Delta Y_t dt + \sqrt{\chi(\alpha)}\nabla dB_t \qquad (27.30)$$

Then, one identifies Y_t as a generalized Ornstein–Uhlenbeck process with characteristics $\mathfrak{A} = \frac{1}{2}\Delta$ and $\mathfrak{B} = \sqrt{\chi(\alpha)}\nabla$.

For the asymmetric case fix as well a function $H \in S(\mathbb{R})$. Then

$$M_t^{N,H} = Y_t^N(H) - Y_0^N(H) - \int_0^t \frac{1}{\sqrt{N}}\sum_{x\in\mathbb{Z}}\nabla^N H\left(\frac{x}{N}\right)W_{x,x+1}^A(\eta_s)ds \quad (27.31)$$

is a martingale with respect to $\tilde{\mathscr{F}}_t = \sigma(\eta_s, s \leq t)$, whose quadratic variation is given by

$$\int_0^t \frac{1}{N^2}\sum_{x\in\mathbb{Z}}\left(\nabla H\left(\frac{x}{N}\right)\right)^2[p\eta(x)(1-\eta(x+1))+q\eta(x+1)(1-\eta(x))]ds. \quad (27.32)$$

Since $\sum_{x\in\mathbb{Z}}\nabla^N H(\frac{x}{N}) = 0$, the integral part of the martingale can be written as:

$$\int_0^t \frac{1}{\sqrt{N}}\sum_{x\in\mathbb{Z}}\nabla^N H\left(\frac{x}{N}\right)\left[W_{x,x+1}^A(\eta_s) - E_{\nu_\alpha}(W_{x,x+1}^A(\eta_s))\right]ds. \qquad (27.33)$$

As we need to write the expression inside last integral in terms of the fluctuation field Y_s^N, we are able to replace $W_{x,x+1}^A(\eta_s) - E_{\nu_\alpha}(W_{x,x+1}^A(\eta_s))$ by $(p-q)\chi'(\alpha)[\eta_s(x) - \alpha]$, with the use of the:

Theorem 27.3. *(G. [4])(Boltzmann–Gibbs Principle) For every local function g, for every $H \in S(\mathbb{R})$ and every $t > 0$,*

$$\lim_{N\to\infty} E_{\nu_\alpha}\left[\int_0^t \frac{1}{\sqrt{N}}\sum_{x\in\mathbb{Z}}H\left(\frac{x}{N}\right)\left\{\tau_x g(\eta_s) - \tilde{g}(\alpha) - \tilde{g}'(\alpha)[\eta_s(x) - \alpha]\right\}ds\right]^2 = 0$$

$$(27.34)$$

where $\tilde{g}(\alpha) = E_{\nu_\alpha}[g(\eta)]$.

Since $\lim_{N\to+\infty} E_{\nu_\alpha}(M_t^{N,H})^2 = 0$ and by the Boltzmann–Gibbs Principle, the limit density field satisfies

27 Simple Exclusion Process: From Randomness to Determinism

$$Y_t(H) = Y_0(H) - \int_0^t Y_s(\mathfrak{C}H)ds, \tag{27.35}$$

where $\mathfrak{C} = v\nabla$ with $v = (p - q)(1 - 2\alpha)$, which in turn means that Y_t satisfies:

$$dY_t = v\nabla Y_t dt. \tag{27.36}$$

In this case we obtain a simple expression for Y_t given by $Y_t(H) = Y_0(T_t H)$ with $T_t H(u) = H(u + vt)$, which is the semigroup associated to $\mathfrak{C}$.

For $t \geq 0$, let $\mathscr{F}_t$ be the σ-algebra on $D([0, T], \mathscr{H}_{-k})$ generated by $Y_s(H)$ for $s \leq t$ and H in $S(\mathbb{R})$. Restricted to $\mathscr{F}_0$, Q is a Gaussian field with covariance given by

$$E_Q(Y_0(G)Y_0(H)) = \chi(\alpha) < G, H > . \tag{27.37}$$

I remark here that if one takes $\alpha = 1/2$ then $v = 0$ and as a consequence $Y_t(H) = Y_0(H)$, which means that there is no temporal evolution of the density fluctuation field, so in order to have some non trivial temporal evolution we have to speed up the process in a longer time scale. It is shown in [4] that until the time scale $tN^{4/3}$ the same behavior is observed. Nevertheless it is conjectured by Spohn in [10] that this same behavior is expected until the time scale $tN^{3/2}$.

27.4.2 Non-Equilibrium Case

Here I start by stating the Central limit theorem for the empirical measure starting from a Bernoulli product measure of varying parameter for the SSEP. Fix a profile $\rho_0 : \mathbb{R} \to [0, 1]$ and denote by $\nu_{\rho_0(\cdot)}$ the product measure on $\{0, 1\}^{\mathbb{Z}}$ such that for a site $x \in \mathbb{Z}$:

$$\nu_{\rho_0(\cdot)}(\eta(x) = 1) = \rho_0(x/N). \tag{27.38}$$

Let $k \in \mathbb{N}$ and define $\mathscr{H}_k$ as above. Let $\rho_t(x) = E_{\nu_{\rho_0(\cdot)}}[\eta_t(x)]$.

Define the density fluctuation field acting on functions $H \in \mathscr{H}_k$ as

$$Y_t^N(H) = \frac{1}{\sqrt{N}} \sum_{x \in \mathbb{Z}} H\left(\frac{x}{N}\right)(\eta_t(x) - \rho_t(x)). \tag{27.39}$$

Let Q_N be the probability measure on $D(\mathbb{R}_+, \mathscr{H}_{-k})$ induced by the density fluctuation Y^N and $\nu_{\rho_0(\cdot)}$. It was shown by Galves, Kipnis and Spohn the following:

Theorem 27.4. *Fix $k \geq 4$. Let $\eta(\cdot)$ be the SSEP evolving on the time scale tN^2 and starting from $\nu_{\rho_0(\cdot)}$. Let Q be the probability measure concentrated on $C(\mathbb{R}^+, \mathscr{H}_{-k})$ corresponding to the Ornstein–Uhlenbeck process Y_t with mean zero and covariance given by*

402 P. Gonçalves

$$E_Q\left[Y_t(H)Y_s(G)\right] = \int_{\mathbb{R}} (T_{t-s}H)G\chi_s du - \int_0^s \int_{\mathbb{R}} (T_{t-r}H)(T_{s-r}G)\{\partial_r \chi_r - \Delta\chi_r\}dudr$$

(27.40)

for $0 \le s < t$ and G, H in $\mathcal{H}_k$. In this expression $(T_t)_t$ denotes the semigroup associated to the Laplacian and χ_s for the function $\chi(s,u) = \rho(s,u)(1 - \rho(s,u))$. Then, the sequence $(Q_N)_N$ converges to Q.

On the other hand the Central Limit Theorem for the empirical measure for the TASEP was shown by Rezakhanlou in [9]. The idea of the proof is to consider the TASEP as a growth model, ie the configuration space consists of functions h: $0 \le h(i+1) - h(i) \le 1$ for all $i \in \mathbb{Z}$. With rate one, each $h(i)$ increases by one unit provided that the resulting configuration does not leave the configuration space; otherwise the growth is suppressed. The Central Limit Theorem is established for $\rho_N(x,t) = \frac{1}{N}h([xN, tN])$. Assuming initially that the probability law of $\rho_N(x,0)$ is the same as $g(x) + \sqrt{1/N} B(x) + o(\sqrt{1/N})$ for a continuous function g (piecewise convex) and a continuous random process $B(\cdot)$, then at later times $\rho_N(x,t)$ can be stochastically represented as $\bar{\rho}(x,t) + \sqrt{1/N}Z(x,t) + o(\sqrt{1/N})$ where $\bar{\rho}$ is the unique solution of the corresponding Hamilton–Jacobi equation and $Z(x,t)$ is a random process that is given by a variational expression involving $B(\cdot)$. For more general initial conditions the problem is still open.

27.5 Superposition of Both Dynamics

In this section I consider a superposition of both dynamics defined above. This process is called Weakly Asymmetric Simple Exclusion (WASEP) and its generator, denoted by $\mathscr{L}_W$, is given by:

$$\mathscr{L}_W = \mathscr{L}_S + \frac{1}{N}\mathscr{L}_A,$$

(27.41)

with $\mathscr{L}_S$ and $\mathscr{L}_A$ defined as above.

Suppose that the asymmetric part of the generator is given with totally asymmetric jumps to the right. Starting this process from a sequence of measures $(\mu_N)_N$ associated to a profile $\rho_0(\cdot)$, under the *parabolic time scale* tN^2,

$$\pi^N_{tN^2} \xrightarrow[N \to +\infty]{} \rho(t,u)du$$

(27.42)

in $\mu^N S^W_N(t)$-probability, where $\rho(t,u)$ is a weak solution of the Burgers equation with viscosity

$$\partial_t \rho(t,u) + \nabla F(\rho(t,u)) = \frac{1}{2}\Delta\rho(t,u)$$

(27.43)

Here S^W_N is the semigroup associated to the generator $\mathscr{L}_W$ and $F(\rho) = \rho(1-\rho)$.

27 Simple Exclusion Process: From Randomness to Determinism

On the other hand for the equilibrium Central Limit theorem for the empirical measure for the process speeded up by tN^2 it follows that the density fluctuation field defined as above, converges to a generalized Ornstein–Uhlenbeck process, ie the limit density fluctuation field is the solution of

$$dY_t = (1 - 2\alpha)\nabla Y_t dt + \frac{1}{2}\Delta Y_t dt + \sqrt{\alpha(1 - \alpha)}\nabla dW_t, \tag{27.44}$$

where W_t is a Brownian motion.

For more general initial conditions I refer the interested reader to [1].

References

1. De Masi, A., Presutti, E., Scacciatelli, E.: The weakly asymmetric simple exclusion process. Ann. Inst. Henri Poincaré. **25**, 1–38 (1989)
2. Ferrari, P., Kipnis, C.: Second class particles in the rarefation fan. Ann. Int. Henri Poincaré. (Sect. B) **31**, 143–154 (1995)
3. Ferrari, P., Kipnis, C., Saada, H.: Microscopic structure of travelling waves in for the simple asymmetric exclusion process. Ann. Probab. **31**(1), 226–244 (1991)
4. Gonçalves, P.: Central limit theorem for a tagged particle in asymmetric simple exclusion, Stoch. Proc. Appl. **118**, 474–502 (2008)
5. Kipnis, C., Landim, C.: Scaling limits of interacting particle systems. Grundlehren der Mathematischen Wissenschaften [Fundamental Principles of Mathematical Sciences], 320. Springer-Verlag, Berlin (1999)
6. Liggett, T.: Interacting Particle Systems. Springer, NY (1985)
7. Ravishankar, K.: Fluctuations from the hydrodynamic limit for the symmetric simple exclusion in $\mathbb{Z}^d$. Stoch. Proc. Appl. **42**, 31–37 (1992)
8. Rezakhanlou, F.: Hydrodynamic limit for attractive particle systems on $\mathbb{Z}^d$. Commun. Math. Physics **140**, 417–448 (1991)
9. Rezakhanlou, F.: A central limit theorem for the asymmetric simple exclusion process. Annales de l'Institute Henri Poincaré (PR -38), 437–464 (2002)
10. Spohn, H.: Large scale Dynamics of Interacting Particles. Springer (1991)

Chapter 28
Universality in PSI20 fluctuations

Rui Gonçalves, Helena Ferreira, and Alberto A. Pinto

Abstract We consider the α re-scaled PSI20 daily index positive returns $r(t)^{\alpha}$ and negative returns $(-r(t))^{\alpha}$ called, after normalization, the α positive and negative fluctuations, respectively. We use the Kolmogorov–Smirnov statistical test as a method to find the values of α that optimize the data collapse of the histogram of the α fluctuations with the truncated Bramwell–Holdsworth–Pinton (BHP) probability density function (pdf) f_{BHP} and the truncated generalized log-normal pdf f_{LN} that best approximates the truncated BHP pdf. The optimal parameters we found are $\alpha_{BHP}^{+} = 0.48$, $\alpha_{BHP}^{-} = 0.46$, $\alpha_{LN}^{+} = 0.50$ and $\alpha_{LN}^{-} = 0.49$. Using the optimal $\alpha's$ we compute the analytical approximations of the pdf of the normalized positive and negative PSI20 index daily returns $r(t)$. Since the BHP probability density function appears in several other dissimilar phenomena, our result reveals a universal feature of the stock exchange markets.

A.A. Pinto (✉)
LIAAD-INESC Porto LA e Departamento de Matemática, Faculdade de Ciências, Universidade do Porto, Rua do Campo Alegre, 687, 4169-007, Porto, Portugal
and
Centro de Matemática e Departamento de Matemática e Aplicações, Escola de Ciências, Universidade do Minho, Campus de Gualtar, 4710-057 Braga, Portugal
e-mail: aapinto@fc.up.pt

R. Gonçalves
LIAAD-INESC Porto LA e Secção de Matemática, Faculdade de Engenharia, Universidade do Porto, R. Dr. Roberto Frias s/n, 4200-465 Porto, Portugal
e-mail: rjasg@fe.up.pt

H. Ferreira
LIAAD-INESC Porto LA, Porto, Portugal
and
Centro de Matemática da Universidade do Minho, Campus de Gualtar, 4710-057 Braga, Portugal
e-mail: helenaisafer@gmail.com

M.M. Peixoto et al. (eds.), *Dynamics, Games and Science I*, Springer Proceedings in Mathematics 1, DOI 10.1007/978-3-642-11456-4_28,
© Springer-Verlag Berlin Heidelberg 2011

28.1 Introduction

The modeling of the time series of stock prices is a main issue in economics and finance and it is of vital importance in the management of large portfolios of stocks [5–7,9,17,20–22,24,26,32,33]. Here we study the PSI20 index (see also [11,12,28]). The PSI20, an acronym of Portuguese Stock Index, is a benchmark stock market index of companies that trade on Euronext Lisbon, the main stock exchange of Portugal. The index tracks the prices of the twenty listings with the largest market capitalization and share turnover in the PSI Geral, the general stock market of the Lisbon exchange. It is one of the main national indices of the pan-European stock exchange group Euronext alongside Brussels (BEL20), Paris (CAC 40) and Amsterdam (AEX). Let $Y(t)$ be the PSI20 index adjusted close value at day t. We define the *PSI20 index daily return* on day t by

$$r(t) = \frac{Y(t) - Y(t-1)}{Y(t-1)}$$

We define the α *re-scaled PSI20 daily index positive returns* $r(t)^{\alpha}$, for $r(t) > 0$, that we call, after normalization, the α *positive fluctuations*. We define the α *re-scaled PSI20 daily index negative returns* $(-r(t))^{\alpha}$, for $r(t) < 0$, that we call, after normalization, the α *negative fluctuations*. We analyze separately the α positive and α negative daily fluctuations that can have different statistical and economic natures due, for instance, to the leverage effects (see, for example, [1, 2, 18, 19, 23]). Our aim is to find the values of α that optimize the data collapse of the histogram of the α positive and negative fluctuations to the Bramwell–Holdsworth–Pinton (truncated BHP) probability density function (pdf) f_{BHP} truncated to the support range of the data (see Appendix A and Bramwell et al. [4]). Our approach is to apply the Kolmogorov–Smirnov (K–S) statistic test as a method to find the values of α that optimize the data collapse. We observe that the P values of the Kolmogorov–Smirnov test vary continuously with α. The highest P values $P_{BHP}^{+} = 0.95\ldots$ and $P_{BHP}^{-} = 0.77\ldots$ of the Kolmogorov–Smirnov test for the positive and negative fluctuations are attained for the α values $\alpha_{BHP}^{+} = 0.48\ldots$ and $\alpha_{BHP}^{-} = 0.46\ldots$, respectively. Using this data collapse we do a change of variable that allows us to compute the analytical approximations

$$f_{BHP,PSI20,+}(x) = 5.71x^{-0.52} f_{BHP}(24.3x^{0.48} - 2.04)$$
$$f_{BHP,PSI20,-}(x) = 4.80x^{-0.54} f_{BHP}(21.0x^{0.46} - 2.01)$$

of the pdf of the normalized positive and negative PSI20 index daily returns in terms of the BHP pdf f_{BHP}. We also find, using the K–S statistic test, the generalized log-normal pdf f_{LN} that best approaches the BHP pdf f_{BHP} in the support range $[-3, 9]$ (see Appendix B and Bramwell et al. [4]). As before, our aim is to find the values of α that optimize the data collapse of the histogram of the α positive and negative fluctuations to the generalized log-normal (truncated LN) pdf f_{LN} truncated to the

28 Universality in PSI20 fluctuations 407

support range of the data (see Appendix B and Bramwell et al. [4]). Again, we apply the K–S statistic test as a method to find the values of α that optimize the data collapse. We observe that the P values of the K–S statistic test vary continuously with α. The highest P values $P_{LN}^{+} = 0.88\ldots$ and $P_{LN}^{-} = 0.85\ldots$ of the Kolmogorov–Smirnov test for the positive and negative fluctuations are attained for the values $\alpha_{LN}^{+} = 0.50\ldots$ and $\alpha_{LN}^{-} = 0.49\ldots$, respectively. Using this data collapse, we do a change of variable that allows us to compute the analytical approximations

$$f_{LN,PSI20,+}(x) = 5.56x^{-0.50} f_{LN}(25.73x^{0.50} - 1.96)$$
$$f_{LN,PSI20,-}(x) = 5.93x^{-0.51} f_{LN}(22.88x^{0.49} - 1.89)$$

of the probability density functions of the normalized positive and negative PSI20 index daily returns in terms of the LN pdf f_{LN}. We observe that

$$P_{LN}^{+} < P_{BHP}^{+} \quad \text{and} \quad P_{LN}^{-} > P_{BHP}^{-}.$$

Similar results have been observed for some stock indices, exchange rates, commodity prices and energy sources (see [11–13, 16, 28]). Since the BHP probability density function appears in several other dissimilar phenomena (see, for example, [8, 10, 14–16, 31]), our result reveals a universal feature of the stock exchange markets. Furthermore, these results lead to the construction of a new qualitative and quantitative econophysics model for the stock market based on the two-dimensional spin model (2dXY) at criticality (see [30]). We also obtain similar results for different indices and different time scales and for other time series like comodity prices, exchange rates and bio-energy prices (see [13,28]). Our results lead to a new stochastic differential equation model for the stock exchange market indices (see [27]) that, in particular, gives a better understanding of the stock exchange crises (see [29]).

28.2 Positive PSI20 Index Daily Returns

Let T^{+} be the set of all days t with positive returns, i.e.

$$T^{+} = \{t : r(t) > 0\}.$$

Let $n^{+} = 1218$ be the cardinal of the set T^{+}. The α re-scaled PSI20 daily index positive returns are the returns $r(t)^{\alpha}$ with $t \in T^{+}$. Since the total number of observed days is $n = 2481$, we obtain that $n^{+}/n = 0.49$. The mean μ_{α}^{+} of the α re-scaled PSI20 daily index positive returns is given by

$$\mu_{\alpha}^{+} = \frac{1}{n^{+}} \sum_{t \in T^{+}} r(t)^{\alpha} \tag{28.1}$$

The *standard deviation* σ_α^+ of the α re-scaled PSI20 daily index positive returns is given by

$$\sigma_\alpha^+ = \sqrt{\frac{1}{n^+} \sum_{t \in T^+} r(t)^{2\alpha} - (\mu_\alpha^+)^2}. \qquad (28.2)$$

We define the α *positive fluctuations* by

$$r_\alpha^+(t) = \frac{r(t)^\alpha - \mu_\alpha^+}{\sigma_\alpha^+} \qquad (28.3)$$

for every $t \in T^+$. Hence, the α *positive fluctuations* are the normalized α re-scaled $PSI20$ daily index positive returns. Let L_α^+ be the *smallest* α positive fluctuation, i.e.

$$L_\alpha^+ = \min_{t \in T^+} \{r_\alpha^+(t)\}.$$

Let R_α^+ be the *largest* α positive fluctuation, i.e.

$$R_\alpha^+ = \max_{t \in T^+} \{r_\alpha^+(t)\}.$$

28.2.1 Data Collapse to a Truncated BHP

We denote by $F_{\alpha,+}$ the *probability distribution of the α positive fluctuations*. Let the *truncated BHP probability distribution* $F_{BHP,\alpha,+}$ be given by

$$F_{BHP,\alpha,+}(x) = \frac{F_{BHP}(x)}{F_{BHP}(R_\alpha^+) - F_{BHP}(L_\alpha^+)}$$

where F_{BHP} is the BHP probability distribution (see Appendix A and Bramwell et al. [4]). We apply the K–S statistic test to the null hypothesis claiming that the probability distributions $F_{\alpha,+}$ and $F_{BHP,\alpha,+}$ are equal. The Kolmogorov–Smirnov P value P_α^+ is plotted in Fig. 28.1. We observe that $\alpha_{BHP}^+ = 0.48\ldots$ is the point where the P value $P_{\alpha_{BHP}^+}^+ = 0.95\ldots$ attains its maximum. We note that

$$\mu_{\alpha_{BHP}^+}^+ = 0.084\ldots, \sigma_{\alpha_{BHP}^+}^+ = 0.041\ldots, L_{\alpha_{BHP}^+}^+ = -1.972\ldots \text{ and } R_{\alpha_{BHP}^+}^+ = 6.071\ldots$$

It is well-known that the Kolmogorov–Smirnov P value P_α^+ decreases with the distance

$$D_{\alpha,+} = \| F_{\alpha,+} - F_{BHP,\alpha,+} \|$$

between $F_{\alpha,+}$ and $F_{BHP,\alpha,+}$. In Fig. 28.1, we plot

$$D_{\alpha_{BHP}^+,+}(x) = \left| F_{\alpha_{BHP}^+,+}(x) - F_{BHP,\alpha_{BHP}^+,+}(x) \right|$$

28 Universality in PSI20 fluctuations

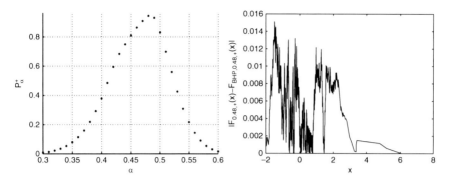

Fig. 28.1 *Left*: The Kolmogorov–Smirnov P value P_α^+ for values of α in the range $[0.3, 0.6]$; *Right*: The map $D_{0.48,+}(x) = |F_{0.48,+}(x) - F_{BHP,0.48,+}(x)|$

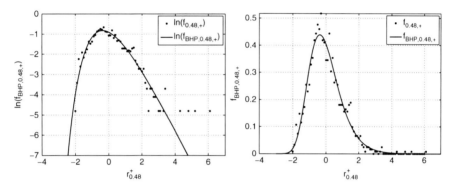

Fig. 28.2 *Left*: The histogram of the α_{BHP}^+ positive fluctuations with the truncated BHP pdf $f_{BHP,0.48,+}$ on top, in the semi-log scale; *Right*: The histogram of the α_{BHP}^+ positive fluctuations with the truncated BHP pdf $f_{BHP,0.48,+}$ on top

and we observe that $D_{\alpha_{BHP}^+,+}(x)$ attains its maximum value 0.0151 for the α^+ positive fluctuations below the mean of the probability distribution. In Fig. 28.2, we show the data collapse of the histogram $f_{\alpha_{BHP}^+,+}$ of the α_{BHP}^+ positive fluctuations to the truncated BHP pdf $f_{BHP,\alpha_{BHP}^+,+}$.

Given that the probability distribution of the α_{BHP}^+ positive fluctuations $r_{\alpha_{BHP}^+}^+(t)$ is approximated by $F_{BHP,\alpha_{BHP}^+,+}$, the pdf of the PSI20 daily index positive returns $r(t)$ is approximated by (see [11])

$$f_{BHP,PSI20,+}(x) = \frac{\alpha_{BHP}^+ x^{\alpha_{BHP}^+ - 1} f_{BHP}\left(\left(x^{\alpha_{BHP}^+} - \mu_{\alpha_{BHP}^+}^+\right)/\sigma_{\alpha_{BHP}^+}^+\right)}{\sigma_{\alpha_{BHP}^+}^+ \left(F_{BHP}\left(R_{\alpha_{BHP}^+}^+\right) - F_{BHP}\left(L_{\alpha_{BHP}^+}^+\right)\right)}.$$

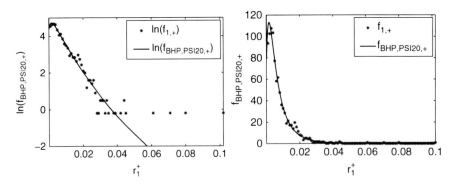

Fig. 28.3 *Left*: The histogram of the fluctuations of the positive returns with the pdf $f_{BHP,PSI20,+}$ on top, in the semi-log scale; *Right*: The histogram of the fluctuations of the positive returns with the pdf $f_{BHP,PSI20,+}$ on top

Hence, we get

$$f_{BHP,PSI20,+}(x) = 5.71 x^{-0.52} f_{BHP}(24.3 x^{0.48} - 2.04).$$

In Fig. 28.3, we show the data collapse of the histogram of the positive returns to our proposed theoretical pdf $f_{BHP,PSI20,+}$.

28.2.2 Data Collapse to a Truncated log-normal

We denote by $F_{\alpha,+}$ the *probability distribution of the α positive fluctuations*. Let the truncated log-normal probability distribution $F_{LN,\alpha,+}$ be given by

$$F_{LN,\alpha,+}(x) = \frac{F_{LN}(x)}{F_{LN}(R_\alpha^+) - F_{LN}(L_\alpha^+)}$$

where F_{LN} is the log-normal probability distribution (see Appendix B and Bramwell et al. [4]). We apply the K–S statistic test to the null hypothesis claiming that the probability distributions $F_{\alpha,+}$ and $F_{LN,\alpha,+}$ are equal. The Kolmogorov–Smirnov P value P_α^+ is plotted in Fig. 28.4. We observe that $\alpha_{LN}^+ = 0.50\ldots$ is the point where the P value $P_{\alpha_{LN}^+}^+ = 0.88\ldots$ attains its maximum. We note that

$$\mu_{\alpha_{LN}^+}^+ = 0.076\ldots, \quad \sigma_{\alpha_{LN}^+}^+ = 0.039\ldots, \quad L_{\alpha_{LN}^+}^+ = -1.905\ldots \text{ and } R_{\alpha_{LN}^+}^+ = 6.258\ldots$$

It is well-known that the Kolmogorov–Smirnov P value P_α^+ decreases with the distance

$$D_{\alpha,+} = \|F_{\alpha,+} - F_{LN,\alpha,+}\|$$

28 Universality in PSI20 fluctuations

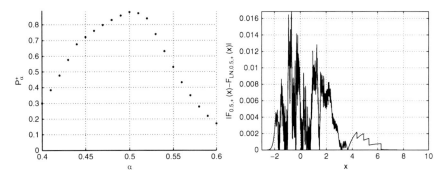

Fig. 28.4 *Left*: The Kolmogorov–Smirnov P value P_α^+ for values of α in the range $[0.4, 0.6]$; *Right*: The map $D_{0.50,+}(x) = |F_{0.50,+}(x) - F_{LN,0.50,+}(x)|$

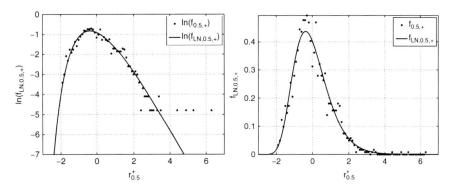

Fig. 28.5 *Left*: The histogram of the α_{LN}^+ positive fluctuations with the truncated log-normal pdf $f_{LN,0.50,+}$ on top, in the semi-log scale; *Right*: The histogram of the α_{LN}^+ positive fluctuations with the truncated log-normal pdf $f_{LN,0.50,+}$ on top

between $F_{\alpha,+}$ and $F_{LN,\alpha,+}$. In Fig. 28.4, we plot

$$D_{\alpha_{LN}^+,+}(x) = \left| F_{\alpha_{LN}^+,+}(x) - F_{LN,\alpha_{LN}^+,+}(x) \right|$$

and we observe that $D_{\alpha_{LN}^+,+}(x)$ attains its maximum value 0.0168 for the α_{LN}^+ positive fluctuations below the mean of the probability distribution. In Fig. 28.5, we show the data collapse of the histogram $f_{\alpha_{LN}^+,+}$ of the α_{LN}^+ positive fluctuations to the truncated LN pdf $f_{LN,\alpha^+,+}$.

Given that the probability distribution of the α_{LN}^+ positive fluctuations $r_{\alpha_{LN}^+}^+(t)$ is approximated by $F_{LN,\alpha_{LN}^+,+}$, the pdf of the PSI20 daily index positive returns $r(t)$ is approximated by (see [11])

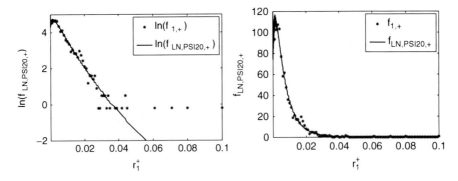

Fig. 28.6 *Left*: The histogram of the fluctuations of the positive returns with the pdf $f_{LN,PSI20,+}$ on top, in the semi-log scale; *Right*: The histogram of the fluctuations of the positive returns with the pdf $f_{LN,PSI20,+}$ on top

$$f_{LN,PSI20,+}(x) = \frac{\alpha_{LN}^+ x^{\alpha_{LN}^+ - 1} f_{LN}\left(\left(x^{\alpha_{LN}^+} - \mu_{\alpha_{LN}^+}^+\right)/\sigma_{\alpha_{LN}^+}^+\right)}{\sigma_{\alpha_{LN}^+}^+ \left(F_{LN}\left(R_{\alpha_{LN}^+}^+\right) - F_{LN}\left(L_{\alpha_{LN}^+}^+\right)\right)}.$$

Hence, we get

$$f_{LN,PSI20,+}(x) = 6.56 x^{-0.50} f_{LN}(25.73 x^{0.50} - 1.96).$$

In Fig. 28.6, we show the data collapse of the histogram of the positive returns to our proposed theoretical pdf $f_{LN,PSI20,+}$.

28.3 Negative PSI20 Index Daily Returns

Let T^- be the set of all days t with negative returns, i.e.

$$T^- = \{t : r(t) < 0\}.$$

Let $n^- = 1158$ be the cardinal of the set T^-. Since the total number of observed days is $n = 2481$, we obtain that $n^-/n = 0.47$. The α *re-scaled PSI20 daily index negative returns* are the returns $(-r(t))^\alpha$ with $t \in T^-$. We note that $-r(t)$ is positive. The *mean* μ_α^- of the α re-scaled PSI20 daily index negative returns is given by

$$\mu_\alpha^- = \frac{1}{n^-} \sum_{t \in T^-} (-r(t))^\alpha \qquad (28.4)$$

28 Universality in PSI20 fluctuations 413

The *standard deviation* σ_α^- of the α re-scaled PSI20 daily index negative returns is given by

$$\sigma_\alpha^- = \sqrt{\frac{1}{n^-} \sum_{t \in T^-} (-r(t))^{2\alpha} - (\mu_\alpha^-)^2}. \tag{28.5}$$

We define the α *negative fluctuations* by

$$r_\alpha^-(t) = \frac{(-r(t))^\alpha - \mu_\alpha^-}{\sigma_\alpha^-} \tag{28.6}$$

for every $t \in T^-$. Hence, the α *negative fluctuations* are the normalized α re-scaled PSI20 daily index negative returns. Let L_α^- be the *smallest* α negative fluctuation, i.e.

$$L_\alpha^- = \min_{t \in T^-} \{r_\alpha^-(t)\}.$$

Let R_α^- be the *largest* α negative fluctuation, i.e.

$$R_\alpha^- = \max_{t \in T^-} \{r_\alpha^-(t)\}.$$

28.3.1 Data Collapse to a Truncated BHP

We denote by $F_{\alpha,-}$ the *probability distribution of the α negative fluctuations*. Let the *truncated BHP probability distribution* $F_{BHP,\alpha,-}$ be given by

$$F_{BHP,\alpha,-}(x) = \frac{F_{BHP}(x)}{F_{BHP}(R_\alpha^-) - F_{BHP}(L_\alpha^-)}$$

where F_{BHP} is the BHP probability distribution. We apply the K–S statistic test to the null hypothesis claiming that the probability distributions $F_{\alpha,-}$ and $F_{BHP,\alpha,-}$ are equal. The Kolmogorov–Smirnov P *value* P_α^- is plotted in Fig. 28.7. Hence, we observe that $\alpha_{BHP}^- = 0.46\ldots$ is the point where the P value $P_{\alpha_{BHP}^-}^- = 0.77\ldots$ attains its maximum. We note that

$$\mu_{\alpha_{BHP}^-}^- = 0.095\ldots \; \sigma_{\alpha_{BHP}^-}^- = 0.048\ldots \; L_{\alpha_{BHP}^-}^- = -1.930\ldots \text{ and } R_{\alpha_{BHP}^-}^- = 5.232\ldots$$

It is well-known that the Kolmogorov–Smirnov P value P_α^- decreases with the distance

$$D_{\alpha,-} = \|F_{\alpha,-} - F_{BHP,\alpha,-}\|$$

between $F_{\alpha,-}$ and $F_{BHP,\alpha,-}$. In Fig. 28.7, we plot

$$D_{\alpha_{BHP}^-,-}(x) = \left| F_{\alpha_{BHP}^-,-}(x) - F_{BHP,\alpha_{BHP}^-,-}(x) \right|$$

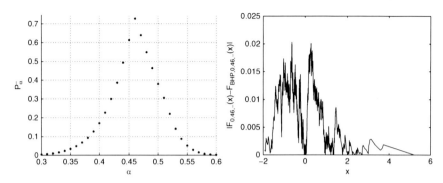

Fig. 28.7 *Left*: The Kolmogorov–Smirnov P value P_α^- for values of α in the range $[0.3, 0.6]$; *Right*: The map $D_{0.46,-}(x) = |F_{0.46,-}(x) - F_{BHP,0.46,-}(x)|$

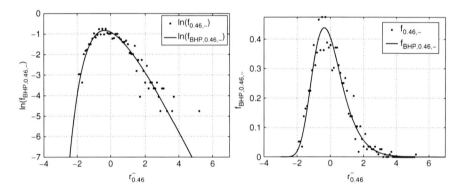

Fig. 28.8 *Left*: The histogram of the α_{BHP}^- negative fluctuations with the truncated BHP pdf $f_{BHP,0.46,-}$ on top, in the semi-log scale; *Right*: The histogram of the α_{BHP}^- negative fluctuations with the truncated BHP pdf $f_{BHP,0.46,-}$ on top

and we observe that $D_{\alpha_{BHP}^-,-}(x)$ attains its maximum value 0.0202 for the α_{BHP}^- negative fluctuations below the mean of the probability distribution. In Fig. 28.8, we show the data collapse of the histogram $f_{\alpha_{BHP}^-,-}$ of the α_{BHP}^- negative fluctuations to the truncated BHP pdf $f_{BHP,\alpha_{BHP}^-,-}$.

Given that the probability distribution of the α_{BHP}^- negative fluctuations $r_{\alpha_{BHP}^-}^-(t)$ is approximated by $F_{BHP,\alpha_{BHP}^-,-}$, the pdf of the *PSI20* daily index (symmetric) negative returns $-r(t)$, with $T \in T^-$, is approximated by (see [11])

$$f_{BHP,PSI20,-}(x) = \frac{\alpha_{BHP}^- x^{\alpha_{BHP}^- 1} f_{BHP}\left(\left(x^{\alpha_{BHP}^-} - \mu_{\alpha_{BHP}^-}^-\right)/\sigma_{\alpha_{BHP}^-}^-\right)}{\sigma_{\alpha_{BHP}^-}^- \left(F_{BHP}\left(R_{\alpha_{BHP}^-}^-\right) - F_{BHP}\left(L_{\alpha_{BHP}^-}^-\right)\right)}.$$

Hence, we get

$$f_{BHP,PSI20,-}(x) = 4.80 x^{-0.54} f_{BHP}(21.0 x^{0.46} - 2.0)$$

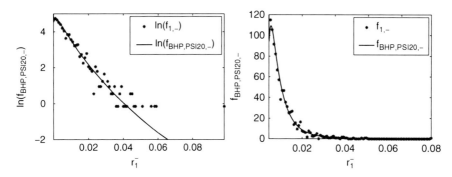

Fig. 28.9 *Left*: The histogram of the negative returns with the pdf $f_{BHP,PSI20,-}$ on top, in the semi-log scale; *Right*: The histogram of the negative returns with the pdf $f_{BHP,PSI20,-}$ on top

In Fig. 28.9, we show the data collapse of the histogram of the negative returns to our proposed theoretical pdf $f_{BHP,PSI20,-}$.

28.3.2 Data Collapse to a Truncated log-normal

We denote by $F_{\alpha,-}$ the *probability distribution of the α negative fluctuations*. Let the *truncated log-normal probability distribution* $F_{LN,\alpha,-}$ be given by

$$F_{LN,\alpha,-}(x) = \frac{F_{LN}(x)}{F_{LN}(R_\alpha^-) - F_{LN}(L_\alpha^-)}$$

where F_{LN} is the log-normal probability distribution. We apply the K–S statistic test to the null hypothesis claiming that the probability distributions $F_{\alpha,-}$ and $F_{LN,\alpha,-}$ are equal. The Kolmogorov–Smirnov P *value* P_α^- is plotted in Fig. 28.10. Hence, we observe that $\alpha_{LN}^- = 0.49\ldots$ is the point where the P value $P_{\alpha_{LN}^-}^- = 0.85\ldots$ attains its maximum. We note that

$$\mu_{\alpha_{LN}^-} = 0.083\ldots \quad \sigma_{\alpha_{LN}^-} = 0.044\ldots \quad L_{\alpha_{LN}^-} = -1.833\ldots \text{ and } R_{\alpha_{LN}^-} = 5.462\ldots$$

It is well-known that the Kolmogorov–Smirnov P value P_α^- decreases with the distance

$$D_{\alpha,-} = \|F_{\alpha,-} - F_{LN,\alpha,-}\|$$

between $F_{\alpha,-}$ and $F_{LN,\alpha,-}$. In Fig. 28.10, we plot

$$D_{\alpha_{LN}^-,-}(x) = \left|F_{\alpha_{LN}^-,-}(x) - F_{LN,\alpha_{LN}^-,-}(x)\right|$$

and we observe that $D_{\alpha_{LN}^-,-}(x)$ attains its maximum value 0.0179 for the α^- negative fluctuations below the mean of the probability distribution. In Fig. 28.11, we

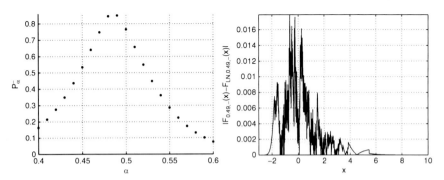

Fig. 28.10 *Left*: The Kolmogorov–Smirnov P value P_α^- for values of α in the range $[0.4, 0.6]$; *Right*: The map $D_{0.49,-}(x) = |F_{0.49,-}(x) - F_{LN,0.49,-}(x)|$

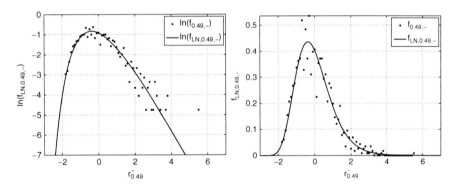

Fig. 28.11 *Left*: The histogram of the α_{LN}^- negative fluctuations with the truncated log-normal pdf $f_{LN,0.49,-}$ on top, in the semi-log scale; *Right*: The histogram of the α_{LN}^- negative fluctuations with the truncated log-normal pdf $f_{LN,0.49,-}$ on top

show the data collapse of the histogram $f_{\alpha_{LN}^-,-}$ of the α_{LN}^- negative fluctuations to the truncated BHP pdf $f_{LN,\alpha_{LN}^-,-}$.

Given that the probability distribution of the α_{LN}^- negative fluctuations $r_{\alpha_{LN}^-}^-(t)$ is approximated by $F_{LN,\alpha_{LN}^-,-}$, the pdf of the *PSI20* daily index (symmetric) negative returns $-r(t)$, with $T \in T^-$, is approximated by (see [11])

$$f_{LN,PSI20,-}(x) = \frac{\alpha_{LN}^- x^{\alpha^--1} f_{LN}\left(\left(x^{\alpha_{LN}^-} - \mu_{\alpha_{LN}^-}^-\right)/\sigma_{\alpha_{LN}^-}^-\right)}{\sigma_{\alpha_{LN}^-}^-\left(F_{LN}\left(R_{\alpha_{LN}^-}^-\right) - F_{LN}\left(L_{\alpha_{LN}^-}^-\right)\right)}.$$

Hence, we get

$$f_{LN,PSI20,-}(x) = 5.93 x^{-0.51} f_{LN}(22.88 x^{0.49} - 1.89)$$

28 Universality in PSI20 fluctuations

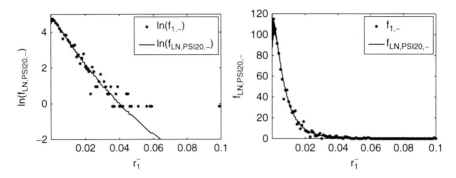

Fig. 28.12 *Left*: The histogram of the fluctuations of the positive returns with the pdf $f_{LN,PSI20,-}$ on top, in the semi-log scale; *Right*: The histogram of the fluctuations of the positive returns with the pdf $f_{LN,PSI20,-}$ on top

In Fig. 28.12, we show the data collapse of the histogram of the negative returns to our proposed theoretical pdf $f_{LN,PSI20,-}$.

28.4 Conclusions

We computed the analytical approximations of the pdf of the normalized PSI20 index daily positive and negative returns in terms of the truncated BHP pdf and of the truncated LN pdf. We showed the data collapse of the histogram of the positive and negative returns to our proposed theoretical pdfs.

28.5 Appendix A: The BHP Pdf

The universal nonparametric BHP pdf was discovered by Bramwell, Holdsworth and Pinton [3]. The *BHP probability density function (pdf)* is given by

$$f_{BHP}(\mu) = \int_{-\infty}^{\infty} \frac{dx}{2\pi} \sqrt{\frac{1}{2N^2} \sum_{k=1}^{N-1} \frac{1}{\lambda_k^2}} e^{ix\mu\sqrt{\frac{1}{2N^2}\sum_{k=1}^{N-1}\frac{1}{\lambda_k^2}}}$$

$$\cdot e^{-\sum_{k=1}^{N-1}\left[\frac{ix}{2N}\frac{1}{\lambda_k} - \frac{i}{2}\arctan\left(\frac{x}{N\lambda_k}\right)\right]} \cdot e^{-\sum_{k=1}^{N-1}\left[\frac{1}{4}\ln\left(1 + \frac{x^2}{N^2\lambda_k^2}\right)\right]}$$

(28.7)

where the $\{\lambda_k\}_{k=1}^{L}$ are the eigenvalues, as determined in [4], of the adjacency matrix. It follows, from the formula of the BHP pdf, that the asymptotic values for large deviations, below and above the mean, are exponential and double

418 · R. Gonçalves et al.

exponential, respectively (in this article, we use the approximation of the BHP pdf obtained by taking $L = 10$ and $N = L^2$ in (28.7)). As we can see, the BHP distribution does not have any parameter, except the mean that is normalize to 0 and the standard deviation that is normalized to 1.

28.6 Appendix B: The Log-Normal Pdf

Let $f(x; \theta, \mu, \sigma)$ be the generalized log-normal pdf

$$f(x; \theta, \mu, \sigma) = \frac{1}{\sigma(x - \theta)\sqrt{2\pi}} \exp(-\frac{1}{2\sigma^2}(\ln(x - \theta) - \mu)^2$$

normalized by

$$E(\theta, \mu, \sigma) = \exp(\mu + \frac{\sigma^2}{2}) - \theta = 0$$
$$Var(\theta, \mu, \sigma) = \exp(\sigma^2 - 1) \exp(2\mu + \sigma^2) = 1.$$

Our aim is to find the values $(\theta^*, \mu^*, \sigma^*)$ that optimize the data collapse of the pdf $f(x; \theta, \mu, \sigma)$ with the BHP pdf $f_{BHP}(x)$ in the support range $[-3, 9]$. We found the optimal values

$$\theta^* = -3.6737, \quad \mu^* = 1.2620 \quad \text{and} \quad \sigma^* = 0.2682$$

that maximize the P value $P^* = 0.99999$ of the Kolmogorov–Smirnov (K–S) statistic test. Hence, we define

$$f_{LN}(x) = f(x; \theta^*, \mu^*, \sigma^*).$$

Bramwell et al. [4], using a different optimizing method, also present a generalized log-normal pdf very close to the BHP pdf f_{BHP} and to the pdf $f_{LN}(x)$ in the support range $[-3, 9]$.

Acknowledgements We thank Nico Stollenwerk, Jason Gallas, Peter Holdsworth, Imre Janosi and Henrik Jensen for showing us the relevance of the Bramwell–Holdsworth–Pinton distribution. This work was presented in PODE09, EURO XXIII, Encontro Ciência 2009, ICDEA2009 and in the International Congress of Mathematicians 2010. This work was highlighted in the article [25] after being presented in ICM 2010. We thank LIAAD-INESC Porto LA, Calouste Gulbenkian Foundation, PRODYN-ESF, FEDER, POFC, POCTI and POSI by FCT and Ministério da Ciência e da Tecnologia, and the FCT Pluriannual Funding Program of the LIAAD-INESC Porto LA and of the Research Centre of Mathematics of the University of Minho, for their financial support.

References

1. Andersen, T.G., Bollerslev, T., Frederiksen, P., Nielse, M.: Continuos-Time Models, Realized Volatilities and Testable Distributional Implications for Daily Stock Returns Preprint (2004)
2. Barnhart, S.W., Giannetti, A.: Negative earnings, positive earnings and stock return predictability: An empirical examination of market timing. J. Empirical Finance **16**, 70–86 (2009)
3. Bramwell, S.T., Holdsworth, P.C.W., Pinton, J.F., Universality of rare fluctuations in turbulence and critical phenomena, *Nature*, **396**, 552–554 (1998)
4. Bramwell, S.T., Fortin, J.Y., Holdsworth, P.C.W., Peysson, S., Pinton, J.F., Portelli, B., Sellitto, M.: Magnetic Fluctuations in the classical XY model: the origin of an exponential tail in a complex system. Phys. Rev E **63**, 041106 (2001)
5. Chowdhury, D., Stauffer, D.: A generalized spin model of financial markets. Eur. Phys. J. **B8**, 477–482 (1999)
6. Cont, R., Potters, M., Bouchaud, J.P., Scaling in stock market data: stable laws and beyond. *Scale Invariance and Beyond*. Eds: B. Dubrulle, F. Graner, D. Sornette. Proceedings of the CNRS Workshop on Scale Invariance, Springer, Berlin (1997)
7. Cutler, D.M., Poterba, J.M., Summers, L.H.: What Moves Stock Prices? J. Portfolio Manag. **15**(3), 4–12 (1989)
8. Dahlstedt, K., Jensen, H.J.: Universal fluctuations and extreme-value statistics. J. Phys. A Math. Gen. **34**, 11193–11200 (2001)
9. Gabaix, X., Parameswaran, G., Plerou, V., Stanley, E.: A theory of power-law distributions in financial markets. Nature **423**, 267–270 (2003)
10. Gonçalves, R., Ferreira, H., Pinto, A.A., Stollenwerk, N.: Universality in nonlinear prediction of complex systems. Special issue in honor of Saber Elaydi. J. Differ. Equ. Appl. **15**(11 & 12, 1067–1076 (2009)
11. Gonçalves, R., Ferreira, H., Pinto, A.A., Stollenwerk, N.: Universality in the Stock Exchange Market. J. Differ. Equ. Appl., **17**(6), 35–39 (2011)
12. Gonçalves, R., Ferreira, H., Stollenwerk, N., Pinto, A.A.: Universal fluctuations of the AEX index, *Physica A: Statistical Mechanics and its Applications* **389**, 4776–4784 (2010)
13. Gonçalves, R., Ferreira, H., Jenkins, S., Pinto, A.A.: Universality in energy sources (submitted)
14. Gonçalves, R., Pinto, A.A.: Negro and Danube are mirror rivers. Special issue Dynamics & Applications in honor of Mauricio Peixoto and David Rand, *Journal of Difference Equations and Applications* **16**(12), 1491–1499 (2010)
15. Gonçalves, R., Pinto, A.A., Stollenwerk, N.: Cycles and universality in sunspot numbers fluctuations. Astrophys. J. **691**, 1583–1586 (2009)
16. Gonçalves, R., Pinto, A.A.: Universality in the stock exchange. arXiv:0810.2508v1 [q-fin.ST] (2008)
17. Gopikrishnan, P., Meyer, M., Amaral, L., Stanley, H.: Inverse cubic law for the distribution of stock price variations. Eur. Phys. J. B **3**, 139–140 (1998)
18. Kantz, H., Schreiber, T.: Nonlinear Time Series Analysis. Cambridge University Press, Cambridge (1997)
19. Landau, L.D., Lifshitz, E.M.: *Statistical Physics*, Vol. 1, Oxford, Pergamon Press (1980)
20. Lillo, F., Mantegna, R.: Statistical properties of statistical ensembles of stock returns. Int. J. Theor. Appl. Finance **3**, 405–408 (2000)
21. Lillo, F., Mantegna, R.: Ensemble Properties of securities traded in the Nasdaq market. Physica A **299**, 161–167 (2001)
22. Mandelbrot, B.: The variation of certain speculative prices. J. Bus. **36**, 392–417 (1963)
23. Mandelbrot, B.: Fractals and Scaling in Finance. Springer, Berlin (1997)
24. Mantegna, R., Stanley, E.: Scaling behaviour in the dynamics of a economic index. Nature **376**, 46–49 (2001)
25. Mudur, G.S.: Maths for movies, medicine & markets. The Telegraph Calcutta, India, 20/09/2010
26. Pagan, A.: The econometrics of financial markets. J. Emp. Finance **3**, 15–102 (1996)

27. Pinto, A.A.: A stochastic differential equation model for the stock market indices (in preparation)
28. Pinto, A.A., Gonçalves, R., Ferreira, H.: Universal fluctuations of the Dow Jones (submitted)
29. Pinto, A.A.: Stock exchange crises (in preparation)
30. Pinto, A.A.: A qualitative and quantitative Econophysics stock market model (submitted)
31. Pinto, A.A.: Game theory and duopoly models, Interdisciplinary Applied Mathematics, vol. 1-237. Springer, New York (2011)
32. Plerou, V., Amaral, L., Gopikrishnan, P., Meyer, M., Stanley, E.: Universal and nonuniversal properties of cross correlations in financial time series, Phys. Rev. Lett. **83**(7), 1471–1474 (1999)
33. Stanley, H., Plerou, V., Gabix, X.: A Statistical physics view of financial fluctuations: Evidence for scaling and universality. Physica A **387**, 3967–3981 (2008)

Chapter 29
Dynamical Systems with Nontrivially Recurrent Invariant Manifolds

Viacheslav Grines and Evgeny Zhuzhoma

Abstract The goal of this article to give exposition of results demonstrating deep interrelation between topological classification of Dynamical Systems with nontrivially recurrent invariant manifolds and topological classification of standard objects existing on ambient manifold. One can see how the purely topological constructions, very pathological at first glance, appear naturally in Dynamical Systems.

29.1 Topological Classification Flows, Foliations and Two-Webs by Means Geodesic Laminations on Hyperbolic Surfaces

29.1.1 Introduction to the Method

Historical remarks. The idea to study two-dimensional dynamical systems and surface foliations applying nonlocal asymptotic properties of orbits and leaves is due to A. Weil and D.V. Anosov (see also the historical comments in [6]–[13, 24, 114]). In the 1960s, D.V. Anosov put forth the concept that the key to the classification of dynamical systems and foliations on M^2 is a study of arrangement of "infinite" simple curves on M^2 and of the asymptotic behavior of lifts of these curves to the universal covering plane Δ with the use of the absolute S_∞. Especially this approaching turned up effective for dynamical systems with nontrivially recurrent motions and nontrivially recurrent invariant manifolds (the most known of such dynamical systems are pseudo-Anosov homeomorphisms, Anosov and

---

V. Grines (✉)
Nizhny Novgorod State University, 23 Gagarin Ave, Nizhny Novgorod, 603950, Russia
e-mail: vgrines@yandex.ru

E. Zhuzhoma
Nizhny Novgorod State Pedagogical University, 1 Ulyanova Str., Nizhny Novgorod, 603600, Russia
e-mail: zhuzhoma@mail.ru

M.M. Peixoto et al. (eds.), *Dynamics, Games and Science I*, Springer Proceedings in Mathematics 1, DOI 10.1007/978-3-642-11456-4_29,
© Springer-Verlag Berlin Heidelberg 2011

DA diffeomorphisms), and foliations with nontrivially recurrent leaves, see [16]–[23, 63, 64]. Such approach sometimes is called the Anosov–Weil Theory which generally considers asymptotic properties of simple curves lifted to an universal covering, and their "deviation" from the lines of constant geodesic curvature that have the same asymptotic direction.

Aranson and Grines [19] and Markley [99] was first who fruitfully applied properties of the hyperbolic (Lobachevsky) geometry to prove that a nontrivially recurrent trajectory of any flow on M^2 has a co-asymptotic geodesic. As a consequence, given any quasiminimal set that contains such a trajectory, one can construct a special geodesic lamination, a geodesic framework. This geodesic framework contains all information about a global topological structure of the quasiminimal set. Levitt [94] used similar geodesic laminations to get the Whitehead classification of surface foliations.

The main concepts. Definitions of dynamical systems, foliations and 2-webs require only the existence of differential structures on supporting manifolds. These differential structures are usually enough, if we consider just local properties of orbits or leaves. But if we study nonlocal properties, we often apply additional structures (for example, algebraical, geometrical, etc.). Here, we use geometrical structures to construct special geodesic laminations, so-called geodesic frameworks.

Recall that geodesic laminations were introduced by Thurston [132, 133] to provide a completion for the space of simple closed curves on M. Ever after, they occur in various problems in low-dimensional topology and geometry as a successful tool to attack these problems. But mainly one considers geodesic laminations endowed with the additional structure of a transverse measure (measured laminations). Here, we apply geodesic laminations without a preferred transverse measure just to obtain a significant topological and dynamical information about surface dynamical systems (with nontrivially recurrent orbits and invariant manifolds), and foliations (with nontrivially recurrent leaves).

To consider a nonlocal asymptotic behavior of orbits, invariant manifolds or leaves, one has to lift these objects to a universal covering space to look a limit set "at infinity". Let us give the more precise definitions. To simplify matters, we restrict ourselves by closed orientable hyperbolic surfaces. Recall that a *hyperbolic surface* $M^2 = M$ is a Riemannian 2-manifold whose universal covering space is the hyperbolic (Lobachevsky) plane, which we'll consider as the unit disk $\Delta = \{z \in \mathbb{C} : |z| < 1\}$ endowed with the Poincare metric of the constant curvature -1. The circle $S_\infty = \partial \Delta = (|z| = 1)$ is called a *circle at infinity* or *absolute*. It is known that a given closed orientable hyperbolic surface M^2, there exists a Fuchsian group Γ of orientation-preserving isometries acting freely on Δ such that $\Delta/\Gamma \cong M^2$. The natural projection $\pi : \Delta \to \Delta/\Gamma$ is a universal covering map which induces a Riemannian structure on M^2. Geodesics of Δ are the circular arcs orthogonal to S_∞ (we suppose that any geodesic is complete with the ideal endpoints in S_∞).

To explain how geodesics and geodesic laminations appear, let us give a formal definition of asymptotic direction for a simple curve. A curve l is *semi-infinite*, if it is an image of $[0; \infty)$ under continuous injective map $[0; \infty) \to M$ that is called

29 Dynamical Systems with Nontrivially Recurrent Invariant Manifolds

a parametrization of the curve. Thus, any semi-infinite curve is endowed with an injective parametrization $[0; \infty) \to l, t \to l(t)$. A curve is *simple*, is it has no self-intersections.

Let $l = \{l(t), t \geq 0\}$ be a semi-infinite simple curve on M, and let $\bar{l}$ be its lifting to Δ. Suppose that $\bar{l}$ tends to precisely one point $\sigma \in S_\infty$ as $t \to \infty$ in the Euclidean metric on the closed disk $\Delta \cup S_\infty$. In this case, we shall say that the curve $\bar{l}$ has an *asymptotic direction* determined by the point σ (we also shall sometimes say that l has an asymptotic direction, and the point σ is *reached* by the curve $\bar{l}$).

Now let $l = \{l(t), t \in R\}$ be an infinite simple curve on M, and let $\bar{l}$ be its lifting to Δ. Here we assume that l is endowed with an injective parametrization $(-\infty; +\infty) \to l$. Suppose that $\bar{l}$ has the asymptotic directions determined by the points σ^+ and σ^- as $t \to +\infty$ and $t \to -\infty$ respectively. If $\sigma^+ \neq \sigma^-$, there exists the geodesic $\bar{g}(\bar{l})$ with the ideal endpoints σ^+, σ^- oriented from σ^- to σ^+. This geodesic $\bar{g}(\bar{l})$ is said to be *co-asymptotic* for $\bar{l}$. The geodesic $\pi(\bar{g}(\bar{l})) = g(l)$ is said to be *co-asymptotic* for l. It can be shown that $g(l)$ has no (transversal) self-intersections. Hence the topological closure of $g(l)$ is a geodesic lamination [50].

Here, we represent many old and some new results on surface dynamical systems and foliations from a "geodesic" point of view. The most results we revisit here are reformulated in a form different from original one. We suggest that this representation based on a purely geometrical object opens new investigations in the theory of surface dynamical systems and foliations. Now we give main definitions.

Rational and irrational points. Let $\Delta/\Gamma \cong M$ be a hyperbolic orientable surface. The group Γ consists of linear-fractional maps that homeomorphically transform the closed disk $\Delta \cup S_\infty$ onto itself. Since M is closed, every isometry $\gamma \in \Gamma$ is a hyperbolic transformation having two fixed points $\gamma^+, \gamma^- \in S_\infty$. A point $\sigma \in S_\infty$ is called *rational* if $\sigma = \gamma^\pm$ for some $\gamma \in \Gamma, \gamma \neq id$. Any point of the set $IR = S_\infty - \cup_{\gamma \in \Gamma}\{\gamma^+, \gamma^-\}$ is called *irrational*.

Local laminations. The motivation for the definition of local lamination is the theorem of Ordinary Differential Equations that says that trajectories of smooth differential equation locally looks like parallel straight lines beyond of singularities. For simplicity, we give the definition of a local lamination for a surface M^2 without boundary, $\partial M^2 = \emptyset$. As usual, one assume that the Euclidean plane R^2 is equipped with Cartesian coordinates (x, y). By a C^0 diffeomorphism we mean a homeomorphism. Fix integers number $0 \leq l \leq r \leq \infty$.

Let $\mathscr{M} \subset M^2$ be a subset of M^2 (which may coincide with M^2) that contains some closed subset $S \subset \mathscr{M}$. Suppose $\mathscr{M}$ is a union $S \bigcup_\alpha L_\alpha$, where L_α are pairwise disjoint C^r-smooth simple curves (α runs through some set of indices). We say that the family $\{L_\alpha\}$ forms a $C^{r,l}$ *local lamination* if, for any point $P \in \mathscr{M} - S$, there exist a neighborhood $U(P)$ of P, and a C^l diffeomorphism $\psi : U(P) \to R^2, \psi(P) = (0,0)$, such that any connected component of the intersection $U(P) \cap L_\alpha$ (provided that this intersection is nonempty) is mapped by ψ onto the line $y = const$ and the restriction $\psi|_{U(P) \cap L_\alpha}$ is a C^r diffeomorphism onto its image, Fig. 29.1. Roughly speaking, $\{L_\alpha\}$ is a family pairwise disjoint simple curves locally homeomorphic to a family of parallel straight lines. We call the family $\{L_\alpha\} = \mathscr{D}$ a $C^{r,l}$ local lamination with the set of singularities $S \overset{\text{def}}{=} Sing(\mathscr{D})$.

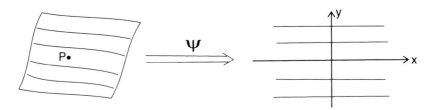

Fig. 29.1

The set $\mathcal{M} = \left(\bigcup_\alpha L_\alpha\right) \cup S$ is called a *support of the local lamination* $\mathcal{D}$. The curves L_α are called *leaves*. Each point of the set $Sing\,(\mathcal{D})$ is called a *singularity*. A point that is not a singularity is called *regular*.

The neighborhoods $U(P)$ where $P \in \mathcal{M} - Sing\,(\mathcal{D})$ are called *neighborhoods with the structure of a linear local lamination*, and the diffeomorphisms ψ are called *rectifying diffeomorphisms*. The pre-image $\psi^{-1}([-1;+1] \times [-1;+1])$ is called a *closed trivially foliated box*. The interior of this box is an *open trivially foliated box*. Actually, foliated boxes are neighborhoods with the structure of a linear local lamination, but they are bounded in part by transversal segments. Given any leaf L_α, a connected component of the intersection of L_α with an open trivially foliated box is called a *local leaf*. Local leaves form a base of topology on each L_α. We call this topology the *intrinsic (or interior) topology of the leaf* L_α. Taking in mind this topology, we speak on compactness of a leaf or that a leaf is homeomorphic to some 1-dimensional manifold, for example, R, S^1 and so on.

Let Σ be a segment (the image of the unit interval $[0;1]$ under an embedding of $[0;1]$ into M^2) through the regular point P. If there is the rectifying diffeomorphisms $\psi: U(P) \to R^2$ such that $\Sigma \subset U(P)$ and ψ maps Σ into the line $x = 0$, then Σ is called *locally transversal segment* at P. The segment Σ is called *transversal* if it is locally transversal at each point of $\mathcal{M} \cap \Sigma$. A closed simple curve C is called a *closed transversal* if every arc-wise part of C is a transversal segment. A local lamination is of *Cantor type* if $\mathcal{M} \cap \Sigma$ is a Cantor set on Σ.

The concept of a local lamination generalizes the classical concepts of lamination and foliation. If $\mathcal{M}$ is closed and $Sing\,(\mathcal{D}) = \emptyset$, then $\mathcal{D}$ is called a $C^{r,l}$ *lamination*. An important example of a lamination is a geodesic lamination. Note that a local $C^{r,l}$ lamination without singularities is not always a lamination. If $\mathcal{M} = M^2$, then $\mathcal{D}$ is called a $C^{r,l}$ *foliation*. One may say that a local lamination with singularities is a "foliation" (with singularities) on a some subset. If this subset is closed and there are no singularities, then we obtain a lamination. If this subset coincides with a manifold (and there may be some singularities), then the local lamination is a foliation. It follows from the aforesaid that the concept of a local lamination is a quite general concept, which includes, as particular cases, the concepts of lamination and foliation.

Geodesic laminations. A *geodesic lamination* is a nonempty collection of mutually disjoint *simple* (i.e. with no transversal self-intersections) geodesics whose union is a closed subset of M. Denote by $\mathscr{L}(M) = \mathscr{L}$ the set of geodesic laminations on M. A simplest geodesic lamination is any union of simple pairwise disjoint closed geodesics. A few complicated example of a geodesic lamination one gets by adding to a simplest geodesic lamination a finite collection of non-closed geodesics that spirally tend (in both directions) to closed geodesics. Such a lamination is called a *trivial geodesic lamination*. Note that the non-closed geodesics belonging to trivial geodesic laminations are isolated i.e., any point on such a geodesic has a neighborhood that intersects with the geodesic lamination only along a unique arc of the geodesic containing this point. Let us denote the family of trivial geodesic laminations by Λ_{triv}. Obviously every lift of a geodesic from any trivial geodesic lamination has rational ideal endpoints. Thus, it is natural to call a geodesic lamination *nontrivial* if it contains a non-closed geodesic that is non-isolated in the geodesic lamination. Any nontrivial geodesic lamination contains a continual set of non-closed geodesics each of which is nontrivially recurrent (self-limiting) i.e., the intrinsic topology on the geodesic does not coincide with the topology induced by the topology of surface (see, for example, [18, 50] which contain a proof). To construct an example of a nontrivial geodesic lamination one can take a simple nontrivially recurrent geodesic g then the topological closure $clos\ (g)$ will be a nontrivial geodesic lamination. However by definition, a nontrivial geodesic lamination can contain, in general, closed and, as well as, isolated geodesics. A lamination is said to be *strongly nontrivial* if it consists of non-isolated geodesics. Every geodesic of such lamination is nontrivially recurrent. Nontrivially recurrent geodesic has an irrational asymptotic direction i.e., every lift of a geodesic from a strongly nontrivial geodesic lamination has irrational ideal endpoints (see for, example, [12, 18, 50]). A lamination is *minimal* if it contains no proper sub-laminations.

Denote by Λ set of all strongly nontrivial and minimal geodesic lamination and for $G \in \Lambda$ apply abbreviation **sntm** geodesic lamination. By definition each leaf of any **sntm** geodesic lamination G is non closed recurrent and dense in G. A lamination G on M is said to be *irreducible* if any closed geodesic on M intersects G.

If $G \in \Lambda$ is irreducible then we call it *irrational* and denote by $\Lambda^{irr} \subset \Lambda$ set of all irrational lamination. It is easy to understand that for any irrational lamination $G \in \Lambda^{irr}$ the compliment $M \setminus G$ consists of finite many domains such that each of them is an immersion of open disk.

For $G \in \Lambda$ choose and fix some orientation on every geodesic from G. This orientations are said to be *compatible* if, for any geodesic $l \in G$ and any point $m \in l$, there exists a transversal segment Σ through m endowed with a normal orientation such that the intersection indices of all geodesics from G (intersecting Σ) with Σ are equal. A geodesic lamination is called *orientable* (*non-orientable* if its geodesics admit (do not admit) compatible orientations. Denote by Λ_{or} (Λ_{non}) the set of orientable (respectively, non-orientable) **sntm** geodesic laminations from Λ then $\Lambda = \Lambda_{or} \cup \Lambda_{non}$.

Let G be a geodesic lamination on M. Clearly, the pre-image $\pi^{-1}(G) = \overline{G}$ is a local geodesic lamination on Δ. Denote by $G(\infty) \subset S_\infty$ the set of points reached by the lamination $\overline{G}$. In other words, $G(\infty)$ is the set of ideal endpoints of all geodesics from $\overline{G}$. If $\mathscr{L}$ is a family of all geodesic laminations on M, $\mathscr{L}(\infty)$ denotes the union of all sets $G(\infty)$ where $G \in \mathscr{L}$.

Generalized mapping class group GM is the quotient *Homeo* $(M)/Homeo_0 (M)$, where *Homeo* (M) is the group of homeomorphisms of M and *Homeo$_0$* (M) is the subgroup of homeomorphisms homotopic to the identity. Given $f \in Homeo (M)$, denote by $[f] \in GM$ the image of f under the natural projection *Homeo* $(M) \rightarrow$ *Homeo* $(M)/Homeo_0 (M)$. It is known that any homeomorphism $f : M \rightarrow M$ induces a one-to-one map $f_* : \mathscr{L} \rightarrow \mathscr{L}$, and any $g \in [f]$ induced the same map $g_* = f_*$ [36,50]. Given $\lambda \in \mathscr{L}$, the family $GM(\lambda) = \{ f_*(\lambda), \ f_* \in GM \}$ is called an *orbit of the geodesic lamination* λ.

Surface foliations and flows. By a *foliation* $\mathscr{F}$ with a set of singularities *Sing* $(\mathscr{F})$ on a surface M we mean a decomposition of $M - Sing (\mathscr{F})$ into pairwise disjoint simple curves l_α locally homeomorphic to a family of parallel straight lines. Any curve l_α is called a *leaf*. Any point of *Sing* $(\mathscr{F})$ is called a *singularity*. Let l be a nonclosed leaf of a foliation $\mathscr{F}$. Any point $x \in l$ divides l into two semileaves, say l^+ and l^-. A semileaf $l^{(\cdot)}$ is called *nontrivially recurrent* if its intrinsic topology does not coincide with the topology of $l^{(\cdot)}$ as a subset of M. A leaf l is said to be *nontrivially recurrent* if both its semileaves are nontrivially recurrent. The topological closure of a nontrivially recurrent semileaf is called a *quasiminimal set*. A foliation is (topologically) transitive if it has a leaf that is dense in M. Every isolated singularity of transitive foliation has at least one separatrix [37].

Following [59], we'll call $\mathscr{F}$ *highly transitive* if the set *Sing* $(\mathscr{F})$ is finite and every leaf of $\mathscr{F}$ is dense in M. The definition of highly transitive foliation admits the existence of so-called fake saddles, i.e. saddles with only two saddle sectors. Clearly that the existence and number of fake saddles does not connect with the topology of the surface M and any fake saddle is an artificial thing. A highly transitive foliation is called *weakly irrational* if it has no fake saddles. A highly transitive foliation $\mathscr{F}$ is said to be *irrational* if $\mathscr{F}$ has no fake saddle and thorns.

Let f^t be a flow on M meaning that $f^t : M \times R \rightarrow M$ is a one-parameter group of homeomorphisms f^t of M. Denote by $l(m) = l$ a trajectory through a point $m \in M$ and by $fix (f^t)$ a set of all fixed points of f^t, where m is a *fixed point* if $l(m) = m$. Due to the local structure of a flow in a neighborhood of regular (i.e., non-fixed) point, the trajectories of f^t form the foliation $\mathscr{F}$ with *Sing* $(\mathscr{F}) = fix (f^t)$. If a given foliation, there is such a flow, the foliation is called *orientable*. In this context, a flow can be considered as an orientable foliation.

Existence of co-asymptotic geodesics. Let l be a nontrivially recurrent trajectory of a flow f^t. Aranson and Grines [19] proved that there exists the co-asymptotic geodesic $g(l)$ showing that the both positive and negative semitrajectories of l have asymptotic directions and this directions are different (i.e. $\alpha(\overline{l}) \neq \omega(\overline{l})$). We give a schematic proof of this fundamental result to demonstrate methods of Hyperbolic Geometry. Since l is a nontrivially recurrent trajectory, there exists a simple closed transversal C such that l intersects C at set of parameter values which is unbounded both above and below. Then $\overline{l}$ intersects transversally the sequence of curves $\overline{C}_1, \ldots,$ $\overline{C}_n, \ldots \in \pi^{-1}(C)$ as $t \rightarrow +\infty$. Since the group Γ is discontinuous, the properties of the hyperbolic plane Δ imply that the topological limit of the sequence $\overline{C}_n$ is a

29 Dynamical Systems with Nontrivially Recurrent Invariant Manifolds   427

unique point, say $\sigma \in S_\infty$. Hence, $\omega(\bar{l}) = \sigma$. Similarly, $\alpha(\bar{l}) \in S_\infty$. Since C is a transversal, $\alpha(\bar{l}) \neq \omega(\bar{l})$.

This Aranson–Grines's result can be generalized as follows. Let l be an infinite simple curve on M that intersects transversally some closed simple curve C infinitely many times. Suppose that there is no loop that is homotopic to zero and formed by an arc of l and arc of C. Then l has the co-asymptotic geodesic $g(l)$ [12]. In particular, any leaf that is not a separatrix of an irrational foliation has a co-asymptotic geodesic. Anosov [6] obtained the following sufficient condition for the existence of asymptotic direction of a semitrajectory, which is the most general condition up now. Let $\bar{l}$ be a lift on $\overline{M}$ of a semitrajectory l of a flow f^t on a closed surface of non-positive Euler characteristic M. Suppose that the set $\mathit{fix}\,(f^t)$ of fixed points is contractible (i.e. there is a continuous map $\varphi : M \times [0, 1] \to M$ such that $\varphi(\cdot, 0) = id$ and $\varphi(\mathit{fix}\,(f^t), 1) = m_0$, where m_0 is some point of M). Then $\bar{l}$ is either bounded or has an asymptotic direction. Anosov [6] proved the last assertion even for an arbitrary set $\mathit{fix}\,(f^t)$ provided that f^t is analytic.

Geodesic frameworks of quasiminimal sets. First, we give the definition of a geodesic framework for a quasiminimal set of a flow. Let Q be a quasiminimal set of a flow f^t. According to [51] the set Q contains continuum of nontrivially recurrent trajectories. Let l be a nontrivially recurrent trajectory from Q. As proved above, there is the co-asymptotic geodesic $g(l)$. One can prove that $g(l)$ has no self-intersections. Therefore, the topological closure $clos\,[g(l)]$ of $g(l)$ is a geodesic lamination (see [18, 50]). This geodesic lamination is independent of the choice of l since, due to the classical Maier's paper [96], any nontrivially recurrent trajectory in the quasiminimal set Q is dense in Q (see the modern proof in [15] and some generalizations in [31]). So the following definition is well defined. The geodesic lamination $clos\,[g(l)] = G(Q)$ is called a *geodesic framework* of Q. One can prove that $G(Q) \in \Lambda^{or}$ that is $G(Q)$ is strongly nontrivial minimal and oriented geodesic lamination. If f^t is transitive, then $Q = M$. In this case, $G(M) = G(f^t)$ is called a *geodesic framework of the flow f^t*.

A similar definition of geodesic framework holds for a quasiminimal set Q of a foliation provided that some nontrivially recurrent leaf from Q has a co-asymptotic geodesic and every nontrivially recurrent leaf from Q is dense in Q. Sufficient conditions of this are in [16, 29, 30]. One can prove that an irrational foliation has a nonempty geodesic framework G, which is an irrational (not necessary oriented) geodesic lamination, $G \in \Lambda^{irr}$.

29.1.2 Foliations and Two-Webs on Hyperbolic Surfaces

Here, we represent some results on topological classification of irrational foliations, nontrivial minimal sets of flows and irrational 2-webs on a hyperbolic orientable closed surface.

29.1.2.1 Irrational Foliations

Recall that two foliations $\mathcal{F}_1$, $\mathcal{F}_2$ on a surface M are *topologically equivalent* if there exists a homeomorphism $h : M \to M$ such that $h(Sing\ (\mathcal{F}_1)) = Sing\ (\mathcal{F}_2)$ and h sends the leaves of $\mathcal{F}_1$ into the leaves of $\mathcal{F}_2$. It is impossible to classify all surface foliations. But if we restrict ourselves to special classes, this problem could be manageable. In general, the classification assumes the following (independent) steps.

1. Find a constructive topological invariant which takes the same values for topologically equivalent foliations.
2. Describe all topological invariants which are admissible, i.e. may be realized in the chosen class of foliations.
3. Find a standard representative in each equivalence class, i.e. given any admissible invariant, one constructs a foliation whose invariant is the admissible one.

An invariant is called *complete* if it takes the same value if and only if two foliations are topologically equivalent. The 'if' part only gives a *relative* invariant.

Invariants fall into three major classes: homology (or cohomology), homotopy, and combinatorial. Poincaré rotation number is most familiar, which carries an interesting arithmetic information, being at the same time homology and homotopy invariant. Combinatorial invariants (exm., Peixoto and Conley-Lyapunov graphs) are good for description of flows without nontrivially recurrent trajectories. Homology and homotopy invariants (exm., fundamental class of Katok [90] and homotopy rotation class of Aranson–Grines [19] respectively) are convenient for description of flows with nontrivially recurrent trajectories. A homotopy invariant that is most related to the Riemannian structure of surface is a geodesic framework. In terms of the geodesic frameworks we can reformulate the Aranson–Grines [19] classification of irrational flows as follows.

Theorem 29.1. *Let f_1^t, f_2^t be two irrational flows on a closed orientable hyperbolic surface M. Then f_1^t, f_2^t are topologically equivalent via a homeomorphism $M \to M$ homotopic to identity if and only if their geodesic frameworks coincide, $G(f_1^t) = G(f_2^t)$.*

Theorem 29.2. *Let f^t be an irrational flow on a closed orientable hyperbolic surface M. Then its geodesic framework $G(f^t)$ is an orientable irrational geodesic lamination, $G(f^t) \in \Lambda_{or} \cap \Lambda^{irr}$.*

Theorem 29.3. *Given any orientable irrational geodesic lamination G on a closed orientable hyperbolic surface M, there is an irrational flow f^t on M such that $G(f^t) = G$.*

Due to Nielsen [110, 111], we see that an irrational orientable geodesic framework is a complete invariant up to the action of the generalized mapping class group GM for irrational flows. Thus an irrational orientable geodesic framework is similar to the Poincare irrational rotation number which is a complete invariant (up to the recalculation with the unimodular integer matrices) for minimal torus flows. Below,

29 Dynamical Systems with Nontrivially Recurrent Invariant Manifolds   429

we'll see that this similarity keeps for the continuity of irrational rotation number under a perturbations of a flow.

Remark that the same results is true for closed non-orientable surfaces of genus ≥ 4 [26]. The similar theorems take place for irrational foliations but one omits the orientability of geodesic framework.

Theorem 29.4. *Let $\mathscr{F}_1$, $\mathscr{F}_2$ be two irrational foliations on a closed orientable hyperbolic surface M. Then $\mathscr{F}_1$, $\mathscr{F}_2$ are topologically equivalent via a homeomorphism $M \to M$ homotopic to identity if and only if their geodesic frameworks coincide, $G(\mathscr{F}_1) = G(\mathscr{F}_2)$.*

Theorem 29.5. *Let $\mathscr{F}$ be a irrational foliation on a closed orientable hyperbolic surface M. Then its geodesic framework $G(\mathscr{F})$ is irrational, $G(\mathscr{F}) \in \Lambda^{irr}$.*

Theorem 29.6. *Given any irrational geodesic lamination G on a closed orientable hyperbolic surface M, there is an irrational foliation $\mathscr{F}$ on M such that $G(\mathscr{F}) = G$.*

Thus, an orbit of irrational geodesic framework is a complete invariant for the class of irrational foliations.

29.1.2.2 Nontrivial Minimal Sets

Let us consider the Aranson–Grines [20] classification of minimal nontrivial sets. Recall that a minimal set of a flow is called *nontrivial (exceptional)* if it is neither a fixed point, nor a closed trajectory, nor the whole surface M. An exceptional minimal set is nowhere dense and consists of continuum nontrivially recurrent trajectories, each being dense in the minimal set. Moreover, an exceptional minimal set is locally homeomorphic to the product of the Cantor set and a segment. The most familiar flow with an exceptional minimal set is the Denjoy flow (first constructed by Poincare [124]) on the torus T^2.

Two minimal sets N_1, N_2 of the flows f_1^t, f_2^t respectively are *topologically equivalent* if there exists a homeomorphism $\varphi : M \to M$ such that $\varphi(N_1) = N_2$ and φ maps the trajectories of N_1 onto the trajectories of N_2.

Let N be an exceptional minimal set. A pair of trajectories l_1, $l_2 \subset N$ is called *special* if there exists a simply connected component Ω of $M \setminus N$ such that the accessible boundary of Ω equals $l_1 \cup l_2$. It is natural to call Ω a *cell of Denjoy*.

Any flow on T^2 with an exceptional minimal set must have special pairs. Conversely, the existence of special pairs on a hyperbolic surface M is artificial. Any flow f^t having an exceptional minimal set with special pairs on M can be mapped by a blow-down operation onto the flow with an exceptional minimal set that has no special pairs. So the first step is a classification of exceptional minimal sets with no special pairs.

Theorem 29.7. *Let N_1, N_2 be exceptional minimal sets with no special pairs of flows f_1^t, f_2^t respectively on a closed orientable hyperbolic surface M. Then N_1, N_2*

are topologically equivalent via a homeomorphism $M \rightarrow M$ homotopic to identity if and only if their geodesic frameworks coincide, $G(N_1) = G(N_2)$. Furthermore, the geodesic framework $G(N)$ of any exceptional minimal set N (possibly, with special pairs) belongs to $\Lambda_o r$ that is an $G(N)$ is orientable **sntm** *geodesic lamination, and vise versa, given any geodesic lamination $G \in \Lambda_{or}$, there is a flow f^t with exceptional minimal set N with no special pairs such that $G(N) = G$. Moreover, let N be an exceptional minimal set of flow f^t on M which has no special pairs. Then there is a flow f_0^t on M with the following properties:*

1. *The geodesic lamination $G(N)$ is an exceptional minimal set of the flow f_0^t.*
2. *Minimal sets N and $G(N)$ are topologically equivalent via a homeomorphism homotopic to the identity.*

We see that the orbit of orientable irrational geodesic lamination is a complete invariant for exceptional minimal sets with no special pairs. In the general case when an exceptional minimal set can have special pairs, we need the notation of marked geodesics as follows. It is easy to see that a cell of Denjoy corresponds to a geodesic which is called *marked*. Such geodesics form a marked subset $G_m(N)$ in a geodesic framework $G(N)$ of an exceptional minimal set N.

Theorem 29.8. *Let N_1, N_2 be exceptional minimal sets of flows f_1^t, f_2^t respectively on a closed orientable hyperbolic surface M. Then N_1, N_2 are topologically equivalent via a homeomorphism $M \rightarrow M$ homotopic to identity if and only if their geodesic frameworks and corresponding marked subsets coincide, $G(N_1) = G(N_2)$, $G_m(N_1) = G_m(N_2)$.*

To solve the part of realization in the classification problem, let us introduce the notion of an interior geodesic in a geodesic framework. Roughly speaking, an interior geodesic is self-limiting from the both sides. More precisely, let g be a geodesic from a geodesic framework $G \in \Lambda$ (it means that G is **sntm** geodesic lamination) and let Σ be a transversal geodesic segment through some point of g. Then the intersection $G \cap \Sigma$ is a Cantor set and thus any open component of $\Sigma - G \cap \Sigma$ is an open interval. If g does not pass through endpoints of open components of $\Sigma - G \cap \Sigma$, then l is called *interior*. This definition does not depend on the choice of Σ.

Theorem 29.9. *Let N be an exceptional minimal set of a flow f^t on a closed orientable hyperbolic surface M. Then the geodesic framework $G(N)$ of N is an orientable* **sntm** *geodesic lamination, $G(N) \in \Lambda_{or}$, with a countable (possibly, finite) marked subset $G_m(N)$ that consists of interior geodesics. The cardinality of $G_m(N)$ equals the cardinality of the set of Denjoy cells. Vise versa, given any* **sntm** *geodesic lamination $G \in \Lambda_{or}$ with a marked subset $G_m \subset G$ consisting of countable set of interior geodesics, there is a flow f^t with exceptional minimal set N such that $G(N) = G$ and $G_m(N) = G_m$.*

Thus the orbit of orientable **sntm** geodesic lamination with marked subset consisting of countable set of interior geodesics is a complete invariant for exceptional minimal sets.

29 Dynamical Systems with Nontrivially Recurrent Invariant Manifolds 431

As to exceptional minimal sets for foliations, let us remark that there are such sets with empty geodesic frameworks (exp., the stable or unstable manifolds of generalized pseudo-Anosov homeomorphism). Therefore we must restrict ourselves by some classes of foliations or special exceptional minimal sets. For example, one can consider foliations with finitely many singularities such that all of them are saddles of negative index, or one can consider so-called widely disposed exceptional minimal sets. In the both cases the similar classification holds just omitting the orientability condition of geodesic frameworks.

Let us introduce the notion of a Denjoy foliation on a hyperbolic surface, which in sense generalizes the notion of Denjoy flow on the torus. A foliation $\mathscr{F}$ whose singular set $Sing\,(\mathscr{F})$ consists of saddles with negative indices is called a *Denjoy foliation* on M if it has a unique exceptional minimal set N satisfying the following conditions: (1) Every component w of $M - N$ is simply connected; (2) every Denjoy cell does not contain singularities; (3) every component w of $M - N$ which is not a Denjoy cell contains a unique saddle of the index that equals the index of w (i.e. a number of separatrices equals a number of leaves which form the accessible boundary of w).

One can show that a geodesic framework of Denjoy foliation is an irrational geodesic lamination with marked subset consisting of countable set of interior geodesics. The classification of Denjoy foliations is word in word the same as for the irrational foliations: the orbit of an irrational geodesic framework with marked subset consisting of countable set of interior geodesics is a complete invariant.

29.1.2.3 Irrational 2-Webs

The web theory is a classical area of geometry and is mainly devoted to solving local problems. However, 2-webs also naturally appear in the theory of dynamical systems on surfaces as pairs of stable and unstable foliations of Smale horseshoes, Anosov diffeomorphisms, pseudo-Anosov homeomorphisms, and diffeomorphisms with Plykin attractors. The topological equivalence of these webs is clearly a necessary condition for the classification up to conjugacy of these diffeomorphisms and homeomorphisms.

2-web on a surface is a pair of foliations such that they have a common singular set and are topologically transversal at all non-singular points. Let us show how a "web" of geodesic frameworks helps to classify so-called irrational 2-webs [25].

2-web is *irrational* if it consists of a pair of irrational foliations. Two 2-webs (F_1, F_2) and (F_1', F_2') on M are *topologically equivalent* if there is a homeomorphism $f : M \to M$ which maps the foliations $F_i, i = 1, 2$, to the corresponding foliations F_i'.

Theorem 29.10. *Two irrational 2-webs (F_1, F_2) and (F_1', F_2') on a closed orientable hyperbolic surface M are topologically equivalent via a homeomorphism $M \to M$ homotopic to identity if and only if their geodesic frameworks coincide, $G(F_1) = G(F_1'),\ G(F_2) = G(F_2')$.*

Let (F_1, F_2) be an irrational 2-web. Recall that every geodesic framework $G(F_i)$, $i = 1, 2$, is an irrational geodesic lamination and hence, the set $M \setminus G(F_i)$ consists of finitely sided convex polygons whose sides are (complete) geodesics with ideal vertices. Moreover, the pair of geodesic frameworks $(G(F_1), G(F_2))$ has the following properties:

1. The sets $M \setminus G(F_i)$, $i = 1, 2$, have the same number of connected components which equal to the number of (common) singularities of the foliations F_i.
2. For each connected component $D_1 \subset M \setminus G(F_1)$ there is exactly one connected component $D_2 \subset M \setminus G(F_2)$ such that one can lift D_1 and D_2 to geodesic polygons $d_1, d_2 \subset \Delta$ respectively with alternating vertices on S_∞.

Two transversal geodesic frameworks $(G(F_1), G(F_2))$ are called *compatible* if conditions (1) and (2) above are satisfied.

Theorem 29.11. *For any irrational 2-web (F_1, F_2) on M, the geodesic frameworks $(G(F_1), G(F_2))$ are transversal and form a compatible pair of irrational geodesic laminations. Conversely, any such pair uniquely (up to a homeomorphism homotopic to identity) determines an irrational 2-web on M.*

29.1.3 Properties of Geodesic Frameworks

We see that a geodesic frameworks is often a complete invariant or an essential part of complete invariant for important classes of surface foliations and dynamical systems. Therefore, it is natural to study carefully properties of geodesic frameworks.

29.1.3.1 Deviations

One of the important aspect of the Anosov–Weil theory is a deviation of a foliation from its geodesic framework. This aspect is especially nutty for irrational foliations (including flows) and exceptional minimal sets because its geodesic frameworks are complete invariants. Let us give definitions.

Suppose a semi-infinite continuous curve $\bar{l} = \{\bar{l}(t), t \geq 0\}$ has the asymptotic direction $\sigma \in S_\infty$. Take one of the oriented geodesics, say $\bar{g}$, with the same positive direction σ (i.e. σ is one of the ideal endpoints of $\bar{g}$). Such geodesic $\bar{g}$ is called a *representative* of σ. Let $d(t) = \bar{d}(\bar{l}(t), \bar{g})$ be the Poincare distance between $\bar{l}(t)$ and $\bar{g}$. If there is a constant $k > 0$ such that $d(t) \leq k$ for all $t \geq 0$, we'll say that $\bar{l}$ has a *bounded deviation property*. The following theorems was proved in [24].

Theorem 29.12. *Let f^t be a flow with finitely many fixed points on a closed hyperbolic surface M. Let $\bar{l}$ be a semitrajectory of the covering flow $\bar{f}^t$ on Δ. Suppose that $\bar{l}$ has an asymptotic direction. Then $\bar{l}$ has the bounded deviation property.*

Theorem 29.13. *Let F be a foliation on a closed hyperbolic surface M. Suppose that all singularities of F are topological saddles. Let $\overline{L}$ be either a generalized or ordinary leaf of the covering foliation $\overline{F}$. Then $\overline{L}$ has an asymptotic direction and the bounded deviation property.*

After Theorems 29.12, 29.13, it is natural to study the "width" of surface flows and foliations with respect to its geodesic frameworks. Put by definition,

$$\overline{d}_{\overline{L}} = sup_{\overline{m} \in \overline{L}} \overline{d}(\overline{m}, \overline{g}(\overline{L})).$$

Theorem 29.14. *Let F be a foliation on a closed hyperbolic surface M. Suppose that all singularities of F are topological saddles; then*

$$\sup\{\overline{d}_{\overline{L}}\} < \infty,$$

where $\overline{L}$ ranges over the set of all generalized and ordinary leaves of the covering foliation $\overline{F}$.

This theorem means the uniformity of deviations of leaves from a geodesic framework of foliation. The supremum above is called a *deviation of foliation from its geodesic framework*. As a consequence, we see that the deviation of irrational foliation from its geodesic framework is finite. It is the interesting problem to study the influence of this deviation on dynamical properties of foliation. One can prove that a deviation of exceptional minimal set from its geodesic framework is also finite. Note that an analytic flow can have a continuum set of fixed points. Nevertheless the strong smoothness allows to prove the following result [32].

Theorem 29.15. *If f^t is an analytic flow on a closed hyperbolic orientable surface M, then any semitrajectory of f^t with an asymptotic direction has the bounded deviation property.*

For flat closed surfaces (torus and Klein bottle), a similar theorem was proved by Anosov [7, 9].

29.1.3.2 Dynamics and Absolute

In this section we show how some properties of points of S_∞ influence on dynamical properties of flows and foliations. In particular, the first theorem says that if a foliation (or flows) with a finite set of singularities reaches an irrational point, then the foliation has a quasiminimal set.

Recall that $\Lambda(\infty) \subset S_\infty$ ($\Lambda^{irr}(\infty) \subset \Lambda(\infty)$) is a set of points reached by pre-images of geodesics from all **sntm** geodesic laminations belonging to Λ (from all irrational geodesic laminations belonging to Λ^{irr}). Denote by $\Lambda_{or}(\infty) \subset S_\infty$ the set of points reached by pre-images of geodesics from all orientable **sntm** geodesic laminations (belonging to Λ).

Theorem 29.16. *Let $\mathcal{F}$ be a foliation with finitely many singularities on a closed orientable hyperbolic surface M. If $\mathcal{F}$ has a semi-leaf with an irrational direction determined by a point $\sigma \in S_\infty$, then $\sigma \in \Lambda(\infty)$. Moreover, $\mathcal{F}$ has a quasiminimal set (in particular, $\mathcal{F}$ has nontrivially recurrent leaves). If $\mathcal{F}$ is orientable and has a quasiminimal set, then $\mathcal{F}$ has a non-empty geodesic framework which reaches a point from $\Lambda_{or}(\infty)$.*

Theorem 29.17. *Let $\mathcal{F}$ be an orientable foliation with a finitely many singularities on M. If its geodesic framework $G(\mathcal{F})$ reaches a point from $\Lambda(\infty) - \Lambda^{irr}(\infty)$, then $\mathcal{F}$ is not highly transitive and there is a homotopically nontrivial closed curve that is not intersected by any nontrivially recurrent leaf. If $G(\mathcal{F})$ reaches a point from $\Lambda^{irr}(\infty)$, then $\mathcal{F}$ has an irreducible quasiminimal set (i.e. any nontrivially homotopic closed curve on M intersects this quasiminimal set). Moreover, $\mathcal{F}$ is either highly transitive or can be obtained from a highly transitive foliation by a blow-up operation of at least countable set of leaves and by the Whitehead operation. In the last case, when $\mathcal{F}$ is not highly transitive, $\mathcal{F}$ has a unique nowhere dense quasiminimal set.*

Take a geodesic framework $G \in \Lambda$ (that is G is **sntm** geodesic lamination). Then $\pi^{-1}(G) = \overline{G}$ is a local geodesic lamination on the hyperbolic plane Δ. A point $\sigma \in \overline{G}(\infty)$ is a *point of first kind* if there is only one geodesic of $\overline{G}$ with the endpoint σ. Otherwise, σ is called a *point of second kind*. One can prove that this definition does not depend on the choosing of $G \in \Lambda^{irr}$. The following theorem shows that the type of asymptotic direction reflects certain "dynamical" properties of foliation [31].

Theorem 29.18. *Let $\mathcal{F}$ be an irrational foliation on M and let l^+ be a positive semi-leaf of $\mathcal{F}$ such that its lifting $\overline{l}^+$ to Δ has the asymptotical direction $\sigma \in S_\infty$. Then $\sigma \in \Lambda^{irr}(\infty)$. Moreover,*

1. *If σ is a point of first kind then l^+ belongs to a nontrivially recurrent leaf.*
2. *If σ is a point of second kind then l^+ belongs to an α-separatrix of some saddle singularity of $\mathcal{F}$.*

One can reformulate above theorem for flows replacing $\Lambda(\infty)$ by $\Lambda_{or}(\infty)$ and $\Lambda^{irr}(\infty)$ by $\Lambda_{or}^{irr}(\infty) = \Lambda^{irr}(\infty) \cap \Lambda_{or}$.

Put by definition, $\Lambda^{irr}(\infty) \cap \Lambda_{non}(\infty) = \Lambda_{non}^{irr}(\infty)$. The set $\Lambda_{non}^{irr}(\infty)$ is dense and has zero Lebesgue measure on S_∞. One holds the following sufficient condition of the existence of continuum fixed points set for flows.

Theorem 29.19. *Suppose a flow f^t on M reaches a point from $\Lambda_{non}^{irr}(\infty)$. Then f^t has a continual set of fixed points. Furthermore, f^t has neither nontrivially recurrent semitrajectories nor closed transversals nonhomotopic to zero.*

29.1.3.3 Absolute and Smoothness

In this section we show that some points of S_∞ attained by C^∞ flows prevent to be analytic for this flows. Recall that $\sigma \in S_\infty$ is called a *point achieved by f^t*

29 Dynamical Systems with Nontrivially Recurrent Invariant Manifolds   435

if there is a positive (or negative) semitrajectory $l^{\pm}$ of f^t such that some covering semitrajectory $\bar{l}^{\pm}$ for $l^{\pm}$ has the asymptotic direction defined by σ. Sometimes we'll say that f^t reaches σ.

Denote by A_{fl}, A_∞, $A_{an} \subset S_\infty$ the sets of points achieved by all topological, C^∞, and analytic flows respectively. Due to the remarkable result of Anosov [7], $A_{fl} = A_\infty$. Obviously, $A_{an} \subset A_\infty$. It follows from the following theorem that $A_\infty - A_{an} \neq \emptyset$ [17, 32]–[34].

Theorem 29.20. *There exists a continual set $U(M) \subset A_\infty$ such that given any C^∞ flow f^t that reaches a point from $U(M)$, is not analytic. The set $U(M)$ is dense and has zero Lebesgue measure on S_∞.*

One can present explicitly a set that belongs to $U(M)$. Namely, one can prove that the points attained by geodesics of non-orientable irrational geodesic laminations are in $U(M)$,

$$\Lambda_{non}(\infty) \subset A_\infty - A_{an}.$$

Starting with Theorem 29.20, one can deduce that the set of points attained by analytic flows contains the points attained by the simple closed geodesics and all irrational points of A_{an} attained by geodesics of orientable weakly irrational geodesic laminations,

$$\Lambda_{triv}(\infty) \subset A_{an} \subset \Lambda_{triv}(\infty) \cup \Lambda_{or}(\infty).$$

29.1.3.4   On Continuity and Collapse of Geodesic Frameworks

There is a deep theory on the dependence of Poincare rotation number for circle diffeomorphisms [10,80]. For the class of transitive circle diffeomorphisms, a Poincare rotation number is a complete invariant of conjugacy. Well known that a transitive circle diffeomorphism has an irrational rotation number that depends continuously on perturbations of the diffeomorphism in C^1 topology (even C^0 topology). Similar results hold for rotation numbers of minimal flows on the torus.

A complete topological invariant of strongly irrational foliations (in particular, flows) is a strongly irrational geodesic framework. Since the set of geodesic laminations can be endowed with a structure of Hausdorff topological space [41, 50, 133], it is natural to study the dependence of geodesic frameworks on perturbations of foliations on M.

Recall that a geodesic lamination is called *rational* if it does not contain nontrivially recurrent geodesics. Note that a rational geodesic lamination necessary contains closed geodesics. Moreover, any geodesic of such a lamination has a rational asymptotic direction. A rational geodesic lamination is called *strongly rational* if it consists of only closed geodesics. Actually, a strongly rational geodesic lamination is a simplest one.

A geodesic framework of an irrational flow is irrational and orientable. This geodesic framework is an analog of irrational rotation number of minimal torus flows. The following results generalize ones of [27].

Theorem 29.21. *Let f^t be an irrational C^1-flow induced by a vector field $v \in X^1(M)$ on a closed orientable hyperbolic surface M. Suppose that all fixed points of f^t are hyperbolic saddles. Let U be a neighborhood of the geodesic framework $G(f^t)$ of f^t. Then there is a neighborhood $O^1(v)$ of v in the space $X^1(M)$ of all C^1-vector fields such that any flow g^t generated by $w \in O^1(v)$ has a non-empty geodesic framework $G(g^t)$ that belongs to U.*

Theorem 29.21 is similar to the assertion that an irrational rotation number of a minimal torus flow depends continuously on perturbations of the flow in the space of C^1-flows.

According to Pugh's C^1 Closing lemma [125], given a vector field v with non-trivially recurrent trajectories, there is a vector field w arbitrary close to v in the space $X^1(M)$ such that w has a periodic trajectory that is nonhomotopic to zero. As a consequence we get a so-called "instability" of irrational geodesic framework, which is similar to the instability of an irrational rotation number (given a torus vector field with irrational Poincaré rotation number, there is an arbitrary close vector field with rational rotation number).

Theorem 29.22. *Let f^t be an irrational C^1-flow induced by a vector field $v \in X^1(M)$ on a closed orientable hyperbolic surface M. Suppose that all fixed points of f^t are hyperbolic saddles. Then for any neighborhood U of the geodesic framework $G(f^t)$ and any neighborhood $O^1(v)$ of v in the space $X^1(M)$ of C^1-vector fields there is a flow g^t generated by $w \in O^1(v)$ such that the geodesic framework $G(g^t)$ is strongly rational and belongs to U.*

As far as rational geodesic frameworks is concerned, then there are examples both of continuous and discontinuous dependence on parameters of a flow. A simplest example for continuous dependence of a rational geodesic framework gives a Morse–Smale flow, which obviously has a rational geodesic framework. Its geodesic framework does not vary under small perturbations of the flow because any Morse–Smale flow is structurally stable. Two theorems below describe virtual scenario of the destruction of a rational geodesic framework.

Theorem 29.23. *On a closed hyperbolic orientable surface M there is a one-parameter family of C^∞ flows f_μ^t which depends continuously on the parameter $\mu \in [0; 1]$ and such that the following conditions are satisfied:*

1. *For all $\mu \in [0; 1)$ the flow f_μ^t has an irrational geodesic framework $G(f_\mu^t) \neq \emptyset$ which does not depend on the parameter μ, $G(f_0^t) = G(f_\mu^t)$.*
2. *The flow f_1^t has a rational geodesic framework $G(f_1^t)$.*
3. *There is a neighborhood U of $G(f_1^t)$ such that $G(f_\mu^t) \notin U$ as $\mu \in [0; 1)$.*

Theorem 29.24. *On a closed hyperbolic orientable surface M there is a one-parameter family of C^∞-flows f_μ^t which depends continuously on the parameter $\mu \in [0; 1]$ such that the following conditions are satisfied:*

1. *For all $\mu \in [0; 1]$ the flow f_μ^t has a rational geodesic framework $G(f_\mu^t) \neq \emptyset$ which does not depend on the parameter μ as $\mu \in [0; 1)$.*
2. *There is a neighborhood U of $G(f_1^t)$ such that $G(f_\mu^t) \notin U$ as $\mu \in [0; 1)$.*

29 Dynamical Systems with Nontrivially Recurrent Invariant Manifolds

Discontinuity of a rational geodesic framework is not surprising, since there are flows on torus (and the Klein bottle) with rational rotation number which varies in a "jump-like" fashion under arbitrarily small perturbations [100, 101].

We now formulate a theorem on the existence of one bifurcation of a geodesic framework which is similar to the 'blue-sky catastrophe' bifurcation of flow and corresponds to a certain family of flows.

Theorem 29.25. *On a closed hyperbolic orientable surface M there is a one-parameter family of C^∞ flows f_μ^t which depends continuously on the parameter $\mu \in [0; 1)$ such that the following conditions are satisfied:*

1. For all $\mu \in [0; 1)$ the flow f_μ^t has a strongly rational geodesic framework $G(f_\mu^t) \neq \emptyset$.
2. The lengths of closed geodesics in $G(f_\mu^t)$ tend uniformly to infinity as $\mu \to 1$.
3. $G(f_1^t) = \emptyset$.

A bifurcation described in Theorem 29.25 we will call a *collapse of geodesic framework*.

The following theorem gives some information on a set of fixed points of a flow under which a collapse of the geodesic framework takes place.

Theorem 29.26. *Let f_μ^t be a one-parameter family of C^∞-flows which depends continuously on the parameter $\mu \in [0; 1]$ on a closed hyperbolic orientable surface M. Assume that:*

1. For all $\mu \in [0; 1)$ the flow f_μ^t has a strongly rational geodesic framework $G(f_\mu^t) \neq \emptyset$.
2. The lengths of closed geodesics in $G(f_\mu^t)$ tend uniformly to infinity as $\mu \to 1$.
3. $G(f_1^t) = \emptyset$.

Then the flow f_1^t has infinitely many fixed points.

29.2 Nontrivial Basic Sets with Recurrent Invariant Manifolds

Introduction. For applications, the most important invariant sets (including basic sets) of dynamical system are attractors which at a first step can be divided into the following three groups:

1. Trivial attractors (periodic attracting isolated orbits, sinks).
2. Nontrivial attractors which are sub-manifolds topologically embedded in M.
3. Attractors which do not belong to the above two groups (sometimes such attractors are called strange).

Lorentz attractors and expanding attractors belong to the third and obviously the most complicated (but the most interesting) group. Among attractors with a uniform hyperbolic structure, the most attention was paid to expanding attractors introduced by Williams [136, 139]. The reason is a natural connection of an expanding attractor with quasi 2-webs consisting of laminations and transversal foliations (see below the precise statements). In 1975, Newhouse [109] introduced wildly embedded zero-dimensional basic sets that are locally homeomorphic to the product of standard Cantor set and Antoine necklace. After that, Robinson and Williams [128], Bothe [44], Isaenkova and Zhuzhoma [85] constructed different types wildly embedded basic sets. We briefly describe these examples below. Using geodesic frameworks introducing in the previous section, we exhibit some classification results.

29.2.1 Examples of Expanding Attractors and Wild Basic Sets

First, let us give some definitions. An invariant set Λ is called *attracting* if a neighborhood U of Λ such that $clos\, f(U) \subset U$. The neighborhood U is attracting one. A closed invariant set Λ is called *attractor* if there is an attracting neighborhood U of Λ such that $\cap_{i \geq 0} f^i(U) = \Lambda$ and the restriction $f|_\Lambda$ is transitive. Recall briefly the notion of *topological dimension of a set*. This definition is given inductively. A set Λ has topological dimension zero provided for each point $p \in \Lambda$, there is an arbitrarily small neighborhood U of p such that $\partial(U) \cap \Lambda = \emptyset$. (It is not always possible to take U as a ball, as the example of Antoine's necklace shows.) Then, inductively, a set Λ is said to have dimension $n \geq 1$ provided for each point $p \in \Lambda$, there is an arbitrarily small neighborhood U of p such that $\partial(U) \cap \Lambda$ has dimension $n - 1$. See [53] for a more complete discussion of topological dimension. A nontrivial hyperbolic attractor Λ is *expanding* if the topological dimension of Λ equals the dimension of a fiber E_Λ^u, $\dim E_x^u = \dim \Lambda$ $(x \in \Lambda)$.

Note that for any attractor Λ (not necessary expanding), the inclusion $W^u(x) \subset \Lambda$ holds for any point $x \in \Lambda$ [118, 139]. Therefore, $\dim \Lambda \geq \dim E_\Lambda^u$. It follows that if one assumes the topological dimension of Λ equals zero, then $\dim E_\Lambda^u = 0$. Hence, Λ must be an isolated periodic attracting orbit, i.e. a trivial attractor. Thus a topological dimension of an expanding attractor more or equals one. Obviously, $\dim \Lambda < \dim M$, otherwise Λ have to coincide with M. So,

$$1 \leq \dim \Lambda \leq \dim M - 1.$$

The most familiar expanding attractors are: (1) Smale solenoid; (2) DA-attractor (a nontrivial attractor of a DA-diffeomorphism); (3) Plykin attractor. We describe schematically these examples (with others) below. Later on for simplicity, we'll suppose that an expanding attractor is connected (otherwise, one can take a connected component and some iteration f^k under whose this component is invariant).

29.2.1.1 Smale-Bothe Solenoids

A solenoid was first independently introduced by Vietoris [135] in 1927 and by Van Danzig in 1930. They considered solenoids from different points of view (see

Introduction in [134]). One of the definition of solenoid is the intersection of a nested sequence of solid tori $T_1 \supset T_1 \supset \ldots, T_i \supset \ldots$, where T_{i+1} is wrapped around inside T_i longitudinally p_i times in a smooth fashion without folding back [1], Fig. 29.2.

In Topology, the solenoid $\cap_{i \geq 0} T_i$ gives the example of 1-dimensional connected set that is circular chainable but can not be embedded into a surface [38, 39]. Recall that a *chain* is a finite collection of open sets $d_1, \ldots, d_k$ such that d_i intersects d_j if and only if $|i - j| = 1$. If all d_i are of diameter less that ε, the chain is called an ε-chain. A set is *circular chainable*, if for each positive number ε it can be covered by a circular ε-chain.

In Topological Dynamics, solenoids was first introduced in [107] as the example of a locally disconnected minimal set consisting of almost periodic trajectories of a flow. Special flows with solenoidal invariant sets was considered in [86]. In Hyperbolic Dynamics, solenoids were introduced by Smale in his celebrated paper [130] as hyperbolic attractors by the following way. Let $N = D^2 \times S^1$ be a solid torus, where the circle S^1 and disk D^2 are endowed with the usual coordinates, $S^1 = [0; 1]/(0 \sim 1)$, $D^2 = \{(x; y) \,|\, x^2 + y^2 \leq 1\}$. Let $f : N \to N$ be a D^2-level preserving embedding, that is $f(D^2 \times \{t\}) \subset D^2 \times \{pt\}$ for $\forall t \in S^1 = R^1$ (mod 1)) and $p \geq 2$, such that $f(\{\cdot\} \times S^1)$ is a p-string braid and the radius of $f(D^2 \times \{\cdot\})$ is $\frac{1}{p^2}$. Geometrically, f can be described as an expanding map of degree p in the S^1 direction and a strong contraction in the D^2 direction. The image is thinner across in the D^2 direction by a factor $\frac{1}{p^2}$, see Fig. 29.3b for $p = 2$. By construction,

$$\Lambda = \bigcap_{k \geq 0} f^k(N)$$

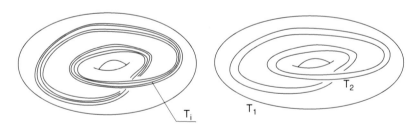

Fig. 29.2

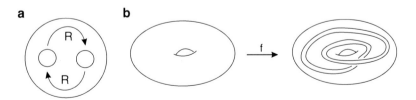

Fig. 29.3

is a one-dimensional expanding attractor which is called *Smale solenoid*. The careful description of Smale's construction can be found in many books on Dynamical Systems, see exm. [91, 127].

Gibbons [60] proved that there is an extension of f to a diffeomorphism of the 3-sphere S^3 such that the non-wandering set of $f : S^3 \to S^3$ consists of Λ and one-dimensional contracting repeller that is a solenoid for f^{-1} (actually in [60], the construction of such diffeomorphisms $S^3 \to S^3$ is presented in more general assumptions). Moreover, by construction, there is a loop of tangencies of stable and unstable manifolds such that this loop is C^1 stable.

We represent the more general construction of a so-called pure solenoid due to Bothe [43]. For the 2-torus T^2 that is a boundary of the solid torus $N = D^2 \times S^1$, one can choose the representatives of generates of $\pi_1(T^2)$: the meridian μ which is homotopy to zero in N but non-homotopy to zero in T^2 and the longitude λ which has the index of intersection $+1$ with μ. Let β be a monotone knot in *int N* and $N_1 \subset int\ N$ a solid torus corresponding to β ("fat" knot). In the similar way, one can define the meridian μ_1 and longitude λ_1 for the torus $T_1^2 = \partial N_1$. There is a diffeomorphism $f : N \to N_1$ such that

$$f(D^2 \times \{t\}) \subset D^2 \times \{\varepsilon pt\}, \quad \forall t \in S^1 = \mathrm{R}^1 \quad (\mathrm{mod}\ 1),$$

$$(f|_{T^2})_* (\lambda) = \lambda_1^j \mu_1^m, \quad (f|_{T^2})_* (\mu) = \mu_1^\delta$$

for some $\varepsilon = \pm 1$, $\delta = \pm 1$, $j \geq 2$, and $m \in \mathrm{Z}$. The classical Smale example corresponds to $\varepsilon = 1$, $\delta = 1$, and $m = 0$. Then

$$\bigcap_{i \geq 0} f^i(N) \overset{\mathrm{def}}{=} \Lambda_{\beta,m,\varepsilon,\delta}$$

is a one-dimensional expanding attractor.

29.2.1.2 DA-Attractor

A diffeomorphism with a DA-attractor is obtained by a so-called Smale surgery performed on a codimension one Anosov automorphism of the n-torus T^n. For this reason, it is called the *Derived from Anosov diffeomorphism* or *DA-diffeomorphism*. It was first introduced by Smale [130].

Take a codimension one Anosov automorphism $A : T^n \to T^n$ with one-dimensional stable splitting and codimension one unstable splitting. Let p_0 be a fixed point of A. One can carefully insert instead of point p_0 a tiny ball with a source P_0 and two saddle type fixed points, see Fig. 29.4, in such a way that we get a diffeomorphism, say $f : T^n \to T^n$, with a codimension one expanding attractor

$$\Lambda = T^n - W^u(P_0)$$

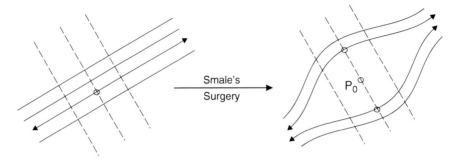

Fig. 29.4

which is called a *DA-attractor*. Smale surgery is similar to a Poincaré–Denjoy blowing up operation producing a Denjoy foliation from a minimal foliation on the 2-torus.

29.2.1.3 Plykin Attractor and Lakes of Wada

The construction of Lakes of Wada was first published in 1917 by the Japanese mathematician Kunizo Yoneyama [141], who credited the discovery to his teacher Takeo Wada. The Lakes of Wada are formed by starting with an open unit square of dry land (homeomorphic to the plane), and then digging 3 lakes according to the following rule: on day $n = 1, 2, 3, \ldots$ extend consequently each lake so that it passes within r_n distance an of all remaining dry land, where $r_1, \ldots, r_n, \ldots$ is some sequence of positive real numbers tending to 0, see in Fig. 29.5 the digging at first day. This should be done so that after a finitely many days the remaining dry land has connected interior, and each lake is open. After an infinite number of days, the three lakes are still disjoint connected open sets, and the remaining dry land is the boundary of each of the 3 lakes. Obviously, this construction can start with any $k \geq 3$ lakes. Moreover, identifying the boundary of the unit square, one can thought of the Wada construction on a 2-sphere.

Wada lakes naturally appear in the example of one-dimensional expanding attractor on the two-sphere S^2 by Plykin [119]. Note that Smale-Bothe solenoids and DA-attractors are orientable expanding attractors while the Plykin attractor is non-orientable. Starting with this example, one can construct a diffeomorphism of any closed surface with a codimension one non-orientable expanding attractor. Bearing in mind these examples, we call a codimension one non-orientable expanding attractor a *Plykin attractor*.

Let us give the sketch of modern construction of Plykin attractor. We start with an arbitrary dimension $n \geq 2$. Denote by $J : T^n \to T^n$ the involution $\mathbf{x} \to -\mathbf{x}$ (mod 1) which has 2^n fixed points $v_1, \ldots, v_{2^n}$. Take the DA-diffeomorphism $g : T^n \to T^n$ with the codimension one orientable expanding attractor Ω_g such that g commutes with J and has the fixed points $v_1, \ldots, v_{2^n}$. Denote by $q : T^n \to T^n/J$

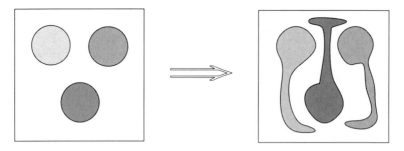

Fig. 29.5

the natural projection which is a double branched covering with the branch points $v_1, \ldots, v_{2^n}$. It is not hard to see that for $n = 2$ the quotient space T^2/J is a 2-sphere. Since $g(-\mathbf{x}) = -g(\mathbf{x})$, g induces the diffeomorphism $f : S^2 \to S^2$ with the one-dimensional expanding attractor $q(\Omega_g)$. If $n \geq 3$, the quotient space T^n/J is not a manifold. Therefore the branch points must be removed to get a codimension one expanding attractor on the open manifold $M = q(T^n - \cup_{i=1}^{2^n} v_i)$.

Note that Plykin's diffeomorphism $f : S^2 \to S^2$ is structurally stable and contains necessarily so-called 1-bunches. Roughly speaking, a 1-bunch corresponds to a component of $S^2 - \Lambda$ whose accessible boundary consists of a unique (one-dimensional) unstable manifold of a periodic point.

In [28], the exact upper estimate for numbers of codimension one expanding attractors of surface diffeomorphisms was obtained. This estimate depends on a genus of a surface and a number of 1-bunches.

Fokkink and Oversteegen [56] proved that any lamination of Cantor type in $\mathbb{R}^2$ and S^2 can be obtained by a lake Wada construction. Moreover, they proved that such a lamination has at least four complementary domain. This corresponds to the estimation by Plykin [119] for the number of complementary domains of expanding attractor on S^2.

29.2.1.4 Newhouse Basic Sets of Antoine Necklace Type

Newhouse [109] introduced Antoine necklace in Hyperbolic Dynamical Systems. Let us recall the construction by Antoine [14] who presented the first example of so-called wild embedding. In the interior of the solid torus T_1, form a set T_2 which is the union of a finite collection of solid tori linked in cyclic order as indicated in Fig. 29.6a. The components C_i of T_2 are indicated schematically by circles. The number of components of T_2 is $k \geq 4$. Figure 29.6b shows what any three successive components of T_2 looks like. Inductively, given a set T_n which is the union of k^{n-1} disjoint solid tori, for each component C_i of T_n let ϕ_i be a similarity $T_1 \to C_i$, that is, a contraction, and let $T_{n+1} = \cup \phi_i(T_1)$. This gives a descending sequence T_1, $T_2, \ldots$. We define

$$\mathscr{C} = \bigcap T_n.$$

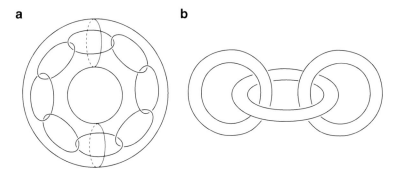

Fig. 29.6 The finite collection T_2 of solid tori linked in cyclic order (**a**); three successive components of T_2 (**b**)

The set $\mathscr{C}$ is called *Antoine necklace*.

In Topology, the main property of the Antoine necklace is that the complement $R^3 - \mathscr{C}$ to the Antoine necklace is not simply connected while $\mathscr{C}$ is zero-dimensional. Moreover, if K is any standard "middle-third" Cantor set on a line in R^3, there are no homeomorphisms $R^3 \to R^3$ that takes K onto $\mathscr{C}$.

Newhouse [109] constructed the structurally stable diffeomorphism $S^4 \to S^4$ with the basic set (transitive closed and hyperbolic) locally homeomorphic to the product of classical Cantor set and Antoine necklace. Thus, the basic set is zero-dimensional and locally the product of Cantor-type sets. We omit the precise description because below one give the generalization of Newhouse's construction.

29.2.1.5 Robinson–Williams Attractors

In [128], there was constructed two homeomorphic expanding attractors of codimension three with the same dynamics but embedded in different way in manifolds.

The first construction is similar to Smale's construction of a solenoid. Let $A : T^2 \to T^2$ be given by the matrix

$$A = \begin{pmatrix} 3 & 1 \\ 1 & 2 \end{pmatrix}, \quad \det A = 5, \quad \lambda_{1,2} = \frac{1}{2}(5 \pm \sqrt{5}).$$

Define $g : T^2 \times D^3 \to T^2 \times D^3$ as follows

$$(x, y, r) \to (A(x, y), \frac{1}{4}r + \frac{1}{2}\exp 2\pi i x)$$

where $(x, y) \in T^2$, $r \in D^3$, and $\exp 2\pi i x$ is thought of as a vector lying in $R^2 \times \{0\} \subset R^3$. Any 3-ball $(x_0, y_0) \times D^3$ intersects $g(T^2 \times D^3)$ in five disjoint 3-balls of radius $\frac{1}{4}$. We see that $\Lambda_g = \cap_{n \geq 0} g^n(T^2 \times D^3)$ is an expanding attractor which intersects $(x_0, y_0) \times D^3$ in a Cantor set.

The second construction. Let $R_t : S^1 \times D^2 \to S^1 \times D^2$ be given by $R_t(\theta, w) = (\theta + t, w)$. This is a rotation of solid torus around S^1 direction. Let $\varphi : S^1 \times D^2 \to S^1 \times D^2$ be an embedding *into* such that the solid toruses $R_{\frac{i}{5}} \varphi(S^1 \times D^2)$, $i = 0, 1, 2, 3, 4$, are disjoint and link, i.e. form an Antoine's necklace configuration. At last, define $f : T^2 \times S^1 \times D^2 \to T^2 \times S^1 \times D^2$ by $f(x, y, z) = (A(x, y), R_x \circ \varphi(z))$ where $(x, y) \in T^2$, $z \in S^1 \times D^2$. Any solid torus $(x_0, y_0) \times S^1 \times D^2$ intersects $f(T^2 \times S^1 \times D^2)$ in a chain of solid tori with two succeeding tori linked. Therefore, $\Lambda_f = \cap_{n \geq 0} f^n(T^2 \times S^1 \times D^2)$ is an expanding attractor which intersects $(x_0, y_0) \times S^1 \times D^2$ in a zero-dimensional but wildly embedded Antoine's necklace.

Since both $f|_{\Lambda_f} : \Lambda_f \to \Lambda_f$ and $g|_{\Lambda_g} : \Lambda_g \to \Lambda_g$ are modelled on $A : T^2 \to T^2$, they are conjugate. However there is no a homeomorphism from a neighborhood of Λ_f to a neighborhood of Λ_g taking Λ_f to Λ_g.

29.2.1.6 NRW-Attractors[1]

Using Newhouse's and Robinson–Williams's technics, Isaenkova and Zhuzhoma [85] constructed the diffeomorphism

$$f : D^2 \times T^2 \to D^2 \times T^2$$

with 1-dimensional expanding attractor locally homeomorphic to the product of R and Antoine's necklace. Let R be the rotation of N along its axis such that T_n is invariant under R for any n, see Fig. 29.3a. Embed the solid torus $N = D^2 \times S^1$ in R^3, and represent 4-dimensional manifold $D^2 \times T^2$ as $N \times [0; 1] \subset \mathrm{R}^4$ with $N \times \{0\}$, $N \times \{1\}$ identified by R. The next Fig. 29.7 depicts $N \times \{0\}$ as a subset of R^3. The vertical direction is to be thought of as R^3 while the horizontal direction may be thought of as R. Note that $N \times [0; 1]/R$ is diffeomorphic to $D^2 \times T^2$, since R is a homotopy trivial mapping. Let $G_1, \dots, G_k$ be solid tori that form the Antoine configuration T_2, where each G_i is linked with neighbors G_{i-1}, G_{i+1}. Let $\psi_i N \to G_i$ be a contraction, $|D\psi_i| \leq \lambda < 1$. Take the k-fold covering space $N \times [0; k]/(N \times \{0\} \sim N \times \{k\})$ for $N \times S^1$, where the group of deck transformations generated by the mapping $(z, t) \longmapsto (R^{-1}(z), t + 1)$. Define the diffeomorphism $f_i : N \times [0, 1] \to G_i \times [0; k]$ by $f_i(z, t) = (\psi_i(z), kt - k - ki)$. This diffeomorphism is a covering for the diffeomorphism $f : N \times S^1 \to: N \times S^1$ that has a 1-dimensional expanding attractor locally homeomorphic to the product of a line and Antoine necklace.

29.2.2 Williams Construction

Let Λ be an expanding attractor. Since $\dim \Lambda = \dim W^u(x)$ for any $x \in \Lambda$, $\dim(\Lambda \cap W^s(x)) = 0$. Moreover, one can prove that the intersection $\Lambda \cap W^s(x)$ is a Cantor set [118, 136, 139]. It follows from the theorem on the continuous dependence

---

[1] Abbreviation NRW is Formed by First Letters of the Names Newhouse, Robinson and Williams.

29 Dynamical Systems with Nontrivially Recurrent Invariant Manifolds

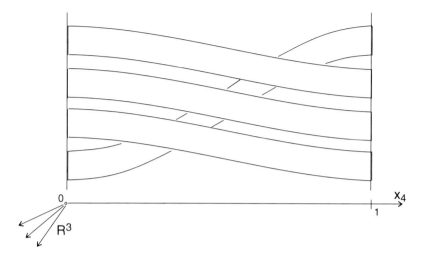

Fig. 29.7

of stable and unstable manifolds on initial conditions (see [82, 127, 130]) that an n-dimensional expanding attractor is locally homeomorphic to the product of an n-ball and a Cantor set, $\Lambda \stackrel{loc}{\simeq} D^n \times C$. As a matter of principle, the product $D^n \times C$ can be widely embedded in an ambient manifold M. However Bothe [42] has shown that this is not a case. He has proved the following basic observation:

Given any expanding attractor Λ and any point $x \in \Lambda$, there is a neighborhood V of x and a homeomorphism

$$\varphi : D^{codim \Lambda} \times D^{\dim \Lambda} \to V$$

such that

$$\varphi^{-1}(V \cap \Lambda) = (\text{Cantor set}) \times D^{\dim \Lambda}.$$

Moreover, every $\varphi(D^{codim \Lambda} \times \{\cdot\})$ belongs to a stable manifold passing through some point of the attractor Λ.

Roughly speaking, an expanding attractor is a Cantor type lamination formed by the unstable manifolds (as leaves) passing through the points of the expanding attractor, and this lamination has a transversal foliation in some neighborhood.

Let N be a compact neighborhood of an expanding attractor Λ. Following Williams, put $x \sim y$ iff the points $x, y \in N$ belong to the same component of $N \cap W^s(z)$ for some point $z \in \Lambda$. Williams proved that the neighborhood can be chosen so that:

- The quotient space $N/\sim \stackrel{def}{=} K$ is a branched manifold
- The following commutative diagram holds

$$f(N) \xleftarrow{f} N$$
$$\downarrow \subset$$
$$N \qquad \downarrow q$$
$$\downarrow q$$
$$K \xleftarrow{g} K$$

where $q : N \to N/\sim$ is a quotient map and $g : K \to K$ is the expansion induced by f. Branched manifolds are smooth manifolds with certain singularities which in the 1-dimensional case are branch points as shown in Fig. 29.8.

Let Σ be an *inverse limit* of

$$K \xleftarrow{g} K \xleftarrow{g} \cdots \xleftarrow{g} K \xleftarrow{g} \cdots$$

i.e. Σ is the set of sequences $(x_0, \ldots, x_i, \ldots)$ with $x_i = g(x_{i+1})$ where K is a branched n-manifold, g is an expansion, and $h : \Sigma \to \Sigma$ is a *shift* defined as follows:

$$h(x_0, x_1, \ldots) = (g(x_0), x_0, x_1, \ldots).$$

Suppose now that (1) $NW(g) = K$; (2) given any $z \in K$, there is a neighborhood U of z and $j \in N$ such that $g^j(U)$ is an n-cell. In this case, Σ is called an *n-solenoid*. Williams [139] proved the following theorem.

Theorem 29.27. *Let Λ be an n-dimensional expanding attractor of a diffeomorphism f. Then the restriction $f|_\Lambda$ of f on Λ is conjugate to the shift map h of an n-solenoid. Vise versa, given a shift map $h : \Sigma \to \Sigma$ of the n-solenoid Σ, there is a manifold M and a diffeomorphism f of M such that f has an n-dimensional expanding attractor Λ and $f|_\Lambda$ is conjugate to h.*

Idea of the proof. The diagram above induces the diagram

$$\vdots \qquad \vdots \qquad \vdots$$
$$\downarrow \subset \qquad \downarrow \subset \qquad \downarrow \subset \quad \cdots$$
$$f^2(N) \xleftarrow{f} f(N) \xleftarrow{f} N$$
$$\downarrow \subset \qquad \downarrow \subset$$
$$f(N) \xleftarrow{f} N$$
$$\downarrow \subset$$
$$N \qquad \downarrow q \qquad \downarrow q$$
$$\downarrow q$$
$$K \xleftarrow{g} K \xleftarrow{g} K \xleftarrow{g} \cdots$$

Each vertical inverse limit is the intersection $\Lambda = \cap_{k \geq 0} f^k(N)$. The horizontal inverse limit yields Σ with the shift h. Therefore this diagram induces a map $R : \Lambda \to \Sigma$ such that the following commutative diagram holds:

29 Dynamical Systems with Nontrivially Recurrent Invariant Manifolds

Fig. 29.8

$$\begin{array}{ccc} \Lambda & \xleftarrow{f|_\Lambda} & \Lambda \\ \downarrow R & & \downarrow R \\ \Sigma & \xleftarrow{h} & \Sigma \end{array}$$

where R is defined by $R(x) = (q(x), qf^{-1}(x), qf^{-2}(x), \ldots)$. One can prove that R is a homeomorphism.

The converse statement means that given a shift map $h : \Sigma \to \Sigma$ of the n-solenoid Σ, there is a manifold M and a diffeomorphism f of M such that f has an n-dimensional expanding attractor Λ and $f|_\Lambda$ is conjugate to h. For simplicity, we sketchily represent here Williams's construction for 1-solenoids [136].

Let K be a 1-dimensional branched manifold and $g : K \to K$ an expansion that determines a 1-solenoid Σ with shift map h, see Fig. 29.9 where the branched manifold $K = A \cup B \cup C$ has two branch points and empty boundary. The expanding immersion g is defined by

$$A \to -B + A + B, \quad B \to C - B + A, \quad C \to B + C - B,$$

where $+(-)$ denotes composition with the (reverse) path. One can check that all points of K are nonwandering and each point of K has a neighborhood whose image under g is an arc (this example is from [136]).

Certainly that K can be smoothly embedded in the sphere S^3 (we identify K with this embedding). Then K has a neighborhood M_0 that is a fiber bundle over K with a fiber disk. The mapping $g : K \to K \subset M_0$ can be approximated by a smooth embedding φ so that $\varphi(x)$ and $g(x)$ lie in the same fiber. We may suppose that φ sends fibers into fibers, φ is a contraction on each fibers, and φ maps various fibers apart, Fig. 29.10.

We can assume $S^3 \subset S^4$. Using arguments of [61] where one proved that two locally flat embedding of S^1 in S^4 are isotopic, one can prove that φ is isotopic to the identity on S^4. Hence there is a diffeomorphism $f : S^4 \to S^4$ such that $f|_{M_0} = \varphi$. Define $M_i = \varphi^i(M_0)$, $\Lambda = \cap_{i \geq 0} M_i$. Then Λ is an 1-dimensional expanding attractor such that $f|_\Lambda$ is conjugate to g.

Theorem 29.27 give rises to the natural questions: (1) what interrelation exists between two n-solenoids corresponding to the same Λ?; (2) when two n-solenoids corresponding to different expanding attractors are conjugate? Such type questions were considered in [138, 139].

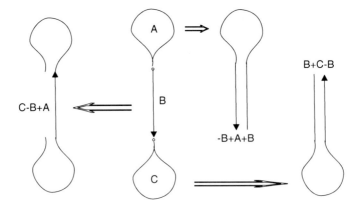

Fig. 29.9

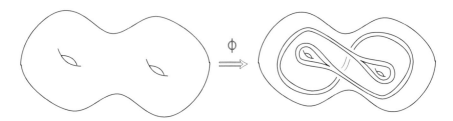

Fig. 29.10

Anderson and Putnam [2] proved that the dynamics of wide class of the substitutions on the space of tilings are conjugated to shifts of n-solenoids, see also [140].

Let us mention some results concerning homology and cohomology groups. A Riemannian metric on an ambient manifold makes each unstable manifold $W^u(x) \subset \Lambda$ into a complete Riemannian manifold. Denote by $B(x, r)$ the r-ball in $W^u(x)$ with the center x. If x is a periodic point, $W^u(x)$ admits a uniformly expanding self-diffeomorphism. Hence, the growth of volume $B(x, r)$ is dominated by a polynomial. This result was used by Plante [117] to prove that an orientable codimension one expanding attractor defines a nontrivial element of the homology group $H_1(M)$. Hence, $H_1(M) \neq 0$. Sullivan and Williams [131] proved that the real Čech homology of orientable expanding attractor Λ in its top dimension is nontrivial and finite-dimensional, $\check{H}_{\dim \Lambda}(\Lambda, \mathrm{R}) \neq 0$. Farrell and Jones [54] constructed the orientable 2-dimensional expanding attractor Λ with $\check{H}_1(\Lambda, \mathrm{R}) = 0$ such that Λ is not the total space of a fiber bundle with a manifold for a base space and a Cantor set for fiber. Other interesting examples can be found in [87].

29 Dynamical Systems with Nontrivially Recurrent Invariant Manifolds 449

29.2.3 Dimension One Expanding Attractors

The problem of classification of the dynamics formed by iterations of maps reduces to the problem of (topological) conjugacy for maps itself that generate corresponding dynamical systems. Solving this problem, it is natural to study the conjugacy for restrictions of maps under consideration to their invariant sets. If a class of diffeomorphisms under consideration has nontrivial basic sets (exm., expanding attractors), there are two ways to do that. The first way is to ask, when the restriction of two maps to their basic sets conjugate? Following [45], we shall call such basic sets *intrinsically conjugate* or *intrinsically equivalent*. The corresponding classification is called an *intrinsic classification*. This type of classification was obtained by Williams [139] for expanding attractors. The second way is to ask, when two diffeomorphisms are conjugate in some neighborhoods of their basic sets. Such basic sets are called *neighbor conjugate*. If the neighborhoods are whole manifolds, the basic sets are called (simply) *conjugate*. Obviously, if basic sets are neighbor conjugate, they are intrinsically conjugate. One can say that the intrinsical classification describes dynamics on basic sets thyself, while the classification under a neighbor conjugacy takes in mind additionally an embedding of basic sets into manifolds. Therefore, the second type of classification is stronger than the first one. Robinson and Williams [128] constructed two diffeomorphisms f and g of 5-dimensional manifolds with 2-dimensional expanding attractors Λ_f and Λ_g respectively such that $f|_{\Lambda_f} : \Lambda_f \to \Lambda_f$ is conjugate to $g|_{\Lambda_g} : \Lambda_g \to \Lambda_g$ but there is not even a homeomorphism from a neighborhood of Λ_f to a neighborhood of Λ_g taking Λ_f to Λ_g (see Robinson–Williams examples above). Taking into account the NRW-attractors considered above, N. Isaenkova and E. Zhuzhoma [85] proved that given any $d \geq 3$, there are compact d-manifolds M^d, N^d and diffeomorphisms $f : M^d \to M^d$, $g : N^d \to N^d$ with 1-dimensional expanding attractors Λ_f and Λ_g respectively such that $f|_{\Lambda_f}, g|_{\Lambda_g}$ are intrinsically conjugate but are not neighbor conjugate. Below we consider topological classifications under a neighbor conjugacy or conjugacy.

29.2.3.1 Bothe's Classification of Pure Solenoids

Let $f : M^3 \to M^3$ be a diffeomorphism of compact 3-manifold M^3 with a one-dimensional expanding attractor Λ. Following Bothe [43], we call Λ a *pure solenoid*, if the basin $B(\Lambda)$ of Λ contains a (closed) solid torus N such that $\Lambda \subset N$, $f(N) \subset int\ N$, and f maps the central circle of N to a monotone nontrivial knot in N. In 1983, Bothe [43] proved the following theorem (here a 3-sphere is considered as a particular case of a lens space, $S^3 = L_{1,0}$).

Theorem 29.28. *Let $f : M^3 \to M^3$ be a diffeomorphism of a closed 3-manifolds M^3. Suppose that f has a pure solenoid Λ. Then M^3 can be represented as a connected sum $M^3 = L_{p,q}\#M_1$, $p > 0$, with a lens space summand $P \subset M^3$ (this means that the boundary ∂P is a 2-sphere and there is an open 3-ball B*

in $L_{p,q}$ such that P is homeomorphic to $L_{p,q} - B$) such that Λ is contained in int P. Moreover, given any lens space $L_{p,q}$, $p > 0$, there is a diffeomorphism $f : L_{p,q} \to L_{p,q}$ with a pure solenoid.

Idea of the proof. Recall that if M^2 be a connected compact 2-sided surface properly embedded in M^3, then M^2 is said to be *compressible* if either M^2 bounds a 3-ball, or there is an essential, simple closed curve on M^2 which bounds a disk in M^3. Otherwise, M^2 is said to be *incompressible*.

By definition of a pure solenoid, there is a solid torus N such that $\Lambda \subset N$, $f(N) \subset int\ N$. Note that since f is a global homeomorphism, $M^3 - int\ N$ and $M^3 - int\ f(N)$ are homeomorphic. Suppose that the 2-torus $\partial N \stackrel{\text{def}}{=} T^2$ is an incompressible in $M^3 - int\ N$. Since the central axis of N is mapped by f to a nontrivial knot which is a central axis of $f(N)$, T^2 and $f^{-1}(T^2)$ are not parallel in $M^3 - int\ N$. It follows that there is the infinite sequence

$$T^2, f^{-1}(T^2), \ldots, f^{-i}(T^2), \ldots \subset M^3 - int\ N$$

of disjoint non-parallel incompressible 2-tori. This contradicts to Haken Finiteness Theorem (see, exm., [79]). Hence, T^2 is compressible in $M^3 - int\ N$. This means that there is a properly embedded disc $(D, \partial D) \subset (M^3 - int\ N, T^2)$ such that ∂D is an essential circle in T^2. It follows that $M^3 - int\ N =$ (solid torus)#M_1 and M^3 is obtained as a conglutination of N and solid torus)#M_1 along the boundaries of the solid toruses.

The idea of second part of the theorem on the existence of a diffeomorphism $f : L_{p,q} \to L_{p,q}$ with a pure solenoid, for simplicity, we demonstrate for the 3-sphere $S^3 = L_{1,0}$, which can be obtained by identifying two solid tori N_1 and N_2 along their common boundary $\partial N_1 = \partial N_2$ that is a torus, Fig. 29.11.

Consider two links (C_1, β_2), (C_2, β_1) in S^3, where $C_1, \beta_1 \in N_1$ and $C_2, \beta_2 \in N_2$. Let us show schematically that there is a diffeotopy $\varphi_t : S^3 \to S^3$ such that $\varphi_1(C_1) = \beta_1$, $\varphi_1(\beta_2) = C_2$, $\varphi_0 = id$. Since the identification $\partial N_1 \to \partial N_2$ takes

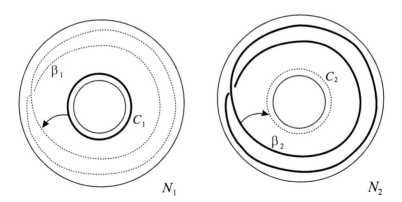

Fig. 29.11

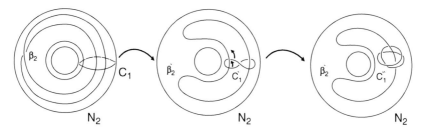

Fig. 29.12

a longitude to a meridian and vice versa, C_1 can be transform to a curve in N_2, see left part of Fig. 29.12. Then one can pull $C_1 \to C_1' \to C_1''$ as indicated in Fig. 29.12. At the same time, one can pull $\beta_2 \to \beta_2'$. Finally, deform $C_1'' \to \beta_1$, $\beta_2' \to C_2$.

Denote by $T(k)$ a tubular neighborhood of a simple closed curve k. The diffeotopy φ_t induces a diffeotopy of the tubular neighborhoods $T(C_1) \to T(\beta_1)$, $T(\beta_2) \to C_2$. One can assume that φ_1 preserves disk structure. Taking in mind that there are diffeomorphisms $N_1 \leftrightarrow T(C_1)$, $N_2 \leftrightarrow T(C_2)$ preserving disk structure, we see that φ_1 induces a Smale diffeomorphism $N_1 \to T(\beta_1)$ that can be extended to a diffeomorphism of S^3. □

Theorem 29.28 means that in sense all pure solenoids can be obtained, up to conjugacy, from diffeomorphisms of lens spaces. Theorem 29.28 was rediscovered in [88].

Recall that Bothe [43] introduced a model diffeomorphism $f_{\beta,m,\varepsilon,\delta} : N \to N$ with a one-dimensional expanding attractor $\Lambda_{\beta,m,\varepsilon,\delta}$, where $f(D^2 \times \{t\}) \subset D^2 \times \{\varepsilon p t\}$, $(f|_{T^2})_*(\lambda) = \lambda_1^j \mu_1^m$, $(f|_{T^2})_*(\mu) = \mu_1^\delta$. By Theorem 29.28, $f_{\beta,m,\varepsilon,\delta}$ can be extended to a diffeomorphism of some lens space $L_{r,s} \to L_{r,s}$. Denote by $B_{\beta,m,\varepsilon,\delta}$ the basin of $\Lambda_{\beta,m,\varepsilon,\delta}$, $B_{\beta,m,\varepsilon,\delta} = B(\Lambda_{\beta,m,\varepsilon,\delta})$.

Let us consider the classification of pure solenoids by Bothe [43].

Theorem 29.29. *Let $f : M^3 \to M^3$ be a diffeomorphism of a closed 3-manifolds M^3. Suppose that f has a pure solenoid Λ with the basin $B(\Lambda)$. Then the restriction $f|_{B(\Lambda)}$ of f on $B(\Lambda)$ is conjugate to the restriction $f_{\beta,m,\varepsilon,\delta}|_{B_{\beta,m,\varepsilon,\delta}}$ of some model diffeomorphism $f_{\beta,m,\varepsilon,\delta}$ on the basin $B_{\beta,m,\varepsilon,\delta} = B(\Lambda_{\beta,m,\varepsilon,\delta})$ of $\Lambda_{\beta,m,\varepsilon,\delta}$.*

Let Λ_i be an attractors of a diffeomorphism f_i, $i = 1, 2$. Following [45] (see also [46]) we say that the attractors Λ_1, Λ_2 are *basin equivalent* if there is a homeomorphism $\varphi : B(\Lambda_1) \to B(\Lambda_2)$ such that

$$f_2 \circ \varphi|_{B(\Lambda_1)} = \varphi \circ f_1|_{B(\Lambda_1)}.$$

Theorem 29.29 says that a pure solenoid in a 3-manifold is basin equivalent to a model pure solenoid reducing the problem of classification to the classification of model pure solenoids. This classification done in [43].

29.2.3.2 Bunches of Expanding Attractors

Dimension one expanding attractors on a surface are codimension one attractors. Let us give some definitions for such attractors for any dimension ≥ 1. Let Λ be a codimension one expanding attractor on closed n-manifold M^n, $n \geq 2$. Then Λ consists of $(n-1)$-dimensional unstable manifolds W_x^u, $x \in \Lambda$ (thus, dim $E_\Lambda^s = 1$) and locally homeomorphic to the product of $(n-1)$-dimensional Euclidean space and a Cantor set of an interval [118,139]. Recall that any Cantor set may be obtained after deleting from the interval the countable set of disjoint open intervals, called adjacent intervals. Each endpoint of an adjacent interval is called a *boundary point* of the Cantor set. An unstable manifold W_x^u passing through a boundary point of the Cantor set is called *boundary*. Union of all boundary unstable manifolds form so-called *accessible boundary* of Λ from $M^n - \Lambda$. One can prove that there are only finitely many boundary unstable manifolds each from which passes through a periodic point [62, 63, 78, 119].

The boundary unstable manifolds of Λ split into a finite number of so-called bunches as follows, Fig. 29.13. The pairwise disjoint unstable manifolds $W^u(p_1), \ldots, W^u(p_k)$ is said to be a k-*bunch* if there are points $x_i \in W^u(p_i)$ and arcs

$$[(x_i, y_i)]_\emptyset^s, \quad y_i \in W^u(p_{i+1}), \quad 1 \leq i \leq k, \text{ where } p_{k+1} = p_1, y_k \in W^u(p_1),$$

and there are no $(k+1)$-bunches containing the given one. The boundary periodic points $p_1, \ldots, p_k$ are called *associated*.

The main difference between $n = 2$ and $n \geq 3$ is that an expanding attractor on M^2 can have k-bunches for any $k \in \mathbb{N}$ (k is even if the expanding attractor is orientable) while for $n \geq 3$, an expanding attractor on M^n can have only 1- and 2-bunches (only 2-bunches if the expanding attractor is orientable). It is a reason to consider codimension one expanding attractors on surfaces and codimension one expanding attractors on manifolds of higher dimension (≥ 3) separately.

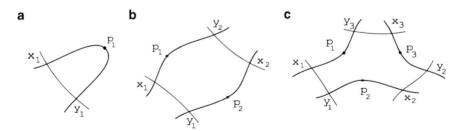

Fig. 29.13 (**a**) 1-bunch, (**b**) 2-bunch, (**c**) 3-bunch

29 Dynamical Systems with Nontrivially Recurrent Invariant Manifolds 453

29.2.3.3 Grines–Plykin–Zhirov Classification of Surface Attractors

Let $f : M \to M$ be a diffeomorphism with uniformly hyperbolic non-wandering set $NW(f) = clos\, Per(f)$, where M is a closed orientable surface of genus $g \geq 0$. Let Ω be a one-dimensional basic set of f. Then Ω is either an attractor or a repeller [118]. If Ω is an attractor, the unstable manifold of Ω belongs to Ω. Thus any one-dimensional attractor of f is nontrivial expanding attractor. Here we consider a little more general case considering one-dimensional expanding attractors, which are not necessary connected set.

The problem of the topological classification requires to find necessary and sufficient conditions for the existence the homeomorphism that conjugates restrictions of diffeomorphisms to their expanding attractors. To be precise, let Ω and Ω' be expanding attractors of diffeomorphisms f and f', respectively. When there is a homeomorphism $g: M \to M$ such that

$$g(\Omega) = \Omega', \quad f'|_\Omega = gfg^{-1}|_{\Omega'}.$$

The first results in solving this problem were obtained by the Grines [62]–[64] for orientable attractors. Recall that a nontrivial basic set Ω is called orientable if for any point $x \in \Omega$ and fixed positive numbers α, β the intersection index of manifolds $W_\alpha^s(x)$ and $W_\beta^u(x)$ is the same at all points of intersection, where $W_\alpha^s(x) = \{y \in W^s(x)|l(x, y) < \alpha\}$ and $W_\beta^u(x) = \{y \in W^u(x)|l(x, y) < \beta\}$ (l is a metric on $W^s(x)$ and $W^u(x)$). The generalization of the orientability of basic set is a widely disposition which is defined as follows. A nontrivial basic set Ω is called *widely disposed* if there is no null-homotopic loop formed by arcs (segments) of stable and unstable manifolds of a point from Ω. The results above by Grines were generalized by Plykin [119, 122] to the cases of widely disposed expanding attractors on a orintable surface of genus ≥ 1 and expanding attractors with bunches of degree no greater than two on a orientable surface of genus ≥ 0. Then Grines and Plykin [74] obtained the topological classification of widely disposed expanding attractors on non-orientable surfaces.

For arbitrary one-dimensional expanding attractors on an orientable surface M of genus ≥ 0, the above stated problem was completely solved by the Grines and Kalai [68, 70, 71] (see also reviews [21, 22]) reducing the topological classification to the algebraic classification of the generalized hyperbolic automorphisms of fundamental groups for canonical supports of expanding attractors. Here our exposition follows to [65] and [68].

Let us recall that a periodic point $p \in \Omega$ is called *boundary* if one of the connected components of $W^s(p) \setminus p$ does not intersect Ω. According to [63] and [119], there are finitely many of boundary periodic points. For a boundary periodic point p, denote by $W_\emptyset^s(p)$ a component of $W^s(p) \setminus p$ that does not intersect Ω. If q is a saddle periodic point and $x, y \in W^\sigma(q)$, $\sigma \in \{s, u\}$, then we denote by $[x, y]^\sigma$, $[x, y)^\sigma$, $(x, y]^\sigma$, and $(x, y)^\sigma$ connected arcs in the manifold $W^\sigma(q)$ with the endpoints x, y. We recall that any nontrivial basic set Ω can be represented as a finite union $\Omega_1 \cup \cdots \cup \Omega_m$ of connected closed subsets ($m \geq 1$), which are called C-dense components of Ω, where $f^m(\Omega_i) = \Omega_i$, $f(\Omega_i) = \Omega_{i+1}$, ($\Omega_{m+1} = \Omega_1$), and for each point $x \in \Omega_i$, $i \in \{1, \dots, m\}$, the set $W^\sigma(x) \cap \Omega_i$ is dense in Ω_i, $\sigma \in \{s, u\}$ [3, 47]). It follows from [62]–[64], [119]–[122], that the accessible boundary from inside of the set $M \setminus \Omega_i$ uniquely falls into a finite number $R(\Omega)$ of bunches. Each bunch C is a union of r_C unstable manifolds $W^u(p_1) \cup \dots \cup W^u(p_{r_C})$ of boundary periodic points $p_1, \dots, p_{r_C}$ with the following property: there exists a sequence of points $x_1, \dots, x_{2r_C}$ such that

(1) x_{2i-1} and x_{2i} belong to distinct connected components of the set $W^u(p_i) \setminus p_i$.
(2) $x_{2i+1} \in W^s(x_{2i})$ (we set $x_{2r_C+1} = x_1$).
(3) $(x_{2i}, x_{2i+1})^s \cap \Omega = \emptyset$, $i = \overline{1, r_C}$.
(4) The curve $L^u_{2i} \cup (x_{2i}, x_{2i+1})^s \cup L^u_{2i+1}$ is the accessible from inside boundary of the domain D_i that is an immersion of the open disk into the surface M, where L^u_{2i}, (L^u_{2i+1}) is a connected component of the set $W^u(x_{2i}) \setminus x_{2i}$ $(W^u(x_{2i+1}) \setminus x_{2i+1})$ that does not contain the point p_i (p_{i+1}), we set $L^u_{2r_C+1} = L^u_1$, Fig. 29.13.

Lemma 29.1. *Let Ω be an attractor of an A-diffeomorphism f consisting of m C-dense components $\Omega_1, \dots, \Omega_m$ each of which contains a collection $\hat{C}_i$ consisting of $R(\Omega)$ bunches of degree r_C for $C \in \hat{C}_i$. Then there exist a neighborhood V of the set Ω that is the union of m neighborhoods V_i of components Ω_i, a compact submanifold N_Ω that is the union of m compact two-dimensional submanifolds $N_1, \dots, N_m$ with the boundary, and a diffeomorphism f_Ω of the submanifold N_Ω such that*

(1) $\Omega_i \subset V_i \subset N_i$.
(2) $f_\Omega|_V = f|_V$.
(3) Every submanifold N_i has $R(\Omega)$ boundary components, is of genus $q \geq 0$ and of a negative Euler characteristic $\chi(N_i) = 2 - 2q - R(\Omega)$ (the numbers q and $R(\Omega)$ are uniquely determined by Ω).
(4) The set $N_\Omega \setminus (\Omega \cup \partial N_\Omega)$ consists of wandering points of the diffeomorphism f_Ω and is the union of $m R(\Omega)$ disjoint domains that are immersions of the open annulus into the manifold M. The accessible from inside boundary of each of such domains consists exactly of one bunch C of the set Ω and of one boundary component ∂N_Ω of the manifold N_Ω containing exactly r_C saddle periodic points and exactly r_C source periodic points of the diffeomorphism f_Ω.

The submanifold N_Ω is called the *canonical support*, and the pair (N_Ω, f_Ω) is called the *canonical form of the attractor* Ω, Fig. 29.14.

Fix a number $i \in \{1, \dots, m\}$. Since the Euler characteristic of the submanifold N_i is negative, it follows from the Nielsen theory [113] that there is a discrete group F of hyperbolic isometries of the hyperbolic plane Δ and a subset $H_F \subset \Delta$ such that F is isomorphic to the fundamental group of N_i and the quotient set H_F / F is homeomorphic to N_i. We denote by π_i the natural projection $H_F \Rightarrow N_i$. By [113], every element $\gamma \in F$ (different from the identity) has exactly two fixed points lying

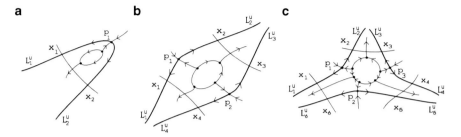

Fig. 29.14

on $S_\infty = \partial \Delta$. Such points are said to be rational. By [113], the closure E_F of all rational points is a Cantor set on S_∞, $E_F = S_\infty \setminus \cup_{k=1}^{k=\infty}(\alpha_k, \beta_k)$, where (α_k, β_k) are adjacent intervals of E_F. Moreover, a geodesic $l_k \subset \Delta$ whose ideal endpoints are α_k and β_k belongs to H_F, and its image under π_i is one of the boundary components of N_i. Let $\tilde{E} = E_F \cup (\cup_{k=1}^{k=\infty} l_k)$. The set $\tilde{E}$ is homeomorphic to a circle and is the boundary of the set $\widetilde{U} = \mathrm{int}\ H_F$ which, is homeomorphic to the open disk and is a universal covering for $\mathrm{int}\ N_i$.

Denote by f_i the restriction of the diffeomorphism f_Ω^m to N_i. Let $\bar{f}_i : H_F \to H_F$ be the covering diffeomorphism for f_i Then $\pi_i \bar{f}_i = f_i \pi_i$. The mapping $\bar{f}_i$ induces the automorphism $\bar{f}_{i*}$ of the group F onto itself by the formula $\bar{f}_{i*}(\gamma) = \bar{f}_i \gamma \bar{f}_i^{-1}$, $\gamma \in F$. According [113] (Sect. 2, p. 9), $\bar{f}_{i*}$ induces a unique homeomorphism $\bar{f}_i^*$ of the set E_F onto itself, and the diffeomorphism $\bar{f}_i$ is uniquely extended to the set E_F and coincides with the homeomorphism $\bar{f}_i^*$ on E_F. If $\bar{\bar{f}}_i$ is a mapping of H_F onto itself covering the diffeomorphism f_i and different from $\bar{f}_i$, then there exists an element $\alpha \in F$ such that $\bar{\bar{f}}_i = \alpha \bar{f}_i$. The mapping $\bar{\bar{f}}_i$ induces the automorphism $A_\alpha \bar{f}_{i*}$, where A_α is the inner automorphism of the group F given by the formula $A_\alpha(\beta) = \alpha \beta \alpha^{-1}$, $\beta \in F$. Thus, to each C-dense component Ω_i of Ω, we get an automorphism of F, which is defined up to an inner automorphism. The pair consisting of the group F and the automorphism $\tau_i = \bar{f}_{i*}$ is denoted by $(F, \tau_i)_{\Omega_i}$ and called *algebraic representation* of Ω_i. An automorphism τ_i is called a *generalized hyperbolic automorphism* if, for any $n \neq 0$ and $\beta, \gamma \in F$ such that $\gamma \neq qid$ and $\pi_i(l_\gamma)$ is not a boundary component of the submanifold N_i, the condition $\beta \tau_i^n(\gamma) \beta^{-1} \neq \gamma$ holds.

Let f, f' be diffeomorphisms whose non-wandering sets contain expanding attractors Ω, Ω' consisting of C-dense components $\Omega_1, \ldots, \Omega_m$ and $\Omega'_1, \ldots, \Omega'_m$ respectively. The algebraic representations $(F, \tau_i)_{\Omega_i}$ and $(F', \tau'_i)_{\Omega'_i}$ are said to be *algebraically adjoint* if there exists an isomorphism $\psi : F \to F'$ such that $\tau'_i = \psi \tau_i \psi^{-1}$.

The following theorems was proved in [68, 70, 71].

Theorem 29.30. *The automorphism $\bar{f}_{i*}$ is a generalized hyperbolic automorphism.*

Theorem 29.31. *For the existence of a homeomorphism $g : N_\Omega \to N'_{\Omega'}$ such that $g(\Omega) = \Omega'$ and $f'|_\Omega = gfg^{-1}|_{\Omega'}$, it is necessary and sufficiently that there exist algebraically adjoint algebraic representations $(F, \tau_i)_{\Omega_i}$ and $(F', \tau'_i)_{\Omega'_i}$ for some coding of C-dense components of the sets Ω and Ω' and for some $i \in \{1, \dots, m\}$.*

Thus the problem of topological classification of expanding attractors is reduced to the algebraic classification of the generalized hyperbolic automorphisms of fundamental groups of supports of attractors. In the paper [67] was obtained the problem of realization of one-dimenional attractors of A-diffeomorphisms of surfaces by means construction of hyperbolic homeomorphisms possessed by pair strongly nontrivial transversal geodesic laminations and in the paper [66] was considered the deviation property for invariant manifolds of points belonging to widely disposition attractors from appropriate geodesics (see also [69] for more detailed information).

In a combinatorial language, the classification problem of such automorphisms was obtained by Zhirov [142]–[147]. The classification problem itself is supplemented by the enumeration problem. The latter is posed as follows. It is required to find a representative for each class of topological conjugacy of attractors with a bounded complexity. Here the complexity is understood both in topological and dynamical senses. The topological complexity of an expanding attractor is characterized by the structure of the accessible boundary of the complement of the attractor. The dynamical complexity is characterized by the topological entropy of the restriction of diffeomorphism on the attractor which is assumed to by bounded. Zhirov proved that under these assumptions there exists only a finite number of topological conjugacy classes which means that the enumeration problem is well posed. The main innovation of the approach developed in [142]–[147] is the solving all considered problems by means of finite algorithms.

The solution of the conjugacy and enumeration problems is based on the combinatorial method. It consists in assigning to expanding attractor a finite set of parameters, which is called a code. The definition of the code is formulated on the basis of the geometric construction of so-called fundamental manifold Π of this attractor. The latter is a concept related to the concept of support in the sense of Grines–Kalai. Π is a surface with boundary and with a partition into a finite number of so-called bands (a band partition, whereby Π is called a band surface) that behave in a certain way under the action of f. The situation is similar to the Markov partitions for the Anosov diffeomorphisms of a 2-torus: Π is an analogue of the torus with cuts along some intervals of unstable manifolds of fixed point and the bands are analogues of the elements of the Markov partition. The code describes the partition of Π into bands (and characterizes their disposition) and the action of f on the latter. The basic property of the code is its absolute invariance with respect to the conjugacy of diffeomorphisms on neighborhoods of attractors, which means that the diffeomorphisms are conjugate if and only if some of their codes coincide. The enumeration problem is also solved by means of codes.

29 Dynamical Systems with Nontrivially Recurrent Invariant Manifolds 457

29.2.3.4 Structural Stable Diffeomorphisms with Expanding Attractors

The pattern of an expanding attractor Λ on M^2 is organized by a few of the unstable curves it contains: the *boundary unstable curves*. These are the unstable curves of boundary periodic points. Let $p \in \Lambda$ be a boundary periodic point where Λ is an expanding attractor of a structurally stable diffeomorphism $f : M^2 \to M^2$. Then one can prove that the set $\overline{W^s_\emptyset(p)} \setminus (W^s_\emptyset(p) \cup p)$ consists of exactly one periodic point which is a source [65]. An important corollary of this fact is the following theorem [65, 66].

Theorem 29.32. *If a structurally stable diffeomorphism $f : M^2 \to M^2$ has an expanding attractor, then f has also a periodic source (isolated repelling periodic orbit).*

Let us now define a class $S(M^2)$ for which it is possible to obtain a complete topological invariants similar to Peixoto's distinguished graphs. We say that an A-diffeomorphism f is in $S(M^2)$ if:

- The non-wandering set of f is a union of expanding attractors, contracting repellers and isolated periodic orbits.
- There are only a finite set of heteroclinic orbits belonging to the intersection of stable and unstable manifolds (separatrices) of isolated periodic saddle type points.

A graph is defined as follows. The vertices of this graph correspond to isolated periodic points, C-dense components of expanding attractors and contracting repellers, and heteroclinic domains. The edges of this graph correspond to separatrices of isolated periodic points and boundary periodic points of the one-dimensional basic sets. The graph is endowed with some additional structures. Among them there are automorphisms of the fundamental groups of supports of C-dense components of attractors and repellers, and heteroclinic permutations, describing the topology of the intersection of stable and unstable manifolds of isolated saddle type periodic points. One can prove that the graph is a complete invariant of conjugacy for diffeomorphisms of the class $S(M^2)$.

29.2.4 Codimension One Expanding Attractors

Let $f : T^n \to T^n$ be a diffeomorphism of the n-torus T^n, $n \geq 3$, and Λ a codimension one orientable expanding attractor of f. Following ideas of Newhouse [108], Grines and Zhuzhoma [75] (see also [76]) proved that there is a neighborhood $U(\Lambda)$ of Λ such that (1) $U(\Lambda) \subset W^s(\Lambda)$; (2) $T^n - U(\Lambda)$ consists of a finitely many n-balls $B_1, \ldots, B_m$; (3) $f^{-j}(B_i) \subset B_i$ for some $j \in \mathbb{N}$ and any $1 \leq i \leq m$. It means that f looks like a DA-diffeomorphism up to dynamics in the balls B_i where it globally remains the dynamics of a periodic repelling point. This implies the following crucial statement.

Theorem 29.33. *Suppose $f : T^n \to T^n$ is a diffeomorphism with a codimension one orientable expanding attractor Λ. Then the automorphism*

$$f_* : H_1(T^n, R^n) \to H_1(T^n, R^n)$$

of the first homology group $H_1(T^n, R^n) \simeq R^n$ has no eigenvalues of absolute value 1 (i.e., f_* is hyperbolic). Moreover, there is a codimension one Anosov diffeomorphism $A \overset{\text{def}}{=} A(f) : T^n \to T^n$ and a continuous map $h : T^n \to T^n$ homotopic to identity such that

1. $h(\Lambda) = T^n$.
2. $A \circ h|_\Lambda = h \circ f|_\Lambda$.
3. $T^n - \Lambda$ *is a union of finitely many n-cells w_i such that given any w_i, $h(w_i) = W_A^u(p_i)$ where p_i is a periodic point of A.*

Idea of the proof is based on the following remarkable result of Franks [57]. Recall that a diffeomorphism $C : M^n \to M^n$ a π_1-*diffeomorphism* if, given any homeomorphism $g : K \to K$ of a compact CW complex and any map $h : K \to M^n$ such that $C_* h_* = h_* g_*$, there is a unique base-point-preserving map $h' : K \to M^n$, homotopic to h, such that $C \circ h' = h' \circ g$. Due to Theorem 29.33, f_* is hyperbolic. Take an algebraic automorphism $A(f) : T^n \to T^n$ such that $f_* = A(f)_*$. By Proposition 2.1 in [57], $A(f)_*$ is a π_1-diffeomorphism. Then there exists a map $h : T^n \to T^n$, homotopic to the identity, such that $h \circ f = A(f) \circ h$. qed

By definition, put

$$P(f, h) = \{x \in T^n | h^{-1}(x) \text{ contains at least two points}\}.$$

The following theorem was proved in [75,76] and it follows from Theorem 29.33 and Arov's Theorem 2 [35] which states that an ergodic automorphism of T^n is linear.

Theorem 29.34. *Suppose $f_1, f_2 : T^n \to T^n$ are diffeomorphisms having orientable expanding attractors of codimension one Λ_1 and Λ_2 respectively. Then there exists a homeomorphism $\varphi : T^n \to T^n$ such that*

$$\varphi(\Lambda_1) = \Lambda_2 \text{ and } f_2|_{\Lambda_2} = \varphi f_1 \varphi^{-1}|_{\Lambda_2}$$

if and only if there is a linear map $\psi : T^n \to T^n$ such that

$$\psi \circ h_1 = h_2 \circ \psi \text{ and } \psi(P(f_1, h_1)) = P(f_1, h_1),$$

where $h_i : T^n \to T^n$ $(i = 1, 2)$ are continuous maps homotopic to the identity and such that $h_i \circ f_i = A_i \circ h_i$, $(f_i)_* = (A_i)_*$.

Note that for $n = 2$ Theorems 29.33, 29.34 was proved in [64]. It immediately follows from theorem 29.34 that $f_1|_{B(\Lambda_1)}$ is conjugate to $f_1|_{B(\Lambda_1)}$ by a homotopy trivial homeomorphism iff f_1 and f_2 are obtained from the same, up to conjugacy,

29 Dynamical Systems with Nontrivially Recurrent Invariant Manifolds 459

codimension one Anosov diffeomorphism $A : T^n \to T^n$ ($n \geq 2$) by the Smale surgery in the same set of finitely many periodic points (in the complements of expanding attractors, f_1 and f_2 can have different dynamics). Formally, a conjugacy invariant of $f|_{B(\Lambda)}$ is A, $f_* = A_*$, with a finite set of periodic points of A. In 1980, Plykin [120] (see also [122]) extended the result by Grines and Zhuzhoma [75, 149] and got the following description of a basin of codimension one expanding attractor (orientable and non-orientable) in any manifold.

Theorem 29.35. *Let Λ be a codimension one expanding attractor of $f \in$ Diff (M^n), $n \geq 3$. Then*

For orientable expanding attractor, the basin $B(\Lambda)$ is homeomorphic to T^n minus a finitely many points $\{p_1, \ldots, p_k\}$. Moreover, there is an extension of $f|_{B(\Lambda)}$ to a homeomorphism $T^n \to T^n$ which conjugates to a DA-diffeomorphism.
For non-orientable expanding attractor, the basin $B(\Lambda)$ is double covered by T^n minus a finitely many points.

Idea of the proof. Take a compact neighborhood $U(\Lambda)$ of the orientable Λ such that $U(\Lambda)$ belongs to $B(\Lambda)$ and the boundary of $U(\Lambda)$ is a union of a finitely many $(n-1)$-spheres. Gluing this spheres by balls B_1, we get a closed manifold M_1^n. Since Λ is orientable, the lamination formed by $W^u(x)$, $x \in \Lambda$, can be extended to a codimension one foliation by planes to M_1^n. Novikov [116] and Hsiang–Wall [84] theorems imply that M_1^n homeomorphic to T^n for $n \geq 5$. For the cases $n = 3$ and $n = 4$, one needs some dynamical reasons according to [57]. For the non-orientable Λ, one can prove the existence of a 1-bunch to construct the corresponding involution (see details in [122]). □

This theorem implies that various diffeomorphisms with an orientable expanding attractors of codimension one are constructed from DA-diffeomorphisms of T^n. Due to Theorem 29.35, manifolds admitting such diffeomorphisms have a decomposition into a special connected sum (that is similar to the decomposition in Theorem 29.28 for diffeomorphisms with solenoids).

Theorem 29.36. *Suppose $f : M^n \to M^n$ is a diffeomorphism having an orientable expanding attractor Λ of codimension one ($n \geq 3$). Then M^n can be represented as a connected sum $M^n = T^n \# M_1$ with a torus summand $T \subset M^n$ (this means that the boundary ∂T is an $(n-1)$-sphere and there is an open n-ball B in T^n such that T is homeomorphic to $T^n - B$) such that Λ is contained in int T.*

Similar statements hold for non-orientable codimension one expanding attractors. It follows from Seifert-van Kampen theorem that if M^n admits a codimension one expanding attractor (orientable or non-orientable), then the fundamental group $\pi_1(M^n)$ contains a subgroup isomorphic to the integer lattice Z^n [121, 149]. This generalizes [93], where one proved that if a closed n-manifold M^n, $n \geq 3$, admits a codimension one expanding attractor (orientable or non-orientable), then M^n has a nontrivial fundamental group.

Taking into account Theorems 29.34 and 29.35 allows Plykin [120, 122] to get a complete invariant of conjugacy for codimension one expanding attractors. For an orientable expanding attractor Λ, the complete invariant the same with the case $\Lambda \subset T^n$. For the non-orientable Λ, we have to add the involution $\theta : T^n \to T^n$ with a finitely many fixed points $Fix\ \theta$ such that $A \circ \theta = \theta \circ A$ and $Fix\ \theta \subset P$, $\theta(P) = P$, where $P = \{p_1, \ldots, p_k\}$.

Let us consider the additional restriction for f being structurally stable. The authors [77, 78] proved that f must be almost DA-diffeomorphism of the n-torus.

Theorem 29.37. *Suppose f is a structurally stable diffeomorphism of a closed n-manifold M^n ($n \geq 3$) and Λ is a codimension one orientable expanding attractor of f. Then:*

1. *M^n is homotopy equivalent to the n-torus T^n. If $n \neq 4$, then M^n is homeomorphic to T^n.*
2. *The spectral decomposition of f consists of Λ, and a finite nonzero number of repelling periodic orbits of index n, and a finite number (maybe zero) of periodic saddle orbits of unstable index $n - 1$.*

Plykin informed us (personal communication) that from Theorem 29.35 one can deduce that M^4 is also homeomorphic to T^4. The crucial step to prove Theorem 29.37 is the following statement.

Theorem 29.38. *Let f be an A-diffeomorphism of a closed n-manifold M^n ($n \geq 3$), and Λ an orientable expanding attractor of codimension one. Suppose $\Omega \neq \Lambda$ is a nontrivial basic set of f such that $W^u(\Omega) \cap W^s(\Lambda) \neq \emptyset$. Then f is not structurally stable.*

Idea of the proof. Taking into account the Mañé-Robinson theorem [98, 126], it is sufficient to prove that f does not satisfy the strong transversality condition. Therefore we can assume that Morse's index of every basic set Θ with $W^u(\Theta) \cap W^s(\Lambda) \neq \emptyset$ is not less than $n - 1$. Suppose the theorem is not correct; then any stable manifold $W^s(x)$, $x \in \Lambda$, intersects transversally any unstable manifold $W^u(y)$, $y \in NW(f)$. By the condition of the theorem and the paper [83], $W^s(z) \cap W^u(z') \neq \emptyset$ for some points $z \in \Lambda$, $z' \in \Omega$. Hence the unstable manifold $W^u(z')$ is either $(n - 1)$-dimensional or n-dimensional. If $W^u(z')$ is n-dimensional, then z' is a periodic point, and so Ω is trivial. Thus, $W^u(z')$ is $(n - 1)$-dimensional. Since Ω is nontrivial, $W^u(z')$ "must return" providing a tangency, see Fig. 29.15. This contradiction with [98, 126] concludes the proof. $\qquad\square$

Theorem 29.37 allows to classify, up to conjugacy, structurally stable diffeomorphisms with orientable expanding attractors of codimension one on the torus T^n. For these diffeomorphisms we introduce the complete invariant of conjugacy, a data set, as follows. Given any pair (p, q) of associated boundary periodic points, we assign the number $n(p, q) \in \mathbb{N}$ of repelling periodic points that are inside the characteristic sphere S_{pq} corresponding to the 2-bunch $B = W^u(p) \cup W^u(q)$. This number is well defined, because it does not depend on the choice of a characteristic sphere. Obviously, $n(p, q) = n(f^m(p), f^m(q))$ for any $m \in \mathbb{Z}$. Therefore we can assign the number $n(p, q) \overset{\text{def}}{=} n(O(p, q))$ to the pair of orbits $O(p)$, $O(q)$ of the points p, q.

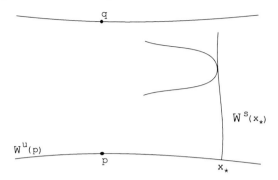

Fig. 29.15

Let $\{O(p_i, q_i)\}_{i=1}^{k}$ be the pairs of orbits of associated boundary points, and let $\{n(O(p_i, q_i))\}_{i=1}^{k}$ be the set of corresponding natural numbers defined above. To each periodic orbit $h(O(p_i)) = h(O(q_i))$ of $A(f)$, we assign the number $n(O(p_i, q_i))$. The collection $\{h(O(p_i)), n(O(p_i, q_i))\}_{i=1}^{k}$, denoted by $\mathscr{D}(f, h)$, is called the *data set* of f, $\mathscr{D}(f, h) = \{h(O(p_i)), n(O(p_i, q_i))\}_{i=1}^{k}$.

Let A be an arbitrary codimension one hyperbolic automorphism of T^n, and let $\{O_j\}_{j=1}^{r}$ be any finite family of periodic orbits of A. To any orbit O_j, let us assign an arbitrary natural number $n_j \in \mathbb{N}$. The collection $\{O_j, n_j\}_{j=1}^{r}$ is called an *admissible data set of the automorphism* A. Note that by Theorem 29.37, a data set of f is admissible whenever f is structurally stable. Suppose $\{O_j^1, n_j^1\}_{j=1}^{r_1}$ and $\{O_j^2, n_j^2\}_{j=1}^{r_2}$ are admissible data sets of codimension one hyperbolic automorphisms A_1 and A_2 respectively. These data sets are called *equivalent* if $r_1 = r_2 \stackrel{\text{def}}{=} r$ and there is an affine transformation $\psi : T^n \to T^n$ (i.e., the composition of an automorphism and translation, $\psi = Ax + \xi$) such that

$$\psi \circ A_1 = A_2 \circ \psi, \quad \psi\left(\bigcup_{j=1}^{r} O_j^1\right) = \bigcup_{j=1}^{r} O_j^2, \quad n(\psi(O_j)) = n(O_j), \quad 1 \leq j \leq r.$$

Theorem 29.39. *Suppose $f_1, f_2 : T^n \to T^n$ are structurally stable diffeomorphisms having orientable expanding attractors of codimension one Ω_1 and Ω_2 respectively. Then f_1 and f_2 are conjugate if and only if the data sets $\mathscr{D}(f_1, h_1)$ and $\mathscr{D}(f_2, h_2)$ are equivalent, where $h_i : T^n \to T^n$ ($i = 1, 2$) are continuous maps homotopic to the identity and such that $h_i \circ f_i = A_i \circ h_i$, $(f_i)_* = (A_i)_*$.*

Theorem 29.40. *Let A be a codimension one hyperbolic automorphism of T^n such that the stable manifolds of A are one-dimensional. Given an admissible data set $\{O_j, n_j\}_{j=1}^{r}$ of A, there is a structurally stable diffeomorphism $f : T^n \to T^n$ having an orientable expanding attractor of codimension one and such that $\mathscr{D}(f, h) = \{O_j, n_j\}_{j=1}^{r}$, where $h : T^n \to T^n$ is a continuous map homotopic to the identity with $h \circ f = A \circ h$ and $f_* = A_*$.*

The proof of Theorem 29.39 follows the proving of Theorem 29.34 (see details in [78]). As to Theorem 29.40, f is constructed by the 'surgery operation' described by Smale [130] and Williams [137] for the 2-torus (the neat construction is in [127]), and [58] for the 3-torus, and [122] for any n-torus, $n \geq 2$.

Let us consider the question on the existence of structurally stable diffeomorphisms with Plykin attractors i.e., codimension one non-orientable expanding attractors. Due to [119], such diffeomorphisms exist on any closed surfaces. As to odd-dimensional closed manifolds, the answer is negative. The following theorem was proved in [102] for dim $M = 3$ and arbitrary dim $M = 2m + 1$ in [104].

Theorem 29.41. *Let* $f : M \rightarrow M$ *be a structurally stable diffeomorphism of a closed* $(2m + 1)$-*manifold* M, $m \geq 1$. *Then* f *does not contain Plykin attractors.*

Idea of the proof. The crucial step is the following lemma.

Lemma 29.2. *Let* $f : M \rightarrow M$ *be an A-diffeomorphism of a closed* $(2m + 1)$-*manifold* M, $m \geq 1$. *If the spectral decomposition of* f *contains a Plykin attractor, then* M *is non-orientable.*

This lemma is not true for even-dimensional manifold (see remark after theorem 29.40). After Lemma 29.2, one use Theorem 29.37 and properties of double coverings. qed

After Theorem 29.41, it is natural to consider the existence of Plykin attractors on closed d-manifolds, $d \geq 3$, including $d = 2m + 1$. Recall that an A-diffeomorphism f is Ω-*stable* if all diffeomorphisms C^1-close to f preserve the structure of $NW(f)$. One can prove the following theorem [104].

Theorem 29.42. *Given any* $d \geq 3$, *there is an* Ω-*stable A-diffeomorphism* f *of closed* d-*manifold* M *such that* f *has a codimension one non-orientable expanding attractor.*

By construction, the manifolds M^{2m}, M^{2m+1} in Theorem 29.40 are orientable and non-orientable respectively. The question on existence of Plykin attractors for structurally stable diffeomorphisms of even-dimensional manifolds (except dim $M = 2$) is still open.

Note that Plykin [123] considered the question when two codimension one expanding attractors are topologically homeomorphic thyself (with no a commutative diagram) and have got the topological classification of such attractors. The more general case (Denjoy minimal sets of codimension one foliations) but in more abstract formulations was considered in [148].

29.2.5 Surface Basic Sets and Hirsch Problem

In 1969, Hirsch [81], posed the following question. Let $f : N \rightarrow N$ be a diffeomorphism of a manifold N and $M \subset N$ a hyperbolic invariant set. Whether the restriction $f|_M$ of f to M is Anosov if M is a closed manifold? Following Mañé

29 Dynamical Systems with Nontrivially Recurrent Invariant Manifolds 463

[97], this restriction $f|_M : M \to M$ we call a *quasi-Anosov diffeomorphism*. In his doctoral thesis, Mane [97] proved that $f : M \to M$ is quasi-Anosov if and only if any of the following conditions hold: (1) the set $\{||Df^n(x)\mathbf{v}||\} : n \in \mathbb{Z}$ is unbounded for all non-zero vectors $\mathbf{v} \in T_x M$ (actually, $||Df^n(x)\mathbf{v}|| \to \infty$ either for $n \to +\infty$ or for $n \to -\infty$); (2) $T_x W^s(x) \cap T_x W^u(x) = \{0\}$ for all $x \in M$; (3) $W^s(x)$ has the same dimension for all periodic points $x \in M$. Mañé proved that a quasi-Anosov diffeomorphism f is Anosov if and only if f is structurally stable. Now, the Hirsch problem means the description of the topology of M and dynamics of f.

In 1976, Franks and Robinson [58] constructed the example of a quasi-Anosov diffeomorphism $f : M^3 \to M^3$ that is not Anosov as essentially follows. They consider a diffeomorphism $F : T^3 \to T^3$ with a codimension one DA-attractor, where T^3 is a 3-torus. The converse F^{-1} has a codimension one contracting repeller. Then one cuts suitable neighborhoods of fixed points (F has an isolated source, and F^{-1} has an isolated sink), and carefully glues together along their boundaries so that the stable and unstable manifolds intersect quasi-transversally. The diffeomorphisms F, F^{-1} form the quasi-Anosov f of M^3, where M^3 is a connected sum $T^3 \sharp T^3$. Note that the non-wandering set of f consists of orientable expanding attractor and contracting repeller. Medvedev and Zhuzhoma [103] constructed the similar example of a quasi-Anosov diffeomorphism the non-wandering set of whose consists of non-orientable expanding attractor and contracting repeller. Before the gluing together two copies of T^3, one can perform a quotient of T^3 by the involution $\theta : x \to -x \, mod \, 1$. The complete decision of Hirsch problem for 3-manifolds was obtained by Fisher and Rodriguez Hertz [55].

Theorem 29.43. *Let $f : M^3 \to M^3$ be a quasi-Anosov diffeomorphism of a closed 3-manifold M^3. If M^3 is orientable, then*

$$M^3 = T_1 \sharp \cdots \sharp T_k \sharp H_1 \sharp \cdots \sharp H_r$$

is the connected sum of $k \geq 1$ tori $T_i = T^3$ and $r \geq 0$ handles $H_j = S^2 \times S^1$. If M^3 is non-orientable, then

$$M^3 = \tilde{T}_1 \sharp \cdots \sharp \tilde{T}_k \sharp H_1 \sharp \cdots \sharp H_r$$

is the connected sum of $k \geq 1$ tori quotiented by involutions $\tilde{T}_i = T^3/\theta$ and $r \geq 0$ handles $H_j = S^2 \times S^1$.

Theorem 29.44. *Let $f : M^3 \to M^3$ be a quasi-Anosov diffeomorphism of a closed 3-manifold M^3. Then*

1. *The non-wandering set $NW(f)$ consists of finitely many codimension one expanding attractors, codimension one contracting repellers, and isolated periodic hyperbolic points.*
2. *For each attractor $\Lambda_f \subset NW(f)$, there exists a DA-diffeomorphism $g : T^3 \to T^3$ with a codimension one expanding attractor Λ_g and a finite set Q of isolated periodic sources such that the restriction of f to its basin of attraction $W^s(\Lambda_f)$*

is conjugate to the restriction of g to the punctured torus $T^3 \setminus Q$ quotiented by a map $\vartheta : T^3 \to T^3$. If M^3 is orientable, ϑ is the identity map. If M^3 is non-orientable, ϑ is the involution θ. An analogous result holds for the repellers of $NW(f)$.

In [129], one proved that a quasi-Anosov diffeomorphism of 3-torus is Anosov. If a quasi-Anosov diffeomorphism $f : M^4 \to M^4$ of 4-manifold M^4 is not Anosov, then the fundamental group $\pi_1(M^4)$ contains a subgroup isomorphic to Z^6.

The generalization of Hirsch problem was proposed by V. Grines who suggested consider nontrivial basic sets that belong to topologically embedded invariant sub-manifolds. The first natural step is to study basic sets that belong to embedded surfaces in a 3-manifold. Following [72, 73], we call a basic set $\mathscr{B}$ of diffeomorphism $f : M^3 \to M^3$ a *surface basic set* if $\mathscr{B}$ belongs to an f-invariant closed surface $M^2_{\mathscr{B}}$ topologically embedded in the 3-manifold M^3. The f-invariant surface $M^2_{\mathscr{B}}$ is called a *supporting surface* for $\mathscr{B}$. By definition, a supporting surface is not necessary connected. But it is obviously that there is some power of diffeomorphism f for which every surface basic set has connected supporting surface.

Let us recall some notation on topological embedding. For $1 \le m \le n$, we presume Euclidean space R^m to be included naturally in R^n as the subset whose final $(n - m)$ coordinates each equals 0. Let $e : M^m \to N^n$ be an embedding of closed m-manifold M^m in the interior of n-manifold N^n. One says that $e(M^m)$ is *locally flat* at $e(x)$, $x \in M^m$, if there exists a neighborhood $U(e(x)) = U$ and a homeomorphism $h : U \to R^n$ such that

$$h(U \cap e(M^m)) = R^m \subset R^n.$$

Otherwise, $e(M^m)$ is *wild* at $e(x)$. When manifolds M^m, N^n are triangulable the notion of flatness is closely related to the notion of tameness. Let M be a triangulable set in R^n. If there is a homeomorphism $h : R^n \to R^n$ such that $h(M)$ is a polygon, then M is *timely embedded*. Otherwise, M is *wild*.

Obviously, that the topological dimension of surface basic set is no more than two. Grines, Medvedev and Zhuzhoma [72, 73] described completely dynamics of 2-dimensional surface basic sets.

Theorem 29.45. *Let $f : M^3 \to M^3$ be an A-diffeomorphism with the surface two-dimensional basic set $\mathscr{B}$ and $M^2_{\mathscr{B}}$ is a supporting surface for $\mathscr{B}$. Then $\mathscr{B} = M^2_{\mathscr{B}}$, and there is a number $k \ge 1$ such that $M^2_{\mathscr{B}}$ is a union $M^2_1 \cup \cdots \cup M^2_k$ of disjoint tamely embedded surfaces such that every M^2_i is homeomorphic to the 2-torus T^2. Moreover, there is $k \ge 1$ such that the restriction f^k to M^2_i ($i \in \{1, \ldots, k\}$) is conjugate to Anosov automorphism of T^2.*

Note that the supporting 2-torus can be essentially non-smoothly embedded in a 3-manifold [89]. Medvedev and Zhuzhoma [106] proved the following theorem that says in sense that a surface diffeomorphism can be embedded in 3-dimensional diffeomorphism such that the embedding surface becomes a wildly embedded attracting invariant surface.

29 Dynamical Systems with Nontrivially Recurrent Invariant Manifolds

Theorem 29.46. *Let g be a homotopy trivial Ω-stable diffeomorphism of closed orientable surface M^2. Suppose g has an isolated sink. Then given any closed 3-manifold M^3, there is the Ω-stable diffeomorphism $f : M^3 \to M^3$ such that:*

1. There is a wildly embedded surface $M \subset M^3$ homeomorphic to M^2 such that M is an invariant and attracting set of f.
2. The restriction $f|_M$ conjugates g.

If g is homotopy nontrivial, the above assertion holds for $M^3 = M^2 \times S^1$.

As a consequence, in contrast with 2-dimensional basic sets, there are one-dimensional and zero-dimensional nontrivial surface basic sets with wild supporting surfaces.

Acknowledgements The research was partially supported by RFFI grants 08-01-00547, 11-01-00730 and grant of government of Russian Federation No 11.G34.31.0039. The authors are grateful to D. Anosov, V. Medvedev, and R. Plykin for useful discussions.

References

1. Aarts, J.M., Fokkink, R.J.: The classification of solenoids. Proc. Am. Math. Soc. **111**, 1161–1163 (1991)
2. Anderson, J.E., Putnam, I.F.: Topological invariants for substitution tillings and their associated C^*-algebra. Ergod. Theroy Dyn. Syst. **18**, 509–537 (1998)
3. Anosov, D.V.: On one class of invariant sets of smooth dynamical systems. Proceedings of the International Conference "Nonlinear Oscillations". Qualitative Methods, Kiev, vol. 2, pp. 39–45 (1970)
4. Anosov, D.V.: Structurally stable systems. Tr. Mat. Inst. im. V.A. Steklova Akad. Nauk SSSR, **169**, 59–93(1985); Engl. transl. Proc. Steklov Inst. Math. **169**, 61–95 (1985)
5. Anosov, D.V.: Basic Concepts. Elementary Theory, in Dynamical Systems-1 (VINITI, Moscow, 1), Itogi Nauki Tekh., Ser.: Sovrem. Probl. Mat., Fundam. Napravl., 156–204 (1985); Engl. transl. in Dynamical Systems I (Springer, Berlin, 1988), Encycl. Math. Sci. 1
6. Anosov D.V.: On the behavior of trajectories, on the Euclidian and Lobachevsky plane, covering trajectories of flows on closed surfaces, I. Izvestia Acad. Nauk SSSR, Ser. Mat. **51(1)**, 16-43 (1987) (in Russian); *Transl. in*: Math. USSR, Izv. 30(1988)
7. Anosov D.V.: On the behavior of trajectories, on the Euclidian and Lobachevsky plane, covering the trajectories of flows on closed surfaces, II. Izvestia Acad. Nauk SSSR, Ser. Mat. **52**, 451-478 (1988) (in Russian); *Transl. in*: Math. USSR, Izvestia. **3(3)**, 449–474 (1989)
8. Anosov D.V.: On the behavior of trajectories, in the Euclidian and Lobachevsky plane, covering the trajectories of flows on closed surfaces, III. Izvestiya Ross. Akad. Nauk, Ser. Mat. **59(2)**, 63-96 (1995) (in Russian); **MR** 97c:58134.
9. Anosov, D.V.: Flows on closed surfaces and behavior of trajectories lifted to the universal covering plane. J. Dyn. Control Syst. **1(1)**, 125–138 (1995)
10. Arnold, V.: Small denominators I. Mapping of the circle onto itself. Izvestia AN SSSR, ser. matem. **25**, 21–86 (1961) (Russian); MR 25#4113
11. Anosov D.V., Solodov V.V.: Hyperbolic Sets, in Dynamical Systems-9 (VINITI, Moscow), Itogi Nauki Tekh., Ser.: Sovrem. Probl. Mat., Fundam. Napravl. **66**, 12–99 (1991); Engl. transl. in Dynamical Systems IX (Springer, Berlin, 1995), Encycl. Math. Sci. 66.
12. Anosov, D.V., Zhuzhoma, E.: Asymptotic behavior of covering curves on the universal coverings of surfaces. Trudi MIAN, **238**, 5–54 (2002) (in Russian)
13. Anosov, D.V., Zhuzhoma, E.: Nonlocal asymptotic behavior of curves and leaves of laminations on covering surfaces. Proc. Steklov Inst. Math. **249**, 221 (2005)

14. Antoine, L.: Sur l'homéomorphisme de deux figures et leurs voisinages. J. Math. Pure et Appl. **4**, 221–325 (1921)
15. Aranson, S., Belitsky, G., Zhuzhoma, E.: Introduction to Qualitative Theory of Dynamical Systems on Closed Surfaces. Trans. Math. Monogr. Am. Math. Soc. **153** (1996)
16. Aranson, S., Bronshtein, I., Nikolaev, I., Zhuzhoma, E.: Qualitative theory of foliations on closed surfaces. J. Math. Sci. **90**(3), 2111–2149 (1998)
17. Aranson, S., Gorelikova, I., Zhuzhoma, E.: The influence of the absolut on the local and smooth properties of foliations and homeomorphisms with invariant foliations on closed surfaces. Dokl. Math. **64**, 25–28 (2001)
18. Aranson, S., Grines, V.: Dynamical systems with minimal entropy on two-dimensional manifolds. Selecta Math. Sovi. **2**(2), 123–158 (1982)
19. Aranson, S., Grines, V.: On some invariant of dynamical systems on 2-manifolds (necessary and sufficient conditions for topological equivalence of transitive dynamical systems). Math. USSR Sb. **19**, 365–393 (1973)
20. Aranson, S., Grines, V.: On the representation of minimal sets of flows on 2-manifolds by geodesic lines. Math. USSR Izv. **12**, 103–124 (1978)
21. Aranson, S., Grines, V.: Topological classification of cascades on closed two-dimensional manifolds. Usp. Mat. Nauk. **45**(4), 3–32 (1990). Engl. transl. in: Russ. Math. Surv. **45**(1), 1–35 (1990)
22. Aranson, S., Grines, V.: Cascades on surfaces. Dynamical Systems-9. VINITI. Moscow. (1991)), Itogi Nauki Tekh., Ser.: Sovrem. Probl. Mat., Fundam. Napravl. **66**, 148–187; Engl. transl. Dynamical Systems IX (Springer, Berlin, 1995), Encycl. Math. Sci. **66**, 141–175
23. Aranson, S., Grines, V.: Cascades on Surfaces. Encyclopaedia of Mathematical Sciences. Dyn. Syst. 9, **66**, 141–175 (1995)
24. Aranson, S., Grines, V., Zhuzhoma, E.: On Anosov-Weil problem. Topology **40**, 475-502 (2001)
25. Aranson, S., Grines, V., Kaimanovich, V.: Classification of supertransitive 2-webs on surfaces. J. Dyn. Control Syst. **9**(4), 455–468 (2003)
26. Aranson, S., Telnyh, I., Zhuzhoma, E.: Transitive and highly transitive flows on closed nonorientable surfaces. Mat. Zametki **63**(4), 625–627 (1998) (Russian)
27. Aranson, S., Medvedev, V., Zhuzhoma, E.: On continuity of geodesic frameworks of flows on surfaces. Russ. Acad. Sci. Sb Math. **188**, 955–972 (1997)
28. Aranson, S., Plykin, R., Zhirov, A., Zhuzhoma, E.: Exact upper bounds for the number of one-dimensional basic sets of surface A-diffeomorphisms. J. Dyn. Control. Syst. **3**(1), 1–18 (1997)
29. Aranson, S., Zhuzhoma, E.: Quasiminimal sets of foliations and one-dimensional basic sets of axiom A diffeomorphisms of surfaces. Dokl. Akad. Nauk, **330**(3), 280–281 (1993) (Russian); Transl. in Russ. Acad. Sci. Dokl. Math. **47**(3), 448–450 (1993)
30. Aranson, S., Zhuzhoma, E.: On structure of quasiminimal sets of foliations on surfaces. Mat. Sb. **185**(8), 31–62 (1994) (Russian); Transl. in Russ. Acad. Sci. Sb. Mat. **82**, 397–424 (1995)
31. Aranson, S., Zhuzhoma, E.: Maier's theorems and geodesic laminations of surface flows. J. Dyn. Control. Syst. **2**(4), 557–582 (1996)
32. Aranson S., Zhuzhoma E.: On asymptotic directions of analytic surface flows. Preprint. Institut de Recherche Mathématiques de Rennes, CNRS, 68 (2001)
33. Aranson, S., Zhuzhoma, E.: Circle at infinity influences on the smoothness of surface flows. In: Walczak, P., et al. Proceedings of *Foliations: Geometry and Dynamics*, held in Warsaw, May 29 – June 9, 2000, pp. 185–195. World Scientific, Singapore (2002)
34. Aranson, S., Zhuzhoma, E.: Nonlocal properties of analytic flows on closed orientable surfaces. Proc. Steklov Ins. Math. **244**, 2–17 (2004)
35. Arov, D.: On topological conjugacy of automorphisms and shifts of compact commutative groups. Uspekhi Mat. Nauk **18**(5), 133–138 (1963) (in Russian)
36. Beardon, A.F.: The geometry of discrete groups. Springer, Berlin (1983)
37. Bendixon, I.: Sur les courbes definies par les equations differentielles. Acta Math. **24**, 1–88 (1901)

38. Bing, R.H.: A simple closed curve is the only homogeneous bounded plane continuum that contains an arc. Can. J. Math. **12**, 209–230 (1960)
39. Bing, R.H.: Embedding circle-line continua in the plane. Canadian J. Math. **14**, 113–128 (1962)
40. Blankenship, W.: Generalization of a construction by Antoine. Ann. Math. **53**, 276–297 (1951)
41. Bonahon, F., Xiaodong, Z.: The metric space of geodesic laminations on a surface: I. Geom. Topol. **8**, 539–564 (2004)
42. Bothe, H.: The ambient structure of expanding attractors, I. Local triviality, tubular neighborhoods. Math. Nachr. **107**, 327–348 (1982)
43. Bothe, H.: The ambient structure of expanding attractors, II. Solenoids in 3-manifolds. Math. Nachr. **112**, 69–102 (1983)
44. Bothe, H.: Transversaly wild expanding attractors. Math. Nachr. **157**, 25–49 (1992)
45. Bothe, H.: How hyperbolic attractors determine their basins. Nonlinearity **9**, 1173–1190 (1996)
46. Bothe, H.: Strange attractors with topologically simple basins. Topol. Appl. **114**, 1–25 (2001)
47. Bowen, R.: Periodic points and measures for axiom A diffeomorphisms. Trans. Am. Math. Soc. **154**, 337–397 (1971)
48. Brown, M.: Locally flat embeddings of topological manifolds. Ann. Math. **75**, 331–341 (1962)
49. Cantrell, J.: n-frames in Euclidean k-space. Proc. Am. Math. Soc. **15**(4), 574–578 (1964)
50. Casson, A.J., Bleiler, S.A.: Automorphisms of Surfaces after Nielsen and Thurston. London Mathematical Society Student Texts, Cambridge University Press, Cambridge (1988)
51. Cherry, T.M.: Topological properties of the solutions of odinary differntial equation. Am. J. Math. **59**, 957–982 (1937)
52. van Danzig, D.: Über topologisch homogene Kontinua. Fund. Math. **14**, 102–105 (1930)
53. Edgar, G.: Measure, Topology, and Fractal Geometry. Springer, NY (1990)
54. Farrell, F., Jones, L.: New attractors in hyperbolic dynamics. J. Diff. Geom. **15**, 107–133 (1980)
55. Fisher, T., Rodriguez Hertz, M.: Quasi-Anosov diffeomorphisms of 3-manifolds. Trans. Am. Math. Soc. **361**(7), 3707–3720 (2009)
56. Fokkink, R., Oversteegen, L.: The geometry of laminations. Fund. Math. **151**, 195–207 (1996)
57. Franks, J.: Anosov diffeomorphisms. In: Global Analisys, Proceedings of the Symposia in Pure Mathematics, vol. 14, pp. 61–94. AMS, Providence, RI (1970)
58. Franks, J., Robinson, C.: A quasi-Anosov diffeomorphism that is not Anosov. Trans. Am. Math. Soc. **223**, 267–278 (1976)
59. Gardiner, C.J.: The structure of flows exhibiting nontrivial recurrence on two-dimensional manifolds. J. Diff. Equ. **57**(1), 138–158 (1985)
60. Gibbons, J.: One-dimensional basic sets in the three-sphere. Trans. Am. Math. Soc. **164**, 163–178 (1972)
61. Gluck, H.: Unknotting S^1 in S^4. Bull. Am. Math. Soc. **69**(1), 91–94 (1963)
62. Grines, V.: On topological equivalence of one-dimensional basic sets of diffeomorphisms on two-dimensional manifolds. Uspekhi Mat. Nauk **29**(6), 163–164 (1974) (in Russian)
63. Grines, V.: On topological conjugacy of diffeomorphisms of a two-dimensional manifold onto one-dimensional orientable basic sets I. Trans. Mosc. Math. Soc. **32**, 31–56 (1975)
64. Grines, V.: On topological conjugacy of diffeomorphisms of a two-dimensional manifold onto one-dimensional orientable basic sets II. Trans. Mosc. Math. Soc. **34**, 237–245 (1977)
65. Grines, V.: On topological classification of structurally stable diffeomorphisms of surfaces with one-dimensional attractors and repellers. Mat. Sb. **188**(4), 57–94 (1997)
66. Grines, V.: Structural stability and asymptotic behavior of invariant manifolds of A-diffeomorphisms of surfaces. J. Dyn. Control Syst. **3**(1), 91–110 (1997)
67. Grines, V.: Representation of one-dimenional attractors of A-diffeomorphisms of surfaces by hyperbolic homeomorphisms. Mat. Zametki **62**, 76–87 (1997)

68. Grines, V.: Topological classification of one-dimensional attractors and repellers of A-diffeomorphisms of surfaces by means of automorphisms of fundamental groups of supports. J. Math. Sci. **95**, 2523–2545 (1999)
69. Grines, V.: On topological classification of A-diffeomorphisms of surfaces. J. Dyn. Control Syst. **6**(1), 97–126 (2000)
70. Grines, V., Kalai, Kh.: On topological equivalence of diffeomorphisms with nontrivial basic sets on two-dimensional manifolds. Cor'kii Math. Sb. 40–48 (1988)
71. Grines, V., Kalai, Kh.: Topological classification of basic sets without pairs of conjugate points of A-diffeomorphisms of surfaces. Gorky State Univ. Press. Dep. VINITI Feb. 10, 1137–1188 (1988) (Russian)
72. Grines, V., Medvedev, V., Zhuzhoma, E.: On two-dimensional surface attractors and repellers on 3-manifolds. Prépublication. Univrsité de Nantes, Laboratoire de Mathématiques Jean-Leray, UMR 6629, 1–16 (2004)
73. Grines, V., Medvedev, V., Zhuzhoma, E.: On two-dimensional surface attractors and repellers on 3-manifolds. Math. Notes **78**, 757–767 (2005)
74. Grines, V., Plykin, R.: Topological Classification of Amply Situated Attractors of A-Diffeomorphisms of Surfaces. Methods of Qual. Theory Diff. Equ. Relat Topics., AMS transl. Ser. 2. **200**, 135–148 (2000)
75. Grines, V., Zhuzhoma, E.: The topological classification of orientable attractors on an n-dimensional torus. Russ. Math. Surv. **34**, 163–164 (1979)
76. Grines, V., Zhuzhoma, E.: Necessary and sufficient conditions for the topological equivalence of orientable attractors on n-dimensional torus. Diff. and Integr. Equat. Sb. of Gorky University, pp. 89–93 (1981) (in Russian)
77. Grines, V., Zhuzhoma E.: On structurally stable diffeomorphisms with expanding attractors or contracting repellers of codimension one. Dokl. Akad. Nauk **374**(6), 735–737 (2000)
78. Grines, V., Zhuzhoma, E.: On structurally stable diffeomorphisms with codimension one expanding attractors. Trans. Am. Math. Soc., **357**, 617–667 (2005)
79. Hempel, J.: 3-manifolds. Annals Math. Studies, vol. 86. Princeton University Press, Princeton (1976)
80. Hermann, M.R.: Sur la conjuguaison différentiable des difféomorphismes du cercle à des rotations, Publ. Math. IHES **49**, 2–233 (1979)
81. Hirsch, M.: Invariant subsets of hyperbolic sets. Lect. Notes Math. **206**, 126–135 (1969)
82. Hirsch, M., Pugh, C.: Stable manifolds and hyperbolic sets. In: Global Analysis, Proceedings of the Symposium, vol. 14, pp. 133–163. AMS, Providence, RI (1970)
83. Hirsch, M., Palis, J., Pugh, C., Shub, M.: Neighborhoods of hyperbolic sets. Invent. Math. **9**, 121–134 (1970)
84. Hsiang, W.C., Wall, C.T.C.: On homotopy tori, II. Bull. Lond. Math. Soc. **1**, 341–342 (1969)
85. Isaenkova, N., Zhuzhoma, E.: Classification of one-dimensional expanding attractors. Math. Notes **86**, 333–341 (2009)
86. Ittai, K.: Strange attractors of uniform flows. Trans. Am. Math. Soc. **293**, 135–159 (1986)
87. Jones, L.: Locally strange hyperbolic sets. Trans. Am. Math. Soc. **275**, 153–162 (1983)
88. Jiang, B., Ni, Y., Wang, S.: 3-manifolds that admit knotted solenoids. ArXiv: math. GT/0403427. (accepted for publ. in Trans. of the Amer. Math. Soc.; electronically published on 27.02.2004).
89. Kaplan, J., Mallet-Paret, J., Yorke, J.: The Lapunov dimension of nonwhere diffferntiable attracting torus. Ergod. Theory Dyn. Syst. **4**, 261–281 (1984)
90. Katok, A.: Invariant measures of flows on oriented surface, Dokl. Akad. Nauk SSSR, **211**, 775–778(1973) (Russian); Transl. in: Sov. Math. Dokl. **14**, 1104–1108 (1973)
91. Katok, A., Hasselblatt, B.: A First Course in Dynamics. Cambridge University Press, Cambridge (2203)
92. Keldysh, L.: Topological embeddings in euclidean space. Moscow. Nauka. Proceedings of Math. Inst. V. A. Steklova., vol. 81. (1966)
93. Kollmer, H.: On hyperbolic attractors of codimension one. Lect. Notes Math. **597**, 330–334 (1977)

94. Levitt, G.: Foliations and laminations on hyperbolic surfaces. Topology **22**, 119–135 (1983)
95. Langevin, R.: Quelques nouveaux invariants des difféomorphismes de Morse-Smale d'une surface. Annales de l'institut Fourier, Grenoble. **43**, 265–278 (1993)
96. Maier, A.: Trajectories on the closed orientable surfaces. Mat. Sbornik. **54**, 71–84 (1943)
97. Mañé, R.: Quasi-Anosov diffeomorphisms and hyperbolic manifolds. Trans. Am. Math. Soc. **229**, 351–370 (1977)
98. Mañé, R.: A proof of C^1 stability conjecture. Publ. Math. IHES **66**, 161–210 (1988)
99. Markley, N.G.: The structure of flows on two-dimensional manifolds. These. Yale University (1966)
100. Medvedev, V.: On a new type of bifurcations on manifolds, Math. USSR, Sb. Math. **41**, 487–492 (1982)
101. Medvedev, V.: On the 'blue-sky catastrophe' bifurcation on two-dimensional manifolds. Math. Notes **51**, 118–125 (1992)
102. Medvedev, V., Zhuzhoma, E.: Structurally stable diffeomorphisms have no codimension one Plykin attractors on 3-manifolds. In: Proceedings of *Foliations: Geometry and Dynamics* (held in Warsaw, May 29 – June 9, 2000), pp. 355–370. World Scientific, Singapore, (2002)
103. Medvedev, V., Zhuzhoma, E.: On non-oreintable two-dimensional basic sets on 3-manifolds. Sb. Math. **193**, 896–888 (2002)
104. Medvedev, V., Zhuzhoma, E.: On the existence of codimension one non-orientable expanding attractors. J. Dyn. Control Syst. **11**, 405–411 (2005)
105. Medvedev, V., Zhuzhoma, E.: Global dynamics of Morse-Smale systems. Proc. Steklov Inst. Math. **261**, 112–135 (2008)
106. Medvedev, V., Zhuzhoma, E.: Surface basic sets with widely embedded support surfaces. Matem. Zametki **85**, 356–372 (2009)
107. Nemytskii, V., Stepanov, V.: Qualitative Theory of Differential Equations. Princeton University Press, Princeton, NJ (1960)
108. Newhouse, S.: On codimension one Anosov diffeomorphisms. Am. J. Math. **92**, 761–770 (1970)
109. Newhouse, S.: On simple arcs between structurally stable flows. Lect. Notes Math. **468**, 262–277 (1975)
110. Nielsen, J.: Uber topologische Abbildungen gesclosener Flächen, Abh. Math. Sem. Hamburg Univ., Heft 3, pp. 246–260 (1924)
111. Nielsen, J.: Untersuchungen zur Topologie der geshlosseenen zweiseitigen Flächen. I, Acta Math. **50**, 189–358 (1927), II - Acta Math. **53**, 1–76 (1929), III - Acta Math. **58**, 87–167 (1932)
112. Nielsen, J.: Die Structure periodischer Transformation von Flachen. Det. Kgl. Dansk Videnskaternes Selskab. Math.-Phys. Meddelerser. 15 (1937)
113. Nielsen, J.: Surface transformation classes of algebraically finite type. Det. Kgl. Dansk Videnskaternes Selskab. Math.- Phys. Meddelerser **21**, 1–89 (1944)
114. Nikolaev, I., Zhuzhoma, E.: Flows on 2-dimensional manifolds. Lecture Notes in Mathematic, vol. 1705. Springer, Berlin (1999)
115. Nitecki, Z.: Differentiable Dynamics. MIT, Cambridge (1971)
116. Novikov, S.P.: Topology of foliations. Trans. Mosc. Math. Soc. **14**, 268–304 (1965)
117. Plante, J.: The homology class of an expanded invariant manifold. Lect. Notes Math. **468**, 251–256 (1975)
118. Plykin, R.V.: On the topology of basic sets of Smale diffeomorphisms. Math. USSR Sb. **13**, 297–307 (1971)
119. Plykin, R.V.: Sources and sinks of A-diffeomorphisms of surfaces. Math. USSR Sb. **23**, 223–253 (1974)
120. Plykin, R.V.: On hyperbolic attractors of diffeomorphisms. Usp. Math. Nauk. **35**, 94–104 (1980) (Russian)
121. Plykin, R.V.: On hyperbolic attractors of diffeomorphisms (non-orientable case). Usp. Math. Nauk **35**, 205–206 (1980) (Russian)
122. Plykin, R.V.: On the geometry of hyperbolic attractors of smooth cascades. Russ. Math. Surv. **39**, 85–131 (1984)

123. Plykin, R.V.: On a problem of topological classification of strange attractors of dynamical systems. Usp. Math. Nauk **57**, 123–166 (2002) (Russian)
124. Poincare, H.: Sur les courbes définies par les equations differentielles. J. Math. Pure Appl. **2**, 151–217 (1886)
125. Pugh, C.: The closing lemma. Ann. Math. **89**, 956–1009 (1967)
126. Robinson C.: Structural stability of C^1 diffeomorphisms. J. Diff. Equ. **22**, 28–73 (1976)
127. Robinson, C.: Dynamical Systems: stability, symbolic dynamics, and chaos. Studies in Advanced Mathematic, 2nd edn. CRC, Boca Raton (1999)
128. Robinson, C., Williams, R.: Classification of expanding attractors: an example. Topology **15**, 321–323 (1976)
129. Rodriguez, H.J., Ures, R., Vieitez, J.: On manifolds supporting quasi-Anosov diffeomorphisms. C. R. Acad. Sci. Paris, Ser. I **334**, 321–323 (2002)
130. Smale, S.: Differentiable dynamical systems. Bull. Am. Math. Soc. **73**, 747–817 (1967)
131. Sullivan, D., Williams, R.: On the homology of attractors. Topology **15**, 259–262 (1976)
132. Thurston, W.: The geometry and topology of three-manifolds. Lecture notes. Princeton University, Princeton (1976–1980)
133. Thurston, W.: On the geometry and dynamics of diffeomorphisms of surfaces. Bull. Am. Math. Soc. **19**, 417–431 (1988)
134. Takens, F.: Multiplications in solenoids as hyperbolic attractors. Topol. Appl. **152**, 219–225 (2005)
135. Vietoris, L.: Über den höheren Zusammenhahg kompakter Räume und Klasse von zusammenhangstreuen Abbildungen. Math. Ann. **97**, 454–472 (1927)
136. Williams, R.F.: One-dimensional non-wandering sets. Topology **6**, 473–487 (1967)
137. Williams, R.F.: The 'DA' maps of Smale and structural stability In: Global Analysis, Proceedings of the Symposium on Pure Mathematics, vol. 14, pp. 329–334. AMS, Providence, RI (1970)
138. Williams, R.F.: The 'DA' maps of Smale and structural stability. In: Global Analysis, Proceedings of the Symposium on Pure Mathematics, vol. 14, pp. 341–361. AMS, Providence, RI (1970)
139. Williams, R.: Expanding attractors. Publ. Math. IHES **43**, 169–203 (1974)
140. Williams, R.: The Penrose, Amman, and DA tiling spaces are Cantor set fiber bundles. Ergod. Theroy Dyn. Syst. **18** (1988)
141. Yoneyama, K.: Theory of Continuous Set of Points. Tohoku Math. J. **12**, 43–158 (1917)
142. Zhirov, A.Yu.: Enumeration of hyperbolic attractors on orientanle surfaces and applications to pseudo-Anosov homeomorphisms. Russ. Acad. Sci. Dokl. Math. **47**, 683–686 (1993)
143. Zhirov, A.Yu.: Hyperbolic attractors of diffeomorphisms of orientable surfaces. Part 1. Coding, classification and coverings. Mat. Sb. **185**, 3–50 (1994)
144. Zhirov, A.Yu.: Hyperbolic attractors of diffeomorphisms of orientable surfaces. Part 2. Enumeration and application to pseudo-Anosov diffeomorphisms. Mat. Sb. **185**, 29–80 (1994)
145. Zhirov, A.Yu.: Hyperbolic attractors of diffeomorphisms of orientable surfaces. Part 3. A classification algorithm. Mat. Sb. **186**, 59–82 (1995)
146. Zhirov, A.Yu.: Complete combinatorial invariants for conjugacy of hyperbolic attractors of diffeomorphisms of surfaces. J. Dyn. Control Syst. **6**, 397–430 (2000)
147. Zhirov, A.Yu.: Combinatorics of one-dimensional hyperbolic attractors of surface diffeomorphisms. Tr. Steklov Math. Ins. **244**, 143–215 (2004) (Russian)
148. Zhuzhoma, E.: Topological classification of foliations which are defined by a one dimensional Pfaff forms on the n-dimensional torus. Diff. Equ. **17**, 898–904 (1981)
149. Zhuzhoma, E.: Orientable basic sets of codimension 1. Sov. Math. Izv. VUZ **26**(5), 17–25 (1982)

Chapter 30
Some Recent Results on the Stability of Endomorphisms

J. Iglesias, A. Portela, and A. Rovella

Abstract This work aims to provide a short description of some old and new results on the theory of stability and related concepts for discrete dynamical systems. The emphasis is posed on noninvertible maps.

30.1 General Definition

The problem of describing the structural stability is central in the theory of dynamical systems, not only by its theoretical interest but also to validate dynamical models appearing in other diverse contexts. Roughly speaking, stability of a system means that small changes in the model will not modify some determined features of the initial model in a qualitative sense. This obviously requires formal definitions of the words *small*, *features* and *qualitative*, so different scenarios will supply corresponding applications.

We give a general definition of stability as follows. Let be given a topology τ on a space X, an equivalence relation $\equiv$ on a space Y and an operator $\Phi : X \to Y$. In the above discussion, X will be the space of models, τ gives sense to proximity between models, the space Y will be the feature of the model we are interested in, and Φ associates to each model a determined feature. With this correspondances in mind, we give a general definition of what may be understood as *stability*.

Definition 30.1. A point $f \in X$ is said $(\tau, \equiv, \Phi)$-stable if there exists a neighborhood U of f such that $\Phi(f) \equiv \Phi(g)$ for every $g \in U$.

J. Iglesias (✉) and A. Portela
IMERL, Facultad de Ingeniería, Universidad de la República, J. Herrera y Reissig 565, Montevideo 11300, Uruguay
e-mail: jorgei@fing.edu.uy, aldo@fing.edu.uy

A. Rovella
Centro de Matemática Facultad de Ciencias, Universidad de la República, Iguá 4225, Montevideo 11400, Uruguay
e-mail: leva@cmat.edu.uy

M.M. Peixoto et al. (eds.), *Dynamics, Games and Science I*, Springer Proceedings in Mathematics 1, DOI 10.1007/978-3-642-11456-4_30,
© Springer-Verlag Berlin Heidelberg 2011

This simple definition includes most global concepts known, as C^r stability, Ω stability, inverse limit stability and ergodic stability, for discrete and continuous dynamical systems, as well as local concepts, as Lyapounov stability and stability of solutions of partial differential equations. For example, if M is a smooth manifold, X is the space of class C^r maps of M endowed with its usual C^r topology, $Y = X$, Φ is the identity, and $\equiv$ is conjugacy or topological equivalence (i.e, $f \equiv g$ if and only if there exists a homeomorphism h such that $fh = hg$), then we find the more traditional concept named C^r structural stability. Throughout this work we will be mostly interested in this one and some other closely related kind of stabilities, and not on the wide scopes of the general definition. The study of the structural stability of maps was perhaps a little bit left aside by the dynamicists since 1987, when R. Mañé published the proof of the long standing conjecture of C^1 stability for diffeomorphisms [17]. However, for noninvertible maps, C^1 structural stability was not characterized yet, and there exist examples of maps that for some $r > 1$, are C^r but not C^1 structurally stable. The theory for continuous time system ran parallel with the discrete one, but we will mainly restrict to discrete dynamics. We also focus on topological and not on metric aspects, so stable ergodicity, a concept that involves very diverse tools, will not be mentioned.

In Sect. 30.2 we present the more usual definitions and those results that seem fundamental for the theory. Sections 30.3–30.5 are devoted to the classical results: diffeomorphisms, Anosov endomorphisms, expanding maps. The remaining sections are reserved to recent results and open problems.

30.2 Notations

In this section we fix notations and explain the basic theory of hyperbolicity and its relation with stability. The whole section may be dropped by those acquainted with the notions of hyperbolicity and invariant manifolds and with the differences between the definitions for invertible and for noninvertible maps.

When M and N are smooth manifolds, the space of C^r maps from M to N equipped with the standard topology will be denoted by $C^r(M, N)$; if $M = N$, then $C^r(M) = C^r(M, N)$. The set of C^r diffeomorphisms of M is denoted by $D^r(M)$.

When dealing with a noninvertible map f in M, the inverse limit of the pair (M, f) is defined as the set

$$\tilde{M}_f = \{\underline{z} = (z_n) \in M^{\mathbb{Z}} \ : \ f(z_{n-1}) = z_n\}.$$

$\tilde{M}_f$ is always compact and shift-invariant. If π_0 is the projection onto the 0-coordinate, then $\pi_0 F = f\pi_0$, where F denotes the restriction of the shift to $\tilde{M}_f$. Consider the space H of homeomorphisms of subspaces of $M^{\mathbb{Z}}$;

30 Some Recent Results on the Stability of Endomorphisms 473

Definition 30.2. Let $\equiv$ be the relation of conjugacy on H and define $\Phi:C^r(M) \to H$ by $\Phi(f) = F$. Then f is said C^r inverse stable if it is $(C^r, \equiv, \Phi)$ stable, according to Definition 30.1.

In other words, a map f is inverse stable if the set of full orbits is preserved under small perturbations.

The hyperbolicity is one of the fundamental concepts in the characterizations of different types of stabilities of maps. A compact, invariant set K is said hyperbolic for f if there exists a continuous invariant bundle E^s defined on K and constants $C > 0$ and $0 < \lambda < 1$, such that the following conditions hold:

1. For any $n > 0$ it holds that $|T_x f^n(v)| \leq C\lambda^n |v|$ for every $v \in E_x^s$.
2. If $[Tf]$ denotes the map induced by Tf on the quotient $TM|_{E^s}$, then the norm of $[Tf]^n$ is greater than $1/(C\lambda^n)$.

It can be proved that if K is hyperbolic for f then, for every $\underline{z} \in \tilde{K}_f := \{\underline{z} \in \tilde{M}_f : z_n \in K \ \forall n \in \mathbb{Z}\}$, there exists a subspace $E_{\underline{z}}^u$ of $T_{z_0}(M)$ such that $E_{\underline{z}}^u \oplus E_{z_0}^s = T_{z_0}M$, where $z_0 = \pi_0(\underline{z})$.

The set K is expanding for f if it is hyperbolic with $E^s = \{0\}$.

The nonwandering set of f is denoted by Ω_f as usual. For noninvertible maps, there are two definitions for Axiom A.

Definition 30.3. A map $f \in C^r(M)$ satisfies the weak Axiom A if the nonwandering set of f is hyperbolic and equal to the closure of periodic points of f. It is well known that for an Axiom A map there exists a decomposition of Ω_f into disjoint, closed, forward invariant transitive sets, called basic pieces.

The map f is Axiom A if it is weak Axiom A and every basic piece Λ is expanding or injective (i.e. the restriction of f to Λ is one to one).

A map $f \in C^r(M)$ is Anosov if M is a hyperbolic set for f. For example, the linear map

$$A = \begin{pmatrix} n & 1 \\ 1 & 1 \end{pmatrix}$$

induces a weak Axiom A map f on the two torus if $n > 2$, and an Axiom A diffeomorphism if $n = 2$.

The following example is due to Przytycki [25]:

Example 30.1. A weak Axiom A map f of the two torus none of whose perturbations is Axiom A, and whose attracting basic piece is not injective.

Consider $f_1(z) = z^2$ defined on S^1, f_2 a diffeomorphism of the circle with a fixed attractor a and a fixed repeller r. Then $f = (f_1, f_2)$ is a weak Axiom A map of the two torus with an attracting basic piece $S^1 \times \{a\}$. The restriction of f to it is not injective. Any perturbation of it has an attracting basic piece which is not injective.

The definition of Axiom A was first given for diffeomorphisms by S.Smale. Weak Axiom A is defined for noninvertible maps and was introduced by F. Przytycki [24]. From these results it follows that the nonwandering set of a weak Axiom A map can be decomposed in a finite union of transitive sets, called basic pieces. Say that

$\Lambda_1 >> \Lambda_2$ for basic pieces Λ_1 and Λ_2 when there exists an orbit $\{x_n\}$ such that $x_{-n} \to \Lambda_1$ and $x_n \to \Lambda_2$. The Axiom A is said to satisfy the no cycles condition if the relation defined above has no cycles.

The theory of invariant manifolds for Axiom A maps was developed by several authors, mainly Palis, Hirsch, Pugh and Schub [8,9]. We introduce the notations of invariant manifolds. There are some differences when the map is not invertible. If f is an Axiom A map, then there exists $\epsilon > 0$ such that for every $x \in \Omega_f$, the set

$$W_\epsilon^s(x; f) = \{y \in M : d(f^n(x), f^n(y)) \xrightarrow[n \to +\infty]{} 0$$
$$\text{and} \quad d(f^n(x), f^n(y)) < \epsilon \; \forall n > 0\}$$

is an embedded manifold (called local stable manifold) having the same dimension as $E^s(x)$ and tangent to $E^s(x)$ at the point x. Here d denotes the distance in the manifold M. On the other hand, local unstable sets are defined in $\tilde{M}_f$, as

$$W_\epsilon^u(\underline{x}; F) = \{\underline{y} \in \tilde{M}_f : d(x_n, y_n) \xrightarrow[n \to -\infty]{} 0 \; \text{and} \; d(x_n, y_n) < \epsilon \; \forall n < 0\}.$$

It holds that π_0 projects the unstable set of $\underline{x}$ to an embedded manifold in M that is tangent to $E^u(\underline{x})$ at x_0.

The Strong Transversality is a necessary condition for C^r structural stability. It says that stable and unstable sets may intersect transversally. It is restricted to Axiom A maps. To explain this condition, we must also take care of the fact that a point $x \in M$ may have different preorbits converging to the nonwandering set. On the other hand, for every $x \in M$ there exists a unique $y \in \Omega_f$ such that $d(x_n, y_n) \to 0$ as $n \to +\infty$. Note that for every $\underline{x} = (x_n)_{\{n \in \mathbb{Z}\}} \in \tilde{M}_f$ it holds that there exists a unique basic piece Λ such that $x_n \to \Lambda$ when $n \to -\infty$. Moreover, there exists a unique $\underline{y} \in \Omega(F)$ such that $d(x_n, y_n) \to 0$ as $n \to -\infty$. The following invariance holds: $F(W_\epsilon^u(\underline{y}; F)) \supset W_\epsilon^u(F(\underline{y}); F)$. This implies that for every $\underline{x} \in \tilde{M}_f$ there exists a subspace $E^u(\underline{x}) \subset T_{x_0} M$, defined as

$$Tf_{x_{-k}}^k (T_{x_{-k}} (\pi_0(W_\epsilon^u(F^{-k}(\underline{y}); F))),$$

where k is such that $x_{-k} \in \pi_0(W_\epsilon^u(F^{-k}(\underline{y}); F))$. Define

$$E^u(x) = \bigcap_{\pi_0^{-1}(x)} E^u(\underline{x}).$$

Definition 30.4. An Axiom A map f satisfies the Strong Transversality condition if

1. For every $x \in M$, the subspaces $E^u(\underline{x})$ are in general position as $\underline{x}$ varies in $\pi_0^{-1}(x)$.
2. For every $x \in M$, $n > 0$ and $y \in \Omega_f$ such that $f^n(x) \in W_\epsilon^s(y)$ it holds that

$$Tf^n(E^u(x)) \oplus T_{f^n(x)}(W_\epsilon^s(y)) = T_{f^n(x)}(M).$$

30 Some Recent Results on the Stability of Endomorphisms 475

The first condition says that there exists a finite number of different $E^u(\underline{x})$ as $\underline{x}$ varies in $\pi_0^{-1}(\underline{x})$, and that the sum of the codimensions of the $E^u(\underline{x})$ equals the codimension of $E^u(x)$. The Strong Transversality condition implies that whenever the unstable set of a basic piece Λ_1 intersects another basic piece Λ_2, then Λ_1 must be an expanding piece. This property has been stated by Przytycki as a necessary condition for C^1 stability.

One of the fundamental tools concerning stability is the C^1 closing lemma.

Theorem 30.1. *Let f be a C^1 endomorphism of a manifold M. Given a point $x \in \Omega_f$ and a C^1 neighborhood $\mathcal{U}$ of f, there exists a map $g \in \mathcal{U}$ such that x is a periodic point of g. Moreover, there exists a residual subset $\mathcal{R}$ of $C^1(M)$ such that the set of periodic points of f is dense in Ω_f for every $f \in \mathcal{R}$.*

The first version of this theorem, valid for diffeomorphisms, was proved by C. Pugh [26]. Two further versions, that hold for endomorphisms without critical points the first, and with a finite number of critical points the second, were obtained by L. Wen in [34] and [35]. The general version stated is recent, [31]. The question if a similar result holds in topologies C^r with $r > 1$ remains open, and is a central problem in the theory.

Definition 30.5. A map f is C^r-Ω stable if it is $(X, \equiv, \Phi)$-stable, where $X = C^r(M)$, $Y = \{f_{|\Omega_f} : f \in C^r(M)\}$, Φ is the restriction operator and $\equiv$ is conjugacy.

The comprehension of the nexus between Ω stability and hyperbolicity has been one of the main goals in the theory. Roughly speaking, that hyperbolicity implies Ω stability was solved for diffeomorphisms in the middle seventies. Smale proved the Ω stability of Axiom A diffeomorphisms with the no cycles property [33]. This theorem was the extended to noninvertible maps by Przytycki [25]:

Theorem 30.2. *If f is a weak Axiom A map without critical points, then*

1. *f is C^1 inverse Ω stable if and only there are no cycles.*
2. *f is C^1-Ω stable if and only if f is Axiom A and there are no cycles.*

From theorem above and the example 1 it follows the no density of the omega stability.

The other direction was and still is very hard. Adapting the proof given by Mañé of the stability conjecture to noninvertible maps, the last result in this direction was obtained by Aoki, Moriyasu and Sumi (see [1]):

Theorem 30.3. *Let $f \in C^1(M)$ be an interior point of the set of maps all of whose periodic points are hyperbolic. Assume that f has no critical points in the nonwandering set.*
Then the nonwandering set of f has a hyperbolic structure.

This implies that if f is C^1-Ω stable (and critical points are wandering) then Ω_f is a hyperbolic set and the closing lemma implies that the set of periodic points of f is dense in the nonwandering set. Moreover, the Theorem of Przytycki implies that f is Axiom A.

Problem 30.1. Prove the same result without any assumption on the singular set.

30.3 Diffeomorphisms

This is a very brief history concerning the stability of C^1 diffeomorphisms, a problem satisfactory solved. When beginning the sixties, Morse and Smale proposed a model of simple dynamics: a Morse Smale map has finitely many nonwandering hyperbolic points and uniformly transverse intersections of stable and unstable manifolds. J. Palis [21] proved the structural stability of Morse Smale systems, while M. Peixoto [23] showed that the class of Morse Smale vectorfields are dense in compact manifolds of dimension two. It was thought that these classes of maps would characterize C^1 structural stability. Afterwards, beautiful examples given by Anosov and Smale showed that maps with an infinite number of nonwandering points can also be structurally stable. The theory of dynamical systems had a great impulse since these examples appear, and after the definition of hyperbolicity and Axiom A maps were given by Smale, and the theory of invariant manifolds was developed, it was conjectured by Palis and Smale [22] that the C^1 structurally stable maps are those satisfying the Axiom A and the Strong Transversality condition (Definition 30.4). It was first proved that the conditions were sufficient for C^1 stability: This part of the problem was solved by Robbin ([29], 1974) for C^2 topology and then by Robinson ([30], 1976) in C^1 topology. In the meanwhile a great step towards the proof of the necessity was the C^1 closing lemma of Pugh. Joining this with the theorems of Kupka and Smale about the genericity of maps whose periodic points are hyperbolic and its invariant manifolds satisfy the transversality condition, and the proof of Franks [6] of the necessity of hyperbolicity of the periodic points, the problem left was to prove the necessity of the hyperbolicity of the nonwandering set. This was finally solved by R. Mañé in [17], 1987, after more than 10 years of big efforts and progresses had been obtained by several mathematicians. The main problem remaining is:

Problem 30.2. For $r > 1$, characterize C^r structural stability for diffeomorphisms.

Concerning this problem, E. Pujals has recently written an interesting survey [27].

30.4 Expanding Maps and Anosov Endomorphisms

A smooth map f of a manifold M is called expanding if M is an expanding set for f. Clearly such map cannot be a diffeomorphism unless the manifold is noncompact. M. Shub [32] showed the stability of expanding maps in 1969. An Anosov endomorphism is a map for which the whole manifold is a hyperbolic set. It was proved by F. Przytycki [25] and independently by R. Mañé and C. Pugh [18] that an Anosov endomorphism is C^1 stable if and only if it is an expanding map or a diffeomorphism. It is also due to Przytycki [24] the proof that every Anosov endomorphism without singularities is inverse limit stable. Further extensions of this result were obtained by Ikeda [10], who showed the inverse stability of Anosov maps with critical points, and by J. Quandt [28], who has considered also maps defined in Banach manifolds.

30 Some Recent Results on the Stability of Endomorphisms

30.5 Geometric Stability

We make a digression here to consider some geometric aspects of maps with singularities and a non dynamical concept of stability, that involves maps between different manifolds. This is intended to describe some of the geometrical features of maps, in a sense that as will be seen below, is closely related to the dynamics when the map is an endomorphism exhibiting critical points.

Definition 30.6. Say that a map f between smooth manifolds M and N is C^r-*geometrically stable* if it is $(C^r(M, N), \equiv, \Phi)$ stable where Φ is the identity operator and $\equiv$ is geometric equivalence: $f \equiv g$ if and only if there exist maps $\varphi \in D^1(M)$ and $\psi \in D^1(N)$ such that $f\varphi = \psi g$.

This concept was introduced by R. Thom; as the maps φ and ψ involved in the equivalence relation are diffeomorphisms, the set of critical values of f and g are diffeomorphic, as well as the set of critical values. It was conjectured by R. Thom, and proved by J. Mather in the middle sixties, that this concept is equivalent to the so called infinitesimal stability, that we proceed to define.

Let $f : M \to N$ and Z be a C^∞ vectorfield along f, that is, $Z : M \to TN$ satisfies $Z(x) \in T_{f(x)}N$. The map f is infinitesimally stable if there exist C^∞ vectorfields $X \in \mathbb{X}(M)$ and $Y \in \mathbb{X}(N)$ such that $Z = Tf \circ X - Y \circ f$. This is not a concept of stability since is given as a property of a map, and does not mention any behavior of its perturbations. The following is a beautiful characterization of geometric stability: A map f is geometrically stable if and only if it is infinitesimally stable.

This was conjectured by R. Thom; the proof was obtained some years later by J. Mather. The concept of geometric stability has a local version: a map f is locally geometrically stable at a point p if there exists a neighborhood U of p such that the restriction of f to U is geometrically stable. Of course that the interest appears when p is a critical point of f. When f is locally geometrically stable at p we say that p is a nondegenerate critical point. It is well known that for an open and dense set of $C^\infty(M, N)$ maps every critical point is nondegenerate. Actually, more than this can be said:

Theorem 30.4. *Given manifolds M and N, there exists an integer r depending just on the dimensions of M and N and an open and dense subset $\mathcal{G}_r(M, N)$ of $C^r(M, N)$ such that every critical point of any map in $\mathcal{G}_r(M, N)$ is nondegenerate.*

However, geometrically stable maps are not dense in general (see [7]). Assume that a map $f \in C^\infty(M)$ has no degenerate critical points. Then the set of points $z \in S(f)$ such that the kernel of the tangent map of f at z has dimension one is either empty or a codimension one submanifold, denoted $S_1(f)$. The set of points in $S_1(f)$ such that the kernel of Tf_z is tangent to $S_1(f)$ is either empty or a submanifold of codimension two, and is denoted by $S_{11}(f)$ or $S_{12}(f)$. If $S_{12}(f)$ is not empty, then the set $S_{13}(f)$ of points in $z \in S_{12}(f)$ such that the kernel of Tf_z is tangent to $S_{12}(f)$ is either empty or a submanifold of codimension three. By recurrence one can define $S_{1^n}(f)$ for every positive n. These are called Morin singularities.

Assume that f has nonempty $S_{1^n}(f)$. Then f is C^{n+1} but not C^n geometrically stable.

Returning to dynamics, fix a C^r endomorphism f of a manifold M; there is another concept, also called infinitesimal stability, introduced by Robbin [29], (who was inspired in Moser [20]) in his proof of the C^2 structural stability of diffeomorphisms: a map f is infinitesimally stable if given a vectorfield Z along f, there exists a vectorfield X on M such that $Z = Tf(X) - X(f)$. It was proved by R.Mañé that for C^1 topology this is equivalent to structural stability when f is a diffeomorphism [16]. The problem if this generalizes to C^r diffeomorphisms or to general endomorphisms is still open.

30.6 Maps with Nonempty Critical Set

To understand the stability of maps with nonempty critical set we introduce the following fact:

Claim. *For generic maps, topological equivalence implies geometric equivalence.*

To explain this fact, note that the maps $f(x) = x^3$ and $g(x) = x/2$ are locally topologically equivalent, but not locally geometrically equivalent at 0.

Let $\mathcal{G}_r(M) = \mathcal{G}_r(M, M)$, as defined in Theorem 30.4 of the previous section. Assume that maps f and g in $\mathcal{G}_r(M)$ are topologically conjugate. Then there exist C^1 diffeomorphisms φ and ψ such that $f\psi = \varphi g$, which means geometrical equivalence. We do not have a reference for the proof of this assertion.

This implies a fundamental fact:

Claim. *Structural stability implies geometric stability.*

This can be precisely stated as follows: if $f \in C^k(M)$ is C^k structurally stable, then it is C^k geometrically stable.

Indeed, if f is not C^k geometrically stable, then one can find two perturbations that are generic and not geometrically equivalent.

For example, the above fact implies that if f is C^1 structurally stable, then the critical set of f must be empty. In general, a C^k structurally stable map cannot have Morin singularities of type S_{1^k}.

It may be rather easy to construct Morin singularities S_{1^k} in any manifold of dimension k.

Problem 30.3. In any manifold of dimension k there exists a C^{k+1} structurally stable map that is not C^k structurally stable.

Example 30.2. A map of the circle with singularities satisfying the Axiom A is C^2 structurally stable if there are no critical relations (meaning that no critical point is periodic and the orbit of a critical point does not meet another critical point), but it cannot be C^1 stable.

Example 30.3. A C^2 structurally stable map in the k-sphere.

Let $\varphi : I \to I$ be a generic C^2 map defined in the interval $I = [1, 2]$ such that $\varphi(1) = 1$, $\varphi(2) = 2$, and φ has exactly two critical points $a < b$. It can also be assumed that φ can be extended as a C^2 map in $[0, +\infty)$ taking $\varphi(x) = x$ whenever $x \notin I$. The genericity assumption implies $\varphi(a) > \varphi(b)$. Let S^k denote the k-dimensional sphere, identify $\mathbb{R}^k$ with $S^{k-1} \times [0, +\infty)$ and define the auxiliary map $\psi : S^{k-1} \times [0, +\infty) \to S^{k-1} \times [0, +\infty)$ by $\psi(x, y) = (x, \varphi(y))$. If B is an arbitrary open ball in $\mathbb{R}^k$ and α is a diffeomorphism from $\mathbb{R}^k$ to B, then define $\psi_B = \alpha \psi \alpha^{-1}$. Note that ψ_B can be extended as the identity in $S^k \setminus B$. To construct the example, let F be a diffeomorphism: the north pole-south pole in S^k, and let B be a ball in S^k such that $F^n(B) \cap B = \emptyset$ for every $n > 0$. Then the map f defined as $f = F\psi$ is a generic C^2 map whose critical set consists in the union of two S^{k-1} spheres. Moreover, it is clear that f is C^2 structurally stable.

Example 30.4. A C^1-Omega stable map f that satisfies the Strong Transversality condition and no perturbation of f is structurally stable.

Let f be an Axiom A diffeomorphism of S^2 whose basic pieces are fixed repellers and a Plykin attractor. Again, let B be a wandering ball and define $f = F\psi_B$ (ψ_B as in the above example). Then f is an Axiom A map and is C^1-Ω stable. The stable manifolds of points in the attractor are one dimensional, and cover the complement of the repellers. Then there exists at least four points of tangency between stable manifolds and the two circles of critical points (see Fig. 30.1). On one hand, standard arguments allow us to prove that for generic C^r perturbations of f the stable manifold of every periodic is transverse to S_f. On the other hand, it is possible to construct a perturbation for which a periodic point of the attractor has a tangency with S_f: these two maps cannot be topologically equivalent.

Observe that until now the examples of structurally stable maps belong to one of these classes: diffeomorphisms, expanding, one dimensional, finite nonwandering

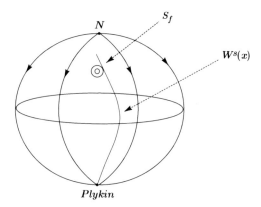

Fig. 30.1 The tangencies between stable manifolds and the two circles of critical points

set. In 2008, another class of examples were given in [12], showing that certain maps of the Riemann sphere are C^3 but not C^2 structurally stable:

Theorem 30.5. *If R is a hyperbolic rational map of the Riemann sphere, then there exist arbitrary small C^∞ perturbations of R that are C^3 structurally stable. Moreover, no perturbation of R is C^2 structurally stable.*

When the perturbations are restricted to the set of holomorphic maps, then we have the following theorem of Mañé et al. [19]:

Theorem 30.6. *If R is a hyperbolic rational map of the sphere without critical relations, then R is H structurally stable, meaning $(\tau, \equiv, \Phi)$-stable, where τ is the topology of uniform convergence in the space of meromorphic maps of the Riemann sphere, $\equiv$ is conjugacy, and Φ the identity operator.*
Moreover, H structurally stable maps are dense in the space of rational maps.

Problem 30.4. Does H structural stability imply hyperbolicity of the Julia set?

Theorem 30.5 is intended to show examples of structurally stable maps having nonempty critical set and nontrivial nonwandering set, in dimension greater than one. A set of sufficient conditions for C^∞ stability that complements the previous examples was recently found by P. Berger [3], in the following statement:

Theorem 30.7. *Let f be a C^∞ weak Axiom A endomorphism of a compact manifold M such that every basic piece is expanding or an attractor. Denote by R the union of the expanding sets, by A the union of the attractors, and by $\hat{\Omega}$ the union of the preimages of Ω. Assume, in addition, that the following conditions hold:*

1. *There are no singularities in $\hat{\Omega}$.*
2. *The restriction of f to $M \setminus \hat{\Omega}$ is C^∞ infinitesimally stable.*
3. *The map f is transverse to the stable foliation, that is: for any $y \in A$, for every z in a local stable manifold $W^s_{loc}(y)$, for any $n \geq 0$ and every point $x \in f^{-n}(\{z\})$, we have:*
$$Tf^n(T_x M) + T_z(W^s_{loc}(y)) = T_z M$$
Then f is C^∞ structurally stable.

Example 30.4 above satisfies the first and second but not the third hypothesis. The proof of the theorem relies on two very different branches: on one hand, it uses some of the techniques introduced by J. Mather in his proof of the equivalence between geometric stability and infinitesimal stability, and on the other the generalization of the theory of invariant manifolds to the case of endomorphisms, that was obtained by P. Berger [2]. The introduction of item 2 is a major contribution of this result. Note that this hypothesis is imposed on the set of wandering orbits, so it has no dynamical meaning, it is intended to control the geometrical aspects of critical sets; on the other side, the hyperbolicity controls the nonwandering orbits. This result motivates some interesting questions, pointing to a characterization of the C^r structural stability.

Problem 30.5. Does a similar statement hold for Axiom A maps with saddle type basic pieces?

Problem 30.6. Try to give a characterization of C^r structural stability for Axiom A maps whose basic pieces are attracting or expanding and there are no critical points in $\hat{\Omega}$.

The following was posed in [18]:

Problem 30.7. Does C^r structural stability imply that the set of critical points has its future orbit disjoint to the nonwandering set?

If a diffeomorphism f is C^1 structurally stable, then any iterate f^k is also C^1 structurally stable, because one can use the characterization by Axiom A plus Strong Transversality. However, the geometry of the singular sets may produce unexpected results. We will use below that for generic maps in dimension two, the critical set is a one dimensional manifold.

Example 30.5. A C^∞ structurally stable map f such that f^2 is not C^∞ structurally stable.

Begin with $p(z) = \lambda(2z^3 - 3z^2) + 1$ where λ is a real negative small parameter. Considering its restriction to the real axis, one can see that $p(0) = 1$, that $p'(0) = p'(1) = 0$ and that there is an attracting fixed point $a > 1$. Moreover, the Julia set of the polynomial p is hyperbolic, because both critical points are attracted to a. The map p itself is not stable because the critical points are degenerate and one of them belongs to the orbit of the other. Using the genericity of maps with nondegenerate critical points, one can produce a C^∞ perturbation f of p such that the following conditions hold:

1. The critical set of f consists of two small circles C_0 and C_1, the first close to 0 and the other close to 1.
2. The image $f(C_0)$ intersects C_1 transversally (see Fig. 30.2).
3. The forward iterates of $f(C_0) \cup C_1$ are pairwise disjoint and converge to the fixed attractor a.
4. f and p are Ω equivalent.

It follows that f satisfies the hypothesis of Theorem 30.7, so it is C^∞ structurally stable. The critical set of f^2 contains $C_0 \cup f^{-1}(C_1)$: this cannot be a one dimensional manifold, hence f^2 cannot be geometrically stable, and by the claim at the beginning of the section follows that it is not structurally stable.

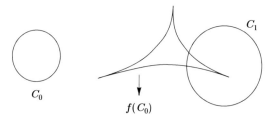

Fig. 30.2 The image $f(C_0)$ intersects C_1 transversally

In the spirit of the fundamental fact stated at the beginning of this section, there exists another approach to the problem of dealing with stability of maps with critical points. In [13], another concept of stability was defined: A map f is said C^r stable modulo singular sets if there exists a C^r neighborhood $\mathcal{U}$ of f such that, if two maps g_1 and g_2 in $\mathcal{U}$ are geometrically conjugate, then they are also topologically conjugate. The following result was proved:

Theorem 30.8. *If a map f is C^1 stable modulo singular sets, then f satisfies the (strong) Axiom A, has no critical points in the nonwandering set and each component of S_f must be contained in the basin of a periodic attractor.*

On the other hand, it can be proved as in [14] that an Axiom A endomorphism whose basic pieces are expanding or attracting, has no critical points in $\hat{\Omega}$ and satisfies the non critical relations property, is C^1 stable modulo singular sets.

This concept is intended to understand when a map is stable regardless of the obstructions introduced by the presence of wandering singularities. Examples of maps satisfying these hypothesis are certain hyperbolic rational maps in the Riemann sphere. Indeed, a hyperbolic rational map is Axiom A and its basic pieces are expanding or attracting; if moreover, there are no critical relations, then the map is C^1 stable modulo singular sets.

There exist some other results involving stability in restricted contexts. For example, Franke has shown that generic contracting maps are structurally stable in a topology that depends on the dimension of the ambient manifold, [5]. Ikegami considered deformed horseshoes to give examples of the nondensity of Axiom A endomorphisms, see [11].

The characterization of Ω stability is another interesting problem still open. It was shown by Przytycki that a weak Axiom A map without singularities is C^1-Ω stable (Definition 30.5) if and only if it is strong Axiom A without cycles. Under the hypothesis of weak Axiom A, he also showed the inverse stability of the nonwandering set. By Theorem 30.3 it follows that a C^1 map without critical nonwandering points, C^1-Ω stability is equivalent to Axiom A plus the no cycles condition. Mañé and Pugh gave an example of a C^1-Ω stable map whose critical set persistently intersects the nonwandering set. Then some conditions were found sufficient for a map having nonwandering critical points to be C^1-Ω stable, [4]. It was also proved that these were necessary conditions for C^1-Ω stability in manifolds of dimension two.

Problem 30.8. Find necessary and sufficient conditions for a map f to be C^1-Ω stable.

We finish this section with a brief comment on inverse stability (Definition 30.2). It was shown by Przytycki that Anosov endomorphisms without critical points are always C^1 inverse stable [25]. When the critical set does not intersect the nonwandering set, Theorem 30.3 implies that weak Axiom A is a necessary condition for C^1 inverse stability. The problem of completely characterize inverse stability seems difficult when there exist saddle type basic pieces intersecting the critical sets. However, if Problem 30.1 is solved by the affirmative, then weak Axiom A

30 Some Recent Results on the Stability of Endomorphisms 483

will be necessary for C^1 inverse stability, and it seems plausible that the following problem has a solution.

Problem 30.9. Weak Axiom A plus Strong Transversality is equivalent to C^1 inverse stability for maps without saddle type basic pieces.

30.7 Maps Without Singularities

In this section we consider stability for maps without critical points. As was stated above, the C^1-Ω stability was already characterized. It was also explained that C^1 stability implies the absence of critical points. By virtue of the Theorems 30.3 and [25] already stated, C^1 structural stability implies strong Axiom A. Moreover, the Strong Transversality condition is also necessary for C^1 structural stability [15]. To characterize C^1 structural stability of arbitrary C^1 maps it remains:

Problem 30.10. Prove that the above conditions are sufficient for C^1 stability.

This was solved in manifolds of dimension two [15].

It is not easy to find examples of structurally stable maps.

Claim: If a product F is structurally stable, then F is a diffeomorphism or an expanding map.

Let $F = (f, g)$. Then f and g are structurally stable and one can assume that f is not a diffeomorphism and g is not expanding. If f is a noninvertible structurally stable map, then there exists a basic piece Λ that is backward invariant ($f^{-1}(\Lambda) = \Lambda$): this follows from the Strong Transversality condition. As g is not expanding, then at least one basic piece of g is an attractor. But then the product of f and g has a basic piece that is not expanding nor injective, thus contradicting the Axiom A.

Example 30.6. The first example of a C^1 structurally stable noninvertible nonexpanding map is in dimension one, and was found by M. Shub [32].

Let $f(z) = z^2$ be defined in the unit circle, and define a new map g equal to f outside a neighborhood of 1, and such that the fixed point 1 is modified to be an attractor. This should be named a *derived from expanding* map. The resulting endomorphism has degree equal to two, its nonwandering set consists of an attracting fixed point and a Cantor expanding set (see Fig. 30.3). Examples in higher dimensions are harder to find.

Example 30.7. A class of examples were obtained in [14]; Let f be the endomorphism of $S^1 \times S^2$ defined by the formula:

$$f(z, w) = (z^2, z/2 + w/3),$$

where S^2 was identified with the extended complex plane.

Fig. 30.3 A derived from expanding map endomorphism

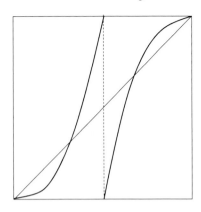

The nonwandering set is the union of two basic pieces: One of them is an attracting solenoid A obtained as the intersection of the forward images of $S^1 \times \mathbb{D}$, where $\mathbb{D}$ is the unit disc. The other is the expanding basic piece $R := S^1 \times \{\infty\}$. The basin of attraction of A is equal to the complement of R. This simple formula hides an interesting particularity of the map. Note that the map has degree equal to two, and that the restriction of f to A is injective, so the set $A' := f^{-1}(A) \setminus A$ must be nonempty and contained in the (connected) basin of A, but f has no singularities. Further preimages of A' are disjoint, contained in the basin of A, and convergent to R.

Example 30.8. The last example is due to Przytycki [25]. It is an Axiom A map satisfying the Strong Transversality condition.

Let f be the map of Example 30.6, and consider the product $F = f \times g$, where g is a diffeomorphism of the circle with an attractor a and a repeller b. This map, as was explained above, is not Axiom A because it has a basic piece that is neither expanding nor injective. The map is modified to make the saddle type basic piece injective (see [25], page 76).

References

1. Aoki, N., Moriyasu, K., Sumi, N.: C^1 maps having hyperbolic periodic points. Fundam. Math. **169**, 1–49 (2001)
2. Berger, P.: Persistence of stratifications of normally expanded laminations. Preprint (2007)
3. Berger, P.: Structural stability of attractor-repellor endomorphisms with singularities. Preprint (2008)
4. Delgado, J., Romero, N., Rovella, A., Vilamajó, F.: Omega stability for maps with nonwandering critical points. Fundam. Math. **193**, 23–35 (2007)
5. Franke, J.E.: Structural stability of smooth contracting endomorphisms on compact manifolds. Trois études en Dynamique Qualitative, Asterisque, No. 31, Soc. Math. France, Paris, pp. 141–188, (1976)
6. Franks, J.: Necessary conditions for stability of diffeomorphisms. Trans. Am. Math. Soc. **158**, 301–308, (1971)

7. Golubitsky, M., Guillemin, V.: Stable mappings and their singularities. Graduate Texts in Mathematics, vol. 14. Springer, New York (1973)
8. Hirsch, M., Palis, J., Pugh, C., Shub, M.: Neighborhoods of hyperbolic sets. Invent. Math. **9**, 121–134 (1970)
9. Hirsch, M., Pugh, C., Shub, M.: Invariant manifolds. Lecture Notes in Mathematics, vol. 583. Springer, Berlin (1977)
10. Ikeda, H.: Ω-inverse limit stability theorem. Trans. Am. Math. Soc. **349**, 2183–2200 (1996)
11. Ikegami, G.: Nondensity of Ω-stable endomorphisms and rough Ω-stabilities for endomorphisms, pp. 52–91. Dynamical Systems (Santiago, 1990)
12. Iglesias, J., Portela, A., Rovella, A.: Structurally stable perturbations of polynomials in the Riemann sphere. Ann. Inst. H. Poincar Anal. Non Linaire **25**(6), 1209–1220 (2008)
13. Iglesias, J., Portela, A., Rovella, A.: Stability modulo singular sets. Fund. Math. **204**(2), 155–175 (2009)
14. Iglesias, J., Portela, A., Rovella, A.: $C^1 C^1$ stable maps: examples without saddles. Fund. Math. 208 (2010), no. 1, 23–33, 37Cxx (37Dxx)
15. Iglesias, J., Portela, A., Rovella, A.: C^1 stability of endomorphisms on two dimensional manifolds. Preprint (2009)
16. Mañé, R.: On infinitesimal and absolute stability of diffeomorphisms. Dynamical systems–Warwick 1974 (Proc. Sympos. Appl. Topology and Dynamical Systems, Univ. Warwick, Coventry, 1973/1974; presented to E. C. Zeeman on his fiftieth birthday), pp. 151–161. Lecture Notes in Math., vol. 468, Springer, Berlin (1975)
17. Mañé, R.: A proof of the C^1 stability conjecture. (English) Publ. Math., Inst. Hautes Étud. Sci. **66**, 161–210 (1987)
18. Mañé, R., Pugh, C.: Stability of endomorphisms. Lecture Notes in Math., vol. 468, pp. 175–184. Springer, Berlin (1975)
19. Mañé, R., Sad, P., Sullivan, D.: On the dynamics of rational maps. Annales Scientifiques de L'E.N.S. 4e séries tome 16, N° 2, pp. 193–217 (1983)
20. Moser, J.: On a theorem of Anosov. J. Diff. Equ. **5**, 411–440 (1969)
21. J.Palis. On Morse-Smale dynamical systems. Topology **8**, 385–404 (1968)
22. Palis, J., Smale, S.: Structural stability theorems. In: Global Analysis, Proceedings of the Symposiums in Pure Mathematical, XIV. (1970)
23. Peixoto, M.M.: Structural stability on two-dimensional manifolds. Topology **1**, 101–120 (1962)
24. Przytycki, F.: Anosov endomorphisms. Stud. Math. **58**(3), 249–285 (1976)
25. Przytycki, F.: On Ω-stability and structural stability of endomorphisms satisfying Axiom A. Stud. Math. LX., 61–77 (1977)
26. Pugh, C.: An improved closing lemma and a general density theorem. Am. J. Math. **89**, 1010–1021 (1967)
27. Pujals, E.: Some simple questions related to the C^r stability conjecture. Nonlinearity **21**(11) 233–237 (2008)
28. Quandt, J.: On inverse limit stability for maps. J. Diff. Equ. **79**(2), 316–339 (1989)
29. Robbin, J.W.: A structural stability theorem. Ann. Math. (2) **94**, 447–493 (1971)
30. Robinson, C.: Structural stability of C^1 diffeomorphisms. J. Diff. Equ., **22**, 28–73 (1976)
31. Rovella, A., Sambarino, M.: The C^1 closing lemma for generic C^1 endomorphism. Annales de l'Institut Henri Poincare (C) Non Linear Analysis. **27**(6), 1461–1469 (2010)
32. Shub, M.: Endomorphisms of compact differentiable manifols. Amer. J. Math. **91**, 175–199 (1969)
33. Smale, S.: The Ω-stability theorem. In: Global Analysis, Proceedings of the Symposium in Pure Mathematical, XIV (1970)
34. Wen, L.: The C^1 closing lemma for endomorphisms with finitely many singularities. Proc. Am. Math. Soc. **114**(1), 217–223 (1992)
35. Wen, L.: The C^1 closing lemma for non-singular endomorphisms. Ergod. Theory. Dyn. Syst., **11**, 393–412 (1991)

Chapter 31
Differential Rigidity and Applications in One-Dimensional Dynamics

Yunping Jiang

Abstract In this survey, I summarize some work towards understanding of differential rigidity and smooth conjugacy in one-dimensional dynamics. In particular, I focus on those dynamical systems that have critical points and on those dynamical systems that have only $C^{1+\alpha}$, $0 < \alpha < 1$, smoothness.

31.1 Introduction

Mostow's rigidity theorem says that two closed hyperbolic 3-manifolds are isometrically equivalent if and only if they are homeomorphically equivalent [13]. A closed hyperbolic 3-manifold can be viewed as the quotient space of a Kleinian group acting on the upper-half 3-space. So a homeomorphic equivalence between two closed hyperbolic 3-manifolds can be lifted to a homeomorphism of the upper-half 3-space preserving group actions. It has a continuous extension to the boundary of upper-half 3-space and this boundary can be viewed as the extended complex plane $\overline{\mathbb{C}}$. This continuous extension is quasiconformal on the boundary, which we view as the extended complex plane $\overline{\mathbb{C}}$. Since a quasiconformal homeomorphism of the complex plane is absolutely continuous and since ϕ is equivariant in the sense that $\phi \circ \gamma \circ \phi^{-1}$ is a Möbius transformation for every transformation γ in the Kleinian group that covers the 3-manifold, it is not possible for ϕ to have an invariant line field, and one concludes that ϕ must be a Möbius transformation.

The parallel situation for a closed hyperbolic Riemann surface is more complicated. Such a surface can be viewed as upper half plane $\mathbb{H}$ factored by a finitely generated Fuchsian group whose limit set is the extended real axis, $\overline{\mathbb{R}} = \mathbb{R} \cup \{\infty\}$. Just as for 3-manifolds, a homeomorphic equivalence between two such surfaces can

---

Y. Jiang
Department of Mathematics, Queens College of the City, University of New York,
New York, USA
and
Department of Mathematics, The Graduate Center, CUNY, New York, USA
e-mail: yunping.jiang@qc.cuny.edu

M.M. Peixoto et al. (eds.), *Dynamics, Games and Science I*, Springer Proceedings
in Mathematics 1, DOI 10.1007/978-3-642-11456-4_31,
© Springer-Verlag Berlin Heidelberg 2011

be also lifted to a homeomorphism of the upper half plane preserving group actions, and this homeomorphism extends to the boundary, which in this case is $\overline{\mathbb{R}}$. The induced boundary homeomorphism ϕ defined on $\overline{\mathbb{R}}$ is quasisymmetric. But there is a new complication that arises because quasisymmetric homeomorphisms are not necessarily absolutely continuous. One can view this complication as what makes possible the theory of deformations of such groups, that is, Teichmüller theory. Nonetheless a version of Mostow's rigidity theorem is still available.

Theorem 31.1. *(Mostow [13]). Suppose $X = \mathbb{H}/\Gamma_X$ and $Y = \mathbb{H}/\Gamma_Y$ are two closed hyperbolic Riemann surfaces covered by finitely generated Fuchsian groups Γ_X and Γ_Y of finite analytic type. Suppose ϕ from $\overline{\mathbb{R}}$ to $\overline{\mathbb{R}}$ induces the isomorphism by $\gamma \mapsto \phi \circ \gamma \circ \gamma^{-1}$. Then ϕ is a Möbius transformation if and only if it is absolutely continuous.*

A stronger version of Mostow's rigidity theorem is proved by Tukia in [19].

Theorem 31.2. *(Tukia [19]). Suppose $X = \mathbb{D}/\Gamma_X$ and $Y = \mathbb{D}/\Gamma_Y$ are two closed hyperbolic Riemann surfaces with the same hypotheses as in the previous theorem. Then ϕ is a Möbius transformation if and only if it is differentiable at one radial limit point with non-zero derivative.*

For a proof of this theorem from the dynamical systems point of view, we refer the reader to [3].

In the 1980s Sullivan set up a dictionary between Kleinian groups and one-dimensional dynamical systems. In his classes at the CUNY Graduate Center between 1986 and 1989 the following theorems were presented, [17].

Theorem 31.3. *(Shub [14]). Suppose f and g are two C^1 orientation preserving circle expanding endomorphisms of the same degree. Then f and g are topologically conjugate by an orientation preserving homeomorphism ϕ.*

Moreover,

Theorem 31.4. *(Shub–Sullivan [15]). Suppose f and g are two analytic orientation preserving circle expanding endomorphisms of the same degree. Then the conjugacy ϕ is analytic if and only if it is absolutely continuous.*

This theorem can be generalized to the $C^{1+\alpha}$ case for any $0 < \alpha \leq 1$. We say f is $C^{1+\alpha}$ if it is differentiable and its derivative f' is a Hölder continuous with Hölder exponent α. That is, there is a number $K > 0$ such that

$$|f'(x) - f'(y)| \leq K|x - y|^{\alpha}, \quad \forall \, x, \, y.$$

In particular, for $\alpha = 1$, we say f is C^{1+1}, or equivalently $C^{1+Lipschitz}$ if f' is Lipschitz, that is if there is a number $K > 0$ such that

$$|f'(x) - f'(y)| \leq K|x - y|, \quad \forall \, x, \, y.$$

31 Differential Rigidity and Applications in One-Dimensional Dynamics

Theorem 31.5. *Suppose f and g are $C^{1+\alpha}$ orientation preserving circle expanding endomorphisms of the same degree for any $0 < \alpha \leq 1$ (see Sect. 31.2.4). Then the conjugacy ϕ is $C^{1+\alpha}$ if and only if it is absolutely continuous.*

Moreover,

Theorem 31.6. *Suppose f and g are two $C^{1+\alpha}$ orientation preserving circle expanding endomorphisms of the same degree for any $0 < \alpha \leq 1$ (see Sect. 31.2.4). Then the conjugacy ϕ is $C^{1+\alpha}$ if and only if all of the eigenvalues of f and g at corresponding periodic points are identical.*

See for example [2, 5–9, 11] for proofs of Theorems 5 and 6. Theorem 6 is a consequence of Theorem 5 because if eigenvalues of f and g are equal at all corresponding periodic points, then one can use Markov partitions for f and g to prove the conjugating map ϕ is bi-Lipschitz. A bi-Lipschitz homeomorphism is absolutely continuous. The following analogue to Tukia's theorem was presented by Sullivan [17].

Theorem 31.7. *(Sullivan [17]). Suppose f and g are two analytic orientation preserving circle expanding endomorphisms of the same degree. Then the conjugacy ϕ is analytic if and only if it is differentiable at one point with non-zero derivative.*

The argument in the outline of the proof given by Sullivan in [17] can easily be generalized to the $C^{1+Lipschitz}$ case.

Theorem 31.8. *Suppose f and g are two $C^{1+Lipschitz}$ orientation preserving circle expanding endomorphisms of the same degree. Then the conjugacy ϕ is $C^{1+Lipschitz}$ if and only if it is differentiable at one point with nonzero derivative.*

However, the argument cannot be used for the $C^{1+\alpha}$ case for $0 < \alpha < 1$. Let me first give the outline of the proof of Theorem 31.8 and show why it cannot be used in the $C^{1+\alpha}$ case.

Outline of the proof of Theorem 31.8. Suppose h is differentiable at a point x_0 on the circle. Then

$$h(x) = h(x_0) + h'(x_0)(x - x_0) + o(|x - x_0|)$$

for x close to x_0 and suppose

$$g \circ h = h \circ f.$$

Consider $\{x_n = f^n(x_0)\}_{n=0}^{\infty}$. Let $0 < a < 1$ be a real number. Consider the interval $I_n = (x_n, x_n + a)$. Let $J_n = (x_0, z_n)$ be an interval such that

$$f^n : J_n \to I_n$$

is a C^{1+1} diffeomorphism. Let $f^{-n} : I_n \to J_n$ denote its inverse. Since f is expanding, the length $|J_n| \to 0$ as $n \to \infty$. Similarly, we have

$$g_n : h(J_n) \to h(I_n)$$

is a C^{1+1} diffeomorphism. Let $g^{-n} : h(I_n) \to h(J_n)$ be its inverse. Then

$$h(x) = g^n \circ h \circ f^{-n}(x), \quad x \in I_n.$$

Let

$$\alpha_n(x) = \frac{x - x_0}{x_n - x_0} : J_n \to (0, 1)$$

and

$$\beta_n(x) = \frac{x - h(x_0)}{h(x_n) - h(x_0)} : h(J_n) \to (0, 1).$$

Then

$$h(x) = (g_n \circ \beta_n^{-1}) \circ (\beta_n \circ h \circ \alpha_n^{-1}) \circ (\alpha_n \circ f^{-n})(x), x \in I_n.$$

The key estimate comes from the following distortion lemma (refer to, for example, [11]).

Lemma 31.1. *(The Lipschitz case). There is a constant $C > 0$ independent of n and any inverse branches of f^n and g^n such that*

$$\left| \log \left| \frac{(f^{-n})'(x)}{(f^{-n})'(y)} \right| \right| \leq C |x - y|, \text{ for all } x \text{ and } y \text{ in } I_n$$

and

$$\left| \log \left| \frac{(g^{-n})'(x)}{(g^{-n})'(y)} \right| \right| \leq C |x - y|, \text{ for all } x \text{ and } y \text{ in } h(I_n).$$

From this distortion property, one can conclude that $g^n \circ \beta_n^{-1}$ and $\alpha_n \circ f^{-n}$ are sequences of bi-Lipschitz homeomorphisms with a uniform Lipschitz constant. Therefore, they have convergent subsequences. Without loss of generality, let us assume that these two sequences themselves are convergent. The map $\beta_n \circ h \circ \alpha_n^{-1}$ converges to a linear map.

Since the unit circle is compact and all I_n have fixed length a, there is a subsequence I_{n_i} of intervals such that $\cap_{i=1}^{\infty} I_{n_i}$ contains an interval I of positive length. Without loss of generality, let us assume that $\cap_{n=1}^{\infty} I_n$ contains an interval I of positive length. Thus h is a bi-Lipschitz homeomorphism on I. Since f and g are expanding, this implies h is bi-Lipschitz on the whole unit circle, and so Theorem 31.8 follows from Theorem 5. This completes the proof.

This argument cannot be used if we assume only that f and g are $C^{1+\alpha}$ for some $0 < \alpha < 1$. The reason is that in this situation we only have the following distortion estimate (refer to, for example, [11]).

Lemma 31.2. *(The Hölder case). Suppose f and g are $C^{1+\alpha}$ for $0 < \alpha < 1$. Then there is a constant $C > 0$ independent of n and any inverse branches of f^n and g^n such that*

$$\left| \log \frac{|(f^{-n}(x))'|}{|(f^{-n}(y))'|} \right| \leq C |x - y|^{\alpha}, \text{ for all } x \text{ and } y \in I_n$$

31 Differential Rigidity and Applications in One-Dimensional Dynamics

and

$$\left| \log \frac{|(g^{-n}(x))'|}{|(g^{-n}(y))'|} \right| \le C|x-y|^\alpha, \text{ for all } x \text{ and } y \in h(I_n).$$

Therefore, $g^n \circ \beta_n^{-1}$ and $\alpha_n \circ f^{-n}$ are only sequences of α-Hölder homeomorphisms with a uniform Hölder constant. Since we cannot conclude that h is bi-Lipschitz, for this case the assumption must be slightly modified. We will mention a related result in the next section (see Theorem 15).

31.2 Quasi-Hyperbolic One-Dimensional Maps

I have studied the differential rigidity problem and the smooth conjugacy problem in one-dimensional dynamics with critical points and in the $C^{1+\alpha}$ case since 1986. The first work for one-dimensional maps with critical points is for generalized Ulam-von Neumann transformations in my PhD dissertation [4]. Soon this work was extended to geometrically finite one-dimensional maps in [5]. And later, it is extended to quasi-hyperbolic one-dimensional maps in [6]. In this section, I would like to give a brief description of this work. All results mentioned below are distributed in several papers. I will not indicate them one by one. The reader who is interested in more details can go to papers [2,4–10] and the book [11].

31.2.1 Quasi-Hyperbolicity

Let M be the interval $[0,1]$ or the unit circle $\mathbb{R}/\mathbb{Z}$. Let $f : M \to M$ be a piecewise C^1 map. A point $c \in M$ is said to be singular if either $f'(c)$ does not exist or $f'(c)$ exists but $f'(c) = 0$. A singular point c is said to be *power law* if there is an interval $(c - \tau_c, c + \tau_c)$, $\tau_c > 0$, such that the restrictions of f to $(c - \tau_c, c)$ and to $(c, c + \tau_c)$ are C^1 and such that there is a real number $\gamma = \gamma(c) \ge 1$ such that the limits

$$\lim_{x \to c-} \frac{f'(x)}{|x-c|^{\gamma-1}} = B_- \quad \text{and} \quad \lim_{x \to c+} \frac{f'(x)}{|x-c|^{\gamma-1}} = B_+$$

exist and are non-zero. The number γ is called the exponent at c.

For a power law singular point c, let

$$r_{c,-}(x) = \frac{f'(x)}{|x-c|^{\gamma-1}}, \quad x \in (c - \tau_c, c)$$

and

$$r_{c,+}(x) = \frac{f'(x)}{|x - c|^{\gamma - 1}}, \quad x \in (c, c + \tau_c).$$

A singular point c is called critical if $\gamma > 1$.

Let SP denote the set of all singular points and let CP denote the set of all critical points. Let $PSO = \cup_{i=1}^{\infty} f^i(SP)$ be the set of post-singular orbits.

Remark 31.1. The exponent is C^1-invariant meaning that if f and g are C^1 conjugated maps (i.e., there is a C^1 diffeomorphism h such that $h \circ f = g \circ h$), then the exponents of f and g are the same at corresponding power law critical points.

In the rest of this survey, I always assume that f satisfies the following two conditions:

(1) SP is finite (could be empty).
(2) Every singular point in SP is power law type.

Let $\tau > 0$ be a real number. For every critical point $c \in CP$, let $U_c = [c - \tau, c + \tau]$. Suppose τ is small enough so that different U_c are disjoint. Define

$$U = U(\tau) = \cup_{c \in CP} U_c \quad \text{and} \quad V = V(\tau) = \overline{M \setminus U(\frac{\tau}{2})}.$$

Denote

$$U_{c,-} = U_c \cap (c - \tau, c) \quad \text{and} \quad U_{c,+} = U_c \cap (c, c + \tau).$$

Definition 31.1 (Hölder Continuity). We call the map f $C^{1+\alpha}$ for some $0 < \alpha \leq 1$ if there is $\tau > 0$ such that

(i) f on every component of $M \setminus SP$ is C^1 and the derivative f' is α-Hölder continuous.
(ii) Every $r_{c,\pm}|U_{c,\pm}$ is α-Hölder continuous.

Definition 31.2 (Chain of intervals). A sequence of intervals $\{I_i\}_{i=0}^n$ is called a chain (with respect to f) if

(a) $I_i \subset M \setminus SP$ for all $0 \leq i \leq n$
(b) $f : I_i \to I_{i+1}$ is a C^1-diffeomorphism for every $0 \leq i \leq n - 1$
(c) Either $I_i \subseteq V(\tau)$ for all $0 \leq i \leq n - 1$ or the last interval $I_n \subseteq U(\tau)$ (but in later case, some I_i, $0 < i < n - 1$, may not be contained in $U(\tau)$ or $V(\tau)$)

A chain $\mathscr{I} = \{I_i\}_{i=0}^n$ is said to be regulated if either $I_i \subseteq V$ or $I_i \subseteq U$ for all $0 \leq i \leq n$.

Definition 31.3. [Quasi-Hyperbolicity] The map f is said to be quasi-hyperbolic if

(1) f is $C^{1+\alpha}$ for some $0 < \alpha \leq 1$.
(2) $\overline{PSO} \cap U(\tau) = \emptyset$ for some number $\tau > 0$; and
(3) there are two constants $C > 0$ and $0 < \mu < 1$ such that for any chain $\{I_i\}_{i=0}^n$, $|I_0| \leq C\mu^n |I_n|$.

31 Differential Rigidity and Applications in One-Dimensional Dynamics 493

Let $\tau > 0$ be a fixed small number in the rest of this survey. Suppose f and g are two quasi-hyperbolic maps. We say f and g are topologically conjugate if there is a homeomorphism h from M onto itself such that

$$f \circ h = h \circ g.$$

31.2.2 Geometrical Finiteness

In this section we define a subspace of quasi-hyperbolic one-dimensional maps which we call geometrically finite.

Let $SO = \cup_{n=0}^{\infty} f^n(SP)$ be the union of singular orbits of f. If SO is non-empty and finite, let $\eta_1 = \{I_0, \cdots, I_{k-1}\}$ be the set of the closures of intervals in $M \setminus SO$, then (f, η_1) has Markovian property. This means that

(a) $I_0, \ldots, I_{k-1}$ have pairwise disjoint interiors.
(b) The union $\cup_{i=0}^{k-1} I_i$ of all intervals in η_1 is M.
(c) The restriction $f : I \to f(I)$ for every interval I in η_1 is homeomorphic.
(d) The image $f(I)$ of every interval I in η_1 is the union of some intervals in η_1.

We call η_1 a *Markov partition*.

Let $g_i = (f|I_i)^{-1}$ be the inverse of $f : I_i \to f(I_i)$ for each $I_i \in \eta_1$. A sequence $w_n = i_0 \cdots i_{n-1}$ of 0's, $\cdots$, $(k-1)$'s is called *admissible* if the domain $f(I_{i_l})$ of g_{i_l} contains $I_{i_{l+1}}$ for all $0 \le l < n - 1$. For an admissible sequence $w_n = i_0 \cdots i_{n-1}$ of 0's, $\cdots$, $(k-1)$'s, we can define $g_{w_n} = g_{i_0} \circ \cdots g_{i_{n-1}}$ and $I_{w_n} = g_{w_n}(f(I_{i_{n-1}}))$. Let η_n be the set of the intervals I_{w_n} for all admissible sequences of length n. It is also a Markov partition of M respect to f. We call it the nth-partition of M induced from (f, η_1). Let κ_n be the maximum of the lengths of intervals in η_n.

Definition 31.4 (Geometrical Finiteness). We call f geometrically finite if

(i) The set of singular orbits SO is non-empty and finite.
(ii) No critical point is periodic.
(iii) There are constants $C > 0$ and $0 < \mu < 1$ such that $\kappa_n \le C\mu^n$ for all $n > 0$.

A point $p \in M$ is called periodic of period k if $f^i(p) \ne p$ for all $0 < i < k$ but $f^k(p) = p$. When $k = 1$, we also call it fixed. For a periodic point p of period k, $e_p = (f^k)'(p)$ is called the eigenvalue of f at p. Then p is called attractive if $|e_p| < 1$; parabolic if $|e_p| = 1$; expanding if $|e_p| > 1$. A critical point c of a C^2 map is called non-degenerate if $f'(c) = 0$ and $f''(c) \ne 0$.

We now give two examples of geometrically finite maps. The first example is a consequence of Theorem 6.3, pp. 261–262 in [12].

Example 31.1. A C^2 map f with only non-degenerate critical points such that PSO and SP are both finite and $PSO \cap SP = \emptyset$ and all periodic points are expanding.

The Schwarzian derivative of a C^3 map h is defined as

$$S(h) = \frac{h'''}{h'} - \frac{3}{2}\left(\frac{h''}{h'}\right)^2.$$

We say that h has negative Schwarzian derivative if $S(h)(x) < 0$ for all x. Singer (see [16]) proved that if f is C^3 and has negative Schwarzian derivative, then the immediate basin of every attractive or parabolic periodic orbit contains at least one critical orbit. Therefore, if f has negative Schwarzian derivative and if $PSO \cap SP = \emptyset$ and if $\overline{PSO}$ contains neither attractive nor parabolic periodic points, then all periodic points of f are expanding. The map f is said to be preperiodic if for every singular point c, $f^m(c)$ is an expanding periodic point for some integer $m \geq 1$. Then $\overline{PSO} = PSO$ contains neither attractive nor parabolic periodic points. A special case of Example 31.1 is that

Example 31.2. A preperiodic C^3 map f having negative Schwarzian derivative.

If two geometrically finite one-dimensional maps f and g is topologically conjugate by a homeomorphism h, then the conjugacy h is quasisymmetric (see [10] or [11, pp88–91, Sect. 3.5] for a proof). This implies that h is a Hölder continuous function.

31.2.3 Generalized Ulam-von Neumann Transformations

A special subspace of geometrically finite maps is generalized Ulam-von Neumann transformation, which is the first class I have studied in this direction.

Definition 31.5 (Generalized Ulam-von-Neumann Transformations). Suppose $M = [-1, 1]$. We call f a *generalized Ulam-von Neumann transformation* if

(1) f is geometrically finite with only one singular point 0
(2) $f(-1) = f(1) = -1$ and $f(0) = 1$
(3) $f|[-1, 0]$ is increasing and $f|[0, 1]$ is decreasing

One example of a generalized Ulam-von Neumann transformation is $f(x) = 1 - 2|x|^\gamma$ for $\gamma \geq 1$. Another one is $f(x) = -1 + 2\cos(\pi x/2)$. If f is a generalized Ulam-von Neumann transformation, let $I_0 = [-1, 0]$ and $I_1 = [0, 1]$. We then have that $f(I_0) = f(I_1) = M$. Thus $\eta_0 = \{I_0, I_1\}$ is a Markov partition. The post-singular orbit $PSO = \cup_{i=1}^{\infty} f^i(0)$ is $\{-1, 1\}$.

Any two generalized Ulam-von Neumann transformations f and g are topologically conjugate, the conjugacy h is quasisymmetric (see [4] or [11, pp. 88–91, Sect. 3.5] for a proof). This implies that h is a Hölder continuous function.

31.2.4 $C^{1+\alpha}$ Circle Endomorphisms

Another class of quasi-hyperbolic one-dimensional maps is orientation-preserving circle coverings.

31 Differential Rigidity and Applications in One-Dimensional Dynamics 495

Let $M = \mathbb{R}/\mathbb{Z}$ be the unit circle. Suppose f is an orientation-preserving circle covering. Suppose f is C^1. Then f is called a circle expanding endomorphism if there are constants $C > 0$ and $\mu > 1$ such that

$$(f^n)'(x) \geq C\mu^n, \quad x \in M, \quad n \geq 1.$$

An example of an expanding circle endomorphism is $x \mapsto dx \pmod 1$, where $d > 1$ is an integer. If the topological degree of f is d, then f is topologically conjugate to $x \mapsto dx \pmod 1$. Thus any two C^1 circle expanding endomorphisms f and g of the same degree are topologically conjugate.

An expanding circle endomorphism f is called $C^{1+\alpha}$ for some $0 < \alpha \leq 1$ if f is C^1 and its derivative f' is α-Hölder continuous. If both f and g are $C^{1+\alpha}$ circle expanding endomorphisms of the same degree, then the conjugacy h is quasisymmetric (see [11, pp. 88–91, Sect. 3.5] or [2] for a proof). This implies that h is a Hölder continuous function.

A $C^{1+\alpha}$ circle expanding endomorphism for $0 < \alpha \leq 1$ is quasi-hyperbolic and has no singular points.

31.3 Differential Rigidity and Applications

We use Leb to mean the Lebesgue measure on M. Our map f (or g) in this section satisfies three more techniqical conditions:

(1) $\overline{PSO}$ has measure zero, i.e., Leb$(\overline{PSO}) = 0$.
(2) The set $\overline{PSO}$ is not an attractor. More precisely, there is an open neighborhood $\overline{PSO} \subset W \neq M$ such that for any point $p \in M$ either f^n falls into $\overline{PSO}$ eventually (i.e., $\{f^n(p)\}_{n=N}^{\infty} \subseteq \overline{PSO}$ for some $N > 0$) or it leaves W infinitely often (i.e., there is a subsequence $\{f^{\circ n_i}(p)\}_{i=1}^{\infty} \subseteq M \setminus W$).
(3) The map f is mixing, that is, for any intervals $I, J \subset M$, there is an integer $n \geq 0$ such that $f^n(J) \supseteq I$.

The last two conditions are invariant under topological conjugacy. The condition (3) implies that the dynamical system $\{f^n\}_{n=0}^{\infty}$ cannot be decomposed.

Denote $M_0 = M \setminus \overline{PSO}$. For any point $p \in M$, define the backward orbit of p to be the countable set

$$BO(p) = \cup_{n=0}^{\infty} f^{-n}(p)$$

Suppose that f and g are two quasi-hyperbolic maps conjugate by h, that is,

$$h \circ f = g \circ h.$$

Then from the equation, if h is differentiable at $p \in M_0$, it is necessarily differentiable at all points in $BO(p)$.

Definition 31.6. We call h differentiable at $p \in M_0$ with uniform bound if there are a small neighborhood Z of p and a constant $C > 0$ such that

$$C^{-1} \leq |h'(q)| \leq C, \quad q \in BO(p) \cap Z.$$

We have the following theorem in one-dimensional dynamics (see [6]).

Theorem 31.9. *Suppose that f and g are quasi-hyperbolic maps conjugate by h. Then h restricted to M_0 is C^1 if and only if h is differentiable at one point $p \in M_0$ with uniform bound.*

The key distortion estimation in the proof of Theorem 31.9 is the following distortion result. The reader can refer to [6, pp. 369–371] for a complete proof of this lemma. He can also find a complete proof of this lemma in the geometrically finite case in [11, pp. 82–97, Sect. 3.4]. The estimation of the distortion for a dynamical system is always important in the study of dynamical systems. Chapters 1, 2, and 3 of [11] contain a general discussion of distortion properties.

We use $d(\cdot, \cdot)$ to mean the distance between two points or two sets.

Lemma 31.3 ($C^{1+\text{Hölder}}$**-Denjoy-Koebe Type Distortion Lemma**). *Suppose f is a quasi-hyperbolic map. Then there are constants $C, D > 0$ such that for any regulated chain $\mathscr{I} = \{I_i\}_{i=0}^n$ and for all x and y in I_0,*

$$\left| \log \left(\frac{|(f^n)'(x)|}{|(f^n)'(y)|} \right) \right| \leq C |x_n - y_n|^{\frac{\alpha}{\gamma}} + D \frac{|x_n - y_n|}{d(\{x_n, y_n\}, \overline{PSO})},$$

where $x_n = f^n(x)$ and $y_n = f^n(y)$ and where $0 < \alpha \leq 1$ is the Hölder exponent and $\gamma \geq 1$ is the maximum of all the exponents of power law critical points.

Consider the conjugacy $h(x) = \frac{2}{\pi} \arcsin(x)$ between $f(x) = 1 - 2x^2$ and $g(x) = 1 - 2|x|$ on $[-1, 1]$. The maps h and h^{-1} are both C^1 on $(-1, 1)$. But h' is not uniformly continuous because the exponents of f and g at 0 are different. Note that the exponent at a singular point is invariant under C^1 conjugacy. Furthermore, we have the following improvement of Theorem 31.9 (see [6]).

Theorem 31.10. *Suppose f and g and h are the same as those in Theorem 31.9. Suppose all the exponents of f and g at the corresponding singular points are the same. Then h restricted to the closure of every interval of M_0 is a $C^{1+\beta}$ diffeomorphism for some $0 < \beta \leq 1$ if and only if h is differentiable at one point in M_0 with uniform bound.*

We can use the equality of eigenvalues of f and g at corresponding periodic points to verify the condition, differentiable at one point with uniform bound, in Theorems 31.9 and 31.10 (see [6]).

Lemma 31.4. *Suppose f and g and h are those in Theorem 31.9. If h is differentiable at a point p in M_0 with non-zero derivative and if there is an open interval*

31 Differential Rigidity and Applications in One-Dimensional Dynamics 497

Y about p such that the absolute values of the eigenvalues of f and g at periodic points in Y and at corresponding periodic points in $h(Y)$ are the same, then h is differentiable at p with uniform bound.

The above lemma combining with Theorems 31.9 and 31.10 give us the following results (see [6]).

Corollary 31.1. *Suppose f and g and h are those in Theorem 31.9. Then $h|M_0$ is C^1 if and only if h is differentiable at one point p in M_0 with non-zero derivative and the absolute values of the eigenvalues of f and g at periodic points in a small neighborhood Y of p and at corresponding periodic points in $h(Y)$ are the same.*

Corollary 31.2. *Suppose f and g and h are those in Theorem 31.9. Suppose all the exponents of f and g at the corresponding singular points are also the same. Then h restricted on the closure of every interval of M_0 is a $C^{1+\beta}$ diffeomorphism for some fixed $0 < \beta \leq 1$ if and only if h is differentiable at one point p in M_0 with non-zero derivative and the absolute values of the eigenvalues of f and g at periodic points in a small neighborhood Y of p and at corresponding periodic points in $h(Y)$ are the same.*

The following rigidity result is now can be obtained from Theorems 31.9 and 31.10 and Corollaries 31.1 and 31.2 (see [6]).

Corollary 31.3. *Suppose f and g and h are those in Theorem 31.9. Then $h|M_0$ is C^1 if and only if there is a small interval Y of M such that $h|Y$ is absolutely continuous. Furthermore, if all the exponents of f and g at the corresponding singular points are also the same, then h restricted on the closure of every interval of M_0 is $C^{1+\beta}$ for some fixed $0 < \beta \leq 1$ if and only if there is a small interval Y of M such that $h|Y$ is absolutely continuous.*

The reader can find the detailed proofs of Theorems 31.9 and 31.10 and Corollaries 31.1, 31.2, and 31.3 in [6]. We give an outline of the proof of Theorem 31.9 in the next section.

Previously, we have proved Theorem 31.9 for geometrically finite one-dimensional maps. In [18], Sullivan defined scaling functions for Cantor sets on the line and used them in the study of differentiable structures on Cantor sets. Furthermore, in [4,5,7,11], we defined the scaling function S_f for every geometrically finite one-dimensional map f by using Markov partitions. We have proved that the scaling function S_f exists for any geometrically finite one-dimensional map f and is a function defined on the dual symbolic space. Moreover, if f has no critical point, then the scaling function S_f is Hölder continuous; and if f has critical points, then the scaling function S_f is discontinuous with certain jump discontinuity. Using scaling functions and Theorem 31.9, we studied the smooth conjugacy between two geometrically finite one-dimensional maps in [7,8] as follows.

Theorem 31.11. *Suppose f and g and h are two geometrically finite one-dimensional maps. Then $h|M_0$ is C^1 if and only if $S_f = S_g$. Furthermore, if* *all the exponents of f and g at the corresponding singular points are also the same, then h restricted on the closure of every interval of M_0 is $C^{1+\beta}$ for some fixed $0 < \beta \le 1$ if and only if $S_f = S_g$.*

The space of generalized Ulam-von Neumann transformations is the first class of one-dimensional maps with critical points which we studied in this direction (refer to [2, 4, 6, 9, 11]).

Theorem 31.12. *Suppose that f and g are two generalized Ulam-von Neumann transformations on $[-1, 1]$. Suppose h is the conjugacy from f to g. Then $h|(-1, 1)$ is C^1 if and only if h is differentiable at one point $p \in (-1, 1)$ with uniform bound. If the exponents of f and g at the singular point 0 are also the same, denoted as $\gamma \ge 1$ and both f and g are $C^{1+\alpha}$ for some $0 < \alpha \le 1$, then h is $C^{1+\alpha/\gamma}$ if and only if h is differentiable at one point $p \in (-1, 1)$ with uniform bound.*

Theorem 31.13. *Suppose that f and g and h are the same as in Theorem 31.12. Then $h|(-1, 1)$ is C^1 if and only if h is absolutely continuous. If the exponents of f and g at the singular point 0 are also the same, denoted as $\gamma \ge 1$ and both f and g are $C^{1+\alpha}$ for some $0 < \alpha \le 1$, then h is $C^{1+\alpha/\gamma}$ if and only if h is absolutely continuous.*

Theorem 31.14. *Suppose that f and g and h are the same as in Theorem 31.12. Then $h|(-1, 1)$ is C^1 if and only if eigenvalues of f and g at corresponding periodic points in $(-1, 1)$ are the same. If the exponents of f and g at the singular point 0 are also the same, denoted as $\gamma \ge 1$ and both f and g are $C^{1+\alpha}$ for some $0 < \alpha \le 1$, then h is $C^{1+\alpha/\gamma}$ if and only if eigenvalues of f and g at corresponding periodic points in $[-1, 1]$ are the same.*

For the $C^{1+\alpha}$ circle expanding endomorphisms case for $0 < \alpha < 1$, we have that

Theorem 31.15. *Suppose f and g are $C^{1+\alpha}$ circle expanding maps of the same degree for some $0 < \alpha < 1$. Suppose h is the topological conjugacy from f to g, that is, $f \circ h = h \circ g$. Then h is $C^{1+\alpha}$ if and only if h is differentiable at one point p on the circle with uniform bound.*

31.4 Outline of the Proof of Theorem 31.9

The "only if" part of Theorem 31.9 is obvious. We outline the proof of the "if part" of Theorem 31.9 in [6] (see also [7, 8]) by the following steps, in which we use Lemma 31.3 repeatedly.

Suppose f and g are $C^{1+\alpha}$ for $0 < \alpha \le 1$. Suppose the conjugacy h is differentiable with uniform bound at p. Let Z be an open interval about p in Definition 31.6. Let $\Lambda = \cup_{n=0}^{\infty} f^{-n}(\overline{PSO})$. Then $\mathrm{Leb}(\Lambda) = 0$. Let Ω be the set of all self-recurrent points in $M \setminus \Lambda$ of f.

Step 1: The set Ω has full Lebesgue measure in M.

Step 2: $h|Z$ is bi-Lipschitz.

31 Differential Rigidity and Applications in One-Dimensional Dynamics

This step is important in the proof. As we pointed out in the introduction, we need to develop a new technique to prove this step as follows. Let Ψ_1 be the set of intervals of $f^{-1}(Z)$ contained in Z. Inductively, let Ψ_n be the set of intervals of $f^{-n}(Z)$ contained in $Z \setminus \left(\cup_{I \in \Psi_{n-1}} I \right)$. Because f is mixing, there are infinitely many integers n such that Ψ_n is non-empty.

Suppose Ψ_n is non-empty. Then for any interval $I \in \Psi_n$, $f^n : I \to Z$ and $g^n : h(I) \to h(Z)$ are $C^{1+\alpha}$-diffeomorphisms. Moreover we have that

$$\frac{|h(I)|}{|I|} = \frac{|(f^n)'(x)|}{|(g^n)'(h(y))|} \frac{|h(Z)|}{|Z|}$$

for some $x, y \in I$. Without loss of generality, we assume that $\{f^k(I)\}_{k=0}^n$ and $\{g^k(h(J))\}_{k=0}^n$ are regulated chains.

Let $q \in I \subseteq Z$ be the preimage of p under $f^n : I \to Z$. Then

$$\frac{(f^n)'(q)}{(g^n)'(h(q))} = \frac{h'(q)}{h'(p)}.$$

So we have that a constant $C_1 > 0$ such that

$$C_1^{-2} \le \frac{|(f^n)'(q)|}{|(g^n)'(h(q))|} \le C_1^2.$$

Therefore,

$$C_1^{-2} \frac{|(f^n)'(x)|}{|(f^n)'(q)|} \frac{|(g^n)'(h(q))|}{|(g^n)'(h(y))|} \frac{|h(Z)|}{|Z|} \le \frac{|h(I)|}{|I|}$$

$$\le C_1^2 \frac{|(f^n)'(x)|}{|(f^n)'(q)|} \frac{|(g^n)'(h(q))|}{|(g^n)'(h(y))|} \frac{|h(Z)|}{|Z|}.$$

Applying Lemma 31.3, there is a constant $C_2 > 0$ such that

$$C_2^{-1} \le \frac{|(f^n)'(x)|}{|(f^n)'(q)|} \le C_2 \quad \text{and} \quad C_2^{-1} \le \frac{|(g^n)'(h(q))|}{|(g^n)'(h(y))|} \le C_2.$$

So we have a constant $C_3 > 0$ such that

$$C_3^{-1} \le \frac{|h(I)|}{|I|} \le C_3.$$

Suppose $x < y$ are in Z. Let $\Psi_1(x, y)$ be the set of intervals of $f^{-1}(Z)$ contained in $[x, y]$. Inductively, let $\Psi_n(x, y)$ be the set of intervals of $f^{-n}(Z)$ contained in $[x, y] \setminus \left(\cup_{I \in \Psi_n} I \right)$. Then

$$\cup_{n=1}^{\infty} \cup_{I \in \Psi_n(x,y)} I$$

is the union of pairwise disjoint intervals and its closure is $[x, y]$. Let

$$A = [x, y] \setminus \left(\cup_{n=1}^{\infty} \cup_{I \in \Psi_n(x,y)} I \right).$$

Since every point $z \neq x, y$ in A is not self-recurrent, we have that $\mathrm{Leb}(A) = 0$. Hence

$$\mathrm{Leb}\left(\cup_{n=1}^{\infty} \cup_{I \in \Psi_n(x,y)} I \right) = \sum_{n=1}^{\infty} \sum_{I \in \Psi_n(x,y)} |I| = [x, y].$$

Similarly,

$$\mathrm{Leb}\left(\cup_{n=1}^{\infty} \cup_{I \in \Psi_n(x,y)} h(I) \right) = \sum_{n=1}^{\infty} \sum_{I \in \Psi_n(x,y)} |h(I)| = h([x, y]).$$

The additive formula implies that

$$C_3^{-1} \leq \frac{|h(x) - h(y)|}{|x - y|} \leq C_3.$$

Therefore, $h|Z$ is bi-Lipschitz.

The next step is to promote from the bi-Lipschitz property to the uniformly continuity property for the derivative h' on a subset of Z with full Lebesgue measure.

Since $h|Z$ is bi-Lipschitz, h' exists a.e. in Z and is integrable. Since $(h|Z)'(x)$ is measurable, $h|Z$ is a homeomorphism, and $\mathrm{Leb}(\Lambda) = 0$, we can find a point p_0 in $Z \setminus \Lambda$ and a subset E_0 containing p_0 such that

(1) $h|Z$ is differentiable at every point in E_0
(2) p_0 is a density point of E_0
(3) $h'(p_0) \neq 0$
(4) The derivative $h'|E_0$ is continuous at p_0.

We know there is a subsequence $\{f^{n_k}(p_0)\}_{k=1}^{\infty} \subseteq M \setminus W$ converging to a point q_0 in $M \setminus W$. Let $I_0 = (a, b)$ be an open interval about q_0 such that $C_4 = d(\overline{I_0}, \overline{PSO}) > 0$. There is a sequence of interval $\{I_k\}_{k=1}^{\infty}$ such that $p_0 \in I_k \subseteq Z$ and $f^{n_k} : I_k \to I_0$ is a $C^{1+\alpha}$ diffeomorphism. Without loss of generality, we may assume that $\{I_{l,k} = f^l(I_k)\}_{l=0}^{n_k}$ is a regulated chain for every $k \geq 1$. Then $\mathrm{Leb}(I_k)$ goes to zero as k tends to infinity.

For any positive integer s, there is an integer $N_s > 0$ such that

$$\frac{\mathrm{Leb}(E_0 \cap I_k)}{\mathrm{Leb}(I_k)} \geq 1 - \frac{1}{s}$$

for all $k > N_s$. Let $E_k = f^{n_k}(E_0 \cap I_k)$. Then h is differentiable at every point in E_k and there is a constant $C_5 > 0$ such that

31 Differential Rigidity and Applications in One-Dimensional Dynamics

$$\frac{\mathrm{Leb}(E_k \cap I_0)}{\mathrm{Leb}(I_0)} \geq 1 - \frac{C_5}{s}$$

for all $k > N_s$ because $\{f^{n_k}|I_k\}_{k=1}^{\infty}$ have uniformly bounded distortion. Let $E = \bigcap_{s=1}^{\infty} \bigcup_{k > N_s} E_k$. Then E has full measure in I_0 and h is differentiable at every point in E with non-zero derivative.

Step 3: $h'|E$ is uniformly continuous.

For any x and y in E, let z_k and w_k be the preimages of x and y under the diffeomorphism $f^{n_k} : I_k \to I_0$. Then z_k and w_k are in E_0. From $h \circ f = g \circ h$, we have that

$$h'(x) = \frac{(g^{n_k})'(h(z_k))}{(f^{n_k})'(z_k)} h'(z_k)$$

and

$$h'(y) = \frac{(g^{n_k})'(h(w_k))}{(f^{n_k})'(w_k)} h'(w_k).$$

So

$$\left| \log \left(\frac{h'(x)}{h'(y)} \right) \right| \leq \left| \log \left| \frac{(g^{n_k})'(h(z_k))}{(g^{n_k})'(h(w_k))} \right| \right| + \left| \log \left| \frac{(f^{n_k})'(w_k)}{(f^{n_k})'(z_k)} \right| \right| + \left| \log \left(\frac{h'(z_k)}{h'(w_k)} \right) \right|$$

Applying Lemma 31.3 to both f and g, we can find a constant $C_6 > 0$ such that

$$\left| \log \left| \frac{(f^{n_k})'(w_k)}{(f^{n_k})'(z_k)} \right| \right| \leq C_6 |x - y|^{\alpha}$$

and

$$\left| \log \left| \frac{(g^{n_k})'(h(z_k))}{(g^{n_k})'(h(w_k))} \right| \right| \leq C_6 |h(x) - h(y)|^{\alpha}$$

for all $k \geq 1$. Therefore,

$$\left| \log \left(\frac{h'(x)}{h'(y)} \right) \right| \leq C_6 \left(|x - y|^{\alpha} + |h(x) - h(y)|^{\alpha} \right) + \left| \log \left(\frac{h'(z_k)}{h'(w_k)} \right) \right|$$

for all $k \geq 1$. Since $h'|E_0$ is continuous at p_0, the last term in the last inequality tends to zero as k goes to infinity. Hence

$$\left| \log \left(\frac{h'(x)}{h'(y)} \right) \right| \leq C_6 \left(|x - y|^{\alpha} + |h(x) - h(y)|^{\alpha} \right).$$

This means that $h'|E$ is uniformly continuous.

The step 3 implies that $h|I_0$ is actually C^1. Then the last step is easy to get by using the mixing condition.

Step 4: $h|M_0$ is C^1.

502 Y. Jiang

Acknowledgements This survey article is written under an invitation by Professor Alberto Pinto in the occasion of the 60th birthday of Professor David Rand. The author would like to use this opportunity to thank Professor Rand for kindness help when he just graduated from the CUNY Graduate Center. Professors Rand and Pinto have been worked in this direction and made a contribution to this direction. The reader who is interested in this direction and their work can go to their survey articles in this volume (see also [1]). The author would like to use this opportunity to thank Professor Dennis Sullivan and Professor Fred Gardiner for their lectures and for their support and help for many years during this research. The author's work in this direction has been partially supported by grants from NSF and awards from PSC-CUNY and grants from AMSS and MCM at Chinese Academy of Sciences.

References

1. Ferreira, F., Pinto, A.A.: Explosion of smoothness from a point to everywhere for conjugacies between diffeomorphisms on surfaces. Ergod. Theory Dynam. Syst. **23**, 509–517 (2003)
2. Jiang, Y.: Teichmüller structures and dual geometric Gibbs type measure theory for continuous potentials. Preprint.
3. Jiang, Y.: Function model of the Teichmüller space of a closed hyperbolic Riemann surface. Preprint.
4. Jiang, Y.: Generalized Ulam-von Neumann transformations. Ph.D. Thesis, Graduate School of CUNY and UMI publication (1990)
5. Jiang, Y.: Dynamics of Certain Smooth One-Dimensional Mappings. I: The $C^{1+\alpha}$-Denjoy-Koebe Distortion Lemma; II. Geometrically Finite One-Dimensional Mappings; III. Scaling Function Geometry; IV. Asymptotic Geometry of Cantor Sets. IMS preprint series, 91-1a, 91-1b, 91-12a, 91-12b
6. Jiang, Y.: Differentiable rigidity and smooth conjugacy. Ann. Acad. Sci. Fenn. Math. **30**, 361–383 (2005)
7. Jiang, Y.: Smooth classification of geometrically finite one-dimensional maps. Trans. Am. Math. Soc., **348**(6), 2391–2412 (1996)
8. Jiang, Y.: On rigidity of one-dimensional maps. Contemp. Math., AMS Series, **211**, 319–431 (1997)
9. Jiang, Y.: On Ulam-von Neumann transformations. Comm. Math. Phys. **172**(3), 449–459 (1995)
10. Jiang, Y.: Geometry of geometrically finite one-dimensional maps. Comm. Math. Phys. **156**(3), 639–647 (1993)
11. Jiang, Y.: Renormalization and Geometry in One-Dimensional and Complex Dynamics. Advanced Series in Nonlinear Dynamics, vol. 10, xvi+309 pp. World Scientific Publishing Co. Pte. Ltd., River Edge, NJ, ISBN 981-02-2326-9 (1996)
12. de Melo, W., van Strien, S.: One-Dimensional Dynamics. Springer, Berlin (1993)
13. Mostow, D.: Strong rigidity of locally symmetric spaces. Ann. of Math. Stud., vol. 78. Princeton, NJ (1972)
14. Shub, M.: Endomorphisms of compact differentiable manifolds. Am. J. Math. **91**, 129–155 (1969)
15. Shub, M., Sullivan, D.: Expanding Endomorphisms of the circle revisited. Ergod. Theroy Dynam. Syst. **5**, 285–289 (1987)
16. Singer, D.: Stable orbits and bifurcations of maps of the interval. SIAM J. Appl. Math. **35**, 260–267 (1978)
17. Sullivan, D. : Class notes at the CUNY Graduate from 1986–1990
18. Sullivan, D.: Differentiable structures on fractal-like sets determined by intrinsic scaling functions on dual Cantor sets. The Mathematical Heritage of Hermann Weyl. A.M.S. Proc. Symp. Pure Math. **48**, 15–23 (1987)
19. Tukia, P.: Differentiability and rigidity of Möbius groups. Invent. Math. **82**, 557–578 (1985)

Chapter 32
Minimum Regret Pricing of Contingent Claims in Incomplete Markets

C. Kountzakis, S.Z. Xanthopoulos, and A.N. Yannacopoulos

Abstract In this paper we propose a contingent claim pricing scheme between two counterparties in an incomplete one period market. According to our approach the two counterparties of a non-marketed contingent claim select a pair of pricing kernels, in order to agree on a common price, by minimizing their joint regret function, which quantifies the departure from their initial beliefs. The joint regret function is a convex combination of entropy-like or norm-dependent functionals. The relevant optimization problem is posed in terms of a partially finite convex programming problem in the space of pricing kernels.

32.1 Introduction

A main strand of the existing literature on incomplete markets has been devoted to the problem of pricing contingent claims, provided that the basic market structure is given (see for example [1] and references within). The main findings of this literature is that in an incomplete market the price map (pricing kernel) is not unique, therefore leading to a whole range of arbitrage free prices for a contingent claim. By now there are well established methods to find this range of prices, however, it is still unclear how one of these prices is eventually selected. Clearly, additional criteria are needed in order to select a particular price out of the range of all possible prices.

C. Kountzakis (✉) and S.Z. Xanthopoulos
Department of Statistic and Actuarial-Financial Mathematics, University of the Aegean, Karlovassi, 83200 Samos, Greece
e-mail: chrkoun@aegean.gr, sxantho@aegean.gr

A.N. Yannacopoulos
Department of Statistics, Athens University of Economics and Business, Patission 76, 10434 Athens, Greece
e-mail: ayannaco@aueb.gr

M.M. Peixoto et al. (eds.), *Dynamics, Games and Science I*, Springer Proceedings in Mathematics 1, DOI 10.1007/978-3-642-11456-4_32,
© Springer-Verlag Berlin Heidelberg 2011

One such criterion that has been proposed in the literature, is the minimization of an entropy- like measure. Typically, a pricing kernel may be interpreted as a probability measure -usually called an "Arrow–Debreu measure"- under which the price of any European claim is the expectation of its discounted payoff. According to this criterion, the pricing kernel that is selected is the one corresponding to the Arrow–Debreu measure Q that is closer to the "true" statistical measure P on the states of the world, that is the one that minimizes a Kuhlback-Leibner like entropy measure $\mathscr{H}(Q, P)$. This suggestion has been supported by utility pricing arguments, which are related with the entropy minimization problem via duality (see for example in [2–4] and references within etc.).

The aim of the paper is to contribute towards this strand of literature by proposing alternative criteria for the pricing of the claim. While one may consider the entropy minimization approach as a game of the economic agent against nature, we introduce an approach that involves the interaction of agents among themselves, so that they are led to the adoption of a single price. Our approach, without refuting the entropy minimization approach, complements it with alternative scenarios, that widen the scope of the theory. Our criteria are based on the concept of update of beliefs of the two counterparties involved in the buying and selling of a contingent claim, as introduced in [5].

We choose to work within the framework of a one period economy with finitely many primary assets and infinitely many states of the world. Two agents are willing to enter as counterparties into a new contract representing a European type contingent claim. If the claim is not replicable by the primary assets then a multitude of arbitrage free prices exists. Only one of these arbitrage free prices will eventually be realised, provided that the two agents are willing to conclude the transaction. We will present a scenario on the mechanism that leads to the realization of this single price. In brief it goes as follows: We suppose that each counterparty has an initial belief about the probability distribution of the future states of the world and based on this belief chooses a reference pricing kernel that prices the contract in a way that is consistent to the underlying market of primary assets. We further assume however that this belief is not rigid, in the sense that each counterparty is "willing" to review her beliefs by observing the other. As they review their beliefs, the two counterparties employ new pricing kernels for their pricing, thus departing from their initial reference kernels. This departure from the initial pricing kernels depends on the relative bargaining power of the interacting agents and is done with some reluctance which may cause regret. We suggest that the two counterparties will reach a unique commonly accepted price by minimizing their joint regret.

The rest of the paper is organized as follows: In the next section we fix ideas and notation of the general mathematical framework that we will need. In Sect. 32.3 we introduce our model and set up the corresponding primal optimization problem. In Sect. 32.4 we prove theorems on the existence of solution to the primal optimization problem. In Sect. 32.5 we formulate the dual problem and show its equivalence to the primal optimization problem. This allows us to offer an economic interpretation of our approach within a portfolio optimization framework. These ideas are further discussed through the more concrete examples of quadratic and entropic regret

32 Minimum Regret Pricing in Incomplete Markets

functions. Finally, in Sect. 32.6 we offer an interpretation of the price volatility of the contingent claim, by allowing for randomness to the relative bargaining power of the interacting agents. The Appendix gives a summary of standard notions and results from the theory of partially ordered linear spaces which are used in the paper.

32.2 Basic Concepts and Notation

In this section we present the general mathematical framework that is needed for our pricing model. The material here is heavily based on [6,7] where one can look for more details.

We consider a one-period financial markets model. There is one time period defined by two points t_0 and t_1. At time t_0 there is no uncertainty about the prevailing state of the world but there is uncertainty concerning the state of the world that will prevail at time t_1. This uncertainty is represented by a probability space $(\Omega, \mathcal{F}, \mu)$ where Ω denotes the set of the possible states of the world at time t_1 and μ denotes the relevant *statistical measure*. Uncertainty is resolved at time t_1.

All conceivable t_1-consumptions constitute a *payoff space* E, considered as a partially ordered Banach subspace of $\mathbb{R}^\Omega$. We denote by E_+ the positive cone of E. In general E will be some reflexive $L^p(\Omega, \mathcal{F}, \mu)$, endowed with the usual partial ordering: $x \geq y$ if and only if the set $\{\omega \in \Omega : x(\omega) \geq y(\omega)\}$ is a set lying in $\mathcal{F}$ of μ-probability 1. $(\Omega, \mathcal{F}, \mu)$ is considered a σ-finite probability space.

The market consists of finitely many primitive contracts, indexed $1, \ldots, J$, with x_j denoting the payoff of the j-th primitive contract. Without loss of generality, the payoffs of the primitive contracts are assumed to be linearly independent and positive (i.e. elements of E_+).

It is assumed that investing in the primitive contracts can take place in real quantities and that short sales are unlimited. Thus, the *portfolio space* Θ, is some subspace of the function space $\mathbb{R}^J$ and it is considered to be a normed linear space.

The *payoff operator* $R : \Theta \to E$, is defined as a linear, bounded and one -to -one operator. Its range $R(\Theta)$, is called the *asset span*.

According to the one-period model, the markets open at t_0, investors choose portfolios $\theta \in \Theta$ at t_0, while at t_1 they enjoy the payoff $R(\theta)$.

Let M denote the closure of $R(\Theta)$. If $E = M$, the market of the primitive contracts is called *complete*, otherwise it is called *incomplete*.

The cone $C = \{\theta \in \Theta | R(\theta) \in E_+\}$ is called the *portfolio dominance cone*.

The primitive contracts have initial prices given by a *contract price vector* q which can be considered as an element of Θ^*, the dual of Θ. For a portfolio $\theta \in \Theta$ and a price vector $q \in \Theta^*$, the price of the portfolio at time t_0 is $q(\theta) =_\Theta \langle \theta, q \rangle_{\Theta^*}$, where $\langle \Theta, \Theta^* \rangle$ denotes the *portfolio-price duality*.

A contract price q is an *arbitrage-free price* if it is a strictly positive functional of the portfolio dominance cone C, that is if $q(\theta) > 0$, for any θ such that $R(\theta) \in C \setminus \{0\}$.

A *pricing kernel* is an element $p \in E^*$, which is a strictly positive functional of E_+ such that $q(\theta) = p(R(\theta))$, where q is an arbitrage-free price.

506 C. Kountzakis et al.

The set of pricing kernels of the market is denoted by $\mathcal{Q}$ and it is non-empty when q is arbitrage-free.

32.3 Update of Beliefs and the Pricing Optimization Problem

We consider an incomplete market, the structure of which is known to all participating agents. Two agents A_1 and A_2 are willing to enter as counterparties into a new (indivisible) contract that represents a non-replicable European contingent claim. Since the market is incomplete, there is a whole range of non-arbitrage prices for this claim, so they face the problem of agreeing on just one of them.

The initial beliefs (together with the risk preferences) of each agent about the distribution characteristics of the future states of the world, influence the choices of the non arbitrage prices at which they would be willing to trade the claim. Thus, suppose that agent A_i, $(i = 1, 2)$ chooses a pricing kernel $\xi_i \in \mathcal{Q}$ according to which she performs her pricing. We will treat the ξ_i as "beliefs" of the agents concerning the future states of nature. Because of the incompleteness of the market, it is generally unlikely that the two agents with initial beliefs ξ_1 and ξ_2 will quote a commonly agreed price for the claim.

Suppose now that the agents want to reach an agreement, as they feel that it is mutually beneficial to exchange this contingent claim. Assume furthermore, that the two agents are not rigid about their beliefs, that is they are willing to adopt a different $\xi_i \in \mathcal{Q}$ which may lead to price agreement. However, their initial beliefs ξ_1^R and ξ_2^R respectively, serve as reference beliefs. Eventhough an agent may change her mind from her initial belief, this happens with some reluctance. The deviation from the reference beliefs may be quantified by *regret functions* $\mathfrak{R}_i : \mathcal{Q} \times \mathcal{Q} \to \mathbb{R}^+$, $i = 1, 2$, which have the property that $\mathfrak{R}_i$ increases as the "deviation" of ξ_i from the reference ξ_i^R increases.

Thus, we suggest that the agents will choose ξ_1, ξ_2 respectively so that they agree on a common price for the claim, but at the minimum possible common regret. In other words they share the regret between them so as to agree on a common price. The sharing rule is given by a convex combination of the individual regret functions.

Therefore the commonly accepted price for the contingent claim with payoff c, is obtained as the solution of an optimization problem in the space of pricing kernels, of the form

$$\min_{\xi_1, \xi_2 \in \mathcal{Q}} \quad \lambda_1 \mathfrak{R}_1(\xi_1, \xi_1^R) + \lambda_2 \mathfrak{R}_2(\xi_2, \xi_2^R)$$

$$\text{subject to} \tag{32.1}$$

$$\langle c, \xi_1 \rangle = \langle c, \xi_2 \rangle,$$

where $\mathcal{Q}$ denotes the set of the pricing kernels of the market given an arbitrage-free price q for the primitive assets, $\lambda_i \in (0, 1)$ for $i = 1, 2$ and $\lambda_1 + \lambda_2 = 1$.

The finally agreed price -if the problem has a solution- depends on (λ_1, λ_2). An interesting question is what λ_1, λ_2 represent in the model and how their values can

32 Minimum Regret Pricing in Incomplete Markets

be chosen. In general we may say that $\lambda_i, i = 1, 2$ represent some sort of relative bargaining power of the agents. In the last section of the paper we will treat these λ as random variables and we will see how this treatment may offer an explanation for price volatility.

One last point that remains unclear is how to quantify the deviation of one pricing kernel from the other. The following suggestions for the modelling of this deviation will be considered in the sequel:

Type I regret functionals: Norm like regret functionals

Since $\mathcal{Q}$ is a subset of $L^q(\Omega, \mathcal{F}, \mu)$, which is taken to be endowed with its usual norm $|| \cdot ||_q$ then a possible representation for $\mathfrak{R}_i$ would be $\mathfrak{R}_i(\xi_i, \xi_i^R) = \phi_i(|| \xi_i - \xi_i^R ||_q)$ where ϕ_i is an increasing and convex real valued function.

The function ϕ_i is increasing because the two counterparties are reluctant to deviate from their reference pricing kernels, hence the regret for adopting a different belief on the state space is increasing with respect to the distance of this pricing kernel from the reference kernel. The convexity of $\phi_i, i = 1, 2$ reflects the usual conditions assumed in economics, related to risk aversion arguments.

Type II regret functionals: Entropic like functionals

A second type of possible regret functionals can be entropic like functionals. We assume that $\mathfrak{R}_i, i = 1, 2$ are functionals of the form

$$\mathfrak{R}_i(\xi_i, \xi_i^R) = \mathbb{E}_{Q_i^R} \left[\frac{dQ_i}{dQ_i^R} \ln(\frac{dQ_i}{dQ_i^R}) \right]$$

namely entropy like functionals that measure the deviation of the risk -neutral measure Q_i whose Radon–Nikodym derivative is defined via the pricing kernel ξ_i from the risk -neutral measure Q_i^R whose Radon–Nikodym derivative is defined via the reference pricing kernel ξ_i^R. A direct computation indicates that

$$\mathfrak{R}_i(\xi_i, \xi_i^R) = \int_\Omega \frac{\xi_i}{\xi_i^R} ln(\frac{\xi_i}{\xi_i^R}) dQ_i^R,$$

while the same integral can be written in the form

$$\mathfrak{R}_i(\xi_i, \xi_i^R) = (1 + r) \int_\Omega \xi_i ln(\frac{\xi_i}{\xi_i^R}) dP,$$

with respect to the statistical measure. The function $\mathfrak{R}_i(\xi_i, \xi_i^R)$ which is actually the entropy $I(Qi, Q_i^R)$ of the corresponding risk neutral measure Q_i with respect to Q_i^R is well defined for any $\xi_i \in \mathcal{Q}$ since Q_i and Q_i^R are equivalent measures since they are equivalent to P.

Then the agents choose a pair of pricing kernels (ξ_1, ξ_2) such that the "distance" of this choice from their initial reference pair of pricing kernels (ξ_1^R, ξ_2^R), as quantified by the entropy, is minimized. This is an important generalization to the standard entropy minimization method according to which all agents in an imcomplete market, choose the pricing kernel ξ^* that corresponds to the risk neutral measure that minimizes the entropy with regard to the statistical measure. In other words, we can view the standard method as a game that the agents play against nature. Of course the nature's beliefs are rigid and her reference pricing kernel is the density of the statistical measure.

In our approach we consider the interaction between the two agents in the choice of the risk neutral measure through comparison of beliefs and preferences, rather than simply the comparison of the risk neutral meaures with the statistical measure.

32.4 Existence of Solution to the Optimization Problem

In this section we prove theorems about the existence of solution to the primal optimization problem for both types of regret funtions that we introduced.

The problem will be considered in some non-empty uniformly lower bounded set of pricing kernels $\mathcal{Q}_a = \{\xi \in \mathcal{Q} | \xi(\omega) \geq a$ for any $\omega \in \Omega\}$ of the market M for some $a > 0$. Considering the problem in $\mathcal{Q}_a$ means that the prices of the contingent commodities in the complete market cannot be lower than a certain level, an assumption which is reasonable from the economic point of view. We remind that the *contingent commodity* associated with the state $\omega \in \Omega$ is a commodity whose every unit is consumed if the state ω occurs. Considering the problem in $\mathcal{Q}_a$ is consistent with the one-period model of financial markets, since if $\xi \geq a1, a > 0$ with respect to the usual partial ordering in L^q we have that $\int_{\Omega} \xi dP = \frac{1}{1+r} \geq a$. Hence, it suffices to determine some a that satisfies the last inequality.

Throughout this paragraph we assume that $E = L^p(P)$ (or $E = \ell^p(P)$) and that $\mathcal{Q} \subset E^*_{++}$ where by E^*_{++} we denote the set of strictly positive functionals on E_+. This is a convex subset of E^*_+.

32.4.1 Reformulation of the Optimization Problem

We now rewrite this optimization problem in a more compact form. Define the map $A : E^* \times E^* \to \mathbb{R}^{2J+1}$ as follows:

$$(f_1, f_2) \mapsto \begin{bmatrix} (x_1, 0) \\ (x_2, 0) \\ \vdots \\ (x_J, 0) \\ (0, x_1) \\ (0, x_2) \\ \vdots \\ (0, x_J) \\ (c, -c) \end{bmatrix} \quad (f_1, f_2) = \begin{bmatrix} \langle x_1, f_1 \rangle \\ \langle x_2, f_1 \rangle \\ \vdots \\ \langle x_J, f_1 \rangle \\ \langle x_1, f_2 \rangle \\ \langle x_2, f_2 \rangle \\ \vdots \\ \langle x_J, f_2 \rangle \\ \langle c, f_1 \rangle - \langle c, f_2 \rangle \end{bmatrix}$$

32 Minimum Regret Pricing in Incomplete Markets

and the joint regret function $g : \mathcal{Q} \times \mathcal{Q} \to \mathbb{R}$ where

$$g(\xi_1, \xi_2) = \lambda_1 \Re_1(\xi_1, \xi_1^R) + \lambda_2 \Re_2(\xi_2, \xi_2^R).$$

Then the optimization problem takes the following form:

$$\min_{\xi_1, \xi_2 \in E_{++}^*} g(\xi_1, \xi_2)$$
$$\text{subject to} \tag{32.2}$$
$$A(\xi_1, \xi_2) \in b + \{0\}$$

where $b = (q_1, q_2, \ldots, q_J, q_1, q_2, \ldots, q_J, 0)^T \in \mathbb{R}^{2J+1}$.

The following lemmas will be useful.

Lemma 32.1. *The following hold:*

1. The adjoint of the map A is:

$$A^* : \mathbb{R}^{2J+1} \to E \times E,$$

where

$$y = (y_1, y_2, \ldots, y_{2J+1}) \mapsto$$

$$A^*(y) = \sum_{j=1}^{J} y_j(x_j, 0) + \sum_{j=J+1}^{2J} y_j(0, x_{j-J}) + y_{2J+1}(c, -c).$$

2. $P = \{0\}$ is a polyhedral cone.

Proof. (1) Follows from a simple calculation.

(2) Note that P can be expressed as follows: $P = \{x \in \mathbb{R}^{2J+1} | x = \lambda 0, \lambda \geq 0\}$, i.e as a finitely generated cone in $\mathbb{R}^{2J+1}$. $\square$

Lemma 32.2. *The set $\mathcal{Q}$ of the pricing kernels of the market given an arbitrage-free price q for the primitive assets, is a convex set.*

Proof. If $\xi, \xi' \in \mathcal{Q}$ then $q(\theta) = \langle R(\theta), \xi \rangle, q(\theta) = \langle R(\theta), \xi' \rangle$ for the price q and for any portfolio θ, where R is the payoff operator of the market. Hence for any real number $a \in (0, 1)$ we obtain $q(\theta) = a(q(\theta)) + (1 - a)(q(\theta)) = a\langle R(\theta), \xi \rangle + (1 - a)\langle R(\theta), \xi' \rangle = \langle R(\theta), a\xi + (1 - a)\xi' \rangle$ for any portfolio θ. $\square$

Lemma 32.3. *The joint regret function has the following properties:*

1. g is a convex function.
2. g is continuous, hence lower semicontinuous and closed.
3. g is a monotonically regular function.

Proof. (1) Immediate since each $\mathfrak{R}_i, i = 1, 2$ is a convex function (The convexity of $\mathfrak{R}_i$ follows from a straight forward calculation taking into account that ϕ_i are increasing and convex).

(2) The form of g guarantees that it is continuous, hence *lower semicontinuous* in the sense of [8] and for this reason it is closed, according to the relevant remark met in [8, p. 88].

(3) Because of the fact that $E^* = L^q(\Omega, \mathscr{F}, P)$ (or $E^* = \ell^q(\Omega, \mathscr{F}, P)$) being partially ordered by its usual ordering is a Banach lattice, g is a *monotonically regular function*, according to the Corollary 6.2 in [9, p. 53]. $\square$

32.4.2 Regret Functions of Type I

The results here treat the case of regret functions of Type I.

Theorem 32.1. *Assume that*

1. $E = L^p(\mu)$ *(or $E = \ell^p(\mu)$).*
2. *The functions ϕ_1, ϕ_2 are both equal to the function $\phi(x) = x^q, x \in [0, \infty)$, where $q > 1$ is the conjugate coefficient of p.*
3. *The statistical measure μ is such that $\mu(\omega) > 0$ for any $\omega \in \Omega$.*

Then a solution to the optimization problem (32.2) exists.

Proof. We only give the proof for the case where $E = \ell^p(P)$. The case where $E = L^p(P)$ proceeds similarly, with a few minor changes sketched briefly in Remark 32.1.

The proof follows along the lines of the proof of the Theorem 2.1 in [2] but takes advantage of the reflexivity of the space $\ell^p(P)$ if $p > 1$.

Since the set of pricing kernels $\mathscr{D}$ is non-empty, the set

$$\{(\xi_1, \xi_2) \in \mathscr{D}_a \times \mathscr{D}_a | \langle c, \xi_1 \rangle = \langle c, \xi_2 \rangle\}$$

is non-empty, hence

$$\inf\{g(\xi_1, \xi_2) | A(\xi_1, \xi_2) = b, (\xi_1, \xi_2) \in \mathscr{D}_a \times \mathscr{D}_a\} < +\infty.$$

By assumption 2 the joint regret function assumes the form

$$g(\xi_1, \xi_2) = \lambda_1 \|\xi_1 - \xi_1^R\|_q^q + \lambda_2 \|\xi_2 - \xi_2^R\|_q^q,$$

where $\lambda_1, \lambda_2 \in (0, 1)$ with $\lambda_1 + \lambda_2 = 1$.

There exists a sequence $(\xi_{1,n}, \xi_{2,n})_{n \in \mathbb{N}} \subseteq \mathscr{D}_a \times \mathscr{D}_a$ such that

$$g(\xi_{1,n}, \xi_{2,n}) \downarrow \inf\{g(\xi_1, \xi_2) | A(\xi_1, \xi_2) = b, (\xi_1, \xi_2) \in \mathscr{D}_a \times \mathscr{D}_a\}.$$

32 Minimum Regret Pricing in Incomplete Markets

Note that this sequence, considered in the space $\ell^q(P) \oplus \ell^q(P)$ endowed with the $\|.\|_2$ norm, is bounded. This is true since for $i = 1, 2$:

$$\|\xi_{i,n} - \xi_i^R\|_q^q \le \frac{1}{\lambda_i} g(\xi_{1,n}, \xi_{2,n}) \le \frac{1}{\lambda_i} g(\xi_{1,1}, \xi_{2,1}),$$

Note that $d(x) = x^{\frac{1}{q}}$ is increasing, so that

$$\|\xi_{1,n} - \xi_1^R\|_q^2 + \|\xi_{2,n} - \xi_2^R\|_q^2 \le (\frac{1}{\lambda_1} g(\xi_{1,1}, \xi_{2,1}))^{\frac{2}{q}} + (\frac{1}{\lambda_2} g(\xi_{1,1}, \xi_{2,1}))^{\frac{2}{q}}.$$

Since $\ell^q(P) \oplus \ell^q(P)$ endowed with the $\|.\|_2$ norm is reflexive, there exists a weakly convergent subsequence with weak limit denoted as (ξ_1, ξ_2). Then Mazur's theorem (for the statement, see Theorem 2.5.16 and Corollary 2.5.19 in [11]) implies the existence of a sequence of convex combinations of the terms of $(\xi_{1,n}, \xi_{2,n})_{n\in\mathbb{N}}$ which converges to (ξ_1, ξ_2) in the norm topology of the space $\ell^q(P) \oplus \ell^q(P)$. Let us denote by $(\tilde{\xi}_{1,n}, \tilde{\xi}_{2,n}), n \in \mathbb{N}$ the terms of this sequence. The dual space of $\ell^q(P) \oplus \ell^q(P)$ endowed with the $\|.\|_2$ norm is $\ell^p(P) \oplus \ell^p(P)$ endowed with the $\|.\|_2$ norm. Let e_ω^* the linear functional defined by $e_\omega^*(\omega) = 1$ while $e_\omega^*(\omega') = 0$ if $\omega' \ne \omega$. Since $(e_\omega^*, 0)$ and $(0, e_\omega^*)$ belong to $\ell^p(P) \oplus \ell^p(P)$ for any $\omega \in \Omega$, we get that

$$\langle (\tilde{\xi}_{1,n}, \tilde{\xi}_{2,n}), (e_\omega^*, 0) \rangle \to \langle (\xi_1, \xi_2), (e_\omega^*, 0) \rangle = \xi_1(\omega).$$

Similarly,

$$\langle (\tilde{\xi}_{1,n}, \tilde{\xi}_{2,n}), (0, e_\omega^*) \rangle \to \langle (\xi_1, \xi_2), (0, e_\omega^*) \rangle = \xi_2(\omega).$$

For any $n \in \mathbb{N}$, $\tilde{\xi}_{1,n}, \tilde{\xi}_{2,n}$ are convex combinations of pricing kernels, therefore positive functionals of $\ell_+^p(P)$. This together with $e_\omega^* \in \ell_+^p(P)$ implies that $\tilde{\xi}_{1,n}(\omega), \tilde{\xi}_{2,n}(\omega) \ge 0$ for any n, hence $\xi_1(\omega), \xi_2(\omega) \ge 0$. The last condition implies that ξ_1, ξ_2 are positive functionals of $\ell_+^p(P)$, since the set $\{e_\omega^*, \omega \in \Omega\} \subseteq \ell_+^p(P)$ is a Schauder basis for this space, being actually a positive basis for it (see [7, p. 498]. By the definition of the norm in $\ell^q(P) \oplus \ell^q(P)$ and the above convergence results, we have for $i = 1, 2$

$$\lim_n \|\tilde{\xi}_{i,n} - \xi_i\|_q = 0,$$

or else

$$\lim_n \|\tilde{\xi}_{i,n} - \xi_i^R\|_q = \|\xi_i - \xi_i^R\|_q \quad \text{and} \quad \lim_n \|\tilde{\xi}_{i,n} - \xi_i^R\|_q^q = \|\xi_i - \xi_i^R\|_q^q.$$

We may suppose that there exist natural numbers $k_1 < k_2 < k_3 < \ldots < k_n < \ldots$ such that

$$\tilde{\xi}_{1,n} = \sum_{k=k_n}^{k_{n+1}-1} \lambda_k^{(n)} \xi_{1,k}$$

where $\lambda_k^{(n)} \in [0, 1]$, $\sum_{k=k_n}^{k_{n+1}-1} \lambda_k^{(n)} = 1$. Hence

$$\|\tilde{\xi}_{1,n} - \xi_1^R\|_q = \| \sum_{k=k_n}^{k_{n+1}-1} \lambda_k^{(n)} \xi_{1,k} - \sum_{k=k_n}^{k_{n+1}-1} \lambda_k^{(n)} \xi_1^R \|_q = \| \sum_{k=k_n}^{k_{n+1}-1} \lambda_k^{(n)} (\xi_{1,k} - \xi_1^R) \|_q$$

$$\leq \sum_{k=k_n}^{k_{n+1}-1} \lambda_k^{(n)} \|\xi_{1,k_n} - \xi_1^R\|_q \leq \|\xi_{1,k_n} - \xi_1^R\|_q.$$

Note that $(\xi_{1,k_n})_{n\in\mathbb{N}}$ is a subsequence of $(\xi_{1,n})_{n\in\mathbb{N}}$ and $\phi(x) = x^q$ is an increasing function on $[0, \infty)$, hence we obtain that

$$\lim_n \|\tilde{\xi}_{1,n} - \xi_1^R\|_q^q = \|\xi_1 - \xi_1^R\|_q^q \leq \lim_n \|\xi_{1,k_n} - \xi_1^R\|_q^q.$$

Similarly,

$$\lim_n \|\tilde{\xi}_{2,n} - \xi_2^R\|_q^q = \|\xi_2 - \xi_2^R\|_q^q \leq \lim_n \|\xi_{2,k_n} - \xi_2^R\|_q^q.$$

Hence

$$g(\xi_1, \xi_2) = \lim_n g(\tilde{\xi}_{1,n}, \tilde{\xi}_{2,n}) = \lim_n \lambda_1 \|\tilde{\xi}_{1,n} - \xi_1^R\|_q^q + \lambda_2 \|\tilde{\xi}_{2,n} - \xi_2^R\|_q^q$$

$$\leq \lim_n g(\xi_{1,k_n}, \xi_{2,k_n}) = \lim_n \lambda_1 \|\xi_{1,k_n} - \xi_1^R\|_q^q + \lambda_2 \|\xi_{2,k_n} - \xi_2^R\|_q^q$$

$$= \inf\{g(\xi_1, \xi_2) | A(\xi_1, \xi_2) = b, (\xi_1, \xi_2) \in \mathcal{Q}_a \times \mathcal{Q}_a\}.$$

We need to show that $\langle \xi_1, x_j \rangle = q_j, \langle \xi_2, x_j \rangle = q_j$ for any $j = 1, 2, \ldots, J$ in order to verify that ξ_1, ξ_2 have the properties of a pricing kernel. Since $(\xi_{1,n})_{n\in\mathbb{N}}$, $(\xi_{2,n})_{n\in\mathbb{N}}$ converge weakly in the $\sigma(\ell^q(P), \ell^p(P))$ -topology and since $x_j \in \ell^p(P)$ for any $j = 1, 2, \ldots, J$,

$$\langle \xi_{1,n}, x_j \rangle = q_j \rightarrow \langle \xi_1, x_j \rangle,$$
$$\langle \xi_{2,n}, x_j \rangle = q_j \rightarrow \langle \xi_2, x_j \rangle.$$

It remains to show the strict positivity of the vectors ξ_1, ξ_2 which in needed to assure that they constitute a solution to the optimization problem. For this we need to consider the problem in $\mathcal{Q}_a$. Then by the weak convergence of $(\xi_{1,n})_{n\in\mathbb{N}}, (\xi_{2,n})_{n\in\mathbb{N}}$ whose elements lie in $\mathcal{Q}_a$ we get that $\xi_1(\omega) \geq a, \xi_2(\omega) \geq a$ for any ω. $\square$

Remark 32.1. Note that $L_+^q(P)$ is norm -closed. The proof of Theorem 32.1 also holds in a general non-atomic L^p space, because the set $\{d \in L^q(P) | d \geq a\mathbf{1}\} = a\mathbf{1} + L_+^q$ is a closed set, since the cone is so. Hence the proof is the same in this case.

32 Minimum Regret Pricing in Incomplete Markets

The joint regret function employed in the Corollary below is in some sense a generalization for the case of joint regret functions which are not convex combinations of the regret funtions of the sole agents.

Corollary 32.1. *Assume that the agents report the joint regret function*

$$g(\xi_1, \xi_2) = max\{\phi_1(\|\xi_1 - \xi_1^R\|_q), \phi_2(\|\xi_2 - \xi_2^R\|_q)\}$$

with $\phi_1 = \phi_2 = \phi$ and $\phi(x) = x, x \in [0, \infty)$ and $\xi_1, \xi_2 \in \mathcal{Q}_a$.
Then, the optimization problem (32.2) has a solution.

Proof. The proof follows along the same lines of the proof of Theorem 32.1. First, note that since the set of desired pricing kernels $\mathcal{Q}_a$ is non-empty, the set

$$\{(\xi_1, \xi_2) \in \mathcal{Q}_a \times \mathcal{Q}_a | \langle c, \xi_1 \rangle = \langle c, \xi_2 \rangle\}$$

is non-empty, hence

$$\inf\{g(\xi_1, \xi_2) | A(\xi_1, \xi_2) = b, (\xi_1, \xi_2) \in \mathcal{Q}_a \times \mathcal{Q}_a\} < +\infty.$$

By the form of the joint regret function, we conclude that there exists a sequence

$$(\xi_{1,n}, \xi_{2,n})_{n \in \mathbb{N}} \subseteq \mathcal{Q}_a \times \mathcal{Q}_a$$

such that

$$g(\xi_{1,n}, \xi_{2,n}) \downarrow \inf\{g(\xi_1, \xi_2) | A(\xi_1, \xi_2) = b, (\xi_1, \xi_2) \in \mathcal{Q}_a \times \mathcal{Q}_a\}.$$

This sequence, considered in the space $\ell^q(P) \oplus \ell^q(P)$ endowed with the $\|.\|_2$ norm, is bounded. Indeed,

$$\|\xi_{1,n} - \xi_1^R\|_q \le g(\xi_{1,n}, \xi_{2,n}) \le g(\xi_{1,1}, \xi_{2,1}),$$
$$\|\xi_{2,n} - \xi_2^R\|_q \le g(\xi_{1,n}, \xi_{2,n}) \le g(\xi_{1,1}, \xi_{2,1}).$$

Hence, $g(\xi_{1,n}, \xi_{2,n}) \le 2g(\xi_{1,1}, \xi_{2,1})$. Since $\ell^q(P) \oplus \ell^q(P)$ endowed with the $\|.\|_\infty$ norm is reflexive, a weakly convergent subsequence of this sequence exists. The rest of the proof remains the same as the proof of Theorem 32.1. $\square$

32.4.3 Regret Functionals of Type II

The existence of solution to the minimization problem holds also in the case where the regret functional is of type II, i.e. an entropy like functional (see also Sect. 32.5.4). The next result covers the case of such regret functions when $E = L^2(\Omega, \mathcal{F}, P)$, with $(\Omega, \mathcal{F}, P)$ a σ-finite probability space.

514   C. Kountzakis et al.

Theorem 32.2. *Suppose $E = L^2(\Omega, \mathcal{F}, P)$ is partially ordered by a closed cone B which contains L_+^2 with non-empty interior (so that the risk neutral measures of the market are well-defined). Also suppose that $1 \in M$ is such an interior point and the corresponding set of pricing kernels $\mathcal{Q}$ is such that the subset $\mathcal{Q}_a$ is non-empty for some $a > 0$. Assume that the regret functionals $\mathfrak{R}_i$, $i = 1, 2$ of the seller and the buyer of the claim are entropic like functionals. Then a solution to the minimization problem exists.*

Proof. The joint regret function of the agents is

$$g(\xi_1, \xi_2) = \lambda_1 \mathfrak{R}(\xi_1, \xi_1^R) + \lambda_2 \mathfrak{R}(\xi_2, \xi_2^R).$$

Note that if we suppose that there exist pricing kernels $\xi_1, \xi_2 \in \mathcal{Q}_a$ such that $\xi_1(c) = \xi_2(c)$, then

$$\gamma := \inf_{(\xi_1, \xi_2) \in \mathcal{G}} g(\xi_1, \xi_2) \neq +\infty,$$

where $\mathcal{G} = \{(\xi_1, \xi_2) \in \mathcal{Q}_a \times \mathcal{Q}_a | \xi_1(c) = \xi_2(c)\}$. Also, $\gamma \neq -\infty$ because for any two risk neutral measures $Q_1, Q_2, I(Q_1, Q_2) \geq 0$ holds, where $I(Q_1, Q_2)$ denotes the relative entropy of Q_1 with respect to Q_2.

Consider a minimizing sequence of pairs of pricing kernels $(\xi_{1,n}, \xi_{2,n})_{n \in \mathbb{N}} \subseteq \mathcal{G}$, namely a sequence such that $g(\xi_{1,n}, \xi_{2,n}) \downarrow \gamma$. Note that since I is an interior point of the cone B, then according to the Proposition 3.8.12 in [12] I is a uniformly monotonic functional of B^0. Every pricing kernel ξ of the market M is an element of the base defined by $(1 + r)I$ on B^0, since $\langle 1, \xi \rangle = \frac{1}{1+r}$. Hence every element of $\mathcal{Q}_a$ is bounded with respect to the $\|.\|_2$-norm of $L^2(\Omega, \mathcal{F}, P)$. Namely, the sequences $(\xi_{i,n})_{n \in \mathbb{N}}$, $i = 1, 2$ are both bounded with respect to the $\|.\|_2$-norm of $L^2(\Omega, \mathcal{F}, P)$. But this implies the boundedness of the sequence $(\xi_{1,n}, \xi_{2,n})_{n \in \mathbb{N}}$ in the space $L^2 \times L^2$ for example under the ℓ^1 norm. Since the space L^2 is reflexive and the space $L^2 \times L^2$ is so, there is a weakly convergent subsequence of the sequence $(\xi_{1,n}, \xi_{2,n})_{n \in \mathbb{N}}$. Let us denote this subsequence by $(\xi_{1,n}, \xi_{2,n})_{n \in \mathbb{N}}$ and its weak limit by (ξ_1, ξ_2). The convergence $g(\xi_{1,n}, \xi_{2,n}) \downarrow \gamma$ remains, while since Q_a is weakly closed in L^2 and the kernel of the functional $(c, -c)$ is also a weakly closed set in $L^2 \times L^2$, the weak limit (ξ_1, ξ_2) of the above subsequence is a pair of pricing kernels such that $\xi_1(c) = \xi_2(c)$ -or else an element of $\mathcal{G}$. From Mazur's theorem we take that there is a sequence of convex combinations of the terms of $(\xi_{1,n}, \xi_{2,n})_{n \in \mathbb{N}}$ denoted by $(\tilde{\xi}_{1,n}, \tilde{\xi}_{2,n})_{n \in \mathbb{N}}$ such that $(\tilde{\xi}_{1,n}, \tilde{\xi}_{2,n})_{n \in \mathbb{N}}$ converges to (ξ_1, ξ_2) in the norm of $L^2 \times L^2$ we had considered. But the last convergence in $L^2 \times L^2$ implies that

$$\tilde{\xi}_{i,n} \to \xi_i, \quad \text{in } L^2(\Omega, \mathcal{F}, P), \quad i = 1, 2,$$

and using Hölder inequality we deduce that

$$\tilde{\xi}_{i,n} \to \xi_i, \quad \text{in } L^1(\Omega, \mathcal{F}, P), \quad i = 1, 2.$$

Also, the last two convergence implications lead to the following convergence results:

$$\tilde{\xi}_{i,n}\frac{dP}{dQ_R^i} \to \xi_i\frac{dP}{dQ_R^i}, \quad \text{in } L^1(\Omega,\mathscr{F},Q_R^i), \quad i=1,2$$

From Theorem 12.6 in [7] there are subsequences of

$$\left(\tilde{\xi}_{i,n}\frac{dP}{dQ_R^i}\right)_{n\in\mathbb{N}}, i=1,2$$

denoted again by

$$\left(\tilde{\xi}_{i,n}\frac{dP}{dQ_R^i}\right)_{n\in\mathbb{N}}, i=1,2$$

and functions $g_i \in L^2(\Omega,\mathscr{F},Q_R^i) \subseteq L^1(\Omega,\mathscr{F},Q_R^i), i=1,2$ respectively, such that $|\tilde{\xi}_{i,n}\frac{dP}{dQ_R^i}| \le g_i$, Q_R^i -a.e. (almost everywhere), such that $\tilde{\xi}_{i,n}\frac{dP}{dQ_R^i}(\omega) \to \xi_i\frac{dP}{dQ_R^i}(\omega)$, Q_R^i-a.e. in Ω $(i=1,2)$. Since $\frac{dQ_R^i}{dP}(\omega) > 0$, P-a.e., this also implies $\frac{dP}{dQ_R^i}(\omega) > 0$, Q_R^i-a.e., $i=1,2$ because the measures P, Q_R^i are equivalent. Since $\xi = \frac{1}{1+r}\frac{dQ}{dP}$, for any pricing kernel of the market M and the corresponding risk -neutral measure Q, we get $\frac{dP}{dQ} = (1+r)\frac{1}{\xi}$, namely $|\frac{\tilde{\xi}_{i,n}}{\xi_R^i}| \le (1+r)g_i$, Q_R^i -a.e. $(i=1,2)$. Hence

$$\frac{\tilde{\xi}_{i,n}}{\xi_R^i}(\omega) \to \frac{\xi_i}{\xi_R^i}(\omega),$$

Q_R^i-a.e. in Ω $(i=1,2)$. Since the function $h(x) = x\log x, x > 0$ is continuous in its domain, we get that

$$\frac{\tilde{\xi}_{i,n}}{\xi_R^i}(\omega)\log\frac{\tilde{\xi}_{i,n}}{\xi_R^i}(\omega) \to \frac{\xi_i}{\xi_R^i}(\omega)\log\frac{\xi_i}{\xi_R^i}(\omega),$$

Q_R^i-a.e. in Ω $(i=1,2)$. Since $\log x \le x - 1$, for any $x > 0$ we get $x\log x \le x^2 - x < x^2$ for any $x > 0$. Hence the terms sequences $(\frac{\tilde{\xi}_{i,n}}{\xi_R^i}\log\frac{\tilde{\xi}_{i,n}}{\xi_R^i})_{n\in\mathbb{N}}, i=1,2$ are dominated by the terms of the sequences of Q_R^i- integrable functions

$$\left(\left(\frac{\tilde{\xi}_{i,n}}{\xi_R^i}\right)^2\right)_{n\in\mathbb{N}}.$$

516   C. Kountzakis et al.

But the last sequences are dominated by the functions $((1 + r)g_i)^2$, respectively. Let us verify the last argument that $\left(\left(\frac{\tilde{\xi}_{i,n}}{\xi_R^i}\right)^2\right)_{n\in\mathbb{N}}$ are sequences of Q_R^i- integrable functions, $i = 1, 2$.

Since for every $\xi \in \mathcal{Q}_a$ we have $\xi(\omega) \geq a$, P-a.e., this inequality holds Q-a.e. Hence $\frac{1}{\xi(\omega)} \leq \frac{1}{a}$, Q-a.e., where Q is the risk neutral measure of the market M defined by ξ. Hence in our case,

$$
\int_\Omega \left(\frac{\tilde{\xi}_{i,n}}{\xi_R^i}\right)^2 dQ_R^i = \int_\Omega (\tilde{\xi}_{i,n})^2 \frac{1}{(\xi_R^i)^2} \frac{dQ_R^i}{dP} dP
$$

$$
= \int_\Omega (\tilde{\xi}_{i,n})^2 \frac{1}{(\xi_R^i)^2} \xi_R^i (1 + r) dP = (1 + r) \int_\Omega (\tilde{\xi}_{i,n})^2 \frac{1}{\xi_R^i} dP
$$

$$
\leq (1 + r) \int_\Omega (\tilde{\xi}_{i,n})^2 \frac{1}{a} dP = (1 + r) \frac{1}{a} \|\tilde{\xi}_{i,n}\|_2 < +\infty.
$$

Hence by the Lebesgue Dominated Convergence Theorem, we get

$$
\int_\Omega \frac{\tilde{\xi}_i}{\xi_R^i} \log \frac{\tilde{\xi}_i}{\xi_R^i} dQ_R^i = \lim_n \int_\Omega \frac{\tilde{\xi}_i}{\xi_R^i} \log \frac{\tilde{\xi}_{i,n}}{\xi_R^i} dQ_R^i,
$$

or else

$$
I(Q^i, Q_R^i) = \lim_n I(\tilde{Q}_n^i, Q_R^i), \ i = 1, 2
$$

where Q_i is the risk neutral measure of the market M corresponding to the pricing kernel ξ_i and $\tilde{Q}_n^i$ is the risk neutral measure of the market M which corresponds to the pricing kernel $\tilde{\xi}_{i,n}$, $i = 1, 2$. From the weak convergence of a sequence of convex combinations of $(\xi_{1,n}, \xi_{2,n})_{n\in\mathbb{N}}$, we get that

$$
(\tilde{\xi}_{1,n}, \tilde{\xi}_{2,n}) \in conv(\{(\xi_{1,n}, \xi_{2,n}), (\xi_{1,n+1}, \xi_{2,n+1}), \ldots\}),
$$

where $conv(A)$ denotes the convex hull of the set A. From the convexity of the relative entropy and the fact that $g(\xi_{1,n}, \xi_{2,n})$ is a decreasing sequence of real numbers, we obtain that $I(\tilde{Q}_n^i, Q_R^i) \leq I(Q_n^i, Q_R^i)$ for any $n \in \mathbb{N}$. Finally, if λ is the bargaining power of the buyer of the claim c, we get

$$
g(\xi_1, \xi_2) = \lambda I(Q^1, Q_R^1) + (1 - \lambda)I(Q^2, Q_R^2) = \lim_n \left(\lambda I(\tilde{Q}_n^1, Q_R^1) + (1 - \lambda)I(\tilde{Q}_n^2, Q_R^2)\right)
$$

$$
\leq \lim_n \left(\lambda I(Q_n^1, Q_R^1) + (1 - \lambda)I(Q_n^2, Q_R^2)\right) = \gamma.
$$

The last inequality implies that the infimum of the values of g over the set $\mathcal{G}$ is actually a minimum and by this way, we verified that the minimization problem of g over $\mathcal{G}$ has a solution. $\qquad\square$

Remark 32.2. The assumption about the cone B mentioned in the statement of the last Theorem is valid if Ω is finite and in this case we take an affirmative answer

32 Minimum Regret Pricing in Incomplete Markets

about the existence of solution to the primal optimization problem with entropy-like regret functionals for the agents. If Ω is infinite, we wonder whether such a cone exists.

32.5 The Dual Problem

In this section we formulate the dual problem of (32.2) and show their equivalence. This allows us to offer an economic interpretation of our approach within a portfolio optimization framework. Finally, we further discuss these ideas in the more concrete examples of quadratic and entropic regret functions.

32.5.1 Duality Approach

The following duality result plays a crucial role in our work

Proposition 32.1. *Let* $b = (q_1, q_2, \ldots, q_J, q_1, q_2, \ldots, q_J, 0)^T \in \mathbb{R}^{2J+1}$, $P = \{0\}$
and $K = E_+^*$.

If the primal problem (32.2) admits a solution, then the solution of this problem is equivalent to the solution of the following dual problem

$$\max_{y \in \mathbb{R}^{2J+1}} \quad b \cdot y - g^*(\psi)$$

$$\text{subject to} \tag{32.3}$$

$$\psi - A^*(y) \in K^0, y \in P^0,$$

$$\psi \in E \oplus E,$$

where g^* *is the convex conjugate of* g, K^0 *is the dual cone of* E_+^*, *which is actually* E_+. *The dual cone of* $P = \{0\}$ *is the whole space* $\mathbb{R}^{2J+1}$.

Proof. Note that $E = L^p(P)$ or $E = \ell^p(P)$ with $1 < p < +\infty$. The proof that follows is given explicitly for $E = \ell^p(P)$ spaces but it also holds for $E = L^p(P)$, thanks to the definition of the quasi-relative interior given in [13, p. 28].

The original problem (32.1), or better its equivalent form (32.2), resembles the primal *conical convex model* problem studied by [9] (p. 53 op. cit.) due to the properties satisfied by the joint regret function g (see Lemma 32.3), with one important difference; in the aforementioned paper the primal variable is allowed to take values over a whole cone K, while in the present work ξ_1, ξ_2 must be strictly positive and continuous linear functionals of the positive cone E_+. Within the framework of $\ell^p(P)$ spaces, the set of the strictly positive, continuous linear functionals of the positive cone ℓ_+^p is the set $\{x \in \ell_+^q \mid x_n > 0 \text{ for any } n \in \mathbb{N}\}$. This is the quasi-relative interior $qri(\ell_+^q)$ of the cone ℓ_+^q, according to the terminology used in [13], since $\ell^p(P)$ is the L^p space with respect to a probability measure on $\mathbb{N}$. From the continuity of the function g we have that

$$\inf\{g(\xi_1,\xi_2)|A(\xi_1,\xi_2)=b, \xi_1,\xi_2 \in \ell_+^q\} = \inf\{g(\xi_1,\xi_2)|A(\xi_1,\xi_2)=b, \xi_1,\xi_2 \in qri(\ell_+^q)\}.$$

Therefore, according to Theorem 6.3 in [9] solving the primal conical convex model problem (32.2), is equivalent to solving the dual problem. $\qquad\square$

It is interesting to note that the dual problem is a finite dimensional problem (an optimization problem in $\mathbb{R}^{2J+1}$) which is of course easier to handle than the primal problem which is in general infinite dimensional. Therefore the duality result given in Proposition 32.1 is important in its own right as it allows us to substitute the primal problem with a much easier problem. Furthermore, as we shall see in Sect. 32.5.2 the dual problem has an interesting economic interpretation in terms of a portfolio optimization framework.

The following technical comments concerning Proposition 32.1 are due now.

Remark 32.3. As indicated by the proof of Theorem 6.3 of [9] a necessary and sufficient condition such that the pair $(\bar{\xi}_1,\bar{\xi}_2)$ of pricing kernels is a solution to the primal problem is the existence of some $\bar{\psi} \in \partial g(\bar{\xi}_1,\bar{\xi}_2)$ and of some $\bar{\lambda} \in \mathbb{R}^{2J+1}$ such that $\bar{\psi} - A^*(\bar{y}) \in E_+ \oplus E_+$ and $\bar{\psi} - A^*(\bar{y})(\bar{\xi}_1,\bar{\xi}_2) = 0$. But since $(\bar{\xi}_1,\bar{\xi}_2)$ is a strictly positive functional of $\ell_+^p \oplus \ell_+^p$ it follows that $\bar{\psi} - A^*(\bar{y})$ is the zero element. Hence, the solution of the primal problem reduces to finding some $\bar{y} \in \mathbb{R}^{2J+1}$ such that $A^*(\bar{y}) \in \partial g(\bar{\xi}_1,\bar{\xi}_2)$ or else $(\bar{\xi}_1,\bar{\xi}_2) \in (\partial g)^{-1}(A^*(\bar{y}))$. According to [8, p.103] since g is a lower semicontinuous and proper convex function, the map $(\partial g)^{-1}$ and the subdifferential ∂g^* of the convex conjugate of g coincide. Hence finding a solution to the primal problem is equivalent to finding some $\bar{y} \in \mathbb{R}^{2J+1}$ such that there is a pair $(\bar{\xi}_1,\bar{\xi}_2)$ of pricing kernels such that $(\bar{\xi}_1,\bar{\xi}_2) \in \partial g^*(A^*(\bar{y}))$.

32.5.2 An Economic Interpretation of the Dual Problem

In this paragraph we propose an interesting economic interpretation of the dual problem, in terms of a utility maximization problem.

Let us consider $-g^*$, where g^* is the convex conjugate of g, as a utility function expressing the preferences of the buyer and the seller. $-g^*$ satisfies the properties that a utility function must satisfy as follows from the properties of the convex conjugate as well as the reflexivity of the space $E \oplus E$. If we consider the function g to be defined over the whole space $E^* \oplus E^*$, then the conjugate of g is defined all over the whole space $E \oplus E$, which may be interpreted as consisting of pairs of "consumption" bundles for the buyer and the seller respectively. Hence, by definition,

$$g^*(x) = \sup_{\xi \in E^* \oplus E^*} \{\langle x, \xi \rangle - g(\xi)\}, \quad \forall x \in E \oplus E,$$

and this can be interpreted as the maximum of the valuation of the "consumption" x under the pair of pricing kernels ξ, minus the regret corresponding to it for any such ξ.

Then the pair of pricing kernels $\bar{\xi} = (\bar{\xi}_1, \bar{\xi}_2)$, which solves the primal problem, may be interpreted as the marginal utility for the seller and the buyer respectively. This marginal utility is calculated on portfolios consisting of the primitive assets and a position on the non marketed claim (short position on the claim for the seller and long position on the claim for the buyer if $\bar{y}_{2J+1} \geq 0$).

We can therefore provide an economic interpretation of the dual problem (32.3) as a portfolio optimization problem. The dual variable $y \in \mathbb{R}^{2J+1}$ can be interpreted as the composition of a joint portfolio for the buyer and the seller of the claim consisting of the primary traded assets available in the market and the shares of the non-marketed claim involved in the transaction. The term $b \cdot y$ can be considered as the cost at $t = 0$ of this portfolio, whereas $A^*(y)$ can be considered as the payoff of this portfolio at $t = 1$ in the different possible states of the world. By considering $-g^*$ as a utility function of the payoff of the portfolio, one may easily see that the dual problem (32.3) is equivalent to finding this portfolio that maximizes the utility of the payoff of the portfolio y, but at the minimum initial cost. Therefore, the dual problem is equivalent to finding the cheapest portfolio in the market that can maximize the utility of the payoff at $t = 1$. This situation is reminiscent of what is known in the case of choosing the pricing kernel such that the entropy of the pricing kernel with respect to the statistical measure of the market is minimized. It is well known [2] that in this case the dual problem is the problem of finding the cheapest portfolio that maximizes an exponential expected utility. What we observe here is a generalization of this situation, for more general utility functions and more importantly $-g^*$ may be interpreted as a utility function that takes into account the preferences of both agents involved in this transaction, towards willingness to deviate from the reference measure, or in other words takes into account the regret functions of both agents.

32.5.3 The Example of Quadratic Regret Functions

An interesting special case is the case where the regret function is given as the linear combination of two norms in an appropriately chosen Hilbert space. This is the case of quadratic regret functions which we treat in some detail because is an interesting generalization of the important notion of projection pricing introduced in [10]. The optimization problem (32.1), which is crucial to the pricing of the contingent claim in the approach we propose in the present paper, may be viewed through such a framework if we decide to choose E to be a Hilbert space H.

As an example we may consider $H = \ell^2(P)$ endowed with its usual inner product.

520   C. Kountzakis et al.

Within the functional context described above, if we suppose that $\phi_1 = \phi_2 = \phi$ with $\phi(x) = \frac{1}{2}x^2$ then the optimization problem (32.1) takes the following form:

$$\min_{(\xi_1,\xi_2)\in\mathscr{Q}\oplus\mathscr{Q}} \quad \frac{1}{2}\|\xi_1 - \xi_1^R\|^2 + \frac{1}{2}\|\xi_2 - \xi_2^R\|^2$$

$$\text{subject to} \tag{32.4}$$

$$(\xi_1,\xi_2) \in (\xi_1^R,\xi_2^R) + ker(x^j,-x^j), \quad j = 1,2,..,J$$

$$(\xi_1,\xi_2) \in ker(c,-c)$$

where $(c,-c),(x^j,-x^j), j = 1,2,\ldots,J$ are taken as continuous linear functionals of $H \oplus H$, and ker denotes the kernel of the functional.

In the next proposition we illustrate how to use the general duality results of Proposition 32.1 to obtain a solution to problem (32.4) by treating the finite dimensional equivalent dual problem, which admittedly is easier to handle than the original problem.

Proposition 32.2. *The dual of problem (32.4) assumes the form*

$$\max_{y\in\mathbb{R}^{2J+1}} \quad b \cdot y - g^*(\psi)$$

$$\text{subject to} \tag{32.5}$$

$$\psi - A^*(y) \in \ell_+^2 \oplus \ell_+^2,$$

$$\psi \in \ell^2 \oplus \ell^2,$$

Proof. Immediate by a direct application of Proposition 32.1.   □

It is interesting to note that problem (32.5) corresponds to a Markowitz type portfolio optimization problem, thanks to Lemma 32.1.

Proposition 32.3. *Suppose that $\lambda_1 = \lambda_2 = \frac{1}{2}$, $1 \in M$, and $c \in \ell^\infty(P)$. Suppose furthermore that the payoffs of the primary assets x^j, $j = 1,2,\ldots,J$ are such that their Gram matrix G, defined by $G := (G_{ij}) = (\langle x^i,x^j \rangle)$, is invertible. Define $u = (u_1,\ldots,u_J) \in \mathbb{R}^J$, with $u_j = \langle c,x^j \rangle$, $j = 1,\ldots,J$, and $v = -2y_{2J+1}G^{-1} \cdot u$, and assume that $\sum_{j=1}^J v_j x^j \in \ell^\infty(P)$. Then, the optimization problem (32.5) has a solution.*

Proof. Let $(\overline{\xi}_1,\overline{\xi}_2)$ be a candidate for the minimizer. Since g is Gateaux differentiable the necessary and sufficient conditions for solution via the dual problem of (32.1) (see Remark 32.3) give that $\psi = (\lambda_1(\overline{\xi}_1 - \xi_1^R), \lambda_2(\overline{\xi}_2 - \xi_2^R))$ and the constraints take the form

$$\lambda_1(\overline{\xi}_1 - \xi_1^R) - \sum_{j=1}^J y_j x^j - y_{2J+1}c \in \ell_+^2, \tag{32.6}$$

$$\lambda_2(\overline{\xi}_2 - \xi_2^R) - \sum_{j=J+1}^{2J} y_j x^{j-J} + y_{2J+1}c \in \ell_+^2, \tag{32.7}$$

32 Minimum Regret Pricing in Incomplete Markets

and

$$\lambda_1\|\bar{\xi}_1\|^2 - \lambda_1\langle \xi_1^R, \bar{\xi}_1 \rangle - \sum_{j=1}^{J} y_j\langle x^j, \bar{\xi}_1 \rangle - y_{2J+1}\langle c, \bar{\xi}_1 \rangle$$

$$+ \lambda_2\|\bar{\xi}_2\|^2 - \lambda_2\langle \xi_2^R, \bar{\xi}_2 \rangle - \sum_{j=J+1}^{2J} y_j\langle x^{j-J}, \bar{\xi}_2 \rangle + y_{2J+1}\langle c, \bar{\xi}_2 \rangle = 0.$$

(32.8)

We now define the function $F : \mathbb{R}^{2J+1} \times \ell_+^2 \times \ell_+^2 \to \mathbb{R}$ as

$$F(y_1, y_2, \dots, y_{2J+1}, \bar{\xi}_1, \bar{\xi}_2) = \lambda_1\|\bar{\xi}_1\|^2 - \lambda_1\langle \xi_1^R, \bar{\xi}_1 \rangle - \sum_{j=1}^{J} y_j\langle x^j, \bar{\xi}_1 \rangle - y_{2J+1}$$

$$\langle c, \bar{\xi}_1 \rangle + \lambda_2\|\bar{\xi}_2\|^2 - \lambda_2\langle \xi_2^R, \bar{\xi}_2 \rangle - \sum_{j=J+1}^{2J} y_j\langle V^{j-J}, \bar{\xi}_2 \rangle + y_{2J+1}\langle c, \bar{\xi}_2 \rangle.$$

Let us assume that the reference kernels ξ_1^R, ξ_2^R are such that $\langle c, \xi_1^R \rangle < \langle c, \xi_2^R \rangle$. In order to solve the dual problem, we have to find the zeros of F which satisfy the conditions (32.6),(32.7), with the requirement that $(\bar{\xi}_1, \bar{\xi}_2)$ is a feasible element of the primal problem's constraints set.

Suppose, for the moment, the existence of real numbers

$$y_1, y_2, \dots, y_J, y_{J+1}, \dots, y_{2J}, y_{2J+1}$$

such that for $\lambda_1, \lambda_2 \neq 0$ we have that

$$\bar{\xi}_1 = \frac{1}{\lambda_1}(\lambda_1 \xi_1^R + \sum_{j=1}^{J} y_j x^j + y_{2J+1}c), \tag{32.9}$$

$$\bar{\xi}_2 = \frac{1}{\lambda_2}(\lambda_2 \xi_2^R + \sum_{j=1}^{J} y_{J+j} x^j - y_{2J+1}c) \tag{32.10}$$

are strictly positive functionals of ℓ_+^2 (recall that ξ_1^R, ξ_1^R are strictly positive functionals of the cone ℓ_+^2). For this choice of $y_k, k = 1, \dots, 2J + 1$ we have

$$F(y_1, y_2, \dots, y_{2J+1}, \bar{\xi}_1, \bar{\xi}_2) = 0.$$

Substituting the candidate solution (32.9–32.10) into the constraint $\langle c, \bar{\xi}_1 \rangle = \langle c, \bar{\xi}_2 \rangle$ we obtain

$$y_{2J+1} = \frac{\langle c, \xi_2^R \rangle - \langle c, \xi_1^R \rangle + \frac{1}{\lambda_2} \sum_{j=1}^{J} y_{J+j} \langle c, x^j \rangle - \frac{1}{\lambda_1} \sum_{j=1}^{J} y_j \langle c, x^j \rangle}{(\frac{1}{\lambda_1} + \frac{1}{\lambda_2}) \|c\|_2^2}.$$

(32.11)

We now turn to the constraints that $\bar{\xi}_1, \bar{\xi}_2$ must be such that they give the same prices q_j for the primitive assets x^j. These constraints become

$$\left\langle \frac{1}{\lambda_1} \sum_{j=1}^{J} y_j x^j, x^i \right\rangle - \left\langle \frac{1}{\lambda_2} \sum_{j=1}^{J} y_{J+j} x^j, x^i \right\rangle = -y_{2J+1} \left(\frac{1}{\lambda_1} + \frac{1}{\lambda_2} \right) \langle c, x^i, \rangle,$$

(32.12)

where $i = 1, \dots, J$. Equations (32.11) and (32.12) is system of $J + 1$ linear equations with $2J + 1$ unknowns, y_j, $j = 1, \dots, 2J + 1$.

In the special case where $\lambda_1 = \lambda_2 = \frac{1}{2}$ this assumes the compact form

$$G \cdot (a_1 - a_2) = -2y_{2J+1}u,$$

where $a_1 = (y_1, y_2, \dots, y_J)^T$, $a_2 = (y_{J+1}, \dots, y_{2J})^T$ and

$$y_{2J+1} = \frac{1}{4\|c\|^2} \left(\langle c, \xi_2^R \rangle - \langle c, \xi_1^R \rangle \right).$$

Since G is invertible $a_1 - a_2 = -4y_{2J+1}G^{-1} \cdot u$. This specifies the difference of the y_k, $k = 1, \dots, 2J$ as well as y_{2J+1}.

It remains to check the assumption that $(\bar{\xi}_1, \bar{\xi}_2)$, defined as in (32.9–32.10) with a choice for y_j, $j = 1, \dots, 2J + 1$, compatible with the constraints, are strictly positive functionals of ℓ_+^2. Denote by v the vector $-2y_{2J+1}G^{-1} \cdot u$. By setting $a_{1j} = a_{2j} + v_j$, on account of $a_1 - a_2 = -2y_{2J+1}G^{-1} \cdot u$, we have to specify a vector $a_2 \in \mathbb{R}^J$ such that $\sum_{j=1}^{J} a_{2j}x^j - y_{2J+1}c$ and $\sum_{j=1}^{J} (a_{2j} + v_j)x^j + y_{2J+1}c$ are positive elements of $\ell^2(P)$. If $\geq$ denotes the usual partial ordering of $\ell^2(P)$, a_2 must be such that $\sum_{j=1}^{J} a_{2j}x^j \geq y_{2J+1}c$ and $\sum_{j=1}^{J} a_{2j}x^j \geq -\sum_{j=1}^{J} v_j x^j - y_{2J+1}c$. It suffices to find a_2 such that $\sum_{j=1}^{J} a_{2j}x^j \geq (-\sum_{j=1}^{J} v_j x^j - y_{2J+1}c) \vee y_{2J+1}c$, where $\vee$ denotes the pointwise supremum of sequences. If $\sum_{j=1}^{J} v_j x^j \in \ell^\infty(P)$ then the last element given by a supremum of sequences is an element of $\ell^\infty(P)$, denoted by s.

The vector a_2 of coefficients can be specified if we find some asset $d \in M$ which super-replicates s, a fact which is assured if $1 \in M$. Then $d = \sum_{j=1}^{J} d_j x^j$ with $d \geq s$. Note that $\|s\|_\infty 1 \geq s$. Hence d could be equal to $\|s\|_\infty 1$. If $1 = \sum_{j=1}^{J} y_j^0 x^j$, then $a_1 = \|s\|_\infty (y_1^0, y_2^0, \dots, y_J^0)$ is the portfolio we need. $\square$

The price obtained for the non marketed contingent claim c by Proposition 32.3, is in general obtained through the inverse of the Gram matrix. This is simply the correlation matrix of the primitive assets of the economy. Therefore, the optimization

32 Minimum Regret Pricing in Incomplete Markets

problem, leading to the determination of the price of the continget claim, resembles both formally and in concept the finite dimensional Markowitz problem, and the proposed price is similar to that obtained through the CAPM pricing procedure. Furthermore, our approach may be seen as a generalization of the projection pricing approach proposed in (Luenberger [10]) in the case of more than one agents.

Remark 32.4. The choice $\lambda_1 = \lambda_2 = \frac{1}{2}$ was made for the sake of simplicity. All the arguments can be generalized for any choice of $\lambda_1, \lambda_2 \in (0, 1)$ with similar qualitative results.

32.5.4 The Example of Entropy Like Regret Functions

We will see here that the dual problem leads to a portfolio maximization problem for expected utility of the exponential type. The utility of the two agents will be a linear combination of two exponential utilities with risk aversion coefficients properly adjusted by the use of the weights λ_1, λ_2 that are present in the regret function.

For simplicity we work in the framework of the sequence spaces $\ell^p(P)$ but the result holds for more general $L^p(P)$ spaces (see Remark 32.5).

Lemma 32.4. Let $g(\xi_1, \xi_2) = \lambda_1 I(Q_1, Q_1^R) + \lambda_2 I(Q_2, Q_2^R)$, where I is the relative entropy.
If the interest rate r between the period 0 and the period 1 is equal to zero, then

$$g^*(x) = \sum_{n=1}^{\infty} \xi_1^R(n) \exp\left(\frac{1}{\lambda_1} x_1(n) - 1\right) P_n + \sum_{n=1}^{\infty} \xi_2^R(j) \exp\left(\frac{1}{\lambda_2} x_2(n) - 1\right) P_n$$

$$(32.13)$$

Proof. Let $\{\xi_i(n)\}_{n=1}^{\infty}$, $i = 1, 2$ be the pricing kernels of the two agents and $\{\xi_i^R(n)\}_{n=1}^{\infty}$, $i = 1, 2$ be the corresponding reference pricing kernels. Then the regret function of the agent i, $(i = 1, 2)$ is the relative entropy of the risk neutral measure Q_i with respect to Q_i^R or else

$$\mathfrak{R}_i(\xi_i, \xi_i^R) = (1+r) \sum_{n=1}^{\infty} \xi_i(j)(\ln(\xi_i(j)) - \ln(\xi_i^R(n))) P_n, \quad i = 1, 2,$$

where the sequence $(P_n)_{n\in\mathbb{N}}$ of real numbers such that $P_n > 0$ for any state n denotes the statistical measure. Let $x = (x_1, x_2)$ be a joint consumption bundle of the agents.
By definition $g^*(x) = g^*(x_1, x_2)$ is

$$g^*(x) = \sup_{(\xi_1, \xi_2)} \{\langle x, \xi \rangle - g(\xi)\}$$

$$= \sup_{(\xi_1, \xi_2)} \left\{ \sum_{i=1}^{2} \sum_{n=1}^{\infty} \xi_i(n) x_i(n) P_n - \sum_{i=1}^{2} \lambda_i \sum_{n=1}^{\infty} (1+r) \xi_i(n)(\ln(\xi_i(n)) - \ln(\xi_i^R(n))) P_n \right\}.$$

524 C. Kountzakis et al.

We now take first order conditions with respect to $\xi_i(j)$, $j = 1, 2, \ldots$, $i = 1, 2$. This gives that

$$\ln(\xi_i^*(n)) = \frac{1}{\lambda_i(1+r)} x_i(n) - 1 + \ln(\xi_i^R(n))$$

Substituting back into the function to be maximized we obtain after proper simplifications and supposing that $r = 0$ that

$$g^*(x) = \sum_{n=1}^{\infty} \xi_1^R(n) \exp\left(\frac{1}{\lambda_1} x_1(n) - 1\right) P_n + \sum_{n=1}^{\infty} \xi_2^R(n) \exp\left(\frac{1}{\lambda_2} x_2(n) - 1\right) P_n$$

(32.14)

Therefore, the conjugate function corresponds to the sum of expected exponential utility functions with risk aversion coefficients that are related to the weights used in the calculation of the convex combination of the entropies. □

Remark 32.5. The result of Lemma 32.4 can readily be generalized for the framework of $L^P(P)$ spaces. Then, at least at a formal level, the first order conditions become

$$\ln(\xi_i^*(\omega)) = \frac{1}{\lambda_i(1+r)} x_i(\omega) - 1 + \ln(\xi_i^R(\omega)), \quad i = 1, 2$$

almost surely in ω (where now the states of the world are thought of as a continuum) and so

$$g^*(x) = \mathbb{E}_{Q_1^R}\left[\exp\left(\frac{1}{\lambda_1} x_1 - 1\right)\right] + \mathbb{E}_{Q_2^R}\left[\exp\left(\frac{1}{\lambda_2} x_2 - 1\right)\right]$$

The formal arguments can be turned into rigorous ones using standard arguments.

Proposition 32.4. *In the case of entropic regret function, the dual of problem (32.2) assumes the form*

$$\max_{y \in \mathbb{R}^{2J+1}} \quad b \cdot y - g^*(\psi)$$

$$\text{subject to} \qquad\qquad\qquad (32.15)$$

$$\psi - A^*(y) \in \ell_+^q \oplus \ell_+^q,$$

$$\psi \in \ell^q \oplus \ell^q,$$

where $g^*(\psi)$ is the exponential utility function given in (32.13).

Proof. The proof is a straightforward application of Lemma 32.4 and Proposition 32.1. □

32 Minimum Regret Pricing in Incomplete Markets 525

It is easily seen that the solution of the dual problem is equivalent to the problem of maximization of an exponential utility function of the final wealth of a portfolio, in a market with a finite number of assets. The only difference, from the standard setting of this problem, is that the utility function is now the sum of the expected utility functions of two agents, each calculated using the reference measure of the agent. Although the solution of this problem does not admit an explicit form (in contrast to the quadratic regret case), the solvability of the dual problem follows by standard arguments and is omitted.

32.6 Regret Volatility and Dispersion of Bargaining Power

The weights λ_1, λ_2 in the joint regret function can be interpreted as bargaining power indicators for the counterparties of the claim c. Since there is not enough information on the exact values of these weights, we may consider them as random variables. But then the minimum regret price of the claim is also a random variable with variance that can be interpreted as the claims price volatility.

The above informal discussion may be turned into a more rigorous argument as follows:

Consider the measurable space $(\Delta_2, \mathscr{B}(\Delta_2))$ where $\mathscr{B}(\Delta_2)$ is the σ-algebra of the Borel sets of Δ_2, where the simplex Δ_2 is considered as a topological space endowed with the induced usual metric topology of $\mathbb{R}^2$.

Let us fix the regret functions ϕ_1, ϕ_2 and the reference pricing kernels ξ_1^R, ξ_2^R. Then, for each $(\lambda_1, \lambda_2) \in \Delta_2$ we associate a joint regret function $\phi(\lambda_1, \lambda_2) : \mathscr{Q}_a \times \mathscr{Q}_a \to \mathbb{R}$ where

$$\phi(\lambda_1, \lambda_2)(\xi_1, \xi_2) = \lambda_1 \phi_1(\|\xi_1 - \xi_1^R\|) + \lambda_2 \phi_2(\|\xi_2 - \xi_2^R\|).$$

Clearly, $\phi(\lambda_1, \lambda_2)$ is continuous, since ϕ_1, ϕ_2 are. For $(\lambda_1, \lambda_2) \in \Delta_2$ the constraints set of the problem of the joint minimization of the regret is the non-empty set

$$\{(\xi_1, \xi_2) \in \mathscr{Q}_a \times \mathscr{Q}_a | \xi_1(c) = \xi_2(c)\}.$$

Hence we define the constant correspondence $z : \Delta_2 \to 2^{\mathscr{Q}_a \times \mathscr{Q}_a}$ by

$$z(\lambda_1, \lambda_2) = \{(\xi_1, \xi_2) \in \mathscr{Q}_a \times \mathscr{Q}_a | \xi_1(c) = \xi_2(c)\}.$$

Clearly z is lower hemicontinuous. This correspondence indicates that the set of the constraints for the problem of the minimization of the joint regret function is the same for every $(\lambda_1, \lambda_2) \in \Delta_2$.

Let now $Gr(z)$ denote the graph of z. We define a unified regret function $F : Gr(z) \to \mathbb{R}$ with

$$F((\lambda_1, \lambda_2), (\xi_1, \xi_2)) = -\left(\lambda_1 \phi_1(\|\xi_1 - \xi_1^R\|) + \lambda_2 \phi_2(\|\xi_2 - \xi_2^R\|)\right).$$

The function F is continuous is clearly lower semicontinuous. Also, for any $(\lambda_1, \lambda_2) \in \Delta_2$, the value function

$$m(\lambda_1, \lambda_2) = \sup_{\{(\xi_1, \xi_2) \in \mathcal{Q}_a \times \mathcal{Q}_a \mid \xi_1(c) = \xi_2(c)\}} -(\lambda_1 \phi_1(\|\xi_1 - \xi_1^R\|) + \lambda_2 \phi_2(\|\xi_2 - \xi_2^R\|))$$

for $(\lambda_1, \lambda_2) \in \Delta_2$ is well defined and lower semicontinuous due to Lemma 16.29 in [7]. Therefore, m is Borel measurable. Hence, by assigning a probability to each pair (λ_1, λ_2) of bargaining power weights, we may consider the probability space $(\Delta_2, \mathcal{B}(\Delta_2), P)$ and the volatility of the minimum regret adopted by counterparties of the claim is the variance of the random variable $-m$:

$$V(\phi_1, \phi_2, \xi_1^R, \xi_2^R) = \int_{\Delta_2} (m(\lambda_1, \lambda_2))^2 dP(\lambda_1, \lambda_2) - \left(\int_{\Delta_2} m(\lambda_1, \lambda_2) dP(\lambda_1, \lambda_2) \right)^2.$$

It remain to verify when the variance of m is well-defined.

For the second integral: $\|\xi_1 - \xi_1^R\| > 0, \|\xi_2 - \xi_2^R\| > 0$, for (ξ_1, ξ_2) belonging to the solution set of the optimization problem. Consequently, $\phi_1(\|\xi_1 - \xi_1^R\|) > \phi_1(0)$ and $\phi_2(\|\xi_2 - \xi_2^R\|) > \phi_2(0)$. Assuming $\phi_1(0) = 0, \phi_2(0) = 0$, we obtain $-(\lambda_1 \phi_1(\|\xi_1 - \xi_1^R\|) + \lambda_2 \phi_2(\|\xi_2 - \xi_2^R\|)) \leq 0$ and $m(\lambda_1, \lambda_2) \leq 0$ for each $(\lambda_1, \lambda_2) \in \Delta_2$. Hence the second integral exists since it is the integral of an upper bounded, Borel measurable function.

For the first integral: Clearly, $m^2(\lambda_1, \lambda_2)$ is Borel measurable, being the square of a Borel measurable function. Notice that if $\phi_1(x) = \phi_2(x) = x, x \in [0, \infty)$ and the payoff space $L^p(P)$ (or $\ell^p(P)$) is partially ordered by a closed cone B such that I is an interior point of it, then $\mathcal{Q}_a$ is bounded. If $M > 0$ is such a bound, then $0 \leq g(\xi_1, \xi_2) \leq \|\xi_1 - \xi_1^R\| + \|\xi_2 - \xi_2^R\| \leq 4M$. Hence $-4M \leq -g(\xi_1, \xi_2) \leq 0$ and $-4M \leq m(\lambda_1, \lambda_2) \leq 0$. Thus, $m^2(\lambda_1, \lambda_2) \leq 16M^2$ which means that $m^2(\lambda_1, \lambda_2)$ is upper bounded, therefore its integral with respect to P exists.

Appendix

Partially Ordered Linear Spaces

In this paragraph, we provide standard definitions and results from the theory of partially ordered linear spaces which are used in this paper. Let E be a (normed) linear space. A set $C \subseteq E$ satisfying $C + C \subseteq C$ and $\lambda C \subseteq C$ for any $\lambda \in \mathbb{R}_+$ is called *wedge*. A wedge for which $C \cap (-C) = \{0\}$ is called *cone*. A pair $(E, \geq)$ where E is a linear space and $\geq$ is a binary relation on E satisfying the following properties:

(a) $x \geq x$ for any $x \in E$ (reflexive)
(b) If $x \geq y$ and $y \geq z$ then $x \geq z$, where $x, y, z \in E$ (transitive)

32 Minimum Regret Pricing in Incomplete Markets

(c) If $x \geq y$ then $\lambda x \geq \lambda y$ for any $\lambda \in \mathbb{R}_+$ and $x + z \geq y + z$ for any $z \in E$, where $x, y \in E$ (compatible with the linear structure of E)

is called *partially ordered linear space*. The binary relation $\geq$ in this case is a *partial ordering* on E. The set $P = \{x \in E \,|\, x \geq 0\}$ is called *(positive) wedge* of the partial ordering $\geq$ of E. Given a wedge C in E, the binary relation $\geq_C$ defined as follows:

$$x \geq_C y \iff x - y \in C,$$

is a partial ordering on E, called *partial ordering induced by C on E*. If the partial ordering $\geq$ of the space E is *antisymmetric*, namely if $x \geq y$ and $y \geq x$ implies $x = y$, where $x, y \in E$, then P is a cone.

E' denotes the linear space of all linear functionals of E, while E^* is the norm dual of E^*, in case where E is a normed linear space.

Suppose that C is a wedge of E. A functional $f \in E'$ is called *positive functional* of C if $f(x) \geq 0$ for any $x \in C$. $f \in E'$ is a *strictly positive functional* of C if $f(x) > 0$ for any $x \in C \setminus C \cap (-C)$. A linear functional $f \in E'$ where E is a normed linear space, is called *uniformly monotonic functional* of C if there is some real number $a > 0$ such that $f(x) \geq a \|x\|$ for any $x \in C$. In case where a uniformly monotonic functional of C exists, C is a cone. $C^0 = \{f \in E^* \,|\, f(x) \geq 0$ for any $x \in C\}$ is the *dual wedge of C in E^**. Also, by C^{00} we denote the subset $(C^0)^0$ of E^{**}. It can be easily proved that if C is a closed wedge of a reflexive space, then $C^{00} = C$. If C is a wedge of E^*, then the set $C_0 = \{x \in E \,|\, \hat{x}(f) \geq 0$ for any $f \in C\}$ is the *dual wedge of C in E*, where $\hat{\ } : E \to E^{**}$ denotes the natural embedding map from E to the second dual space E^{**} of E. Note that if for two wedges K, C of E, $K \subseteq C$ holds, then $C^0 \subseteq K^0$.

If C is a cone, then a set $B \subseteq C$ is called *base* of C if for any $x \in C \setminus \{0\}$ there exists a unique $\lambda_x > 0$ such that $\lambda_x x \in B$. The set $B_f = \{x \in C \,|\, f(x) = 1\}$ where f is a strictly positive functional of C is the *base of C defined by f*. B_f is bounded if and only if f is uniformly monotonic. If B is a bounded base of C such that $0 \notin \overline{B}$ then C is called *well-based*. If C is well-based, then a bounded base of C defined by a $g \in E^*$ exists. If $E = C - C$ then the wedge C is called *generating*, while if $E = \overline{C - C}$ it is called *almost generating*. If C is generating, then C^0 is a cone of E^* in case where E is a normed linear space. Also, $f \in E^*$ is a uniformly monotonic functional of C if and only if $f \in int C^0$, where $int C^0$ denotes the norm-interior of C^0. If E is partially ordered by C, then any set of the form $[x, y] = \{r \in E \,|\, y \geq_C r \geq_C x\}$ where $x, y \in C$ is called *order-interval of E*. If E is partially ordered by C and for some $e \in E$, $E = \cup_{n=1}^{\infty} [-ne, ne]$ holds, then e is called *order-unit* of E. If E is a normed linear space, then if every interior point of C is an order-unit of E. If E is moreover a Banach space and C is closed, then every order-unit of E is an interior point of C. The partially ordered vector space E is a *vector lattice* if for any $x, y \in E$, the supremum and the infimum of $\{x, y\}$ with respect to the partial ordering defined by P exist in E. In this case $\sup\{x, y\}$ and $\inf\{x, y\}$ are denoted by $x \vee y$, $x \wedge y$ respectively. If so, $|x| = \sup\{x, -x\}$ is the *absolute value* of x and if E is also a normed space such that $\| |x| \| = \|x\|$ for any $x \in E$, then E is called *normed lattice*. Finally, if E is a partially ordered Banach space whose positive cone is E_+, if E has a Schauder basis $(e_n)_{n\in\mathbb{N}}$, this basis is called *positive basis* if and only if $E_+ = \{x = \sum_{n=1}^{\infty} \lambda_n e_n | \lambda_n \geq 0, n \in \mathbb{N}\}$.

References

1. Karatzas, I., Shreve, S.: Methods of Mathematical Finance Applications of Mathematics, vol. 39. Springer, New York (1998)
2. Fritelli, M.: The minimal entropy martingale measure and the valuation property in incomplete markets. Math. Finance **10**, 39–52 (2000)
3. Bellini, F. Frittelli, M. : On the existence of minimax martingale measures. Math. Finance **12**, 1–21 (2002)
4. Föllmer, H., Schied, A.: Stochastic Finance: An Introduction in Discrete Time. Walter de Gruyter, Berlin (2002)
5. Xanthopoulos, S., Yannacopoulos, A.N. : Scenarios for the price determination in incomplete markets. Int. J. Theor. Appl. Finance **5**, 415–445 (2008)
6. Aliprantis, C.D., Brown, D.J., Polyrakis, I.A., Werner, J.: Portfolio dominance and optimality in infinite security markets. J. Math. Econ. **30**, 347–366 (1998)
7. Aliprantis, C.D., Border, K.C.: Infinite Dimensional Analysis, a Hitchhiker's Guide, 2nd edn. Springer, New York (1999)
8. Barbu, V., Precupanu, T.: Convexity and Optimization in Banach Spaces. D.Riedel Publishing Company, Bucharest (1986)
9. Borwein J.M., Lewis, A.S.: Partially finite convex programming, Part II: Explicit lattice models. Math. Program. **57**, 49–83 (1993)
10. Luenberger, D.G.: Projection pricing. J. Optim. Theory Appl. **109**, 1–25 (2001)
11. Megginson, R.E.: An Introduction to Banach Space Theory. Springer, New York (1998)
12. Jameson, G.: Ordered Linear Spaces, Lecture Notes in Mathematics, vol. 141. Springer, Berlin (1970)
13. Borwein, J.M., Lewis, A.S.: Partially finite convex programming, Part I: Quasi relative interiors and duality theory. Math. Program. **57**, 15–48 (1993)

Chapter 33
A Class of Infinite Dimensional Replicator Dynamics

D. Kravvaritis, V. Papanicolaou, T. Xepapadeas, and A.N. Yannacopoulos

Abstract We introduce a class of infinite dimensional replicator dynamics in the form of nonlinear and non local integrodifferential equations. We study the properties of the steady state of the equation and their connections with Nash equilibria of the game as well as the global stability of the steady state using techniques from the theory of variational inequalities and infinite dimensional dynamical systems.

33.1 Introduction

Evolutionary game dynamics is a major part of modern game theory, which tries to explain the how a population of players updates their strategies in the course of a game according to the strategies success. As Hofbauer and Sigmund put it [1], strategies with high payoff will spread through the population either through learning, imitation or inheriting strategies, according to a feedback loop, the dynamics of which will determine the long time behavior of the game. The subject of evolutionary game theory, which has found interesting applications in biology and economics, is exactly the dynamics of this feedback loop.

One particularly popular update dynamics scheme is the replicator dynamics scheme. According to this scheme the population updates their strategies by adjusting the logarithmic rate of change of the population density in proportion to the

---

D. Kravvaritis (✉) and V. Papanicolaou
Department of Mathematics, National Technical University, Athens, Greece
e-mail: dkrav@math.ntua.gr, papanico@math.ntua.gr

T. Xepapadeas
Department of International and European Economic Studies, Athens University of Economics
and Business, Athens, Greece
e-mail: xepapad@aueb.gr

A.N. Yannacopoulos
Department of Statistics, Athens University of Economics and Business,
Patission 76, 10434 Athens, Greece,
e-mail: ayannaco@aueb.gr

M.M. Peixoto et al. (eds.), *Dynamics, Games and Science I*, Springer Proceedings
in Mathematics 1, DOI 10.1007/978-3-642-11456-4_33,
© Springer-Verlag Berlin Heidelberg 2011

529

530 D. Kravvaritis et al.

difference between the actual payoff for a given strategy profile and the average payoff.

33.2 Set Up of the Model

Consider a symmetric two player game where each of the players has a continuous strategy space S, whose members are denoted by x, and assume that each player may adopt a mixed strategy modeled by a probability measure on S, absolutely continuous with respect to the Lebesgue measure with density $u(x)$, considered an element of a suitable Hilbert space H with inner product $(\cdot, \cdot)$. The game is characterized by its payoff functional, which is defined through a bilinear form $a(u, v) = \int_S (Au)(x)u(x)dx$ modeling the payoff of player 1 if she plays strategy $u(x)$ while player 2 plays strategy $v(x)$. The bilinear form, following standard arguments defines a payoff operator A acting in the space of densities. We will assume that the bilinear form a is continuous and coercive.

The replicator dynamics is based on the assumption that players update their strategies by comparing payoffs of particular strategies with average payoffs if both players adopt the same mixed strategy. We assume that the density is a function of time, $u(t, x)$ the evolution of this mixed strategy follows the law

$$\frac{\partial u}{\partial t} = (Au - (u, Au))u \tag{33.1}$$

This general equation covers a large number of interesting applications, including infinite dimensional matrix games, the general infinite dimensional replicator dynamics of Oessler and Riedel [2] etc.

In this paper we restrict attention to measures with density $u(t, x)$ so that replicator dynamics become

$$\frac{\partial u}{\partial t}(t, x) = \left\{ \int_S f(x, y)u(t, y)dy - \int_S \int_S f(z, y)u(t, y)u(t, z)dydz \right\} u(t, x)$$

This is a non local integrodifferential equation, which allows us to dispose of the boundedness assumption for the payoff kernel f and assume that in general $\int_S f(x, y)u(t, y)dy =: Au(t, x)$ where A is in general an unbounded operator e.g. A can be a differential operator (f is a singular kernel). In this case the replicator dynamics equation reduces to a nonlocal nonlinear parabolic PDE.

33.3 Nash Equilibria, Steady States and Their Stability

Definition 33.1. In the case of symmetric games a strategy u is a Nash equilibrium if $(w, Au) \leq (u, Au), \ \forall w \in H$.

This is equivalent to the variational inequality $(u - w, Au) \geq 0, \ \forall w \in K$, where the K is the subset of H such that $K = \{u \in H \mid u \geq 0, \int_S u\, dx = 1\}$. Based

33 A Class of Infinite Dimensional Replicator Dynamics

on a classical result of Stampachia [3] on variational inequalities and a technical lemma concerning closedness and convexity of the set K, this leads to the following characterization of Nash equilibria:

Proposition 33.1. *There exist Nash equilibria for the infinite dimensional game and are given by the solution of the minimization problem* $\inf_{u \in K} a(u, u)$ *or by the solution of the nonlinear, (nonlocal elliptic) equation* $A u - (u, A u) = 0$ *i.e. the steady state of the replicator dynamics.*

The existence of a steady state is guaranteed by the following proposition which is based on the Lax–Milgram lemma [3].

Proposition 33.2. *There exists a unique stationary state of the infinite dimensional replicator dynamics equation (33.1) that can be constructed through the auxiliary Poisson type equation* $A\phi = C$ *with appropriate choice of the constant C.*

Using the well known results concerning the solutions of Poisson equation in various domains one may obtain a number of qualitative results as well as analytic forms of solutions for the steady state of the game.

The following proposition shows the existence of a Lyapunov functional for the infinite dimensional replicator dynamics equation (33.1).

Proposition 33.3. *Suppose $u = u(t, x)$ satisfies the infinite dimensional replicator dynamics equation. Then the quantity* $E(t) = \left\| (-A)^{1/2} u \right\|^2$ *satisfies $E'(t) \leq 0$, for all t for which $u(t, x)$ exists.*

Based on Proposition 33.3 and interpolation arguments we have the following stability result:

Proposition 33.4. *Classical solutions of the infinite dimensional replicator equation are globally stable, in the sense that they converge to Nash equilibria.*

33.4 Examples from Economic Theory

33.4.1 Congestion Games

Assume that the strategy space is $\mathbf{Z}$, and u_i is the fraction of players that play strategy i. Let us assume further that congestion effects are present, i.e. the player at i benefits when more players prefer nearby strategies $i + 1$ and $i - 1$. This assumption is a reasonable modelling assumption for games concerning limited resources, e.g. fisheries games, bandwidth management in networks etc. It is also relevant for common pool resource games with spatial structure where an agent's strategy depends on actions taken by agents in his/her immediate neighborhood. In this case the payoff operator may be e.g. of the form $A u = (u_{i+1} - u_i) + (u_{i-1} - u_i)$ which in the continuous limit this becomes the Laplacian $A u = \Delta u$. Operators of this form are coercive and continuous therefore falling exactly in the general framework of the present work.

33.4.2 New Economic Geography

Let S denote a geographical space, $x \in S$ a location (site) on this space and $u(x,t)$ the density of firms located in site x at time t. Desirability of a location is reflected in the market potential, defined as:

$$P(x,t) = \int_S h(D_{xz}) u(z,t) \, dz,$$

where D_{xz} denotes distance between location x and all other locations $z \in S$. The function $h(D_{xz})$ incorporates both centripetal agglomerative forces and dispersing centrifugal forces. One plausible specification for the function $h(D_{xz})$ provided by [4] is the case where h is the linear combination of two exponential functions, each of which models the effects of centripetal and centrifugal forces respectively.

The basic behavioral assumption, that firms immigrate towards locations with market potential above the average, can be modelled by the replicator dynamics

$$\frac{\partial u(x,t)}{\partial t} = u(x,t)[P(x,t) - \int_S P(x,t) u(x,t) \, dx]. \tag{33.2}$$

This is a problem of the general form studied here but A is an integral operator. This may present problems concerning the coercivity of the operator A, unless the kernel is singular. In the case of singular kernels it is feasible to reduce formally the above equation to a nonlocal nonlinear partial differential equation.

Acknowledgements The authors wish to acknowledge the partial support of a National Technical University of Athens Caratheodory Research Grant.

References

1. Hofbauer, J., Sigmund, K.: Evolutionary game dynamics. Bul. AMS **40**, 479–519 (2003)
2. J. Oechssler, J., Riedel, F.: Evolutionary dynamics on infinite strategy spaces. Econ. Theory. **17**, 141–162 (2001)
3. Brezis, H.: Analyse Fonctionelle, Théorie et Applications. Masson, Paris (1983)
4. Krugman, P.: The Self-organizing Economy. Blackwell, London (1996)

Chapter 34
Kinetic Theory for Chemical Reactions Without a Barrier

Gilberto M. Kremer and Ana Jacinta Soares

Abstract A new model of the BE for binary reactive mixtures is here proposed with the aim of describing symmetric reversible reactions without a barrier, assuming appropriate reactive cross sections without activation energy and introducing suitable improvements in the elastic and reactive collision terms. The resulting model assures the correct balance equations and law of mass action, as well as good consistency properties for what concerns equilibrium and entropy inequality. Moreover the non-equilibrium effects induced by the chemical reaction on the distribution function are explicitly determined in a flow regime of slow chemical reaction.

34.1 Introduction

Reactions without a barrier have become an attractive subject of theoretical and experimental studies due to the central role of such reactions in combustion phenomena and many other processes with relevance in astrophysics, organic chemistry, enzymology, chemical physics and biophysics [6, 7]. The so called no-barrier theory has been used in many areas of chemistry, providing quantitative information about the reaction mechanism as well as useful guidance in chemical investigations. Mathematical and modeling approaches for reactions without barriers can help to improve the research on this subject, providing some useful informations about the kinetics of the reactions.

On the other hand, the Boltzmann equation (BE) has been extensively used in the scientific literature to treat chemically reactive systems, after the first studies conducted by Prigogine and co-workers [9] and several others, as documented for

---

A.J. Soares (✉)
Department of Mathematics, University of Minho, 4710-057 Braga, Portugal
e-mail: ajsoares@math.uminho.pt

G.M. Kremer
Department of Physics, University of Paraná, Curitiba, Brazil
e-mail: kremer@fisica.ufpr.br

M.M. Peixoto et al. (eds.), *Dynamics, Games and Science I*, Springer Proceedings in Mathematics 1, DOI 10.1007/978-3-642-11456-4_34,
© Springer-Verlag Berlin Heidelberg 2011

534 G.M. Kremer and A.J. Soares

example in [5, 10] and bibliography therein cited. However, in the literature of the BE extended to chemically reacting gases, only few works consider reactive processes without a barrier and the corresponding collision terms are restricted to some particular chemical regimes for which no significant changes are needed at the model level [10].

In this paper we propose a kinetic model of the BE for a binary mixture undergoing elastic and reactive scattering, adopting an appropriate model of reactive cross section without barriers [2, 11]. At the microscopic scale, the key idea is to modify the collision terms, introducing probability coefficients that describe the possibility of a pair of molecules to collide through an elastic mechanism or a reactive process. This improvement accounts for the fact that a great number of reactive interactions corresponds to a small number of elastic collisions, and vice-versa. In this sense, the considered model represents a new approach, since almost all existent papers introduce reactive cross sections with activation energy and control the presence of both elastic and reactive collisions assuming rather large activation energies and a steric factor which reduce the number of reactive collisions.

34.2 The Model Equations

We consider a binary mixture of constituents $\alpha = A, B$ with binding energies ϵ_A and ϵ_B and equal molecular masses, $m_A = m_B = m$, undergoing the reversible reaction $A + A \rightleftharpoons B + B$. Gas molecules can collide through a binary elastic process which preserves momentum and kinetic energy,

$$m\mathbf{c}_\alpha + m\mathbf{c}_\beta = m\mathbf{c}'_\alpha + m\mathbf{c}'_\beta, \quad \frac{mc_\alpha^2}{2} + \frac{mc_\beta^2}{2} = \frac{mc_\alpha'^2}{2} + \frac{mc_\beta'^2}{2}, \tag{34.1}$$

as well through a reactive process which preserves momentum and total energy,

$$m\mathbf{c}_A + m\mathbf{c}_{A_1} = m\mathbf{c}_B + m\mathbf{c}_{B_1}, \quad 2\epsilon_A + \frac{m(c_A^2 + c_{A_1}^2)}{2} = 2\epsilon_B + \frac{m(c_B^2 + c_{B_1}^2)}{2}. \tag{34.2}$$

In the above equations, $(\mathbf{c}_\alpha, \mathbf{c}_\beta)$ and $(\mathbf{c}'_\alpha, \mathbf{c}'_\beta)$ denote pre and post collisional velocities, whereas $(\mathbf{c}_A, \mathbf{c}_{A_1})$ and $(\mathbf{c}_B, \mathbf{c}_{B_1})$ refer to the velocities of reactants and products of the forward reaction, respectively. We use the sub-index 1 to distinguish two molecules of the same species.

34.2.1 Boltzmann Equation

The gas mixture is characterized in the phase space by the one-particle distribution functions $f(\mathbf{x}, \mathbf{c}_\alpha, t)$, $\alpha = A, B$, with $f_\alpha d\mathbf{x} d\mathbf{c}_\alpha$ denoting the number of

34 Kinetic Theory for Chemical Reactions Without a Barrier 535

α-particles in the volume element $dx d\mathbf{c}_\alpha$ around position x and velocity $\mathbf{c}_\alpha$, at time t. We introduce $f = \{f_\alpha, f_\beta\}$. In absence of external body forces, we propose the following system of two Boltzmann equations for the distribution functions

$$\frac{\partial f_\alpha}{\partial t} + c_i^\alpha \frac{\partial f_\alpha}{\partial x_i} = \mathcal{Q}_\alpha^E(\underline{f}) + \mathcal{Q}_\alpha^R(\underline{f}), \quad \alpha = A, B,$$

where the first term on the r.h.s. refers to contributions from elastic collisions and the second one to contributions from chemical interactions. They are given by

$$\mathcal{Q}_\alpha^E(\underline{f}) = \sum_{\beta=A}^{B}(1 - \chi_\beta \delta_{\alpha\beta}) \int \left[f_\alpha' f_\beta' - f_\alpha f_\beta \right] g_{\beta\alpha} \sigma_{\alpha\beta} d\Omega_{\alpha\beta} d\mathbf{c}_\beta \tag{34.3}$$

$$\mathcal{Q}_\alpha^R(\underline{f}) = \int \left[\chi_\beta f_\beta f_{\beta_1} \sigma_\beta^* \frac{g_\beta^2}{g_\alpha^2} - \chi_\alpha f_\alpha f_{\alpha_1} \sigma_\alpha^* \right] g_\alpha d\Omega_\beta d\mathbf{c}_{\alpha_1}, \quad \alpha = A, B. \tag{34.4}$$

In the reactive term, it is implicit that the index $\beta = A, B$ is always different from the index α, since the chemical reaction predicts that the reactants and products are of different species. Moreover, $g_{\alpha\beta}$ stands for relative velocity of an elastic collision and g_A, g_B for those of the reactants and products involved in chemical interactions. The symbols $d\Omega_{\alpha\beta}$ and $d\Omega_\beta$ represent elements of solid angles which characterize the scattering processes $\sigma_{\alpha\beta}$ is the differential elastic cross section, σ_α^* and σ_β^* are differential reactive cross sections for forward and backward reactions.

We have introduced probability coefficients χ_A and χ_B relative to the AA and BB encounters, in order to account for the possibility of a pair of molecules of the same species interact through an elastic collision or a reactive process. This represents an improvement in the kinetic theory of chemically reacting gases, suitable to treat chemical reactions without activation energy.

34.2.2 Differential Cross Sections

For what concerns the differential cross sections of elastic encounters, we adopt the simple model of rigid spheres, namely $\sigma_{\alpha\beta} = d^2/4$, where d is the molecular diameter.

On the other hand, for reactive cross sections we have to adopt a suitable model to describe chemical reactions without a barrier, so that such a model should not consider an activation energy. In particular, if we consider an attractive potential energy of the form $V(r) = -K/r^n$, we can choose the following reactive model (see [2, 11]),

$$\sigma_\alpha^* = \frac{d_r^2}{4} \left(\frac{m g_\alpha^2}{4k T_0} \right)^{n-1/2}, \tag{34.5}$$

where T_0 is a characteristic temperature, d_r a reactive collision diameter and the exponent n may range from $-3/2$ to $1/2$. The values $n = 1/2, n = 0$ and $n = 1/6$ stand for hard-spheres reactions, ion-molecule reactions and reactions of neutral species, respectively.

34.3 The Macroscopic Equations

It is well known that the Boltzmann equations (34.3) give a detailed microscopic picture which is not necessary in general for applied studies involving the model. Instead of these equations, one has to consider the corresponding macroscopic ones, which are time-space evolution equations for mean quantities, as density, momentum and energy.

34.3.1 Transfer Equations

These equations result from (34.3), when one multiplies the equation by an arbitrary function, say $\psi_\alpha \equiv \psi(x, c_\alpha, t)$, and then integrate the resulting equation over all velocities c_α. It results the so called transport equation for constituent α

$$
\frac{\partial}{\partial t} \int \psi_\alpha f_\alpha d\mathbf{c}_\alpha + \frac{\partial}{\partial x_i} \int \psi_\alpha c_i^\alpha f_\alpha d\mathbf{c}_\alpha - \int \left(\frac{\partial \psi_\alpha}{\partial t} + c_i^\alpha \frac{\partial \psi_\alpha}{\partial x_i} \right) f_\alpha d\mathbf{c}_\alpha
$$

$$
= \sum_{\beta=A}^{B} (1 - \chi_\beta \delta_{\alpha\beta}) \int (\psi_\alpha' - \psi_\alpha) f_\alpha f_\beta g_{\beta\alpha} \sigma_{\alpha\beta} d\Omega_{\alpha\beta} d\mathbf{c}_\beta d\mathbf{c}_\alpha
$$

$$
- \frac{\nu_\alpha}{2} \int (\psi_\alpha + \psi_{\alpha_1}) \chi_A \left[\frac{\chi_B}{\chi_A} f_B f_{B_1} - f_A f_{A_1} \right] \sigma_A^\star g_A d\Omega d\mathbf{c}_{A_1} d\mathbf{c}_A. \qquad (34.6)
$$

The corresponding transport equation for the mixture is obtained by summing (34.6) over all constituents, yielding

$$
\frac{\partial}{\partial t} \sum_{\alpha=A}^{B} \int \psi_\alpha f_\alpha d\mathbf{c}_\alpha + \frac{\partial}{\partial x_i} \sum_{\alpha=A}^{B} \int \psi_\alpha c_i^\alpha f_\alpha d\mathbf{c}_\alpha - \sum_{\alpha=A}^{B} \int \left(\frac{\partial \psi_\alpha}{\partial t} + c_i^\alpha \frac{\partial \psi_\alpha}{\partial x_i} \right) f_\alpha d\mathbf{c}_\alpha
$$

$$
= \frac{1}{4} \sum_{\alpha,\beta=A}^{B} (1 - \chi_\beta \delta_{\alpha\beta}) \int (\psi_\alpha + \psi_\beta - \psi_\alpha' - \psi_\beta') \left[f_\alpha' f_\beta' - f_\alpha f_\beta \right] g_{\beta\alpha} \sigma_{\beta\alpha} d\Omega_{\beta\alpha} d\mathbf{c}_\beta d\mathbf{c}_\alpha
$$

$$
+ \frac{1}{2} \int (\psi_A + \psi_{A_1} - \psi_B - \psi_{B_1}) \chi_A \left[\frac{\chi_B}{\chi_A} f_B f_{B_1} - f_A f_{A_1} \right] \sigma_A^\star g_A d\Omega d\mathbf{c}_{A_1} d\mathbf{c}_A.
$$

$$
(34.7)
$$

34 Kinetic Theory for Chemical Reactions Without a Barrier

34.3.2 Macroscopic Fields

For the considered gas mixture, the macroscopic picture is described by seven scalar fields namely the partial particle number densities n_α, the partial internal energies density $\varrho_\alpha \varepsilon_\alpha$ (with $\alpha = A, B$) and the velocity of the mixture v_i. In terms of the one-particle distribution function f_α these fields are defined by

$$n_\alpha = \int f_\alpha d\mathbf{c}_\alpha = \frac{\varrho_\alpha}{m} \quad \text{with} \quad n = \sum_{\alpha=A}^{B} n_\alpha \quad \text{and} \quad \varrho = \sum_{\alpha=A}^{B} \varrho_\alpha, \tag{34.8}$$

$$v_i = \frac{1}{\varrho} \sum_{\alpha=A}^{B} \int m c_i^\alpha f_\alpha d\mathbf{c}_\alpha, \tag{34.9}$$

$$\varrho_\alpha \varepsilon_\alpha = \int \frac{m}{2} \xi_\alpha^2 f_\alpha d\mathbf{c}_\alpha, \tag{34.10}$$

where $\xi_i^\alpha = c_i^\alpha - v_i$ is the molecular peculiar velocity of constituent α. We assume that both constituents have the same temperature T, which is the temperature of the mixture, so that its internal energy density of constituent α is in fact given by $\varrho_\alpha \varepsilon_\alpha = 3 n_\alpha k T / 2$.

34.3.3 Field Equations

The balance equations for the fields (34.8–34.10) are obtained from the transfer equations (34.6) and (34.7) with pertinent choices of ψ_α. More in detail (34.6) with $\psi_\alpha = 1$ or $\psi_\alpha = m \xi_\alpha^2 / 2$, and (34.7) with $\psi_\alpha = m c_i^\alpha$ lead to the equations for particle number densities, momentum and internal energies in the form

$$\frac{\partial n_\alpha}{\partial t} + \frac{\partial}{\partial x_i} \left(n_\alpha u_i^\alpha + n_\alpha v_i \right) = \tau_\alpha \tag{34.11}$$

$$\frac{\partial \varrho v_i}{\partial t} + \frac{\partial}{\partial x_j} \left(p_{ij} + \varrho v_i v_j \right) = 0, \tag{34.12}$$

$$\frac{3}{2} \frac{\partial n_\alpha k T}{\partial t} + \frac{\partial}{\partial x_i} \left(q_i^\alpha + \frac{3}{2} n_\alpha k T v_i \right) - \frac{\varrho_\alpha}{\varrho} u_i^\alpha \frac{\partial p_{ij}}{\partial x_j} + p_{ij}^\alpha \frac{\partial v_j}{\partial x_i} = \zeta_\alpha. \tag{34.13}$$

In the above equations, τ_α is the rate of reaction and ζ_α the production term of internal energy density of constituent α, given by

$$\tau_\alpha = \int \left[\chi_\beta f_\beta f_{\beta_1} \sigma_\beta^\star \frac{g_\beta^2}{g_\alpha^2} - \chi_\alpha f_\alpha f_{\alpha_1} \sigma_\alpha^\star \right] g_\alpha d\Omega_\beta d\mathbf{c}_{\alpha_1} d\mathbf{c}_\alpha, \quad \text{with} \quad \tau_B = -\tau_A, \tag{34.14}$$

$$\zeta_\alpha = \sum_{\beta=A}^{B} (1 - \chi_\beta \delta_{\alpha\beta}) \int \frac{1}{2} m(\xi_\alpha'^2 - \xi_\alpha^2) f_\alpha f_\beta g_{\beta\alpha} \sigma_{\beta\alpha} d\Omega_{\beta\alpha} d\mathbf{c}_\beta d\mathbf{c}_\alpha$$

$$+ \int \frac{m}{4} (\xi_\alpha^2 + \xi_{\alpha_1}^2) \left[\chi_\beta f_\beta f_{\beta_1} \sigma_\beta^\star \frac{g_\beta^2}{g_\alpha^2} - \chi_\alpha f_\alpha f_{\alpha_1} \sigma_\alpha^\star \right] g_\alpha d\Omega_\beta d\mathbf{c}_{\alpha_1} d\mathbf{c}_\alpha.$$

$$(34.15)$$

Moreover, u_i^α, q_i^α and p_{ij}^α are the diffusion velocity, heat flux and pressure tensor of each constituent, defined by

$$u_i^\alpha = \frac{1}{n_\alpha} \int \xi_i^\alpha f_\alpha d\mathbf{c}_\alpha, \qquad q_i^\alpha = \int \frac{1}{2} m \xi_\alpha^2 \xi_i^\alpha f_\alpha d\mathbf{c}_\alpha, \qquad p_{ij}^\alpha = \int m \xi_i^\alpha \xi_j^\alpha f_\alpha d\mathbf{c}_\alpha,$$

with

$$\sum_{\alpha=A}^{B} \varrho_\alpha u_i^\alpha = 0, \qquad q_i = \sum_{\alpha=A}^{B} q_i^\alpha, \qquad p_{ij} = \sum_{\alpha=A}^{B} p_{ij}^\alpha.$$

34.4 Consistency of the Model

The model here proposed has good consistency properties for what concerns conservation laws, chemical exchange rates, trend to equilibrium and entropy production. These features are confirmed trough the following results whose detailed proofs are here omitted for sake of brevity.

Theorem 34.1. *The elastic collision terms (34.3) are such that*

$$\int_{\mathbb{R}^3} \mathcal{Q}_\alpha^E(\underline{f}) d\mathbf{c}_\alpha = 0, \quad \alpha = A, B. \tag{34.16}$$

Proof. The proof follows from the definition of the elastic collision terms resorting to the usual symmetry properties of the gain and loss contributions. $\qquad\square$

The result of Theorem 34.1 is well known in the literature of the Boltzmann equation for inert mixtures (see, for example [3]). It means that elastic collisions do not modify the concentration of each constituent, as expected.

Theorem 34.2. *The elastic and reactive collision terms (34.3) and (34.4) are such that*

$$\sum_{\alpha=A}^{B} \int_{\mathbb{R}^3} \psi_\alpha(\mathbf{c}_\alpha) \left(\mathcal{Q}_\alpha^E(\underline{f}) + \mathcal{Q}_\alpha^R(\underline{f}) \right) d\mathbf{c}_\alpha = 0, \tag{34.17}$$

for $\psi_\alpha = m$, $\psi_\alpha = m c_\alpha$, $\psi_\alpha = \epsilon_\alpha + \frac{1}{2} m c_\alpha^2$, $\alpha = A, B$.

34 Kinetic Theory for Chemical Reactions Without a Barrier

Proof. The proof follows from the definition of the elastic and reactive collision terms and uses similar symmetry techniques to those referred in the proof of the previous result. □

The result of Theorem 34.2 states that elastic and reactive collision terms are consistent with the physical conservation laws for mass, momentum and total energy of the whole mixture.

Theorem 34.3. *The reactive collision terms (34.4) are such that*

$$\int_{\mathbb{R}^3} \mathcal{Q}_A^R(\underline{f}) dc_A = -\int_{\mathbb{R}^3} \mathcal{Q}_B^R(\underline{f}) dc_B. \tag{34.18}$$

Proof. The proof follows the same line of arguments. In particular one changes the reactant variables with the corresponding ones of the products. □

Theorem 34.3 states that reactive collision terms assure the correct chemical exchange rates for the considered chemical process.

Theorem 34.4. *The following statements are equivalent:*

(a) The distribution functions are Maxwellian, that is

$$f_\alpha^M = n_\alpha \left(\frac{m}{2\pi k T} \right)^{3/2} \exp\left(-\frac{m(c_\alpha - u)^2}{2k T} \right), \quad \alpha = A, B, \tag{34.19}$$

with the particle number densities n_α restricted to de mass action law

$$\frac{2(\epsilon_A - \epsilon_B)}{k T} = \ln \left(\frac{\chi_A n_A^2}{\chi_B n_B^2} \right); \tag{34.20}$$

(b) $\mathcal{Q}_A^E(\underline{f}) = 0$ and $\mathcal{Q}_A^R(\underline{f}) = 0$;

(c) $\mathcal{Q}_A^E(\underline{f}) + \mathcal{Q}_A^R(\underline{f}) = 0$.

Proof. The proof proceeds showing that (a)⇒(b)⇒(c)⇒(a). The first implication comes straightforward from the substitution of expressions (34.19) and (34.20) into the definitions (34.3) and (34.4) of the elastic and reactive collision terms. The second implication is trivially satisfied. For the last implication one has to use again the symmetry properties and the well known inequality $(1 - x) \ln x \le 0$ for all $x > 0$. □

Theorem 34.4 defines the equilibrium solutions to the reactive system (34.3) of Boltzmann equations. More in detail, expressions (34.19), alone, define a mechanical equilibrium solution, whereas expressions (34.19) and (34.20), together, define a mechanical, thermal and chemical equilibrium solution.

540   G.M. Kremer and A.J. Soares

Theorem 34.5. *The elastic and reactive collision terms (34.3) and (34.4) are such that*

$$\sum_{\alpha=A}^{B} \int \left(\mathcal{Q}_\alpha^E\,(\underline{f}) + \mathcal{Q}_\alpha^R(\underline{f}) \right) \ln(\sqrt{\chi_\alpha}\, f_\alpha) d\mathbf{c}_\alpha \le 0. \tag{34.21}$$

Proof. If we multiply each equation of system (34.3) by $\ln\left(\sqrt{\chi_\alpha} f_\alpha\right)$, then integrate over the velocity $\mathbf{c}_\alpha$ and take the sum over all constituents, we obtain

$$\sum_{\alpha=A}^{B} \int \left(\mathcal{Q}_\alpha^E\,(\underline{f}) + \mathcal{Q}_\alpha^R(\underline{f}) \right) \ln(\sqrt{\chi_\alpha}\, f_\alpha) d\mathbf{c}_\alpha = \Sigma_E + \Sigma_R, \tag{34.22}$$

where Σ_E and Σ_R are the entropy production terms due to elastic scattering and chemical reactions, respectively given by

$$\Sigma_E = \frac{k}{4} \sum_{\alpha=A}^{B} \sum_{\beta=A}^{B} (1 - \chi_\beta \delta_{\alpha\beta}) \int f_\alpha' f_\beta'$$

$$\times \left(1 - \frac{f_\alpha f_\beta}{f_\alpha' f_\beta'} \right) \ln \left(\frac{f_\alpha f_\beta}{f_\alpha' f_\beta'} \right) g_{\beta\alpha}\sigma_{\beta\alpha} d\Omega_{\beta\alpha} d\mathbf{c}_\beta d\mathbf{c}_\alpha, \tag{34.23}$$

$$\Sigma_R = \frac{k}{2} \int \chi_B f_B f_{B_1} \left(1 - \frac{\chi_A f_A f_{A_1}}{f_B f_{B_1}} \right) \ln \left(\frac{\chi_A f_A f_{A_1}}{f_B f_{B_1}} \right) \sigma_A^\star g_A d\Omega d\mathbf{c}_{A_1} d\mathbf{c}_A. \tag{34.24}$$

The expressions on the r.h.s of (34.23) and (34.24) define negative semi-definite quantities, thanks again to the inequality $(1 - x) \ln x \le 0$.   □

The result expressed in Theorem 34.5 means that elastic and reactive collision terms assure a positive entropy production, since it is equivalent to the inequality

$$\frac{\partial}{\partial t} (\varrho\eta) + \frac{\partial}{\partial x_i} (\phi_i + \varrho\eta v_i) \ge 0, \tag{34.25}$$

where

$$\varrho\eta = -k \sum_{\alpha=A}^{B} \int f_\alpha \ln \left(\sqrt{\chi_\alpha} f_\alpha \right) d\mathbf{c}_\alpha, \quad \phi_i = -k \sum_{\alpha=A}^{B} \int f_\alpha \xi_i^\alpha \ln \left(\sqrt{\chi_\alpha} f_\alpha \right) d\mathbf{c}_\alpha, \tag{34.26}$$

are the entropy density and its flux, respectively.

34 Kinetic Theory for Chemical Reactions Without a Barrier 541

34.5 The Non-Equilibrium Distribution Function

In the literature of the Boltzmann equation extended to chemically reacting gases, the determination of the non-equilibrium distribution function containing the effects induced by the chemical reaction constitutes a fundamental topic. See, for example, paper [10] and references cited therein. In fact, the chemical reaction induces a perturbation of the local equilibrium, disturbing the molecular velocity distribution function from its Maxwellian form. As a consequence, important qualitative changes of the system properties occur.

The deviations induced by the chemical reaction on the distribution functions have been explicitly computed in the paper [8], using the Chapman–Enskog method [4] combined with Sonine polynomial representation of the distribution functions. A chemical regime of slow processes has been considered, meaning that the reaction is close to its initial stage. In this case, reactive collisions are less frequent than elastic encounters and the chemical relaxation time is larger than the elastic one. This means that reactive collision terms and material time derivatives, $\mathscr{D} = (\partial/\partial t) + v_i \partial/\partial x_i$, are of the same order whereas the gradients of the fields are of successive order. The Boltzmann equation (34.3) can then be re-written as

$$\mathscr{D} f_\alpha + \lambda \xi_i^\alpha \frac{\partial f_\alpha}{\partial x_i} - \int \left[\chi_\beta f_\beta f_{\beta_1} \sigma_\beta^\star \frac{g_\beta^2}{g_\alpha^2} - \chi_\alpha f_\alpha f_{\alpha_1} \sigma_\alpha^\star \right] g_\alpha d\Omega_\beta d\mathbf{c}_{\alpha_1} \qquad (34.27)$$

$$= \frac{1}{\lambda} \sum_{\beta=A}^{B} (1 - \chi_\beta \delta_{\alpha\beta}) \int \left[f_\alpha' f_\beta' - f_\alpha f_\beta \right] g_{\beta\alpha} \sigma_{\alpha\beta} d\Omega_{\alpha\beta} d\mathbf{c}_\beta,$$

where λ is a formal parameter of the order of the Knudsen number [4]. We then insert the expansions

$$f_\alpha = f_\alpha^M + \lambda f_\alpha^{(1)} + \lambda^2 f_\alpha^{(2)} + \cdots \quad \text{and} \quad \mathscr{D} = \mathscr{D}^{(0)} + \lambda \mathscr{D}^{(1)} + \lambda^2 \mathscr{D}^{(2)} + \cdots$$
$$(34.28)$$

into the Boltzmann equations (34.27) and equate equal powers of λ. We then obtain the integral equations for f_α^M and $f_\alpha^{(1)}$ in the form [8],

$$\sum_{\beta=A}^{B} (1 - \chi_\beta \delta_{\alpha\beta}) \int \left[f_\alpha^{M\prime} f_\beta^{M\prime} - f_\alpha^M f_\beta^M \right] g_{\beta\alpha} \sigma_{\alpha\beta} d\Omega_{\alpha\beta} d\mathbf{c}_\beta = 0, \qquad (34.29)$$

$$\mathscr{D}^{(0)} f_\alpha^M - \int \left[\chi_\beta f_\beta^M f_{\beta_1}^M \sigma_\beta^\star \frac{g_\beta^2}{g_\alpha^2} - \chi_\alpha f_\alpha^M f_{\alpha_1}^M \sigma_\alpha^\star \right] g_\alpha d\Omega_\beta d\mathbf{c}_{\alpha_1} \qquad (34.30)$$

$$= \sum_{\beta=A}^{B} (1 - \chi_\beta \delta_{\alpha\beta}) \int \left[f_\alpha^{(1)\prime} f_\beta^{M\prime} + f_\alpha^{M\prime} f_\beta^{(1)\prime} - f_\alpha^{(1)} f_\beta^M - f_\alpha^M f_\beta^{(1)} \right] g_{\beta\alpha} \sigma_{\alpha\beta} d\Omega_{\alpha\beta} d\mathbf{c}_\beta.$$

The solution of the integral equation (34.29) is the Maxwellian distribution function given by expression (34.19), with partial number densities n_α completely uncorrelated, since no chemical equilibrium condition is involved. Concerning (34.30), since the field gradients are absent, one admits that its solution is a small deviation from the Maxwellian distribution, expressed in terms of Sonine polynomials as

$$f_\alpha^{(1)} = f_\alpha^M \left[a_1^\alpha \left(\frac{3}{2} - \frac{m\xi_\alpha^2}{2kT} \right) + a_2^\alpha \left(\frac{15}{8} - \frac{5m\xi_\alpha^2}{4kT} + \frac{m^2\xi_\alpha^4}{8k^2T^2} \right) \right], \qquad (34.31)$$

where $\xi_i^\alpha = c_i^\alpha - v_i$ is the peculiar velocity of constituent α, and a_1^α and a_2^α are scalar coefficients to be determined using the Chapman–Enskog method. See the details reported in paper [8]. We have obtained $a_1^A = a_1^B = 0$ and a_2^A, a_2^B given by

$$a_2^A = \frac{2\chi_A x_A}{15x_A^2\chi_A + 16x_A\chi_A - 31} \left(\frac{T}{T_0} \right)^{n-\frac{1}{2}} \Gamma\left(n + \frac{3}{2}\right) n(n-1) \left(\frac{d_r}{d} \right)^2, \quad (34.32)$$

$$a_2^B = \frac{x_A(x_A\chi_A - 1)}{1 - x_A} a_2^A, \qquad (34.33)$$

where $x_A = n_A/n$ and $x_B = 1 - x_A$ represent the molar fraction of the constituents. Expressions (34.31) with coefficients $a_1^\alpha = 0$ and coefficients a_2 given by (34.32) and (34.33) completely determine the non-equilibrium distribution function $f_\alpha^{(1)}$, in the considered slow chemical regime. The consequent non-equilibrium effects induced on the macroscopic properties of the reacting mixture can then be evaluated as well. This study is reported in paper [8], where some numerical simulations have been implemented, showing satisfactory results that are in agreement with experimental predictions.

Furthermore, other different chemical regimes are investigated in a paper in preparation in view of studying transport properties, see [1]. Some details about the approximating procedure employed to determine the non-equilibrium distribution function are reported in that paper.

Acknowledgements The paper is partially supported by Brazilian Research Council (CNPq), by Minho University Mathematics Centre (CMAT-FCT) and by Project FCT – PTDC/MAT/ 68615/2006.

References

1. Alves, G.M., Kremer, G.M., Marques Jr., W., Soares, A.J.: A Kinetic Model for Chemical Reactions without Barriers: Transport Coefficients and Eigenmodes (submitted)
2. Brouard, M.: Reaction Dynamics. Oxford University Press, Oxford (1998)
3. Cercignani, C.: The Boltzmann Equation and its Application. Springer, New York (1988)
4. Chapman, S., Cowling, T.G.: The Mathematical Theory of Non-uniform Gases. Cambridge University Press, Cambridge (1970)
5. Giovangigli, V.: Multicomponent Flow Modeling. Birkhäuser, Boston (1999)

34 Kinetic Theory for Chemical Reactions Without a Barrier

6. Guthrie, J.P., Pitchko, V.: Reactions of carbocations with water and azide ion: calculation of rate constants from equilibrium constants and distortion energies using No-Barrier Theory. J. Phys. Org. Chem. **17**, 548–559 (2004)
7. Hessler, J.P.: New empirical rate expressions for reactions without a barrier: Analysis of the reaction of CN with O_2. J. Chem . Phys. **111**, 4068–4076 (1999)
8. Kremer, G.M., Soares, A.J.: A Kinetic Model for Chemical Reactions without Barriers. Rarefied Gas Dynamics, Ed. Akashi Abe, AIP Conference Proceedings 1084, New York, 105–110 (2009)
9. Prigogine, I., Mahieu, M.: Sur la perturbation de la distribution de Maxwell par des reactions chimiques en phase gazeuse. Physica **16**, 51–64 (1950)
10. Shizgal, B.D., Chikhaoui, A.: On the use temperature parameterized rate coefficients in the estimation of non-equilibrium reaction rates. Phys. A **365**, 317–332 (2006)
11. Stiller, W.: Arrhenius Equation and Non-Equilibrium Kinetics. BSB Teubner Text, Leipzig (1989)

Chapter 35
Dynamical Gene-Environment Networks Under Ellipsoidal Uncertainty: Set-Theoretic Regression Analysis Based on Ellipsoidal OR

Erik Kropat, Gerhard-Wilhelm Weber, and Selma Belen

Abstract We consider dynamical gene-environment networks under ellipsoidal uncertainty and discuss the corresponding set-theoretic regression models. Clustering techniques are applied for an identification of functionally related groups of genes and environmental factors. Clusters can partially overlap as single genes possibly regulate multiple groups of data items. The uncertain states of cluster elements are represented in terms of ellipsoids referring to stochastic dependencies between the multivariate data variables. The time-dependent behaviour of the system variables and clusters is determined by a regulatory system with (affine-) linear coupling rules. Explicit representations of the uncertain multivariate future states of the system are calculated by ellipsoidal calculus. Various set-theoretic regression models are introduced in order to estimate the unknown system parameters. Hereby, we extend our *Ellipsoidal Operations Research* previously introduced for gene-environment networks of strictly disjoint clusters to possibly overlapping clusters. We analyze the corresponding optimization problems, in particular in view of their solvability by interior point methods and semidefinite programming and we conclude with a discussion of structural frontiers and future research challenges.

E. Kropat (✉)
Institute for Theoretical Computer Science, Mathematics and Operations Research, Universität der Bundeswehr München, Werner-Heisenberg-Weg 39, 85577 Neubiberg, Germany
e-mail: erik.kropat@unibw.de

G.-W. Weber
Institute of Applied Mathematics, Middle East Technical University, 06531 Ankara, Turkey
and
Faculty of Economics, Business and Law, University of Siegen, Siegen, Germany
and
Center for Research on Optimization and Control, University of Aveiro, Aveiro, Portugal
and
Faculty of Science, Universiti Teknologi Malaysia (UTM), Skudai, Malaysia
e-mail: gweber@metu.edu.tr

S. Belen
CAG University, Yenice-Tarsus, 33800 Mersin, Turkey
e-mail: sbelen@cag.edu.tr

M.M. Peixoto et al. (eds.), *Dynamics, Games and Science I*, Springer Proceedings in Mathematics 1, DOI 10.1007/978-3-642-11456-4_35,
© Springer-Verlag Berlin Heidelberg 2011

545

35.1 Introduction

The development of microarray technologies enabled researchers in genetics to monitor the expression values of thousands of genes simultaneously. The availability of such huge data sets challenged bioinformatics and mathematics and led to the development of new methods for knowledge discovery in functional genomic data sets. Many concepts from data mining and statistical analysis were applied in order to reveal the cellular processes involved. In particular, *clustering techniques* were used for an identification of functionally related groups of genes. Among them were, for example, techniques as *k-means* [43, 78], *hierarchical clustering* [14, 22, 43], *self-organizing maps* [32, 40, 48], *principle component analysis* [59, 75], *singular value decomposition* [6, 59] and *support vector machines* [12, 31]. However, these methods often resulted in *disjoint clusters* and in many applications such a *hard clustering* is too strict because of the quality of the data or the presence of outliers and errors. In addition, a single gene (or a group of genes) can have a regulating effect on various clusters of genes, as for example in context with the identification of synexpression groups and the analysis of synexpression control networks [33]. *Fuzzy-clustering* [53] or other methods resulting in a partially overlapping cluster decomposition can alleviate the effects of noise-prone data and can lead to a more flexible representation of interconnections between groups of data. Although these methods – exact or flexible – proved to be sufficient in identifying, e.g., damaged or cancerous genes and groups of regulating genetic items, they are nevertheless considered as *static methods* which do not shed any light on the time-dependent behaviour of the genetic network. *Time-series analysis* [23] or related approaches can be applied to forecast the time-dependent states of the expression values. In our studies [16, 63, 69], we demonstrate the way how we develop a time-discrete dynamics whose parameters we identified and how we use the given data in order to test the goodness of the regression. Since the time-discrete dynamics can be gained by various kinds of discretization schemes, the aforementioned comparison also helps to test the quality of these schemes and rank them. It turned out that 3rd order Heun's method has a smoothing effect with respect to the prediction, which is regarded to be very good and 'natural' in the (natural) context of gene-environment networks, and that it leads to a faster convergence to equilibrium points of the dynamics.

In ways such as aforementioned, the gained time-discrete dynamics supports the prediction. As we explained in [55, 67, 72], we can also in further ways use these dynamics for a texting of our model, i.e., of the quality of data fitting. Actually, in various application contexts it is known or at least considered to be guaranteed, that the 'expression levels' of state variables are staying in bounded intervals [1, 2, 4, 56]. If, however, our discrete dynamics emerged in some direction in an unbounded kind, then we could conclude that this dynamics, e.g., it parameters identified by us, cannot be accepted. In such a case, the hypothesis of the model has to be rejected and, within our entire learning processes, improvements in the model structure be made and the parameter estimation restarted.

Here, we combine methods from clustering theory and dynamic systems under the presence of errors and uncertainty. For an analysis of the interconnections between clusters of genes and environmental factors and a prediction of their future states we study *gene-environment networks under ellipsoidal uncertainty*. In contrast to other models of *genetic systems*, these networks capture and assess the regulating effects of additional environmental factors. In general, gene-environment networks consist of two major groups of data items – the *genes* (or proteins and other molecules) and the so-called *environmental factors*, which stand for other cell components like toxins, transcription factors etc. that often play an important but nevertheless underestimated role in the regulatory system. We note that beside genetic systems many other examples from life sciences and systems biology refer to *gene-environment networks*. Among them are, e.g., *metabolic networks* [10, 41, 70], *immunological networks* [20], networks in *toxicogenomics* [34], but also *social-* and *ecological networks* [19]. We refer to [18, 24, 26, 42, 49–51, 54, 76, 77] for applications, practical examples and numerical calculations. Recent studies on gene-environment networks focussed on errors and uncertainty. The potential deviation from measurement values and predictions of *each gene* was measured in terms of error intervals by imposing bounds on each variable. Various regression models have been developed and studied with the help of *generalized Chebychev approximation* and, equivalently to that approximation, *semi-infinite* and even *generalized semi-infinite optimization* [58, 60, 62, 64, 65, 67, 70, 71, 73]. However, error intervals referring to single variables do not reflect correlations of the multivariate data within specific clusters of genetic and environmental items. In our approach we apply clustering techniques for an identification of functionally related groups of genes (or environmental factors) commonly exerting influence on other groups of genes and/or environmental items [15, 44]. In particular, we focus on *possibly overlapping clusters* and by this we further extend the approach from [27], where a strict subdivision of data was assumed. Each cluster stands for a group of correlated data items. In order to measure data uncertainty, the multivariate state of genes (or environmental factors) in a cluster will be represented in terms of *ellipsoids*. These uncertainty sets refer to stochastic representations of errors and are directly related to Gaussian distributions and the corresponding covariance matrices. Error ellipsoids are considered as more flexible than error intervals where stochastic dependencies among any two of the errors made in the measurement of expression values and environmental levels are not taken into account explicitly. However, the two approaches are related, because any confidence ellipsoid can be inscribed into a sufficiently large and suitably oriented parallelpipe or, in reverse, it can be contained in such a paraxial set.

The dynamics of the uncertain (ellipsoidal) states of genes and environmental factors are represented by a time-dependent *regulatory model*. The coupling rules of this model are based on *ellipsoidal calculus* and they determine the interactions between the various clusters. With this model, predictions of the future states can be calculated explicitly. In addition, we introduce an *iterative procedure* for calculating the centers and shape matrices of the ellipsoidal states. The parameter constellation of the regulatory model refers to the topology and the degree of connectivity of the underlying gene-environment network. For an estimation of the unknown parameters, various set-theoretic regression models are introduced. These models are heavily effected by the cluster decomposition and the overlap of clusters. The associated objective functions of the regression models compare the ellipsoidal predictions of the regulatory model and the results from microarray experiments and environmental measurements, however, in a set-theoretic sense. They depend on the distance of cluster centers and on nonnegative criteria functions which measure, e.g., the sum of squares of semiaxes (which corresponds to the trace of the configuration matrix) or the length of the largest semiaxes (which corresponds to the eigenvalues of the shape matrix). We note that semi-definite programming and interior point methods can be applied for solution.

In general, gene-environment networks comprise thousands of genes and additional factors. For this reason, the underlying network has a high number of branches. Often the connections between the clusters are weak so that the related contribution to the system is negligible. In order to reduce complexity we delete weak connections. This could be achieved by introducing bounds on the number of incoming branches. By imposing such additional constraints we obtain mixed integer regression problems. Since these constraints are very strict, we turn to a further relaxation that could be achieved by replacing binary constraints with continuous constraints leading to regression models based on continuous optimization.

The Chapter is organized as follows: In Sect. 35.2, we review basic facts about ellipsoidal calculus required for a representation of the dynamic states of multivariate noise prone data. In Sect. 35.3, the time-dependent regulatory model for overlapping groups of genes and environmental factors under ellipsoidal uncertainty is introduced. In addition, we provide an algorithm that allows to calculate predictions of future states of groups of genes and environmental items in terms of centers and shape matrices of the corresponding ellipsoids. In Sect. 35.4, we turn to a set-theoretic regression analysis for parameter estimation of the dynamic (ellipsoidal) system. Various regression models are introduced and we discuss their solvability by means of semi-definite programming. Finally, we address a reduction of complexity by network rarefication in Sect. 35.5, where we further extend the dynamic model and discuss related mixed integer approximation and a relaxation based on continuous optimization.

35.2 Ellipsoidal Calculus

The time-dependent multivariate states of the gene-environment network under consideration will be represented in terms of ellipsoidal sets. Predictions of the future ellipsoidal states are calculated with a time-discrete model based on *ellipsoidal calculus*. Here, we shortly review the basic operations of ellipsoidal calculus such as *sums, intersections (fusions)* and *affine-linear transformations* of ellipsoids. The family of ellipsoids in $\mathbb{R}^p$ is closed with respect to affine-linear transformations but neither the sum nor the intersection is generally ellipsoidal, so both must be approximated by ellipsoidal sets.

35 Dynamical Gene-Environment Networks Under Ellipsoidal Uncertainty

35.2.1 Ellipsoidal Descriptions

An *ellipsoid* in $\mathbb{R}^p$ will be parameterized in terms of its center $c \in \mathbb{R}^p$ and a symmetric non-negative definite *configuration (or shape) matrix* $\Sigma \in \mathbb{R}^{p \times p}$ as

$$\mathscr{E}(c, \Sigma) = \{\Sigma^{1/2} u + c \mid \|u\| \leq 1\},$$

where $\Sigma^{1/2}$ is any matrix square root satisfying $\Sigma^{1/2}(\Sigma^{1/2})^T = \Sigma$. When Σ is of full rank, the non-degenerate ellipsoid $\mathscr{E}(c, \Sigma)$ may be expressed as

$$\mathscr{E}(c, \Sigma) = \{x \in \mathbb{R}^p \mid (x - c)^T \Sigma^{-1}(x - c) \leq 1\}.$$

The eigenvectors of Σ point in the directions of principal semiaxes of $\mathscr{E}$. The lengths of the semiaxes of the ellipsoid $\mathscr{E}(c, \Sigma)$ are given by $\sqrt{\lambda_i}$, where λ_i are the eigenvalues of Σ for $i = 1, \ldots, p$. The volume of the ellipsoid $\mathscr{E}(c, \Sigma)$ is given by $\mathrm{vol}\,\mathscr{E}(c, \Sigma) = V_p \sqrt{\det(\Sigma)}$, where V_p is the volume of the unit ball in $\mathbb{R}^p$, i.e.,

$$V_p = \begin{cases} \dfrac{\pi^{p/2}}{(p/2)!}, & \text{for even } p, \\[2ex] \dfrac{2^p \pi^{(p-1)/2}((p-1)/2)!}{p!}, & \text{for odd } p. \end{cases}$$

35.2.2 Affine Transformations

The family of ellipsoids is closed with respect to *affine transformations*. Given an ellipsoid $\mathscr{E}(c, \Sigma) \subset \mathbb{R}^p$, matrix $A \in \mathbb{R}^{m \times p}$ and vector $b \in \mathbb{R}^m$ we get $A\mathscr{E}(c, \Sigma) + b = \mathscr{E}(Ac + b, A\Sigma A^T)$. Thus, ellipsoids are preserved under affine transformation. If the rows of A are linearly independent (which implies $m \leq p$), and $b = 0$, the affine transformation is called *projection* [30].

35.2.3 Sums of K Ellipsoids

Given K bounded ellipsoids of $\mathbb{R}^p$, $\mathscr{E}_k = \mathscr{E}(c_k, \Sigma_k), k = 1, \ldots, K$, their *geometric (Minkowksi) sum* $\mathscr{E}_1 + \mathscr{E}_1 = \{z_1 + z_2 \mid z_1 \in \mathscr{E}_1, \ z_2 \in \mathscr{E}_2\}$ is not generally an ellipsoid. However, it can be tightly approximated by parameterized families of external ellipsoids. We adapt the notion of the minimal trace ellipsoid from [17] and introduce the outer ellipsoidal approximation $\mathscr{E}(\sigma, P) = \oplus_{k=1}^K \mathscr{E}_k$ containing the sum $\mathscr{S} = \sum_{k=1}^K \mathscr{E}_k$ of ellipsoids which is defined by

$$\sigma = \sum_{k=1}^K c_k$$

and

$$P = \left(\sum_{k=1}^{K} \sqrt{\mathrm{Tr}\, \Sigma_k} \right) \left(\sum_{k=1}^{K} \frac{\Sigma_k}{\sqrt{\mathrm{Tr}\, \Sigma_k}} \right).$$

35.2.4 Intersection of Ellipsoids

The intersection of two ellipsoids is generally not an ellipsoid. For this reason we replace this set by the outer ellipsoidal approximation of minimal volume and adapt the notion of *fusion* of ellipsoids from [45]. Given two non-degenerate ellipsoids $\mathscr{E}(c_1, \Sigma_1)$ and $\mathscr{E}(c_2, \Sigma_2)$ in $\mathbb{R}^p$ with $\mathscr{E}(c_1, \Sigma_1) \cap \mathscr{E}(c_2, \Sigma_2) \neq \emptyset$ we define an ellipsoid

$$\mathscr{E}_\lambda(c_0, \Sigma_0) := \{ x \in \mathbb{R}^p \mid \lambda (x - c_1)^T \Sigma_1^{-1} (x - c_1)$$
$$+ (1 - \lambda)(x - c_2)^T \Sigma_2^{-1} (x - c_2) \leq 1 \},$$

where $\lambda \in [0, 1]$. The ellipsoid $\mathscr{E}_\lambda(c_0, \Sigma_0)$ coincides with $\mathscr{E}(c_1, \Sigma_1)$ and $\mathscr{E}(c_2, \Sigma_2)$ for $\lambda = 1$ and $\lambda = 0$, respectively. In order to determine a tight external ellipsoidal approximation $\mathscr{E}_\lambda(c_0, \Sigma_0)$ of the intersection of $\mathscr{E}(c_1, \Sigma_1)$ and $\mathscr{E}(c_2, \Sigma_2)$, we introduce

$$\mathscr{X} := \lambda \Sigma_1^{-1} + (1 - \lambda) \Sigma_2^{-1}$$

and

$$\tau := 1 - \lambda(1 - \lambda)(c_2 - c_1)^T \Sigma_2^{-1} \mathscr{X}^{-1} \Sigma_1^{-1} (c_2 - c_1).$$

The ellipsoid $\mathscr{E}_\lambda(c_0, \Sigma_0)$ is given by the center

$$c_0 = \mathscr{X}^{-1} (\lambda \Sigma_1^{-1} c_1 + (1 - \lambda) \Sigma_2^{-1} c_2)$$

and shape matrix

$$\Sigma_0 = \tau \mathscr{X}^{-1}.$$

The *fusion* of $\mathscr{E}(c_1, \Sigma_1)$ and $\mathscr{E}(c_2, \Sigma_2)$, whose intersection is a nonempty bounded region, is defined as the ellipsoid $\mathscr{E}_\lambda(c_0, \Sigma_0)$ for the value $\lambda \in [0, 1]$ that minimizes its volume [45]. The fusion of $\mathscr{E}(c_1, \Sigma_1)$ and $\mathscr{E}(c_2, \Sigma_2)$ is $\mathscr{E}(c_1, \Sigma_1)$, if $\mathscr{E}(c_1, \Sigma_1) \subset \mathscr{E}(c_2, \Sigma_2)$; or $\mathscr{E}(c_2, \Sigma_2)$, if $\mathscr{E}(c_2, \Sigma_2) \subset \mathscr{E}(c_1, \Sigma_1)$; otherwise, it is $\mathscr{E}_\lambda(c_0, \Sigma_0)$ defined as above where λ is the only root in $(0, 1)$ of the following polynomial of degree $2p - 1$:

$$\tau (\det \mathscr{X}) \, \mathrm{Tr}\, (\mathrm{co}(\mathscr{X})(\Sigma_1^{-1} - \Sigma_2^{-1})) - p (\det \mathscr{X})^2$$
$$\times (2c_0^T \Sigma_1^{-1} c_1 - 2c_0^T \Sigma_2^{-1} c_2 + c_0^T (\Sigma_2^{-1} - \Sigma_1^{-1}) c_0 - c_1^T \Sigma_1^{-1} c_1 + c_2^T \Sigma_2^{-1} c_2) = 0.$$

Here, $\mathrm{co}(\mathscr{X})$ denotes the matrix of cofactors of $\mathscr{X}$. Since $\mathscr{X}^{-1} = \mathrm{co}(\mathscr{X}) / \det \mathscr{X}$, we represent this polynomial as

$$\tau (\det \mathscr{X})^2 \, \mathrm{Tr} \, (\mathscr{X}^{-1}(\Sigma_1^{-1} - \Sigma_2^{-1})) - p(\det \mathscr{X})^2$$
$$\times \, (2c_0^T \Sigma_1^{-1} c_1 - 2c_0^T \Sigma_2^{-1} c_2 + c_0^T (\Sigma_2^{-1} - \Sigma_1^{-1})c_0 - c_1^T \Sigma_1^{-1} c_1 + c_2^T \Sigma_2^{-1} c_2) = 0.$$

We note that it is also possible to define an inner ellipsoidal approximation. The method of finding the internal ellipsoidal approximation of the intersection of two ellipsoids is described in [57].

35.3 Gene-Environment Systems Under Ellipsoidal Uncertainty

35.3.1 Clusters of Gene-Environment Data

Various approaches from clustering and classification can be applied to analyze the structure of gene-environment networks. In this way, certain groups of genes and environmental factors can be identified which exert a more or less regulating influence on other groups of data items. Usually these groups cannot be divided unambiguously since a single gene can have a regulating effect on various clusters of genes and, thus, belongs to different clusters. In addition, the quality of the available data sets may not be sufficient for an identification of disjoint groups. For this reason, we assume that in the preprocessing step of clustering a number of *overlapping* clusters of genes and environmental factors can be identified. Such a partition can be achieved for example with one of the many variants of *fuzzy-c-means clustering* [53]. The specific gene-environment network under consideration consists of n genes and m environmental factors, where the vector $\mathbb{X} = (\mathbb{X}_1, \ldots, \mathbb{X}_n)^T$ denotes the expression values of the genes and the vector $\mathbb{E} = (\mathbb{E}_1, \ldots, \mathbb{E}_m)^T$ stands for the values of the environmental factors. The set of genes is divided in R overlapping clusters $C_r \subset \{1, \ldots, n\}$, $r = 1, \ldots, R$ and the set of all environmental items is divided in S overlapping clusters $D_s \subset \{1, \ldots, m\}$, $s = 1, \ldots, S$. We note that the paper [27] focussed on disjoint clusters assuming a strict sub-division of the variables where the relations $C_{r_1} \cap C_{r_2} = \emptyset$ for all $r_1 \neq r_2$ and $D_{s_1} \cap D_{s_2} = \emptyset$ for all $s_1 \neq s_2$ are fulfilled.

The *(crisp) states* of the elements of these clusters are given by subsets of the vectors $\mathbb{X}$ and $\mathbb{E}$. That means, we assign a $|C_r|$-subvector X_r of $\mathbb{X}$ to each cluster of genes which is given by the indices of C_r. Similarly, E_s is a $|D_s|$-subvector of $\mathbb{E}$ given by the indices of D_s.

For a representation of the *uncertain states* of the aforementioned clusters we identify the clusters with error ellipsoids. That means, the vectors X_r represent ellipsoidal states of the genes in cluster C_r given by the ellipsoid $\mathscr{E}(\mu_r, \Sigma_r) \subset \mathbb{R}^{|C_r|}$ and E_s represent the ellipsoidal states of the environmental items in cluster D_s given by the ellipsoids $\mathscr{E}(\rho_s, \Pi_s) \subset \mathbb{R}^{|D_s|}$. The ellipsoid $\mathscr{E}(\mu_r, \Sigma_r)$ is characterized by $|C_r| + |C_r|^2$ coefficients and the ellipsoid $\mathscr{E}(\rho_s, \Pi_s)$ is determined by $|D_s| + |D_s|^2$ variables. The number of coefficients can be reduced by assuming symmetric shape matrices what refers to specific correlation of the data variables. We note that ellipsoids can be identified with intervals if clusters are singletons. It is also possible that some of the variables are exactly known. In this situation, the ellipsoids $\mathscr{E}(\mu_r, \Sigma_r)$ and $\mathscr{E}(\rho_s, \Pi_s)$ are flat. However, as we are interested in approximations we can avoid this by imposing lower bounds on the semiaxes lengths or an artificial extension in the corresponding coordinate directions of length $\varepsilon > 0$. Similarly, degenerate or needle-shaped ellipsoids can be avoided by imposing upper bounds on the extension of the semiaxes.

35.3.2 The Linear Model

In this section, we introduce a dynamic model that allows to predict the time-dependent (ellipsoidal) states of the clusters in the gene-environment regulatory network. This model is based on four types of cluster interactions and regulating effects:

(GG) genetic cluster regulates genetic cluster
(EG) environmental cluster regulates genetic cluster
(GE) genetic cluster regulates environmental cluster
(EE) environmental cluster regulates environmental cluster.

As shown in Sect. 35.3.1, each cluster corresponds to a functionally related group of genes or environmental factors and the uncertain states of these clusters are represented in terms of parameterized ellipsoids

$$X_r = \mathscr{E}(\mu_r, \Sigma_r) \subset \mathbb{R}^{|C_r|}, \quad E_s = \mathscr{E}(\rho_s, \Pi_s) \subset \mathbb{R}^{|D_s|}.$$

With the ellipsoidal calculus introduced in Sect. 35.2, the dynamics and interactions between the various clusters of genetic and environmental items is given by the linear model

$$\left.\begin{aligned} X_j^{(\kappa+1)} &= \xi_{j0} + \left(\bigoplus_{r=1}^{R} \mathscr{A}_{jr}^{GG} X_r^{(\kappa)} \right) \oplus \left(\bigoplus_{s=1}^{S} \mathscr{A}_{js}^{EG} E_s^{(\kappa)} \right) \\ E_i^{(\kappa+1)} &= \zeta_{i0} + \left(\bigoplus_{r=1}^{R} \mathscr{A}_{ir}^{GE} X_r^{(\kappa)} \right) \oplus \left(\bigoplus_{s=1}^{S} \mathscr{A}_{is}^{EE} E_s^{(\kappa)} \right) \end{aligned}\right\} \quad (EC)$$

with $\kappa \geq 0$ and $j = 1, 2, \ldots, R, i = 1, 2, \ldots, S$. The system (EC) is defined by (affine) linear coupling rules, what implies that all future states of genetic and environmental clusters are ellipsoids themselves. In particular, the sums $\bigoplus_{r=1}^{R} \mathscr{A}_{jr}^{GG} X_r^{(\kappa)}$

and $\bigoplus_{s=1}^{S} \mathscr{A}_{js}^{EG} E_s^{(\kappa)}$ describe the *cumulative effects* of all genetic and environmental clusters exerted on the elements of cluster C_j in a set theoretic or ellipsoidal

35 Dynamical Gene-Environment Networks Under Ellipsoidal Uncertainty 553

sense. In the same way, the (ellipsoidal) sums $\bigoplus_{r=1}^{R} \mathscr{A}_{ir}^{GE} X_r^{(\kappa)}$ and $\bigoplus_{s=1}^{S} \mathscr{A}_{is}^{EE} E_s^{(\kappa)}$ refer to the additive genetic and environmental effects on cluster D_i. The *degree of connectivity* between the individual clusters is given by the (unknown) interactions matrices $\mathscr{A}_{jr}^{GG} \in \mathbb{R}^{|C_j| \times |C_r|}$, $\mathscr{A}_{js}^{EG} \in \mathbb{R}^{|C_j| \times |D_s|}$, $\mathscr{A}_{ir}^{GE} \in \mathbb{R}^{|D_i| \times |C_r|}$, and $\mathscr{A}_{is}^{EE} \in \mathbb{R}^{|D_i| \times |D_s|}$. These matrices are in turn sub-matrices of the general interaction matrices $\mathscr{A}^{GG} \in \mathbb{R}^{n \times n}$, $\mathscr{A}^{EG} \in \mathbb{R}^{n \times m}$, $\mathscr{A}^{GE} \in \mathbb{R}^{m \times n}$, $\mathscr{A}^{EE} \in \mathbb{R}^{m \times m}$. In case of disjoint clusters, the aforementioned sub-matrices constitute distinct building blocks of the interaction matrices [27,28]. For overlapping clusters, the structure of the interaction matrices is much more complicated; it reflects the cluster structure and the sub-matrices are partly composed of the same elements. This also holds for the intercepts $\xi_{j0} \in \mathbb{R}^{|C_j|}$ and $\zeta_{i0} \in \mathbb{R}^{|D_i|}$ which are partly overlapping sub-vectors of the vectors $\xi_0 = (\xi_{10}, \dots, \xi_{n0})^T \in \mathbb{R}^n$ and $\zeta_0 = (\zeta_{10}, \dots, \zeta_{m0})^T \in \mathbb{R}^m$, respectively. We note that the initial values of the linear system (EC) can be defined by the first genetic and environmental measurements, i.e., $X_j^{(0)} = \overline{X}_j^{(0)}$ and $E_i^{(0)} = \overline{E}_i^{(0)}$.

The unknown parameters of the linear model (EC) have to be determined by a regression analysis based on ellipsoidal data sets. Since all matrices and vectors of (EC) are parts of the general interaction matrices and intercepts, $n^2 + 2nm + m^2 + n + m = (n + m)^2 + n + m$ unknown parameters have to be determined. We will provide more details on regression analysis in Sect. 35.4.

Remark (Gene-environment networks). The clusters of genes and environmental factors can be considered as the nodes of a so-called *gene-environment network*. Such a network usually consists of a high number of branches what refers to the inherent connections between the clusters. The branches between the nodes (or clusters) are weighted by the matrices and intercept vectors of the linear coupling rules of model (EC) and the nodes are weighted by the time-dependent ellipsoidal states of the clusters. Hereby, network analysis and concepts from discrete mathematics become applicable and features like connectedness, cycles and shortest paths can be investigated [25].

35.3.3 Algorithm

The regulatory system (EC) allows to predict the ellipsoidal states of genes and environmental factors with the set-theoretic calculus introduced in Sect. 35.2. In order to avoid set-valued calculations we propose to determine the centers and shape matrices of the predictions $X_j^{(\kappa+1)}$ and $E_s^{(\kappa+1)}$ of (ellipsoidal) genetic and environmental cluster states by an iterative procedure. Throughout this section we assume $\kappa \geq 0$. The states of the genetic clusters C_j, $j = 1, 2, \dots, R$, are given by the ellipsoids

$$X_j^{(\kappa+1)} = \mathscr{E}\left(\mu_j^{(\kappa+1)}, \Sigma_j^{(\kappa+1)}\right)$$

with center

$$\mu_j^{(\kappa+1)} = \xi_{j0} + \sum_{r=1}^{R} A_{jr}^{GG} \mu_r^{(\kappa)} + \sum_{s=1}^{S} A_{js}^{EG} \rho_s^{(\kappa)}$$

and shape matrix

$$\Sigma_j^{(\kappa+1)} = \left(\sqrt{\mathrm{Tr}\,\mathscr{G}_j^{(\kappa)}} + \sqrt{\mathrm{Tr}\,\mathscr{H}_j^{(\kappa)}} \right) \cdot \left(\frac{\mathscr{G}_j^{(\kappa)}}{\sqrt{\mathrm{Tr}\,\mathscr{G}_j^{(\kappa)}}} + \frac{\mathscr{H}_j^{(\kappa)}}{\sqrt{\mathrm{Tr}\,\mathscr{H}_j^{(\kappa)}}} \right),$$

where

$$\mathscr{G}_j^{(\kappa)} = \left(\sum_{r=1}^{R} \sqrt{\mathrm{Tr}\,G_{jr}^{GG}} \right) \cdot \left(\sum_{r=1}^{R} \frac{G_{jr}^{GG}}{\sqrt{\mathrm{Tr}\,G_{jr}^{GG}}} \right),$$

$$\mathscr{H}_j^{(\kappa)} = \left(\sum_{s=1}^{S} \sqrt{\mathrm{Tr}\,H_{js}^{EG}} \right) \cdot \left(\sum_{s=1}^{S} \frac{H_{js}^{EG}}{\sqrt{\mathrm{Tr}\,H_{js}^{EG}}} \right)$$

and

$$G_{jr}^{GG} = A_{jr}^{GG} \Sigma_r^{(\kappa)} (A_{jr}^{GG})^T, \quad H_{js}^{EG} = A_{js}^{EG} \Pi_s^{(\kappa)} (A_{js}^{EG})^T.$$

Similarly, the states of the environmental cluster D_i, $i = 1, 2, \ldots, S$, can be represented in terms of ellipsoids

$$E_i^{(\kappa+1)} = \mathscr{E}\left(\rho_i^{(\kappa+1)}, \Pi_i^{(\kappa+1)} \right)$$

with center

$$\rho_i^{(\kappa+1)} = \zeta_{i0} + \sum_{r=1}^{R} A_{ir}^{GE} \mu_r^{(\kappa)} + \sum_{s=1}^{S} A_{is}^{EE} \rho_s^{(\kappa)}$$

and shape matrix

$$\Pi_i^{(\kappa+1)} = \left(\sqrt{\mathrm{Tr}\,\mathscr{M}_i^{(\kappa)}} + \sqrt{\mathrm{Tr}\,\mathscr{N}_i^{(\kappa)}} \right) \cdot \left(\frac{\mathscr{M}_i^{(\kappa)}}{\sqrt{\mathrm{Tr}\,\mathscr{M}_i^{(\kappa)}}} + \frac{\mathscr{N}_i^{(\kappa)}}{\sqrt{\mathrm{Tr}\,\mathscr{N}_i^{(\kappa)}}} \right),$$

where

$$\mathscr{M}_i^{(\kappa)} = \left(\sum_{r=1}^{R} \sqrt{\mathrm{Tr}\,M_{ir}^{GE}} \right) \cdot \left(\sum_{r=1}^{R} \frac{M_{ir}^{GE}}{\sqrt{\mathrm{Tr}\,M_{ir}^{GE}}} \right),$$

$$\mathscr{N}_i^{(\kappa)} = \left(\sum_{s=1}^{S} \sqrt{\mathrm{Tr}\,N_{is}^{EE}} \right) \cdot \left(\sum_{s=1}^{S} \frac{N_{is}^{EE}}{\sqrt{\mathrm{Tr}\,N_{is}^{EE}}} \right)$$

and

$$M_{ir}^{GE} = A_{ir}^{GE} \Sigma_r^{(\kappa)} (A_{ir}^{GE})^T, \quad N_{is}^{EE} = A_{is}^{EE} \Pi_s^{(\kappa)} (A_{is}^{EE})^T.$$

35.4 Regression Analysis Under Ellipsoidal Uncertainty

35.4.1 The Regression Problem

The linear model (EC) depends on $(n + m)^2 + n + m$ unknown parameters which define the system dynamics but also the strength of the interconnections between the genetic and environmental clusters. In this section, we introduce our main regression model for an estimation of these parameters and, thus, of the entries of the interaction matrices $\mathscr{A}_{jr}^{GG}, \mathscr{A}_{js}^{EG}, \mathscr{A}_{ir}^{GE}, \mathscr{A}_{is}^{EE}$ as well as the intercepts ξ_{j0} and ζ_{i0} for overlapping clusters. Since the model (EC) is based on ellipsoidal sets, a set-theoretic regression analysis has to be established. The input data is given in terms of ellipsoidal genetic and environmental observations

$$\overline{X}_r^{(\kappa)} = \mathscr{E}\left(\overline{\mu}_r^{(\kappa)}, \overline{\Sigma}_r^{(\kappa)}\right) \subset \mathbb{R}^{|C_r|}, \quad \overline{E}_s^{(\kappa)} = \mathscr{E}\left(\overline{\rho}_s^{(\kappa)}, \overline{\Pi}_s^{(\kappa)}\right) \subset \mathbb{R}^{|D_s|},$$

with $r = 1, 2, \ldots, R$, $s = 1, 2, \ldots, S$ and $\kappa = 0, 1, \ldots, T$ which are taken at sampling times $t_0 < t_1 < \ldots < t_T$. These measurements have to be compared with the first T predictions of the model (EC) given by

$$\widehat{X}_j^{(\kappa+1)} = \mathscr{E}\left(\widehat{\mu}_j^{(\kappa+1)}, \widehat{\Sigma}_j^{(\kappa+1)}\right) := \xi_{j0} + \left(\bigoplus_{r=1}^R \mathscr{A}_{jr}^{GG} \overline{X}_r^{(\kappa)}\right) \oplus \left(\bigoplus_{s=1}^S \mathscr{A}_{js}^{EG} \overline{E}_s^{(\kappa)}\right),$$

$$\widehat{E}_i^{(\kappa+1)} = \mathscr{E}\left(\widehat{\rho}_i^{(\kappa+1)}, \widehat{\Pi}_i^{(\kappa+1)}\right) := \zeta_{i0} + \left(\bigoplus_{r=1}^R \mathscr{A}_{ir}^{GE} \overline{X}_r^{(\kappa)}\right) \oplus \left(\bigoplus_{s=1}^S \mathscr{A}_{is}^{EE} \overline{E}_s^{(\kappa)}\right),$$

with $j = 1, 2, \ldots, R$, $i = 1, 2, \ldots, S$ and $\kappa = 0, 1, \ldots, T - 1$.

In our set-theoretic regression, we try to maximize the overlap of the predictions and measurement values (both ellipsoids). For this reason, we introduce the ellipsoidal approximation of the intersection given by

$$\Delta X_r^{(\kappa)} := \widehat{X}_r^{(\kappa)} \cap \overline{X}_r^{(\kappa)} \quad \text{and} \quad \Delta E_s^{(\kappa)} := \widehat{E}_s^{(\kappa)} \cap \overline{E}_s^{(\kappa)},$$

with $r = 1, 2, \ldots, R$, $s = 1, 2, \ldots, S$ and $\kappa = 1, \ldots, T$, where $\cap$ denotes the fusion of ellipsoids introduced in Sect. 35.2.3. In addition, the centers of the ellipsoids are adjusted, so that their squared distance

$$\left\| \widehat{\mu}_r^{(\kappa)} - \overline{\mu}_r^{(\kappa)} \right\|_2^2 \quad \text{and} \quad \left\| \widehat{\rho}_s^{(\kappa)} - \overline{\rho}_s^{(\kappa)} \right\|_2^2$$

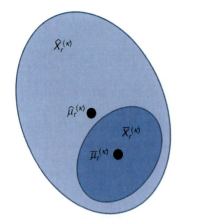
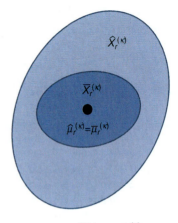

Fig. 35.1 Overlap of ellipsoids: The intersections of the two ellipsoids $\widehat{X}_r^{(\kappa)}$ and $\overline{X}_r^{(\kappa)}$ have the same geometrical size with the same measure of fusions on the left and the right side. On the right side, the centers $\widehat{\mu}_r^{(\kappa)}$ and $\overline{\mu}_r^{(\kappa)}$ are adjusted in order to minimize the difference between the centers of ellipsoids

becomes minimized (cf. Fig. 35.1). This leads us to the following regression problem:

$$(R) \quad \text{Maximize} \quad \sum_{\kappa=1}^{T} \Bigg\{ \sum_{r=1}^{R} \bigg[\big\| \Delta X_r^{(\kappa)} \big\|_* - \big\| \widehat{\mu}_r^{(\kappa)} - \overline{\mu}_r^{(\kappa)} \big\|_2^2 \bigg] \\ + \sum_{s=1}^{S} \bigg[\big\| \Delta E_s^{(\kappa)} \big\|_* - \big\| \widehat{\rho}_s^{(\kappa)} - \overline{\rho}_s^{(\kappa)} \big\|_2^2 \bigg] \Bigg\}.$$

Here, $\| \cdot \|_*$ denotes a measure that reflects the geometrical size of the intersections (fusions) and we assume that $\| \Delta X_r^{(\kappa)} \|_* = 0$, if $\Delta X_r^{(\kappa)} = \emptyset$ and $\| \Delta E_s^{(\kappa)} \|_* = 0$, if $\Delta E_s^{(\kappa)} = \emptyset$. There exist various measures related to the shape of the intersections, e.g., the *volume* (which corresponds to the ellipsoid matrix determinant), the *sum of squares of semiaxes* (which corresponds to the trace of the configuration matrix) or the *length of the largest semiaxes* (which corresponds to the eigenvalues of the shape matrix). For further details on geometrical (ellipsoidal) measures and the related regression problems we refer to [28].

35.4.2 Variants of the Regression Problem

In this section, we introduce specific formulations of the regression model (R). As mentioned above, the objective function of this model depends on a measure of the geometrical size of the intersections (fusions) $\Delta X_r^{(\kappa)}$ and $\Delta E_s^{(\kappa)}$ which is related to the corresponding shape matrices. In general, nonnegative-valued criteria functions

$\psi(\mathcal{E}(0, Q))$ defined on the set of all nondegenerate ellipsoids can be applied to measure the size of a p-dimensional ellipsoid $\mathcal{E}(0, Q)$. These functions are monotonous by increasing with respect to inclusion, i.e., $\psi(\mathcal{E}_1) \leq \psi(\mathcal{E}_2)$ if $\mathcal{E}_1 \subseteq \mathcal{E}_2$. Such measures are, e.g.,

(a) *The trace of Q*,

$$\psi_T(\mathcal{E}(0, Q)) := \operatorname{Tr} Q = \lambda_1 + \ldots + \lambda_p,$$

where λ_i are the eigenvalues of Q (i.e., $\operatorname{Tr} Q$ is equal to the sum of the squares of the semiaxes),

(b) *The trace of square of Q*,

$$\psi_{TS}(\mathcal{E}(0, Q)) := \operatorname{Tr} Q^2,$$

(c) *The diameter*,

$$\psi_{Dia}(\mathcal{E}(0, Q)) := \operatorname{diam}(\mathcal{E}(0, Q)) := d,$$

where

$$\max\{\lambda_i \in \mathbb{R} \mid i = 1, \ldots, p\} = \left(\frac{d}{2}\right)^2,$$

so that $d/2$ is the radius of the smallest p-dimensional ball that includes $\mathcal{E}(0, Q)$.

For further details on criteria functions we refer to [29], p. 101. The measures stated above lead to different representations of the regression problem (R) and in the following sections we study them in more detail.

Remark (Representation of fusions). For numerical calculations and an estimation of parameters of the regression problem (R), explicit representations of the fusions $\Delta X_r^{(\kappa)}$ and $\Delta E_s^{(\kappa)}$ are required. In the following we explain how these representations can be calculated with the ellipsoidal calculus of Sect. 35.2:

The fusion $\Delta X_r^{(\kappa)} = \widehat{X}_r^{(\kappa)} \cap \overline{X}_{C_r}^{(\kappa)}$ is an ellipsoid $\mathcal{E}(\Delta\mu_r^{(\kappa)}, \Delta\Sigma_r^{(\kappa)})$ with center

$$\Delta\mu_r^{(\kappa)} = \left[\mathscr{X}_r^{(\kappa)}\right]^{-1} \left(\lambda\left[\widehat{\Sigma}_r^{(\kappa)}\right]^{-1}\widehat{\mu}_r^{(\kappa)} + (1-\lambda)\left[\overline{\Sigma}_r^{(\kappa)}\right]^{-1}\overline{\mu}_r^{(\kappa)}\right)$$

and shape matrix

$$\Delta\Sigma_r^{(\kappa)} = \xi_r^{(\kappa)}\left[\mathscr{X}_r^{(\kappa)}\right]^{-1},$$

where

$$\mathscr{X}_r^{(\kappa)} := \lambda\left[\widehat{\Sigma}_r^{(\kappa)}\right]^{-1} + (1-\lambda)\left[\overline{\Sigma}_r^{(\kappa)}\right]^{-1}$$

and

$$\xi_r^{(\kappa)} := 1 - \lambda(1-\lambda)\big(\overline{\mu}_r^{(\kappa)} - \widehat{\mu}_r^{(\kappa)}\big)^T \big[\overline{\Sigma}_r^{(\kappa)}\big]^{-1}\big[\mathscr{X}_r^{(\kappa)}\big]^{-1}\big[\widehat{\Sigma}_r^{(\kappa)}\big]^{-1}\big(\overline{\mu}_r^{(\kappa)} - \widehat{\mu}_r^{(\kappa)}\big).$$

The parameter λ is the only root in $(0,1)$ of the following polynomial of degree $2|C_r| - 1$:

$$\xi_r^{(\kappa)}\big(\det \mathscr{X}_r^{(\kappa)}\big)^2 \operatorname{Tr}\left(\big[\mathscr{X}_r^{(\kappa)}\big]^{-1}\big(\big[\widehat{\Sigma}_r^{(\kappa)}\big]^{-1} - \big[\overline{\Sigma}_r^{(\kappa)}\big]^{-1}\big)\right) - |C_r|\big(\det \mathscr{X}_r^{(\kappa)}\big)^2$$

$$\times \left(2\big[\Delta\mu_r^{(\kappa)}\big]^T\big[\widehat{\Sigma}_r^{(\kappa)}\big]^{-1}\widehat{\mu}_r^{(\kappa)} - 2\big[\Delta\mu_r^{(\kappa)}\big]^T\big[\overline{\Sigma}_r^{(\kappa)}\big]^{-1}\overline{\mu}_r^{(\kappa)}\right.$$

$$+ \big[\Delta\mu_r^{(\kappa)}\big]^T\big(\big[\overline{\Sigma}_r^{(\kappa)}\big]^{-1} - \big[\widehat{\Sigma}_r^{(\kappa)}\big]^{-1}\big)\Delta\mu_r^{(\kappa)} - \big[\widehat{\mu}_r^{(\kappa)}\big]^T\big[\widehat{\Sigma}_r^{(\kappa)}\big]^{-1}\widehat{\mu}_r^{(\kappa)}$$

$$+ \big[\overline{\mu}_r^{(\kappa)}\big]^T\big[\overline{\Sigma}_r^{(\kappa)}\big]^{-1}\overline{\mu}_r^{(\kappa)}\right) = 0.$$

Similarly, the fusion $\Delta E_s^{(\kappa)} = \widehat{E}_s^{(\kappa)} \cap \overline{E}_s^{(\kappa)}$ is an ellipsoid $\mathscr{E}\big(\Delta\rho_s^{(\kappa)}, \Delta\Pi_s^{(\kappa)}\big)$ with center

$$\Delta\rho_s^{(\kappa)} = \big[\mathscr{Y}_s^{(\kappa)}\big]^{-1}\big(\lambda\big[\widehat{\Pi}_s^{(\kappa)}\big]^{-1}\widehat{\rho}_s^{(\kappa)} + (1-\lambda)\big[\overline{\Pi}_s^{(\kappa)}\big]^{-1}\overline{\rho}_s^{(\kappa)}\big)$$

and shape matrix

$$\Delta\Pi_s^{(\kappa)} = \eta_s^{(\kappa)}\big[\mathscr{Y}_s^{(\kappa)}\big]^{-1},$$

where

$$\mathscr{Y}_s^{(\kappa)} := \lambda\big[\widehat{\Pi}_s^{(\kappa)}\big]^{-1} + (1-\lambda)\big[\overline{\Pi}_s^{(\kappa)}\big]^{-1}$$

and

$$\eta_s^{(\kappa)} := 1 - \lambda(1-\lambda)\big(\overline{\rho}_s^{(\kappa)} - \widehat{\rho}_s^{(\kappa)}\big)^T\big[\overline{\Pi}_s^{(\kappa)}\big]^{-1}\big[\mathscr{Y}_s^{(\kappa)}\big]^{-1}\big[\widehat{\Pi}_s^{(\kappa)}\big]^{-1}\big(\overline{\rho}_s^{(\kappa)} - \widehat{\rho}_s^{(\kappa)}\big).$$

The parameter λ is the only root in $(0,1)$ of the following polynomial of degree $2|D_s| - 1$:

$$\eta_s^{(\kappa)}\big(\det \mathscr{Y}_s^{(\kappa)}\big)^2 \operatorname{Tr}\left(\big[\mathscr{Y}_s^{(\kappa)}\big]^{-1}\big(\big[\widehat{\Pi}_s^{(\kappa)}\big]^{-1} - \big[\overline{\Pi}_s^{(\kappa)}\big]^{-1}\big)\right) - |D_s|\big(\det \mathscr{Y}_s^{(\kappa)}\big)^2$$

$$\times \left(2\big[\Delta\rho_s^{(\kappa)}\big]^T\big[\widehat{\Pi}_s^{(\kappa)}\big]^{-1}\widehat{\rho}_s^{(\kappa)} - 2\big[\Delta\rho_s^{(\kappa)}\big]^T\big[\overline{\Pi}_s^{(\kappa)}\big]^{-1}\overline{\rho}_s^{(\kappa)}\right.$$

$$+ \big[\Delta\rho_s^{(\kappa)}\big]^T\big(\big[\overline{\Pi}_s^{(\kappa)}\big]^{-1} - \big[\widehat{\Pi}_s^{(\kappa)}\big]^{-1}\big)\Delta\rho_s^{(\kappa)} - \big[\widehat{\rho}_s^{(\kappa)}\big]^T\big[\widehat{\Pi}_s^{(\kappa)}\big]^{-1}\widehat{\rho}_s^{(\kappa)}$$

$$+ \big[\overline{\rho}_s^{(\kappa)}\big]^T\big[\overline{\Pi}_s^{(\kappa)}\big]^{-1}\overline{\rho}_s^{(\kappa)}\right) = 0.$$

35 Dynamical Gene-Environment Networks Under Ellipsoidal Uncertainty

35.4.2.1 The Trace Criterion

We now turn to the specific formulations of the regression problem (R). The first criterion is based on the *traces of the shape matrices* of the fusions $\Delta X_r^{(\kappa)}$ and $\Delta E_s^{(\kappa)}$. The geometrical size of these ellipsoids is measured in terms of their (squared) lengths of semiaxes and, thus, the traces of the shape matrices $\Delta \Sigma_r^{(\kappa)}$ and $\Delta \Pi_s^{(\kappa)}$:

$$
(R_{Tr}) \qquad \text{Maximize} \qquad \sum_{\kappa=1}^{T} \left\{ \sum_{r=1}^{R} \left[\mathrm{Tr}\left(\Delta \Sigma_r^{(\kappa)}\right) - \sum_{j=1}^{|C_r|} \left(\widehat{\mu}_{r,j}^{(\kappa)} - \overline{\mu}_{r,j}^{(\kappa)}\right)^2 \right] \right.
$$
$$
\left. + \sum_{s=1}^{S} \left[\mathrm{Tr}\left(\Delta \Pi_s^{(\kappa)}\right) - \sum_{i=1}^{|D_s|} \left(\widehat{\rho}_{s,i}^{(\kappa)} - \overline{\rho}_{s,i}^{(\kappa)}\right)^2 \right] \right\}.
$$

The trace of the shape matrix of an ellipsoid is equal to the sum of the squares of the semiaxes. For this reason, the regression problem takes the form

$$
(R'_{Tr}) \qquad \text{Maximize} \qquad \sum_{\kappa=1}^{T} \left\{ \sum_{r=1}^{R} \sum_{j=1}^{|C_r|} \left[\lambda_{r,j}^{(\kappa)} - \left(\widehat{\mu}_{r,j}^{(\kappa)} - \overline{\mu}_{r,j}^{(\kappa)}\right)^2 \right] \right.
$$
$$
\left. + \sum_{s=1}^{S} \sum_{i=1}^{|D_s|} \left[\Lambda_{s,i}^{(\kappa)} - \left(\widehat{\rho}_{s,i}^{(\kappa)} - \overline{\rho}_{s,i}^{(\kappa)}\right)^2 \right] \right\},
$$

where $\lambda_{r,j}^{(\kappa)}$ and $\Lambda_{s,i}^{(\kappa)}$ are the eigenvalues of $\Delta \Sigma_r^{(\kappa)}$ and $\Delta \Pi_s^{(\kappa)}$, respectively.

35.4.2.2 The Trace of the Square Criterion

When we measure the size of an ellipsoid by the *traces of the square* of its shape matrix, we obtain the following regression problem:

$$
(R_{TS}) \qquad \text{Maximize} \qquad \sum_{\kappa=0}^{T} \left\{ \sum_{r=1}^{R} \left[\mathrm{Tr}\left(\Delta \Sigma_r^{(\kappa)}\right)^2 - \sum_{j=1}^{|C_r|} \left(\widehat{\mu}_{r,j}^{(\kappa)} - \overline{\mu}_{r,j}^{(\kappa)}\right)^2 \right] \right.
$$
$$
\left. + \sum_{s=1}^{S} \left[\mathrm{Tr}\left(\Delta \Pi_s^{(\kappa)}\right)^2 - \sum_{i=1}^{|D_s|} \left(\widehat{\rho}_{s,i}^{(\kappa)} - \overline{\rho}_{s,i}^{(\kappa)}\right)^2 \right] \right\}.
$$

35.4.2.3 The Diameter Criterion

The maximal extension of the fusions can be used to define a further regression model. Here, the *diameter* of the ellipsoids $\Delta X_r^{(\kappa)}$ and $\Delta E_s^{(\kappa)}$ (or the size of the smallest balls which include the fusions) is used in the objective function:

(R_{Dia}) Maximize

$$\sum_{\kappa=0}^{T} \left\{ \sum_{r=1}^{R} \left[\text{diam}\big(\mathscr{E}\big(0, \Delta\Sigma_r^{(\kappa)}\big)\big) - \sum_{j=1}^{|C_r|} \big(\widehat{\mu}_{r,j}^{(\kappa)} - \overline{\mu}_{r,j}^{(\kappa)}\big)^2 \right] \right.$$
$$\left. + \sum_{s=1}^{S} \left[\text{diam}\big(\mathscr{E}\big(0, \Delta\Pi_s^{(\kappa)}\big)\big) - \sum_{i=1}^{|D_s|} \big(\widehat{\rho}_{s,i}^{(\kappa)} - \overline{\rho}_{s,i}^{(\kappa)}\big)^2 \right] \right\}.$$

An equivalent formulation of (R_{Dia}) can be given in terms of the eigenvalues of $\Delta\Sigma_r^{(\kappa)}$ and $\Delta\Pi_s^{(\kappa)}$:

(R'_{Dia}) Maximize

$$\sum_{\kappa=1}^{T} \left\{ \sum_{r=1}^{R} \left[2 \cdot \sqrt{\lambda_r^{(\kappa)}} - \sum_{j=1}^{|C_r|} \big(\widehat{\mu}_{r,j}^{(\kappa)} - \overline{\mu}_{r,j}^{(\kappa)}\big)^2 \right] \right.$$
$$\left. + \sum_{s=1}^{S} \left[2 \cdot \sqrt{\Lambda_s^{(\kappa)}} - \sum_{i=1}^{|D_s|} \big(\widehat{\rho}_{s,i}^{(\kappa)} - \overline{\rho}_{s,i}^{(\kappa)}\big)^2 \right] \right\}$$

with $\lambda_r^{(\kappa)} := \max\{\lambda_{r,j}^{(\kappa)} \mid j = 1, \ldots, |C_r|\}$ and $\Lambda_s^{(\kappa)} := \max\{\Lambda_{s,i}^{(\kappa)} \mid i = 1, \ldots, |D_s|\}$. As the objective function of (R'_{Dia}) is nonsmooth with well-understood max-type functions [60–62] but not Lipschitz-continuous, we also introduce the additional regression problem

(R''_{Dia}) Maximize

$$\sum_{\kappa=0}^{T} \left\{ \sum_{r=1}^{R} \left[\lambda_r^{(\kappa)} - \sum_{j=1}^{|C_r|} \big(\widehat{\mu}_{r,j}^{(\kappa)} - \overline{\mu}_{r,j}^{(\kappa)}\big)^2 \right] \right.$$
$$\left. + \sum_{s=1}^{S} \left[\Lambda_s^{(\kappa)} - \sum_{i=1}^{|D_s|} \big(\widehat{\rho}_{s,i}^{(\kappa)} - \overline{\rho}_{s,i}^{(\kappa)}\big)^2 \right] \right\}$$

as an alternative proposal.

35.4.3 Optimization Methods

In this section, we summarize solution methods for the regression models of the previous subsections. The objective functions of these volume-related programming problems depend on, e.g., the eigenvalues of symmetric positive semidefinite shape matrices $\Delta\Sigma_r^{(\kappa)}$ and $\Delta\Pi_s^{(\kappa)}$ as well as the distance of the centers $\Delta\mu_r^{(\kappa)}$ and $\Delta\rho_s^{(\kappa)}$ of the fusions $\Delta X_r^{(\kappa)}$ and $\Delta E_s^{(\kappa)}$. For this reason, methods from semidefinite programming [11] can be applied. In particular, the regression model (R'_{Tr}) refers to sums of all eigenvalues of the shape matrices $\Delta\Sigma_r^{(\kappa)}$ and $\Delta\Pi_s^{(\kappa)}$. These objective functions can also be considered as positive semidefinite representable functions ([8], p. 80) and *interior point methods* can applied [35–37, 39]. Alternatively, associated *bilevel problems* can be introduced which could be solved by *gradient methods*. In fact,

35 Dynamical Gene-Environment Networks Under Ellipsoidal Uncertainty

in [38] structural frontiers of conic programming are discussed with other optimization methods compared, and future applications in machine learning and data mining prepared. However, we would like to underline that in the areas regression and classification of statistical learning (cf. e.g., [21, 46]), our optimization based methods provided and further promise very good and competitive results [13, 24, 52, 66, 74].

35.5 Network Rarefication Based on Mixed Integer Programming and Continuous Programming

The gene-environment network of clusters defined by the interaction matrices of the linear model (EC) is usually highly-interconnected. The regression analysis of the previous sections allows an identification of the corresponding degrees of connectivity of all branches between the clusters, although its actual influence is small and negligible. In addition, a particular cluster is generally influenced by a limited (not too high) number of regulating genetic and environmental clusters and usually it regulates only a small number of clusters. For this reason, we introduce a method for diminishing the number of branches with the aim of network rarefication during the regression process. This goal could be achieved by introducing upper bounds on the indegrees and outdegrees of nodes of the gene-environment network. In other words, the number of clusters regulating a specific genetic or environmental cluster in our network as well as the number of clusters regulated by a particular cluster has to be bounded. We impose *binary constraints* in order to decide whether or not there is a connection between two clusters and by this we obtain a *mixed-integer optimization problem*. After this model is provided, we will pass to *continuous optimization* and introduce a model with *continuous constraints*. This is because of the exclusive nature of binary constraints that could even destroy the connectivity of the gene-environment network.

35.5.1 Mixed Integer Regression Problem

Given two clusters A, B we use the notation $A \leftarrow B$ if cluster A is regulated by cluster B and $A \not\leftarrow B$ if cluster A is not regulated by cluster B. Now, we define the Boolean matrices

$$\chi_{jr}^{GG} = \begin{cases} 1, & \text{if } C_j \leftarrow C_r \\ 0, & \text{if } C_j \not\leftarrow C_r, \end{cases} \qquad \chi_{js}^{EG} = \begin{cases} 1, & \text{if } C_j \leftarrow D_s \\ 0, & \text{if } C_j \not\leftarrow D_s, \end{cases}$$

$$\chi_{ir}^{GE} = \begin{cases} 1, & \text{if } D_i \leftarrow C_r \\ 0, & \text{if } D_i \not\leftarrow C_r, \end{cases} \qquad \chi_{is}^{EE} = \begin{cases} 1, & \text{if } D_i \leftarrow D_s \\ 0, & \text{if } D_i \not\leftarrow D_s, \end{cases}$$

indicating whether or not pairs of clusters in our regulatory network are directly related. If two clusters are not related, the corresponding parts of the matrices A^{GG}, A^{GE}, A^{EG}, A^{EE} have zero entries in case of a disjoint cluster decomposition.

The *indegree* of the genetic cluster C_j in our regulatory network is defined with respect to all genetic and environmental clusters by

$$\deg(C_j)_{in}^{GG} := \sum_{r=1}^{R} \chi_{jr}^{GG} \quad \text{and} \quad \deg(C_j)_{in}^{EG} := \sum_{s=1}^{S} \chi_{js}^{EG},$$

where $j \in \{1, \ldots, R\}$. By this, the indegrees $\deg(C_j)^{GG}$ and $\deg(C_j)^{EG}$ count the number of genetic and environmental clusters which regulate cluster C_j. The *overall indegree* of the genetic cluster C_j is defined by

$$\deg(C_j)_{in} := \deg(C_j)_{in}^{GG} + \deg(C_j)_{in}^{EG}.$$

Similarly, for $i \in \{1, \ldots, S\}$ the *indegree* of cluster D_i with respect to the environmental clusters and the genetic clusters is given by

$$\deg(D_i)_{in}^{GE} := \sum_{r=1}^{R} \chi_{ir}^{GE} \quad \text{and} \quad \deg(D_i)_{in}^{EE} := \sum_{s=1}^{S} \chi_{is}^{EE}.$$

In this way, the indegrees $\deg(D_i)^{GE}$ and $\deg(D_i)^{EE}$ count the number of genetic and environmental clusters which regulate cluster D_i. The *overall indegree* of cluster D_i is defined by

$$\deg(D_i)_{in} := \deg(D_i)_{in}^{GE} + \deg(D_i)_{in}^{EE}.$$

In the same way, bounds on the outdegree, i.e., the number of outgoing branches can be introduced. Firstly, binary values are defined to determine whether or not there is an outgoing connection:

$$\zeta_{jr}^{GG} = \begin{cases} 1, & \text{if } C_r \leftarrow C_j \\ 0, & \text{if } C_r \not\leftarrow C_j, \end{cases} \qquad \zeta_{js}^{EG} = \begin{cases} 1, & \text{if } D_s \leftarrow C_j \\ 0, & \text{if } D_s \not\leftarrow C_j, \end{cases}$$

$$\zeta_{ir}^{GE} = \begin{cases} 1, & \text{if } C_r \leftarrow D_i \\ 0, & \text{if } C_r \not\leftarrow D_i, \end{cases} \qquad \zeta_{is}^{EE} = \begin{cases} 1, & \text{if } D_s \leftarrow D_i \\ 0, & \text{if } D_s \not\leftarrow D_i. \end{cases}$$

Now, the *outdegree* of genetic cluster C_j with respect to all genetic and environmental clusters can be expressed as

$$\deg(C_j)_{out}^{GG} := \sum_{r=1}^{R} \zeta_{jr}^{GG} \quad \text{and} \quad \deg(C_j)_{out}^{GE} := \sum_{s=1}^{S} \zeta_{js}^{GE},$$

where $j \in \{1, \ldots, R\}$. The outdegrees $\deg(C_j)_{out}^{GG}$ and $\deg(C_j)_{out}^{EG}$ count the number of genetic and environmental clusters *regulated by* cluster C_j. The *overall outdegree* of genetic cluster C_j is given by

$$\deg(C_j)_{out} := \deg(C_j)_{out}^{GG} + \deg(C_j)_{out}^{EG}.$$

The *outdegree* of environmental cluster D_i with respect to the environmental clusters and the genetic clusters is defined by

$$\deg(D_i)_{out}^{GE} := \sum_{r=1}^{R} \zeta_{ir}^{GE} \quad \text{and} \quad \deg(D_i)_{out}^{EE} := \sum_{s=1}^{S} \zeta_{is}^{EE},$$

where $i \in \{1, \ldots, S\}$. The outdegrees $\deg(D_i)_{out}^{GE}$ and $\deg(D_i)_{out}^{EE}$ count the number of genetic and environmental clusters *regulated by* cluster D_i. The *overall outdegree* of cluster D_i is

$$\deg(D_i)_{out} := \deg(D_i)_{out}^{GE} + \deg(D_i)_{out}^{EE}.$$

As mentioned above, we introduce upper bounds on the indegrees and the outdegrees of the nodes (clusters) with the aim of network rarefication. These values have to be given by the practitioner and they can depend on any a priori information. Including these additional constraints, we obtain the following *mixed integer optimization problem*:

$$
(MI1) \left\{
\begin{aligned}
\text{Maximize} \quad & \sum_{\kappa=1}^{T} \left\{ \sum_{r=1}^{R} \left\| \Delta X_r^{(\kappa)} \right\|_* - \left\| \widehat{\mu}_r^{(\kappa)} - \overline{\mu}_r^{(\kappa)} \right\|_2^2 \right. \\
& \left. + \sum_{s=1}^{S} \left\| \Delta E_s^{(\kappa)} \right\|_* - \left\| \widehat{\rho}_s^{(\kappa)} - \overline{\rho}_s^{(\kappa)} \right\|_2^2 \right\} \\[1em]
\text{subject to} \quad & \deg(C_j)_{in}^{GG} \leq \alpha_j^{GG}, \ j = 1, \ldots, R \\
& \deg(C_j)_{in}^{EG} \leq \alpha_j^{EG}, \ j = 1, \ldots, R \\
& \deg(D_i)_{in}^{EG} \leq \alpha_i^{EG}, \ i = 1, \ldots, S \\
& \deg(D_i)_{in}^{EE} \leq \alpha_i^{EE}, \ i = 1, \ldots, S \\[1em]
& \deg(C_j)_{out}^{GG} \leq \beta_j^{GG}, \ j = 1, \ldots, R \\
& \deg(C_j)_{out}^{EG} \leq \beta_j^{EG}, \ j = 1, \ldots, R \\
& \deg(D_i)_{out}^{GE} \leq \beta_i^{GE}, \ i = 1, \ldots, S \\
& \deg(D_i)_{out}^{EE} \leq \beta_i^{EE}, \ i = 1, \ldots, S.
\end{aligned}
\right.
$$

In model *(MI1)*, individual bounds on the indegrees and outdegrees of each genetic and environmental cluster are imposed what allows us to control the connectivity of the gene-environment network. This approach is an extension of the

564 E. Kropat et al.

regression problems with bounds on the indegrees in [27, 28]. Similar mixed-integer problems for an analysis of gene-environment networks based on interval arithmetics were presented in [64, 65, 67].

In model *(MI1)* the number of connections of each cluster with genetic and environmental clusters are considered separately. In a further step, we can combine these bounds and impose restrictions on the total number of all ingoing or outgoing branches of each cluster:

$$
(MI2)\begin{cases}
\text{Maximize} & \displaystyle\sum_{\kappa=1}^{T}\left\{\sum_{r=1}^{R}\left\|\Delta X_r^{(\kappa)}\right\|_* - \left\|\widehat{\mu}_r^{(\kappa)} - \overline{\mu}_r^{(\kappa)}\right\|_2^2 \right. \\
& \left. + \displaystyle\sum_{s=1}^{S}\left\|\Delta E_s^{(\kappa)}\right\|_* - \left\|\widehat{\rho}_s^{(\kappa)} - \overline{\rho}_s^{(\kappa)}\right\|_2^2\right\} \\[2ex]
\text{subject to} & \deg(C_j)_{in} \leq \gamma_j, \ j = 1, \ldots, R \\[1ex]
& \deg(D_i)_{in} \leq \delta_i, \ i = 1, \ldots, S \\[2ex]
& \deg(C_j)_{out} \leq \varepsilon_j, \ j = 1, \ldots, R \\[1ex]
& \deg(D_i)_{out} \leq \varphi_i, \ i = 1, \ldots, S.
\end{cases}
$$

35.5.2 Continuous Programming

As mentioned above, the binary constraints in *(MI1)* and *(MI2)* can lead to restrictions of the network connectivity. For this reason, continuous optimization is applied for a relaxation of *(MI1)* by replacing the binary variables χ_{jr}^{GG}, χ_{js}^{EG}, χ_{ir}^{GE} and χ_{is}^{EE} with real variables P_{jr}^{GG}, P_{js}^{EG}, P_{ir}^{GE}, $P_{is}^{EE} \in [0, 1]$, which is also interpretable as probabilities (we refer to [47] for optimization models with probabilistic constraints). These variables should linearly depend on the corresponding elements of $\mathscr{A}_{jr}^{GG}$, $\mathscr{A}_{js}^{EG}$, $\mathscr{A}_{ir}^{GE}$, $\mathscr{A}_{is}^{EE}$.

The real-valued *indegree* of cluster C_j in our regulatory network with respect to the genetic and environmental clusters are now defined by

$$
\deg(C_j)_{in}^{GG} := \sum_{r=1}^{R} P_{jr}^{GG}\left(\mathscr{A}_{jr}^{GG}\right) \quad \text{and} \quad \deg(C_j)_{in}^{EG} := \sum_{s=1}^{S} P_{js}^{EG}\left(\mathscr{A}_{js}^{EG}\right),
$$

respectively. Similarly, the real-valued *indegree* of cluster D_i is given by

$$
\deg(D_i)_{in}^{GE} := \sum_{r=1}^{R} P_{ir}^{GE}\left(\mathscr{A}_{ir}^{GE}\right) \quad \text{and} \quad \deg(D_i)_{in}^{EE} := \sum_{s=1}^{S} P_{is}^{EE}\left(\mathscr{A}_{is}^{EE}\right).
$$

35 Dynamical Gene-Environment Networks Under Ellipsoidal Uncertainty

In the same way, we can adapt the outdegrees of clusters by replacing the binary variables ζ_{jr}^{GG}, ζ_{js}^{EG}, ζ_{ir}^{GE} and ζ_{is}^{EE} with real variables Q_{jr}^{GG}, Q_{js}^{EG}, Q_{ir}^{GE}, $Q_{is}^{EE} \in [0,1]$ linearly depending on the corresponding elements of $\mathscr{A}_{jr}^{GG}$, $\mathscr{A}_{js}^{EG}$, $\mathscr{A}_{ir}^{GE}$, $\mathscr{A}_{is}^{EE}$. Now, the real-valued *outdegrees* of cluster C_j with respect to the genetic and environmental clusters are defined by

$$\deg(C_j)_{out}^{GG} := \sum_{r=1}^{R} Q_{jr}^{GG}(\mathscr{A}_{jr}^{GG}) \quad \text{and} \quad \deg(C_j)_{out}^{EG} := \sum_{s=1}^{S} Q_{js}^{EG}(\mathscr{A}_{js}^{EG}),$$

respectively. Similarly, the real-valued *outdegree* of cluster D_i is given by

$$\deg(D_i)_{out}^{GE} := \sum_{r=1}^{R} P_{ir}^{GE}(\mathscr{A}_{ir}^{GE}) \quad \text{and} \quad \deg(D_i)_{out}^{EE} := \sum_{s=1}^{S} Q_{is}^{EE}(\mathscr{A}_{is}^{EE}).$$

When we replace the strict binary constraints of the mixed-integer problem *(MI1)* with the aforementioned 'soft constraints', we obtain the following *continuous programming problem*:

$$(C1) \begin{cases} \text{Maximize} & \sum_{\kappa=1}^{T} \left\{ \sum_{r=1}^{R} \left\| \Delta X_r^{(\kappa)} \right\|_* - \left\| \widehat{\mu}_r^{(\kappa)} - \overline{\mu}_r^{(\kappa)} \right\|_2^2 \right. \\ & \left. + \sum_{s=1}^{S} \left\| \Delta E_s^{(\kappa)} \right\|_* - \left\| \widehat{\rho}_s^{(\kappa)} - \overline{\rho}_s^{(\kappa)} \right\|_2^2 \right\} \\ \\ \text{subject to} & \sum_{r=1}^{R} P_{jr}^{GG}(\mathscr{A}_{jr}^{GG}) \leq \alpha_j^{GG}, \ j = 1, \dots, R \\ & \sum_{s=1}^{S} P_{js}^{EG}(\mathscr{A}_{js}^{EG}) \leq \alpha_j^{EG}, \ j = 1, \dots, R \\ & \sum_{r=1}^{R} P_{ir}^{GE}(\mathscr{A}_{ir}^{GE}) \leq \alpha_i^{GE}, \ i = 1, \dots, S \\ & \sum_{s=1}^{S} P_{is}^{EE}(\mathscr{A}_{is}^{EE}) \leq \alpha_i^{EE}, \ i = 1, \dots, S \\ \\ & \sum_{r=1}^{R} Q_{jr}^{GG}(\mathscr{A}_{jr}^{GG}) \leq \beta_j^{GG}, \ j = 1, \dots, R \\ & \sum_{s=1}^{S} Q_{js}^{EG}(\mathscr{A}_{js}^{EG}) \leq \beta_j^{EG}, \ j = 1, \dots, R \\ & \sum_{r=1}^{R} Q_{ir}^{GE}(\mathscr{A}_{ir}^{GE}) \leq \beta_i^{GE}, \ i = 1, \dots, S \\ & \sum_{s=1}^{S} Q_{is}^{EE}(\mathscr{A}_{is}^{EE}) \leq \beta_i^{EE}, \ i = 1, \dots, S. \end{cases}$$

We can also combine these real-valued restrictions on the total number of all ingoing or outgoing branches of each cluster and so we obtain a relaxation of model *(M2)* in terms of the following *continuous programming problem*:

$$(C2) \begin{cases} \text{Maximize} & \sum_{\kappa=1}^{T} \left\{ \sum_{r=1}^{R} \left\| \Delta X_r^{(\kappa)} \right\|_* - \left\| \widehat{\mu}_r^{(\kappa)} - \overline{\mu}_r^{(\kappa)} \right\|_2^2 \right. \\ & \quad \left. + \sum_{s=1}^{S} \left\| \Delta E_s^{(\kappa)} \right\|_* - \left\| \widehat{\rho}_s^{(\kappa)} - \overline{\rho}_s^{(\kappa)} \right\|_2^2 \right\} \\ \\ \text{subject to} & \sum_{r=1}^{R} P_{jr}^{GG}\left(\mathscr{A}_{jr}^{GG}\right) + \sum_{s=1}^{S} P_{js}^{EG}\left(\mathscr{A}_{js}^{EG}\right) \leq \gamma_j, \ j = 1, \ldots, R \\ & \sum_{r=1}^{R} P_{ir}^{GE}\left(\mathscr{A}_{ir}^{GE}\right) + \sum_{s=1}^{S} P_{is}^{EE}\left(\mathscr{A}_{is}^{EE}\right) \leq \beta_i, \ i = 1, \ldots, S \\ \\ & \sum_{r=1}^{R} Q_{jr}^{GG}\left(\mathscr{A}_{jr}^{GG}\right) + \sum_{s=1}^{S} Q_{js}^{EG}\left(\mathscr{A}_{js}^{EG}\right) \leq \gamma_j, \ j = 1, \ldots, R \\ & \sum_{r=1}^{R} Q_{ir}^{GE}\left(\mathscr{A}_{ir}^{GE}\right) + \sum_{s=1}^{S} Q_{is}^{EE}\left(\mathscr{A}_{is}^{EE}\right) \leq \delta_i, \ i = 1, \ldots, S. \end{cases}$$

35.6 Conclusion

We considered dynamical gene-environment networks under ellipsoidal uncertainty. The hidden interconnections and regulating effects of possibly overlapping clusters of genetic and environmental items are revealed by a new set-theoretic regression methodology. By this, we further extend our *Ellipsoidal Operations Research* introduced in [27] that was based on our and our colleagues studies on regulatory systems in computational biology and life sciences with data uncertainty. In these studies, interval arithmetics was applied to express various kinds of errors and uncertainty. Here, we further extend this approach by considering stochastic dependencies between groups of genes and environmental factors. Furthermore, the clusters of data items are not strictly divided as in [27] and they can overlap what is motivated by the fact that a single gene or environmental factor can influence more than one other gene. The representation of data uncertainty in terms of ellipsoids is more flexible than the error intervals of single variables. In particular, ellipsoids are directly related to covariance matrices. For this reason, the often used Gaussian random noise refers to our ellipsoidal approach. However, Gaussian random distributions are often considered as simplifications. In future works, we will extend our regression models based on ellipsoidal uncertainty and we will focus on set-theoretic approaches with semi-algebraic sets and approximations of convex or

35 Dynamical Gene-Environment Networks Under Ellipsoidal Uncertainty

non-convex error sets [7,9]. This new perception will be combined with refined optimization methods that offer a new perspective for the analysis of regulatory systems under uncertainty.

Acknowledgements The authors express their cordial gratitude to Professor Alberto Pinto for inviting us to contribute with our chapter to this distinguished book in honour of Prof. Dr. Mauricio Peixoto and Prof. Dr. David Rand.

References

1. Alparslan Gök, S.Z.: Cooperative interval games. PhD Thesis, Institute of Applied Mathematics, Middle East Technical University, Ankara, Turkey (2009)
2. Alparslan Gök, S.Z., Branzei, R., Tijs, S.: Convex interval games. Preprint at IAM, Middle East Technical University, Ankara, Turkey, and Center for Economic Research, Tilburg University, The Netherlands (2008)
3. Alparslan Gök, S.Z., Branzei, R., Tijs, S.: Airport interval games and their Shapley value. Operations Research and Decisions, **2**, 9–18 (2009)
4. Alparslan Gök, S.Z., Miquel, S., Tijs, S.: Cooperation under interval uncertainty. Math. Methods Oper. Res. **69**, 99–109 (2009)
5. Alparslan Gök, S.Z., Weber, G.-W.: Cooperative games under ellipsoidal uncertainty. In: The Proceedings of PCO 2010, 3rd Global Conference on Power Control and Optimization, Gold Coast, Queensland, Australia, Feb 2–4, 2010 (ISBN: 978-983-444483-1-8)
6. Alter, O., Brown, P.O., Botstein, D.: Singular value decomposition for genome-wide expression data processing and modeling. PNAS. **97**(18), 10101–10106 (2000)
7. Benedetti, R.: Real algebraic and semi-algebraic sets. Hermann, Ed. des Sciences et des Arts, Paris (1990)
8. Ben-Tal, A.: Conic and robust optimization. Lecture notes (2002)
   Available at http://iew3.technion.ac.il/Home/Users/morbt.phtml.
9. Bochnak, J., Coste, M., Roy, M.-F.: Real algebraic geometry. Springer, New York (1998)
10. Borenstein, E., Feldman, M.W.: Topological signatures of species interactions in metabolic networks. J. Comput. Biol. **16**(2), 191–200 (2009). doi: 10.1089/cmb.2008.06TT
11. Boyd, S., Vandenberghe, L.: Convex optimization. Cambridge University Press, Cambridge (2004)
12. Brown, M.P.S., Grundy, W.N., Lin, D., Cristianini, N., Sugnet, C.W., Furey, T.S., Ares, M., Haussler, D.: Knowledge-based analysis of microarray gene expression data by using support vector machines. PNAS **97**(1), 262–267 (2000)
13. Büyükbebeci, E.: Comparison of MARS, CMARS and CART in predicting default probabilities for emerging markets. MSc. Term Project Report/Thesis in Financial Mathematics, at IAM, METU, Ankara, August 2009
14. Chipman, H., Tibshirani, R.: Hybrid hierarchical clustering with applications to microarray data. Biostatistics **7**(2), 286–301 (2006)
15. Mol, C. De, Mosci, S., Traskine, M., Verri, A.: A Regularized Method for selecting nested groups of relevant genes from microarray data. J. Comput. Biol. **16**(5), 677–690 (2009)
16. Defterli, Ö., Fügenschuh, A., Weber, G.-W.: New discretization and optimization techniques with results in the dynamics of gene-environment networks. In: The proceedings of PCO 2010, 3rd Global Conference on Power Control and Optimization, Gold Coast, Queensland, Australia, Feb 2–4, 2010 (ISBN: 978-983-44483-1-8)
17. Durieu, P., Walter, É., Polyak, B.: Multi-input multi-output ellipsoidal state bounding. J. Optim. Theory Appl. **111**(2), 273–303 (2001)
18. Gebert, J., Lätsch, M., Quek, E.M.P., Weber, G.-W.: Analyzing and optimizing genetic network structure via path-finding. J. Comput. Technol. **9**(3), 3–12 (2004)

19. Gökmen, A., Kayalgil, S., Weber, G.-W., Gökmen, I., Ecevit, M., Sürmeli, A., Bali, T., Ecevit, Y., Gökmen, H., DeTombe, D.J.: Balaban valley project: improving the quality of life in rural area in Turkey. Int. Sci. J. Methods Models Complex. **7**(1) (2004)
20. Harris, J.R., Nystad, W., Magnus, P.: Using genes and environments to define asthma and related phenotypes: applications to multivariate data. Clin. Exp. Allergy **28**(1), 43–45 (1998)
21. Hastie, T., Tibshirani, R., Friedman, J.: The elements of statistical learning. Springer, New York (2001)
22. Herrero, J., Valencia, A., Dopazo, J.: A hierarchical unsupervised growing neural network for clustering gene expression patterns. Bioinformatics **17**(2), 126–136 (2001)
23. Hooper, S.D., Boué, S., Krause, R., Jensen, L.J., Mason, C.E., Ghanim, M., White, K.P., Furlong, E.E.M., Bork, P.: Identification of tightly regulated groups of genes during Drosophila melanogaster embryogenesis. Mol. Syst. Biol. **3**, 72 (2007)
24. Işcanoğlu, A., Weber, G.-W., Taylan, P.: Predicting default probabilities with generalized additive models for emerging markets. Invited lecture, Graduate Summer School on New Advances in Statistics, METU (2007)
25. Kropat, E., Pickl, S., Rössler, A., Weber, G.-W.: On theoretical and practical relations between discrete optimization and nonlinear optimization. In: Special issue Colloquy Optimization – Structure and Stability of Dynamical Systems (at the occasion of the colloquy with the same name, Cologne, October 2000) of Journal of Computational Technologies, vol. 7, pp. 27–62 (2002)
26. Kropat, E., Weber, G.-W., Akteke-Öztürk, B.: Eco-finance networks under uncertainty. In: Herskovits, J., Canelas, A., Cortes, H., Aroztegui, M. (eds.) Proceedings of the International Conference on Engineering Optimization (ISBN 978857650156-5, CD), EngOpt 2008, Rio de Janeiro, Brazil (2008)
27. Kropat, E., Weber, G.-W., Rckmann, J.-J.: Regression analysis for clusters in gene-environment networks based on ellipsoidal calculus and optimization. Dynamics of Continuous, Discrete and Impulsive Systems, Series B: Applications & Algorithms 17(5), 639-657 (2010). In the special issue in honour of Professor Alexander Rubinov.
28. Kropat, E., Weber, G.-W., Pedamallu, C.S.: Regulatory networks under ellipsoidal uncertainty optimization theory and dynamical systems. To appear in the book on "Data Mining", D. Holmes, ed., Springer.
29. Kurzhanski, A.B., Vályi, I.: Ellipsoidal Calculus for Estimation and Control. Birkhäuser, Boston (1997)
30. Kurzhanski, A.A., Varaiya, P.: Ellipsoidal Toolbox Manual. EECS Department, University of California, Berkeley (2008)
31. Lee, Y., Lin, Y., Wahba, G.: Multicategory support vector machines: theory and application to the classification of microarray data and satellite radiance data. J. Am. Stat. Assoc. **99**, 67–81 (2004)
32. Mahony, S., McInerney, J.O., Smith, T.J., Golden, A.: Gene prediction using the self-organizing map: automatic generation of multiple gene models. BMC Bioinform **5**, 23 (2004). doi:10.1186/1471-2105-5-23
33. Marvanova, M., Toronen, P., Storvik, M., Lakso, M., Castren, E., Wong, G.: Synexpression analysis of ESTs in the rat brain reveals distinct patterns and potential drug targets. Mol. Brain Res. **104**(2), 176–183 (2002)
34. Mattes, W.B., Pettit, S.D., Sansone, S.A., Bushel, P.R., Waters, M.D.: Database development in toxicogenomics: issues and efforts. Environ. Health Perspect. **112**(4), 495–505 (2004)
35. Nemirovski, A.: Five lectures on modern convex optimization. C.O.R.E. Summer School on Modern Convex Optimization (2002). Available at http://iew3.technion.ac.il/Labs/Opt/opt/LN/Final.pdf
36. Nemirovski, A.: Lectures on modern convex optimization, Israel Institute of Technology (2002). Available at http://iew3.technion.ac.il/Labs/Opt/opt/LN/Final.pdf
37. Nemirovski, A.: Interior point polynomial time algorithms in convex programming. Lecture Notes (2004). Available at https://itweb.isye.gatech.edu
38. Nemirovski, A.: Modern convex optimization. Lecture at PASCAL Workshop, Thurnau, Germany, March 16–18 (2005)

35 Dynamical Gene-Environment Networks Under Ellipsoidal Uncertainty

39. Nesterov, Y.E., Nemirovskii, A.S.: Interior Point Polynomial Algorithms in Convex Programming. SIAM, Philadelphia (1994)
40. Nikkila, J., Törönen, P., Kaski, S., Venna, J., Castrén, E., Wong, G.: Analysis and visualization of gene expression data using self-organizing maps. Neural Netw. **15**(8–9), 953–966 (2002)
41. Partner, M., Kashtan, N., Alon, U.: Environmental variability and modularity of bacterial metabolic network. BMC Evol. Biol. **7**, 169 (2007). doi:10.1186/1471-2148-7-169
42. Pickl, S.: Der τ-value als Kontrollparameter – Modellierung und Analyse eines Joint-Implementation Programmes mithilfe der dynamischen kooperativen Spieltheorie und der diskreten Optimierung. Thesis, Darmstadt University of Technology, Department of Mathematics (1998)
43. Quackenbush, J.: Computational analysis of microarray data. Nat. Rev. Genet. **2**, 418–427 (2001)
44. Rivolta, M.N., Halsall, A., Johnson, C.M., Tones, M.A., Holley, M.C.: Transcript profiling of functionally related groups of genes during conditional differentiation of a mammalian cochlear hair cell line. Genome Res. **12**, 1091–1099 (2002)
45. Ros, L., Sabater, A., Thomas, F.: An ellipsoidal calculus based on propagation and fusion. IEEE Trans. Syst. Man Cybern. B Cybern. **32**(4), 430–442 (2002)
46. She, Y.: Sparse regression with exact clustering. PhD Thesis, Department of Statistics, Stanford University, USA (2008)
47. Shapiro, A., Dentcheva, D., Ruszczynski, A.: Lectures on Stochastic Programming: Modeling and Theory. MOS-SIAM Series on Optimization 9, Philadelphia (2009), ISBN 978-0898716-87-0
48. Tamayo, P., Slonim, D., Mesirov, J., Zhu, Q., Kitareewan, S., Dmitrovsky, E., Lander, E.S., Golub, T.R.: Interpreting patterns of gene expression with self-organizing maps: methods and application to hematopoietic differentiation. PNAS **96**, (6), 2907–2912 (1999)
49. Taştan, M.: Analysis and prediction of gene expression patterns by dynamical systems, and by a combinatorial algorithm. MSc Thesis, Institute of Applied Mathematics, METU, Turkey (2005)
50. Taştan, M., Ergenç, T., Pickl, S.W., Weber, G.-W.: Stability analysis of gene expression patterns by dynamical systems and a combinatorial algorithm. In: HIBIT – Proceedings of International Symposium on Health Informatics and Bioinformatics, Turkey '05, pp. 67–75. Antalya, Turkey, 2005
51. Taştan, M., Pickl, S.W., Weber, G.-W.: Mathematical modeling and stability analysis of gene-expression patterns in an extended space and with Runge-Kutta discretization. In: Proceedings of Operations Research 2005, pp. 443–450. Springer, Bremen, Sept 2005
52. Taylan, P., Weber, G.-W., Beck, A.: New approaches to regression by generalized additive models and continuous optimization for modern applications in finance, science and techology. In: Burachik, B., Yang, X. (guest eds.) The special issue in honour of Prof. Dr. Alexander Rubinov, Optimization, vol. 56, 5–6, 1–24 (2007)
53. Thomas, B., Raju, G., Sonam, W.: A modified fuzzy c-means algorithm for natural data exploration. World Acad. Sci. Eng. Technol. **49** (2009)
54. Uğur, Ö., Pickl, S.W., Weber, G.-W., Wünschiers, R.: Operational research meets biology: An algorithmic approach to analyze genetic networks and biological energy production. Preprint no. 50, Institute of Applied Mathematics, METU, 2006. Submitted for the special issue of *Optimization* at the occasion of the 5th Ballarat Workshop on Global and Non-Smooth Optimization: Theory, Methods, and Applications (2006)
55. Uğur, Ö., Pickl, S.W., Weber, G.-W., Wünschiers, R.: An algorithmic approach to analyze genetic networks and biological energy production: an introduction and contribution where OR meets biology. Optimization **58**(1), 1–22 (2009)
56. Uğur, Ö., Weber, G.-W.: Optimization and dynamics of gene-environment networks with intervals. In the special issue at the occasion of the 5th Ballarat Workshop on Global and Non-Smooth Optimization: Theory, Methods and Applications, Nov 28–30, 2006, of J. Ind. Manag. Optim., vol. 3(2), 357–379 (2007)
57. Vazhentsev, A.Y.: On internal ellipsoidal approximations for problems of control synthesis with bounded coordinates. J. Comput. Syst. Sci. Int. **39**(3) (2000)

570   E. Kropat et al.

58. Vázques, F.G., Rückmann, J.-J., Stein, O., Still, G.: Generalized semi-infinite programming: a tutorial. J. Comput. Appl. Math. **217**(2), 394–419 (2008)
59. Wall, M., Rechtsteiner, A., Rocha, L.: Singular Value Decomposition and Principal Component Analysis. In: Berrar, D.P., Dubitzky, W., Granzow, M. (eds.) A Practical Approach to Microarray Data Analysis, pp. 91–109, Kluwer, Norwell, MA (2003)
60. Weber, G.-W.: Charakterisierung struktureller Stabilität in der nichtlinearen Optimierung. In: Bock, H.H., Jongen, H.T., Plesken, W.: (eds.) Aachener Beiträge zur Mathematik 5. Augustinus publishing house (now: Mainz publishing house) Aachen (1992)
61. Weber, G.-W.: Minimization of a max-type function: Characterization of structural stability. In: Guddat, J., Jongen, H.Th., Kummer, B., Nožička, F. (eds.) Parametric Optimization and Related Topics III, pp. 519–538. Peter Lang publishing house, Frankfurt a.M., Bern, New York (1993)
62. Weber, G.-W.: Generalized semi-infinite optimization and related topics. In: Hofmannn, K.H., Wille, R. (eds.) Research and Exposition in Mathematics, vol. 29, Heldermann Publishing House, Lemgo (2003)
63. Weber, G.-W., Alparslan-Gök, S.-Z., Defterli, O., Kropat, E.: Modeling, Inference and Optimization of Regulatory Networks Based on Time Series Data. Preprint at Institute of Applied Mathematics, Middle East Technical University, Ankara, Turkey, submitted to *European Journal of Operational Research* (EJOR)
64. Weber, G.-W., Alparslan-Gök, S.Z., Dikmen, N.: Environmental and life sciences: gene-environment networks – optimization, games and control – a survey on recent achievements. In: DeTombe, D. (guest ed.) Invited paper, in the special issue of Journal of Organisational Transformation and Social Change, vol. **5**(3), pp. 197–233 (2008)
65. Weber, G.-W., Alparslan-Gök, S.Z., Söyler, B.: A new mathematical approach in environmental and life sciences: gene-environment networks and their dynamics. Environ. Model. Assess. **14**(2), 267-288 (2009)
66. Weber, G.-W., Batmaz, I., Köksal, G., Taylan, P., Yerlikaya-Özkur, F.: CMARS: A new contribution to nonparametric regression with multivariate adaptive regression splines supported by continuous optimisation. Preprint at IAM, METU, submitted to Advances in Computational and Applied Mathematics (in reviewing)
67. Weber, G.-W., Kropat, E., Akteke-Öztürk, B., Görgülü, Z.-K.: A survey on OR and mathematical methods applied on gene-environment networks. Special Issue on Innovative Approaches for Decision Analysis in Energy, Health, and Life Sciences of Central European Journal of Operations Research (CEJOR) at the occasion of EURO XXII 2007 (Prague, Czech Republic, July 8–11, 2007), vol. 17(3), 315–341 (2009)
68. Weber, G.-W., Kropat, E., Tezel, A., Belen, S.: Optimization applied on regulatory and eco-finance networks – survey and new developments. In: Fukushima, M., Kelley, C.T., Qi, L., Sun, J., Ye, Y. (guest eds.) Pac. J. Optim., vol. **6**(2), 319–340, Special Issue in memory of Professor Alexander Rubinov (2010)
69. Weber, G.-W., Özögür-Akyüz, S., Kropat, E.: A review on data mining and continuous optimization applications in computational biology and medicine. Embryo Today, Birth Defects Research (Part C), **87**, 165–181 (2009)
70. Weber, G.-W., Taylan, P., Alparslan-Gök, S.-Z., Özöğür, S., Akteke-Öztürk, B.: Optimization of gene-environment networks in the presence of errors and uncertainty with Chebychev approximation. *TOP*, the Operational Research journal of SEIO (Spanish Statistics and Operations Research Society) vol. 16(2), 284–318 (2008)
71. Weber, G.-W., Tezel, A.: On generalized semi-infinite optimization of genetic networks. TOP **15**(1), 65–77 (2007)
72. Weber, G.-W., Tezel, A., Taylan, P., Soyler, A., Çetin, M.: Mathematical contributions to dynamics and optimization of gene-environment networks. In: Pallaschke, D., Stein, O. (guest eds.) Special Issue: In Celebration of Prof. Dr. Dr. Hubertus Th. Jongen's 60th Birthday, of Optimization, vol. 57(2), pp. 353–377 (2008)
73. Weber, G.-W., Uğur, Ö., Taylan, P., Tezel, A.: On optimization, dynamics and uncertainty: a tutorial for gene-environment networks. In: the Special Issue Networks in Computational Biology of Discrete Appl. Math., vol. 157(10), pp. 2494–2513 (2009)

35 Dynamical Gene-Environment Networks Under Ellipsoidal Uncertainty

74. Yerlikaya, F.: A new contribution to nonlinear robust regression and classification with MARS and its applications to data mining for quality control in manufacturing. Thesis, Middle East Technical University, Ankara, Turkey (2008)
75. Yeung, K.Y., Ruzzo, W.L.: Principal component analysis for clustering gene expression data. Bioinformatics **17**(9), 763–774 (2001)
76. Yılmaz, F.B.: A mathematical modeling and approximation of gene expression patterns by linear and quadratic regulatory relations and analysis of gene networks. MSc Thesis, Institute of Applied Mathematics, METU, Ankara, Turkey (2004)
77. Yılmaz, F.B., Öktem, H., Weber, G.-W.: Mathematical modeling and approximation of gene expression patterns and gene networks. In: Fleuren, F., den Hertog, D., Kort, P. (eds.) Operations Research Proceedings, pp. 280–287 (2005)
78. Zhang, A.: Advanced Analysis of Gene Expression Microarray Data. World Scientific Pub. Co. Ltd., Singapore (2006)

Chapter 36
Strategic Interaction in Macroeconomic Policies: An Outline of a New Differential Game Approach

T. Krishna Kumar

Abstract Increasing globalization has created a realization by most countries that macroeconomic policy coordination is useful. However, the economic models advanced so far for this purpose do not adequately address the major questions that are of interest for policy coordination. The need for policy coordination arises from differences in resource endowments and differences in savings and investment that have implications for long run growth. The models that are currently in use are only of a comparative static nature based on open IS-LM framework with dynamics brought in as disequilibrium dynamics. It is suggested here that dynamic multi-sectoral multi-regional input output model with current input and capital input matrices may be used along with compatible spatial price equilibrium relationships and error correcting disequilibrium dynamics as the model. Some suggestions are offered to weaken the sufficiency conditions for the existence of a solution to Linear Quadratic Differential Game and some economic examples are discussed.

36.1 Introduction

Until the great depression in 1929–1933 planning and economic policy was confined to national economies, with exchange rates between national currencies being flexible. After the world depression John Maynard Keynes emphasized the problems associated with beggar thy neighbor policies of the pre-depression years and how they led to the depression. He extended the logic of his general theory to the world economy and identified the need to have a Global Bank which performs the

---

T. Krishna Kumar
Samkhya Analytica India Pvt. Ltd., Bangalore, India
and
Economic Analysis Unit, Indian Statistical Institute, Bangalore, India
and
Indian Institute of Management, Bangalore, India
and
3 Graystone Court, Naperville, IL 60565, USA
e-mail: tkkumar@gmail.com

M.M. Peixoto et al. (eds.), *Dynamics, Games and Science I*, Springer Proceedings
in Mathematics 1, DOI 10.1007/978-3-642-11456-4_36,
© Springer-Verlag Berlin Heidelberg 2011

functions of a central bank of a country, but at the global level, viz. controlling the money supply through its credit policies, foreign exchange reserves, and the exchange rates. This had led to the creation of World Bank (International Bank for Reconstruction and Development), and the International Monetary Fund, the first entrusted with the international developmental credit flows and the latter with controlling the exchange rates and providing short term credit for maintaining reasonable levels of foreign exchange reserves.[1] The Fixed exchange rate regime thus replaced the flexible exchange rate regime. The national fiscal and monetary policies were left to be decided by national governments.

During the last four decades, with more trade interactions, and greater capital mobility between countries, the transmission mechanism of shocks from country to country has been quick and of significant magnitude. This has lead to changes in policies of IMF and WB. The switch back to flexible exchange rate regime from fixed exchange rate regime is one such major change. There are also systemic reactions to the new flexible exchange rate regime and global competition in trade. These are in terms of new trade agreements such as NAFTA and monetary unions such as European Union with a common currency Euro and its own European Bank. Although there is some attempt to devise methods for macroeconomic policy coordination, such attempts are either limited to countries within the European Union or happened to be pure academic exercises which adapted existing theoretical open macroeconomic models to study the issues of policy coordination. What is needed is to develop new methods to suit this class of problems rather than apply an existing theoretical tool. It is necessary to make them more suitable for actual adoption at least as guides to macroeconomic policy coordination.

What I attempt to do in this paper are: (a) present a critical review of the existing literature on macroeconomic policy coordination, (b) emphasize the limitations of that literature and suggest ways of making them more useful and operational, (c) identify an operational approach, and suggest a few research problems on differential games arising from this problem, and (d) finally give an illustrative operational example of differential game solution to strategic trade policy.

The plan of the paper is as follows. Section 36.2 presents a brief critical review of the existing literature. Section 36.3 states the problem of macroeconomic policy coordination and identifies a new operational approach. Section 36.4 elaborates on the structure of a LQDG and its solutions. It also suggests a few interesting research problems in the theory of differential games. Section 36.5 illustrates the operational usefulness of the model by presenting a numerical example of a strategic trade policy.

---

[1] There are other ulterior motives for the major super power, the United States of America. For a fuller discussion on the political economy see my earlier paper (Kumar (see [24])).

36.2 Review of Literature

Before I go into the review of literature I shall present a broader overall scenario of evolution of that literature. Macroeconomic policy coordination requires dealing with two or more open economy macro models. In the seventies and up to mid-eighties the macroeconomic policy coordination models used were static and basically open economy IS-LM models. In eighties and beyond the models used have been dynamic but confined to simple post-Keynesian dynamic models, where to open IS-LM model some features of short run dynamics are added (These are Models of Mundell–Fleming or Dornbush variety). In other words, these models are dynamic only in terms of making general price level and nominal wages sticky and then specifying a Walrasian market adjustment mechanism or other short run dynamics, such as adding a Phillips curve type relation. When each country's model used variables that linked the country with the rest of the world, and when such linking variables are *assumed* to be exogenous, the policy prescription required only an optimization model and no policy coordination is either anticipated or suggested.

The need for macroeconomic policy coordination came mainly from four sources. First, with the breakdown of Bretton Woods's statutes and the introduction of flexible exchange rate regime, economists recalled the interwar period of flexible exchange rates with the associated beggar thy neighbor policies and their tragic consequences. Second, the Lucas critique and the issue of credibility of economic policy called into question the simple optimization. This brought rational expectations and game theoretic approach into the literature on coordination of monetary and fiscal policies (see Kydland [25]). Third, the creation of the European Union (EU or European Monetary Union, EMU) with 11 asymmetric European countries coming under one central European Bank and one common monetary policy raised interesting questions regarding coordinating the fiscal policies of the 11 countries with the common monetary policy. Finally, the financial meltdown of September 2008 and the prolonged recession that followed in USA raised concerns to contain it within United States and to contain its adverse impact on other countries through macroeconomic policy coordination. The real missing link still is a need for macroeconomic policy coordination through an institutional mechanism such as the World Bank and IMF, implying drawing up new statutes for these organizations to facilitate macroeconomic policy coordination.

36.2.1 Macroeconomic Policy Coordination Within a Country

Kydland was the first to use differential game theory in macroeconomic policy in his doctoral dissertation at Carnegie-Mellon University in 1973.[2] Kydland (see [25])

---

[2] One may see Tabellini, [41], for an excellent exposition of Kydland's contributions to macroeconomic policy.

shows the inferiority of the non-cooperative Nash equilibrium compared to a cooperative solution. He interprets the policy maker as the dominant player and the typical individual citizen as the non-dominant player. He shows (Kydland see [26]) that in such a game with a dominant player the open loop control policy is time-inconsistent and attributes it to lack of credibility. He also establishes that in order to get time consistency one must go for a feedback control policy. Lucas [28], and Kydland and Prescott (see [27]) are other papers on the same topic that use the concept of rational expectations and argue for the advantage of rule-based policies to create rational expectations equilibrium solution.

There is considerable literature on the interplay between monetary policy and fiscal policy and arguments are often advanced for an independent monetary authority. One can apply the above differential game framework with linear differential equation system with quadratic objective functions, and the monetary authority and the fiscal authority being the two players of the game. It can then be argued that if the monetary authority is truly independent and if one obtains the non-cooperative solution with each policy maker having his own objective function, the corresponding Nash equilibrium is Pareto inferior to a cooperative solution. Using differential games Petit established this result for Italy (Petit, see [35]).

There is a similarity between the policy coordination in European Monetary Union and that of states within a country. The EMU has a common monetary policy and all the 11 member countries have their own fiscal policies. There is a need to put restrictions on large deficit spending by some of the countries as they will have serious adverse repercussions otherwise on other countries. Now referring to a single country for a comparison, the states within a country have their own fiscal policies with their own budgets and taxing powers, while they have a common fiscal policy at the country level and a common monetary policy by an independent central bank. There seems to be a need for fiscal policy coordination between the states of a country, such as need for such coordination between the countries within EMU.

Aarle et al. (see [1]) examine the impact of fiscal policies of member countries with their own labor market distortions on the stability and growth of EMU. They identify the need for coordination such as the stability and growth pact (SGP) that EMU has. They use a differential game model, with Mundell–Fleming type of model. Aarle et al. (see [2]) examine the coalition formation in EMU. The analogy in a single country is, although there may not be any asymmetries in labor market rigidities, there could be asymmetries in public infrastructure that create repercussions of a state's fiscal policy on other states.

36.2.2 International Policy Coordination

Literature on international policy coordination dates back to the late 1960s and 1970s, and two of the most noted studies in this area are by Cooper (see [9]) and Hamada (see [18]). Cooper (see [9]) examines economic policy formulation in an open economy by allowing international capital movements. The paper observes that with international capital movement and increase in the interdependencies between the countries, effectiveness of decentralized policies declines, and coordination of policies between different countries becomes compelling. The paper assumes a two-country model. He specifies the targets of economic policy as the level of unemployment and the rate of economic growth. For instruments of economic policy he chooses government expenditures or open market operations, which are controlled by the nation's economic authorities, and which in turn influence the values taken by the target variables. The effectiveness of policy formulation is measured by the speed at which the target variables are restored to their target levels, in the presence of interdependencies within the countries.

Hamada (see [18]) uses a game theoretic formulation to explain the same problem. Hamada's study became the pioneering work in using economic policy games to explain the gains from coordination. Hamada (see [18]) highlights the importance of monetary interdependencies between different countries while examining the gains from policy coordination. He constructs an n-country game in which the monetary policy of each country is conducted in such a way as to maximize the objective function of its monetary authority, the primary objectives being price stability and balance of payment equilibrium. In the process, he also shows that if there are differences in national preferences concerning inflation and balance of payment policies, they affect the realized outcome of the world inflation rate. While this study assumes a fixed exchange rate, Hamada (see [19]) extends this analysis for flexible exchange rate.

Cooper (see [10]) gives a detailed exposition of economic interdependence between different countries. He defines economic interdependence as a multidimensional economic transaction between two countries, or a country and the rest of the world. Following Hamada, many studies were carried out using multiple-country, macroeconomic policy games, with countries maximizing their respective welfare functions. The welfare functions of the countries, or in other words the objective functions, defined the strategic positions of the countries. Corden (see [11]) defines a case of bilateral monopoly between two governments, whose objective functions are to manage the aggregate nominal demand of its own country. Under flexible exchange rate system he shows that non-cooperative solutions will have deflationary biases relative to cooperative policies. But this remains valid only in the short run as it does not take into account the effects of expectation formation on inflation and unemployment in the later periods. Most of these studies use static macroeconomic models.

Other studies on strategic policy coordination include Canzoneri and Gray (see [6]), Currie and Levine (see [13]), Kehoe (see [22]), Ploeg (see [37]). In general these studies show that when authorities ignore interdependence, the solutions will not be efficient and conclude that when authorities cooperate the result would be Pareto superior. While there are studies suggesting cooperation as a superior strategy, there are other studies which show that there are no clear benefits of international cooperation. Studies like Oudiz and Sachs (see [34]) use a dynamic game model to show possible time inconsistency in the solution. Thus they bring out the importance of credibility of the policies of the players. Frankel and Rockett (see

[16]) suggest that in order to maximize the gains from cooperation, policymakers often come out with incorrect models of policy coordination. This happens primarily because different governments subscribe to different economic philosophies. Lack of knowledge of the true model leads to movement of the target variables in the wrong direction and hence lowers equilibrium rates.

Obstfeld (see [33]) and Rogoff (see [38]) provide an excellent review of some of the models used for policy coordination with Mundell–Fleming–Dornbush type models. The models invariably had one spatial equilibrium condition for the output and another spatial equilibrium condition for the prices, and one more for the capital flows or exchange rate. While Mudell–Fleming model is an open IS-LM model Dornbush included the assumption of sticky prices for wages, and introduced disequilibrium dynamics into the model. This post Keynesian assumption was difficult to uphold among macroeconomists those days by anyone other than Dornbush (see Rogoff's Mundell–Fleming lecture (see [38]). While these models are labeled the first generation models of macroeconomic policy coordination, a second generation of models for macroeconomic policy coordination appeared in the literature. These were extensions of the post-Keynesian models in which micro foundations were added by assuming monopolistic competition, individual representative agents with utility functions, etc. But they retained the short run dynamics and ignored the truly dynamic long run growth aspects. For a review one may see Conzoneri et al. (see [6]).

These models of macroeconomic policy coordination are dynamic only by introduction of disequilibrium dynamics. They are not truly dynamic. In other words these models only capture short run dynamics or business fluctuations and they are devoid of capturing the medium and long term trends in the pursuit and evasion games being played between nations.

Let us examine what the major issues facing countries in the European Monetary Union or between US and China, and US and India, or India and China are. In EMU it is the asymmetric growth scenarios in the member countries, particularly between the two old countries East and West Germany, now united into a united Germany but still having growth differential in the two regions. It is the differential growths in US and China, triggered by labor market distortions and state capitalism in China. These distortions had created a long term growth differential between the countries along with a huge trade deficit for US. Are not the present trade deficits problems related to some long term positions the two countries have taken? Can such positions be reversed? Where are these growth issues addressed in the existing models of macroeconomic policy coordination? Did we not talk about export-led growth of the emerging economies just a few years ago? Where is the issue of export-led growth and the associated policy instruments or strategies such as special economic (exporting) zones (SEZs) in these macroeconomic policy coordination models? Do we not see that a major part of misalignment of the exchange rate of a country with purchasing power parity arises because of market distortions in one or both trading countries that create significant differences in prices of non-traded goods? Where is the distinction made between traded and non-traded goods in these models? It is to address some of these issues that we present an alternative modeling strategy for macroeconomic policy coordination in the next section.

36.3   An Alternate Operational Model

36.3.1   Need for an Alternative Approach

One of the major policy issues facing nations in a global economy today is to shield each country from external shocks, as such shocks can lower growth or cause inflation. These shocks could arise from speculation in output growth, inflation and domestic rate of interest. Such shocks could be transmitted through trade in goods and services, international capital mobility etc. Some of these speculations or expectations arise due to long term trends as well as short run fluctuations. The existing literature was mostly limited to policy coordination between countries of EMU or between US and UK, as in Frankel and Rockett (see [16]). The type of models to use, and the type of problems one faces, will radically be different if the macroeconomic policy coordination is between US and China or between US and India. The policy coordination must deal with growth, investment, and distortions in both product and factor markets.

One may note the recent trends and the political dialogue between USA and China regarding revaluation of the Chinese currency. Likewise one may expect more trade in ICT services between US and India, and one may also expect trade in manufacturing with US shifting from China to India. With these emerging trends the macroeconomic policy coordination between US and China and US and India is not just for academic curiosity but is of great political and economic significance. Macroeconomic policy coordination in such cases must be strategic in nature and the underlying models should incorporate both long run growth as well as short run fluctuations.

In order that a model is policy relevant it must satisfy the following criteria:

1. It must have a model that captures the salient features of economies that are of interest to policy makers, such as how output and prices are determined for traded and not traded goods, and how the general price level is a weighted average of the prices of traded and non-traded goods.
2. It must describe both the long term growth and short run fluctuations.
3. There must be an ongoing continuous activity in each country of specifying the model, estimating it, testing it, validating it, and improving it. It must address the issue of credibility of a policy in terms of its enforceability by having transparency in model specification and its validity. There must also be a mechanism to share information.
4. The policies must not be decided at the beginning of a policy dialogue and frozen at those levels. Instead the policies should be based on feedbacks from the model performance from period to period.

580   T.K. Kumar

5. The model formulation must be such that the solution to the differential game must exist and the solution must be operationally computable given an estimated dynamic model of macroeconomic interactions.
6. There must be an institutional mechanism that will ensure the model assumptions are maintained that include data availability, common model, enforcing the adoption of cooperative solution, imposition of penalties for departing from a coordinated policy etc. (reorganization of IMF and World Bank that is being contemplated may take into account these institutional requirements).

The review of literature given above shows that most of the existing literature on macroeconomic policy coordination does not offer any models that fit into this framework. We see no connection between time series econometric models, such as Vector Auto Regression and Error Correction models, of the macro economies that are estimated, tested, validated and improved on the one hand and the policy models used for macroeconomic policy coordination on the other. We also see little connection between the policy needs and policy models.

36.3.2 A Suggested Alternative Model

In this connection it must be noted rather ironically, that what has been recommended as bad and disposable in a globalizing world, viz. macroeconomic planning and coordination by the state, is quite valuable here. For instance, the long term economic planning in India used Leontieff's dynamic input out model (see Chakravarty [8]), and Eckaus and Parikh (see [14])). Chakravarty and others used closed one country models that were formulated and solved as optimal control problems. Now, in view of different countries choosing policies in their own interests, policies that can have impact on other countries through trade and capital mobility, the policy problem must be formulated as a differential game problem[3]. Macroeconomic policy coordination can benefit immensely from multi-regional, multi-sectoral, and multi-period dynamic input output model. Let me illustrate by specifying one such model in a k-country set up with m sectors producing goods and services. Let us assume that there are a $km \times km$ matrix A of current input requirements per unit output, and a $km \times km$ matrix of requirements of capital inputs determined through a matrix version of an acceleration principle. Let us assume that each country has

---

[3] I recall my reading several working papers and reports on differential games during 1964–1965 while working on my doctoral dissertation. The dissertation was on application of optimal control theory to long term economic planning problem (Kumar (see [23])). The works I came across on differential games then were reports on dynamic games with continuous strategies from the Rand Corporation written by Rufus Isaacs [20] and John Nash [32]. These were on pursuit and evasion differential games Rand undertook for the US Defense department in early fifties. At that time I wondered about possible applications of differential game theory in economics if the decision making involves an underlying dynamic model with a system of differential equations and several decision makers.

a set of commodities that constitute its traded goods, while the rest being the non-traded goods. Let c_t represent a km dimensional vector of final consumption of the m commodities in the k countries. Likewise let us assume that g_t represents km dimensional government demands for the outputs, e_t and i_t represent the km dimensional vectors of exports and imports. We can represent the dynamic Leontieff model as follows:

$$Ax_t + B(x_{t+1} - x_t) + c_t + g_t + e_t - i_t = x_t \qquad (36.1)$$

The above matrix difference equation represent km equations in a km dimensional vector x. The vectors g_t incorporates fiscal policies, the vectors e_t and it incorporate trade policies. With this as the core model one can define other macroeconomic relations such as equations of normal profits with errors. These are equilibrium price relations (derivable from the dual of the dynamic Leontieff equations (see Jorgenson, [21])), the errors representing abnormal profits or losses depending on business cycles etc. To those price equations one can add the Walrasian price adjustment equations for each sector or groups of sectors. One can add another equation that shows the changes in foreign exchange assets, a stock-flow equation coming from a free capital mobility assumption. In order to explain the exchange rate and the interest rate parity one may use two more equations. The model thus consists of a set of simultaneous difference equations depicting dynamic relations, and a set of equations in contemporaneous variables showing equilibrium relations.[4] Several time series macro econometric models such as vector auto-regression and error correction (e.g. Mallick (see [30]), and Mallick and Mohsin (see [31])) or stochastic dynamic general equilibrium models (eg. Bhattarai (see [5]) are built for a single country. Such truly dynamic models can be combined, through transaction linkages between them with country-specific objective functions, to formulate differential game theory models for policy coordination.

We can assume that each country has predetermined target variables for the outputs of the traded and non-traded goods, and for the price level. If desired, the targets could be formulated in terms of nominal values of aggregated sectoral outputs of a few sectors, such as food, shelter etc. The policy instruments could be fiscal, monetary, and trade policies. We may assume that there is a cost of deviating from these targets and such a cost can be reasonably approximated by a quadratic cost function. The macroeconomic policy problem will then turn out to be a dynamic game problem with quadratic cost functions and a system of first order linear difference equations.

---

[4] Thus the model suggested here is similar to VAR-EC models macroeconomists (econometricians) build to study the impulse and response analyses of economic policies. The suggested approach in this paper thus bridges the policy perspective of macroeconomic policy coordination with the macroeconomic policy evaluations countries make using time series econometric models.

36.4 Differential Game Formulation of a Typical Macroeconomic Policy Coordination Problem

36.4.1 Formulation of the Policy Coordination Problem as a Differential Game Problem

For simplicity we shall describe the above problem of macroeconomic policy coordination as a two-nation differential game problem with a linear differential equation system and quadratic cost functions, called briefly as Linear Quadratic Differential Game (LQDG), such as the ones described by Starr and Ho (see [40]), Basar and Olsder (see [2]), and Petit (see [36]) and Kydland (see [25])[5]:

$$\frac{dx}{dt} = Ax + B_1 u_1 + B_2 u_2 \tag{36.2}$$

with $x(0) = x_0$; $x(T) = x_T$, $u_1 \in U_1$ (set of all feasible policies open to player 1) and $u_2 \in U_2$ (set of all feasible policies open to player 2)

$$J_1(u_1, u_2) = \frac{1}{2} x^{'}(T) K_{1T} x(T) + \frac{1}{2} \int_0^T \{x^{'}(T) Q_1 x(T) + u_1^{'}(t) R_{11} u_1^{'}(t)$$
$$+ u_2^{'}(t) R_{12} u_2^{'}(t)\} dt \tag{36.3}$$

$$J_2(u_1, u_2) = \frac{1}{2} x^{'}(T) K_{2T} x(T) + \frac{1}{2} \int_0^T \{x^{'}(T) Q_2 x(T) + u_1^{'}(t) R_{21} u_1^{'}(t)$$
$$+ u_2^{'}(t) R_{22} u_2^{'}(t)\} dt \tag{36.4}$$

Any solution to a game problem such as this depends on several preconditions. These are:

1. Each player knows at time t the state of the system at that time, but does not know the strategy or policy of the other player.
2. Each player knows the strategy space of the other players and their cost functional, and this is part of common knowledge.
3. The system dynamics are known with certainty to both players and that is part of common knowledge.
4. The nature of strategy or policy space (this will be described in some detail below).
5. The nature of solution to the game one is looking for (this will be explained below).

---

[5] Starr and Ho consider a wide class of deterministic differential games. Kydland considers stochastic games but uses the corresponding certainty equivalent deterministic game. The later is possible given the LQDG formulation.

36 Strategic Interaction in Macroeconomic Policies   583

After specifying all these features of a game problem listed above the LQDG problem poses the following questions:

1. Does a solution exist? If so, under what additional conditions does it exist? What are the economic interpretations of those conditions?
2. Is the solution unique? If it is not unique what additional criteria are needed to pick one among a set of solutions?
3. Is the solution stable under perturbations to the system, either in terms of additive shock to the system or in terms of perturbations to the parameters of the dynamic system?
4. How sensitive is the solution to changes in the objective functions (the cost functions)?
5. How should one compute the solution?
6. What economic insights do we get from the solution?
7. Does the interpretation of the solution represent a body of useful knowledge?
8. Can we operationalize and institutionalize the solution in terms of monitoring it and improvising it? (At least as a professionally agreed upon aid to policy advice, even if it is not accepted by the policy makers?)

We must define what we mean by a solution in this game of conflict. There are five different solution concepts. These are:

1. Non-cooperative solution.
2. Minimax solution where each country tries to minimize the maximum cost it could incur under the worst scenario of strategies followed by the other countries.
3. Pareto efficient solution. This may not be unique and one needs further criteria to choose one among many Pareto efficient solutions.
4. Cooperative solution, usually under Nash program where the cooperation depends on the bargaining strength of each country based on what benefits or costs it would incur if it were to go without cooperation.
5. Dominant player or Stackelberg solution.

There are basically two main types of policies, one being more appealing than the other. One is a policy or strategy defined as a function of time determined at the beginning of the planning horizon, with an understanding that once a policy is chosen the country sticks to that policy. Such a strategy space is called an open-loop control policy space. However, it is more appealing to introduce policy instruments that have feedback features and can be adjusted as we go along the time path, monitoring what has been happening to the states of the economies over time. This calls for using two types of differential games, one with open loop control policies and the other with closed loop or feedback policies. Within the closed loop control policies there could still be several other types. The electrical engineers and space engineers designed several automatic control engines or mechanisms that would observe the state of the system and its deviation from the target. They found it useful to measure not only the state of the system at time t, but also the rate of change of the state and how far the state had come or how far it still has to go. Such information was used to control the system. They found that a combination of all these sometimes works better. These feedback or closed loop controls using all the three feedbacks are called proportional control, integrative, and derivative control or a PID control for short.

For a single player optimal control problems it is possible to derive an optimal open loop control and the optimal trajectory of the state, eliminate time from both of them to express the control as a function of the state. This process is called synthesizing the control, leading to a closed loop control strategy. Such synthesis is in general not possible for a 2-person or an n-person non-zero sum differential game. Hence it is possible that an open loop strategy solution does not exist, or even if it exists it will be a nonlinear function of the state (Basar (see [3])). A closed loop solution to the game may, however, exist.

Does each country know the growth models of other countries? What happens if there is asymmetric information on these growth models? When the countries agree to cooperate do they do so sincerely or do they renege on their commitments to cooperate? What is the situation regarding enforcement of cooperative strategies? These are some of the practical issues which the literature on macroeconomic policy coordination does not adequately address. Answers to some of these questions will help us in designing the institutional mechanism or the rules under which the games are played. Frankel and Rocket [16] show that when each country has a different perception of the other's model and both differ from the true model the Nash bargaining solution could move the targets away from the optimal solution with the true models.

36.4.2 Existence of a Solution

One of the reasons why the linear quadratic differential game is chosen is that for any more general differential game problems either the existence of the solution or its uniqueness or its computation could pose problems. Even in this case what is known is a necessary condition that the closed loop control should satisfy, which is the existence of a solution to coupled matrix Riccati differential equations described later. It is known that those coupled Riccati equations may not have a solution in general. If we assume that each player has the information set given in Cruz (see [12]), based on which he would choose his strategy, then there exists a solution, under certain conditions given by Freiling et al. (see [17]). Many authors gave sufficient conditions for the existence of a solution to the coupled matrix Riccati equations. These sufficient conditions are restrictive and new methods can be devised to prove the existence of Nash equilibrium under less restrictive conditions. I shall return to this topic later.

Lukes (see [29]) showed that if a solution to these coupled Riccati equations exists it is unique. Engwerda (see [15]) provides a computational algorithm for computing the solution to coupled matrix Riccati equations. For details one may see Engwerda (see [15]). The literature on differential games regarding the existence of a solution and computation of a solution is quite intricate and could turn off

36 Strategic Interaction in Macroeconomic Policies

a policy maker. These results seem to suggest that in many problems there may not be a solution. But it need not be taken that seriously for reasons explained below.

The problem of existence of solution of a differential game problem can be interpreted in two different ways. First, one may say that the formulation of the problem is correct and the actual situation does not lead to any solution, meaning that whatever strategies countries take there will be some move by some country to disturb and destabilize the situation. There is no convincing reason to say, in that case, that the mathematical specification of the system not admitting a solution is bad. Then one might ask the question what are the possible reasons for the non-existence of the solution. With some examples and counter examples of systems that admit and do not admit the solutions one can get some clear idea on what features are needed in order to have existence of a solution. The mathematical conditions used to derive the existence theorems also give some clues on what is needed to have a solution. One can ask if the system when perturbed will have a solution. Then the direction in which the perturbation is needed to get a solution may also guide us as to how to design or steer the system from an unstable position to a stable position. One may also postulate a kind of proximity theorem that state the solution of an approximate model is an approximate solution. Then even if a given formulation of a differential game does not have a solution one may look for another approximately equivalent problem that admits a solution.

Second, our comprehension of how the system works may not be correct although we know intuitively that the system has a stable solution, or do not have any reason to believe that it has no solution. Then we may look for how to make the system admit a solution by making a better approximation of the system.

The mathematical systems are general descriptions and hence obtaining a solution to a general mathematical problem is asking for a general solution to a general problem. The generality is sometimes so general that there may not be any solution. In that case we are better off taking a less general formulation of the problem that admits a solution and at the same time accounts for a very large number of problems we do encounter. When we run into mathematical difficulties in proving the existence of a solution we may resort to taking a series of specific numerical examples, obtain each country's response function to other country's strategy, from those responses and strategies map the cost fuctionals (as functions of strategies) and see if they have a common point (fixed point), which is the Nash equilibrium. By taking a series of such numerical examples belonging to a class of problems we can generalize the result to that class of problems. With this new kind of computational empirical economics we can enrich our knowledge and even discover new ways of stating and proving existence theorems. I would like to emphasize this empirical approach to what otherwise looks like a formidable mathematical problem. Here I would like to quote Alvin Roth (see [39]):

"However if we do *not* take steps in the direction of adding a solid empirical base to game theory, but instead continue to rely on game theory primarily for conceptual insights (deep and satisfying as these may be), then it is likely that long before a hundred years game theory will have experienced sharply diminishing returns. In this respect, I think the next hundred years will likely bring about a change in the

586   T.K. Kumar

way theoretical and empirical work are related in economics generally, and that, if
not, then the entire discipline of economics may also fail to realize its potential".

Looking at the problem from an intuitive angle the problem involves the following steps:

1. Taking piecewise continuous control variables
2. Plugging them into the differential equations
3. Solving them for the state variable trajectory
4. Plugging the controls and the state variables in the cost functions to check for the
   Nash equilibrium condition

At each one of these steps we are taking variables and transforming them. What
we need is compactness of controls, responses, and cost functional. Compactness
involves boundedness and continuity or absolute continuity. Most of the results in
Nash games draw from the mathematical insights drawn from optimal control theory. In Kumar (see [23]) I proved the existence of a general nonlinear control policy
assuming simple conditions on the control variables that they are bounded and satisfy a Lipschitz condition. I also assumed that the functions are continuous and
differentiable. It is perhaps possible to use the same method of proof and establish
the existence of Nash equilibrium in n-person non-zero sum differential game with
quadratic pay-offs.

36.4.3 The Solution

We assume that the information set for each player are his own control strategies
and control space, the control space of the other player, complete knowledge of the
differential equation system with the initial condition, and the form of his own cost
functional and the cost functional of the other player. We also restrict the control
space for the two players to be the linear feedback or closed loop controls. The
necessary condition for the feedback Nash equilibrium is that:

$$u_1^*(t) = -R_1^{-1}B_1'K_1(t)x(t) \tag{36.5}$$

$$u_2^*(t) = -R_2^{-1}B_1'K_2(t)x(t) \tag{36.6}$$

Where $K_1(t)$ and $K_2(t)$ are solutions of coupled matrix Riccati equations:

$$\frac{dK_1}{dt} = -A'K_1 - K_1A - Q_1 + K_1S_{11}K_1 + K_1S_{22}K_2 - K_2S_{12}K_2 \tag{36.7}$$

$$\frac{dK_2}{dt} = -A'K_2 - K_2A - Q_2 + K_2S_{22}K_2 + K_1S_{11}K_1 - K_1S_{21}K_1 \tag{36.8}$$

$K_1(T) = K_1(T)$ and $K_2(T) = K_2(T)$ and $S_{ij} = B_j R_{jj}^{-1} R_{jj} R_{jj}^{-1} B_j'$ for
$i, j = 1, 2\ i \neq j$.

It may be noted that these coupled matrix Riccati equations are nonlinear. Petit
shows that when the Lagrangian or the Ha miltonian is written and the canonical

36 Strategic Interaction in Macroeconomic Policies   587

equations are written for the system then K_1 and K_2 are the co-state variables or the Lagrangian multipliers. See Petit ([23]) and Freiling et al. (see [17]).

36.5 Differential Game Approach to Strategic Trade Policy

36.5.1 The Strategic Trade Model

We are still in the lookout for good computer software for solving LQDGs for macroeconomic policy coordination with fiscal and monetary policies of two countries. In the mean time we can illustrate the approach suggested here with policy coordination for trade policies between two countries.[6]

We assume that there are two nations called North and South, each being governed by a dynamic Leontiefff model with and without trade links. The model without trade is called Autarky situation and the one with free trade with no quantitative restrictions and tariffs is called the Free Trade situation.[7]

The specification of the model is as follows:

Let $y(t)$ represent the output of country labeled South in period 't' while $z(t)$ the output associated with country North.

We then have the following National Income Identity.

$$y(t) = ay(t) + bDy(t) + c(t) + e(t) - m(t) \qquad (36.9)$$

Where:

D is an operator denoting the first order time derivative,
a is the current input–output coefficient,
b is the incremental capital-output ratio,
c (t) is the final consumption in period't',
e (t) is the exports to country North by the South country, and
m (t) is imports of the South country from the North country.[8]

---

[6] This is an example I had worked out eight years ago using the Maple software 5.1 version. This problem involved numerical solution of first order linear differential equations.

[7] The intercept and the slope parameters of the import and export functions are assumed to be constant in this and the subsequent sections. Later we make them depend on the trade policies, such as imposing quantitative restrictions and import tariffs, to trace the implications of a variety of alternate trade policies.

[8] We assume a composite single commodity. Trade is meaningful only if there are at least two commodities. It is assumed here that there are in fact more than one commodity in the two economies and what are modeled are the aggregate outputs and their growth. The composite price indices in the two countries differ due to differing average costs as reflected in the technical coefficients of the Leontieff model. Comparative advantage and specialization thus become meaningful. One may even say that the composite commodity of one country is different from the composite commodity of the other country.

588 T.K. Kumar

Similarly, we have the following National Income Identity for North Country:

$$z(t) = dz(t) + fDz(t) + g(t) + m(t) - e(t) \tag{36.10}$$

where, d and f are the current and capital input output coefficients of the North country, and $g(t)$ is the final demand or consumption function of the North country.[9]

We assume that the consumption or final demand equations in the two countries are given by:

$$c(t) = c_0 + c_1 y(t) \tag{36.11}$$
$$g(t) = g_0 + g_1 z(t) \tag{36.12}$$

We assume that the export from country South equals the imports of country North, and *vice versa*. We assume the following linear export and import functions:

$$e(t) = e_0 + e_1 z(t) \tag{36.13}$$
$$m(t) = m_0 + m_1 y(t) \tag{36.14}$$

When we substitute expressions (36.12)–(36.14) in (36.9) and (36.10) we get a system of two inhomogeneous differential equations of the first order that are linked to each other if there is trade. By assuming an initial condition we can solve these two equations and determine the dynamic inter-temporal trajectory of the total output or income of each country as a function of time.

We assume that the variables are measured in real terms measured in, say, billions of US Dollars. We also assume that the exchange rate is flexible to generate zero trade balance for each country.

We can specify all the parameters of the model and solve the two differential equations

36.5.2 Comparison of Solutions

We assume the following numerical values for the parameters for deriving the empirical results:

$$a = 0.10; b = 4.0; c_0 = -10; c_1 = 0.75; e_0 = 0 \text{ (autarky)}, = 1.03 \text{ (free trade)};$$
$$e_1 = 0 \text{ (autarky)}, = 0.20 \text{ (free trade) } m_0 = 0 \text{ (autarky)}, = 1 \text{ (free trade)}; \quad (36.15)$$
$$m_1 = 0 \text{ (autarky)}, = 0.35 \text{ (free trade)}; \ d = 0.30; \ f = 3.0; g_0 = -10; g_1 = 0.65$$

---

[9] The technical coefficients "a", "b", "c", and "d" are assumed to differ between the countries. This is equivalent to assuming that the two countries have either unequal access to the latest technology or that the countries choose different technologies depending on their respective factor endowments.

36.5.2.1 Autarky Situation

Assume that there are no exports and imports, i.e., $e(t) = m(t) = 0$. The South Country's structure is now characterized by the following dynamic Leontieff model.

$$4.0^* Dy(t) - 0.15^* y(t) - 10 = 0 \tag{36.16}$$

The North Country's structure is given by the following dynamic Leontieff model:

$$3.0^* Dz(t) - 0.05^* z(t) - 10 = 0 \tag{36.17}$$

Assume the following initial conditions:

$$y(0) = 6, \ z(0) = 10; \tag{36.18}$$

The dynamic path of the two countries, South and North are given by $y(t)$, and $z(t)$ respectively as follows[10]:

$$y(t) = 72.66666667 \ exp(.03750000000t) - 66.66666667 \tag{36.19}$$

$$z(x) = 209.9999999 \ exp(.01666666667t) - 199.9999999 \tag{36.20}$$

These solutions are plotted against time in Fig. 36.1. We assumed a horizon of 15 years for our comparative analysis throughout this study. Throughout this paper the trajectory that starts at 10 refers to the North Country while the one that starts at 6 refers to the South country. It can be noted that the divergence between the two countries widens as time goes on.

36.5.2.2 Free Trade Situation

Let us now assume that South country's exports are given by $e(t)$, which constitutes the imports of North Country. Similarly the imports of South Country and exports of North Country are given by $m(t)$ as given by (36.13) and (36.14) with parameters as specified in (36.16). The structure of South Country is now given by:

$$4.0Dy(t) - .50y(t) - 9.97 + .20z(t) = 0 \tag{36.21}$$

The structure of the North Country is given by:

$$3.0Dz(t) - .25z(t) - 10.03 + .35y(t) = 0 \tag{36.22}$$

---

[10] We used Maple 5.1 software of Maplesoft, Canada for obtaining the numerical solution of a system of ordinary differential equations.

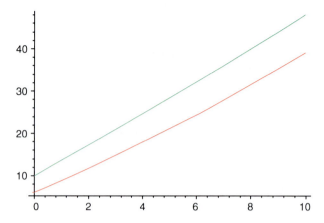

Fig. 36.1 The Trajectories of outputs of the North (*Green*) and the South (*Red*) countries: under autarky situation

The initial conditions are assumed to be the same as those assumed for the Autarky case, viz. $y(0) = 6$, $z(0) = 10$; It can be seen from (36.21) and (36.22) that the growth path of each country depends on the output of the other country.

The solution for the Free Trade situation is given by:

$$y(t) = 3.45933019*10^{-8} \exp(0.1833333333t)$$
$$+ 84.33157892 \exp(0.02499999999t)$$
$$+ .4000000000*10^{-7} \exp(0.1583333333t) - 81.79090915 \quad (36.23)$$

$$z(t) = 168.6631579 \exp(0.02499999999t)$$
$$- 4.03588519 \exp(0.1833333333t) - 154.6272728$$
$$- .2000000001*10^{-8} \exp(-0.1583333333t)$$

$$- .3000000001*10^{-7} \exp(0.1583333333t),$$
$$+ .1000000000*10^{-8} \exp(-0.1583333333t) \quad (36.24)$$

The Free Trade solutions are plotted against time in Fig. 36.2. As can be seen from Figs. 36.1 and 36.2 free trade has benefited the South country at the expense of, the North Country.

Marginal propensity to export by the North country seem to reflect the kind of situation that could prevail when a small (in the sense of GNP) closed economy such as India could be the South country while a large export-seeking large country such as United States is the North country.

Fig. 36.2 The Trajectories of GNPs of the North and the South countries: Case 1 under free trade situation

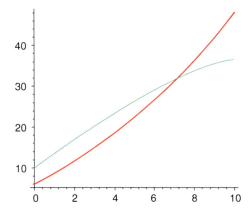

The usual claim made by the free trade proponents, that free trade is beneficial to both countries does not seem to be borne out in this case. We will, however, examine later whether the countries find free trade to be preferable to autarky.

If $y(t)$ is the output in US bill $s, is the time discount factor, and $(0, T)$ is the time horizon then the integral payoff for the plan horizon can be written as:

$$\int_0^T \exp(-\rho t) y(t) dt \qquad (36.25)$$

Optimal growth theory suggests that one must use a Ramsey type objective function that is a strictly concave function of consumption. We assume, however, that what is maximized is the discounted consumption stream. As the consumption function is linear function of income it is a monotonic isomorphic transformation of income, and hence the above pay-off function. We find the numerical values for the integral pay-offs as follows, assuming that the time discount factor is 0.08 and that the plan horizon is 15 years:

Pay-offs for Autarky Situation in US bill $s: (286.42; 223.64) Pay-offs for Free Trade Situation in US bill $s: (227.07; 271.16)

Where the first entry refers to the pay-off to the North and the second entry is the pay-off to the South country.

From this it is obvious that free trade is beneficial to the South and harmful to the North. As a result of free trade the gain to the South is 47.52 and the loss to the North is 59.35. The loss to the North is more than the gain to the South. Which situation will the two countries choose autarky or free trade? We will consider a more general situation of a choice by the two countries of a trade strategy which involves quantitative restrictions and tariffs.

36.5.3 Choosing Optimal Trade Strategies: A Differential Game Formulation

It must be admitted that the model described above is very primitive and highly aggregative. Further the question of trade strategies is made somewhat irrelevant as a result of the new WTO trade regime. There are certain provisions of WTO regime that do permit member countries to use discretionary trade policies. Even violations of the trade regime rules are subject to disputes and dispute resolution mechanisms that take time.

In this section we reformulate the import and export functions with a quantitative restriction dummy variable that shifts the intercept term and a tariff dummy variable that would shift the marginal propensities to export and import.

The following are the possible trade strategies that the two countries are supposed to adopt. Two letters represent the strategy. The first letter represents the level of quantitative restrictions (QR) or quotas, and the second letter represents the level of import tariff (T).

We assume that the quantitative restrictions (QR) imposed by South reduce the intercept of its import function while the tariff (T) reduces its slope or the marginal propensity to import. We assume two levels for each, labeled low (L) and high (H). For the South country QR low equals downward shift of intercept of the import function by 0.02, while QR high equals a downward shift of intercept of the import function by 0.04. When there are no quantitative restrictions the intercept of the import function of South is 1.0. When there is Low tariff in the South country its MPI is 0.30, while a high tariff gives rise to an MPI of 0.20. When there is no tariff at all the MPI for the South country is assumed to be 0.35.

The QR imposed by North at low level results in an intercept of export function of the South country of 1.02 while QR at higher level imposed by the North country would result in an intercept of the South country's exports of 1.01. When there are no quantitative restrictions imposed by the North Country the intercept of the export function of the South country at 1.03. When there is low tariff in North the marginal propensity to export of the South country is 0.15 and when there is high tariff in North the value of this becomes 0.10.

The matrix of strategy pairs is given in Table 36.1 below. NA means that strategy pair is not applicable. There are altogether 26 pairs of strategies. We specify the following values to the parameters:
$a = 0.10$, $b = 4.0$, $c_0 = -10$, $c_1 = 0.75$, $e_0 = 1.03$ (N: no QR), $e_0 = 1.02$ (N:LQR), 1.01(N:HQR); $e_1 = 0.10$(N:HT), 0.15(N:LT), 0.20(N: no-T); $m_0 = 1$ (S: noQR), 0.98 (S:LQR), 0.96 (S:HQR); $m_2 = 0.20$ (S:HT), 0.30 (S:LT), 0.35(S:no-T); $g_0 = -10$; $d = 0.30$; $f = 3.0$; $g_1 = 0.65$.

The integral payoffs for the two countries can be calculated for each of the 26 combinations of the trade strategies. If we assume that the two countries play a non-cooperative game then the optimal trade strategies for the two countries are those that correspond to the Nash-Equilibrium. For each strategy used by a country we determine what the best counter strategy of the other country is. We call these the

36 Strategic Interaction in Macroeconomic Policies

Table 36.1 Matrix of alternate trade strategies of North and South countries

		SOUTH					
		Free trade	LL	LH	HL	HH	Autarky
	Free trade	1	2	3	4	5	NA
	LL	6	7	8	9	10	NA
NORTH	LH	11	12	13	14	15	NA
	HL	16	17	18	19	20	NA
	HH	21	22	23	24	25	NA
	Autarky	NA	NA	NA	NA	NA	26

Table 36.2 Payoff bi-matrix associated with alternate trade strategies of North and South countries

		NORTH					
		Free trade	LL	LH	HL	HH	Autarky
	Free trade	227.07+	268.52+	334.07+	269.38+	334.80+	NA
		271.16*	237.80	184.89	237.10	184.30	NA
	LL	181.96	223.27	289.58	224.04	290.24	NA
		307.66*	274.42	220.90	273.79	220.36	
	LH	145.44	186.17	252.37	186.87	252.97	NA
SOUTH		337.33*	304.53	251.09	303.96	250.60	
	HL	181.54	222.88	289.25	223.65	289.91	NA
		308.00*	274.73	221.17	274.10	220.63	
	HH	145.06	185.82	252.08	186.52	252.67	NA
		337.64*	304.82	251.33	304.25	250.85	
	Autarky	NA	NA	NA	NA	NA	286.42
							223.64

best response functions. We then define the Nash Equilibrium as the set of strategies that are best against the best strategies of the other country.

We present below in Table 36.2 the integral payoffs for all the strategy pairs along with the identification of the best responses of each country against each of the strategies of the other country. The numbers in the top refer to the integral pay-offs of the North Country while the numbers below refer to the integral pay-offs of the South country. The payoffs with " $*$ " are the best response of the South country while the pay-offs with "+" sign are the best responses of the North Country. The trajectories of outputs of the South and North countries associated with these alternate trade strategies are not shown here. Interested readers can contact the author for the Appendix to this article that is not being printed.

From the payoff bi-matrix it is evident that Free Trade is the Nash equilibrium strategy.[11] The two countries would thus choose free trade as the non-cooperative solution. However, it can be noted that if both countries cooperate they can choose

---

[11] This is intuitive. Free trade solution refers to optimization by each country under no trade restrictions. As the objective function itself does not depend on the trade flow an unrestricted maximum is larger than a restricted maximum.

either LH and LH each or HH and HH each which will give more benefits to the North than the loss to the South when compared with the Nash equilibrium, The difference, the gain over the loss is $5.14 billion in LH, LH case, and $5.29 billion in HH, HH case. This gain can be shared between the two countries. Thus there can be cooperative solutions that are better than the non-cooperative solutions.

36.6 Concluding Remarks

This paper reports work in progress. It demonstrates that the existing macroeconomic policy coordination papers use models that use only short run dynamics or fluctuations and are ill-suited to address growth related strategies that must enter into any macroeconomic policy coordination. It gives an illustration of how differential game approach with dynamic Leontieff input output model can be used to determine an optimal trading strategy.

Acknowledgements I thank Puja Guha for her help with part of the literature review.

References

1. Aarle, B.V., Engwerda, J.C., Plasmans, J.E.J., Weeren, A.: Macroeconomic policy interaction under EMU. Open Econ. Rev. **12**, 29–60 (2001)
2. Aarle, B.V., Engwerda, J.C., Plasmans, J.E.J.: Monetary and fiscal policy interactions in EMU: A dynamic game approach. Ann. Oper. Res. **109**, 229–264 (2002)
3. Basar, T.: A counter example in linear quadratic game: the existence of nonlinear Nash solutions, J. Optim. Theory Appl. **14**, 425–430 (1974)
4. Basar, T., Olsder, G.J.: Dynamic Noncoperative Game Theory. Academic, New York (1982)
5. Bhattarai, K.: Econometric and Stochastic General Equilibrium Models for Evaluation of Economic Policies. International Journal of Trade and Global Markets, **4**(2) (2011) Forthcoming
6. Canzoneri, M.B., Gray, J.A.: Monetary policy games and the consequences of non-cooperative behavior. Int. Econ. Rev. **26**(3), 547–564 (1985)
7. Canzoneri, M.B., Robert, E.C., Diba, B.T.: The need for international policy coordination: what's old, what's new, and what's yet to come? J. Int. Econ. **66**, 363–384 (2005)
8. Chakravarty, S.: Optimal programme of capital accumulation in a multisector economy. Econometrica **33**(3), 557–570 (1965)
9. Cooper, R.N.: Macroeconomic policy adjustment in interdependent economies. Q. J. Econ. **83**(1), 1–24 (1969)
10. Cooper, R.N.: Economic interdependence and coordination of economic policies, Vol. 2 of Handbook of International Economics, chapter 23 pp. 1195–1234. Elsevier, Amsterdam (1985)
11. Corden, W.M.: Macroeconomic policy interaction under flexible exchange rates: a two-country model. Economica **52**(205), 9–23 (1985)
12. Cruz, J.B.: Survey of nash and stackelberg equilibrium solutions in dynamic games. Ann. Econ. Soc. Meas. **4**(2), 339–344 (1975)
13. Currie, D., Levine, P.: Macroeconomic policy design in an interdependent world, in 'International Economic Policy Coordination', NBER Chapters, National Bureau of Economic Research, Inc, pp. 228–273 (1985)
14. Eckaus, R.S., Kirit, S.P.: Planning for Growth: Multisectoral, Intertemporal Models Applied to India. MIT, Cambridge, MA (1968)

15. Engwerda, Jacob., LQ Dynamic Optimization and Differential Games. Wiley, England (2005)
16. Frankel, J.A., Rockett, K.E.: International macroeconomic policy coordination when policy-makers do not agree on the true model. Am. Econ. Rev. **78**(3), 318–340 (1988)
17. Friling, G., Jank, G., Abou-Kandil, H.: On global existence of solutions to coupled matrix Riccati equations in closed loop Nash games. IEEE Trans. Autom. Control **41**, 264–269 (1996)
18. Hamada, K.: A strategic analysis of monetary interdependence. J. Polit. Econ. **84**(4), 677–700 (1976)
19. Hamada, K.: Macroeconomic strategy and coordination under flexible exchange rates. In: Dombusch, R., Frankel, J.A. (eds.) International Economic Policy: Theory and Evidence. Johns Hopkins, Baltimore (1979)
20. Isaacs, R.P.: Games of Pursuit, Paper P-257. RAND Corporation. Santa Monica, California (1951)
21. Jorgenson, D.W.: A dual stability theorem. Econometrica **28**(4), 892–899 (1960)
22. Kehoe, P.J.: Coordination of fiscal policies in a world economy. Staff Report 98, Federal Reserve Bank of Minneapolis (1986)
23. Kumar, T.K.: On the existence of an optimal economic policy. Econometrica **37**(2), 600–612 (1969)
24. Kumar, T.K.: Fund-Bank policies of stabilization and structural adjustment: a global and historical perspective. Econ. Polit. Wkly. **28**(17), 815–823 (1993)
25. Kydland, F.: Non-cooperative and dominant player solutions in discrete dynamics games. Int. Econ. Rev. **16**, 312–335 (1975)
26. Kydland, F.: Equilibrium solutions in dynamic dominant-player models. J. Econ. Theory **15**, 307–324 (1977)
27. Kydland, F., Prescott, E.C.: Rules rather than discretion: the inconsistency of optimal plans. J. Polit. Econ. **85**, 473–491 (1977)
28. Lucas, R.: Econometric Policy evaluation: a critique. In: Brunner, K., Meltzer, A.: The Phillips Curve and Labor Markets. Carnegie-Rochester Conference Series on Public Policy, vol. 1, pp. 19–46. American Elsevier, New York (1976)
29. Lukes, D.L.: Equilibrium feedback control in linear games with quadratic cost. SIAM J. Control Optim. **9**, 234–252 (1971)
30. Mallick, S.K.: Policy instruments to avoid output collapse. Appl. Financ. Econ. **16**, 761–776 (2006)
31. Mallick, S.K., Mohsin, M.: On the effects of inflation shocks in a small open economy, Aust. Econ. Rev. **40**(3), 253–266 (2007)
32. Nash, J.: Continuous iteration method for solution of differential games. Rand Corporation. Santa Monica, California, RM-1326 (1954)
33. Obsfeld, M.: International Macroeconomics Beyond Mundell-Fleming Model, Mundell-Fleming Lecture, International Monetary Fund, Nov 2000, Center for International development and Economic Research, University of California, Berkeley, July 1, (2001)
34. Oudiz, G., Sachs, J.: *Macroeconomic policy coordination among the industrial economies*, Brookings Papers on Economic Activity 15 Control Theory and Dynamic Games in Economic Policy Analysis (1984-1), 1–76 (1984)
35. Petit, M.L.: Fiscal and monetary policy coordination: a differential game approach. J. Appl. Econ. (1989)
36. Petit, M.L.: Control Theory and Dynamic Games in Economic Policy Analysis. Cambridge University Press, Cambridge (1990)
37. Ploeg, F.V.D.: International policy coordination in interdependent monetary economies. J. Int. Econ. **25**(1–2), 1–23 (1988)
38. Rogoff, K.: Dornbush's Overshooting Model after Twenty Five Years, IMF Staff Papers, 49 Special Issue (2002)
39. Roth, A.E.: Game theory as a part of empirical economics. Econ. J., January 1991, **101**, 107–114 (1971)
40. Starr, A.W., Ho, Y.C.: Non-zero Sum Differential Games, Journal of Optimization Theory and Applications, **3**, 184–206 (1969)
41. Tabellini, G.: Fynn Kydland and Edward Prescott's contributions to the theory of macroeconomic policy. Scand. J. Econ. **107**(2), 203–216 (2005)

Chapter 37
Renormalization of Hénon Maps

M. Lyubich and M. Martens

Abstract Period doubling cascades are observed at transition to chaos in many models used in the sciences and in physical experiments. These period doubling cascades are very well understood in one-dimensional dynamics. In particular, the microscopic geometrical properties of the attractors do not depend on the actual system, they are universal. Moreover, the attractors of two different maps are smoothly conjugate, they are rigid. Strongly dissipative Hénon maps describe parts of the dynamics of systems close to a homoclinic tangency and are often observed in various models. For these maps the transition to positive entropy also occurs along period doubling cascades. These strongly dissipative Hénon maps can be considered as perturbations of one-dimensional systems. Indeed, some of the universal geometrical properties of the one-dimensional systems are present in the Hénon maps. However, they appear in a much more delicate form: in a probabilistic sense the geometry of the Hénon attractors is the same as their one-dimensional counter part. This phenomenon is revered to as probabilistic universality and rigidity.

37.1 Introduction

Since the universality discoveries, made in the mid-1970s by Feigenbaum [10, 11] and, independently, by Coullet and Tresser [9, 31], these fundamental phenomena have attracted a great deal of attention from mathematicians, pure and applied, and physicists. In particular, universality and the corresponding geometric rigidity of the attractors at the transition from regular behavior to chaotic behavior are central themes in one-dimensional dynamics. Coullet and Tresser conjectured that the universal geometry at transition to chaos in one-dimensional dynamics will also be

---

M. Lyubich (✉) and M. Martens
SUNY, Stony Brook, NY, USA
e-mail: mlyubich@math.sunysb.edu, marco@math.sunysb.edu

M.M. Peixoto et al. (eds.), *Dynamics, Games and Science I*, Springer Proceedings
in Mathematics 1, DOI 10.1007/978-3-642-11456-4_37,
© Springer-Verlag Berlin Heidelberg 2011

observed in higher dimensional systems. This conjecture has been confirmed by many numerical and physical experiments.[1]

A rigorous study of universality and rigidity has been surprisingly difficult and technically sophisticated and so far has only been thoroughly carried out in the case of one-dimensional maps, on the interval or the circle (see [12, 15, 17, 21, 22, 25, 30, 32, 33] and references therein). The study of universality and rigidity is in essence the study of a corresponding renormalization operator. This operator replaces a system by another which describes the original systems on a smaller scale. It acts like a microscope.

A frequently observed transition to chaos in one-dimensional dynamics is the so-called *period doubling cascade to chaos*. The corresponding renormalization operator has a unique hyperbolic fixed point. The dynamics of the renormalization fixed point, which is itself a one-dimensional system, and the behavior of the renormalization operator around this fixed point determine the asymptotic small scale geometry of systems at transition and the asymptotic small scale properties around the boundary of chaos. This explains the observed universality.

A rigorous exploration of universality for dissipative higher dimensional systems was begun in an article by Collet et al. [6]. It is shown in this article that the one-dimensional renormalization fixed point is also a hyperbolic fixed point for renormalization of strongly dissipative higher-dimensional maps close to the one-dimensional renormalization fixed point: this explained the parameter universality observed in families of such systems. A subsequent paper by Gambaudo, van Strien and Tresser [13] demonstrates that, similarly to the one-dimensional situation, infinitely renormalizable two-dimensional maps which are close to the one-dimensional renormalization fixed point have an attracting Cantor set which is, up to topological equivalence, the same as the attractor of the renormalization fixed point.

Observations in physical and numerical experiments indicate that universality and rigidity are also playing a crucial role in higher dimensional dynamics. This survey discusses two-dimensional strongly dissipative Hénon maps at transition to chaos. These are maps which are infinitely renormalizable of period doubling type. Indeed, there is still universality and rigidity but in a much more delicate form.

The Cantor attractors of infinitely renormalizable maps of period doubling type cannot be understood geometrically in terms of their one-dimensional counterpart. Though they lie on a rectifiable curve, the geometry of the Cantor attractors in small scales essentially differs from their one-dimensional counterpart. In fact, a typical map in a family will have *unbounded geometry*. However, almost everywhere the Cantor attractors have the same geometry as their one-dimensional counterpart, so we encounter here a new phenomenon of *probabilistic universality* and *probabilistic rigidity*.

---

[1] This conjecture should be taken with caution as not every transition to chaos is related to a transition in one-dimensional dynamics.

37 Renormalization of Hénon Maps

The topology of infinitely renormalizable maps of period doubling type also differs from the one-dimensional equivalent. These differences come from the bifurcations in the heteroclinic web of the maps in question. The *average Jacobian* of such a map, which is an ergodic theoretical invariant, is closely related to the observed differences in topology and geometry. By changing the average Jacobian one changes the topology of the heteroclinic web and the geometry of the non-universal part of the Cantor attractor.

This survey is based on the series of articles [5, 7, 16, 18, 19]. The results of Sects. 37.3, 37.4, and parts of Sect. 37.5 are generalized by P. Hazard, to maps of more general periodic renormalization types and can be found in [14].

37.2 Unimodal Renormalization

A unimodal map is a smooth map of the interval with only one critical point. The critical point is non-degenerate. A smooth unimodal map $f \in \mathcal{U}$ is *renormalizable* if it contains two disjoint intervals which are exchanged by the map. The two smallest intervals which are exchanged form the first renormalization cycle, $\mathcal{C}_1 = \{I_0^1, I_1^1\}$, where I_0^1 contains the critical point c of f. Let $\mathcal{U}_0$ be the collection of renormalizable maps. The renormalization of $f \in \mathcal{U}_0$ is an affinely rescaled version of the first return map to I_0^1, $f^2 : I_0^1 \to I_0^1$. This defines an operator

$$R_c : \mathcal{U}_0 \to \mathcal{U}$$

Similarly, one can rescale the first return map to I_1^1, the interval which contains the critical value v of f. This defines the second renormalization operator

$$R_v : \mathcal{U}_0 \to \mathcal{U}.$$

The intervals I_0^1 and I_1^1 are called *renormalization domains*.

These renormalization operators are microscopes used to the study the small scale geometry of the dynamics. In particular, $R_c f$ is a unimodal map which describes the dynamics on one scale lower in I_0^1. Similarly $R_v f$ describes the geometry one scale smaller in I_1^1. The strength of renormalization is expressed by the Coullet–Tresser–Feigenbaum Conjecture whose proof has a long history (see [12, 15, 21, 22, 30, 32, 33] and references therein) and was finally obtained in [17].

Theorem 37.1. *There is a unique fixed point f_* of R_c. It is a hyperbolic fixed point with codimension one stable manifold and a one dimensional unstable manifold. The operator R_v also has a unique fixed point f_v. It is also hyperbolic and its stable manifold coincides with the stable manifold of R_c.*

Remark 37.1. The hyperbolicity of the renormalization operators depends on the smoothness class of the unimodal maps. The hyperbolicity holds on the class of $C^{2+\alpha}$ unimodal maps with non degenerate critical point. Compare [8], and [12].

Remark 37.2. The relative length of I_c^1 of f_* in the domain of f_* is called the *universal scaling ratio*. It is denoted by $\sigma < 1$.

A map is infinitely renormalizable if it can be renormalized infinitely many times. That means for each $n \geq 1$, $R^n f \in \mathcal{U}_0$. An infinitely renormalizable map has cycles, pairwise disjoint intervals,

$$\mathscr{C}_n = \{I_i^n | i = 0, 1, 2, \ldots, 2^n - 1\},$$

with $f(I_i^n) = I_{i+1}^n$ and

$$\bigcup \mathscr{C}_{n+1} \subset \bigcup \mathscr{C}_n.$$

This nested sequence of dynamical cycles accumulates on a Cantor set.

$$\mathscr{C} = \bigcap \bigcup \mathscr{C}_n.$$

This Cantor set attracts almost every orbit. It is called the Cantor attractor of the map. The only points whose orbits are not attracted to this Cantor set are the periodic point, of period 2^n, and their stable manifolds. The cycle $\mathscr{C}_n$ is centered around a periodic orbit of length 2^{n+1}. It contains all the periodic orbits of period 2^s with $s \geq n + 1$.

Every small part of the Cantor attractor $\mathscr{C}$ of some infinitely renormalizable map, say within an interval I_i^n of the nth—cycle, can be studied by repeatedly applying one of the renormalization operators R_c or R_v. For each interval in the cycle there is a uniquely defined sequence of length n of choices $w = (c, c, v, c, \ldots, v)$ such that the

$$R_w f = R_c \circ R_c \circ R_v \circ R_c \circ \cdots \circ R_v f$$

describes the dynamics within the given I_i^n. This means that $R_w f$ is an affinely rescaled version of the first return map to I_i^n. Denote the length of a word w by $|w|$. The collection of infinitely renormalizable maps coincides with the stable manifold of the two renormalization operators.

Theorem 37.2. *(Universality) There exists $\rho < 1$ such that for any two infinitely renormalizable maps $f, g \in \mathcal{U}_0$ and any finite word w*

$$dist(R_w f, R_w g) = O(dist(f, g)\rho^{|w|}).$$

The universality means that the Cantor attractors of two infinitely renormalizable maps are asymptotically the same on small scale. However, the actual geometry one observes depends on the place where one zooms in. This universal geometric structure of attractor is far from the well-known middle-third Cantor set, where in every

37 Renormalization of Hénon Maps 601

place one recovers the same geometry. In the Cantor attractor there are essentially no two places with the same asymptotic geometry, [4].

Given two infinitely renormalizable maps $f, g \in \mathcal{U}$, there exists a homeomorphism h between the domains of the two maps which maps orbits to orbits,

$$h \circ f = g \circ h.$$

The maps are conjugated, the homeomorphism is called a conjugation. The dynamics of two conjugated maps are the same from a topological point of view.

Theorem 37.3. *(Rigidity) The conjugation between two infinitely renormalizable maps is differentiable on the attractor.*

If a conjugation is differentiable, it means that on small scale the conjugation is essentially affine. This means that the microscopic geometrical properties of corresponding parts of the attractor are the same. One can deform an infinitely renormalizable map to another infinitely renormalizable map which will deform the geometry on large scale. However, the microscopic structure of the Cantor attractor is not changed: this is the rigidity phenomenon.

The *topology* of the system determines the *geometry* of the system.

This central idea has been rigorously justified in one-dimensional dynamics. It also holds when the systems are not of the period doubling type described above but have topological characteristics which are tame. We will not discuss the most general statement and omit the precise definition of tameness.

37.3 Hénon Renormalization

A smooth map $F: B \to \mathbb{R}^2$, $B = I^h \times I^v$ is called a *Hénon map* if it maps vertical sections of B to horizontal arcs, while the horizontal sections are mapped to parabola-like arcs (i.e., graphs of unimodal functions over the y-axis). Examples of Hénon maps are given by small perturbations of unimodal maps of the form

$$F(x, y) = (f(x) - \varepsilon(x, y), x), \tag{37.1}$$

where $f: I^h \to \mathbb{R}$ is unimodal and ε is small. Note that, in this case, the Jacobian is

$$\text{Jac } F = |\frac{\partial \varepsilon}{\partial y}|.$$

If $\partial \varepsilon / \partial y \neq 0$ then the vertical sections are mapped diffeomorphically onto horizontal arcs, so that F is a diffeomorphism onto a "thickening" of the graph $\Gamma_f = \{(f(x), x)\}_{x \in I^h}$ (Fig. 37.1).

In this case F is a diffeomorphism onto its image which will be briefly called a *Hénon diffeomorphism*. The classical Hénon family is obtained, up to affine

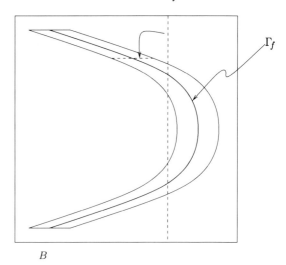

Fig. 37.1 A Hénon-like map

normalization, by letting $f(x)$ be a quadratic polynomial and $\varepsilon(x, y) = by$. A Hénon map with $\varepsilon = 0$ is called a *degenerate Hénon map*. We will mainly discuss *strongly dissipative* maps, e.g. map with a small ε.

Let Ω^h, $\Omega^v \subset \mathbb{C}$ be neighborhoods of I^h and I^v resp. and $\Omega = \Omega^h \times \Omega^v \subset \mathbb{C}^2$. Let $\mathscr{H}_\Omega$ stand for the class of Hénon maps $F \in \mathscr{H}^\omega$ of form (37.1) such that the unimodal map f admits a holomorphic extension to Ω^h and ε admits a holomorphic extension to Ω. The subspace of maps $F \in \mathscr{H}_\Omega$ with $\|\varepsilon\|_\Omega \leq \bar{\varepsilon}$ will be denoted by $\mathscr{H}_\Omega(\bar{\varepsilon})$.

Realizing a unimodal map f as a degenerate Hénon map F_f with $\varepsilon = 0$ yields an embedding of the space of unimodal maps $\mathscr{U}_{\Omega^h}$ into the space of Hénon maps $\mathscr{H}_\Omega$ making it possible to think of $\mathscr{U}_{\Omega^h}$ as a subspace of $\mathscr{H}_\Omega$.

37.3.1 Renormalizable Hénon Maps

An orientation preserving Hénon map is *renormalizable* if it has two saddle fixed points — a *regular* saddle β_0, with positive eigenvalues, and a *flip* saddle β_1, with negative eigenvalues — such that the unstable manifold $W^u(\beta_0)$ intersects the stable manifold $W^s(\beta_1)$ at a single orbit, see Fig. 37.2.

For example, if f is a twice renormalizable unimodal map then a small Hénon perturbation of type (37.1) is a renormalizable Hénon-like map.

Given a renormalizable map F, consider an intersection point $p_0 \in W^u(\beta_0) \cap W^s(\beta_1)$, and let $p_n = F^n(p_0)$. Let D be the topological disk bounded by the arcs of $W^s(\beta_1)$ and $W^u(\beta_0)$ with endpoints at p_0 and p_1. The disk D is invariant under F^2. The map $F^2|D$ is called a *pre-renormalization* of F.

37 Renormalization of Hénon Maps

Fig. 37.2 A renormalizable Hénon map

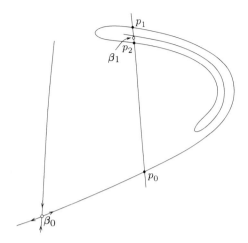

The topological notion of pre-renormalization is not convenient for the analysis of renormalizable maps. The Hénon renormalization operator has three non-conventional aspects. The renormalization domain has a geometric definition, not a topological definition. The renormalization domain is a neighborhood of the *tip*. The tip plays the role of critical value. It will be discussed in more detail. Finally, the *rescalings* are not affine maps, they are carefully chosen diffeomorphisms. A crucial part of the analysis of the Hénon renormalization operator deals with the repeated rescalings: the composed rescalings have a universal asymptotic shape. This asymptotic shape relates the asymptotic geometry of the renormalization to the actual asymptotic small scale geometry of the dynamics of the original map.

37.3.2 The Hénon Renormalization Operator

Hénon maps take vertical lines and map them into horizontal segments. The second iterate of such a map does not have this property. In general one can not find an affine coordinate change such that F^2 after this coordinate change is again a Hénon map. There is essentially only one diffeomorphic coordinate change which does bring F^2 back the Hénon form.

Consider the map $F(x, y) = (f(x) - \varepsilon(x, y), x)$ with the norm of ε small. The second iterate F^2 maps curves of the foliation defined by

$$f(x) - \varepsilon(x, y) = \text{Const}$$

into horizontal segments. The leaf of this foliation through a point (x, y) with x away from the critical point of f is an almost vertical curve. The map is renormalizable if there exists a domain bounded by two curves of the foliation and two

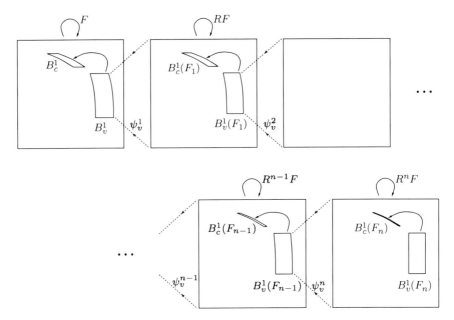

Fig. 37.3 The renormalization microscope

horizontal lines which is mapped into itself by F^2. The renormalization domain will be the smallest domain with this property. Denote this domain by B_v^1, see Fig. 37.3.

The renormalization domain B_v^1 is foliated by almost vertical curves. The diffeomorphism $H : B_v^1 \to \mathbb{R}^2$ defined by

$$H(x, y) = (f(x) - \varepsilon(x, y), y)$$

straightens the leaves, it maps them into vertical straight lines. It maps horizontal lines into horizontal lines, it is a *horizontal diffeomorphism*. The image $H(B_v^1)$ is a rectangle, in fact almost a square. Define $G : H(B_v^1) \to H(B_v^1)$ by

$$G = H \circ F^2 \circ H^{-1}.$$

Let Λ be the dilation which maps $H(B_v^1)$ to a rectangle with unit horizontal size and define the renormalization of F by

$$RF = \Lambda \circ G \circ \Lambda^{-1}.$$

The coordinate change which conjugates RF with $F^2|B_v^1$ is denoted by

$$\psi = (\Lambda \circ H)^{-1}.$$

The renormalization domain of a degenerate map $F_f = (f, x)$ where f is a renormalizable unimodal map with critical point c, $I_0^1 = [f^4(c), f^2(c)]$ and $I_1^1 = [f(c), f^3(c)]$ is

$$B_v^1 = I_1^1 \times I_0^1$$

and the renormalization of F is the degenerate Hénon map corresponding to the unimodal renormalization $R_c f$. Indeed, the Hénon renormalization operator extends the unimodal renormalization operator R_c.

37.3.3 Hyperbolicity of the Hénon Renormalization Operator

Let $\mathscr{I}_\Omega(\bar{\varepsilon})$ denote the subspace of infinitely renormalizable Hénon maps (including degenerate ones) of classes $\mathscr{H}_\Omega(\bar{\varepsilon})$. By the unimodal renormalization theory, the fixed point f_* is a quadratic-like map on some domain $\Omega_* \subset \mathbb{C}$, see e.g., [2] and references therein. Moreover, f_* is a hyperbolic fixed point of R_c in any space $\mathscr{U}_V$ with $V \Subset \Omega_*$. The corresponding degenerate Hénon map is denoted by F_*.

Theorem 37.4. *Assume $\Omega^h \Subset \Omega_*$. Then the map F_* is the hyperbolic fixed point for the Hénon renormalization operator R acting on $\mathscr{H}_\Omega$, with one-dimensional unstable manifold $\mathscr{W}^u(F_*) = \mathscr{W}^u(f_*)$ contained in the space of unimodal maps. Moreover, the differential $DR(F_*)$ has vanishing spectrum on the quotient $T\mathscr{H}_\Omega / T\mathscr{U}_{\Omega^h}$.*

The set $\mathscr{I}_\Omega(\bar{\varepsilon})$ of infinitely renormalizable Hénon maps coincides with the stable manifold

$$\mathscr{W}^s(F_*) = \{F \in \mathscr{H}_\Omega(\bar{\varepsilon}): R^n F \to F_* \text{ as } n \to \infty\},$$

which is a codimension-one real analytic submanifold in $\mathscr{H}_\Omega(\bar{\varepsilon})$.

Corollary 37.1. *For all Ω and $\bar{\varepsilon}$ as above, the intersection of $\mathscr{I}_\Omega(\varepsilon)$ with the Hénon family*

$$F_{a,b}: (x, y) \mapsto (a - x^2 - by, x)$$

is a real analytic curve intersecting transversally the one-dimensional slice $b = 0$ at a_*, the parameter value for which $x \mapsto a - x^2$ is infinitely renormalizable.

37.4 Microscopes for the Invariant Cantor Set

37.4.1 Design of the Microscopes

The set of n-times renormalizable maps is denoted by $\mathscr{H}_\Omega^n(\bar{\varepsilon}) \subset \mathscr{H}_\Omega(\bar{\varepsilon})$. If $F \in \mathscr{H}_\Omega^n(\bar{\varepsilon})$ we use the notation

$$F_n = R^n F.$$

If F is a renormalizable map then its renormalization RF is well defined on the rectangle with unit horizontal size. The coordinate change $\psi = H^{-1} \circ \Lambda^{-1}$ maps this rectangle onto the B_v^1 topological rectangle $A = B_v^1$. If we want to emphasize that some set, say A, is associated with a certain map F we use notation like $A(F)$.

The coordinate change which conjugates $F_k^2 | A(F_k)$ to F_{k+1} is denoted by

$$\psi_v^{k+1} = (\Lambda_k \circ H_k)^{-1} : \mathrm{Dom}(F_{k+1}) \to A(F_k). \tag{37.2}$$

Here H_k is the non-affine part of the coordinate change used to define $R^{k+1} F$ and Λ_k is the corresponding dilation.

Recall that the change of coordinates conjugating the renormalization RF to F^2 is denoted by

$$\psi_v^1 := H^{-1} \circ \Lambda^{-1}.$$

To describe the attractor of an infinitely renormalizable Hénon map we also need the map

$$\psi_c^1 = F \circ \psi_v^1.$$

The subscripts v and c indicate that these maps are associated to the critical *value* and the *critical* point, respectively.

If F is twice renormalizable define similarly, ψ_v^2 and ψ_c^2 be the corresponding changes of variable for RF, and let

$$\psi_{vv}^2 = \psi_v^1 \circ \psi_v^2, \quad \psi_{cv}^2 = \psi_c^1 \circ \psi_v^2, \quad \psi_{vc}^2 = \psi_v^1 \circ \psi_c^2, \quad \psi_{cc}^2 = \psi_c^1 \circ \psi_c^2.$$

For an infinitely renormalizable $F \in \mathscr{I}_\Omega(\epsilon)$ we can proceed this way, and for any $n \geq 0$, we can construct 2^n maps

$$\psi_w^n = \psi_{w_1}^1 \circ \cdots \circ \psi_{w_n}^n, \quad w = (w_1, \ldots, w_n) \in \{v, c\}^n.$$

Consider the domains

$$B_\omega^n = \mathrm{Im} \, \psi_\omega^n.$$

The coordinate changes ψ_ω^n conjugate $R^n F$ to the first return map $F^{2^n} : B_\omega^n \to B_\omega^n$. In this sense they are *microscopes*. The first return maps to the nested domains

$$B_{v^n}^n = \mathrm{Im} \, \psi_{v^n}^n$$

correspond to the renormalizations.

The critical point and critical value of a unimodal map plays a crucial role in its dynamics. The counterpart of the critical value for infinitely renormalizable Hénon maps is the *tip*

$$\{\tau_F\} = \bigcap_{n \geq 1} B_{v^n}^n.$$

Each collection $\{B_\omega^n | \omega \in \{c, v\}^n\}$, $n \geq 1$, consists of pairwise disjoint domains. These collections are called *renormalization cycles*. They are nested, as in the unimodal case. An infinitely renormalizable Hénon map has an invariant Cantor set:

$$\mathscr{O}_F = \bigcap_{n\geq 1} \bigcup_{i=0}^{2^n-1} F^i(B^n_{v^n}) = \bigcap_{n\geq 1} \bigcup_{\omega\in\{v,c\}^n} B^n_\omega.$$

The dynamics on this Cantor set is conjugate to an adding machine. Its unique invariant measure is denoted by μ. The *average Jacobian* is

$$b_F = \exp \int \log \operatorname{Jac} F d\mu.$$

37.4.2 Universality Around the Tip

The convergence of renormalization in the unimodal case is used to study the small scale geometry of the Cantor attractor. This is possible because the coordinate changes used to rescale are affine. In the Hénon case the coordinate changes are not affine. Fortunately, they have a universal asymptotic limit which allows to apply the convergence of renormalization to understand the small scale geometry of the invariant Cantor set.

Let $F \in \mathscr{I}_\Omega(\bar\varepsilon)$ and define

$$\Psi^n_0 = \psi^1_v \circ \psi^2_v \circ \cdots \circ \psi^n_v = \psi^n_{v^n}$$

which is the coordinate change which conjugates the nth-renormalization $R^n F$ to $F^{2^n} : B^n_{v^n} \to B^n_{v^n}$. To describe these maps Ψ^n_0 we will center the coordinate systems around the tips of F and $R^n F$ resp. In these coordinates we introduce the following notation

$$\Psi^n_0 = D^n_0 \circ (\operatorname{id} + S^n_0)$$

where D^n_0 is the derivative of Ψ^n_0 at the tip of $R^n F$ and

$$S^n_0(x, y) = (s^n_0(x, y), 0) = O(\|(x, y)\|^2)$$

near the origin, is the non-linear part of the map. The next Theorem is a crucial tool to study infinitely renormalizable Hénon maps.

Theorem 37.5. *There exists a universal analytic function $v(x)$ and $\rho < 1$ such that the following holds. Given $F \in \mathscr{I}_\Omega(\bar\varepsilon)$ there exists $t_F \asymp -b_F$, a_F, $C_1, C_2 > 0$ such that*

$$D^n_0 \sim \begin{pmatrix} 1 & t_F \\ 0 & 1 \end{pmatrix} \begin{pmatrix} C_1(\sigma^2)^n & 0 \\ 0 & C_2(-\sigma)^n \end{pmatrix}$$

and

$$|x + s^n_0(x, y) - (v(x) + a_F y^2)| = O(\rho^n).$$

Remark 37.3. The coordinate change $\psi^n_{v^n}$ between $R^n F$ and the first return map to $B^n_{v^n}$ around the tip, has well defined asymptotic behavior. In general, the coordinate changes ψ^n_ω between $R^n F$ and the first return map F^{2^n} to B^n_ω do not have such a limit behavior. In fact, there are pieces in the Cantor set where these coordinate changes do degenerate. The study of the general coordinate changes ψ^n_ω and the geometrical consequences is discussed in Sect. 37.5.

A first consequence of the asymptotic behavior of the coordinate change Ψ^n_0 is the following Theorem which describes the universality of the super-exponential convergence of Hénon renormalizations to the unimodal maps, compare Theorem 37.4.

Theorem 37.6. *(Universality) For any* $F \in \mathscr{I}_\Omega(\bar{\varepsilon})$ *with sufficiently small* $\bar{\varepsilon}$, *we have:*

$$R^n F = (f_n(x) - b^{2^n} a(x) y (1 + O(\rho^n)), x),$$

where $f_n \to f_*$ *exponentially fast,* b *is the average Jacobian,* $\rho \in (0, 1)$, *and* $a(x)$ *is a universal function. Moreover,* a *is analytic and positive.*

37.5 Geometry of the Invariant Cantor Set

The strongly dissipative infinitely renormalizable Hénon maps are small perturbations of unimodal maps. Although the invariant Cantor sets in the one-dimensional context are rigid, the geometry of the Cantor set of an infinitely renormalizable Hénon map differs surprisingly from its one-dimensional counterpart. In particular, it can not be understood within the one-dimensional theory.

Theorem 37.7. *Given an infinitely renormalizable Hénon map* F *with* $b_F > 0$, *there are no smooth curves containing* $\mathscr{O}_F$.

The characteristic exponents of the unique invariant measure on $\mathscr{O}_F$ of an infinitely renormalizable Hénon map F with $b_F > 0$, are 0 and $\ln b_F$. The higher dimensional nature of $\mathscr{O}_F$ can be seen more specifically in the following.

Theorem 37.8. *There are no continuous invariant direction fields on* $\mathscr{O}_F$ *when* $b_F > 0$.

Theorem 37.9. *The map* F *is not partially hyperbolic on* $\mathscr{O}_F$ *in the sense that the contracting and neutral line fields corresponding to the characteristic exponents* $\log b_F$ *and* 0 *are discontinuous.*

The universality Theorem 37.6 and the universality of the coordinate changes as described in Theorem 37.5 imply that around the tip one recovers geometrical aspects of the one-dimensional Cantor set. Some of the one-dimensional geometric universality survives in the Hénon maps. However, there are parts in the Cantor set of the Hénon maps whose geometry differs from its one-dimensional counterpart: rigidity does not survive.

The tip of a map $F \in \mathscr{I}_\Omega(\bar{\varepsilon})$ has a stable manifold. This stable manifold is tilted over t_F, see Theorem 37.5, away from the vertical. Although the tilt t_F is very small, it is proportional to $b_F = O(\bar{\varepsilon})$, it has crucial influence on the geometry of the Cantor set. The pieces $B_{v^n}^n$ are very thin parallelograms aligned along the stable manifold of the tip. They are tilted over an angle proportional to b_F, have horizontal size of the order σ^{2n} and vertical size of the order σ^n, see Theorem 37.5. The two pieces $B_{v^n c}^{n+1}, B_{v^{n+1}}^{n+1}$ contained in $B_{v^n}^n$ will be above eachother when

$$b_F \cdot \sigma^n \asymp (\sigma^2)^n.$$

The vertical direction is strongly contracting with a factor of order b_F. One iteration will bring the pieces very close to eachother, relative to their size, see Fig. 37.4. The corresponding pieces of the one-dimensional renormalization fixed point are small curves next to eachother at a distance comparable to their size.

The same distortion phenomenon caused by the tilt happens for the renormalizations. There this distorting effect will be stronger and stronger because the average Jacobian of the renormalizations decays superexponentially. These strong distortions are reflected in the Cantor set of the original map.

This leads to the following Non-Rigidity Theorem.

Theorem 37.10. *(Non-Rigidity) Let F and $\tilde{F}$ be two infinitely renormalizable Hénon maps with average Jacobian b and $\tilde{b}$ resp. Assume $b > \tilde{b}$. Let h be a homeomorphism which conjugates $\tilde{F}|_{\mathcal{O}_{\tilde{F}}}$ and $F|_{\mathcal{O}_F}$ with $h(\tau_{\tilde{F}}) = \tau_F$. Then the Hölder exponent of h is at most $\frac{1}{2}(1 + \ln b / \ln \tilde{b})$.*

In particular, the conjugation between the Cantor set of a unimodal map and the Cantor set of a Hénon map is not smooth.

Corollary 37.2. *Let F be an infinitely renormalizable Hénon map with the average Jacobian $b_F > 0$ and F_0 be a degenerate infinitely renormalizable Hénon map. Let*

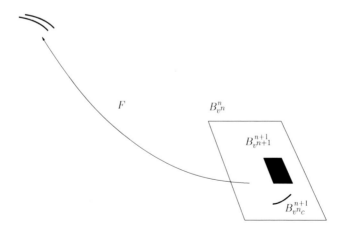

Fig. 37.4 Tilt and unbounded geometry

h be a homeomorphism that conjugates $F_0|_{\mathscr{O}_{F_0}}$ and $F|_{\mathscr{O}_F}$ with $h(\tau_{F_0}) = \tau_F$. Then the Hölder exponent of h is at most $\frac{1}{2}$.

An infinitely renormalizable Hénon map F has *bounded geometry* if

$$\mathrm{diam}(B_{wv}^n) \asymp \mathrm{dist}(B_{wv}^n, B_{wc}^n),$$

for $n \geq 1$ and $w \in \{v, c^{n-1}\}$ and $v \in \{v, c\}$. A slight modified version of this definition would require

$$\mathrm{diam}(B_{wv}^n \cap \mathscr{O}_F) \asymp \mathrm{dist}(B_{wv}^n \cap \mathscr{O}_F, B_{wc}^n \cap \mathscr{O}_F).$$

The one-dimensional renormalization theory relies on the bounded geometry of the Cantor sets, see [20, 25]. This crucial property fails to hold for typical Hénon maps. The following Theorem, see [16], holds for both definitions of bounded geometry:

Theorem 37.11. *There exists G_δ set $G \subset [0, 1]$ of full measure with the following property. Let $F \in \mathscr{I}_\Omega(\bar\varepsilon)$ with sufficiently small $\bar\varepsilon$. The map F does not have bounded geometry if $b_F \in G$.*

Consider a renormalization cycle $\{B_\omega^n \cap \mathscr{O}_F | \omega \in \{v, c\}^n\}$ of an infinitely renormalizable map $F \in \mathscr{I}_\Omega(\bar\varepsilon)$. The non-rigidity theorem implies that the geometry of some of the pieces in this cycle differ from their one-dimensional counterpart. For a typical map the difference can be arbitrary large, see Theorem 37.11.

This phenomenon could restrict tremendously succesfull applications of renormalization in higher dimensions. However, the universal geometrical properties of one-dimensional maps are observed in many higher-dimensional applications. The explanation is that the geometry of most pieces of a renormalization cycle are asymptotically equal to their one-dimensional counterpart. This leads to the notion of *probabilistic universality* and *probabilistic rigidity*.

The precise definition of these probabilistic notions needs some preparation. Consider the degenerate Hénon map $F_* = (f_*(x), x)$, the renormalization fixed point. Observe that for $n \geq 1$ large the pieces B_ω^n are almost straight line segments. The *scaling ratio* of a piece B_ω^n, with $\omega = \omega_0 v \in \{v, c\}^n$ is

$$\sigma_\omega^* = \frac{|\pi_2(B_\omega^n)|}{|\pi_2(B_{\omega_0}^{n-1})|},$$

where π_2 is the projection onto the vertical axis. Notice that $B_{\omega_0}^{n-1}$ is the piece of the previous level containing B_ω^n. The function

$$\omega \mapsto \sigma_\omega^*$$

is called the *universal scaling function*.

Consider an infinitely renormalizable map $F \in \mathscr{I}_\Omega(\bar\varepsilon)$ and a piece B_ω^n. Let us rotate it and then rescale it to horizontal size 1; denote the corresponding linear

Fig. 37.5 Scaling ratios

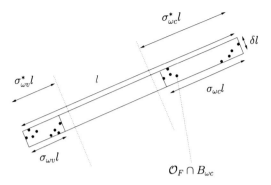

conformal map by A. Choose the map A to obtain minimal numbers $\delta, \sigma_{\omega c}, \sigma_{\omega v} \geq 0$ such that:

(1) $A(B_\omega^n \cap \mathscr{O}_F) \subset [0, 1] \times [0, \delta]$
(2) $A(B_{\omega c}^{n+1} \cap \mathscr{O}_F) \subset [0, \sigma_{\omega c}] \times [0, \delta]$
(3) $A(B_{\omega v}^{n+1} \cap \mathscr{O}_F) \subset [1 - \sigma_{\omega v}, 1] \times [0, \delta]$

where $B_{\omega c}^{n+1}$, and $B_{\omega v}^{n+1}$ are the two pieces of level $n+1$ contained in B_ω^n. We say that B_ω^n is ϵ-*universal* if

$$|\sigma_{\omega c} - \sigma_{\omega c}^*| \leq \epsilon, \quad |\sigma_{\omega v} - \sigma_{\omega v}^*| \leq \epsilon, \quad \text{and} \quad \delta \leq \epsilon.$$

The *precision* of the piece B_ω^n is the smallest $\epsilon > 0$ for which B is ϵ-universal. Let

$$\mathscr{S}^n(\epsilon) \subset \{B_\omega^n\}$$

be the collection of ϵ-universal pieces (Fig. 37.5).

Definition 37.1. The Cantor attractor $\mathscr{O}_F$ of an infinitely renormalizable Hénon map $F \in \mathscr{H}_\Omega(\bar{\epsilon})$ is *universal in probabilistic sense* if there is $\theta < 1$ such that

$$\mu(\mathscr{S}^n(\theta^n)) \geq 1 - \theta^n, \quad n \geq 1.$$

Theorem 37.12. *(Probabilistic Universality) The Cantor attractor $\mathscr{O}_F$ is universal in probabilistic sense.*

Denote the invariant line field of zero characteristic, see [26], by

$$T : \mathscr{O}_F \to \mathbb{P}^1.$$

This line field is not continuous, see Theorem 37.9. However, we can determine sets of arbitrary large measure, with respect to the invariant measure on $\mathscr{O}_F$, on which it is continuous. Namely, for each $N \geq 1$ let

$$X_N = \bigcap_{k \geq N} \mathscr{S}_k(\theta^k),$$

where $\theta < 1$ is given by Theorem 37.12 and notice that

$$\mu(X_N) \geq 1 - O(\theta^N).$$

Let

$$X = \bigcup X_N.$$

Theorem 37.13. *There exists $\beta > 0$ such that the restriction $T \mid X_N$ is β-Hölder*

$$dist(T(x_0), T(x_1)) \leq C_N |x_0 - x_1|^\beta,$$

with $x_0, x_1 \in X_N$.

Theorem 37.14. *The line field T over each X_N consists of β-Hölder tangent lines to $\mathscr{O}_F$. Namely, for each $N \geq 1$ there exists $C_N > 0$ such that*

$$dist(x, T_{x_0}) \leq C_N |x - x_0|^{1+\beta}$$

when $x \in \mathscr{O}_F$, $x_0 \in X_N$.

Remark 37.4. The constants C_N tend to infinity when N becomes large.

Theorem 37.15. *Each set $X_N \subset \mathscr{O}_F$ is contained in a $C^{1+\beta}$-curve.*

Theorem 37.16. *The Cantor set $\mathscr{O}_F$ is contained in a rectifiable-curve.*

Definition 37.2. The attractor $\mathscr{O}_F$ of an infinitely renormalizable Hénon map $F \in \mathscr{H}_\Omega(\bar{\epsilon})$, $\bar{\epsilon} > 0$ small enough, is *rigid in probabilistic sense* if there exists $\beta > 0$ such that for every $\epsilon > 0$ there exists $X \subset \mathscr{O}_F$ with $\mu(X) > 1 - \epsilon$ and such that the restriction $h : X \to h(X)$ of the conjugation $h : \mathscr{O}_F \to \mathscr{O}_{F_*}$, is a $C^{1+\beta}$-diffeomorphism.

Theorem 37.17. *The Cantor attractor $\mathscr{O}_F$ is rigid in probabilistic sense.*

The Hausdorff dimension of a measure μ on a metric space $\mathscr{O}$ is defined as

$$HD_\mu(\mathscr{O}) = \inf_{\mu(X)=1} HD(X).$$

Theorem 37.18. *The Hausdorff dimension is universal*

$$HD_\mu(\mathscr{O}_F) = HD_{\mu_*}(\mathscr{O}_{F_*}).$$

37.6 Topology of the Attractor

The *global attracting set* of a map F is

$$\mathscr{A}_F = \bigcap_{k \geq 0} F^k(\mathrm{Dom}(F))$$

For a discussion on the concept of attractor see [23] and [24]. The dynamics of an infinitely renormalizable map $F \in \mathscr{H}_\Omega(\bar{\epsilon})$, $\bar{\epsilon} > 0$ small enough, is controlled by its periodic orbits $\boldsymbol{\beta}_n$ of period 2^n, $n \geq 0$, and the invariant Cantor set $\mathcal{O}_F$. The periodic orbits are flip saddles.

Theorem 37.19. *Given an infinitely renormalizable Hénon map $F \in \mathscr{I}_\Omega(\bar{\epsilon})$ with $\bar{\epsilon} > 0$ small enough, we have:*

$$\mathscr{A}_F = \overline{W^u(\boldsymbol{\beta}_0)} = \mathcal{O}_F \cup \bigcup_{n \geq 0} W^u(\boldsymbol{\beta}_n).$$

Furthermore, for every point $x \in \mathrm{Dom}(F)$ either $x \in W^s(\boldsymbol{\beta}_n)$ for some $n \geq 0$ or $\omega(x) = \mathcal{O}_F$. The non-wandering set of F is $\Omega_F = \mathscr{P}_F \cup \mathcal{O}_F$.

The second part of Theorem 37.19, concerning the limit sets of points and the non-wandering set, was already obtained in [13].

The topology of the non-wandering set of an infinitely renormalizable Hénon map is as in the degenerate one-dimensional context. However, the attractors $\mathscr{A}_F$ do have a topology which differs from the one-dimensional situation. The topological differences occur in the *heteroclinic web*

$$W = \bigcup_{k \geq 0} W^u(\boldsymbol{\beta}_k) \cup \bigcup_{k \geq 0} W^s(\boldsymbol{\beta}_k).$$

The topology of the heteroclinic web can be changed by changing the average Jacobian. The reason for this lies in the universal geometry observed around the tip. In particular, the rate of accumulation of stable manifolds corresponding to different periodic orbits towards the stable manifold of the tip is universal. Although the average Jacobian is an ergodic theoretical invariant it also controls the accumulation rate towards the tip of the unstable manifolds of periodic points. This geometrical relation between the invariant manifolds and the average Jacobian leads to

Theorem 37.20. *The average Jacobian is a topological invariant.*

The central idea that the topology of the system determines the geometry might still hold for infinitely renormalizable Hénon maps. Maps with different average Jacobian do have different geometry, Theorem 37.10, but also different topology, Theorem 37.20. It is still open whether maps with the same average Jacobian have rigid Cantor attractors.

The heteroclinic web is a countable collection of disjoint curves. A point in the web is called *laminar* if it has a matchbox-neighborhood, a neighborhood homeomorph to $(-1, 1) \times Q$ where $Q \subset [0, 1]$ is a countable set. A heteroclinic tangency is a tangency between some $W^u(\boldsymbol{\beta}_k)$ and $W^s(\boldsymbol{\beta}_n)$.

Remark 37.5. An infinitely renormalizable $F \in \mathcal{H}^n_\Omega(\bar{\varepsilon})$, with $\bar{\varepsilon} > 0$ small enough, can not have homoclinic tangencies, a tangency between $W^u(\boldsymbol{\beta}_k)$ and $W^s(\boldsymbol{\beta}_k)$.

Theorem 37.21. *The heteroclinic web is laminar if there are no heteroclinic tangencies.*

There are examples of infinitely renormalizable maps whose heteroclinic webs do not have laminar points at all, except along finitely many unstable manifolds. Given a periodic orbit $\boldsymbol{\beta}_n$ of F, denote the two exponents by $\lambda^u_n < -1$ and $\lambda^s_n \in (-1, 0)$.

Theorem 37.22. *If the map F has a tangency*

$$W^u(\boldsymbol{\beta}_k) \sqcap W^s(\boldsymbol{\beta}_n),$$

$k < n$, *and*

$$\frac{\ln |\lambda^u_k|}{\ln |\lambda^s_n|} \notin \mathbb{Q}$$

then no point in

$$\mathcal{A}_F \setminus \bigcup_{l<n} W^u(\boldsymbol{\beta}_l)$$

is laminar.

37.7 A Step Towards the Palis Conjecture

A map $F : B \to B$ is *Morse-Smale* if the non-wandering set Ω_F consists of finitely many periodic points, all hyperbolic, and the stable and unstable manifolds of the periodic points are all transversal to each other. The collection $\mathcal{I}^n_\Omega(\bar{\varepsilon}) \subset \mathcal{H}^n_\Omega(\bar{\varepsilon})$ consists of the maps which are exactly n-times renormalizable and have a periodic attractor of period 2^n. The non-wandering set of each map $F \in \mathcal{I}^n_\Omega(\bar{\varepsilon})$, with $\bar{\varepsilon} > 0$ small enough, consists of finitely many periodic points. In particular, a map $F \in \mathcal{I}^n_\Omega(\bar{\varepsilon})$ is Morse-Smale if all its periodic points are hyperbolic and if for every $x, y \in \mathcal{P}_F = \Omega_F$ there are only transverse intersections of $W^u(x)$ and $W^s(y)$.

Theorem 37.23. *Let $\bar{\varepsilon} > 0$ be small enough. The Morse-Smale maps form an open and dense subset of any $\mathcal{I}^n_\Omega(\bar{\varepsilon})$.*

A *Morse-Smale component* is a connected component of the set of non-degenerate Morse-Smale maps in $\mathcal{H}_\Omega(\bar{\varepsilon})$. Morse-Smale maps are structurally stable, see [27]. Two Morse-Smale components in $\mathcal{I}^n_\Omega(\bar{\varepsilon})$ are of different *type* if the maps in the first component are not conjugate to the maps in the other.

Theorem 37.24. *Let $\bar{\varepsilon} > 0$ be small enough. Then for $n \geq 1$ large enough there are countably many Morse–Smale components of different type in $\mathscr{I}_\Omega^n(\bar{\varepsilon})$. The collection of Morse–Smale components in $\mathscr{I}_\Omega^n(\bar{\varepsilon})$ is not locally finite.*

A finitely renormalizable map $F \in \mathscr{H}_\Omega(\bar{\varepsilon})$ is called *hyperbolic* if its non-wandering set can be decomposed as

$$\Omega_F = \Lambda_F \cup P_F,$$

where P_F is a hyperbolic periodic attractor which attracts almost every point and Λ_F a hyperbolic zero-dimensional set. A closed invariant set is hyperbolic if it has an invariant splitting consisting of one stable direction and one unstable direction. See [29] for a general discussion of hyperbolicity and invariant splittings. The map is called *hyperbolic with positive entropy* if Λ_F contains a Cantor set which has positive entropy. The Morse–Smale maps discussed in Theorem 37.24 are hyperbolic.

Theorem 37.25. *Let $\gamma \subset \mathscr{H}_\Omega(\bar{\varepsilon})$ be a smooth curve through $F_0 \in \mathscr{I}_\Omega(\bar{\varepsilon})$, $\bar{\varepsilon} > 0$ small enough, which is transversal to $\mathscr{I}_\Omega(\bar{\varepsilon})$. The hyperbolic maps with positive entropy in γ have positive density in F_0.*

A map $F \in \mathscr{H}_\Omega(\bar{\varepsilon})$ is called *regular* if there exists a periodic attractor which attracts almost every point. It is called *stochastic* if there exists an SRB measure which describes the statistics of almost every orbit. Benedicks and Carleson have shown that the stochastic Hénon maps with a fixed, but small Jacobian, form a set of positive one-dimensional measure, [3]. They discuss Hénon maps in a neighborhood of a specific Misiurewicz unimodal map.

By no means it is a straightforward task to extend their discussion to neighborhoods of unimodal maps with a Collet–Eckmann condition. However, the following result is within reach.

Given a family of unimodal maps, for example the unstable manifold of the period doubling renormalization fixed point. Every Collet–Eckmann map in this family has, for every $\delta > 0$, a neighborhood $U \subset \mathscr{H}_\Omega(\bar{\delta})$ with the following property. Consider a one-parameter family of Hénon maps close enough to the given unimodal family. The fraction of stochastic maps in the part of this Hénon family which crosses the neighborhood U, is at least $1 - \delta$.

Renormalization, the fact that regular and Collet–Eckmann maps have full measure [1], and this extension would prove the following step towards the Palis Conjecture, [28].

Let γ be a curve through an infinitely renormalizable map $F \in \mathscr{I}_\Omega(\bar{\varepsilon})$, $\bar{\varepsilon} > 0$ small enough, which is transversal to $\mathscr{I}_\Omega(\bar{\varepsilon})$. The map F is a Lebesgue density point of the regular and stochastic maps in the curve γ.

37.8 Open Problems

Let us finish with some further questions that naturally arise from the previous discussion.

Problem I:

(1) Prove that F_* is the only fixed point of the Hénon renormalization R, and $R^n F \to F_*$ exponentially for any infinitely renormalizable Hénon map F.

(2) Is it true that the trace of the unstable manifold $\mathcal{W}^u(F_*)$ by the two-parameter Hénon family $F_{c,b} : (x, y) \mapsto (x^2 + c - by, x)$ is a (real analytic) curve γ on which the Jacobian b assumes all values $0 < b < 1$. If so, does this curve converge to some particular point $(c, 1)$ as $b \to 1$?

Problem II:

(1) Is the conjugacy $h: \mathcal{O}_F \to \mathcal{O}_G$ always Hölder?

(2) Can $\mathcal{O}_F$ have bounded geometry when $b_F \neq 0$? If so, does this property depend only on the average Jacobian b_F?

(3) Does the Hausdorff dimension of $\mathcal{O}_F$ depend only on the average Jacobian b_F? (This question was suggested by A. Avila.)

Problem III:

(1) A *wandering domain* is an open set in the basin of attraction of $\mathcal{O}_F$. Do wandering domains exist?

(2) If a map $F \in \mathscr{I}_\Omega(\bar{\varepsilon})$ does not have wandering domains then the union $\mathscr{F}^s$ of all stable manifolds of periodic points is dense in the domain of F. Does there exist $F \in \mathscr{I}_\Omega(\bar{\varepsilon})$ such that $\mathscr{F}^s$ is not laminar even if there are no heteroclinic tangencies?

(3) For $F \in \mathscr{I}_\Omega(\bar{\varepsilon})$ let $\mathscr{F}^s_\tau$ be the union of stable manifolds of the points in the orbit of the tip. Is $\mathscr{F}^s_\tau$ dense in $\mathrm{Dom}(F)$?

Problem IV:

The unique invariant measure on the Cantor attractor $\mathcal{O}_F$ has characteristic exponents 0 and $\ln b_F < 0$. Can the stable characteristic exponent of the tip τ_F differ from $\ln b_F$?

Problem V:

Can we still speak of rigidity of the Cantor attractor $\mathcal{O}_F$?

(1) Are the Cantor attractors rigid within the topological conjugacy classes of maps restricted to a neighborhood of the Cantor attractor?

(2) Prove or disprove that two Cantor attractors $\mathcal{O}_F$ and $\mathcal{O}_{\tilde{F}}$ are smoothly equivalent if and only if they have the same average Jacobian.

Problem VI:

(1) Can different Morse–Smale components

$$MS_1, MS_2 \subset \bigcup_{n \geq 0} \mathscr{I}_\Omega^n(\bar{\varepsilon})$$

  have the same type, that is the maps in MS_1 are conjugate to the maps in MS_2?

(2) As we have shown, the Morse–Smale Hénon maps are dense in the zero entropy region with small Jacobian. Are they dense in the full zero entropy region of dissipative Hénon maps? How about other real analytic families of dissipative two dimensional maps?

(3) The discussion that led to Theorem 37.24 was based on the renormalization structure. However, the non-locally finiteness of the collection of Morse–Smale components might be a more general phenomenon. Study the combinatorics of Morse–Smale components in other real analytic families of dissipative two dimensional maps.

(4) Are the real Morse–Smale Hénon maps from Theorem 37.24 hyperbolic on $\mathbb{C}^2$? To what extent the topology of the real heteroclinic web determines the topology of the corresponding Hénon map on $\mathbb{C}^2$?

Problem VII: Is the convergence of the statistics of the *bad* pieces,

$$\lim_{n \to \infty} \mu(\mathscr{B}_n \setminus \mathscr{S}_n(\varepsilon)) = 0,$$

governed by some sort of universality? This question is related to Problem II on the regularity of the conjugation $h : \mathscr{O}_G \to \mathscr{O}_G$ when $b_F = b_G$.

References

1. Avila, A., Moreira, C.G.: Quasisymmetric robustness of the Collet-Eckmann condition in the quadratic family. Bull. Braz. Math. Soc. **35**(2), 291–331 (2004)
2. Buff, X.: Geometry of the Feigenbaum map. Conf. Geom. Dyn. **3**, 79–169 (1999)
3. Benedicks, M., Carleson, L.: On dynamics of the Hénon map. Ann. Math. **133**, 73–169 (1991)
4. Birkhoff, C., Martens, M., Tresser, C.P.: On the Scaling Structure for Period Doubling. Asterisque **286**, 167–186 (2003)
5. Chandramouli, V.V.M.S.: Universality and non-rigidity, PhD Thesis Stony Brook Dec (2008)
6. Collet, P., Eckmann, J.P., Koch, H.: Period doubling bifurcations for families of maps on $\mathbb{R}^n$. J. Stat. Phys. **25**, 1–15 (1980)
7. de Carvalho, A., Lyubich, M., Martens, M.: Renormalization in the Hénon family, I: universality but non-rigidity. J. Stat. Phys. **121**(5/6), 611–669 (2005)
8. Chandramouli, V.V.M.S., Martens, M., de Melo, W., Tresser, C.P.: Chaotic period doubling. Ergod. Theory Dyn. Syst. **29**, 381–418 (2009)
9. Coullet, P., Tresser, C.: Itération d'endomorphismes et groupe de renormalisation. J. Phys. Colloq. C **539**, C5-25 (1978)

10. Feigenbaum, M.J.: Quantitative universality for a class of non-linear transformations. J. Stat. Phys. **19**, 25–52 (1978)
11. Feigenbaum, M.J.: The universal metric properties of non-linear transformations. J. Stat. Phys. **21**, 669–706 (1979)
12. de Faria, E., de Melo, W., Pinto, A.: Global hyperbolicity of renormalization for C^r unimodal mappings. Ann. Math. **164**(3), 731–824 (2006)
13. Gambaudo, J.-M., van Strien, S., Tresser, C.: Hénon-like maps with strange attractors: there exist C^∞ Kupka-Smale diffeomorphisms on S^2 with neither sinks nor sources. Nonlinearity **2**, 287–304 (1989)
14. Hazard, P.: Hénon Renormalization. PhD Thesis Stony Brook. (2008)
15. Hermann, M.R.: Sur la conjugaision differentiable des diffeomorphismes du cercle. Publ. Math. IHES **49** (1976)
16. Hazard, P., Lyubich, M., Martens, M.: Unbounded geometry for typical infinitely renormalizable Hénon maps. IMS Stony Brook preprint 10-2 and submitted for publication
17. Lyubich, M.: Feigenbaum-Coullet-Tresser Universality and Milnor's Hairiness Conjecture. Ann. Math. **149**, 319–420 (1999)
18. M. Lyubich, M. Martens. Renormalization in the Hénon family, II. The Heteroclinic Web. IMS Stony Brook preprint 08-2 and accepted for publication in Inventiones Mathematicae
19. Lyubich, M., Martens, M.: Renormalization in the Hénon family, III. Probabilistic universality and rigidity. Submitted for publication
20. Martens, M.: Distortion results and invariant cantor sets for unimodal maps. Ergod. Theory Dyn. Syst. **14**, 331–349 (1994)
21. Martens, M.: The periodic points of renormalization. Ann. Math. **147**, 543–584 (1998)
22. McMullen, C.: Renormalization and 3-manifolds which fiber over the circle. Annals of Math. Studies, vol. 135. Princeton University Press, Princeton (1996)
23. Milnor, J.W.: On the concept of attractor. Comm. Math. Phys. **99**, 177–195 (1985)
24. J.W. Milnor. Attractor, Scholarpedia
25. de Melo, W., van Strien, S.: One-Dimensional Dynamics. Springer, Berlin (1993)
26. Oseledec, V.I.: A multiplicative ergodic theorem. Lyapunov characteristic numbers for dynamical systems. Trans. Mosc. Math. Soc. **19**, 197–231 (1969)
27. Palis, J.: On Morse-Smale dynamical systems. Topology **8**, 385–404 (1968)
28. Palis, J.: A global view of dynamics and a Conjecture of the denseness of finitude of attractors. Asterisque **261**, 335–348 (2000)
29. Palis, J., Takens, F.: Hyperbolicity and sensitive chaotic dynamics at homoclinic bifurcations. Cambridge Studies in Advanced Mathematics, vol. 35. Cambridge University Press, Cambridge (1993)
30. Sullivan, D.: Bounds, quadratic differentials, and renormalization conjectures. AMS Centennial Publications. v. 2: Mathematics into Twenty-first Century (1992)
31. Tresser, C., Coullet, P.: Itération d'endomorphismes et groupe de renormalisation. C.R. Acad. Sc. Paris **287A**, 577–580 (1978)
32. Vul, E.B., Sinai, Ya.G., Khanin, K.M.: Feigenbaum universality and the thermodynamical formalism. Russ. Math Surv. **39**(3), 1–40 (1984)
33. Yampolsky, M.: The attractor of renormalization and rigidity of towers of critical circle maps. Comm. Math. Phys. **218**, 537–568 (2001)

Chapter 38
Application of Fractional Calculus in Engineering

J.A. Tenreiro Machado, Isabel S. Jesus, Ramiro Barbosa, Manuel Silva, and Cecilia Reis

Abstract Fractional Calculus (FC) goes back to the beginning of the theory of differential calculus. Nevertheless, the application of FC just emerged in the last two decades. It has been recognized the advantageous use of this mathematical tool in the modelling and control of many dynamical systems. Having these ideas in mind, this paper discusses a FC perspective in the study of the dynamics and control of several systems. The paper investigates the use of FC in the fields of controller tuning, legged robots, electrical systems and digital circuit synthesis.

38.1 Introduction

The generalization of the concept of derivative $D^\alpha[f(x)]$ to non-integer values of α goes back to the beginning of the theory of differential calculus. In fact, Leibniz, in his correspondence with Bernoulli, L'Hôpital and Wallis (1695), had several notes about the calculation of $D^{1/2}[f(x)]$. Nevertheless, the development of the theory of Fractional Calculus (FC) is due to the contributions of many mathematicians such as Euler, Liouville, Riemann and Letnikov [1–3].

The FC deals with derivatives and integrals to an arbitrary order (real or, even, complex order). The mathematical definition of a derivative/integral of fractional order has been the subject of several different approaches [1–3]. For example, the

---

J.T. Machado (✉)
Department of Electrical Engineering, ISEP-Institute of Engineering of Porto, Rua Dr. Antonio Bernardino de Almeida, 4200-072 Porto, Portugal
e-mail: jtm@isep.ipp.pt

I.S. Jesus, R.Barbosa, M. Silva, and C. Reis
Institute of Engineering of Porto, Rua Dr. António Bernardino de Almeida, 4200-072 Porto, Portugal
e-mail: isj@isep.ipp.pt, rsb@isep.ipp.pt, mss@isep.ipp.pt, cmr@isep.ipp.pt

M.M. Peixoto et al. (eds.), *Dynamics, Games and Science I*, Springer Proceedings in Mathematics 1, DOI 10.1007/978-3-642-11456-4_38,
© Springer-Verlag Berlin Heidelberg 2011

Laplace definition of a fractional derivative/integral of a signal $x(t)$ is:

$$D^{\alpha} x(t) = L^{-1} \left\{ s^{\alpha} X(s) - \sum_{k=0}^{n-1} s^k D^{\alpha-k-1} x(t) \Big|_{t=0} \right\} \tag{38.1}$$

where $n - 1 < \alpha \le n$, $\alpha > 0$. The Grünwald–Letnikov definition is given by $(\alpha \in \Re)$:

$$D^{\alpha} x(t) = \lim_{h \to 0} \left[\frac{1}{h^{\alpha}} \sum_{k=0}^{\infty} (-1)^k \frac{\Gamma(\alpha + 1)}{\Gamma(k + 1)\Gamma(\alpha - k + 1)} x(t - kh) \right] \tag{38.2}$$

where Γ is the Gamma function and h is the time increment. Expression (38.2) shows that fractional-order operators are "global" operators having a memory of all past events, making them adequate for modelling memory effects in most materials and systems.

In recent years FC has been a fruitful field of research in science and engineering [1–5]. In fact, many scientific areas are currently paying attention to the FC concepts and we can refer its adoption in viscoelasticity and damping, diffusion and wave propagation, electromagnetism, chaos and fractals, heat transfer, biology, electronics, signal processing, robotics, system identification, traffic systems, genetic algorithms, percolation, modelling and identification, telecommunications, chemistry, irreversibility, physics, control systems, economy and finance.

Bearing these ideas in mind, Sects. 38.2–38.5 present several applications of FC in science and engineering. Finally, Sect. 38.6 draws the main conclusions.

38.2 Tuning of PID Controllers Based on Fractional Calculus

In this section we study a novel methodology for tuning PID controllers such that the response of the compensated system has an almost constant overshoot defined by a prescribed value. The proposed method is based on the minimization of the integral of square error (ISE) between the step responses of a unit feedback control system, whose open-loop transfer function $L(s)$ is given by a fractional-order integrator and that of the PID compensated system [6].

Figure 38.1a illustrates the fractional-order control system that will be used as reference model for the tuning of PID controllers. The corresponding Bode diagrams of amplitude and phase of $L(s)$ are illustrated in Fig. 38.1b.

This choice of $L(s)$ gives a closed-loop system with the desirable property of being insensitive to gain changes. If the gain changes, the crossover frequency ω_c will change, but the phase margin of the system remains PM $= \pi(1 - \alpha/2)$ rad, independently of the value of the gain, as can be seen from the curves of amplitude and phase of Fig. 38.1b. With the order α and the crossover frequency ω_c we can establish the overshoot and the speed of the output response, respectively.

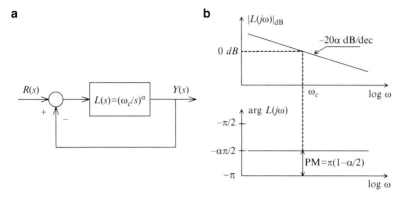

Fig. 38.1 Fractional-order control system with open-loop transfer function $L(s)$ and Bode diagrams of amplitude and phase of $L(j\omega)$ for $1 < \alpha < 2$

The transfer function of the PID controller is:

$$G_c(s) = \frac{U(s)}{E(s)} = K\left(1 + \frac{1}{T_i s} + T_d s\right) \quad (38.3)$$

The design of the PID controller will consist on the determination of the optimum PID set gains (K, T_i, T_d) that minimize J, the integral of the square error (ISE), defined as:

$$J = \int_0^\infty [y(t) - y_d(t)]^2 \, dt \quad (38.4)$$

where $y(t)$ is the step response of the unit feedback control system with the PID controller and $y_d(t)$ is the desired step response of the fractional-order control system of Fig. 38.1a.

To illustrate the effectiveness of proposed methodology we consider the third-order plant transfer function:

$$G_p(s) = \frac{K_p}{(s+1)^3} \quad (38.5)$$

Figure 38.2 shows the step responses and the Bode diagrams of phase of the closed-loop system with the PID for the transfer function $G_p(s)$ for gain variations around the nominal gain ($K_p = 1$) corresponding to $K_p = \{0.6, 0.8, 1.0, 1.2, 1.4\}$, that is, for a variation up to -40% of its nominal value. The system was tuned for $\alpha = 3/2$ (PM $= 45°$), $\omega_c = 0.8$ rad/s. We verify that we get the same desired iso-damping property corresponding to the prescribed (α, ω_c)-values.

We verify that the step responses have an almost constant overshoot independently of the variation of the plant gain around the gain crossover frequency ω_c. Therefore, the proposed methodology is capable of producing closed-loop systems robust to gain variations and step responses exhibiting an iso-damping property.

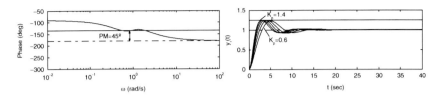

Fig. 38.2 Bode phase diagrams and step responses for the closed-loop system with a PID controller for $G_p(s)$. The PID parameters are $K = 1.9158$, $T_i = 1.1407$ and $T_d = 0.9040$

The proposed method was tested on several cases studies revealing good results. It was also compared with other tuning methods showing comparable or superior results [6].

38.3 Fractional Control of Legged Robots

The present section compares the tuning of Fractional Order (FO) algorithms, applied to the joint control of a walking robot with $n = 6$ legs, equally distributed along both sides of the robot body, each with three rotational joints, $j = \{1, 2, 3\} \equiv$ {hip, knee, ankle} [7].

Regarding the dynamic model for the hexapod body and foot-ground interaction, it is considered the existence of robot body compliance because walking animals have a spine that allows supporting the locomotion with improved stability. The robot body is divided in n identical segments (each with mass $M_b n^{-1}$) and a linear spring-damper system (with parameters defined so that the body behaviour is similar to the one expected to occur on an animal) is adopted to implement the intra-body compliance [7]. The contact of the ith robot feet with the ground is modelled through a non-linear system [7], being the values for the parameters based on the studies of soil mechanics [7].

We evaluate the effect of different PD^μ controller implementations for $G_{c1}(s)$, while G_{c2} is a P controller. The PD^μ $0 < \mu_j \leq 1$ ($j = 1, 2, 3$) algorithm is implemented through a discrete-time 4th-order Padé approximation.

The performance analysis is based on the formulation of two indices measuring the mean absolute density of energy per travelled distance (E_{av}) and the hip trajectory errors (ε_{xyH}) during walking [8]. It is analyzed the system performance of the different PD^μ controller tuning, when adopting a periodic wave gait at a constant forward velocity V_F [7].

To tune the different controller implementations we adopt a systematic method, testing and evaluating several possible combinations of parameters, for all controller implementations. Therefore, we adopt the $G_{c1}(s)$ parameters that establish a compromise in what concerns the simultaneous minimization of E_{av} and ε_{xyH}, and a proportional controller G_{c2} with gain $Kp_j = 0.9(j = 1, 2, 3)$. It is assumed high performance joint actuators (i.e., with almost negligible saturation), having a

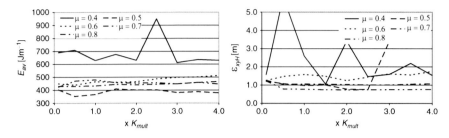

Fig. 38.3 Performance indices E_{av} and ε_{xyH} vs. K_{mult} for the different $G_{c1}(s)$ PD$^\mu$ controller tuning

maximum actuator torque of $\tau_{ijMax} = 400$ Nm. The desired angle between the foot and the ground (assumed horizontal) is established as $\theta_{i3hd} = -15°$. For this case we tune the PD$^\mu$ controllers for values of the fractional order in the interval $+0.1 < \mu_j < +0.9$, establishing $\mu_1 = \mu_2 = \mu_3$.

Since the objective of the walking robots is to walk in natural terrains, we test how the different controllers behave under distinct ground properties.

Considering the previously tuning controller parameters, the values of the ground model parameters are varied simultaneously through a multiplying factor varied in the range $K_{mult} \in [0.1; 4.0]$. This variation for the ground model parameters allows the simulation of the ground behaviour for growing stiffness, from peat to gravel [7]. We conclude that the controller responses are quite similar, meaning that these algorithms are robust to variations of the ground characteristics.

The performance measures *versus* the multiplying factor of the ground parameters K_{mult} are presented on Fig. 38.3.

Analyzing the system performance from the viewpoint of the index E_{av} (Fig. 38.3, left), it is possible to conclude that the best PD$^\mu$ implementation occurs for the fractional order $\mu_j = 0.5$. Moreover, it is clear that the performances of the different controller implementations are almost constant on all range of the ground parameters, with the exception of the fractional order $\mu_j = 0.4$, where E_{av} presents a significant variation.

When the system performance is evaluated in the viewpoint of the index ε_{xyH} (Fig. 38.3, right) we verify that the controller implementations corresponding to the fractional orders $\mu_j = \{0.6, 0.7, 0.8\}$ present the best values. The fractional order $\mu_j = 0.5$ leads to controller implementations with a slightly inferior performance, particularly for values of $K_{mult} > 2.5$. It is also clear on the chart of ε_{xyH} vs. K_{mult} that the fractional order $\mu_j = 0.4$ leads to a controller implementation with a poor performance.

In conclusion, the controllers with $\mu_j = \{0.5, 0.6, 0.7, 0.8\}$ present lower values of the indices E_{av} and ε_{xyH} on almost all range of K_{mult} under consideration. The only exception to this observation occurs for the PD$^\mu$ controller implementation when $\mu_j = 0.5$, that presents slightly higher values of the index ε_{xyH} for values of $K_{mult} > 2.5$. We conclude that the controller responses are quite similar, meaning that these algorithms are robust to variations of the ground characteristics [8].

38.4 Electrical Impedance of Fruits

In an electrical circuit the voltage $u(t)$ and the current $i(t)$ can be expressed as a function of time t:

$$u(t) = U_0 cos(\omega t); \quad i(t) = I_0 cos(\omega t + \phi) \tag{38.6}$$

where U_0 and I_0 are the amplitudes of the signals, ω is the frequency and ϕ is the current phase shift. The voltage and current can be expressed in complex form as:

$$u(t) = Re\left\{U_0 e^{j(\omega t)}\right\}; \quad i(t) = Re\left\{I_0 e^{j(\omega t + \phi)}\right\} \tag{38.7}$$

Consequently, the electrical impedance $Z(j\omega)$ is:

$$Z(j\omega) = \frac{U(j\omega)}{I(j\omega)} = Z_0 e^{j\phi} \tag{38.8}$$

Usually these concepts are applied to inorganic systems leading to the well known resistance, inductance and capacitance fundamental electrical elements [9]. However, the structure of fruits and vegetables have cells that constitute electrical circuits exhibiting a complex behavior. Bearing these facts in mind, in our work we study the electrical impedance for several botanical elements, under the point of view of fractional order systems.

We apply sinusoidal excitation signals $v(t)$, to the botanical system, for several distinct frequencies ω (Fig. 38.4) and the impedance $Z(j\omega)$ is measured based on the resulting voltage $u(t)$ and current $i(t)$. Moreover, we measure the environmental temperature, the weight, the length and width of all botanical elements. This criterion helps us to understand how these factors influence $Z(j\omega)$ [10].

In this study we develop several different experiments for evaluating the variation of the impedance $Z(j\omega)$ with the amplitude of the input signal V_0, for different electrode lengths of penetration inside the element Δ, the environmental temperature T, the weights W and the dimension D. The value of R is changed for each experiment, in order to adapt the values of the voltage and current to the scale of the measurement device.

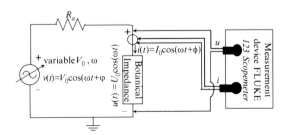

Fig. 38.4 Electrical circuit for the measurement of the botanical impedance $Z(j\omega)$

38 Application of Fractional Calculus in Engineering

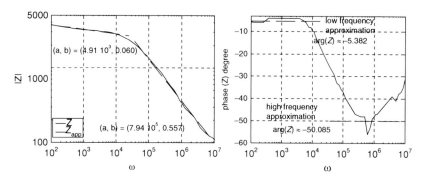

Fig. 38.5 Bode diagrams of the impedance $Z(j\omega)$ for the potato

We start by analyzing the impedance for an amplitude of input signal of $V_0 = 10$ volt, a constant adaptation resistance $R_a = 15$ k ohm, applied to one *Solanum Tuberosum* (potato), with an weight $W = 1.24 \, 10^{-1}$ kg, environmental temperature $T = 26.5$ degree Celsius, dimension $D = 7.97 \, 10^{-2} \times 5.99 \, 10^{-2}$ m, and the electrode length penetration $\Delta = 2.1 \, 10^{-2}$ m.

Figure 38.5 presents the Bode diagrams for $Z(j\omega)$.s The results reveal that the system has a fractional order impedance. In fact, approximating the experimental results in the amplitude Bode diagram through a power function namely by $|Z(j\omega)| = a\omega^{-b}$, we obtain $(a,b) = (4.91 \, 10^3, 0.0598)$, at the low frequencies, and $(a,b) = (7.94 \, 10^5, 0.5565)$, at the high frequencies.

In order to analyze the system linearity we evaluate $Z(j\omega)$ for different amplitudes of input systems, namely, $V_0 = \{5, 15, 20\}$ volt, maintaining constant the adaptation resistance $R_a = 15$ k ohm. The impedance $Z(j\omega)$ has a fractional order and this characteristic does not change significantly with the variation of input signal amplitude. Therefore, we can conclude that this system has a linear characteristic.

In a second experiment, we vary the length Δ of the electrode penetration inside the potato, and we evaluate its influence upon the value of the impedance. Therefore, we adjust the electrode to $\Delta = 1.42 \, 10^{-2}$ m, with $V_0 = 10$ volt and adaptation resistance $R_a = 5$ k ohm, leading to $|Z(j\omega)|$ approximations $(a,b) = (5.48 \, 10^3, 0.0450)$, at the low frequencies, and $(a,b) = (1.00 \, 10^6, 0.5651)$, at the high frequencies. With these results, we conclude that the length of wire inside the potato does not change significantly the values of the fractional orders. Also the linearity was again confirmed.

Similar experiments were developed for several fruits. The results reveal that $Z(j\omega)$ has distinct characteristics according with the frequency range. For low frequencies, the impedance is approximately constant, but for high frequencies, it is clearly of fractional order.

38.5 Circuit Synthesis Using Particle Swarm Optimization

Ornithologists, biologists and psychologists did early research, which led into the theory of particle swarms. In these areas, the term 'swarm intelligence' is well known and characterizes the case when a large number of individuals are able of accomplish complex tasks. Motivated by these facts, some basic simulations of swarms were abstracted into the mathematical field. The usage of swarms for solving simple tasks in nature became an intriguing idea in algorithmic and function optimization. Eberhart and Kennedy were the first to introduce the particle swarm optimization (PSO) algorithm [11, 12], which is an optimization method inspired in the collective intelligence of swarms of biological populations, and was discovered through simplified social model simulation of bird flocking, fishing schooling and swarm theory. In the PSO each particle adjusts its flying according with its own and its companions experiences as described below:

$$v_{id} = v_{id} + c_1\, rand()(p_{id} - x_{id}) + c_2\, Rand()(p_{gd} - x_{id}), \quad x_{id} = x_{id} + v_{id} \qquad (38.9)$$

where c_1 and c_2 are positive constants, $rand()$ and $Rand()$ are two random functions in the range [0,1], $X_i = (x_{i1}, x_{i2}, \dots, x_{iD})$ represents the ith particle, $P_i = (p_{i1}, p_{i2}, \dots, p_{iD})$ is the best previous position (the position giving the best fitness value) of the particle, the symbol g represents the index of the best particle among all particles in the population, and $V_i = (v_{i1}, v_{i2}, \dots, v_{iD})$ is the rate of the position change (velocity) for particle i.

We adopt a PSO algorithm to design combinational logic circuits. A truth table specifies the circuits and the goal is to implement a functional circuit with the least possible complexity. Four sets of logic gates have been defined, as shown in Table 11, being *Gset 2* the simplest one and *Gset 6* the most complex gate set. Logic gate named WIRE means a logical no-operation.

In the PSO scheme the circuits are encoded as a rectangular matrix $\mathbf{A}$ ($row \times column = r \times c$) of logic cells. Three genes represent each cell: $< input1 >< input2 >< gate\ type >$, where $input1$ and $input2$ are one of the circuit inputs, if they are in the first column, or one of the previous outputs, if they are in other columns. The gate type is one of the elements adopted in the gate set. The chromosome is formed with as many triplets as the matrix size demands (*e.g.*, triplets $= 3 \times r \times c$). The initial population of circuits (particles) has a random generation. The initial velocity of each particle is initialized with zero. The following velocities and the new positions are calculated applying (38.9). In this way, each potential solution, called particle, flies through the problem space. For each gene is calculated the corresponding velocity. Therefore, the new positions are as many as the number of genes in the chromosome. If the new values of the input genes result out of range, then a re-insertion function is used. If the calculated gate gene is not allowed a new valid one is generated at random. These particles then have memory and each one keeps information of its previous best position (*pbest*) and its corresponding fitness. The swarm has the pbest of all the particles and the particle with the greatest fitness is called the global best

38 Application of Fractional Calculus in Engineering   627

(*gbest*). Four gate sets were defined: Gset 2 = {AND,XOR,WIRE}, Gset 3 = {AND,OR,XOR,WIRE}, Gset 4 = {AND,OR,XOR,NOT,WIRE} and Gset 6 = {AND,OR,XOR,NOT,NAND,NOR,WIRE}. However, in our case we also use a kind of mutation operator that introduces a new cell in 10% of the population. This mutation operator changes the characteristics of a given cell in the matrix. Therefore, the mutation modifies the gate type and the two inputs, meaning that a completely new cell can appear in the chromosome. To run the PSO we have also to define the number P of individuals to create the initial population of particles. This population is always the same size across the generations, until reaching the solution.

The calculation of the fitness function F_s in (38.10) has two parts, $f_1 = f_1 + 1$ if $\{bit\ i\ of\ Y\} = \{bit\ i\ of\ Y_R\}$, $i = 1, \ldots, f_{10}$ and $f_2 = f_2 + 1$ if gate type = wire, where f_1 measures the functionality and f_2 measures the simplicity. In a first phase, we compare the output Y produced by the PSO-generated circuit with the required values Y_R, according with the truth table, on a bit-per-bit basis. By other words, f_1 is incremented by one for each correct bit of the output until f_1 reaches the maximum value $f_{10} = 2^{ni} \times no$, that occurs when we have a functional circuit. The variables ni and no represent the number of inputs and outputs of the circuit. Once the circuit is functional, in a second phase, the algorithm tries to generate circuits with the least number of gates. This means that the resulting circuit must have as much genes < gate type >≡< wire > as possible. Therefore, the index f_2, that measures the simplicity (the number of null operations), is increased by one (zero) for each wire (gate) of the generated circuit. Consequently, the fitness function F_s is given by:

$$F_s = \begin{cases} f_1, & F_s < f_{10} \\ f_1 + f_2, & F_s \geq f_{10} \end{cases} \tag{38.10}$$

The concept of dynamic fitness function F_d results from an analogy between control systems and the GA case, where we master the population through the fitness function. The simplest control system is the proportional algorithm; nevertheless, there can be other control algorithms, such as, for example, the proportional and the differential scheme. In this line of thought, expression (10) is a static fitness function F_s and corresponds to using a simple proportional algorithm. Therefore, to implement a proportional-derivative evolution the fitness function needs a scheme of the type $F_d = F_s + KD^\mu [F_s]$ [13], where $0 \leq \mu \leq 1$ is the differential fractional-order and $K \in \Re$ is the 'gain' of the dynamical term.

In this study are developed $n = 20$ simulations for each case under analysis. The experiments consist on running the algorithm to generate typical combinational logic circuits, namely a 2-to-1 multiplexer ($M2-1$), a 1-bit full adder ($FA1$), a 4-bit parity checker ($PC4$) and a 2-bit multiplier ($MUL2$). The circuits are generated with the four gate sets and $P = 3,000$, $w = 0.5$, $c_1 = 1.5$ and $c_2 = 2$. We conclude that F_d leads to better results in particular for the $MUL2$ circuit and for the $Av(PT)$. Figure 38.6 presents a comparison between F_s and F_d.

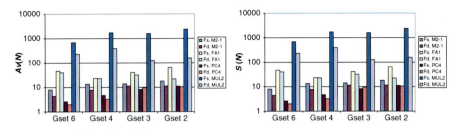

Fig. 38.6 $Av(N)$ and $S(N)$ for the PSO algorithm, $P = 3,000$ using F_s, F_d and $\mu = 0.5$

38.6 Conclusions

We have presented several applications of the FC concepts. It was demonstrated the advantages of using the FC theory in different areas of science and engineering. In fact, this paper studied a variety of different physical systems, namely: tuning of PID controllers using fractional calculus concepts, fractional control of legged robots, electrical impedance of fruits and circuit synthesis using evolutionary algorithms. The results demonstrate the importance of FC in the modelling and control of many systems and motivate for the development of new applications.

References

1. Oldham, K.B., Spanier, J.: The Fractional Calculus. Academic, New York (1974)
2. Miller, K.S., Ross, B.: An Introduction to the Fractional Calculus and Fractional Differential Equations. Wiley, New York (1993)
3. Podlubny, I.: Fractional Differential Equations. Academic, San Diego (1999)
4. Samko, S.G., Kilbas, A.A., Marichev, O.I.: Integrals and Derivatives of the Fractional Order and Some of Their Applications. Nauka and Tekhnika, Minsk (1987)
5. Oustaloup, A.: La Dérivation Non Entière: Théorie, Synthèse et Applications, Editions Hermès, Paris (1995)
6. Barbosa, R.S., Machado, J.A.T., Ferreira, I.M.: Tunning of PID Controller Based on Bode's Ideal Transfer Function. Nonlin. Dyn. **38**(1–4), 305–321 (2004)
7. Silva, M.F., Machado, J.A.T., Jesus, I.S.: Modelling and Simulation of Walking Robots With 3 dof Legs. In: MIC 2006 – The 25th IASTED International Conference on Modelling, Identification and Control. Lanzarote, Spain (2006)
8. Silva, M.F., Machado, J.A.T.: Fractional Order PD^α Joint Control of Legged Robots. Journal of Vibration and Control – Special Issue on Modeling and Control of Artificial Locomotion Systems 12(12), 1483–1501 (2006)
9. Barsoukov, E., Macdonald, J.R.: (2005) Impedance Spectroscopy, Theory, Experiment, and Applications. Wiley, New York
10. Jesus, I.S., Machado, J.A.T., Cunha, J.B., Silva, M.F.: Fractional Order Electrical Impedance of Fruits and Vegetables. Proceedings of the 25th IASTED International Conference on Modeling, Identification and Control, Spain, Feb (2006)
11. Kennedy, J., Eberhart R.C.: Particle Swarm Optimization. Proceedings of the IEEE International Conference Neural Networks: 1942–1948. (1995)

38 Application of Fractional Calculus in Engineering

12. Shi, Y., Eberhart, R.C.: A Modified Particle Swarm Optimizer. Proceedings of the 1998 International Conference on Evolutionary Computation: 69–73. (1998)
13. Reis, C., Machado, J., Cunha, J.: Evolutionary Design of Combinational Circuits Using Fractional-Order Fitness. Proceedings of the Fith EUROMECH Nonlinear Dynamics Conference: 1312–1321. (2005)

Chapter 39
Existence of Invariant Circles for Infinitely Renormalisable Area-Preserving Maps

R.S. MacKay

Abstract Existence of an invariant circle for any orientation-preserving 2D map whose orbit under renormalisation remains forever in a certain bounded subset is proved. The construction dates back to 1984. It was stimulated by a preprint by David Rand doing the same for the dissipative case. To include the general case, notably area-preserving, required a variation on his idea.

39.1 Background

This paper is based on notes that I wrote in April 1984. They were inspired by a preprint by David Rand (eventually published as [1]) in which he proved the analogous result for dissipative annulus maps.

My notes were the starting point for one chapter of the PhD thesis of Nicolai Hoidn whom I supervised from April to September 1984 (eventually published in [2], under his pseudonym), but the idea ended up somewhat obscured under technical analysis there.

The idea was reinvented by Andreas Stirnemann [3], who presented it in the framework of iterated function systems. This made the argument very clear. I was not familiar with the concept back in 1984, but with hindsight one can see it was the right tool, from Fig. 4.4.2.2 of my (1982) PhD thesis (reprinted in [4]), Rand's preprint, and the construction of my notes.

I publish this exposition of my notes now, firstly to put on the record that I had obtained this general existence result back then and secondly to acknowledge David Rand's great influence.

R.S. MacKay
Mathematics Institute, University of Warwick, Coventry CV4 7AL, UK
e-mail: R.S.MacKay@warwick.ac.uk

M.M. Peixoto et al. (eds.), *Dynamics, Games and Science I*, Springer Proceedings
in Mathematics 1, DOI 10.1007/978-3-642-11456-4_39,
© Springer-Verlag Berlin Heidelberg 2011

39.2 Setting

Let (U, T) be a pair of orientation-preserving diffeomorphisms of domains in $\mathbb{R}^2$ to ranges in $\mathbb{R}^2$, which commute on the subset for which both compositions UT, TU are defined. The basic example is (FR, F) where F is a lift to $\mathbb{R}^2$ of an orientation-preserving degree-one map of a cylinder $\mathbb{T} \times \mathbb{R}$ to itself (with $\mathbb{T} = \mathbb{R}/\mathbb{Z}$) (i.e. $F(x+1, y) = F(x, y) + (1, 0)$) and $R(x, y) = (x - 1, y)$ is a deck transformation. The maps U, T do not have to preserve area, nor have twist, but the case with both is the main motivation.

Assumptions. A1. The domains of U and T are assumed to contain a vertical line segment L, without loss of generality in $x = 0$, where both compositions UT, TU are defined and such that $U(L)$ is to the left of L, $T(L)$ to the right of L. The domain of U is assumed to connect L to $T(L)$. The domain of T is assumed to connect L to $U(L)$. See Fig. 39.1.

A2. The horizontal is scaled so that some notion of the horizontal width of the union of the domains of U and T is 1.

The map UT^{-1} takes $T(L)$ to $U(L)$ so the union of the domains of U and T can be considered to be an annulus cut along this line, and the pair (U, T) can be considered as a map of this annulus to an annulus with the same cut, by applying U to points between L and $T(L)$ and T to points between $U(L)$ and L. For points on

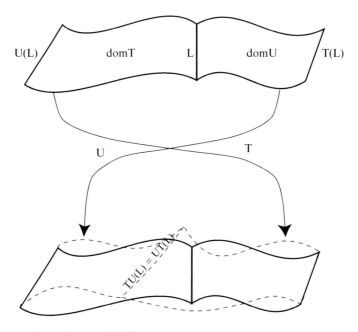

Fig. 39.1 Domains and ranges of U, T

39 Existence of Invariant Circles for Infinitely Renormalisable Area-Preserving Maps

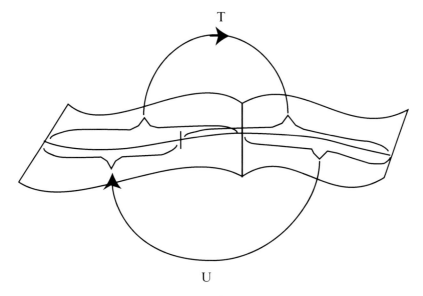

Fig. 39.2 An invariant circle for (U, T)

L we apply both U and T, obtaining two different points, but which are identified under UT^{-1}; their images are the same. Thus we can talk about an orbit segment of a point under (U, T).

39.3 Invariant Circles

Say a curve γ is an *invariant circle* for (U, T) if it joins $U(L)$ to $T(L)$, crosses L once, the image by T of its segment from $U(L)$ to L is defined and contained in γ and the image by U of its segment from L to $T(L)$ is defined and contained in γ. See Fig. 39.2. It is called a "circle" because it becomes a circle under the identification of $T(L)$ with $U(L)$ via UT^{-1}.

An invariant circle has a *rotation number* $\omega \in [0, 1]$, which is the fraction of iterations when U is used.

In my April 1984 notes I restricted attention to the case of *golden circles*, $\omega = \frac{\sqrt{5}-1}{2}$, but I add a section to this paper on the extension to other rotation numbers.

39.4 Renormalisation

If L has a subinterval L' and restrictions of the domains for T and TU can be chosen so that the conditions A1 apply to the pair (TU, T) then choose a map B of the form

$$B(x, y) = \left(\frac{x}{\alpha}, \frac{y}{\beta} - f(x)\right),$$

634 R.S. MacKay

with $\alpha, \beta < 0$, taking L to L' with reversed orientation (thus determining β and $f(0)$), and with $-1/\alpha$ chosen to be the horizontal width of the union of the new domains. Then the pair

$$D(U, T) := (B^{-1}TB, B^{-1}TUB)$$

satisfies the conditions A1 and A2. Note that B preserves the foliation by verticals. The function f is chosen to try to keep the domains for $D(U, T)$ "horizontal". There are various recipes for this, e.g. [4], but the choice will not be important here.

Note that γ is a golden circle for (U, T) iff $B^{-1}\gamma$ is a golden circle for $D(U, T)$. The operator D is called a *renormalisation*.

39.5 Construction

Theorem. *If $(U_n, T_n) := D^n(U_0, T_0)$ are defined for all $n \geq 0$ and lie in the set of (U, T) satisfying A1, A2 and B1–B6:*

B1. $|\beta| \geq \beta_m > 1$
B2. $|\alpha| \geq \alpha_m > 1$
B3. For all $y \in L$, the slopes of the lines from y to $U(0, y)$ and $T(0, y)$ are at most s_m in absolute value
B4. The derivatives of B and UB map vectors of slope at most s_m to vectors of slope at most s_m. For example, the first condition is $\frac{\alpha}{\beta}s_m + |\alpha f'(x)| \leq s_m$
B5. The xx-components of the derivatives of U and T are positive
B6. The derivative of T multiplies the horizontal component of tangent vectors with slope less than s_m by at least $\kappa > 1/\alpha_m$

then (U_0, T_0) has a golden circle.

Actually, we obtain in addition that it is an s_m-Lipschitz graph and that the dynamics on it preserves horizontal order and is conjugate to rotation.

The proof requires the following definition (see Fig. 39.3).

Definition. An orbit segment C for (U, T) is a *cycle* if it forms an s_m-Lipschitz graph, considered as a function y over x, and decomposing it into subsets C_-, C_+ on the left and right of L respectively (including a point on L in both) then $U(C_+)$ is to the left of $T(C_-)$ (with possible overlap only on $UT(L) = TU(L)$) and U, T preserve horizontal order applied to C_+, C_- (ignoring the point which is last in the orbit segment).

Proof. The sequence of line segments $B_0 \ldots B_n(L), n \in \mathbb{Z}_+$, is nested. By B1, it converges to a single point; call it ξ_0.

The idea of the proof is that the closure of the orbit of ξ_0 under (U_0, T_0) is the desired golden circle. This conjecture was already clear to me in 1982, but to prove it required ideas from David Rand's preprint (building up ordered Lipschitz sets and

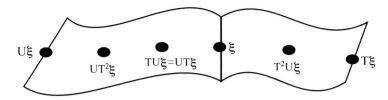

Fig. 39.3 A cycle for (U, T)

checking no gaps), plus a replacement for his dissipativity condition in the control of Lipschitz constants.

Given $n \geq 0$, let $\xi_n = B_{n-1}^{-1} \ldots B_0^{-1} \xi_0$. It is ξ_0 in the n-th coordinate system. Apply the map (U_n, T_n) once to ξ_n, obtaining an orbit segment C_n^n consisting of the three points $U_n(\xi_n), \xi_n, T_n(\xi_n)$ (by our convention about applying both U and T to points of L), ordered from left to right. It is s_m-Lipschitz by B3. Order-preservation is trivial because in the definition we chose to ignore the last point ($U_n(\xi_n)$ and $T_n(\xi_n)$ are identified).

Thus C_n^n is a cycle for (U_n, T_n). Next we use it to make a longer cycle for (U_{n-1}, T_{n-1}) and by induction down to (U_0, T_0).

If C_k^n is a cycle for (U_k, T_k) then we make one for (U_{k-1}, T_{k-1}) by taking $B_{k-1}(C_k^n)$ to be the rescaled cycle with reversed order of points, and concatenating $U_{k-1}B_{k-1}(C_k^n)$, $B_{k-1}(C_k^n)$. The condition B4 guarantees that the points come in the right order and form an s_m-Lipschitz set. Call it C_{k-1}^n. Iterating k from n to 1 produces a cycle C_0^n for (U_0, T_0).

Now take n to infinity. The C_0^n are a nested sequence ($C_0^n \subset C_0^{n+1}$) so the closure C_0^∞ of their union is an s_m-Lipschitz set, invariant under (U_0, T_0) and of golden rotation number.

The only obstacle to C_0^∞ being an invariant circle is that it might have gaps. If there is a gap then it lies between a consecutive pair of $U_0(\xi_0), T_0U_0(\xi_0), \xi_0, T(\xi_0)$, because these form a subcycle of C_0^∞. Call its horizontal length ℓ_0. If the chosen gap lies between $U_0(\xi_0), T_0U_0(\xi_0)$ then take its image by T_0 to obtain a gap between either $T_0U_0(\xi_0), \xi_0$ or $\xi_0, T(\xi_0)$. By condition B6, its horizontal length is at least κ times that of the first one. Thus in any case, we have a gap of horizontal length at least $\kappa\ell_0$. Now apply B_0^{-1} to it to obtain a gap in C_1^∞ between $\xi_1, T_1(\xi_1)$ or $U_1(\xi_1), \xi_1$ respectively. By B2, this gap has horizontal length ℓ_1 at least $\kappa\alpha_m\ell_0$. By induction, C_n^∞ has a gap of horizontal length at least $(\kappa\alpha_m)^n\ell_0$. Since $\kappa\alpha_m > 1$ this eventually exceeds 1, giving a contradiction, because the horizontal length between $U_n(\xi_n)$ and $T_n(\xi_n)$ is of order 1 so there is no room between them for such a long gap. □

Remark. The sequence of pairs of maps $U_{k-1}B_{k-1}, B_{k-1}$ forms a (non-autonomous) iterated function system, of which the invariant circle is the unique pull-back attractor. Thus any seed suffices to construct it.

Remark. The hypotheses hold for F in an open $C^{3+\varepsilon}$-neighbourhood of integrable area-preserving twist maps, thus giving a KAM theorem. More importantly, they

636 R.S. MacKay

appear to hold at the critical fixed point of D (whose existence has finally been proved [5]), so if they can be verified rigorously then all maps on its stable manifold under renormalisation also have a golden circle. This was the main reason for me to develop the idea.

39.6 Extension to Other Rotation Numbers

For rotation number $\omega_0 \in (0, 1)$, define its continued fraction sequence $a_0, a_1, \ldots$ by $a_n = [1/\omega_n]$, $\omega_{n+1} = \frac{1}{\omega_n} - a_n$. Define renormalisation operators

$$D_a(U, T) = (B^{-1}TB, B^{-1}T^a UB)$$

for $a \geq 1$ on those pairs (U, T) for which conditions A1 apply to $(T^a U, T)$ (thus the previous renormalisation $D = D_1$). Then γ is an invariant circle of rotation number ω_0 for (U, T) iff $B^{-1}\gamma$ is one of rotation number ω_1 for $D_{a_0}(U, T)$.

If condition B4 of the Theorem is extended to apply also to $T^{a-1}UB$, $a \geq 1$, then the construction generalises to prove that if $(U_{n+1}, T_{n+1}) = D_{a_n}(U_n, T_n)$ are defined for all $n \geq 0$ and lie in the set satisfying A1, A2 and B1–B6 then (U_0, T_0) has an invariant circle of rotation number ω_0.

If the sequence a_n grows too fast the hypotheses might fail to hold at the conjectured critical points of renormalisation or even on a C^∞-neighbourhood of integrable area-preserving twist maps. Certainly there are Liouville numbers for which the latter occurs. But some growth is permitted, because the B become much more contracting and flattening as $a \to \infty$, compensating for the possible rotation effects of the subsequent composition $T^{a-1}U$. In particular, Hoidn showed that in high enough smoothness classes, the approach gives a KAM theorem for all Diophantine rotation numbers [2], and by considering the effect of the iterated function system on not just Lipschitz constants but also higher derivatives, he deduced a high degree of smooth conjugacy to rotation for the invariant circles.

Acknowledgements I am grateful to David Rand for the interest he has always shown in my work and for the stimulation he has provided to me. And I thank Elaine Greaves-Coelho for preparing the figures for me.

References

1. Rand, D.A.: Existence, non-existence and universal breakdown of dissipative golden invariant tori: III. Invariant circles for mappings of the annulus. Nonlinearity **5**, 681–706 (1992)
2. Haydn, N.: On invariant curves under renormalisation. Nonlinearity **3**, 887–912 (1990)
3. Stirnemann, A.: Renormalization for golden circles. Commun. Math. Phys. **152**, 369–431 (1993)
4. MacKay, R.S.: Renormalisation in area-preserving maps. World Sci Publ Co, River Edge, NJ (1993)
5. Arioli, G., Koch, H.: The critical renormalization fixed point for commuting pairs of area-preserving maps, mp_arc/09-67. Commun. Math. Phys. **295**, 415–429 (2010)

Chapter 40
The Dynamics of Expectations

Wilfredo L. Maldonado and Isabel M.F. Marques

Abstract In this survey we describe the dynamics generated by expectation revisions in intertemporal economic models. Local uniqueness of equilibrium (like stationary states, cycles or sunspot equilibrium) is shown to be a necessary and sufficient condition to obtain stability for the dynamics of expectation revisions. For that reason, such equilibria are called Expectational Stable (E-Stable). Finally, we show how a stationary state which is E-Stable may exhibit a two period cycle bifurcation which is also E-Stable.

40.1 Introduction

In intertemporal economic models, expectations of future values of the state variable have a role in the current decisions of agents. In this way, the learning rule used by them in order to forecast those future values is one of the main subjects of interest among researchers. The convergence of the actual dynamics (generated by the interaction between structural equations and expectations functions) to some rational expectations equilibrium (REE) provides theoretical foundations for the rational expectations hypothesis (Guesnerie [11]).

When agents forecast the value (or distribution of values) of some variable for a future period and the strategies and actions they execute lead to the forecasted value we will say that the system is in a Rational Expectation Equilibrium (REE). However, small deviations in the forecasted value may deviate the current actions and the equilibrium may not be fulfilled. When revisions of expectations allows the system to converge in the long-run to the REE we will say that it is Expectational Stable (E-Stable). All these concepts were introduced by Lucas [14].

W.L. Maldonado (✉) and I.M.F. Marques
Graduate School in Economics, Catholic University of Brasília, SGAN 916, Módulo B, Brasília DF, CEP 70790-160, Brazil
e-mail: wilfredo@pos.ucb.br, isabel.marques@catolica.edu.br

M.M. Peixoto et al. (eds.), *Dynamics, Games and Science I*, Springer Proceedings in Mathematics 1, DOI 10.1007/978-3-642-11456-4_40,
© Springer-Verlag Berlin Heidelberg 2011

In fact, economic agents use a diversity of learning rules that depart from adaptive learning rules, passing by econometric rules and more recently, Bayesian rules of learning. A broad description of learning rules may be found in Evans and Honkapohja [9]. The goal of these learning rules is to obtain convergence to a REE. When a system exhibit equivalence between the E-Stability of a equilibrium and the convergence of a family of learning rules to the REE, we will say that the principle of E-Stability is satisfied.

At this point, a topological feature of the equilibrium is important: the local uniqueness (or determinacy) of it. Intuitively, an equilibrium which has a continuum of equilibria close to it can not be learned. In this way, all the works in this vein depart from the fact that the equilibrium is determinate in order to obtain convergence of learning rules or E-Stability of the equilibrium.

Relationship between E-Stability and local uniqueness of equilibrium in games was studied by Evans and Guesnerie [8]. Applied analysis of these concepts to monetary policy may be found in Bullard and Mitra [3] and Llosa and Tuesta [13].

In this chapter we provide a survey with the main results about determinacy and E-Stability of equilibria in economic dynamic models. We also provide a monopolistic bank model were an E-Stable payment rate bifurcates to a two-period cycle showing the persistence of oscillations in state variables even thought the stability of the stationary state.

The chapter is organized as follows. In Sect. 40.2 we present the basic intertemporal framework to be considered and define the equilibria that may exist in it. To illustrate the concepts we also include two examples corresponding to structures with lagged variables and without them. In Sect. 40.3 the definition of E-Stability is given and the main theorems relating that concept and the local uniqueness of the equilibrium are shown. Finally, in Sect. 40.4 we provide an example of a monopolistic bank model where the stationary payment rate equilibrium has a two-period bifurcation in the perceived-to-actual dynamics of that variable.

40.2 Intertemporal Economic Models and Their Equilibria

In this section we will present the basic (although general) framework in the analysis of intertemporal economic models. Afterward, we will show some explicit examples where equilibria of diverse nature may emerge.

Let $X \subset \mathbb{R}^n$ be the state variable values set of the model. In practice, the state variables may be prices, capital stocks, rates of return, etc. The optimality conditions of agents plus market clearing conditions are summarized in the following (probably) non-linear equations system:

$$Z(x_{t-1}x_t, \mu_{t+1}) = 0; \tag{40.1}$$

where $Z : X \times X \times \mathscr{P}(X) \to \mathbb{R}^n$, x_t is the value of the state variable in period t, μ_{t+1} is the probability distribution of the state variable in period $t+1$. Here, $\mathscr{P}(X)$

40 The Dynamics of Expectations 639

is the set of probability distributions defined on X. Equation (40.1) represents the dynamics of a model with lagged variables (also called model with memory). Simple overlapping generation (OLG) models are models without lagged variables and are also called one-step forward looking models. Structures where agents solve dynamic programming problems or OLG models with delay in production provide examples of models with lagged variables.

Equation (40.1) must be read in the following way: If agents observe $x_{t-1} \in X$ (the state variable value in period $t - 1$) and they have expectations for the next period $t + 1$ defined by $\mu_{t+1} \in \mathscr{P}(X)$ then $x_t \in X$ is the current optimal decision for this variable that equilibrates the markets if and only if $Z(x_{t-1}, x_t, \mu_{t+1}) = 0$. The perfect foresight dynamics is defined by the function[1] $Z(x_{t-1}, x_t, x_{t+1}) = Z(x_{t-1}, x_t, \delta_{x_{t+1}})$ (where δ_z is the Dirac measure concentrated at z).

The following definition classifies the deterministic equilibria that the structure (40.1) may have.

Definition 40.1.
A *steady state* is a $\bar{x} \in X$ such that: $Z(\bar{x}, \bar{x}, \bar{x}) = 0$.
A *perfect foresight equilibrium* is a sequence $(x_t^*)_{t \geq 0} \subset X$ such that for all $t \geq 1$, $Z(x_{t-1}^*, x_t^*, x_{t+1}^*) = 0$.
A *k-cycle* is a perfect foresight equilibrium such that: (a) $x_{t+k}^* = x_t^*$, $\forall t \geq 1$. (b) k is the lowest integer with the property (a).

Existence of cycles in models with lagged variables was proven by several authors (Reichlin [17], Farmer [10], de Vilder [18], Michel and Venditti [16]). Stationary state equilibria and cycles in one-step forward looking models were studied by Grandmont [12] and Azariadis and Guesnerie [2].

When an intertemporal model does not have intrinsic uncertainty (like shocks in production or in preferences) and it supports a truly stochastic probability measure for the future values of the state variable as an equilibrium, we find a very singular phenomenon called a sunspot equilibrium, which we will define at once.

Definition 40.2. A *stationary sunspot equilibrium* (SSE) for the one-step forward looking model $Z(x_t, \mu_{t+1}) = 0$ is a set X_0 and a kernel[2] $Q : X_0 \times \mathscr{B}(X_0) \to [0, 1]$ such that: a) $Q(x, .)$ is truly stochastic for at least one $x \in X_0$ and b) $Z(x, Q(x, .)) = 0$ for all $x \in X_0$.

For models with lagged variables there is a more general definition for that kind of equilibria given by Woodford [19]. Maldonado [15] used a Markovian version of the SSE which is defined next.

---

[1] To keep notation simple, we are using the same notation Z for the function defined on the sets $X \times X \times \mathscr{P}(X)$ and $X \times X \times X$.
[2] It must satisfy that: (*a*) for all $x \in X_0$, $Q(x, .)$ is a probability distribution defined on $\mathscr{B}(X_0)$ which is the set of Borelians defined on X_0, and (*b*) for all $A \in \mathscr{B}(X_0)$, $Q(., A)$ is a measurable function.

Definition 40.3. A *Markovian k stationary sunspot equilibrium* (*k*-SSE) for the model with lagged variables $Z(x_{t-1}, x_t, \mu_{t+1}) = 0$ is a k-tuple $\mathbf{x} = (x^1, \ldots, x^k) \in X^k$ with all of them distinct and k^2 vectors in the interior of S^{k-1} (the $k-1$ dimensional simplex) denoted by m_{ij}; $i = 1, \ldots, k$; $j = 1, \ldots, k$ such that:

$$Z(x^i, x^j, (\mathbf{x}, m_{ij})) = 0, \quad \forall i, j = 1, \ldots, k,$$

where $(\mathbf{x}, m_{ij})$ represents the probability distribution measure with support $x^1, \ldots, x^k$ and with probabilities given by m_{ij} (i.e. $Prob[x^l] = m_{ij}^l$).

Existence of SSE in models without lagged variables results from indeterminate stationary states[3] (Chiappori et al. [4]) or from deterministic cycles (Azariadis and Guesnerie [2]). Araujo and Maldonado [1] showed that deterministic models of that type with complex dynamics also have SSE with stationary probability being absolutely continuous with respect to the Lebesgue measure. In models with memory Woodford [19] concluded that the indeterminacy of the stationary state is a sufficient condition to obtain SSE close to it and Maldonado [15] provided necessary and sufficient conditions for existence of Markovian SSE close to regular cycles.

40.2.1 Some Examples

Now, we are going to describe two specific examples in order to illustrate the framework above. The first one will show a simple OLG model whose dynamics is described by a one-step forward looking equation and exhibit cycles of any order, complex dynamics and sunspot equilibrium with absolute continuous stationary probability. The second example is a model with lagged variables with cycles and multiple stationary states.

40.2.1.1 A Simple Overlapping Generation Model

Let us consider the simple OLG model where the technology is linear and in each period two kind of agents (young and old) coexist in the same proportion. Each young agent supplies x units of labor to produce the unique perishable good in the same quantity (since the technology is linear) and each old agent consumes c units of the good. The agents hold M units of fiat money and the stock of money in circulation is constant for all periods. The utility function is represented by $U(c, x)$, where x is the labor supply when the agent is young and c is the consumption when the agent is old. If $(p_t)_{t \geq 0}$ is a sequence of prices for the good in this economy, each agent will solve:

---

[3] A stationary state equilibrium is called indeterminate if there exists infinitely many other equilibrium paths arbitrarily close to the stationary state.

40 The Dynamics of Expectations

$$\text{Max } U(c_{t+1}, x_t)$$

$$\text{subject to } p_{t+1}c_{t+1} = p_t x_t.$$

If we suppose that in period t the next period price $\tilde{p}_{t+1}$ is a random variable, the agent's problem will be:

$$\text{Max } E[U((p_t/\tilde{p}_{t+1})x_t, x_t)],$$

where the expected value is taken with respect to the probability measure of $\tilde{p}_{t+1}$. In this case the first order condition and the equilibrium equation $\frac{M}{p_t} = x_t$ give us the following equation which defines the equilibria for the model:

$$Z(x_t, \mu_{t+1}) = E_{\mu_{t+1}}[(\tilde{x}_{t+1}/x_t)U_c(\tilde{x}_{t+1}, x_t) + U_x(\tilde{x}_{t+1}, x_t)] = 0,$$

where μ_{t+1} is the probability distribution induced in $\tilde{x}_{t+1} = M/\tilde{p}_{t+1}$. Let us consider the following functional specification: the labor supply $x \in X = [0, 1]$ and the utility function $U(c, x) = Ac - (A/2)c^2 - x$, defined on $[0, 1] \times [0, 1]$. From the first order conditions and the equilibrium equation we obtain the following dynamical system:

$$x_t = Ax_{t+1}(1 - x_{t+1}). \tag{40.2}$$

It is well known that there is a set of values for the parameter A where the dynamics given by (40.2) exhibits cycles of any order and complex behavior in the interval $[0, 1]$. Araujo and Maldonado [1] proved that in this case there also exists a SSE with absolutely continuous stationary probability measure.

40.2.1.2 A Two Sector OLG Model

An OLG model with two sectors was analyzed by de Vilder [18] in order to obtain bifurcations of the stationary state to deterministic cycles. We will describe that model. The utility function of consumers is separable in the current supply of labor and in the future consumption:

$$U(c_{t+1}, l_t) = V_1(l^* - l_t) + V_2(c_{t+1}) = \frac{(l^* - l_t)^{1-\alpha_1}}{1 - \alpha_1} + \frac{c_{t+1}^{1-\alpha_2}}{1 - \alpha_2}. \tag{40.3}$$

The technology uses capital stock and labor aggregating them with a Leontief production function $x_t = \text{Min}\{l_t, k_{t-1}/a\}$. The capital depreciation rate is δ. With all these elements, the optimization problem of the consumer is:

$$\text{Max } V_1(l^* - l_t) + E[V_2(c_{t+1})]$$

$$\text{subject to } p_{t+1}c_{t+1} = w_t l_t.$$

642 W.L. Maldonado and I.M.F. Marques

The first order condition for that problem is

$$l_t V_1'(l^* - l_t) = E[c_{t+1} V_2'(c_{t+1})] \tag{40.4}$$

From efficiency in production we have $x_t = l_t = k_{t-1}/a$. The equilibrium condition in the commodity market requires:

$$\begin{aligned} c_{t+1} &= x_{t+1} - i_{t+1} = x_{t+1} - (k_{t+1} - (1-\delta)k_t) \\ &= x_{t+1} - (ax_{t+2} - (1-\delta)ax_{t+1}) \\ &= (1 + a(1-\delta))x_{t+1} - ax_{t+2} \tag{40.5} \end{aligned}$$

Finally, using the functional form given in (40.3), the equalities (40.5) and $l_t = x_t$ and replacing t by $t-1$ we obtain the following equation governing the dynamics of the production:

$$Z(x_{t-1}, x_t, \mu_{t+1}) = x_{t-1}(l^* - x_{t-1})^{-\alpha_1} - E_{\mu_{t+1}}[((1 + a(1-\delta))x_t - a\tilde{x}_{t+1})^{1-\alpha_2}]. \tag{40.6}$$

40.2.2 Diversity of Equilibria

The basic structure (40.1) (and its particular case with no lagged variables) exhibits several types of equilibria. The simplest one is the stationary state equilibrium which is calculated by solving $Z(\bar{x}, \bar{x}, \bar{x}) = 0$. In fact, the (40.1) depends on several parameters (i.e. $Z(x_{t-1}, x_t, \mu_{t+1}; \theta) = 0$) and we can write the stationary state equilibrium as depending of these parameters. Namely, if $\theta \in \mathbb{R}^J$ represents the J parameters in the model, we will have:

$$Z(\bar{x}(\theta), \bar{x}(\theta), \bar{x}(\theta); \theta) = 0, \ \forall \theta \in \Theta \subset \mathbb{R}^J, \tag{40.7}$$

Analyzing possible bifurcations of the stationary state in (40.7) we may obtain cycles for the model. For example, Grandmont [12] analyze a two-sector OLG economy which leads to a one-step forward looking model where a stationary state may bifurcate into cycles of high order. For the model of Sect. 40.2.1.2, de Vilder [18] proved that there exists an open set of parameter values such that the model (40.6) exhibits a cycle bifurcation of order 11. Maldonado [15] calculated that cycle for the parameter values $a = 3, l^* = 2, \alpha_1 = 0.5, \alpha_2 = 0.77$ and $\delta = 0.1$. The values of that cycle are:

$$x^{(1)} = 0.928; \ x^{(2)} = 1.076; \ x^{(3)} = 1.119; \ x^{(4)} = 0.836; \ x^{(5)} = 0.312;$$
$$x^{(6)} = 0.275; \ x^{(7)} = 0.339; \ x^{(8)} = 0.417; \ x^{(9)} = 0.514; \ x^{(10)} = 0.631;$$
$$x^{(11)} = 0.771.$$

40 The Dynamics of Expectations

An important type of stationary state equilibrium is one called indeterminate, whose definition is given at once.

Definition 40.4. A stationary state equilibrium is called *indeterminate* if the characteristic roots of the linearization of (40.1) have modulus lower than one, namely, if the polynomial[4]:

$$Z_{-1} + Z_0 r + Z_1 r^2 = 0 \tag{40.8}$$

have roots with modulus lower than 1.

When a stationary state is an indeterminate equilibrium for each neighborhood of it we can find a perfect foresight equilibrium converging to that state included in such a neighborhood. This multiplicity of equilibrium is a subject of research because it may demand some sort of coordination among agents or policy intervention in order to attain more efficient equilibria.

In addition to the existence of many deterministic equilibria around an indeterminate stationary state, it is also possible to find truly stochastic equilibria close to it. The following theorems state this result.

Theorem 40.1. *(Chiappori, Geoffard and Guesnerrie [4]) In the one-step forward looking model* $Z(x_t, \mu_{t+1}) = 0$ *if there is an indeterminate stationary state equilibrium* $\bar{x}$ *and the* Z*-function is linear in the second variable (i.e.* $Z(x, \alpha\mu_1 + (1 - \alpha)\mu_2) = \alpha Z(x, \mu_1) + (1 - \alpha)Z(x, \mu_2))$, *then there exists a SSE close to* $\bar{x}$.

For models with lagged variables, although there is not a general theorem like the theorem above, Woodford [19] showed for the specific framework he analyzed that indeterminacy and linearity are also sufficient conditions to obtain SSE.

Finally, Araujo and Maldonado [1] proved that deterministic complex dynamics may arise SSE with a large support. Suppose that for the one-step forward looking model we can explicitly calculate the current state value as a function of its future value, namely there exists a function ϕ such that $Z(\phi(x), x) = 0$ for all x. The function ϕ is called *the backward perfect foresight map* associated to Z. Then we have the following theorem.

Theorem 40.2. *(Araujo and Maldonado [1]) In the one-step forward looking model, if the* Z *function is linear in the second variable and the backward perfect foresight map* ϕ *associated to it is unimodal and has an absolutely continuous invariant measure* μ, *then there exists a SSE whose stationary probability measure is* μ.

In this way, we show how the simple framework (40.1) may support several types of equilibria. In the next section we will discuss the possibility of agents learning some of them by using expectations update and how this is related with the local uniqueness of the equilibria.

---

[4] Z_i is the derivative of $Z(x_{t-1}, x_t, x_{t+1})$ with respect to x_{t+i} for $i = -1, 0, 1$ evaluated at $(\bar{x}, \bar{x}, \bar{x})$.

40.3 Expectational Stability

As we could see in the previous section, the intertemporal economic models may exhibit stationary steady states, cycles, or truly stochastic equilibria. All these equilibria may be seen as parameters that agents want to learn by updating the expectations they have about them. A simple way to state the possibility of learning the equilibria is to test if they are *Expectational Stable* (E-Stable) (Evans [7]). Suppose that the model uses a vector of parameters $\theta \in R^K$ for making the forecast of the future state variable value and when it is done the actual dynamics reveals the correct values of these parameters as being $T(\theta) \in R^K$. A rational expectations equilibrium value of the parameter vector is θ^* such that $T(\theta^*) = \theta^*$. The (REE) defined by θ^* is expectational stable (or weakly E-Stable) if θ^* is locally stable for the dynamics defined by:

$$\frac{d\theta}{d\tau} = T(\theta) - \theta \tag{40.9}$$

Definition 40.5. The REE defined by θ^* is *strongly expectational stable* if it is weakly E-Stable for all overparameterized model of forecast (in a class of parameterizations of the REE).

Definition 40.5 is quite general for parameterizations in dynamic models. In our case, the parameters which define de equilibrium will be the state values themselves. Thus, a k-cycle is strongly E-Stable if the nk-cycle defined from it is weakly stable for all $n \geq 1$.

The possibility of learning an equilibrium should be linked with the uniqueness of it, since multiplicity of equilibria could deviate the updating of expectations from one to another. That "principle" was stated by Evans and Honkapohja [9] and used as a test of the Rational Expectations hypothesis by Guesnerie [11].

Definition 40.6. An equilibrium in called *determinate* if there is a neighborhood which does not contain any other equilibrium for the model.

For instance, for models with no lagged variables ($Z(x_t, \mu_{t+1}) = 0$) a regular[5] k-cycle $(x^1, \ldots, x^k)$ is determinate if and only if $|Z_1(x^1, x^2) \times \ldots \times Z_1(x^k, x^1)| < |Z_0(x^1, x^2) \times \ldots \times Z_0(x^k, x^1)|$. In models with lagged variables it is usual to have the initial value of the state variable as given (i.e. $x_0 \in X$ is given). Therefore it is used the definition of saddle point determinacy. Namely, $\bar{x} \in X$ is saddle point determinate for the model (40.1) if one of the characteristic roots of the linearization $Z_{-1} + Z_0 r + Z_1 r^2 = 0$ has modulus lower than one and the other has modulus greater than one.

Theorem 40.3. *(Evans and Honkapohja [9]) Consider a regular k-cycle $(x^1, \ldots, x^k)$ of the model $Z(x_t, \mu_{t+1}) = 0$. It is determinate if and only if it is strongly E-Stable.*

---

[5] A cycle is regular if $|Z_0(x^j, x^{j+1})| \neq 0$ for $j = 1, \ldots, k - 1$ and $|Z_0(x^k, x^1)| \neq 0$.

40 The Dynamics of Expectations 645

For models with lagged variable we state the theorem below and its proof is given in the appendix.

Theorem 40.4. *Consider a regular stationary state equilibrium $\bar{x}$ of the model with lagged variables $Z(x_{t-1}, x_t, \mu_{t+1}) = 0$. It is saddle point determinate if and only if it is strongly E-Stable.*

The definition of E-Stability given in (40.9) is based in correction of expectations in continuous time. An alternative version may be analogously written in discrete time and then obtaining the same theorems above. In the next section we will use that definition to analyze the E-Stability of the payment rate equilibrium in a micro model of credit.

40.4 Bifurcation of an E-Stable Equilibrium

In this section we provide a micro model of loans where the stationary equilibrium payment rate bifurcates from a E-Stable equilibrium to a two-period cycle equilibrium in the dynamics of the expectations.

The model is as follows. We will consider a loan market model where the borrowers are represented by a single agent and the lender by a monopolistic bank. In this sense, borrowers are interest-rate-takers in the loan market and lenders have market power to decide the interest rate. There are two periods $t = 0, 1$ and in $t = 0$ the borrower demands loans (m) and consumption (c_0). In $t = 1$, the same agent delivers (part of) the debt and also consumes (c_1). In case of the borrower does not deliver his complete obligation, he will suffer a penalty in his utility function. The interest rate on loans are given by $r > 0$. All the terms described above are included in the classical models with possibility of default (Dubey, Geanakoplos and Shubik [6]). Thus, the payoff of the borrower with consumption plan $(c_0, c_1) \in \mathbb{R}^2_+$, loan demand $m \geq 0$ and delivery decision $D \in [0, (1 + r)m]$ is given by:

$$V(c_0, c_1, m, D) = u(c_0) + \delta u(c_1) - \lambda[(1 + r)m - D]. \tag{40.10}$$

The parameter $\lambda > 0$ represents the penalty rate for defaulting and $\delta > 0$ is the discount factor. The budget constrain for the representative agent is defined by the following inequalities:

$$p_0 c_0 \leq p_0 w_0 + m \tag{40.11}$$
$$p_1 c_1 + D \leq p_1 w_1 \tag{40.12}$$
$$0 \leq D \leq (1 + r)m \tag{40.13}$$

In this way, the borrower problem is defined as the maximization of the payoff function (40.10) subject to the restrictions (40.11), (40.12) and (40.13). The demand for loans is the $m^* = m(r)$ component of the solution of that problem.

The lender side of the economy is modeled by a monopolistic bank which decides the value of the interest rate $r > 0$ to be charged for private loans, given the loan demand curve $m = m(r)$ (or its inverse function $r = r(m)$) and the payment rate ρ. Thus, if $c(m)$ is the cost of the monopolist and ρ is the perceived payment rate, its problem is to chose the interest rate for loans in order to maximize:

$$m^* = \text{ArgMax}_{m \geq 0} \, \rho(1 + r(m))m - c(m),$$

where $r^* = r(m(\rho))$ and for simplicity we write $r = r(\rho)$ and $m = m(\rho)$. With that interest rate, the borrower demand $m(\rho)$ and decides to deliver $D = D(r(\rho)) = D(\rho)$. Therefore, the actual payment rate of the market will be:

$$\phi(\rho) = \frac{D(\rho)}{(1 + r(\rho))m(\rho)} \tag{40.14}$$

The function $\phi : [0, 1] \to [0, 1]$ gives us the actual payment rate of the market given the perceived payment rate. A REE is a ρ^* such that $\phi(\rho^*) = \rho^*$. The REE ρ^* is E-Stable if $|\phi'(\rho^*)| < 1$. Let us suppose that the marginal cost of the monopolistic bank is constant (i.e. $c(m) = cm$), then the function ϕ may be considered as parameterized by c. The following proposition (see Devaney [5]) provides conditions to obtain a 2-period cycle bifurcation of the payment rate REE.

Proposition 40.1. *If the function ϕ_c satisfies the following conditions:*

(i) $\phi_c(\rho_c^*) = \rho_c^*$, *for all c*
(ii) *There exists $c_0 > 0$ such that $\phi'_{c_0}(\rho_{c_0}^*) = -1$*
(iii) $\frac{\partial(\phi_c^{(2)})'}{\partial c}|_{c=c_0}(\rho_{c_0}^*) \neq 0$
then, there exists an open interval $I \subset [0, 1]$ containing $\rho_{c_0}^*$ and a function $p : I \to \mathbb{R}_+$ such that for all $\rho \in I$:

$$\phi_{p(\rho)}(\rho) \neq \rho \text{ and } \phi_{p(\rho)}^{(2)}(\rho) = \rho.$$

To illustrate Proposition 1, let us consider the following functional specification: $u(c) = bc - c^2/2$ where $b > 0$; $w_0 = 0$ and $p_0 = 1$. The demand for loans and delivery decisions are give by:

$$m(r) = b - \lambda(1 + \delta)^{-1}(p_1 + 1 + r),$$

$$D(r) = p_1[w_1 - b + \lambda(1 + \delta)^{-1}(p_1 + 1 + r)].$$

Thus, using the first order condition of the monopolist maximization (in this case, Marginal Revenue equal to Marginal Cost divided by ρ) we obtain:

$$m^* = \frac{b}{2} - \frac{\lambda c}{2\rho}; \quad 1 + r^* = \frac{b}{2} + \frac{\lambda c}{2\rho}; \quad D^* = p_1[w_1 - b + \lambda\delta^{-1}p_1]. \tag{40.15}$$

40 The Dynamics of Expectations

Replacing (40.15) in (40.14) we can obtain explicitly the function ϕ and the REE ρ^*:

$$\phi(\rho) = \frac{4k\lambda\rho^2}{b^2\rho^2 - \lambda^2c^2}; \quad \rho^* = \frac{2\lambda k + \sqrt{4\lambda^2k^2 + \lambda^2b^2c^2}}{b^2},$$

where $k = p_1[w_1 - b + \lambda\delta^{-1}p_1]$. Fixing all the parameters but c, we will have the parameterized dynamical system $\rho_{n+1} = \phi_c(\rho_n)$ and the REE $\rho^* = \rho_c^*$. The derivative of the function $\phi_c(\rho)$ evaluated at the REE is:

$$\phi_c'(\rho_c^*) = 1 - \sqrt{1 + \frac{b^2c^2}{4k^2}}.$$

Therefore, there exist some values of c such that $|\phi_c'(\rho_c^*)| < 1$ and the REE is E-stable. For example, we can consider the following parameter specifications: $b = 10$, $w_1 = 10$, $\lambda = 0.7$; $p_1 = 1.1$ and $\delta = 0.99$. For $c = 0.25634$ the stationary REE results $\rho_c^* = 0.03355$, which is E-Stable ($\phi_c'(\rho_c^*) = -0.8013$). On the other hand, if we set $c_0 = 0.29634$ (which results from $\phi_{c_0}'(\rho_{c_0}^*) = -1$) it results a two-period cycle around $\rho_{c_0}^* = 0.03593$ ($\rho_1 = 0.03399$ and $\rho_2 = 0.036$ approximately).

Appendix

Proof. (Theorem 40.4) First, let us characterize the strong E-stability of the steady state. The steady state can be seen as a degenerated cycle of order one, so if we consider the overparameterization of the steady state given by a K-cycle $(x^1, \ldots, x^K)$ the transformation from the perceived to the actual law of motion is given by $T(\theta^1, \ldots, \theta^K) = (x_1(\theta^K, \theta^2), \ldots, x_K(\theta^{K-1}, \theta^1))$ where these functions are defined implicitly by:

$$Z(\theta^{k-1}, x_k(\theta^{k-1}, \theta^{k+1}), \theta^{k+1}) = 0, \quad k = 2, \ldots, K - 1,$$

$$Z(\theta^K, x_1(\theta^K, \theta^2), \theta^2) = 0 \quad \text{and} \quad Z(\theta^{K-1}, x_K(\theta^{K-1}, \theta^1), \theta^1) = 0.$$

Then the K-cycle is E-stable if and only if the matrix $T'(x^1, \ldots, x^K)$ has all its eigenvalues with real part lower than one. Let us denote

$$a_1 = -\frac{Z_0(x^K, x^1, x^2)}{Z_{-1}(x^K, x^1, x^2)}, \quad a_l = -\frac{Z_0(x^{l-1}, x^l, x^{l+1})}{Z_{-1}(x^{l-1}, x^l, x^{l+1})}, \quad l = 2, \ldots, K - 1;$$

$$a_K = -\frac{Z_0(x^{K-1}, x^K, x^1)}{Z_{-1}(x^{K-1}, x^K, x^1)}$$

$$b_1 = -\frac{Z_1(x^K, x^1, x^2)}{Z_{-1}(x^K, x^1, x^2)}, \quad b_l = -\frac{Z_1(x^{l-1}, x^l, x^{l+1})}{Z_{-1}(x^{l-1}, x^l, x^{l+1})}, \quad l = 2, \ldots, K - 1;$$

$$b_K = -\frac{Z_1(x^{K-1}, x^K, x^1)}{Z_{-1}(x^{K-1}, x^K, x^1)},$$

then the matrix $T'(x^1, \ldots, x^K)$ has the following form: For $K \geq 3$ is given by:

$$\begin{pmatrix} 0 & -a_1^{-1}b_1 & 0 \ldots & 0 & 0 & a_1^{-1} \\ a_2^{-1} & 0 & -a_2^{-1}b_2 \ldots & 0 & 0 & 0 \\ \vdots & & & & & \\ 0 & 0 & 0 \ldots & a_{K-1}^{-1} & 0 & -a_{K-1}^{-1}b_{K-1} \\ -a_K^{-1}b_K & 0 & 0 & 0 \ldots & a_K^{-1} & 0 \end{pmatrix}$$

and for $K = 2$:

$$\begin{pmatrix} 0 & a_1^{-1}(1 - b_1) \\ a_2^{-1}(1 - b_2) & 0 \end{pmatrix}.$$

The steady state is strongly E-stable if and only if all overparameterization of it as a cycle is E-stable. It is equivalent to say that the following matrices have all their eigenvalues with a real part lower that one:

$$A_K = \begin{pmatrix} 0 & -b^{-1} & 0 \ldots & 0 & 0 & -ab^{-1} \\ -ab^{-1} & 0 & -b^{-1} \ldots & 0 & 0 & 0 \\ \vdots & & & & & \\ 0 & 0 & 0 \ldots & -ab^{-1} & 0 & -b^{-1} \\ -b^{-1} & 0 & 0 & 0 \ldots & -ab^{-1} & 0 \end{pmatrix}, \quad K \geq 3$$

and for $K = 2$:

$$A_2 = \begin{pmatrix} 0 & -b^{-1}(1 + a) \\ -b^{-1}(1 + a) & 0 \end{pmatrix}.$$

For $K = 2$ the condition for stability of the cycle is that $|a + 1|/|b| < 1$, which is true if and only if $\bar{x}$ is saddle point determinate. For $K \geq 3$, λ is eigenvalue of A_K if and only if $\det((b\lambda)I - M_K) = 0$ where

$$M_K = \begin{pmatrix} 0 & -1 & 0 & 0 \ldots & 0 & 0 & -a \\ -a & 0 & -1 & 0 \ldots & 0 & 0 & 0 \\ 0 & -a & 0 & -1 \ldots & 0 & 0 & 0 \\ \vdots & & & & & & \\ 0 & 0 & 0 & 0 \ldots & -a & 0 & -1 \\ -1 & 0 & 0 & 0 \ldots & 0 & -a & 0 \end{pmatrix}$$

It can be verified that the eigenvalues of M_K are $\{aw_k + \bar{w}_k; k = 0, \ldots, K-1\}$ if K is odd and $\{a\bar{v}_k + v_k; k = 0, \ldots, K-1\}$ if K is even, where w_k and v_k are the K-th roots of 1 and -1 respectively. In any case:

$$\lambda = \frac{a+1}{b}cos(\theta_k) \pm \frac{a-1}{b}sin(\theta_k)\mathrm{i},$$

so the eigenvalues have their real parts lower than one for all K if and only if $|a + 1|/|b| < 1$, which is true if and only if $\bar{x}$ is saddle point determinate. $\square$

Acknowledgements Wilfredo L. Maldonado would like to thank the financial support provided by CNPq (Brazil) through grants 305317/2003-2 and 472178/2006-7.

References

1. Araujo, A., Maldonado, W.: Ergodic chaos, learning and sunspot equilibrium. Econ. Theory **15**, 163–184 (2000)
2. Azariadis, C., Guesnerie, R.: Sunspot and Cycles. Rev. Econ. Stud. **53-5**, 725–737 (1986)
3. Bullard, J., Mitra, K.: Determinacy, learnability and monetary policy inertia. J. Money Credit Bank. **39-5**, 1177–1212 (2007)
4. Chiappori, P.-A., Geoffard, P.-Y., Guesnerie, R.: Sunspot Fluctuations Around a Steady State: The Case of Multidimensional One-Step Forward Looking Economic Models. Econometrica **60-5**, 1097–1126 (1992)
5. Devaney, R.: An Introduction to Chaotic Dynamical Systems. Addison-Wesley Publishing Company, Redwood City (1989)
6. Dubey, P., Geanakoplos, J., Shubik, M.: Default and punishment in general equilibrium. Econometrica **73-1**, 1–37 (2005)
7. Evans, G.: The fragility of sunspots and bubbles. J. Monet. Econ. **23**, 297–317 (1989)
8. Evans, G., Guesnerie, R.: Rationalizability, Strong Rationality, and Expectational Stability. Games Econ. Behav. **5**, 632–646 (1993)
9. Evans, G., Honkapohja, S.: Learning Dynamics. In: Taylor, J.B., Woodford, M. (eds.) Handbook of Macroeconomics vol. I, pp. 449–542. Elsevier Science B.V., Amsterdam (1999)
10. Farmer, R.: Deficit and cycles. J. Econ. Theory **40**, 77–88 (1986)
11. Guesnerie, R.: Theoretical tests of the Rational Expectations hypothesis in economic dynamical models. J Econ. Dyn. Control **17**, 847–864 (1993)
12. Grandmont, J.-M.: On Endogenous Competitive Business Cycles. Econometrica **53**, 995–1045 (1985)
13. Llosa, L.-G., Tuesta, V.: Learning about monetary policy rules when the cost-channel matters. J. Econ. Dyn. Control **33**, 1880–1896 (2009)
14. Lucas, R.: Asset prices in an exchange economy. Econometrica **46**, 1429–1445 (1978)
15. Maldonado, W.L.: Cycles, steady states and susnpot equilibrium in models with memory. Braz. Rev. Econom. **20-2**, 189–199 (2001)
16. Michel, P., Venditti, A.: Optimal growth and cycles in overlapping generations models. Econ. Theory **9**, 511–528 (1997)
17. Reichlin, P.: Equilibrium cycles in an overlapping generations economy with production. J. Econ. Theory **40**, 89–102 (1986)
18. de Vilder, R.: Complicated endogenous business cycles under gross substitutability. J. Econ. Theory **71**, 416–442 (1996)
19. Woodford, M.: Stationary sunspot equilibria in a finance constrained economy. J. Econ. Theory **40**, 128–137 (1986)

Chapter 41
Dynamics on the Circle

W. de Melo

Abstract In this paper we will review several results on the dynamics of circle maps. This includes the theory of circle diffeomorphism where the combinatorial aspects goes back to Poincaré followed by the topological description of the dynamics by Denjoy and the geometric aspects by Herman and Yoccoz. In this case the dynamics is either periodic or quasi-periodic. The dynamics of non-invertible circle maps is much more complicated. The special case of covering maps of the circle is well understood. We will also consider maps with critical points and exhibit parametrized families of such maps that contains essentially all possible dynamical behavior. In the boundary between these two types of dynamical systems we have the critical circle maps also discussed here.

41.1 Definitions and Results

Let $C^r(\mathbb{S}^1), 3 \leq r \leq \omega$ be the space of C^r maps of the circle endowed with the C^r topology. If $r = \omega$ the topology is such that a sequence of real analytic maps f_n converges to a real analytic map f if there exists a neighborhood of the circle in the complex plane such that all maps extend to holomorphic maps in this neighborhood and the sequence of holomorphic extensions converges in the C^0 topology to the holomorphic extension of f. This implies of course that f_n converges to f in the C^k topology for every finite k. However, the space $C^\omega(\mathbb{S}^1)$ is not metrizable and does not satisfy the Baire property which states that a residual set, which is the intersection of a countable number of subsets that are open and dense, is dense. We will also consider smooth families of circle maps, i.e, C^r maps $\Delta \times \mathbb{S}^1 \to \mathbb{S}^1$, where the parameter space Δ is an open subset of some finite dimensional Euclidean space and we endow the space of families with the C^r topology.

W. de Melo
Instituto de Matemática Pura e Aplicada (IMPA), Estrada Dona Castorina, 110, 22460-320 Rio de Janeiro, RJ, Brazil
e-mail: demelo@impa.br

Let $\pi: \mathbb{R} \to \mathbb{S}^1$ be the covering map $\pi(t) = e^{2\pi i t}$. Each circle map f lifts to a map $\tilde{f}: \mathbb{R} \to \mathbb{R}$, i.e $\pi \circ \tilde{f} = f \circ \pi$. We have that $\tilde{f}(t+1) = \tilde{f}(t) + d$ where d is an integer called the *topological degree* of f. Conversely any smooth map of the real line satisfying this property is a lift of a circle map and two different lifts of the same map differs by an integer translation.

We will only consider the subset of maps whose critical points are non-flat, i.e, for each critical point c of such map f, there are smooth local diffeomorphisms ϕ, ψ with $\phi(c) = \psi(f(c)) = 0$ and such that $\psi \circ f \circ \phi^{-1}: x \mapsto x^k$ for some positive integer k. If k is even the critical point is a turning point, otherwise it is an inflection point and the map is a local homeomorphism in a neighborhood of such critical point. For this type of maps we have a very good universal control on the distortion of high iterates that are first return maps to nice intervals, i.e, intervals such that the iterates of the endpoints never return to the interior of the interval. In [29] it is proved that the restriction of the first return map of a small enough nice interval to each component of its domain is either a diffeomorphism of universally bounded distortion or a composition of such diffeomorphism with a map $x \mapsto x^k$ where k depends only on the order of the critical points of the original map. Of course, every non-constant real anaylic maps have this property. Also, for each k there exists a residual set of smooth k parameter family of smooth circle map such that all critical points of each map in one of these families are non-flat.

Definition 41.1. A dynamical dichotomy.

For $f \in C^r(\mathbb{S}^1)$ the *periodic part of the dynamics* is the subset $P(f) \subset \mathbb{S}^1$ such that $x \in P(f)$ if and only if there exists a neighborhood $V \subset \mathbb{S}^1$ of x such that each $y \in V$ is asymptotic to a periodic orbit of f. The complement of $P(f)$, the closed set $Ch(f)$, is the *chaotic* part of the dynamics.

Clearly $P(f)$ is backward invariant and it is almost forward invariant: each connected component of $P(f)$ which does not contain a turning point of f is mapped onto another component. If it contains a turning point it is mapped into the closure of another component (a critical value may be in the boundary of this component).

From the structure theorem proved in [18] each component of $P(f)$ is eventually periodic and the number of periodic components is finite. Also the chaotic part cannot contain a wandering interval, i.e, an interval whose forward orbit is a disjoint sequence of intervals. The chaotic part of the dynamics may contain a periodic interval but in such case the orbit of this periodic interval must contain a turning point. Hence each component of the interior of $Ch(f)$ is eventually periodic and the number of periodic components is finite. From the density of hyperbolic maps in the space of smooth interval maps proved in [15], it follows that for an open and dense set of circle maps $Ch(f)$ has empty interior. If we can prove that for maps f in a dense subset $Ch(f)$ does not contain any critical point we would conclude that the hyperbolic maps are dense in the space of circle maps because, by a theorem of Mañe, [17]. For C^2 maps this is equivalent to saying that all periodic points are hyperbolic and all critical points converge to hyperbolic attractors.

Our space of mappings contains the open subset of circle diffeomorphisms. In the boundary of the space of circle diffeomorphisms we have the so called *critical*

41 Dynamics on the Circle

circle maps, i.e. smooth circle homeomorphisms with a finite number of inflection points. For a C^r, $r \geq 2$, diffeomorphism f either $P(f)$ is the whole circle or it is empty and by a theorem of Poincaré (1895) and Denjoy (1932) f is topologically conjugate to an irrational rigid rotation. By a theorem of Yoccoz, [37] this statement also holds for critical circle maps.

An important example of circle maps is the so called standard family or Arnold family. The standard family is a two parameter family $f_{a,b}$ of circle maps whose lifts $\tilde{f}_{a,b}$ are given by the formula:

$$\tilde{f}_{a,b}(t) = t + a + \frac{b}{2\pi} \sin 2\pi t$$

where $0 \leq a \leq 1$ and $b \geq 0$.

For $b = 0$ we have the family of rigid rotations. If $b < 1$ the maps are circle diffeomorphisms, for $b = 1$ each map is a homeomorphism with a unique turning point and for $b > 1$ each map has two turning points.

This family exhibits a very reach bifurcation diagram which is quite well known in the region $b \leq 1$. In the region $0 < b < 1$ the set of parameter values corresponding to maps that have a hyperbolic periodic attractor is open and dense. It is equal to the union of a countable number of disjoint open strips called Arnolds tongues. Each Arnolds tongue is bounded by a pair of smooth curves that converges to a unique rational value of a as b tends to zero. The complement of the tongues is foliated by smooth curves each one is either in the boundary of a tongue or conveges to an irrational value of a as b tends to zero and all maps in such a curve are topologically conjugate to the same irrational rotation corresponding to the tip of the curve in the a-axis. For (a, b) in a tongue the map $f_{a,b}$ is structurally stable and has a unique attracting periodic point. For a fixed value of $b \in (0, 1)$ let us consider the one parameter family of maps $a \mapsto f_{a,b}$ and let $\rho(a, b)$ be the rotation number of $f_{a,b}$ (the rotation number of a a circle homeomorphism f is equal to the limit $\rho(f) = \lim_{n \to \infty} \frac{\tilde{f}^n(x) - x}{n}$ that exists and is independent of x). It is easy to see that for each fixed b the mapping $a \mapsto \rho(a, b)$ is continuous, monotone, increases from 0 to 1 and is constant and rational in the intersection of the line with each Arnold tongue and this intersection is an interval.

For each $b \in (0, 1]$ let $B(b)$ be the set of values of the parameter a for which the rotation number $\rho(a, b)$ is irrational. For $b < 1$ the closure of $B(b)$ is the bifurcation set since for parameters in the complement of this set the corresponding map is structurally stable. In [11] Herman proved that the bifurcation set $\overline{B(b)}$ is a Cantor set of positive Lebesgue measure. Each map $f_{a,b}$, for $a \in B(b)$ is topologically conjugate to an irrational rotation. This conjugacy may be however very bad. In fact, this Cantor set contains a subset, which is residual, such that for each parameter in this subset the conjugacy between the corresponding map and a rotation is not absolutely continuous and in fact maps a set of zero Lebesgue measure in a set of full Lebesgue measure, [2]. However, if the rotation number is far from rationals, i.e, if ρ satisfies a Diophantine condition

$$|\rho - \frac{p}{q}| \geq \frac{C}{q^{2+\beta}} \qquad \forall p, q \in \mathbb{Z}$$

where $\beta > 0$ and $C > 0$ then, by a very deep rigidity result of Herman, [10], extended by Yoccoz in [38], a smooth diffeomorphism with this rotation number is smoothly conjugate to a rotation.

For each $0 < b < 1$ let $D(b) \subset B(b)$ be the set of values of the parameter a such that the rotation number $\rho(a, b)$ satisfy a Diophantine condition. In [8] it was proved that the Hausdorff dimension of the set $B(b) \setminus D(b)$ is equal to zero. In fact these results hold for generic one parameter families of smooth circle diffeomorphisms: the set of parameter values whose corresponding rotation number is irrational has positive Lebesgue measure and the subset that does not satisfy a Diophantine condition has Hausdorf dimension equal to zero.

For critical circle maps, with a unique critical point, the situation is quite different. For the standard family, the set $B(1)$ of values of the parameter a such that the rotation number $\rho(a, 1)$ is irrational has zero Lebesgue measure, as was established in [33,34]. In fact its Hausdorff dimension is strictly smaller than one and bigger or equal to $\frac{1}{3}$, [9]. Also, two smooth critical circle maps with a unique critical point of the same criticality and the same irrational rotation number are C^1 conjugate as we will discuss in more details below. Another difference is that there are examples of two smooth and even real analytic critical circle maps with a unique critical point of the same criticality and the same rotation number satisfying a Diophantine condition such that the conjugacy is not $C^{1+\alpha}$ for any positive α, see [3,5].

If $b > 1$ the maps $f_{a,b}$ have two turning points and, as was proved in [28], it must have a periodic point for any value of a. For a continuous circle map f of degree 1, the limit $\rho(f, x) = \lim \frac{\tilde{f}^n(y)-y}{n}$ for $\pi(y) = x$ may not exist and when it exists it may depend on x. In general, the set of values of the limit that exists for some x is a closed interval that may degenerate to a point as in the case of orientation preserving circle diffeomorphisms. This is the so called *rotation interval* and for each point a in the rotation interval there exists a compact invariant subset $M_a \subset \mathbb{S}^1$ such that the restriction of f to M_a is minimal, order preserving and $\rho(f, x)$ exists and is equal to a for all $x \in M_a$. In particular for each rational number $\frac{p}{q}$, in the rotation interval, with p, q coprime, there exists a periodic point of f of period q and rotation number $\frac{p}{q}$, see [1]

The standard family is very rich in the sense that any degree one circle map with at most two turning points has essentially the same dynamical behavior as some map in the family. To be more precise, each degree one smooth circle map with at most two turning points is strongly semi-conjugate to some map $f_{a,b}$ in the standard family.

Definition 41.2. A circle map f is *strongly semi-conjugate* to a map g if there exists a continuous, monotone and surjective map $h: \mathbb{S}^1 \to \mathbb{S}^1$ such that $h \circ f = g \circ h$ and for every $y \in \mathbb{S}^1$, $h^{-1}(y)$ is either a unique point or a closed interval entirely contained in $P(f)$.

The above statement is a special case of the following result established in [21].

41 Dynamics on the Circle

Theorem 41.1. *For each integer m let $\mathscr{F}_m$ be the family of $2m$-modal maps of the circle whose lift is given by.*

$$\tilde{f}_\mu(t) = dt + \mu_1 + \mu_{2m}\sin(2\pi m t) + \sum_{j=1}^{m-1}(\mu_{2j}\sin(2\pi j t) + \mu_{2j+1}\cos(2\pi j t)),$$

(41.1)

where $\mu = (\mu_1,\ldots,\mu_{2m}) \in \Delta := \{\mu \in \mathbb{R}^{2m} : \mu_{2m} > 0$ and f_μ is $2m$ − multimodal$\}$ and $d \in \mathbb{Z}$. Then, any smooth circle map with $2n$ turning points is strongly semiconjugate to some $f_\mu \in \mathscr{F}_m$ for some $m \leq n$.

In the above theorem m may be strictly smaller than n. In fact, let us construct a $2n$-modal map g that is strongly semi-conjugate to a $2m$-modal map $f = f_\mu \in \mathscr{F}_m$. We first construct a map f_1 by blowing up the full orbit of a repelling periodic point of f and insert at each point an interval. We get an orbit of intervals that is eventually periodic and in the interior of the corresponding periodic interval we put a periodic attractor. This is a non-essential attractor for the map f_1 in the sense that its immediate basin of attraction does not contain a critical point. Let I be an interval that is not in the forward orbit of this periodic interval but which is mapped into a periodic interval. Let g be a map that coincides with f_1 outside of I and that has $2k$ turning points in I. This map is strongly semi-conjugate to f and has $2(m + k)$ turning points. It is not strongly semi-conjugate to a map in $\mathscr{F}_{m+k}$ because, as it was proved in [21], all these maps have negative Schwarzian derivative and, therefore, do not have non-essential periodic attractors.

Recall that a circle map is hyperbolic if all critical points are in the basin of periodic attractors and the complement of the basin of the periodic attractors is a hyperbolic invariant set: the norm of the derivative of some iterate of the mapping is bigger than one in every point of this invariant set. For a hyperbolic map f the chaotic part of the dynamics, $Ch(f)$ has zero Lebesgue measure, [25].

The family $\mathscr{F}_m$ of the above theorem is the restriction to the circle $\mathbb{S}^1 = \{z \in \mathbb{C}; |z| = 1\}$ of the family of holomorphic endomorphisms of $\mathbb{C} \setminus \{0\}$ defined by $f_\mu(z) = z^d \times \exp(\frac{P_\mu(z)}{z^m})$ where P_μ is a degree $2m$ polynomial whose coefficients are linear combinations of the μ_j. Using strong tools from complex dynamics it was proved in [26] that the set of parameter values $\mu \in \Delta$ such that f_μ is hyperbolic is open and dense in Δ. Furthermore, for μ_0 in this set, the set of $\mu \in \Delta$ such that f_μ is topologically conjugate to μ_0 has at most m connected components.

41.2 Rigidity of Circle Diffeomorphisms and Critical Circle Maps

The proof of Herman's rigidity theorem involves very delicate estimates from real analysis but no complex analysis argument. Recently K. Khanin and A. Teplinsky, [13], found a very simple proof of this theorem that involves only some cross-ratio estimates. In fact they prove that if the diffeomorphis is C^k with $k > \beta + 2$ then the conjugacy to a rotation is $C^{k-1-\beta}$.

The rigidity results for critical circle mappings involve both real analytic estimates and complex dynamics and is related to the behavior of a renormalization operator. To a critical circle mapping f with irrational rotation number we associate a sequence of interval mappings $f_n: J_n \to J_n$ which is the first return mapping of f to the interval J_n. The critical point splits the interval J_n in two intervals I_n, I_{n+1} where the interval I_n returns to J_n after q_{n+1} iterates whereas I_{n+1} returns after q_n iterates. Here $\frac{p_n}{q_n}$ are the convergents of the rotation number of f, i.e, the best rational approximations to the rotation number. By rescaling so that the interval I_n becomes the interval $[0, 1]$ we get a sequence of mappings $R^n(f): [-\lambda_n, 1] \to [-\lambda_n, 1]$ where λ_n is the ratio of the lengths of I_{n+1} and I_n. The critical circle mapping has bounded combinatorics if the ratio $\frac{q_{n+1}}{q_n}$ is uniformly bounded. The first step toward rigidity results for critical circle mappings is the so called real a priori bounds. This uses tools developed by Yoccoz, Herman, Swiatek that involves the control of the distortion of cross-ratios under iteration, see [5]. The real a priori bounds imply that the sequence λ_n is uniformly bounded and that the sequence of renormalized maps lie in a compact set in the C^0 topology. Furthermore, any convergent subsequence is a commuting pair of real analytic maps that have holomorphic extension belonging to the Epstein class of holomorphic pairs, [4].

The next step is the complex a priori bounds: high iterates of a real analytic commuting pair have holomorphic extensions that belong to a compact set of holomorphic commuting pairs. Finally, using McMullen's tools it was proved in [6] the exponential contraction of renormalization for real analytic maps with the same bounded combinatorial type and the $C^{1+\alpha}$ rigidity of these maps.

To study maps with unbounded combinatorial type we first notice that from the real a priori bounds, the renormalized maps of very big renormalization period, i,e, $\frac{q_{n+1}}{q_n}$ big, are very close, in the C^2 topology, to maps that have a parabolic fixed point with bounded second derivative. The next step is to prove the a priori complex bounds for the unbounded case, see [35] and to extend the McMullen's rigidity of towers to include the parabolic towers.

Those were the main new tools used in [36] to get the hyperbolicity or the full limit set of the renormalization operator for real analytic critical circle maps. A different proof of the exponential contraction of the renormalization operator was given in [14] where they prove that the conjugacy between two analytic critical circle maps with the same rotation number is $C^{1+\alpha}$ at the critical point and this implies the exponential contraction of the renormalization operator.

By a careful analysis of the parabolic bifurcation [3] shows the existence real analytic critical circle maps that are not $C^{1+\alpha}$ rigid for any $\alpha > 0$.

However, using a rather precise estimates of the iterates near a saddle-node bifurcation [12] proves that the exponential convergence of the renormalization operator acting on two smooth critical circle maps with the same irrational rotation number implies the existence of a C^1 conjugacy. Hence, combining this with the previous result we get the C^1 rigidity for any real analytic critical circle map with irrational rotation number. This in sharp contrast with the circle diffeomorphism case where,

41 Dynamics on the Circle

in the case of Liouville rotation number, the conjugacy to a rigid rotation may fail to be even quasi-symmetric.

By using an approximation of smooth critical circle maps by real analytic critical circle maps with holomorphic extension to $\mathbb{C} \setminus \{0\}$ the exponential convergence of the renormalization operator also holds for smooth critical circle maps [19] and hence we also have C^1 rigidity for smooth critical circle maps.

There are some other interesting rigidity results in one dimensional dynamics. From [32] we have that if the conjugacy between two smooth expanding circle maps is absolutely continuous then it is smooth. In [28], another type of rigidity is proved for real analytic maps of the circle and of the interval: if the conjugacy between two maps send critical points to critical points of the same order, parabolic periodic points into parabolic periodic points and if the maps have at least one periodic point then the conjugacy is quasi-symmetric. Finally there are many rigidity results for unimodal maps of the intervals that are infinitely renormalizable, see [7, 16, 20, 22–24, 30, 31].

41.3 Open Problems

In this section I will formulate some open problems related to rigidity in dynamics.

- Prove that in the family $f_\mu(z) = z^d \times \exp(\frac{P_\mu(z)}{z^m})$, even if some critical point of f_μ is outside the unit circle, there exists a sequence $\mu_n \to \mu$ such that the restriction of f_{μ_n} to the circle is hyperbolic. This would imply the density of hyperbolicity in the space of smooth circle maps.
- Find all smooth conjugacy invariants of critical circle maps with $n \geq 2$ critical points. Such a map is topologically conjugate to a rigid rotation and so has a unique invariant measure. The ratio of the measures of segments bounded by the critical points are clearly smooth conjugacy invariants.
- A very hard problem: extend the renormalization theory to cover real (non-integer) power law, i.e. for maps of the type $f(x) = \phi(|x|^\alpha)$ where α is a positive real number and ϕ is a smooth interval diffeomorphism. The difficulty here is that we can no longer use the strong tools from complex dynamics.
- Hyperbolicity of renormalization for smooth unimodal maps with unbounded combinatorial type.
- Renormalization of multimodal interval maps, see [27].

References

1. Alsedà, Ll., Llibre, J., Misiurewicz, M.: Combinatorial Dynamics and Entropy in Dimension One, 2nd edn. In: Advanced Series in Nonlinear Dynamics, vol. 5. World Scientific, Singapore (2000)
2. Arnol'd, V.I.: Small denominators. I. Mapping the circle onto itself. Izv. Akad. Nauk SSSR Ser. Mat. **25**, 21–86 (1961)

3. Avila, A.: On the Rigidity of Critical Circle Maps. http://www.proba.jussieu.fr/pageperso/artur/papers.html
4. de Faria, E.: Asymptotic rigidity of scaling ratios for critical circle mappings. Ergod. Theory Dyn. Syst. **19**(4), 995–1035 (1999)
5. de Faria, E., de Melo, W.: Rigidity of critical circle mappings I. J. Eur. Math. Soc. **1**, 339–392 (1999)
6. de Faria, E., de Melo, W.: Rigidity of critical circle mappings II. J. Am. Math. Soc. **13**, 343–370 (2000)
7. de Faria, E., de Melo, W., Pinto, A.A.: Global hyperbolicity of renormalization for C^r unimodal mappings. Ann. Math. **164**, 731–824 (2006)
8. Graczyk, J.: Linearizable circle diffeomorphisms in one-parameter families. Bol. Soc. Bras. Mat. (N.S.) **24**(2), 201–210 (1993)
9. Graczyk, J. Swiątek, G.: Critical circle maps near bifurcation. Comm. Math. Phys. **176**(2), 227–260 (1996)
10. Herman, M.: Sur la conjugaison differentiable des difféomorphismes du cercle a des rotations. Publ.Math. IHES **49**, 5–234 (1979)
11. Herman, M.: Mesure de Lebesgue et nombre de rotation. Geometry and topology (Proc. III Latin Amer. School of Math., Inst. Mat. Pura Aplicada CNPq, Rio de Janeiro, 1976), pp. 271–293. Lecture Notes in Math., vol. 597. Springer, Berlin (1977)
12. Khanin, K., Teplinsky, A.: Robust rigidity for circle diffeomorphisms with singularities. Invent. Math. **169**, 193–218 (2007)
13. Khanin, K., Teplinsky, A.: Herman's theory revisited. Invent. Math. **178**(2), 333–344 (2009)
14. Khmelev, D., Yampolsky, M.: The rigidity problem for analytic critical circle maps. Mosc. Math. J. **6**(2), 317–351 (2006)
15. Kozlovski, O., Shen, W., van Strien, S.: Density of hyperbolicity in dimension one. Ann. Math. (2) **166**(1), 145–182 (2007)
16. Lyubich, M.: Feigenbaum-Coullet-Tresser universality and Milnor's hairiness conjecture. Ann. Math. **149**, 319–420 (1999)
17. Mañé, R.: Hyperbolicity, sinks and measure in one-dimensional dynamics. Comm. Math. Phys. **100**(4), 495–524 (1985)
18. Martens, M., de Melo, W., van Strien, S.: Julia-Fatou-Sullivan theory for real one-dimensional dynamics. Acta Math. **168**(3–4), 273–318 (1992)
19. de Melo, W., Guarinos, P.: Rigidity for smooth critical circle maps. work in progress
20. de Melo, W., Pinto, A.A.: Rigidity of C^2 infinitely renormalizable unimodal maps. Comm. Math. Phys. **208**(1), 91–105 (1999)
21. de Melo, W., Salomo, P.A.S., Vargas, E.: A Full Family of Multimodal Maps of the Circle. Ergod. Theory Dyn. Syst. (in press)
22. McMullen, C.: Complex Dynamics and Renormalization. Annals of Math. Studies, vol. 135. Princeton University Press, Princeton, NJ (1994)
23. McMullen, C.: Renormalization and 3-manifolds which Fiber over the Circle. Annals of Math. Studies, vol. 142. Princeton University Press, Princeton, NJ (1996)
24. de Melo, W., Pinto, A.A.: Smooth conjugacies between C^2 infinitely renormalizable quadratic maps with the same bounded type. Commun. Math. Phys. **208**, 91–105 (1999)
25. de Melo, W., van Strien, S.: One-Dimensional Dynamics. Springer, Berlin (1993)
26. Rempe, L., van Strien, S.: Density of hyperbolicity for classes of real transcendental entire functions and circle maps. arXiv:1005.4627
27. Smania, D.: Phase space universality for multimodal maps. Bull. Braz. Math. Soc. (N.S.) **36**(2), 225–274 (2005)
28. van Strien, S.: Quasi-symmetric rigidity of real analytic one-dimensional maps. work in progress.
29. van Strien, S., Vargas, E.: Real bounds, ergodicity and negative Schwarzian for multimodal maps. J. Am. Math. Soc. **17**(4), 749–782 (2004)
30. Sullivan, D.: Quasiconformal homeomorphisms and dynamics, topology and geometry. Proc. ICM-86, Berkeley, v. II 1216–1228

31. Sullivan, D.: Bounds, quadratic differentials and renormalization conjectures. AMS Centennial Publications, Providence, RI (1992)
32. Shub, M., Sullivan, D.: Expanding endomorphisms of the circle revisited. Ergod. Theory Dyn. Syst. **5**(2), 285–289 (1985)
33. Swiatek, G.: On critical circle homeomorphisms. Bol. Soc. Brasil. Mat. (N.S.) **29**(2), 329–351 (1998)
34. Swiatek, G.: Rational rotation numbers for maps of the circle. Comm. Math. Phys. **119**(1), 109–128 (1988)
35. Yampolsky, M.: Complex bounds for renormalization of critical circle maps. Ergod. Theory. Dyn.Syst. **19**(1), 227–257 (1999)
36. M. Yampolsky, Renormalization horseshoe for critical circle maps. Comm. Math. Phys. 240(1–2), 75–96 (2003)
37. Yoccoz, J.-C.: Il n'y a pas de contre-exemple de Denjoy analytique. C. R. Acad. Sci. Paris **298**, 141–144 (1984)
38. Yoccoz, J.-C.: Conjugaison différentiable des difféomorphismes du cercle dont le nombre de rotation vérifie une condition diophantienne. Ann. Sci. École Norm. Sup. (4) **17**(3), 333–359 (1984)

Chapter 42
Rolling Ball Problems

Waldyr M. Oliva and Gláucio Terra

Abstract A spherical ball of radius δ rests on an oriented surface S embedded in $\mathbb{R}^3$ and has a positive orthonormal frame attached to it. The *states* of the ball are the elements of the 5-dimensional manifold $M = S \times SO(3)$ and a *move* is a smooth path on M corresponding to a rolling of the ball on S without slipping. The moves without slipping or twisting along geodesics are called *pure moves*. Rolling ball problems on S are mainly related to the search of $N(S)$, the minimum number of moves (or moves without twisting, or pure moves) sufficient to reach continuously any final state starting at a given initial state. We mention some results and conjectures relative to the case of a unitary ball ($\delta = 1$) rolling on surfaces of revolution; important cases are: plane, sphere, cylinder and surfaces parallel to Delaunay. The dynamics giving the moves without slipping of the rolling ball problems are nonholonomic, preserve a volume and lead, in certain cases, to the existence of minimal surfaces immersed in M.

42.1 Introduction

The rollings of a spherical ball B of the Euclidean space over a connected smooth surface S embedded in $\mathbb{R}^3$ suggest the study of kinematic (virtual) motions as well as give rise to some special nonholonomic mechanical systems with linear constraints.

The *state* of the ball is given by the pair formed by the point of contact between B and S and by a positive orthonormal frame attached to B. Assume, in this survey,

---

W.M. Oliva
Instituto Superior Técnico, ISR and Departamento de Matemática, Centro de Análise Matemática, Geometria e Sistemas Dinâmicos, Av. Rovisco Pais, 1, 1049-001 Lisbon, Portugal
e-mail: wamoliva@math.ist.utl.pt

G. Terra (✉)
Departamento de Matemática, Instituto de Matemática e Estatística, Universidade de São Paulo, Rua do Matão, 1010, 05508-090 São Paulo, Brazil
e-mail: glaucio.terra@gmail.com, glaucio@ime.usp.br

M.M. Peixoto et al. (eds.), *Dynamics, Games and Science I*, Springer Proceedings
in Mathematics 1, DOI 10.1007/978-3-642-11456-4_42,
© Springer-Verlag Berlin Heidelberg 2011

that B *rolls freely* on S, i.e. $S \cap B$ is the singleton of the contact point. So, the set of all states is identified with the 5-manifold $M = S \times SO(3)$. A *move* is a smooth path on M corresponding to a rolling of B on S without slipping and a *pure move* is a move without twisting along a geodesic of the surface S.

In the present paper, Sect. 42.2 deals with questions related to the kinematical properties of the moves: (a) geometric constraints, (b) the rolling of a ball of radius δ on an analytic surface of revolution and (c) rolling ball problems on S; these are mainly related to the search of $N(S)$, the minimum number of moves (or moves without twisting, or pure moves) sufficient to reach continuously any final state starting at a given initial state. Section 42.3 studies the dynamical properties of the moves; in particular, Sect. 42.3.1 refers to conservation of volume in the nonholonomic system with linear constraints corresponding to the rolling of a homogeneous spherical ball on a surface of revolution with constant mean curvature. As a special case, it is considered the rolling of a ball of radius δ on a surface parallel to a Delaunay, and the conditions on δ for a free rolling are established.

42.2 Kinematic Properties

As we observed in the Introduction, the space of states for the motions of a ball B rolling freely on a surface S is the 5-manifold $M = S \times SO(3)$ where we consider moves (motions of B on S without slliping) and pure moves (moves without twisting along a geodesic of S). We will describe the classic linear geometric constraints, the rolling of the ball B of radius δ on an analytic surface of revolution S and the rolling ball problems on S that is the search of $N(S)$, the minimum number of moves (or moves without twisting, or pure moves) sufficient to reach continuously any final state starting at a given initial state.

42.2.1 The Geometric Constraints

Let S be a smooth connected surface embedded in $\mathbb{R}^3$ and δ the radius of the rolling ball. It is usual to consider on $M = S \times SO(3)$ the following smooth distributions (see [1,4], and also [12]):

- $\mathscr{D}_1$, with constant rank 3, such that the paths tangent to it correspond to motions on S without slipping.
- $\mathscr{D} \subset \mathscr{D}_1$, with constant rank 2, such that the paths tangent to it correspond to motions on S without slipping or twisting.

Concerning the above distributions and assuming that the Gauss curvature of S is distinct everywhere from the curvature $1/\delta^2$ of the ball, we have:

- $\mathscr{D}$ and $\mathscr{D}_1$ are non-integrable.
- $\mathscr{D}$ and $\mathscr{D}_1$ are bracket generating, so they satisfy the hypothesis of the classical Chow's theorem.

42 Rolling Ball Problems

42.2.2 The Rolling of a Ball of Radius δ on an Analytic Surface of Revolution

With analytic data, the following theorems hold:

Theorem 42.1 ([1] and [10]). *For an* analytic *connected surface of revolution $S \subset \mathbb{R}^3$, every point of $S \times SO(3)$ is reachable from any other point of the same manifold by a continuous piecewise analytic path tangent to $\mathscr{D}$ (i.e. corresponding to motions without slipping or twisting).*

Theorem 42.2 ([1]). *Under the same hypothesis, every point of $S \times SO(3)$ is reachable from any other point of the same manifold by a continuous piecewise analytic path whose analytic pieces are* pure moves *(i.e. motions without slipping or twisting along geodesics).*

42.2.3 The Rolling Ball Problem

In the study of the kinematics of *rolling ball problems (RBP)*, one looks, as it is natural, for $N(S)$, the minimum number of moves (or moves without twisting, or pure moves) sufficient to connect, continuously on M, one chosen state to any other state of the ball.

42.2.3.1 The Kendall Problem

We recall briefly the search of $N(\mathbb{R}^2)$, the minimum number of pure moves sufficient to reach any final state of $\mathbb{R}^2 \times SO(3)$, starting from the initial state $(\mathbb{O}, \mathrm{id})$ – the so-called *Kendall Problem*.

Hammersley [8] used strongly the theory of quaternions to prove that $N(\mathbb{R}^2)=3$; in [2] it is also shown a geometric proof for that without the use of quaternions. The two hard steps in the proofs appearing in [2] and [8] are the following propositions:

Proposition 42.1. *Given P_0 in $\mathbb{R}^2$ and $\theta \in \mathbb{S}^1$, we can pass from the initial state (P_0, id) to the final state $(P_0, R_3(\theta))$ doing 3 pure moves, where $R_3(\theta) \in SO(3)$ is the rotation of angle θ around $e_3 = (0,0,1)$.*

Proposition 42.2. *Given P_0 in $\mathbb{R}^2$ and $M \in SO(3)$ with the third column linearly independent with e_3, we can pass from the initial state (P_0, id) to the final state (P_0, M) doing 3 pure moves.*

42.2.3.2 The Rolling of a Unitary Ball on a Sphere of Radius $R > 1$

Let $S = \mathbb{S}^2(R)$ be the sphere of radius $R > 1$ centered at the origin of $\mathbb{R}^3$ and B the unitary ball. As before, it is posed the following question: *How many pure moves*

$N(\mathbb{S}^2(R))$, *rolling outside the sphere, will be sufficient to reach any final state of* $\mathbb{S}^2(R) \times SO(3)$ *starting at* $\left(P_0 = (0,0,1), \mathrm{id}\right)$?

- In [7] it is proved that $3 \leqslant N(\mathbb{S}^2(R)) \leqslant 4$.
- In [2] (an ongoing project), the authors try to prove or disprove the following open question: *is it true that* $N(\mathbb{S}^2(R)) = 3$?
  As in the case of the Kendall Problem, they start by considering two claims analogous to Propositions 42.1 and 42.2. The first claim is already proved, but they are still working on the second one.
- The same questions can be considered when a unitary ball rolls inside a sphere of radius $R > 1$, as suggested in [7] and [10].

42.3 Dynamical Properties

In this section we describe the dynamics of a homogeneous rigid ball rolling on a smooth connected surface S embedded in $\mathbb{R}^3$. To define the d'Alembert–Chetaev dynamics of the RBP, we consider the following smooth data (see [13, 17]):

- $M = S \times SO(3)$, the so-called *configuration space*.
- The *Riemannian metric* **g**, which is obtained by polarization of the kinetic energy $K : M \rightarrow \mathbb{R}$ of the ball.
- The *constraint* (linear), given by the constant rank 3 distribution $\mathscr{D}_1 \subset TM$ (motions of the ball without slipping).

The d'Alembert–Chetaev trajectories of the nonholonomic mechanical system $(M, \mathscr{D}_1, K)$ are given by the equation (see [12, 14, 16] and [18]):

$$\nabla_t \dot{q} = \mathscr{B}_{\mathscr{D}_1}(\dot{q}, \dot{q}) \tag{42.1}$$

where:

- ∇_t denotes the covariant derivative induced by the Levi–Civita connection of $(M, \mathbf{g})$;
- $\mathscr{B}_{\mathscr{D}} : TM \oplus_M \mathscr{D} \rightarrow \mathscr{D}^\perp$ denotes the total second fundamental form of a linear subbundle $\mathscr{D} \subset TM$, given by: $(X, Y) \mapsto P_{\mathscr{D}^\perp} \nabla_X Y$, where $\mathscr{D}^\perp$ is the linear subbundle orthogonal to $\mathscr{D}$ (with respect to g) and $P_{\mathscr{D}^\perp}$ the orthogonal projection.

Equation (42.1) defines a smooth vector field on $\mathscr{D}_1$, the so-called GMA vector field of $(M, \mathscr{D}_1, K)$, whose trajectories are canonical lifts of their projections on M (i.e. a second order vector field).

Concerning the flow of the GMA vector field on $\mathscr{D}_1$, we have:

Theorem 42.3 ([12, 13]). *The GMA flow on $\mathscr{D}_1$ preserves a natural volume or, equivalently, $\mathscr{D}_1^\perp$ is minimal (i.e. the trace of the second fundamental form of $\mathscr{D}_1^\perp$ restricted to $\mathscr{D}_1^\perp \oplus_M \mathscr{D}_1^\perp$ vanishes).*

42 Rolling Ball Problems   665

See [3] for the special case $M = \mathbb{S}^2(R) \times SO(3)$ and [16] for a generalization of the results in [3].

Theorem 42.4 ([12]). *The minimal distribution $\mathscr{D}_1^\perp$ is integrable if, and only if, the surface $\mathscr{C}$ parallel to S, defined by the possible positions of the center of the ball, has constant mean curvature with absolute value $H(\mathscr{C}) = \frac{5}{4\delta}$.*

Example 42.1. Special cases of $\mathscr{C}$ are: *spheres, circular cylinders* and *Delaunay surfaces*, provided that $H(\mathscr{C}) = \frac{5}{4\delta}$.

Corollary 42.1. *The ball of radius δ rolls without slipping inside a sphere of radius R. Then $\mathscr{D}_1^\perp$ is integrable if, and only if, $\delta = \frac{5R}{9}$. In particular, the compact manifold $\mathbb{S}^2(R) \times SO(3)$ admits a 2-dimensional foliation with immersed minimal leaves.*

Remark 42.1. If the ball of radius δ rolls outside $\mathbb{S}^2(R)$, $\mathscr{D}_1^\perp$ is never integrable because, in that case, $H(\mathscr{C}) = \frac{1}{R+\delta}$ cannot be equal to $\frac{5}{4\delta}$.

42.3.1 The Rolling on a Surface Parallel to Delaunay

Since Delaunay surfaces represent all the surfaces of revolution embedded in $\mathbb{R}^3$ with constant mean curvature (except the spheres), it is natural to study, in view of Theorem 42.4, the surfaces S such that their δ-parallel $\mathscr{C}$ are Delaunay. That was done in [16], as summarized below.

In 1841 Delaunay [5] was able to obtain the following description of all surfaces of revolution in $\mathbb{R}^3$ of constant mean curvature (see also [9]). By rolling a given conic section on a line in a plane, and rotating about that line the trace of a focus, one obtains a surface of constant mean curvature in $\mathbb{R}^3$. Conversely, all the surfaces of revolution of constant mean curvature in $\mathbb{R}^3$ (with the exception of spheres) can be described in this way:

Theorem 42.5 (Delaunay [5,9,11]). *For finding the meridian curve of a surface of revolution of which the mean curvature is constant and equal to $(2a)^{-1}$, it must be done by rolling upon the axis of the surface an ellipse (or a hyperbola) such that the major axis (or the transverse axis) is equal to $2a$, and the focus describes the desired curve.*

Delaunay obtained also parametric equations for the surface in terms of positively oriented orthogonal coordinates (x, y) on the plane of the meridian curve, the x coordinates giving the orientation of the axis of rotation of the surface. For the proof he started by integrating the equations for an evolute of the meridian curve and from that he obtained the parametric equations. Therefore these last equations hold only on some intervals on which the evolute can be defined.

Kenmotsu [11] found nice expressions for the meridian curve in terms of the arc length parameter and described, among other interesting results, all complete surfaces of revolution in $\mathbb{R}^3$ with constant mean curvature. In fact, Kenmotsu studied the surfaces of revolution in $\mathbb{R}^3$ with prescribed mean curvature $H(s)$ where s is the arc length of the C^2 meridian curve $(x(s), y(s))$, $s \in I$. The surface of revolution generated by the rotation of that meridian curve, around the x-axis, have the following representation:

$$(x(s), y(s)\cos\theta, y(s)\sin\theta), \quad s \in I \quad \theta \in [0, 2\pi].$$

By the regularity of the surface, we may assume $y(s) > 0$ on the interval I.
So the mean curvature $H(s)$, satisfies,

$$2H(s)y(s) - x'(s) - y(s)[x''(s)y'(s) - x'(s)y''(s)] = 0 \qquad (42.2)$$

where $x'(s)^2 + y'(s)^2 = 1$, $\quad s \in I$.

If the mean curvature is constant and non zero, $H(s) = H \neq 0$, Kenmotsu derived the following expression for the meridian curve $(x(s), y(s)) = X(s; H, B)$, where B is any constant (see [11]):

$$X(s; H, B) = \left(\int_0^s \frac{1 + B\sin 2Ht}{\sqrt{1 + B^2 + 2B\sin 2Ht}} dt, \frac{1}{2|H|}\sqrt{1 + B^2 + 2B\sin 2Hs} \right),$$
$$(42.3)$$

$s \in \mathbb{R}$.

The following interesting remarks are useful (see [11]):

(a) $X(s; -H, B) = X(s; H, -B)$.
(b) $X(s; H, -B) = X(s - \frac{\pi}{2}H; H, B) +$ a constant vector.
(c) $X(s; \lambda H, B) = \frac{1}{\lambda}X(\lambda s; H, B), \lambda > 0$.
   So, without loss of generality, it is enough to consider the cases $B \geq 0$ and $H > 0$.
(d) $X(s; H, 0)$ is the generating line of a circular cylinder.
(e) $X(s; H, 1)$ represents a sequence of continuous half circles over the x-axis which have the radii.
(f) If $0 < B < 1$, $x(s)$ is monotone increasing as $s \to \infty$.
(g) If $B > 1$ $x(s)$ is not monotone.

In the last two cases, $\lim x(s) = \infty$ as $s \to \infty$ because $X(s; H, B)$ is periodic, with period π/H. Moreover, using Delaunay's Theorem 42.5 a simple computation shows that the constants H and B appearing in (42.3) are given by $H = 1/2a$ and $B = e$, where a and e are the semi-axis and the eccentricity of the rolling ellipse or hyperbola. Note that in the case $B > 1$ we obtain points where $x'(s)$ vanishes, that is, the values of $s \in \mathbb{R}$ such that $\sin 2Hs = -\frac{1}{B}$.
From (42.2) one obtains

$$2H = \frac{x'}{y} + x''y' - x'y'' = \frac{1}{a}.$$

42 Rolling Ball Problems

Since $\tau = (x', y')$, $n = (-y', x')$ for (τ, n) positively oriented and $\tau' = k_c n$ (see [6]) we have $x'' y' - x' y'' = -k_c$. And so we can rewrite (42.2) as

$$k_p + k_m = \frac{1}{a}$$

where k_p and k_m are the principal curvatures

$$k_p = \frac{x'}{y},$$

and

$$k_m = -k_c = x'' y' - x' y''.$$

Remark 42.2. Standard computations give us the following expressions

$$k_p = \frac{y^2 \pm b^2}{2ay^2} \tag{42.4}$$

and

$$k_m = \frac{y^2 \mp b^2}{2ay^2} \tag{42.5}$$

where the upper (resp. the lower) sign corresponds to the ellipse of equation $\frac{x^2}{a^2} + \frac{y^2}{b^2} = 1$ (resp. hyperbola of equation $\frac{x^2}{a^2} - \frac{y^2}{b^2} = 1$).

We will show that the surface $S \subset \mathbb{R}^3$ where the sphere rolls is well defined, globally, as an embedding. The surface S is parallel to the given Delaunay surface $C \subset \mathbb{R}^3$ defined by the *locus* of the positions of the center of the sphere when it rolls over S. We consider two types of Delaunay surfaces. For $0 < B < 1$ the meridian curve generating the Delaunay surface is periodic and $x' > 0$. In this case, there is no "loop" in the meridian and we consider $S_\pm$ generated by the meridian curve extending from $x \to -\infty$ to $x \to +\infty$. When $B > 1$, the meridian is still periodic, but the curve extending from $x \to -\infty$ to $x \to +\infty$ presents loops. So to avoid these loops, we consider in the sequel the Delaunay surface generated by rotating the generating curve $(x(s), y(s))$ above over one period and starting at any point where $x' = 0$.

Let us now choose the field of unit normal vectors ξ on the surface C such that at each point ξ coincides with $-n$, where n is the above unit normal to the meridian at that point.

Define S_+ (resp. S_-), the parallel surface to C, by $S_+ = C + \delta \xi$ (resp. $S_- = C - \delta \xi$), with δ to be chosen properly. The condition for S_+ to be an immersion in $\mathbb{R}^3$ is that the parallel distance between S_+ and C, given by the radius of the sphere, be distinct from the reciprocal of the curvature k_c of the meridian generating C. To show that, let us denote by $(\bar{x}, \bar{y})$ a generic point of the meridian of S_+. So

$$(\bar{x}, \bar{y}) = (x, y) + \delta(\xi_x, \xi_y),$$

668   W.M. Oliva and G. Terra

and

$$(\bar{x}', \bar{y}') = (x', y') + \delta(\xi'_x, \xi'_y),$$

where $'$ denotes the derivation with respect to the arc length of the meridian curve of the surface C. From what we saw above one has

$$\xi' = k_c(x', y');$$

then

$$(\bar{x}', \bar{y}') = (x' + \delta k_c x', y' + \delta k_c y') = (1 + \delta k_c)(x', y').$$

So we have immersion if and only if $1 + \delta k_c \neq 0$ along the meridian curve (in fact, we only need to check the previous condition for points where $k_c < 0$). For S_- the condition for immersion is $1 - \delta k_c \neq 0$ and is obtained analogously.

We will prove that the same condition obtained for the immersion is enough to show that we obtain the embedding of the surfaces S_+ and S_-. Start by fixing $\tilde{s} \in \mathbb{R}$ and consider the map $f : s \mapsto \bar{x}(s) - \bar{x}(\tilde{s}) \in \mathbb{R}$. We need to show the injectivity of the function $(x, y) \in C \mapsto (\bar{x}, \bar{y}) \in S_+$ and for that it is enough to show that for any $s \neq \tilde{s}$ we have $\bar{x}(s) \neq \bar{x}(\tilde{s})$. Now assume by contradiction that there exists an $\hat{s} \neq \tilde{s}$ such that $\bar{x}(\hat{s}) = \bar{x}(\tilde{s})$, so we have $f(\hat{s}) = 0$ and also $f(\tilde{s}) = 0$. By the classical Rolle's theorem there exists s_0 strictly between $\hat{s}$ and $\tilde{s}$ such that $f'(s_0) = x'(1 - \delta k_c) = 0$, with $\delta k_c \neq 1$, which is a contradiction because the meridian curve of the Delaunay surface does not have vertical tangents (the same happens with S_-). That proves the following proposition:

Proposition 42.3. *Let k_c be the curvature function of the meridian curve of the Delaunay surface and $\delta > 0$ such that $k_c \delta \neq 1$ (resp. $k_c \delta \neq -1$). Then S_+ (resp. S_-) is an embedded parallel surface to a given Delaunay surface.*

The existence of a "free" rolling motion of a ball of radius $\delta > 0$ over the surface S_+ (resp. S_-) requires that the sphere is not "locked" during its motion over that surface. Here not locked means that the reciprocal of the radius δ of the ball is bigger than the maximum of the absolute values of the principal curvatures of the surface S_+ (resp. S_-) for all points where the ball touches the closure of the set of points of the meridian of the surface S_+ (resp. S_-) in which this meridian is convex.

Using the previous definitions and results, and using the relation between the principal curvatures of the Delaunay surface C and its parallel surfaces $S_\pm$ (see [15])

$$k_i^{S_\pm} = \frac{k_i^C}{1 \mp \delta k_i^C}$$

valid for $\delta < |k_i^C|$; a simple computation then shows (Table 42.1):

Theorem 42.6 ([16]). *The surface S_+ (resp. S_-) parallel to a given Delaunay surface C, as defined above, is well defined and the ball can freely roll over it, if the appropriate condition in the table below is satisfied.*

42 Rolling Ball Problems

Table 42.1 Conditions for the sphere to roll over a surface parallel to a Delaunay surface

	Ellipse	Hyperbola
S_-	$\delta < \begin{cases} a & \text{if } 0 < e < 1/2 \\ \frac{a(1-e)}{e} & \text{if } e \geq 1/2 \end{cases}$	$\delta < a(e-1)$
S_+	$\delta < a(1-e)$	$\delta < \frac{a(e-1)}{e}$

Acknowledgements The first author wishes to thank *Fundação para a Ciência e Tecnologia (Portugal)* for the support through Program POCI 2010/FEDER. The second author wishes to thank *Centro de Análise, Geometria e Sistemas Dinâmicos* at *Instituto Superior Técnico, Lisbon, Portugal*, where he was kindly received during the year 2008 and also *Coordenação de Aperfeiçoamento de Pessoal de Nível Superior (Brazil)* for the support through process 3952-07-0.

References

1. Biscolla, L.M.O.: Controlabilidade do rolamento de uma esfera sobre uma superfície de revolução. Ph.D. Dissertation, IME – USP, São Paulo. (2005)
2. Biscolla, L.M.O., Llibre, J., Oliva, W.M.: The rolling ball problem on the sphere. Ongoing project, Lisbon (2008)
3. Blackall, C.: On volume integral invariants of non-holonomic dynamical systems. Am. J. Math. **63**, 155–168 (1941)
4. Bryant, R.L., Hsu, L.: Rigidity of integral curves of rank 2 distributions. Invent. Math. **114**, 435–461 (1993)
5. Delaunay, C.: Sur la surface de révolution dont la courbure moyenne est constant. Journal de Math'ematiques Pures et Appliquées. Série 1, **6**, 309—320 (1841)
6. do Carmo, M.: Differential Geometry of Curves and Surfaces. Prentice Hall, Englewood Cliffs (1976)
7. Luise M. Frenkel. Resolução do problema de kendall na esfera. Ph.D. Dissertation, IME – USP, São Paulo. (2007)
8. Hammersley, J.M.: Oxford commemoration ball. In: Kingman, J.F.C., Reute, G.E.H. (eds.) Probability, Statistics and Analysis, papers dedicated do David G. Kendall on the occasion of his sixty-fifth birthday, volume 79 of London Math. Society Lecture Notes Series, pp. 112–142. CUP, Cambridge-New York (1983)
9. Hsiang, W.-Y., Yu, W.-C.: A generalization of a theorem of Delaunay. J. Differ. Geom. **16**, 161—177 (1981)
10. Jurdjevic, V.: Geometric Control Theory. Cambridge University Press, Cambridge (1997)
11. Kenmotsu, K.: Surface of revolution with prescribed mean curvature. Tôhoku Math. J. **32**, 147–153 (1980)
12. Kobayashi, M.H., Oliva, W.M.: Nonholonomic systems and the geometry of constraints. Qual. Theory Dyn. Syst. **5**, 247–259 (2004)
13. Kobayashi, M.H., Oliva, W.M.: A note on the conservation of energy and volume in the setting of nonholonomic mechanical systems. Qual. Theory Dyn. Syst. **4**, 383–411 (2004)
14. Kupka, I., Oliva, W.M.: The non-holonomic mechanics. J. Differ. Equ. **169**, 169–189 (2001)
15. Nomizu, K.: Ellie Cartan's work on isoparametric families of hypersurfaces. Proc. Symp. Pure Math. **27**, 191–200 (1975)
16. Oliva, W.M., Kobayashi, M.H., Terra, G.: Anosov flows and invariant measures in constrained mechanical systems. São Paulo Journal of Mathematical Sciences. Proceedings of the 1st IST-IME Meeting. 2(1), 55–76 (2008)
17. Oliva, W.M.: Geometric Mechanics, volume 1798 of Lecture Notes in Mathematics. Springer, Berlin (2002)
18. Terra, G., Kobayashi, M.H.: On classical mechanical systems with non-linear constraints. J. Geom. Phys. **49**(3–4), 385–417 (2004)

Chapter 43
On the Dynamics of Certain Models Describing the HIV Infection

Dayse H. Pastore and Jorge P. Zubelli

Abstract This article concerns some global stability aspects of a class of models introduced by Nowak and Bangham that describe in a fairly successful way the initial phases of the HIV dynamics in the human body as well as some generalizations that take into account mutations. We survey recent results implying that the biologically meaningful positive solutions to such models are all bounded and do not display periodic orbits. For the mutationless cases the dynamics is characterized in terms of certain dimensionless quantities, the so-called basic reproductive rate and the basic defense rate. As a consequence, we infer that the finite dimensional models under consideration cannot account, without further modifications, for the third phase of the HIV infection. We conclude by suggesting a modification that according to our numerical simulations may describe the collapse of the infected patient.

43.1 Introduction

A better understanding of how entire populations of viruses, such as the HIV, interact with immune cells seems to be a key factor in the development of effective long-term therapies or possibly preventive vaccines for deadly diseases such as the acquired immunodeficiency syndrome [11]. Mathematical modeling of the underlying biological mechanisms and a good understanding of the theoretical implications of such models is crucial in this process. Indeed, it helps clarifying and testing assumptions, finding the smallest number of determining factors to explain the biological phenomena, and analyzing the experimental results [1]. Furthermore,

---

D.H. Pastore
CEFET-RJ, Av. Maracanã 229, Rio de Janeiro, RJ 20271-110, Brazil
e-mail: dayse@impa.br

J.P. Zubelli (✉)
IMPA, Est. D. Castorina 110, Rio de Janeiro, RJ 22460-320, Brazil
e-mail: zubelli@impa.br

M.M. Peixoto et al. (eds.), *Dynamics, Games and Science I*, Springer Proceedings
in Mathematics 1, DOI 10.1007/978-3-642-11456-4_43,
© Springer-Verlag Berlin Heidelberg 2011

modeling has already impacted on research at molecular level [11] and important results have been obtained in modeling the virus dynamics for several infections, such as the HIV [10, 15, 17], hepatitis B [7], hepatitis C [8], and influenza [2].

In this work we survey a class of models introduced by Nowak and Bangham in [10] as well as some extensions of these models that take into account mutations. Our main goal is to study the global dynamics of the models. It turns out that in this description two key dimensionless parameters play a crucial role. They are the *basic reproductive ratio* and the *basic defense ratio*.

For the first model under consideration, namely the one that takes into account the infected and uninfected concentrations of CD4+ T cells and the concentration of free HIV in the blood, for any biologically meaningful initial condition one of the following situations will happen: If the basic reproductive ratio is less than one, then eventually the virus is cleared and the disease dies out. If the basic reproductive ratio R_0 is greater than one, then the virus persists on the host approaching a chronic disease steady state. Finally, if $R_0 = 1$ then the two stationary states coincide and the biological solutions approach such state as time goes by. This first model does not consider the immune response provided by the cytotoxic T lymphocytes (CTL). The latter kill cells that are infected with viruses.

For the second model under consideration, namely the one that besides the afore-mentioned variables takes also into account the CTL response concentration, we also characterize the global dynamics according to the values of R_0 and the basic defense rate D_0. If $R_0 < 1$, then eventually the virus is cleared. If $1 < R_0 < 1 + (R_0/D_0)$, then generically the virus persists while the CTL response tends to zero.

We study a third model, also by [10] that besides the above variables takes into account mutations. In this case, if we start with biologically meaningfull initial data in the sense that all coordinates are non-negative, then they remain so for all future times and, furthermore, remain bounded. Recent results in [18] indicate that under mild hypothesis on the model parameters, the equilibria of such systems are globally asymptotically stable. Yet, the full characterization of the nongeneric cases remains open.

We remark in passing that although our focus is primarily HIV, the basic muta-tionless models we are considering may apply to many viral infections besides HIV [11].

The plan for this article goes as follows: In Sect. 43.2 we describe the models under consideration. Three of the models come from those proposed by Nowak and Bangham, while a fourth one involving possibly an arbitrary quantity of virus strains is also discussed. In Sect. 43.3 we present the mathematical statements, as well as some of their proofs, which characterize the long time behavior of the within-host infectious dynamics. In Sect. 43.4 we conclude with a discussion of the results and show some numerical simulations of an invading species illustrating the collapse of the infected individual.

43.2 Methods and Models

We start by recalling the path followed by the within-host HIV infection [11]. First, the HIV enters a T cell. Being a retrovirus, once the HIV is inside the T cell, it makes a DNA copy of its viral RNA. For this process it requires the reverse transcriptase (RT) enzyme. The DNA of the virus is then inserted in the T-cell's DNA. The latter in turn will produce viral particles that can bud off the T cell to infect other ones. Before one such viral particle leaves the infected cell, it must be equipped with *protease*, which is an enzyme used to cleave a long protein chain. Without protease the virus particle is incapable of infecting other T cells.

One of the key characteristics of HIV is its extensive genetic variability. In fact, the HIV seems to be changing continuously in the course of each infection and typically the virus strain that initiates the patient's infection differs from the one found a year ore more after the infection.

In what follows we present four models. Two of them do not take into account mutation, whereas the other ones consider mutation. The difference between the two latter ones is the possibility of mutation on an arbitrary set of strains. This could be a powerful tool in modeling the genetic variability of the HIV within-host variability.

43.2.1 Mutationless Models

Martin Nowak and Charles Bangham in [10] introduced a class of models for the time evolution of the HIV virus in the human organism. The simplest of such models considers the virus, the cells that it attacks, and the infected cells. It is given by

$$\begin{aligned}
\dot{x} &= \lambda - dx - \beta xv, \\
\dot{y} &= \beta xv - ay, \\
\dot{v} &= ky - uv.
\end{aligned} \tag{43.1}$$

Here, the state variables of the system are:

x : Concentration of CD4+ T cells in the blood
y : Concentration of infected CD4+ T cells by the HIV
v : Concentration of free HIV in the blood

The (positive) constants are:

λ : CD4+ T cell supply rate
d : CD4+ T cell death rate
β : Infection rate
a : Death rate of the infected cells
k : Free virus production rate
u : Free virus death rate

The first equation represents the CD4+ T cell rate of change in the blood. Free virus infect healthy cells at a rate proportional to the product of their concentrations, xv. Thus, β is the constant that represents the efficacy of such process. On the other hand, positive cells are produced at a constant rate λ and die at a rate xd.

The second equation concerns the infected cells. They are produced at a rate βxv and perish at a rate ay.

The third equation, represents the free virus dynamics. Infected cells release free virus at a rate proportional to their abundance, y, and free virus are removed from the system at rate uv.

A second model presented by Nowak and Bangham includes the presence of the defense cells in the organism but does not foresee mutation. It is given by:

$$\begin{aligned}
\dot{x} &= \lambda - dx - \beta xv, \\
\dot{y} &= \beta xv - ay - pyz, \\
\dot{v} &= ky - uv, \\
\dot{z} &= cyz - bz.
\end{aligned} \qquad (43.2)$$

Here, the variables and constants are the same ones of System (43.1) and in addition we have:

z : CTL response concentration
p : Infected cells elimination rate by the CTL response
c : CTL reproduction rate
b : CTL death rate

The growth rate of the CTL response concentration in this model is take to be proportional to the product yz of infected cells and virus concentration.

43.2.2 Models with Mutation

The third model introduced by Nowak and Bangham, which now considers mutation, is given by:

$$\begin{aligned}
\dot{x} &= \lambda - dx - x\Sigma_{i=1}^{n}\beta_i v_i, \\
\dot{y}_i &= \beta_i xv_i - ay_i - py_i z_i, \\
\dot{v}_i &= k_i y_i - uv_i, \\
\dot{z}_i &= cy_i z_i - bz_i.
\end{aligned}$$

Here, the index $i \in \{1, \ldots, n\}$ indicates the virus strain (or mutant) and n is the total number of strains. We remark that the only constants that depend on the virus strain are β_i (infection rate for the i-th virus) and k_i (production rate for the i-th virus).

We may assume, without loss of generality, that the virus production rate is a positive constant k independently of the virus strain. This is obtained after changing

43 On the Dynamics of Certain Models Describing the HIV Infection 675

v_i into $k_i v_i / k$ and β_i into $k\beta_i / k_i$ in the previous system. Thus, we get

$$
\begin{aligned}
\dot{x} &= \lambda - dx - x\Sigma_{i=1}^{n}\beta_i v_i, \\
\dot{y}_i &= \beta_i x v_i - a y_i - p y_i z_i, \\
\dot{v}_i &= k y_i - u v_i, \\
\dot{z}_i &= c y_i z_i - b z_i.
\end{aligned}
\tag{43.3}
$$

We shall now present a model that accounts for mutation both in terms of replication ability and escape from immune response. The equations of the model represent rate of change for uninfected cells, infected cells, free virus and CTL response, respectively. The model also simulates the mutation process of the virus.

The fundamental idea here lies in the fact that an integral operator could be used to model in a robust way the multitude of possible genetic variations. Indeed, the genome length of the HIV is of the order of $L = 10^4$ and this *in principle* could encode 4^L different strains [11, Sect. 8.1]. Although obviously most of these strains would not correspond to different viable antigenic responses, it stands to reason that such space could be very large indeed and endowed with a very complex landscape. The different virus strains will be indexed by a parameter $\mu \in \Omega$ where Ω is a set with as little structure as possible. The only structure we require is that it should be a σ-finite measure space. This is motivated by the idea that HIV mutations occur on a very large configuration space. This space, albeit finite, can be modeled by a infinite set in the same spirit of statistical or continuum mechanics.

The model takes the form:

$$
\begin{aligned}
\dot{x} &= \lambda - dx - x\int \beta_\mu v_\mu d\mu, \\
\dot{y}_\mu &= \beta_\mu x v_\mu - a y_\mu - p y_\mu z_\mu, \\
\dot{v}_\mu &= k[(1 - \theta)y_\mu + \theta K[y](\mu)] - u v_\mu, \\
\dot{z}_\mu &= c y_\mu z_\mu - b z_\mu.
\end{aligned}
\tag{43.4}
$$

where $\theta \in [0, 1]$ and the variables y, v and z are functions of the time $t \in [0, \infty)$ and of the virus mutation strain $\mu \in \Omega$. We summarize in Table 43.1 the biological meaning of the variables and parameters occurring in the model.

The mutation process is modeled as follows: Ω is a σ-finite measure space and the integral operator

$$
K[y](\mu) = \int_\Omega K(\mu, \mu')y(\mu')d\mu'
$$

gives the total of viruses that are transformed into strain μ virus.

We assume that K is positive and belongs to $L^1(\Omega \times \Omega)$. We will also assume that

$$
\int_\Omega K(\mu, \mu')d\mu' = \int_\Omega K(\mu', \mu)d\mu' = \overline{K} \in \mathbb{R}, \forall \mu \in \Omega.
\tag{43.5}
$$

Table 43.1 Variables and parameters

Variable	Parameter
x	Uninfected cells in the organism
y_μ	Infected cells with the HIV of strain μ
v_μ	Free HIV of strain μ
z_μ	CTL response that eliminates cells infected by strain μ HIV
λ	Uninfected cells supply rate
d	Uninfected cells death rate
β_μ	Infection rate
a	Infected cells death rate
k	Free virus production rate
u	Free virus death rate
p	Infected cells elimination rate by CTL response
c	CTL reproduction rate
b	CTL death rate

It is natural to request that the total amount of virus, taking into account all strains, to be finite. Thus,

$$\int_\Omega v_\mu d\mu < \infty.$$

Likewise for y_μ and z_μ. It is also natural to require that all such quantities to be bounded almost everywhere in Ω. Thus, we consider the solutions of the system in the space

$$\mathfrak{M} := \mathbb{R} \oplus \left(L^\infty(\Omega, \mathbb{R}^3) \cap L^1(\Omega, \mathbb{R}^3)\right),$$

[14] carried out an analytic study of the integro-differential System (43.4). For such biologically meaningful initial conditions, existence and uniqueness of the solutions were established.

We observe that System (43.4) includes the model of (43.3) as special case if we take Ω as a finite cardinality probability space.

43.3 Results

We will start by describing the stationary solutions of System (43.1) following [14]. We remark that some of the results for the three state-variable systems therein overlap with the comprehensive analysis developed by [4] that used different techniques. They are presented here for the sake of completeness.

It is easily verified that the stationary solutions are

$$X_1^\star = (x_1^\star, y_1^\star, v_1^\star) = \left(\frac{\lambda}{d}, 0, 0\right)$$

43 On the Dynamics of Certain Models Describing the HIV Infection

and

$$X_2^\star = (x_2^\star, y_2^\star, v_2^\star) = \left(\frac{ua}{\beta k}, \frac{k\beta\lambda - uda}{\beta ak}, \frac{k\beta\lambda - uda}{\beta au}\right).$$

The stationary solution $X_1^\star$ corresponds to the absence of the HIV in the organism. On the other hand, the stationary solution $X_2^\star$ corresponds to an equilibrium of infected cells and T cells.

In order to perform the analysis of the infinitesimal behavior of the stationary solutions it is convenient to write the System (43.1) in the form $\dot{X} = F(X)$ where $X = (x, y, v)$ and $F : \mathbb{R}^3 \to \mathbb{R}^3$ is defined by

$$F(X) = \begin{bmatrix} \lambda - dx - \beta xv \\ \beta xv - ay \\ ky - uv \end{bmatrix}.$$

The Jacobian of F takes the form

$$DF(X) = \begin{bmatrix} -d - \beta v & 0 & -\beta x \\ \beta v & -a & \beta x \\ 0 & k & -u \end{bmatrix}.$$

For generic parameters the matrices $DF(X_1^\star)$ and $DF(X_2^\star)$ have nonzero determinant and are hyperbolic points. From the Hartman–Grobman Theorem [6] it follows that the local (infinitesimal) behavior of the system in a neighborhood of the point $X_1^\star$, respectively $X_2^\star$, is determined by the sign of the real part of the eigenvalue of $DF(X_1^\star)$, respectively $DF(X_2^\star)$.

It turns out to be useful to consider what we will call in the sequel *basic reproductive ratio*

$$R_0 := \frac{k\lambda\beta}{dau}.$$

It consists of a dimensionless parameter that considers the ratio of the parameters that contribute to the increase of the variables divided by the parameters that contribute to their depletion. The next result states if R_0 is small, i.e. less than 1, then the equilibrium of infected cells and T cells is unstable while the absence of HIV in the organism is an attractor. The picture is reversed if $R_0 > 1$. More precisely, we have that

Lemma 43.1. *If $R_0 = 1$, then $X_1^\star = X_2^\star$ and $DF(X_1^\star) = DF(X_2^\star)$ possesses two negative eigenvalues and a null one. If $R_0 \neq 1$ the local behavior of the stationary solutions is described according to the following:*

	$R_0 < 1$	$R_0 > 1$
$DF(X_1^\star)$	3 eigenvalues with negative real part (**attractor**)	2 eigenvalues with negative real part and 1 with positive real part (**source**)
$DF(X_2^\star)$	2 eigenvalues with negative real part and 1 with positive real part (**source**)	3 eigenvalues with negative real part (**attractor**)

Remark 43.1. We remark that if $R_0 < 1$ then $X_2^\star$ is not in the biologically relevant domain because two of its components become negative.

We now describe the stationary solutions for System (43.2). Here, we have three stationary points. They are

$$X_1^\star = \left(\frac{\lambda}{d}, 0, 0, 0\right),$$

$$X_2^\star = \left(\frac{ua}{\beta k}, \frac{k\beta\lambda - uda}{\beta ak}, \frac{k\beta\lambda - uda}{\beta au}, 0\right), \text{ and}$$

$$X_3^\star = \left(\frac{\lambda cu}{dcu + \beta kb}, \frac{b}{c}, \frac{kb}{cu}, \frac{\beta\lambda kc - adcu - a\beta kb}{(dcu + \beta kb)p}\right).$$

The stationary solution $X_1^\star$, once again, corresponds to the absence of the HIV in the organism. The stationary solution $X_2^\star$ corresponds, as previously, to a balance of infected and normal cells. The absence of defense cells in the organism ($z = 0$) means that we are back to the previous model. The stationary solution $X_3^\star$ corresponds to a balance between positive, infected, and defense cells. Biologically this point corresponds to the *HIV latency period*, or either, the second phase of the HIV infection.

As in the analysis of the model of (43.1), it will be convenient to write the system in the form $\dot{X} = F(X)$ where now $X = (x, y, v, z)$ and the function $F : \mathbb{R}^4 \to \mathbb{R}^4$ is given by

$$F(X) = \begin{bmatrix} \lambda - dx - x\beta v \\ x\beta v - ay - pyz \\ ky - uv \\ cyz - bz \end{bmatrix}.$$

In the next lemma we collect some information on the infinitesimal behavior of the system in a neighborhood of the stationary points $X_1^\star, X_2^\star$ and $X_3^\star$.

We shall call the constant $D_0 := \frac{c\lambda}{ab}$ the *basic defense rate*. Together with the basic reproductive ratio it is another important dimensionless parameter. It is the ratio of the growth parameters of the immune system and their corresponding death rates. The importance of this constant in our analysis starts with the following:

Lemma 43.2. *If $R_0 = 1$ then $X_1^\star = X_2^\star$ and $DF(X_1^\star)$ has a vanishing eigenvalue. If $R_0 = 1 + \frac{R_0}{D_0}$ then $X_2^\star = X_3^\star$ and $DF(X_2^\star)$ has a vanishing eigenvalue. If $R_0 \neq 1$*

43 On the Dynamics of Certain Models Describing the HIV Infection

and $R_0 \neq 1 + \frac{R_0}{D_0}$, then the infinitesimal behavior of the stationary solutions is described by the following:

	$R_0 < 1$	$1 < R_0 < 1 + \frac{R_0}{D_0}$	$R_0 > 1 + \frac{R_0}{D_0}$
$DF(X_1^*)$	4 eigenvalues with negative real part (**attractor**)	3 eigenvalues with negative real part and 1 with positive real part (**saddle**)	3 eigenvalues with negative real part and 1 with positive real part (**saddle**)
$DF(X_2^*)$	3 eigenvalues with negative real part and 1 with positive real part (**saddle**)	4 eigenvalues with negative real part (**attractor**)	3 eigenvalues with negative real part and 1 with positive real part (**saddle**)
$DF(X_3^*)$	at least 1 eigenvalue with negative real part	at least 1 eigenvalue with negative real part	at least 2 eigenvalues with negative real part

Remark 43.2. As in Model (43.1), the case $R_0 < 1$ leads to X_2^* and X_3^* out of the biologically relevant region. Furthermore, if $R_0 < 1 + (R_0/D_0)$ the stationary point X_3^* is out of the biological range as well. We will show that this range is positively invariant and thus the biologically relevant solutions cannot approach such steady states.

Since the cases where $R_0 = 1$ or $R_0^{-1} + D_0^{-1} = 1$ are nongeneric, we now focus on interpreting the consequences of Lemma 43.2 away from such situations. If $R_0 < 1$, then arbitrary initial conditions (at least close to the equilibrium point) will lead to the clearing of the virus and the disappearence of the infection. It will follow, as a consequence of the results in the next two sections, that this is in fact the case for *arbitrary* positive initial conditions. If $1 < R_0 < 1 + (R_0/D_0)$ then, at least in a neighborhood of the equilibrium point X_2^*, the disease will approach a uniformly persistent state where the CTL response concentration will vanish. In fact, as will be explained in the next two sections, generically the biological solutions will converge to this steady state. See Theorem 43.1.

43.3.1 Boundedness and Positivity

In this paragraph we will answer the following basic question: Will the solutions of Models (43.1) and (43.2) that start from biologically meaningful initial values preserve such property for future times? Here, by biologically meaningful we mean that all coordinates are non-negative and bounded for all times. We split the discussion into two parts, namely, positivity and boundedness.

43.3.1.1 Positivity

Let $\mathbb{R}_+$ denote the set of non-negative real numbers. Obviously, a solution (x, y, v, z) to System (43.1) only admits a biological interpretation if $(x, y, v, z) \in \mathbb{R}_+^4$. As

680   D.H. Pastore and J.P. Zubelli

remarked before, the System (43.2) reduces to System (43.1) if $z = 0$. Hence, we will state all the results for System (43.2).

Proposition 43.1. *Let* $\varphi : [t_0, +\infty) \rightarrow \mathbb{R}^4$ *be a solution of System (43.2). If* $\varphi(t_0) \in \mathbb{R}^4_+$ *then* $\varphi(t) \in \mathbb{R}^4_+$ *for all* $t \in [t_0, \infty)$.

The proof of this result is a straightforward case by case analysis of the behavior of solutions to System (43.2) whenever one of its components vanishes.

The above result also holds for the case that includes mutation given in (43.3).

Proposition 43.2. *Let* $\varphi : [t_0, \infty) \rightarrow \mathbb{R}^{3n+1}$ *be a solution of System (43.3). If* $\varphi(t_0) \in \mathbb{R}^{3n+1}_+$ *then* $\varphi(t) \in \mathbb{R}^{3n+1}_+$ *for all* $t \in [t_0, \infty)$.

43.3.1.2   Boundedness

We already have a lower bound given by Propositions 43.1 and 43.2 for the solutions of Models (43.1) and (43.2) with positive initial values. We now show that the solutions are bounded from above.

We denote by $C_b(I)$ the set of continuous and bounded functions defined on the interval I and taking values in $\mathbb{R}^n$.

Proposition 43.3. *Let* $\varphi : [t_0, \infty) \rightarrow \mathbb{R}^4$ *be a solution of System (43.2). If* $\varphi(t_0) \in \mathbb{R}^4_+$ *then* $\varphi \in C_b[t_0, \infty)$.

Proof. Because of Proposition 43.1, it only remains to prove the existence of an upper bound to the nonnegative solutions of System (43.2).

We start with $x(t)$. Since $\beta, x(t), v(t) \geq 0$ we have from

$$\dot{x} = \lambda - dx - \beta xv \leq \lambda - dx.$$

$$x(t) \leq x(t_0) + \frac{\lambda}{d} \quad \text{for all} \quad t \geq t_0. \tag{43.6}$$

We now go on to prove that $y(t) \in C_b[t_0, \infty)$. From

$$\dot{y} = \beta xv - ay - pzy,$$

since $z(t) \geq 0$ and $y(t) \geq 0$, we have that

$$\dot{y} + ay \leq \beta xv = \lambda - (\dot{x} + xd).$$

Thus,

$$\frac{d}{dt}(ye^{ta}) \leq (\lambda - \frac{d}{dt}(xe^{td})e^{-td})e^{ta}$$

and so

$$\int_{t_0}^{t} \frac{d}{ds}(y(s)e^{sa})ds \leq \int_{t_0}^{t} (\lambda e^{-sd} - \frac{d}{ds}(x(s)e^{sd})e^{s(a-d)})ds.$$

43 On the Dynamics of Certain Models Describing the HIV Infection

Integrating by parts

$$\int_{t_0}^t \frac{d}{dt}(xe^{sd})e^{s(a-d)}ds = x(s)e^{sa}|_{t_0}^t - (a-d)\int_{t_0}^t x(s)e^{sa}ds.$$

Thus,

$$y(t) \le y(t_0)e^{a(t_0-t)} + \frac{\lambda}{a}(1 - e^{a(t_0-t)})$$
$$- \left(x(t) - x(t_0)e^{a(t_0-t)} - (a-d)\int_{t_0}^t x(s)e^{a(s-t)}ds\right). \tag{43.7}$$

Thus, remarking that, for all $t \ge t_0$, $x(t) \ge 0$, $e^{a(t_0-t)} \in [0, 1]$, and $x(t)$ is bounded, we get the boundedness of y. To get more precise bounds we break the analysis into two cases, depending on the sign of $a - d$. If $a - d \le 0$, then (43.7) implies that

$$y(t) \le y(t_0) + \frac{\lambda}{a} + x(t_0) \quad \text{for all} \quad t \ge t_0.$$

If $a - d \ge 0$, then it follows from (43.6) and (43.7) that

$$y(t) \le y(t_0) + \frac{\lambda}{a} + x(t_0) + \frac{(a-d)}{a}\left(\frac{\lambda}{d} + x(t_0)\right)(1 - e^{a(t_0-t)})$$

Thus,

$$y(t) \le y(t_0) + \frac{\lambda}{d} + \left(2 - \frac{d}{a}\right)x(t_0) \quad \text{for all} \quad t \ge t_0$$

Let us now analyze $v(t)$. The equation $\dot{v} = ky - uv$, implies that

$$\frac{d}{dt}(ve^{ut}) = kye^{ut}.$$

Integrating the differential equation, it follows that

$$v(t) = v(t_0)e^{u(t_0-t)} + k\int_{t_0}^t y(s)e^{u(s-t)}ds. \tag{43.8}$$

Since we have already shown that $y \in C_b[t_0, \infty)$ we have

$$v(t) \in C_b[t_0, \infty).$$

Finally, it remains to show that $z(t) \in C_b[t_0, \infty)$. Combining the equations for $\dot{y}$ and $\dot{z}$ in System (43.2) we get

$$\dot{z} + bz = cyz = \frac{c}{p}(\beta vx - \dot{y} - ay).$$

Using the equation $\dot{x} = \lambda - dx - \beta xv$, we have that

$$\dot{z} + bz = \frac{c}{p}(\lambda - dx - \dot{x} - \dot{y} - ay).$$

Hence,

$$z(t) = \left(z(t_0) - \frac{c}{p}\left(\lambda b^{-1} + y(t_0) + x(t_0)\right) \right) e^{b(t_0 - t)} + \frac{c}{p}\left(\lambda b^{-1} - y(t) - x(t)\right)$$
$$+ \frac{c}{p}\left((b - d)\int_{t_0}^{t} x(s)e^{b(s-t)}ds + (b - a)\int_{t_0}^{t} y(s)e^{b(s-t)}ds \right). \tag{43.9}$$

Since x and $y \in C_b[t_0, \infty)$ we have that $z(t) \in C_b[t_0, \infty)$. $\qquad\square$

Proposition 43.4. *Let* $\varphi : [t_0, \infty) \to \mathbb{R}^{3n+1}$ *be a solution of System (43.3). If* $\varphi(t_0) \in \mathbb{R}_+^{3n+1}$ *then* $\varphi \in C_b[t_0, \infty)$.

Proof. We proved in Proposition 43.2 that the components of the solutions to System (43.3) are bounded from below by 0. It remains to show that they have an upper bound.

We shall start by analyzing $x(t)$. Since $x(t) \geq 0$, $v_i(t) \geq 0$ and $\beta_i \geq 0$ for all $t \geq t_0$,

$$\dot{x} = \lambda - dx - x\sum_i^n \beta_i v_i$$

implies that $\dot{x} + dx \leq \lambda$. Thus, as in the proof of the Proposition 43.3, we have that

$$x(t) \leq x(t_0) + \frac{\lambda}{d} \quad \text{for all} \quad t \geq t_0.$$

For the boundedness of $y(t)$, we look at the equation $\dot{y}_i = \beta_i x v_i - a y_i - p y_i z_i$. From Proposition 43.3 we have that

$$\sum_i^n \dot{y}_i + a\sum_i^n y_i \leq x\sum_i^n \beta_i v_i.$$

Let us set $Y(t) := \sum_i^n y_i(t)$, $V(t) = \sum_i^n v_i(t)$ and $Z(t) := \sum_i^n z_i(t)$. Since $x\sum_i^n \beta_i v_i = \lambda - \dot{x} - dx$, we have that

$$\dot{Y}(t) + aY(t) \leq \lambda - \dot{x}(t) - dx(t).$$

As in Proposition 43.3,

$$Y(t) \leq Y(t_0) + \max\left\{\frac{\lambda}{d}, \frac{\lambda}{a}\right\} + \max\left\{1, 2 - \frac{d}{a}\right\} x(t_0) = \overline{Y},$$

that is, $Y(t) \in C_b[t_0, \infty)$. Since $y_i \geq 0$ for all $i = 1, \ldots, n$, we have that $y_i(t) \leq Y(t) \leq \overline{Y}$. Thus, $y_i \in C_b[t_0, \infty)$ for all $i = 1, \ldots, n$.

In the case of v, we have that

$$\dot{V}(t) + uV(t) = k(\theta Y(t) + (1 - \theta)\overline{K}Y(t)) = k(\theta + (1 - \theta)\overline{K})Y(t),$$

because $\sum_{j=1}^{n} K_{i,j} = \overline{K}$. As we saw in the proof of Proposition 43.3,

$$V(t) \leq V(t_0) + \frac{k}{u}(\theta + (1 - \theta)\overline{K})\overline{Y} = \overline{V},$$

that is, $V(t) \in C_b[t_0, \infty)$. We conclude that $v_i(t) \in C_b[t_0, \infty)$ for all $i \in \{1, \ldots, n\}$.
As far as $z_i(t)$ is concerned, using the equation $\dot{z}_i = c y_i z_i - b z_i$, we get

$$\sum_{i}^{n} \dot{z}_i + b\sum_{i}^{n} z_i = \frac{c}{p}\left(x\sum_{i}^{n} \beta_i v_i - \sum_{i}^{n} \dot{y}_i - a\sum_{i}^{n} y_i \right).$$

Thus, as before,

$$\dot{Z}(t) + bZ(t) = \frac{c}{p}(\lambda - dx(t) - \dot{x}(t) - \dot{Y}(t) - aY(t)).$$

The inequality (43.9) is now written as

$$Z(t) = \left(Z(t_0) - \frac{c\lambda}{pb} + \frac{c}{p}Y(t_0) + \frac{c}{p}x(t_0) \right) e^{b(t_0 - t)} + \frac{c\lambda}{pb} - \frac{c}{p}Y(t)$$
$$- \frac{c}{p}x(t) + \frac{c}{p}(b - d)\int_{t_0}^{t} x(s)e^{sb}ds + \frac{c}{p}(b - a)\int_{t_0}^{t} Y(s)e^{sb}ds.$$

So, $Z(t) \leq \overline{Z}$, where $\overline{Z}$ is a constant that depends only on $x(t_0)$ and $Y(t_0)$. It follows that $Z(t) \in C_b[t_0, \infty)$. Consequently, $z_i(t) \in C_b[t_0, \infty)$ for all $i \in \{1, \ldots, n\}$.

43.3.2 Stability of the Equilibrium Points

The global stability of the equilibrium points of Systems (43.1) and (43.2) in the biologically interesting region defined by the positive orthant has attracted several authors. Proofs of these global stability characteristics of the mutationless models (43.1) and (43.2) were given in [5], using Hirsch's theory of competitive differential systems, and more recently by [3] (for system (43.1)) and [4] using Lyapunov functions. More recently, the second author in collaboration with M. Souza in [18] established global stability of the equilibrium points for systems that include those of the form (43.3) under some hypothesis on the corresponding coefficients. This was done by exhibiting suitable Lyapunov functions. As a consequence of such results one can state the following:

Theorem 43.1. *Let* $\varphi : [t_0, \infty) \to \mathbb{R}^4$, $\varphi(t) = (x(t), y(t), v(t), z(t))$, *be a solution of System (43.2) such that* $\varphi(t_0) \in \mathbb{R}_+^4$.

- *If* $R_0 \leq 1$, *then* $\lim_{t \to \infty} \varphi(t) = X_1^\star$.
- *If* $R_0 > 1$ *and* $(y(t_0), v(t_0)) = (0, 0)$, *then* $\lim_{t \to \infty} \varphi(t) = X_1^\star$.
- *If* $1 < R_0 \leq 1 + (R_0/D_0)$ *and* $y(t_0) + v(t_0) \neq 0$ *then* $\lim_{t \to \infty} \varphi(t) = X_2^\star$.
- *If* $R_0 > 1 + (R_0/D_0)$, $z(t_0) = 0$, *and* $y(t_0) + v(t_0) \neq 0$ *then* $\lim_{t \to \infty} \varphi(t) = X_2^\star$.
- *If* $R_0 > 1 + (R_0/D_0)$, $z(t_0) > 0$ *and* $y(t_0) + v(t_0) \neq 0$ *then* $\lim_{t \to \infty} \varphi(t) = X_3^\star$.

The asymptotic behaviour of the solutions of the System (43.1) can be promptly inferred from the result above. One has just to notice that System (43.2) reduces to System (43.1) when restricted to the invariant hypersurface $z = 0$.

Thus, the generic biologically relevant solutions of the models without mutation belong to the basin of attraction of some stationary point of system. This extends a result of [10] who observed this for initial conditions close to the stationary solutions. This also shows that these models do not simulate the last phase of the HIV since the solutions always converge to the absence of virus or the period of latency. In Sect. 43.4 we take up this issue by considering mutation and the invasion of an opportunistic virus numerically.

43.4 Discussion

A number of models for the within-host viral infection by HIV have been proposed and studied by different authors. In particular, a class of three state-variable models was introduced by [16] that modifies the first equation of System (43.1) to a logistic type form. Namely, the first equation takes the form

$$\dot{x} = \lambda - dx + px(1 - x/x_m) - \beta xv, \tag{43.10}$$

A global analysis of both three-dimensional models was performed by [4]. It overlaps consistently with Theorem 43.1. Since they also consider models for which the first equation takes the form (43.10), their models in some situations may give rise to periodic orbits or oscillations. This however, is not the case for our models.

In fact, for the three dimensional models under consideration the solutions of the system eventually enter in the basin of attraction of some stationary point of system. [10] observed this for initial conditions close to the stationary solutions in the mutationless models. Thus the models under consideration do not simulate the last phase of the HIV since the solutions always converge to the absence of virus or to a latency state.

It is well recognized that the HIV does not kill any vital organ [11]. Nevertheless, it destabilizes the immune system leaving the body defenseless to opportunistic virus attacks.

Several mathematical models have been devised to describe the slow decline in the numbers of CD4 cells in the HIV infection and the interaction between HIV and other opportunistic infections [9, 11, 13, 16]. Furthermore, a number of alternative approaches have been proposed to model the third phase of the HIV infection and the onset of AIDS. See for example [11, 12, 19] and references therein. We close this article by considering a model that takes into account the action of an opportunistic virus after the HIV infection. The main point being that of illustrating the potential of the models that include mutation in a general context such as (43.11) and the need for further mathematical inquire into this direction.

The model is given by the following system of equations:

$$
\begin{aligned}
\dot{x} &= \lambda - dx - x \int \beta_\mu v_\mu d\mu - \alpha x v_o \\
\dot{y}_\mu &= \beta_\mu x v_\mu - a y_\mu - p y_\mu z_\mu \\
\dot{v}_\mu &= k[(1 - \theta) y_\mu + \theta K[y](\mu)] - u v_\mu \\
\dot{z}_\mu &= c y_\mu z_\mu - b z_\mu \\
\dot{v}_o &= m v_o - \alpha x v_o - \omega v_o
\end{aligned}
\tag{43.11}
$$

where v_0, the new variable of the system, stands for opportunistic virus. The additional (positive) constants are

α : meeting rate of opportunistic virus with the uninfected cells
m : reproduction rate of the opportunistic virus
ω : death rate of the opportunistic virus

The term that represents the encounter between T cells and the opportunistic virus is $\alpha x v_o$. It appears in the first and in the last equation of the model. The equation for the opportunistic virus has the term $m v_o$ that represents the reproduction of the opportunistic virus. The opportunistic virus infected cells are not considered in this model. The term ωv_o corresponds to the decline of the opportunistic virus. We do not take into account the type of opportunistic virus attacking the organism. The parameter values, the constants and the initial conditions for the opportunistic virus can be found in Table 43.2.

The functions $\beta(\mu)$ and $K(\mu, \mu')$ are taken as Gaussians. The parameters, constants and initial conditions appearing in System (43.4) can be found in Table 43.3.

We have started by simulating the infection using the corresponding solution at hand we used $x(t)$, $y_\mu(t)$, $v_\mu(t)$ and $z_\mu(t)$ as initial conditions for the Model (43.11). The graph of the corresponding solutions are shown as indicated in Table 43.2. The numerical solutions were found using MatLab's function $ode23s$. More information concerning the implementation and validation of the numerical methods to obtain the reported results can be found in [14].

Table 43.2 Numerical experiment list

Number of Strains	m	o	α	$v_o(100)$	Figure
20	3.1	0.01	0.01	10^{-3}	43.1
100	3.1	0.01	0.01	10^{-3}	43.2 and 43.4
100	1.2	1.2	0.1	10^{-3}	43.3

Table 43.3 Parameters, constants, and initial conditions

λ	10	$day^{-1} \times mm^{-3}$	β	2.4×10^{-5}	$day^{-1} \times mm^{-3}$
a	1	$day^{-1} \times mm^{-3}$	p	0.8	$day^{-1} \times mm^{-3}$
c	0.2	$day^{-1} \times mm^{-3}$	d	0.02	day^{-1}
k	360	day^{-1}	u	2.4	day^{-1}
b	1.2	day^{-1}	θ	0.5	
N	20		$x(0)$	10^3	mm^{-3}
$y(\mu, 0)$	0	mm^{-3}	$z(\mu, 0)$	10^{-6}	mm^{-3}
$v_0(0)$	10^{-3}	mm^{-3}	$v(\mu, 0)$	0	mm^{-3}

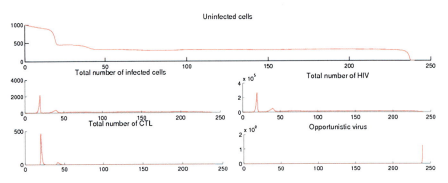

Fig. 43.1 The number of opportunistic virus in the system presents a considerable growth and the number of uninfected cells converged to zero

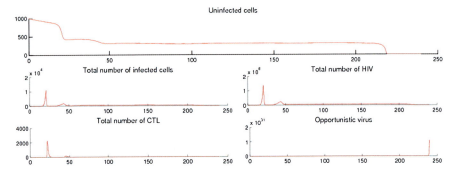

Fig. 43.2 The number of opportunistic virus in the system presents a considerable growth and the number of uninfected cells converged to zero

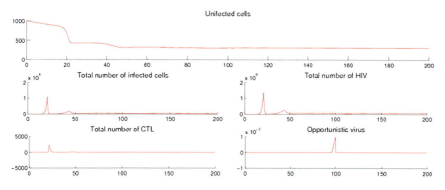

Fig. 43.3 In this simulation, the presence of the opportunistic virus has not caused any change on the equilibrium of the system

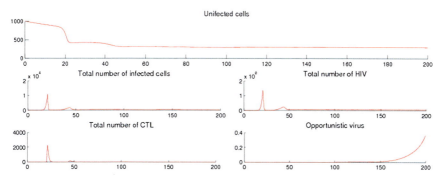

Fig. 43.4 In this simulation the equilibrium between uninfected cells and virus was preserved in the sense that the population of uninfected cells did not go to zero but the number of opportunistic virus grew

In Figs. 43.1 and 43.2 we show simulations where the number of opportunistic virus in the system presents a considerable growth and the number of uninfected cells converged to zero. On the other hand, in Fig. 43.3 the presence of the opportunistic virus has not caused any change on the equilibrium of the system. In Fig. 43.4 the equilibrium of the system was preserved but nevertheless the number of opportunistic virus presented a considerable growth.

From the numerical results presented herein we conclude that the presence of the opportunistic virus may or may not lead the system to an equilibrium state different from the ones of the previous system. Since the human organism is constantly in contact with different kinds of virus this suggests a way to model the post-latency period and the possible collapse of the infected individual.

Acknowledgements JPZ was supported by CNPq under grants 302161/2003-1 and 474085/2003-1. The final version of this work was conducted during the Special Semester on Quantitative Biology Analyzed by Mathematical Methods, October 1st, 2007 -January 27th, 2008, organized by RICAM, Austrian Academy of Sciences.

References

1. Asquith, B., Bangham, C.R.M.: An introduction to lymphocyte and viral dynamics: the power and limitations of mathematical analysis. Proc. R. Soc. Lond. Ser B-Bio. Sci. **270**(1525), 1651–1657 (2003)
2. Bocharov, G.A., Romanyukha, A.A.: Mathematical-model of antiviral immune-response-iii – influenza-a virus-infection. J. Theoret. Biol. **167**(4), 323–360 (1994)
3. Korobeinikov, A.: Global properties of basic virus dynamics models. Bull. Math. Biol. **66**, 879–883 (2004)
4. De Leenheer, P., Smith, H.L.: Virus dynamics: A global analysis. SIAM J. Appl. Math. **63**(4), 1313–1327 (electronic) (2003)
5. Li, M.Y., Muldowney, J.S.: Global stability for the seir model in epidemiology. Math. Biosci. **125**(2):155–164 (1995)
6. Katok, A., Hasselblatt, B.: Introduction to the modern theory of dynamical systems. Cambridge University Press, Cambridge (1995)
7. Marchuk, G.I., Romanyukha, A.A., Bocharov, G.A.: Mathematical-model of antiviral immune-response .2. parameters identification for acute viral hepatitis-b. J. Theoret. Biol. **151**(1), 41–70 (1991)
8. Neumann, A.U., Lam, N.P., Dahari, H., Gretch, D.R., Wiley, T.E., Layden, T.J., Perelson, A.S.: Hepatitis c viral dynamics in vivo and the antiviral efficacy of interferon-alpha therapy. Science **282**(5386), 103–107 (1998)
9. Nowak, M., Anderson, R.M., McLean, R.A., Wolfs, T.F.W., GouDsmit, J., May, R.M.: Antigenic diversity thresholds and the development of AIDS. Science **254** (1991)
10. Nowak, M., Bangham, C.R.M.: Population dynamics of immune responses to persistent viruses. Science **272**, 74–79 (1996)
11. Nowak, M., May, R.M.: Virus Dynamics Mathematical Principles of Immunology and Virology. Oxford University Press, Oxford (2000)
12. Nowak, M.A.: Evolutionary dynamics. Exploring the equations of life, xi 363 p. The Belknap Press of Havard University Press, Cambridge, MA (2006)
13. Nowak, M.A., McMichael, A.J.: How HIV defeats the immune system. Sci. Am. (1995)
14. Pastore, D.H.: The hiv dynamics in the immunological system in the presence of mutation (a dinâmica do hiv no sistema imunológico na presença de mutação). Ph.D. thesis, IMPA (2005)
15. Perelson, A.S., Kirschner, D.E., Deboer, R.: Dynamics of hiv-infection of CD4+ T-cells. Math. Biosci. **114**(1), 81–125 (1993)
16. Perelson, A.S., Nelson, P.W.: Mathematical analysis of HIV-1 dynamics in vivo. SIAM Rev. **41**, 3–44 (1999)
17. Perelson, A.S., Neumann, A.U., Markowitz, M., Leonard, J.M., Ho, D.D.: Hiv-1 dynamics in vivo: Virion clearance rate, infected cell life-span, and viral generation time. Science **271**(5255), 1582–1586 (1996)
18. Souza, M.O., Zubelli, J.P.: Global Stability for a Class of Virus Models with CTL Immune Response and Antigenic Variation. arXiv:0810.4364 (2009)
19. Willensdorfer, M., Nowak, M.A.: Mutation in evolutionary games can increase average fitness at equilibrium. J. Theoret. Biol. **237**(4), 355–362 (2005)

Chapter 44
Tilings and Bussola for Making Decisions

Alberto A. Pinto, Abdelrahim S. Mousa, Mohammad S. Mousa, and Rasha M. Samarah

Abstract We introduce the yes–no decision model, where individuals can make the decision yes or no. We characterize the coherent and uncoherent strategies that are Nash equilibria. Each decision tiling indicates the way coherent and uncoherent Nash equilibria co-exist and change with the relative decision preferences of the individuals for the yes or no decision. There are 289 combinatorial classes of decision tilings, described by the decision bussola, which demonstrates the high complexity of making decision.

44.1 Introduction

The main goal in Planned Behavior or Reasoned Action theories, as developed in the works of Ajzen (see [2]) and Baker (see [3]), is to understand and forecast the way individuals turn intentions into behaviors. Almeida–Cruz–Ferreira–Pinto

---

A.A. Pinto (✉)
LIAAD-INESC Porto LA e Departamento de Matemática, Faculdade de Ciências, Universidade do Porto, Rua do Campo Alegre, 687, 4169-007, Portugal
and
Centro de Matemática e Departamento de Matemática e Aplicações, Escola de Ciências, Universidade do Minho, Campus de Gualtar, 4710-057 Braga, Portugal
e-mail: aapinto@fc.up.pt

A.S. Mousa
LIAAD-INESC Porto LA, Porto, Portugal
and
Departamento de Matemática, Faculdade de Ciências, Universidade do Porto
e-mail: abed11@ritaj.ps

M.S. Mousa and R.M. Samarah
Department of Mathematics, Birzeit University, Birzeit, Palestine
e-mail: moha_mousa@yahoo.com, rasha.abed1@yahoo.com

M.M. Peixoto et al. (eds.), *Dynamics, Games and Science I*, Springer Proceedings in Mathematics 1, DOI 10.1007/978-3-642-11456-4_44,
© Springer-Verlag Berlin Heidelberg 2011

690   A.A. Pinto et al.

(see [1]) created a game theoretical model for reasoned action, inspired by the works of J. Cownley and M. Wooders (see [5]). They studied the way saturation, boredom and frustration can lead to uncoherent (or split or impasse) strategies, and no saturation situations can lead to coherent (or heard or no-split) strategies. Here, we study the yes–no decision model that is a simplified version of the Almeida–Cruz–Ferreira–Pinto decision model. In this model, there are just two possible decisions d that individuals can make. For instance, they have to choose between yes or no, i.e. $d \in \{Yes, No\}$. Each set of economical, educational, political, psychological and social variables gives rise to a decision tiling that indicates all the coherent and no-coherent pure Nash equlibria and also the mixed Nash equilibria in terms of the relative decision preference (taste type) of the individuals for the yes or no decision (see [9, 10]). The yes–no decision model incorporates, in the preference neighbours matrix (crowding type), the preference that an individual has for having other individuals making the same decision as his. The crowding type information gives rise to 289 different combinatorial classes of decision tilings, reflecting the complexity of the yes–no decision model (see [9, 10]). The decision bussola encodes all the information of each combinatorial class of decision tilings and indicates the way small changes in economical educational, political, psychological or social variables can transform one decision tiling, into another thus, creating and annihilating individuals and collective behavior. In this chapter, we survey, in part the work presented in [9, 10].

44.2 Yes–No Decision Model

The *yes–no decision model* has two types $\mathbf{T} = \{t_1, t_2\}$ of individuals $i \in \mathbf{I}$ that have to make one decision $d \in \mathbf{D} = \{Y, N\}$. Let $n_p \geq 1$ be the number of individuals with type t_p.[1] Let $\mathscr{L}$ be the *preference decision matrix* whose *coordinates* w_p^d indicate how much an individual, with type t_p, likes, or dislikes, to make decision d

$$\mathscr{L} = \begin{pmatrix} w_1^Y & w_1^N \\ w_2^Y & w_2^N \end{pmatrix}.$$

The preference decision matrix indicates, for each type, the decision that the individuals prefer, i.e. the individuals taste type (see [1, 4, 5, 9]).

Let $\mathscr{N}_d$ be the *preference neighbors matrix* whose *coordinates* α_{pq}^d indicate how much an individual, with type t_p, likes, or dislikes, that an individual, with type t_q, makes decision d

$$\mathscr{N}_d = \begin{pmatrix} \alpha_{11}^d & \alpha_{12}^d \\ \alpha_{21}^d & \alpha_{22}^d \end{pmatrix}.$$

---

[1] Similarly, we can consider that there is a single individual with type t_p that has to make n_p decisions, or we can, also, consider a mixed model using these two possibilities.

44 Tilings and Bussola for Making Decisions

The preference neighbors matrix indicates, for each type of individuals, whom they prefer, or not, to be with in each decision, i.e. the individuals crowding type (see [1,4,5,9]).

We describe the individuals' decision by a *strategy map* $S : \mathbf{I} \to \mathbf{D}$ that associates to each individual $i \in \mathbf{I}$ its decision $S(i) \in \mathbf{D}$. Let $\mathbf{S}$ be the space of all strategies S. Given a strategy S, let $\mathscr{O}_S$ be the *strategic occupation matrix*, whose coordinates $l_p^d = l_p^d(S)$ indicate the number of individuals, with type t_p, that make decision d

$$\mathscr{O}_S = \begin{pmatrix} l_1^Y & l_1^N \\ l_2^Y & l_2^N \end{pmatrix}.$$

The *strategic occupation vector* $\mathscr{V}_S$, associated to a strategy S, is the vector $(l_1, l_2) = (l_1^y(S), l_2^y(S))$. Hence, l_1 (resp. $n_1 - l_1$) is the number of individuals, with type t_1, that make the decision Y (resp. N). Similarly, l_2 (resp. $n_2 - l_2$) is the number of individuals, with type t_2, that make the decision Y (resp. N). The set $\mathbf{O}$ of all possible *occupation vectors* is

$$\mathbf{O} = \{(l_1, l_2) : 0 \le l_1 \le n_1 \quad \text{and} \quad 0 \le l_2 \le n_2\}.$$

Let $U_1 : \mathbf{D} \times \mathbf{O} \to \mathbb{R}$ the *utility function*, of an individual with type t_1, be given by

$$U_1(Y; l_1, l_2) = \omega_1^Y + \alpha_{11}^Y (l_1 - 1) + \alpha_{12}^Y l_2$$
$$U_1(N; l_1, l_2) = \omega_1^N + \alpha_{11}^N (n_1 - l_1 - 1) + \alpha_{12}^N (n_2 - l_2).$$

Let $U_2 : \mathbf{D} \times \mathbf{O} \to \mathbb{R}$ the *utility function*, of an individuals with type t_2, be given by

$$U_2(Y; l_1, l_2) = \omega_2^Y + \alpha_{22}^Y (l_2 - 1) + \alpha_{21}^Y l_1$$
$$U_2(N; l_1, l_2) = \omega_2^N + \alpha_{22}^N (n_2 - l_2 - 1) + \alpha_{21}^N (n_1 - l_1).$$

Given a strategy $S \in \mathbf{S}$, the *utility* $U_i(S)$, of an individual i with type $t_{p(i)}$, is given by $U_{p(i)}(S(i); l_1^y(S), l_2^y(S))$.

Definition 44.1. A strategy $S^* : \mathbf{I} \to \mathbf{D}$ is a *Nash equilibrium* if, for every individual $i \in \mathbf{I}$ and for every strategy S, with the property that $S^*(j) = S(j)$ for every individual $j \in I \setminus \{i\}$, we have

$$U_i(S^*) \ge U_i(S).$$

Let $x = \omega_1^Y - \omega_1^N$ be the *horizontal relative decision preference* of the individuals with type t_1 and let $y = \omega_2^Y - \omega_2^N$ be the *vertical relative decision preference* of the individuals with type t_2. The *Nash equilibrium domain* $E(S)$ of a strategy S is the set of all pairs (x, y) for which S is a Nash Equilibrium.

Definition 44.2. Let $A_{ij} = \alpha_{ij}^Y + \alpha_{ij}^N$, for $i, j \in \{1, 2\}$, be the coordinates of the *partial threshold order matrix*.

692   A.A. Pinto et al.

As we will show, the partial thresholds encode all the relevant information for the existence of Nash equilibria that are no-coherent strategies.

44.3 Evolutionary Dynamics and Yes–No Decision Models

We implement the *evolutionary deterministic yes–no decision models* as follows (see [9]): Fix an infinite sequence (i_t, d_t), with $t \in \mathbb{N}$, of pairs $(i_t, d_t) \in \mathbf{I} \times \mathbf{D}$ with the property that every pair, contained in $\mathbf{I} \times \mathbf{D}$, occurs in the sequence infinitely often. Given a strategy $S_t : \mathbf{I} \to \mathbf{D}$, at moment t, the strategy $S_{t+1} : \mathbf{I} \to \mathbf{D}$ is defined as follows: (a) $S_{t+1} = S_t | \mathbf{I} \setminus \{i_{t+1}\}$; (b) $S_{t+1}(i_{t+1}) = d_{t+1}$, if i_{t+1} increases its utility by making decision d_{t+1} instead of $S_t(i_{t+1})$ (knowing that $S_{t+1} = S_t | \mathbf{I} \setminus \{i_{t+1}\}$), and $S_{t+1}(i) = S_t(i)$, otherwise. Hence, the Nash equilibria are the fixed points, and vice-versa, of the evolutionary decision deterministic models.

We implement the *evolutionary stochastic yes–no decision models* as follows: Let P be a probability distribution that assigns a positive probability to each pair $(i, d) \in \mathbf{I} \times \mathbf{D}$. Given a strategy $S_t : \mathbf{I} \to \mathbf{D}$, at moment t, we choose randomly a pair (i, D) according to the probability distribution P. The strategy $S_{t+1} : \mathbf{I} \to \mathbf{D}$ is defined as follows: (a) $S_{t+1} = S_t | \mathbf{I} \setminus \{i\}$; (b) $S_{t+1}(i) = d$, if i increases its utility by deciding d instead of $S_t(i)$ (knowing that $S_{t+1} = S_t | \mathbf{I} \setminus \{i\}$), and $S_{t+1}(i) = S_t(i)$, otherwise. Hence, the Nash equilibria are the absorbing states, and vice-versa, of the evolutionary decision stochastic model.

44.4 (Coherent, Coherent) Strategies

A *(coherent, coherent) strategy*[2] is a strategy in which all individuals, with the same type, prefer to make the same decision (see [9]). A *(coherent, coherent) strategy* is described by a map $C : \mathbf{T} \to \mathbf{D}$ that, for every individual i, with type $t_{p(i)}$, indicates its decision $C(p(i))$. Hence, a (coherent, coherent) strategy $C : \mathbf{T} \to \mathbf{D}$ determines an unique strategy $S : \mathbf{I} \to \mathbf{D}$ given by $S(i) = C(p(i))$.

We observe that there are four (coherent, coherent) strategies:

- (Y, Y) *strategy*: all individuals make the decision Y
- (Y, N) *strategy*: all individuals, with type t_1, make the decision Y, and all individuals, with type t_2, make the decision N
- (N, Y) *strategy*: all individuals, with type t_1, make the decision N and all individuals, with type t_2, make the decision Y
- (N, N) *strategy*: all individuals make the decision N

The *horizontal* $H(Y, Y)$ and *vertical* $V(Y, Y)$ *strategic thresholds* of the (Y, Y) strategy are given by

---

[2] or equivalently, *(no-split, no-split) strategy* or *(heard, heard) strategy*.

44 Tilings and Bussola for Making Decisions

Fig. 44.1 (Y,Y) Nash equilibria domain $Q(Y, Y)$

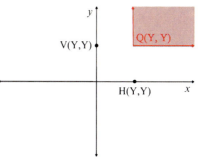

Fig. 44.2 (Y, N) Nash equilibria domain $Q(Y, N)$

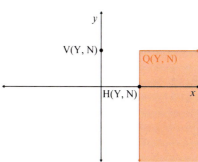

$$H(Y, Y) = -\alpha_{11}^Y(n_1 - 1) - \alpha_{12}^Y n_2 \quad \text{and} \quad V(Y, Y) = -\alpha_{22}^Y(n_2 - 1) - \alpha_{21}^Y n_1.$$

The (Y,Y) *Nash equilibria domain* $Q(Y, Y) = E(Y, Y)$ is the right-upper quadrant (see Fig. 44.1)

$$Q(Y, Y) = \{(x, y) : x \geq H(Y, Y) \text{ and } y \geq V(Y, Y)\}.$$

The horizontal $H(Y, N)$ and vertical $V(Y, N)$ *strategic thresholds* of the (Y, N) strategy are given by

$$H(Y, N) = -\alpha_{11}^Y(n_1 - 1) + \alpha_{12}^N n_2 \quad \text{and} \quad V(Y, N) = \alpha_{22}^N(n_2 - 1) - \alpha_{21}^Y n_1.$$

The (Y, N) *Nash equilibria domain* $Q(Y, N) = E(Y, N)$ is the right-lower quadrant (see Fig. 44.2)

$$Q(Y, N) = \{(x, y) : x \geq H(Y, N) \text{ and } y \leq V(Y, N)\}.$$

The horizontal $H(N, Y)$ and vertical $V(N, Y)$ *strategic thresholds* of the (N, Y) strategy are given by

$$H(N, Y) = \alpha_{11}^N(n_1 - 1) - \alpha_{12}^Y n_2 \quad \text{and} \quad V(N, Y) = -\alpha_{22}^Y(n_2 - 1) + \alpha_{21}^N n_1.$$

Fig. 44.3 (N,Y) Nash equilibria domain $Q(N, Y)$

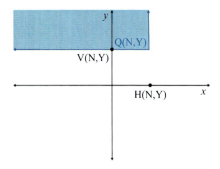

Fig. 44.4 (N,N) Nash equilibria domain $Q(N, N)$

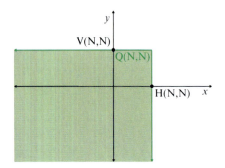

The (N,Y) *Nash equilibria domain* $Q(N, Y) = E(N, Y)$ is the left-upper quadrant (see Fig. 44.3)

$$Q(N, Y) = \{(x, y) : x \leq H(N, Y) \text{ and } y \geq V(N, Y)\}.$$

The horizontal $H(N, N)$ and vertical $V(N, N)$ strategic thresholds of the (N, N) strategy are given by

$$H(N, N) = \alpha_{11}^N(n_1 - 1) + \alpha_{12}^N n_2 \text{ and } V(N, N) = \alpha_{22}^N(n_2 - 1) + \alpha_{21}^N n_1.$$

The (N, N) *Nash equilibria domain* $Q(N, N) = E(N, N)$ is the left-lower quadrant (see Fig. 44.4)

$$Q(N, N) = \{(x, y) : x \leq H(N, N) \text{ and } y \leq V(N, N)\}.$$

The representations of the domains $Q(Y, Y)$, $Q(Y, N)$, $Q(N, Y)$, and $Q(N, N)$ in the plan (x, y) determine the *decision tilings*. Let $U(Y, Y) \subset Q(Y, Y)$, $U(Y, N) \subset Q(Y, N)$, $U(N, Y) \subset Q(N, Y)$, and $U(N, N) \subset Q(N, N)$ be the regions with unique Nash equilibrium. In Fig. 44.5, we represent three decision tilings, (1) with the coherent uniqueness Nash equilibria domains $U(Y, Y)$, $U(Y, N)$, $U(N, Y)$, and $U(N, N)$ colored red, orange, blue and green, respectively, (2) regions without coherent Nash equilibrium colored purple and (3) regions with two, three and four Nash equilibria colored yellow, brown and pink, respectively. In the left tiling, there

44 Tilings and Bussola for Making Decisions

Fig. 44.5 Three examples of strategic thresholds and decision tilings; *left*: $A_{11} < 0$, $A_{12} > 0$, $B_{12} < 0$, $A_{22} < 0$, $A_{21} > 0$, $B_{21} < 0$; *center*: $A_{11} = A_{12} = A_{21} = A_{22} = 0$; *right*: $A_{11} > 0$, $A_{12} < 0$, $B_{12} > 0$, $A_{22} > 0$, $A_{21} < 0$, $B_{21} > 0$

is an unbounded region without coherent Nash equilibrium. In the central tiling, for every relative decision preferences, there is a unique coherent Nash equilibrium, except along the axis, where there are two coherent Nash equilibria, and at the origin, where there are four coherent Nash equilibria. In the right tiling, there are regions with one, two, three and four coherent Nash equilibria.

44.5 (Uncoherent, Coherent) Strategies

An *(uncoherent, coherent) strategy*[3] is a strategy in which all individuals, with type t_2, prefer to make the same decision, but individuals, with type t_1, split between the two decisions Y and N (see [10]). Hence, the (uncoherent, coherent) strategies can be of two types:

- (l, Y) *strategy*: all the individuals, with type t_2, and l individuals, with type t_1, make decision Y, and $n_1 - l$ individuals, with type t_1, make decision N.
- (l, N) *strategy*: l individuals, with type t_1, choose decision Y, but all the individuals, with type t_2, and $n_1 - l$ individuals, with type t_1, choose decision N.

We define the *left horizontal threshold* $H_L(l, Y)$ and the *right horizontal threshold* $H_R(l, Y)$ of the (l, Y) strategy by

$$H_L(l, Y) = -\alpha_{11}^Y(l-1) - \alpha_{12}^Y n_2 + \alpha_{11}^N(n_1 - l)$$

$$H_R(l, Y) = -\alpha_{11}^Y l - \alpha_{12}^Y n_2 + \alpha_{11}^N(n_1 - l - 1).$$

We define the *vertical threshold* $V(l, Y)$ of the (l, Y) strategy by

$$V(l, Y) = -\alpha_{21}^Y l + \alpha_{21}^N(n_1 - l) - \alpha_{22}^Y(n_2 - 1)$$

---

[3] or equivalently, *(split, no-split)* or *(no-heard, heard) strategy*.

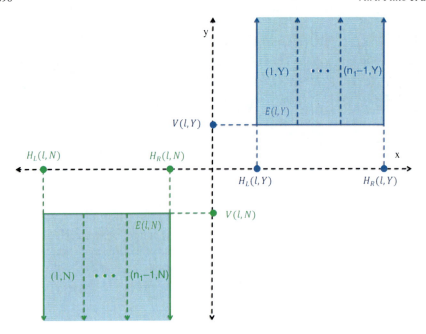

Fig. 44.6 (Uncoherent, coherent) Nash equilibria, $A_{11} < 0$ and $A_{12}n_2 < A_{11}$

The (l, Y) *Nash equilibria domain* $E(l, Y)$ strategy is a Nash Equilibrium if, and only if, $(x, y) \in E(l, Y)$, where

$$E(l, Y) = \{(x, y) : H_L(l, Y) \leq x \leq H_R(l, Y) \text{ and } y \geq V(l, Y)\}.$$

Hence, $E(l, Y)$ is the *Nash Equilibrium domain* of the (l, Y) strategy (see Fig. 44.6).

We define the *left horizontal threshold* $H_L(l, N)$ and the *right horizontal threshold* $H_R(l, N)$ of the (l, N) strategy by

$$H_L(l, N) = -\alpha_{11}^Y(l - 1) + \alpha_{12}^N n_2 + \alpha_{11}^N(n_1 - l)$$

$$H_R(l, N) = -\alpha_{11}^Y l + \alpha_{12}^N n_2 + \alpha_{11}^N(n_1 - l - 1).$$

We define the *vertical threshold* $V(l, N)$ of the (l, N) strategy by

$$V(l, N) = -\alpha_{21}^Y l_1 + \alpha_{21}^N(n_1 - l) + \alpha_{22}^N(n_2 - 1).$$

The (l, N) strategy is a Nash Equilibrium if, and only if, $(x, y) \in E(l, N)$, where

$$E(l, N) = \{(x, y) : H_L(l, N) \leq x \leq H_R(l, N) \text{ and } y \leq V(l, N)\}.$$

Hence, $E(l, N)$ is the *Nash Equilibrium domain* of the (l, N) strategy (see Fig. 44.6).

Since $H_R(l,Y) = H_L(l,Y) - A_{11}$ and $H_R(l,N) = H_L(l,N) - A_{11}$, we have

- If $A_{11} > 0$, there are no (l,Y) and (l,N) Nash equilibria, for every $l \in \{1,\ldots,n_1-1\}$.
- If $A_{11} \leq 0$, there are (l,Y) and (l,N) Nash equilibria, for every $l \in \{1,\ldots, n_1-1\}$.

Hence, the following equalities determine the domains of the (l,Y) Nash equilibria (see Fig. 44.6):

$$H_R(l,Y) = H_L(l+1,Y), H_L(1,Y) = H(N,Y), H_R(n_1-1,Y) = H(Y,Y);$$
$$V(l,N) = V(l+1,N) + A_{21}, V(1,Y) = V(N,Y) - A_{21}, V(n_1-1,Y) = V(Y,Y) + A_{21}.$$

Similarly, the following equalities determine the domains of the (l,N) strategies (see Fig. 44.6):

$$H_R(l,N) = H_L(l+1,N), H_L(1,N) = H(N,N), H_R(n_1-1,N) = H(Y,N);$$
$$V(l,N) = V(l+1,N) + A_{21}, V(1,N) = V(N,N) - A_{21}, V(n_1-1,N) = V(Y,N) + A_{21}.$$

44.6 (Coherent, Uncoherent) Strategies

A *(coherent, uncoherent) strategy*[4] is a strategy in which all individuals, with type t_1, prefer to make the same decision, but individuals, with type t_2, split between the two decisions Y and N (see [10]). Hence, the (coherent, uncoherent) strategies can be of two types:

- (Y,l) *strategy*: all the individuals, with type t_1, and l individuals, with type t_2, make decision Y, and $n_2 - l$ individuals, with type t_2, make decision N.
- (N,l) *strategy*: l individuals, with type t_2, choose decision Y, but all the individuals, with type t_1, and $n_2 - l$ individuals, with type t_2, choose decision N.

We define the *lower vertical threshold* $V_L(Y,l)$ and the *upper vertical threshold* $V_U(Y,l)$ of the (Y,l) strategy by

$$V_L(Y,l) = -\alpha_{22}^Y(l-1) - \alpha_{21}^Y n_1 + \alpha_{22}^N(n_2-l)$$

$$V_U(Y,l) = -\alpha_{22}^Y l - \alpha_{21}^Y n_1 + \alpha_{22}^N(n_2-l-1).$$

We define the *horizontal threshold* $H(Y,l)$ of the (Y,l) strategy by

$$H(Y,l) = -\alpha_{12}^Y l + \alpha_{12}^N(n_2-l) - \alpha_{11}^Y(n_1-1).$$

---

[4] or equivalently, *(no-split, split)* or *(heard, no-heard) strategy*.

The (Y, l) *Nash equilibria domain* $E(Y, l)$ strategy is a Nash Equilibrium if, and only if, $(x, y) \in E(Y, l)$, where

$$E(Y, l) = \{(x, y) : V_L(Y, l) \leq y \leq V_U(Y, l) \text{ and } x \geq H(Y, l)\}.$$

Hence, $E(Y, l)$ is the *Nash Equilibrium domain* of the (Y, l) strategy (see Fig. 44.7).

We define the *lower vertical threshold* $V_L(N, l)$ and the *upper vertical threshold* $V_U(N, l)$ of the (N, l) strategy by

$$V_L(N, l) = -\alpha_{22}^Y (l - 1) + \alpha_{21}^N n_1 + \alpha_{22}^N (n_2 - l)$$

$$V_U(N, l) = -\alpha_{22}^Y l + \alpha_{21}^N n_1 + \alpha_{22}^N (n_2 - l - 1).$$

We define the *horizontal threshold* $H(N, l)$ of the (N, l) strategy by

$$H(N, l) = -\alpha_{12}^Y l + \alpha_{12}^N (n_2 - l) + \alpha_{11}^N (n_1 - 1).$$

The (N, l) *Nash equilibria domain* $E(N, l)$ strategy is a Nash Equilibrium if, and only if, $(x, y) \in E(N, l)$, where

$$E(N, l) = \{(x, y) : V_L(N, l) \leq y \leq V_U(N, l) \text{ and } x \leq H(N, l)\}.$$

Hence, $E(Y, l)$ is the *Nash Equilibrium domain* of the (Y, l) strategy (see Fig. 44.7).

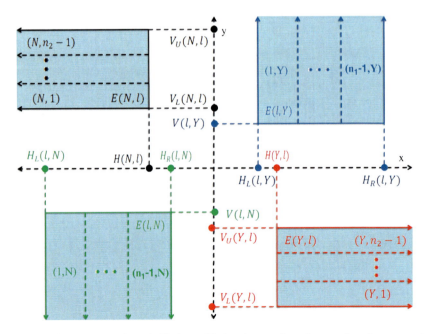

Fig. 44.7 (Coherent, uncoherent) Nash equilibria, $A_{11} < 0$, $A_{22} < 0$, $A_{12} n_2 < A_{11}$ and $A_{21} n_1 > -A_{22}$

44 Tilings and Bussola for Making Decisions

Since $V_U(Y, 1) = V_L(Y, 1) - A_{22}$ and $V_R(N, 1) = V_L(N, 1) - A_{22}$, we have

- If $A_{22} > 0$, there are no (Y, l) and (N, l) Nash equilibria, for every $l \in \{1, \ldots, n_2 - 1\}$.
- If $A_{22} \leq 0$, there are (Y, l) and (N, l) Nash equilibria, for every $l \in \{1, \ldots, n_2 - 1\}$.

Hence, the following equalities determine the domains of the (Y, l) Nash equilibria (see Fig. 44.7):

$$V_U(Y, l) = V_L(Y, l + 1), \quad V_L(Y, 1) = V(Y, N), \quad V_U(Y, n_2 - 1) = V(Y, Y);$$
$$H(Y, l) = H(Y, l + 1) + A_{12}, \quad H(Y, 1) = H(Y, N) - A_{12}, \quad H(Y, n_2 - 1) = H(Y, Y) + A_{12}.$$

Similarly, the following equalities determine the domains of the (N, l) strategies (see Fig. 44.7):

$$V_U(N, l) = V_L(N, l + 1), \quad V_L(N, 1) = V(N, N), \quad V_U(N, n_2 - 1) = V(N, Y);$$
$$H(N, l) = H(N, l + 1) + A_{12}, \quad H(N, 1) = H(N, N) - A_{12}, \quad H(N, n_2 - 1) = H(N, Y) + A_{12}.$$

44.7 (Uncoherent, Uncoherent) Strategies

An *(uncoherent, uncoherent) strategy*[5] is a strategy in which individuals, with type t_1 and type t_2, split between the two decisions Y and N (see [10]).
There are $(n_1 - 1)(n_2 - 1)$ (uncoherent, uncoherent) strategies:

- (l_1, l_2) *strategy*: l_1 individuals, with type t_1, and l_2 individuals, with type t_2, make decision Y, and $n_1 - l_1$ individuals, with type t_1, and $n_2 - l_2$ individuals, with type t_2, make decision N, for $l_1 \in \{1, \ldots, n_1 - 1\}$ and $l_2 \in \{1, \ldots, n_2 - 1\}$.

We define the *left horizontal threshold* $H_L(l_1, l_2)$ and the *right horizontal threshold* $H_R(l_1, l_2)$ of the (l_1, l_2) strategy by

$$H_L(l_1, l_2) = \alpha_{11}^N n_1 + \alpha_{12}^N n_2 + \alpha_{11}^Y - (\alpha_{12}^Y + \alpha_{12}^N)l_2 - (\alpha_{11}^Y + \alpha_{11}^N)l_1$$

$$H_R(l_1, l_2) = \alpha_{11}^N n_1 + \alpha_{12}^N n_2 - \alpha_{11}^N - (\alpha_{12}^Y + \alpha_{12}^N)l_2 - (\alpha_{11}^Y + \alpha_{11}^N)l_1.$$

We define the *down vertical threshold* $V_D(l_1, l_2)$ and the *up vertical threshold* $V_U(l_1, l_2)$ of the (l_1, l_2) strategy by

$$V_D(l_1, l_2) = \alpha_{22}^N n_2 + \alpha_{21}^N n_1 + \alpha_{22}^Y - (\alpha_{21}^Y + \alpha_{21}^N)l_1 - (\alpha_{22}^Y + \alpha_{22}^N)l_2$$
$$V_U(l_1, l_2) = \alpha_{22}^N n_2 + \alpha_{21}^N n_1 - \alpha_{22}^N - (\alpha_{21}^Y + \alpha_{21}^N)l_1 - (\alpha_{22}^Y + \alpha_{22}^N)l_2.$$

---

[5] or equivalently, *(split, split)* or *(no-heard, no-heard) strategy*.

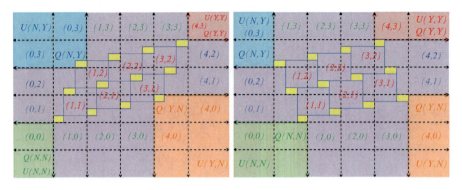

Fig. 44.8 (Uncoherent, uncoherent) Nash equilibria

The (l_1, l_2) strategy is a Nash Equilibrium if, and only if, $(x, y) \in E(l_1, l_2)$, where

$$E(l_1, l_2) = \{(x, y) : H_L(l_1, l_2) \leq x \leq H_R(l_1, l_2) \text{ and } V_D(l_1, l_2) \leq y \leq V_U(l_1, l_2)\}.$$

Hence, $E(l_1, l_2)$ is the *Nash Equilibrium domain* of the (l_1, l_2) strategy (see Fig. 44.8).

Since $H_R(l_1, l_2) = H_L(l_1, l_2) - A_{11}$ and $V_U(l_1, l_2) = V_D(l_1, l_2) - A_{22}$, we have that

- If $A_{11} > 0$ or $A_{22} > 0$, there are no (l_1, l_2) Nash Equilibria, for every $l_1 \in \{1, \ldots, n_1 - 1\}$ and $l_2 \in \{1, \ldots, n_2 - 1\}$.
- If $A_{11} \leq 0$ and $A_{22} \leq 0$, there are (l_1, l_2) Nash Equilibria, for every $l_1 \in \{1, \ldots, n_1 - 1\}$ and $l_2 \in \{1, \ldots, n_2 - 1\}$.

Hence, the following equalities determine the domains of the (l_1, l_2) Nash Equilibria (see Fig. 44.8):

$$H_R(l_1, l_2) = H_L(l_1 + 1, l_2) \text{ and } V_U(l_1, l_2) = V_D(l_1, l_2 + 1).$$

In the left tiling of Fig. 44.8, we have

$$\mathcal{N}_Y = \begin{pmatrix} -1 & -\frac{1}{3} \\ \frac{1}{2} & -1 \end{pmatrix} \text{ and } \mathcal{N}_N = \begin{pmatrix} -1 & -\frac{1}{3} \\ 1 & -1 \end{pmatrix}.$$

The yellow rectangles are regions with two pure Nash equilibria and one mixed Nash equilibrium. In the right tiling of Fig. 44.8, we have

$$\mathcal{N}_Y = \begin{pmatrix} -1 & \frac{1}{3} \\ -\frac{1}{2} & -1 \end{pmatrix} \text{ and } \mathcal{N}_N = \begin{pmatrix} -1 & \frac{1}{3} \\ -1 & -1 \end{pmatrix}.$$

44 Tilings and Bussola for Making Decisions

The yellow rectangles are regions with no pure Nash equilibrium and one mixed Nash equilibrium.

44.8 Bifurcations and Combinatorial Equivalent Tilings

Let $A_{ij} = \alpha_{ij}^Y + \alpha_{ij}^N$, for $i, j \in \{1, 2\}$, be the coordinates of the *partial threshold order matrix*. We observe that

$$H(N, Y) \leq H(Y, Y) \Leftrightarrow A_{11} \leq 0 \Leftrightarrow H(N, N) \leq H(Y, N);$$
$$H(Y, N) \leq H(Y, Y) \Leftrightarrow A_{12} \leq 0 \Leftrightarrow H(N, N) \leq H(N, Y);$$
$$V(N, Y) \leq V(Y, Y) \Leftrightarrow A_{21} \leq 0 \Leftrightarrow V(N, N) \leq V(Y, N);$$
$$V(Y, N) \leq V(Y, Y) \Leftrightarrow A_{22} \leq 0 \Leftrightarrow V(N, N) \leq V(N, Y).$$

Let $B_{11}(n_1, n_2) = A_{11}(n_1 - 1) - A_{12}n_2$, $B_{12}(n_1, n_2) = A_{11}(n_1 - 1) + A_{12}n_2$, $B_{21}(n_1, n_2) = A_{22}(n_2 - 1) + A_{21}n_1$ and $B_{22}(n_1, n_2) = A_{22}(n_2 - 1) - A_{21}n_1$ be the coordinates of the *balanced threshold weight matrix*. We observe that

$$H(N, Y) \leq H(Y, N) \Leftrightarrow B_{11}(n_1, n_2) \leq 0;$$
$$H(N, N) \leq H(Y, Y) \Leftrightarrow B_{12}(n_1, n_2) \leq 0;$$
$$V(N, N) \leq V(Y, Y) \Leftrightarrow B_{21}(n_1, n_2) \leq 0;$$
$$V(Y, N) \leq V(N, Y) \Leftrightarrow B_{22}(n_1, n_2) \leq 0.$$

We say that a decision tiling is *structurally stable*, if all the horizontal and vertical thresholds are pairwise distinct. We say that a decision tiling is *a bifurcation*, if there are, at least, two horizontal thresholds that coincide or there are, at least, two vertical thresholds that coincide (see Figs. 44.9 and 44.10).

We say that a decision tiling is *structurally horizontal (resp. vertical) stable*, if all the horizontal (resp. vertical) thresholds are pairwise distinct. A bifurcation is *horizontally (resp. vertically) single* if, and only if, two horizontal (resp. vertical) thresholds coincide. A bifurcation is *horizontally (resp. vertically) double* if, and only if, two pairs of horizontal (resp. vertical) thresholds coincide. A bifurcation is *horizontally (resp. vertically) degenerated* if all horizontal (resp. vertical) thresholds coincide.

Two decision tilings are *combinatorial equivalent*, if the lexicographic orders of the horizontal and vertical thresholds along the axis are the same in both tilings. The *parameter space PS* is the set

$$PS = \{\underline{\alpha} = (\alpha_{11}^Y, \alpha_{12}^Y, \alpha_{21}^Y, \alpha_{22}^Y, \alpha_{11}^N, \alpha_{12}^N, \alpha_{21}^N, \alpha_{22}^N) \in \mathbb{R}^8\}.$$

The *bifurcation parameter space BPS*

$$BPS = \{\underline{\alpha} \in PS : A_{ij} = 0 \vee B_{ij} = 0, \quad \text{with} \quad i, j \in \{1, 2\}\}$$

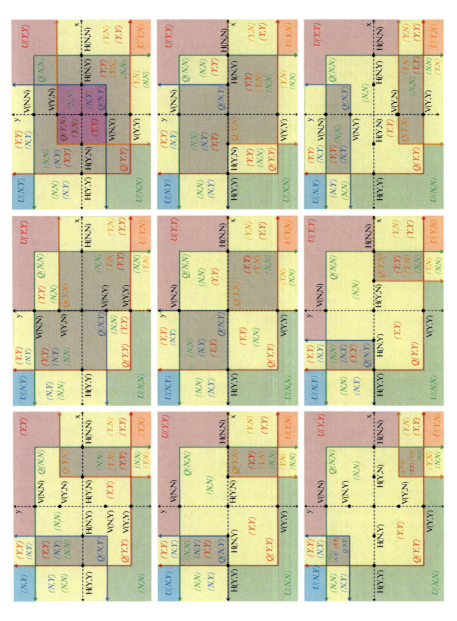

Fig. 44.9 A single horizontal and vertical bifurcations. $A_{11} > 0$ and $A_{22} > 0$; B_{11} and B_{22} changing signs

44 Tilings and Bussola for Making Decisions

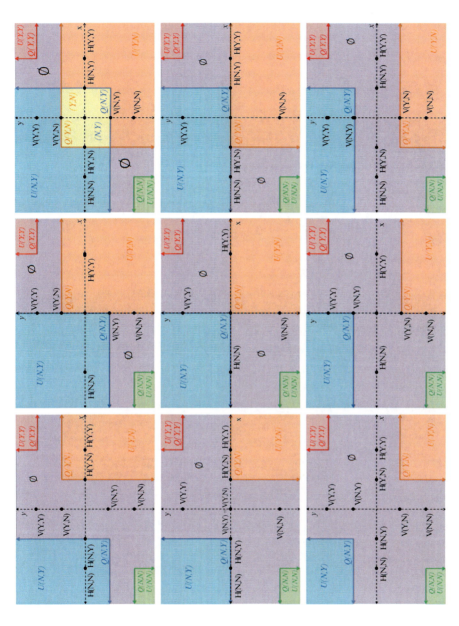

Fig. 44.10 A single horizontal and vertical bifurcations. $A_{11} < 0$ and $A_{22} < 0$; B_{11} and B_{22} changing signs

is the set of all parameters corresponding to bifurcation decision tilings. All parameters, in a same connected component of $PS \setminus BPS$, determine decision tilings that are combinatorial equivalent.

Next we characterize the different orders for the horizontal and vertical thresholds.

Case $H(N, Y) < H(Y, N)$: If $A_{11} < 0$ and $A_{12} > 0$, then

$$H(N, Y) < H(Y, Y) < H(Y, N) \text{ and } H(N, Y) < H(N, N) < H(Y, N).$$

Hence, the horizontal threshold $H(N, Y)$ is the smallest one and the horizontal threshold $H(Y, N)$ is the largest. Therefore, the only indeterminacy to solve this case is the order between the thresholds $H(N, N)$ and $H(Y, Y)$. If $B_{12}(n_1, n_2) < 0$, then $H(N, N) < H(Y, Y)$. If $B_{12}(n_1, n_2) = 0$, then $H(N, N) = H(Y, Y)$. If $B_{12}(n_1, n_2) > 0$, then $H(Y, Y) < H(N, N)$. If $A_{11} = 0$ and $A_{12} > 0$, then $H(N, Y) = H(Y, Y)$ and $H(N, N) = H(Y, N)$ (see Fig. 44.11).

Case $H(Y, Y) < H(N, N)$: If $A_{11} > 0$ and $A_{12} > 0$, then

$$H(Y, Y) < H(N, Y) < H(N, N) \text{ and } H(Y, Y) < H(Y, N) < H(N, N).$$

Hence, the horizontal threshold $H(Y, Y)$ is the smallest one and the horizontal threshold $H(N, N)$ is the largest. Therefore, the only indeterminacy to solve this case is the order between the thresholds $H(N, Y)$ and $H(Y, N)$. If $B_{11}(n_1, n_2) < 0$, then $H(N, Y) < H(Y, N)$. If $B_{11}(n_1, n_2) = 0$, then $H(N, Y) = H(Y, N)$. If $B_{11}(n_1, n_2) > 0$, then $H(Y, N) < H(N, Y)$. If $A_{12} = 0$ and $A_{11} > 0$, then $H(Y, Y) = H(Y, N)$ and $H(N, Y) = H(N, N)$ (see Fig. 44.12).

Case $H(Y, N) < H(N, Y)$: If $A_{11} > 0$ and $A_{12} < 0$, then

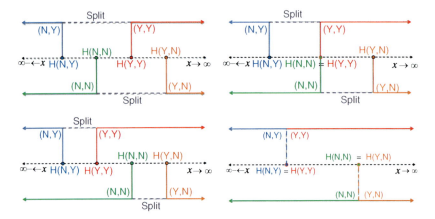

Fig. 44.11 $A_{11} \leq 0$ and $A_{12} > 0$

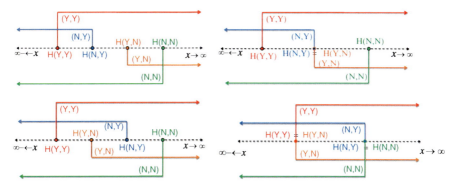

Fig. 44.12 $A_{11} > 0$ and $A_{12} \geq 0$

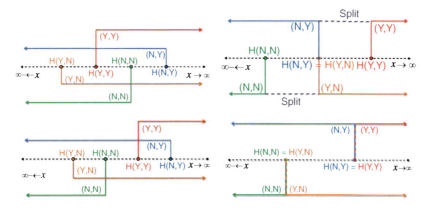

Fig. 44.13 $A_{11} \geq 0$ and $A_{12} < 0$

$$H(Y,N) < H(Y,Y) < H(N,Y) \text{ and } H(Y,N) < H(N,N) < H(N,Y).$$

Hence, the horizontal threshold $H(Y,N)$ is the smallest one and the horizontal threshold $H(N,Y)$ is the largest. Therefore, the only indeterminacy to solve this case is the order between the thresholds $H(Y,Y)$ and $H(N,N)$. If $B_{12}(n_1, n_2) > 0$, then $H(Y,Y) < H(N,N)$. If $B_{12}(n_1, n_2) = 0$, then $H(Y,Y) = H(N,N)$. If $B_{12}(n_1, n_2) < 0$, then $H(N,N) < H(Y,Y)$. If $A_{11} = 0$ and $A_{12} < 0$, then $H(Y,N) = H(N,N)$ and $H(N,Y) = H(Y,Y)$ (see Fig. 44.13).

Case $H(N,N) < H(Y,Y)$: If $A_{11} < 0$ and $A_{12} < 0$, then

$$H(N,N) < H(Y,N) < H(Y,Y) \text{ and } H(N,N) < H(N,Y) < H(Y,Y).$$

Hence, the horizontal threshold $H(N,N)$ is the smallest one and the horizontal threshold $H(Y,Y)$ is the largest. Therefore, the only indeterminacy to solve this case is the order between the thresholds $H(Y,N)$ and $H(N,Y)$. If $B_{11}(n_1, n_2) > 0$, then $H(Y,N) < H(N,Y)$. If $B_{11}(n_1, n_2) = 0$, then $H(Y,N) = H(N,Y)$.

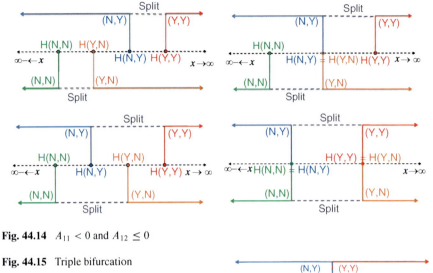

Fig. 44.14 $A_{11} < 0$ and $A_{12} \leq 0$

Fig. 44.15 Triple bifurcation

If $B_{11}(n_1, n_2) < 0$, then $H(N, Y) < H(Y, N)$. If $A_{12} = 0$ and $A_{11} < 0$, then $H(N, N) = H(N, Y)$ and $H(Y, Y) = H(Y, N)$ (see Fig. 44.14).

Case $H_{[(N,N)=(N,Y)=(Y,N)=(Y,Y)]}$ If $A_{12} = 0$ and $A_{11} = 0$, we obtain $H(N, N) = H(N, Y) = H(Y, N) = H(Y, Y)$. Hence, in this case, we have determined all the no-split strategies that are Nash equilibria in terms of the horizontal relative preferences decision x (see Fig. 44.15).

In Fig. 44.16, the thresholds $H(Y, Y)$ (resp. $V(Y, Y)$) are marked by the red dots, the thresholds $H(Y, N)$ (resp. $V(N, Y)$) are marked by the orange dots, the thresholds $H(N, Y)$ (resp. $V(Y, N)$) are marked by the blue dots, and the thresholds $H(N, N)$ (resp. $V(N, N)$) are marked by the green dots. We have four horizontal (resp. vertical) thresholds whose order is determined in each direction of the *bussola*. The way the colored thresholds spiral in the bussola correspond to the way they change with the coordinates of the partial threshold order matrix and with the coordinates of the threshold balanced weight matrix. Hence, a pair (d_1, d_2) of directions in the bussola determine a unique decision tiling, up to combinatorial equivalence, and vice-versa. The russula has the following properties:

- d_1 and d_2 are both in the north side of the bussola if, and only if, there are only (uncoherent, uncoherent) Nash equilibria in the corresponding tiling.
- d_1 is in the north side and d_2 is in the south side of the bussola if, and only if, there are (uncoherent, coherent) Nash equilibria in the corresponding tiling.

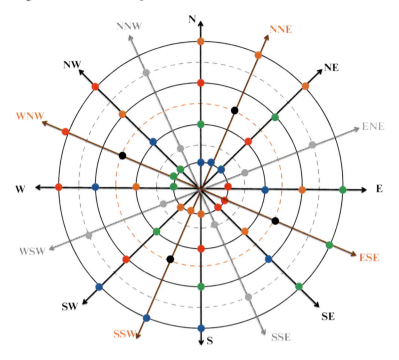

Fig. 44.16 Horizontal (or vertical) decision bussola

- d_1 is in the south side and d_2 is in the north side of the bussola if, and only if, there are (coherent, uncoherent) Nash equilibria in the corresponding tiling.
- d_1 and d_2 are both in the south side of the bussola if, and only if, there are (coherent, coherent) Nash equilibria in the corresponding tiling.

There are 64 combinatorial classes of structurally stable decision tilings and 225 combinatorial classes of bifurcation decision tilings.

44.9 Conclusions

Small changes in the coordinates of the partial threshold order matrix and of the threshold balanced weight matrix, when these coordinates are close to zero, can change their sign and, therefore, alter the order of the horizontal and vertical thresholds. These changes can create and annihilate coherent and uncoherent Nash equilibria giving rise to abrupt changes in individuals and collective behavior.

Acknowledgements This work was presented in the International Congress of Mathematicians ICM 2010, the Second Brazilian Workshop of the Game Theory Society in honor of John Nash and EURO 2010. This work was highlighted in the article [6] after being presented in ICM 2010. We thank LIAAD-INESC Porto LA, Calouste Gulbenkian Foundation, PRODYN-ESF, FEDER,

POFC, POCTI and POSI by FCT and Ministério da Ciência e da Tecnologia, and the FCT Plurian-nual Funding Program of the LIAAD-INESC Porto LA and of the Research Centre of Mathematics of the University of Minho, for their financial support.

References

1. Almeida, L., Cruz, J., Ferreira, H., Pinto, A.A.: Bayesian-Nash equilibria in theory of planned behavior. J. Differ. Equ. Appl., **17**(6) 61–69 (2011)
2. Ajzen, I.: Perceived behavioral control, self-efficacy, locus of control, and the theory of planned behavior. J. Appl. Soc. Psychol. **32**, 665–683 (2002)
3. Baker, S., Beadnell, B., Gillmore, M., Morrison, D., Huang, B., Stielstra, S.: The theory of reasoned action and the role of external factors on heterosexual mens monogamy and condom use. J. Appl. Soc. Psychol. **38**(1), 97–134 (2008)
4. Brida, J., Defesa, M., Faias, M., Pinto, A.A.: Strategic choice in tourism with differentiated crowding types. Econ. Bull., **30**(2), 1509–1515 (2010)
5. Conley, J.P., Wooders, M.H.: Tiebout economies with differential genetic types and endoge-nously chosen crowding characteristics. J. Econ. Theory **98**, 261–294 (2001)
6. Mudur, G.S.: Maths for movies, medicine & markets. The Telegraph Calcutta, India, 20/09/2010. (2010)
7. Pinto, A.A.: Game Theory and Duopoly Models. Interdisciplinary Applied Mathematics Series. Springer, New York (2011)
8. Pinto, A.A., Faias, M., Mousa, A.S.: Resort Pricing and Bankruptcy. Dynamics, Games and Science II. In: Peixoto, M., Pinto, A.A., Rand, D. (eds.) Proceedings in Mathematics series, Springer-Verlag, Chapter 39, 549–555 (2011)
9. Pinto, A.A., Mousa, A.S.: Tilings, Chromosomes and Bussola for Human Evolutionary Dynamical Yes-No Decision Models (submitted)
10. Pinto, A.A., Mousa, A.S.: Uncoherent human decisions (submitted)

Chapter 45
A Hotelling-Type Network

Alberto A. Pinto and Telmo Parreira

Abstract This paper develops a theoretical framework to study spatial price competition in a Hotelling-type network game. Each firm i is represented by a node of degree k_i, where k_i is the number of firm i's direct competitors (neighbors). We investigate price competition à la Hotelling with complete and incomplete information about the network structure. The goal is to investigate the effects of the network structure and of the uncertainty on firms' prices and profits. We first analyze the benchmark case where each firm knows its own degree as well as the rivals' degree. Then, in order to understand the role of information in the price competition network, we also analyze the incomplete information case where each firm knows its type (i.e. number of connections) but not the competitors' type.

45.1 Introduction

Looking at the Hotelling model [14], it's easy to imagine it transposed to a network, where each edge represents a market, disputed by the two firms located at the extreme nodes. In a network, a node can be linked to more than one edge, so it will be disputing as many markets as the number of edges to which it is connected. In each node, a firm establishes a product's selling price that will be used in every market where it is competing. We investigate the effects of the network structure on firms' prices and profits. We assume that each firm's production cost depends

---

A.A. Pinto (✉)
LIAAD-INESC Porto LA e Departamento de Matemática, Faculdade de Ciências, Universidade do Porto, Rua do Campo Alegre, 687, 4169-007, Portugal
and
Centro de Matemática e Departamento de Matemática e Aplicações, Escola de Ciências, Universidade do Minho, Campus de Gualtar, 4710-057 Braga, Portugal
e-mail: aapinto@fc.up.pt

T. Parreira
Universidade de Aveiro, Aveiro, Portugal
e-mail: telmoparreira@ua.pt

M.M. Peixoto et al. (eds.), *Dynamics, Games and Science I*, Springer Proceedings in Mathematics 1, DOI 10.1007/978-3-642-11456-4_45,
© Springer-Verlag Berlin Heidelberg 2011

only upon the degree of the firm's node. We first analyze the benchmark case where every firm knows its node degree and its direct rivals' degree nodes. In this case firms have complete information about their competitors' nodes so they know their competitors production costs. In the case of incomplete information each firm only knows its node degree and the probability distribution of the degrees of the nodes in the network. We determine, explicitly, the Bayesian Nash equilibrium prices and the associated equilibrium expected profits for each firm in the network, as a function of the firm's degree node using the results obtained for the Hotelling model with uncertainty in the production costs of both firms (see [17]).

45.2 Hotelling Model on a Single Line

We assume the buyers of a commodity will be uniformly distributed along a line (normalized to length one). At the two ends of the line there are two firms A and B selling the same commodity with unitary production costs c_A and c_B. No customer has any preference for either seller except on the ground of price plus *transportation cost* t. We will assume that each consumer buys a single unit of the commodity in each unit of time and in each unit of length of the line. Denote A's *price* by p_A and B's *price* by p_B. The point of division x between the regions served by the two entrepreneurs is determined by the condition that at this place it is a matter of indifference whether one buys from A or from B (see Fig. 45.1). Throughout the paper we assume that *production costs* c_A and c_B are such that the demand of both firms is above zero i.e. $|c_A - c_B| \leq t$. We also assume that every consumer is willing to pay at most v, but v is sufficiently high such that every buyer has always the option to buy from both firms (see [1] and [13] for Hotelling and related models)

45.2.1 Hotelling Model with Complete Information

In the complete information case, it is well-known that the *indifference consumer location x* is given by

$$p_A + tx = p_B + t(1 - x).$$

Hence, we have that

$$x = (t + p_B - p_A)/2t.$$

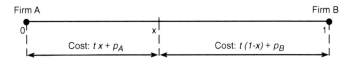

Fig. 45.1 Hotelling's linear city

45 A Hotelling-Type Network

Then, the *profits*, for each firm, are given by

$$\pi_A(p_A, p_B) = (p_A - c_A)x = (p_A - c_A)\left(\frac{1}{2} + \frac{p_B - p_A}{2t}\right);$$
(45.1)

$$\pi_B(p_A, p_B) = (p_B - c_B)(1 - x) = (p_B - c_B)\left(\frac{1}{2} + \frac{p_A - p_B}{2t}\right).$$
(45.2)

From the first order condition (FOC), the *Nash equilibrium prices* are given by

$$p_A^* = t + \frac{2}{3}c_A + \frac{1}{3}c_B;$$

$$p_B^* = t + \frac{2}{3}c_B + \frac{1}{3}c_A.$$

Furthermore, the corresponding profits, for each firm, are

$$\pi_A = \frac{t}{2} + \frac{c_B - c_A}{3} + \frac{(c_B - c_A)^2}{18t} = \frac{(3t + (c_B - c_A))^2}{18t};$$

$$\pi_B = \frac{t}{2} + \frac{c_A - c_B}{3} + \frac{(c_A - c_B)^2}{18t} = \frac{(3t + (c_A - c_B))^2}{18t}.$$

45.2.2 Hotelling Model with Incomplete Information

In this subsection, we consider a symmetric Hotelling model with incomplete information, (for other duopoly models with incomplete information, see [4–10, 16]) where both firms have a specific space of price strategies associated with their production costs, and we compute the corresponding Bayesian Nash equilibrium in prices.

Let the triple (I, Ω, q) represent the (finite, countable or uncountable) set I with σ-algebra Ω and probability measure q, over I. Let $c : I \to \mathbb{R}_0^+$ be a measurable function with finite expected value:

$$E(c) = \int_I c^z dq(z) < \infty;$$

We assume that $dq(z)$ denotes the probability of the *belief* of each firm on the production costs of the other firm.

Theorem 45.1. *The Bayesian Nash equilibrium prices for the symmetric Hotelling game with incomplete information are*

$$p_A^z = p_B^z = t + \frac{c^z + E(c)}{2}.$$

Furthermore, the ex-ante expected profits $\pi^{EA}(c^z)$ are given by

$$\pi^{EA}(c^z) = \frac{(2t - c^z + E(c))^2}{8t}.$$

We observe that the equilibrium prices and the expected profits depends only upon their own production cost, the expected values of the production costs and the transportation costs.

The proof is based on the result presented in [17], where the Bayesian Nash equilibrium prices is also computed for an asymmetric Hotelling model with uncertainty on the production cost of both firms.

45.3 Hotelling-Type Networks

The Hotelling network consists of nodes where the firms are located, and edges normalized with length one, where the consumers are uniformly distributed. We assume that each firm F_A has a *production cost* c_A and that each buyer can shop in either of the firms located at the extreme points of its edge. The customer has no preferences for either firm, except on the grounds of price plus the *transportation cost t* that it is proportional to its distance to the firm. Each firm, represented by a node of *degree k*, competes in k markets with its k *neighbors*, and practices the same competitive price in all the k the markets, i.e. consumers in different edges of the same node pay the same price to the firm in that node plus its transportation cost.

45.3.1 Hotelling's Model in Networks with Complete Information

Let G be a network with N nodes. For every firm F_A, let V_A be the set of all firms that share a common edge with firm F_A, i.e. the direct neighbors of firm F_A. We assume that all players possess complete knowledge of the prevailing network, including the production costs of the firms at all nodes. For each market on an edge with Firms F_A and F_B, the *indifferent buyer* is given by the following distance $x_{A,B}$ of firm A, given by

$$p_A + t x_{A,B} = p_B + t(1 - x_{A,B}). \tag{45.3}$$

Solving equality (45.3), we find the location, $x_{A,B}$ of the indifferent buyer,

$$x_{A,B} = \frac{p_B - p_A + t}{2t}. \tag{45.4}$$

The profit associated with this market for the firm F_A is

$$\pi_{A,B} = (p_A - c_A) x_{A,B} = \frac{1}{2t}(p_A - c_A)(t - p_A) + \frac{1}{2t}(p_A - c_A)p_B. \tag{45.5}$$

45 A Hotelling-Type Network

For each firm F_A, in a node with degree k_A, the *profit function* π_A is the sum of the profits obtained in every market, i.e.

$$\pi_A = \frac{k_A}{2t}(p_A - c_A)(t - p_A) + \sum_{B \in V_A} p_B \frac{(p_A - c_A)}{2t}. \tag{45.6}$$

Lemma 45.1. *For a network with N nodes, the Nash equilibrium prices are the solution of the following linear system of N equations*

$$p_A = \frac{1}{2}\left(t + c_A + \sum_{B \in V_A} \frac{p_B}{k_A}\right). \tag{45.7}$$

Proof. From the FOC, we obtain that

$$\frac{\partial \pi_A}{\partial p_A} = \frac{k_A}{2t}(t - 2p_A + c_A) + \frac{k_A}{2t}\sum_{B \in V_A} \frac{p_B}{k_A} = 0.$$

□

Let us consider the *exponential cost function*

$$c_k = te^{-\frac{k}{10}}$$

that relates the cost of production with the degree of the node. We note that the cost of production decreases with the degree of the node. Using the exponential cost function, for the network given in Fig. 45.2, we compute the prices and the corresponding profits (see Table 45.1). In this case, we can observe some interesting

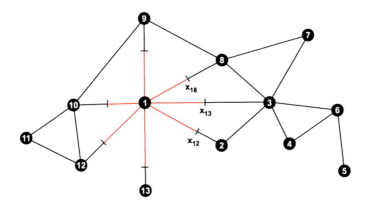

Fig. 45.2 A network example. The *dashes* in the edges represents a possible division of markets in each edge that firm 1 competes and x_{12}, x_{13} and x_{18} are the indifferent consumers at different edges

714   A.A. Pinto and T. Parreira

Table 45.1 Equilibrium prices and profits

Node	Degree:k	c(k)	p(k)	Profit(k)	Profit(k)/k
1	7	0,7047t	1,7661t	3,943t	0,5633t
2	2	0,9048t	1,8405t	0,8755t	0,4378t
3	6	0,7408t	1,7864t	3,2796t	0,5466t
4	2	0,9048t	1,863t	0,918t	0,459t
5	1	0,9512t	1,9035t	0,4534t	0,4534t
6	3	0,8607t	1,8558t	1,4854t	0,4951t
7	2	0,9048t	1,8525t	0,898t	0,449t
8	4	0,8187t	1,8138t	1,9804t	0,4951t
9	3	0,8607t	1,8307t	1,4115t	0,4705t
10	4	0,8187t	1,8224t	2,0148t	0,5037t
11	2	0,9048t	1,868t	0,9276t	0,4638t
12	3	0,8607t	1,8398t	1,4378t	0,4793t
13	1	0,9512t	1,8587t	0,4117t	0,4117t

points: firms at nodes with the same degree can have different equilibrium prices, e.g., nodes 5 and 13 have degree 1, but node 13 has a neighbor of degree 7 whereas the neighbor of node 5 has degree 2; equilibrium prices of the firms at nodes 5 and 13 are $1,9t$ and $1,86t$, respectively, and the profits of the firms at nodes 5 and 13 are $0,45t$ and $0,41t$, respectively. We also note that firms at nodes with higher degree achieve lower prices, but they obtain more profits per market because they attract more clients.

45.3.2 Hotelling's Model in Networks, with Incomplete Information

Let G be a network with finite or infinite set of nodes. Every firm F_A has a production cost $c_k = c_A$ that just depends upon the degree $k = k_A$ of the node. The firms have incomplete information, i.e. every firm F_A knows his own degree k_A but ignores the degree of the other firms. The overall degree distribution is assumed common knowledge. Let $q = \{q_k\}_{k=0}^{\infty}$ be the probability density of the degree of the nodes k in the network, i.e. each q_k denotes the fraction of firms who have k neighbors. Noting that the frequency with which each node of degree k is encountered is proportional to the product kq_k, the corresponding probability density for the degree distribution of a neighboring node i.e., one that is chosen as the neighbor of some randomly selected node (see [11] and [12]) is given by $\tilde{q} = \{\tilde{q}_k\}_{k=0}^{\infty}$, where

$$\tilde{q}_k = \frac{q_k k}{\sum_{k'=0}^{\infty} q_{k'} k'}.$$

45 A Hotelling-Type Network

Given a firm in a node, the *expected production cost* $E(c)$ of a firm in a neighboring node is given by

$$E(c) = \int_I c_k d\tilde{q}_k < \infty.$$

Theorem 45.2. *The Bayesian Nash equilibrium prices, for the Hotelling network with incomplete information, are*

$$p_k = t + \frac{c_k + E(c)}{2}, \tag{45.8}$$

where p_k is the competitive price of a firm located at a node of degree k. Furthermore, the ex-ante expected profits in equilibrium are

$$\pi_k^{EA} = \frac{k(2t - c_k + E(c))^2}{8t}. \tag{45.9}$$

The proof is in [17].

Let $c(k; \pi) = 2t + E(c) - \sqrt{\frac{8t\pi}{k}}$, where π is a given profit.

Corollary 45.1.
$$c_k < c(k; \pi) \Leftrightarrow \pi_k^{EA} > \pi.$$

In Fig. 45.3, the dependence of the cost c_k in the degree node k is shown for several ex-ante expected isoprofits.

By equality (45.6), the ex-post profit of a firm at a node with degree k, given that their neighbor firm costs are $c^1, \dots, c^k$, is

$$\pi_k^{EP}(c^1, \dots, c^k) = (p_k - c_k)\left(\frac{k}{2} + \frac{\sum_{i=1}^k p^i - p_k}{2t}\right), \tag{45.10}$$

where p_k and p^i, $i = 1, \dots, k$ are the Bayesian–Nash equilibrium prices (45.8). Note that $p^i = p_{degree(V_i)}$, where V_i is the i neighbor of the node with degree k.

Remark 45.1. (*static analysis*) For the Hotelling network, with incomplete information, we have that

$$\pi_k^{EP}(c^1, \dots, c^k) - \pi_k^{EA} = \left(\sum_{i=1}^k c^i - kE(c)\right)\left(\frac{1}{4} + \frac{E(c) - c_k}{8t}\right). \tag{45.11}$$

Furthermore,

$$\sum_{i=1}^k c^i < kE(c) \quad \text{if, and only if,} \quad \pi_k^{EP}(c^1, \dots, c^k) < \pi_k^{EA}. \tag{45.12}$$

The proof is in [17].

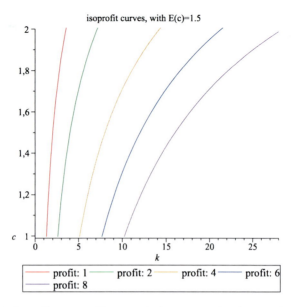

Fig. 45.3 Ex-ante isoprofit curves, with transportation cost $t = 1$ and expected cost $E(c) = 1.5$

We are going to study three different types of networks and we will compare the prices and the profits obtained with complete and with incomplete information: (a) a regular network; (b) A network where the degree distribution is assumed Poisson (studied by Erdös and Rényi [3]) and the network is generated by a mechanism where connectivity is set at random (every possible link is formed with a fixed independent probability) and the framework is stationary (the set of nodes is large but given); and (c) a network where the degree distribution is scale-free (i.e. it is given by a power law). Barabási and Albert [2] have shown that these structures arise in growing environments where new links are again set at random, but with a (linear) bias in favor of nodes that are more highly connected. We will use the following relation

$$c_k = te^{-\frac{k}{10}}$$

between the node's degree and the production cost.

Let the network be regular, where each node has degree k. In this case, the prices and profits obtained with incomplete information are the same as the ones obtained with complete information, because if all firms know the network's distribution, then they know the network. The prices are given by

$$p_k = t + c_k$$

and the profits are given by

$$\pi^k = \frac{kt}{2}.$$

45 A Hotelling-Type Network

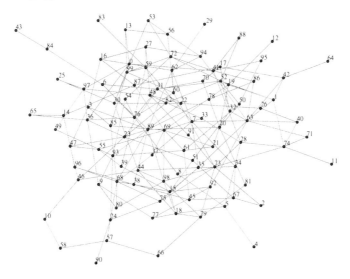

Fig. 45.4 Network with Poisson distribution

Let the network degree be Poisson distributed. The probability that a randomly selected node has degree k is given by

$$f(k; \lambda) = \frac{e^{-\lambda} \lambda^k}{k!} \tag{45.13}$$

where λ is the *average network degree*. In Fig. 45.4, we present a network with 100 nodes generated using a Poisson distribution with average network degree close to 4. In Table 45.2, we show the prices and profits with complete information and the expected prices and profits with incomplete information.

Let the network degree be scale-free. The probability that a randomly selected node has degree k is given by

$$p_k = \frac{k^{-\gamma}}{\zeta(\gamma)} \tag{45.14}$$

where $\gamma \geq 2$ is a parameter that determines the decay of the distribution and $\zeta(\gamma) = \sum_{k=1}^{\infty} k^{-\gamma}$ is the Riemann zeta function. The average degree is given by

$$z(\gamma) = \frac{\zeta(\gamma - 1)}{\zeta(\gamma)}$$

In Fig. 45.5, we present a scale-free network with 100 nodes.

In Table 45.3, we show the prices and profits with complete information and the expected prices and profits with incomplete information.

Table 45.2 Poisson distribution: complete vs. incomplete information. The table relates the prices, profits and profits per market in each node of the network with Poisson degree distribution. p refers to prices and l refers to profits. $E(p)$ and $E(l)$ are the expected prices and profits. The last row of the table are the averages of the respective column values. The differences are computed in absolute value

node	k	c(k)	p	E(p)	p-E(p)	l	E(l)	l-E(l)	l/k	E(l)/k	(l-E(l))/k
0	8	0,449t	1,531t	1,534t	0,0028t	4,68t	4,705t	0,0244t	0,585t	0,588t	0,0031t
1	5	0,607t	1,597t	1,612t	0,0154t	2,453t	2,53t	0,0767t	0,491t	0,506t	0,0153t
2	2	0,819t	1,711t	1,719t	0,0073t	0,797t	0,81t	0,013t	0,398t	0,405t	0,0065t
3	4	0,67t	1,631t	1,644t	0,013t	1,847t	1,897t	0,0502t	0,462t	0,474t	0,0126t
4	1	0,905t	1,788t	1,762t	0,0259t	0,39t	0,367t	0,0225t	0,39t	0,367t	0,0225t
...	...	...	...	...	...	...	...	...	...	...	...
99	6	0,549t	1,58t	1,584t	0,004t	3,187t	3,212t	0,025t	0,531t	0,535t	0,0042t
					0,0149t			0,0479t			0,0141t

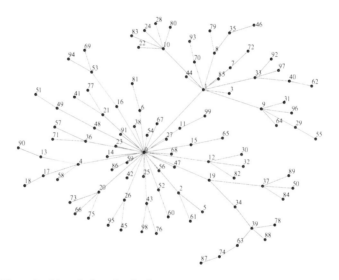

Fig. 45.5 Network with scale-free distribution

45.4 Conclusion

We present the Bayesian Nash equilibrium prices and associated equilibrium expected profits of each firm in a network as a function of the firm's degree node and we found that these prices do not depend on the network distribution, except on their first moments. We computed the price and profit values, in equilibrium, for some classical networks: regular networks, random networks and scale-free networks. Moreover, we compared the prices and profits obtained in the case of

45 A Hotelling-Type Network

Table 45.3 Scale-Free distribution: complete vs. incomplete information

node	k	c(k)	p	E(p)	p-E(p)	l	E(l)	l-E(l)	1/k	E(l)/k	(1-E(l))/k
0	30	0,045t	1,353t	1,321t	0,0324t	25,48t	24,23t	1,2496t	0,849t	0,808t	0,0417t
1	10	0,368t	1,52t	1,48t	0,0402t	6,636t	6,18t	0,4551t	0,664t	0,618t	0,0455t
2	3	0,741t	1,696t	1,666t	0,0299t	1,369t	1,284t	0,0844t	0,456t	0,428t	0,0281t
3	1	0,905t	1,712t	1,748t	0,0358t	0,326t	0,356t	0,0295t	0,326t	0,356t	0,0295t
4	4	0,67t	1,676t	1,631t	0,0447t	2,021t	1,845t	0,1757t	0,505t	0,461t	0,0439t
...	...	...	...	...	...	...	...	...	...	...	...
99	1	0,905t	1,801t	1,748t	0,0532t	0,402t	0,356t	0,0463t	0,402t	0,365t	0,0463t
					0,065t			0,1067t			0,0571t

The table relates the prices, profits and profits per market in each node of the network with Scale-Free degree distribution p refers to prices and l refers to profits. $E(p)$ and $E(l)$ are the expected prices and profits. The last row of the table are the averages of the respective column values. The differences are computed in absolute value.

complete information with the prices and profits obtained in the case of incomplete information.

Acknowledgements This work was presented in the International Congress of Mathematicians ICM 2010, in the Second Brazilian Workshop of the Game Theory Society in honor of John Nash. This work was highlighted in the article [15] after being presented in ICM 2010. We thank Humberto Moreira and Rosa Branca Esteves for their suggestions in a early version of this work. We thank Calouste Gulbenkian Foundation, PRODYN-ESF, FEDER, POFC, POCTI and POSI by FCT and Ministério da Ciência e da Tecnologia, and the FCT Pluriannual Funding Program of the LIAAD-INESC Porto LA and of the Research Centre of Mathematics of the University of Minho, for their financial support.

References

1. Anderson, S., de Palma, A. Thisse, J-F.: Discrete Choice Theory of Product Differentiation. MIT Press, Cambridge, Massachusetts. 1992
2. Barabási, A.-L., Albert, B.: Emergence of scaling in random networks. Science **286**, 509–512 (1999)
3. Erdös, P., Rényi, A.: On Random graphs I. Publicationes Mathematicae Debrecen **6**, 290–297 (1959)
4. Ferreira, F.A., Ferreira, F., Ferreira, M., Pinto, A.A.: Quantity competition in a differentiated duopoly. In: Machado, J.A.T., et als. (eds.) Intelligent Engineering Systems and Computational Cybernetics, pp. 365–374. Springer, New York (2008)
5. Ferreira, F.A., Ferreira, F., Pinto, A.A.: Bayesian price leadership. In: Tas, K., et als. (eds.) Mathematical Methods in Engineering, pp. 359–369. Springer, Dordrecht (2007)
6. Ferreira, F.A., Ferreira, F., Pinto, A.A.: Flexibility in Stackelberg Leadership. In: Machado, J.A.T., et als. (eds.) Intelligent Engineering Systems and Computational Cybernetics, pp. 399–405. Springer, New York (2008)
7. Ferreira, F.A., Ferreira, F., Pinto, A.A.: 'Own' price influences in a Stackelberg leadership with demand uncertainty. Braz. J. Bus. Econ. **8**(1), 29–38 (2008)

8. Ferreira, F., Ferreira, F.A., Pinto, A.A.: Price-setting dynamical duopoly with incomplete information. In: Machado, J.A.T., et als. (eds.) Nonlinear Science and Complexity, pp. 1–7. Springer, Berlin (2009)
9. Ferreira, F. A., Ferreira, F. and Pinto, A. A., Uncertainty on a Bertrand duopoly with product differentiation. In: Machado, J.A.T., et als. (eds.): Nonlinear Science and Complexity, pp. 1–7. Springer, Berlin (2009)
10. Ferreira, F.A., Ferreira, F., Pinto, A.A.: Unknown costs in a duopoly with differentiated products. In: Tas, K., et als. (eds.): Mathematical Methods in Engineering, pp. 371–379. Springer, Dordrecht (2007)
11. Galeotti, A., Goyal, S., Jackson, M., Vega-Redondo, F., Yariv, L.: Network Games. The Review of Economic Studies **77**, 218–244 (2010)
12. Galeotti, A., Vega-Redondo, F.: Complex networks and local externalities: a strategic approach. Forthcoming in International Journal of Economic Theory
13. Graitson, D.: Spatial competition á la Hotelling: a selective survey. The Journal of Industrial Economics. **31**, 11–25 (1982)
14. Hotelling, H.: Stability in Competition. Econ. J. **39**, 41–57 (1929)
15. Mudur, G.S., Maths for movies, medicine & markets. *The Telegraph Calcutta, India*, 20/09/2010
16. Pinto, A.A.: Game Theory and Duopoly Models. Interdisciplinary Applied Mathematics Series. Springer, New York (2010)
17. Pinto, A.A., Parreira, T.: Hotelling model with uncertainty on the production costs and networks (submited)

Chapter 46
The Closing Lemma in Retrospect

Charles Pugh

Voici un fait que je n'ai pu démontrer rigoureusement, mais qui me parait pourtant très vraisemblable. Étant données des équations ... et une solution particulière quelconque de ces équations, on peut toujours trouver une solution périodique (dont la période peut, il est vrai, être très longue), telle que la différence entre les deux solutions soit aussi petite qu'on le veut, pendant un temps aussi long qu'on le veut. D'ailleurs, ce qui nous rend ces solutions périodiques si précieuses, c'est qu'elles sont, pour ainsi dire, la seule brèche par où nous puissions essayer de pénétrer dans une place jusqu'ici réputée inabordable.

– Henri Poincaré, 1892

Abstract This paper presents a discussion of the closing lemma, its origins and development.

46.1 Rowland Hall

On a Friday afternoon in 1963, I was making my way down a staircase in Rowland Hall at Johns Hopkins, when I encountered my advisor, Phil Hartman. He had just returned from a seminar at RIAS[1] where he'd listened to Mauricio Peixoto talk about structural stability. At the time, structural stabilty was a new idea in the West. Hartman mentioned to me a question of Peixoto that he thought I "might find interesting," namely "If a vector field X has a recurrent orbit, can X be perturbed so the recurrent orbit becomes a closed orbit?" The perturbation was to be an additional vector field Δ with small C^1-size. The new vector field $Y = X + \Delta$ should have a

---

C. Pugh
Department of Mathematics, University of Toronto, 40 St. George Street, Toronto, ON, Canada M5S 2E4
e-mail: cpugh@math.utoronto.ca

[1] Research Institute for Advanced Study, the forerunner to the Lefschetz Center for Dynamical Systems at Brown University.

M.M. Peixoto et al. (eds.), *Dynamics, Games and Science I*, Springer Proceedings in Mathematics 1, DOI 10.1007/978-3-642-11456-4_46,
© Springer-Verlag Berlin Heidelberg 2011

722 C. Pugh

closed orbit near the recurrent orbit of X, and this accounts for the name "Closing Lemma."[2] You *close* a recurrent orbit. Peixoto had already proved a version of the Closing Lemma as part of his two dimensional structural stability theorem – *the generic vector field on a compact surface is structurally stable* – but was seeking a more definitive result.

Hartman's crucial advice was this: *Figure out why the obvious construction of Δ fails.* During the following weekend I did exactly that, and saw how to get around the worst errors.

46.2 Two Wrong Proofs

It's easiest to think about the question for vector fields on compact surfaces, and that's where I focused my attention – flows in dimension two.[3] The simplest case is that of constant vector fields X on the torus whose orbits have irrational slope. All the orbits of X are recurrent and they all become closed when the slope is changed from irrational to rational. Not very enlightening, but at least re-assuring. See Fig. 46.1.

In general, a recurrent orbit of a vector field X on a surface crosses a local transversal segment P repeatedly. You think of a point p_0 on P, and you track its successive return points under the X-flow φ as

$$ p_1 = \varphi_{t_1}(p_0), \quad p_2 = \varphi_{t_2}(p_0), \cdots \in P $$

where $0 < t_1 < t_2 < \ldots$. Recurrence implies that the points p_n accumulate at p_0. Then you try to perturb X to $Y = X + \Delta$ so that instead of merely returning near p_0, the Y-orbit returns to p_0 itself. See Fig. 46.2. The intuition is that you want Δ to deform the X-trajectories so p_n moves toward and eventually equals p_0.

To keep everything simple, it's a good idea to think of X near the transversal as trivial. In the right local coordinate system at p_0, P is a segment on the y-axis and X is the constant horizontal vector field $X = \partial/\partial x$. I called such a coordinate system a *flowbox*. See Fig. 46.3. Flowbox coordinates exist at every nonsingular point, i.e., wherever $X(p) \neq 0$.

---

[2] I later learned that Réné Thom was responsible for the concept and statement of the Closing Lemma. He needed the result as a lemma in his proof that the generic ordinary differential equation has no first integrals, i.e. no global smooth functions which are constant on the ODE's orbits. After lengthy discussions, Peixoto convinced Thom that his proof of the Closing Lemma had a major gap. It produced a perturbation that was C^0-small but not C^r-small, and failed to lead to a genericity result in the class of C^r ODE's. Years later, Truman Bewley proved Thom's genericity result, bypassing the Closing Lemma.

[3] In this survey paper, for no special reason, I have chosen to concentrate on vector fields and flows rather than diffeomorphisms.

46 The Closing Lemma in Retrospect

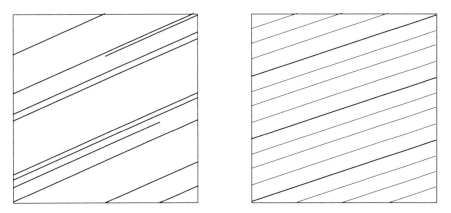

Fig. 46.1 The surface is the torus – the square with opposite edges identified. The orbits of X have irrational slope and are recurrent. The orbits of $Y = X + \Delta$ have rational slope and are closed.

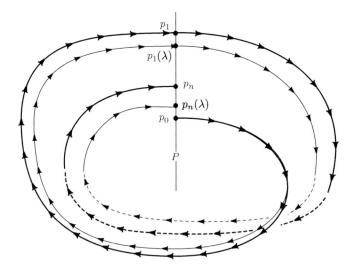

Fig. 46.2 The Y-orbit through p_0 returns to P at $p_n(\lambda)$, which is closer to p_0 than is p_n. The dotted curves indicate orbits traveling on distant parts of the surface where there may be handles permitting them to "cross under" the orbits shown connecting p_0 to p_1 and $p_1(\lambda)$.

Using a flowbox F with coordinates $-1 \leq x \leq 0$ and $-1 \leq y \leq 1$, there are two "obvious" (but incorrect) proofs of the Closing Lemma. Both rely on Peixoto's perturbing vector field

$$\Delta = \epsilon \lambda \beta(x, y)\left(\frac{\partial}{\partial y}\right)$$

where β is a fixed, smooth bump function with support F, ϵ is a positive constant, and $\lambda \in [-1, 1]$ is a parameter. When ϵ is small, Δ has small C^1-size. In fact it has small C^r-size for all r. The nth return of the recurrent X-orbit is p_n. For many

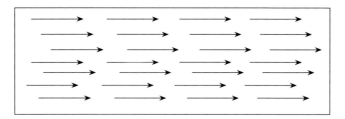

Fig. 46.3 The X-flow in a flowbox is trivial. It is unit speed horizontal translation.

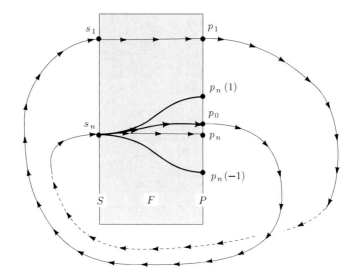

Fig. 46.4 The Intermediate Value Theorem gives a $\lambda \in (-1, 1)$ such that the local $Y(\lambda)$-orbit connects s_n to p_0, apparently creating a closed orbit through p_0.

large n, p_n is very near p_0. The nth return of p_0 with respect to $Y(\lambda) = X + \Delta$ is $p_n(\lambda)$. Orbits take unit time to cross F from its left side $S = \{-1\} \times [-1, 1]$ to its right side $P = \{0\} \times [-1, 1]$. A local $Y(\lambda)$-orbit connects $s_n(\lambda)$ to $p_n(\lambda)$.

When $\lambda = 0$, $s_n(0) = s_n$. When $\lambda = 1$, the local Y-orbit through s_n leaves the flowbox F above p_0 and when $\lambda = -1$, it leaves F below p_0. The Intermediate Value Theorem gives a $\lambda \in (-1, 1)$ such that the local Y-orbit through s_n leaves F exactly at p_0, which apparently implies that the Y-orbit through p_0 is closed. See Fig. 46.4.

The error in this proof involves the points $p_k \in P$ with $0 < k < n$. They are *intermediate intersections* of the recurrent orbit and the transversal, where "intermediate" refers to the time order, not the order along the transversal. The perturbation affects all of them, not merely the last one. It may perfectly well be that the local $Y(\lambda)$-orbit connects s_n to p_0, but $Y(\lambda)$ also changes the orbit from p_0 to S. No longer need the Y-orbit through p_0 cross S at s_n. This simplistic way of connecting s_n to p_0 may *break* the orbit that connects p_0 to s_n.

46 The Closing Lemma in Retrospect

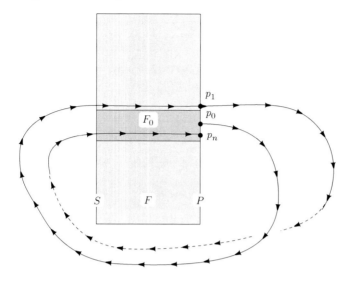

Fig. 46.5 A thin flowbox F_0 eliminates intermediate intersections of the X-orbit through p_0.

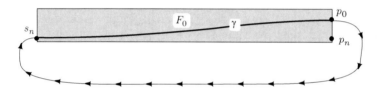

Fig. 46.6 $Y = X + \Delta$ has a closed orbit through p_0.

Now, if there *were* no intermediate intersections, the preceding proof would be OK, and this leads to the second incorrect proof. Take a subflowbox $F_0 \subset F$ thin enough so it is disjoint from the X-orbit between p_0 and s_n. This can be done by looking at a large n so that p_n is closer to p_0 than any p_k with $0 < k < n$, and it solves the intermediate intersection problem. See Fig. 46.5. Since F_0 is free from intermediate intersections, it is called a *free flowbox* and the short transversal segment $P_0 = P \cap F_0$ is called a *free segment*. Scale the perturbing vector field Δ to F_0. Then $s_n = s_n(\lambda)$, and the $Y(\lambda)$-orbit connects s_n to p_0, again apparently producing a closed orbit through p_0. See Fig. 46.6.

The error here is subtler. It involves the relative positions of p_0 and p_n on the free segment $P_0 = P \cap F_0$. You want $Y = X + \Delta$ to connect s_n to p_0, you want Δ to have C^1-size $< \epsilon$, and you want Δ to have support in F_0. At first glance, this seems like no problem at all, and in fact it's how Thom saw things. Basically, as in Fig. 46.6, you just draw a smooth local Y-orbit γ from s_n to p_0 in F_0. Since F_0 is thin, γ is nearly horizontal, so its tangent is nearly equal to X, and $Y - X$ is C^r-small.

The trouble is that merely because the local orbits of one vector field C^r-approximate the local orbits of another, the two vector fields need not be C^r close

Fig. 46.7 The vector field $Y = \partial/\partial x + \Delta$ does not C^1-approximate $X = \partial/\partial x$ despite the fact that all its local orbits are nearly horizontal in the C^∞ sense.

together. More is required. The transverse properties of the local orbit assemblies must also be C^r close together.[4] See Fig. 46.7. The vector field Y defines the Poincaré map $\phi : S \to P$ according to the formula

$$\phi(s) = \varphi_1(s, Y)$$

where $s = (-1, y) \in S$ and $\varphi_t(s, Y)$ is the Y-orbit through s. Because Δ is C^1-small, $\phi'(s)$ must be approximately equal to 1. This is the necessary transverse property of Y.

Since Δ has support in F_0, ϕ sends its left edge S_0 to its right edge P_0, endpoints being sent to endpoints. Since $\phi' \approx 1$, ϕ cannot alter the relative positions of points of S_0 significantly. ϕ sends points near the middle of S_0 to points near the middle of P_0. Now it may very well happen that p_0 lies near the top of P_0 while p_n lies near the bottom, which would make it impossible for ϕ to send s_n to p_0. That's the error in the thin flowbox proof. See Fig. 46.6.

What can be learned from these two wrong proofs? First, it's easy to be misled by one's intuition and optimistic pictures. Second, intermediate intersections are disastrous. They must be avoided at all costs. Third, given a C^1-small perturbation in a thin flowbox, one can hope to estimate its relative or proportional effect on the Poincaré map ϕ across the flowbox. More precisely, if $s = (-1, y) \in S_0$ and $p = (0, y)$ then estimates on the *proportional lift*

$$L(s) = \frac{|\phi(s) - p|}{|S_0|}$$

are possible.

46.3 Good Position

Over that weekend in 1963, I came to a rough understanding of the two wrong proofs, and I realized that a certain amount of proportional progress (in the sense that $p_n(\lambda)$ becomes closer to p_0 relative to the width of a thin flowbox) can be achieved. The proportional progress was on the order of ϵ, *provided that the points p_0, p_n lie in the middle part of the free segment P_0*. On the other hand, if they lie

---

[4] The same issue arises for dynamical foliations. Often, the stable manifold foliation has smooth leaves, but has unavoidably bad properties transverse to the leaves.

46 The Closing Lemma in Retrospect

Fig. 46.8 The intermediate intersection p_k is too close to p_0 so the pair (p_0, p_n) gets replaced by the pair (p_0, p_k).

near the endpoints of P_0 relative to the length of P_0 all is lost: the proportional progress is $\ll \epsilon$. (ϵ is approximately the C^1 size of the perturbation Δ.)

So how to cope? The segment P_0 is dictated by its avoidance of the intermediate intersections. What's to be done if an intermediate intersection p_k actually does occur very near p_0 or p_n in comparison to $|p_0 - p_n|$? For then p_0 or p_n would lie too near the endpoints of P_0. Well, this was my only original idea, and it's not much. Simply replace the bad orbit pair (p_0, p_n) by a sub orbit pair consisting of the too close p_k and the closer of p_0, p_n. See Fig. 46.8. (An *orbit pair* (p, q) has $\varphi_\tau(p) = q$ for some $\tau > 0$.) If the sub orbit pair is still bad (intermediate intersections occur too close to the new pair), repeat the replacement. There are only finitely many intermediate intersections so the process terminates with a good orbit pair (p_i, p_j) having $0 \le i < j \le n$. But is the good pair near the original recurrent point p_0? (If not, making a closed orbit through p_i is worthless.)

To handle this, you have to say how close is too close. You say an orbit pair (p, q) on the transversal P is *good* (or has good *relative position*) when $\varphi_\tau(p) = q$ for some $\tau > 0$ and all intermediate intersections $\varphi_t(p) \cap P$ with $0 < t < \tau$ occur outside the interval

$$P(p,q) = \{y : |y - p| \le 2|p - q|/3\} \cup \{y : |y - q| \le 2|p - q|/3\}.$$

This interval has length $7|p - q|/3$ and is centered at the midpoint of $[p, q]$.

What happens when you replace a bad orbit pair (p, q) with a potentially better sub orbit pair (p', q')? The new pair lies in $P(p, q)$ and

$$|p' - q'| \le \frac{2}{3}|p - q|.$$

This implies that the eventual good orbit pair (p_i, p_j) produced from (p_0, p_n) is not too far from p_0. In fact, its distance from p_0 is dominated by the geometric series

$$\sum_{k=0}^{\infty} \left(\frac{2}{3}\right)^k |p_0 - p_n| = 3|p_0 - p_n|,$$

which is as small as you want.

So I saw how to get good pairs $(p, q) = (p_i, p_j)$ near the original recurrent point p_0, and how to make approximately ϵ proportional progress in pushing one toward the other without breaking the connection between them. So what? You want proportional progress on the order of unity (actually 3/7 would suffice) but you only have proportional progress on the order of ϵ. The answer is simple. Just use N very thin disjoint flowboxes where $N > 1/\epsilon$. Proportional progress of ϵ in each box adds up to proportional progress on the order of unity when you have N flowboxes. This idea of "spreading the perturbation along the recurrent orbit" is roughly the same thing as taking a single, very long and very thin flowbox.

46.4 Cleaning Things Up

There were a few things to clean up. For instance the Poincaré map from the kth thin flowbox F_k to the next one potentially cancels the proportional progress made in F_k. That was worrisome, so I "assumed it away" by restricting to distal flows – flows like the irrational torus flow in which orbits stay approximately parallel. But eventually things turned out to be OK because the Poincaré map is approximately linear, and linear maps preserve proportions. See Sect. 46.5 below.

There were several other issues, but nothing seemed shaky at the time. Lurking in the background was the fact that I had defined the perturbation Δ using flowbox coordinates, and a vector field is once less differentiable than the coordinate system it's defined in. So without realizing it I was assuming X was at least C^2. See Sect. 46.10 below.

In fact I also thought that the perturbation could be made C^r small for all $r \geq 1$. Luckily there was no ArXiv in 1963, and I saw why my technique did not handle the C^r case, $r \geq 2$, before final submission of an announcement. The reason is simple. If Δ has support in a thin, unit time-length flow box F of width w and has C^r size ϵ then its maximum size is ϵw^r. The maximum lift (i.e., the maximum deflection of the local Y-orbits in the y-direction) is at most ϵw^r, so the proportional lift is at most ϵw^{r-1}. To get total proportional lift on the order of unity requires N disjoint flowboxes, and N is a lot larger than $1/\epsilon$ when $r \geq 2$. Specifically, if the flowboxes have widths $w_1, \ldots, w_N$ and the total proportional lift is on the order of unity then we must have

$$\epsilon(w_1^{r-1} + \cdots + w_N^{r-1}) \approx 1.$$

But the area of each box is approximately $w_k \geq w_k^{r-1}$ when $r \geq 2$, so this implies that the total area of the flowboxes exceeds the area of the surface when ϵ is small, contradicting the fact the flowboxes are disjoint.

46 The Closing Lemma in Retrospect

It is amazing that the C^2 Closing Lemma remains an open problem *even for flows on surfaces*. Carlos Gutierrez, Catherine Carroll, and others have C^r Closing Lemma results for special surfaces such as the two-holed torus, but for the general case, nothing!

46.5   Visiting Mauricio in Rio

In 1963, after Mauricio's visit to Baltimore, I corresponded with him about the Closing Lemma and his structural stability paper. He invited me to spend that summer at IMPA in Rio. I remember getting his first letter in its green and yellow Brazilian airmail envelope. Pretty exotic.

At the time, Mauricio was in his forties and IMPA occupied a small building in Botafogo, a district in the South zone of Rio part way between downtown and Copacabana. I shared an office with Ivan Kupka who was Mauricio's student and IMPA's first PhD. During that period, Mauricio was my effective advisor. Mauricio and IMPA molded my outlook on mathematics as much as anything else, and it would not be too far fetched to see the Closing Lemma as the second PhD thesis at IMPA.

One of the biannual events in Brazilian mathematics of the era was the week long meeting at Poços de Caldas, the Colóquio Brasileiro de Matemática. Poços is a small town in the state of Minas Gerais about 300 miles from Rio. It's a kind of a low key hot springs resort, and most of the participants from Rio traveled there by bus.

We were directed to be present in the central bus terminal at 6 am on a Saturday morning for the chartered bus taking us to Poços. By starting well before 5, my wife and I managed to arrive there by public buses (I was too cheap to try for a cab), only getting lost twice. The first lesson confirmed our impression of Brazilian time – our bus waited until 7:30 or so, while the group of mathematicians finished assembling. And also, there was no grousing about the delay. That was the second Brazilian lesson.

Our bus was of the round, school bus variety and was equipped with a many geared manual transmission. It was a bit larger than the local buses in Rio, noted for their floor seating policy that avoided police enforcement of over crowding laws. Ours was packed to capacity and then some. Nobody had to sit on the floor though.

The first part of the journey was easy enough since we followed the main highway between Rio and São Paulo. Then we turned off into the hinterlands near Guaratinguetá. For the next hundred miles, while bumping along over dusty, hilly roads, Mauricio and I talked about structural stability and the Closing Lemma off and on, puzzling about the possible cancellation of proportional progress from one flow box perturbation to the next. We drew pictures of a necklace of thinner and thinner flowboxes, trying to imagine how the flow from one flowbox to the next could work against us.

After a while we realized that this intermediate flow was not our enemy at all. The thinner the flowboxes were, the more nearly linear the Poincaré maps were. And linear maps do not distort proportions. If, after n steps, we had pushed the point $q_n(\lambda)$ half the way from q_n to p_n then the flow from the nth flow box to the $(n+1)^{\text{st}}$ would hardly affect this.

Mauricio and I thought of this as the last brick in the two dimensional proof, and were not even too perturbed when the bus broke down a mile or so out from Poços.

46.6 Flows in Dimension Three

Returning to Johns Hopkins in the 1963–1964 year, I tried to see how to make my proof work in higher dimensions. Existence of the free regions on the transversal (regions without intermediate intersections) is proved in the same way. It's essentially a metric space argument, and you get a good pair (p, q) (which is close to the recurrent point) as before.

The free region $P(p, q)$ for the good pair is peanut shaped: it is the union of two balls $B(p, 3d/4) \cup B(q, 3d/4)$ on a transversal P, where $d = |p - q|$. See Fig. 46.9. So far so good. You can push $q_n(\lambda)$ toward p_n, flowbox after flowbox, with the hope that eventually you get $q_n(\lambda)$ to exit the Nth flowbox at p_N. If you can do this you get a closed orbit through p.

But there is a difficulty. It's the shape of the free regions under the Poincaré maps $\pi_n : P_{n-1} \to P_n$ from one flowbox end-face to the next. The composite $\Pi_n = \pi_n \circ \cdots \circ \pi_1$ can squash the original free region $R_0 = P(p, q)$ exponentially. The nth free region $R_n = \Pi_n(R_0)$ can become quite eccentric. See Fig. 46.10.

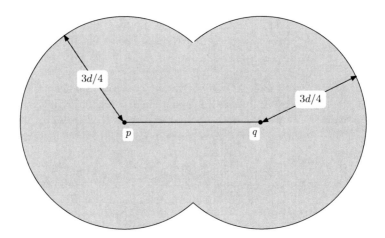

Fig. 46.9 For vector fields in dimension 3, the free region of a good pair (p, q) is a peanut-shaped region on the two dimensional transversal P.

46 The Closing Lemma in Retrospect

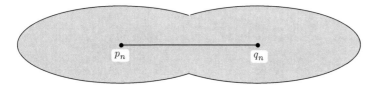

Fig. 46.10 The composite Poincaré map $\Pi_n : P_0 \to P_n$ can squash the original peanut shaped free region.

The amount of progress in pushing q_n toward p_n is limited by the smallest width w_n of the flowbox, and in the squashed case the eccentricity w_n/d_n can be exponentially small, where $d_n = |p_n - q_n|$. If $w_n/d_n \leq \mu^n$ with $0 < \mu < 1$ then the total proportional progress in pushing q_n toward p_n in N flowboxes, with an ϵ-push in each box would be

$$\sum_{n=0}^{N-1} \frac{\epsilon w_n}{d_n} < \sum_{n=0}^{\infty} \frac{\epsilon w_n}{d_n} = \frac{\epsilon}{1-\mu}$$

which is not on the order of unity as $\epsilon \to 0$. This kills the pushing strategy as it stands.

I puzzled a long time over what to do about this, and finally came up with the idea of splitting the closing construction into two cases.

Case 1. *The old, easy case*. There is a subsequence of iterated free regions whose shapes are uniformly similar to the original peanut. Then the previous proof for flows on surfaces works fine. You get a sequence of $N > K/\epsilon$ flowboxes (where K is a constant bounding the amount of distortion of the free regions) and a proportional lift on the order of ϵ/K in each. The total proportional lift is on the order of unity and the perturbed vector field $Y = X + \sum \Delta_n$ has a closed orbit through p.

Case 2. *The new case*. There is no subsequence of iterated free regions with uniformly bounded shape. The Poincaré maps are essentially linear when the flowboxes are thin, so you can take a subsequence of flowboxes F_n along the recurrent orbit in which the linearized distortion tends to infinity. Under a linear map, a square becomes a parallelogram, so the end-face P_n of F_n looks like an eccentric parallelogram whose eccentricity tends to infinity as $n \to \infty$. The shape of the parallelogram converges to the shape of a segment.

There are two ways the shape of a parallelogram can decay as the eccentricity tends to infinity. The angle between the sides can tend to zero, or the ratio of one edge to another can tend to zero. The former case reduces to the latter by choosing the original parallelogram carefully.[5] The free regions do look like vertically squashed peanuts.

---

[5] Here is a place that I should have used ellipses instead of parallelograms. (Later, Jiehua Mai did exactly this.) For a sequence of ellipses whose eccentricity tends to infinity is easier to analyze than a sequence of parallelograms. I more or less took the limit of the "major axes of the parallelograms" without realizing it.

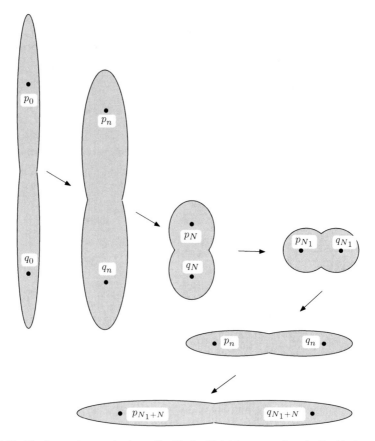

Fig. 46.11 The free regions are horizontally thin for $N+1$ iterates and vertically thin for a further $N+1$ iterates.

Now comes the trick. By choosing a new metric on the original transversal P_0, and choosing a good pair (p,q) with respect to the new metric, the free region $R_0 \subset P_0$ can be made to look like (with respect to the original metric) a peanut squashed horizontally, and the shape of R_n remains horizontally thin for $N \approx 1/\epsilon$ iterates of the Poincaré maps. Then for some $N_1 > N$ and $N_1 \leq n \leq N + N_1$, the shapes of the R_n are again thin vertically. See Fig. 46.11. With this picture you can gradually push q_n toward p_n.

The idea is to concentrate always on the thin direction, for you can make good progress in the thin direction. During the first N iterates, when R_n is horizontally thin, you push q_n horizontally, so that eventually the horizontal coordinate of $q_N(\lambda)$ agrees with the horizontal coordinate of p_N. Then you wait until the free region is vertically thin. You do nothing for $N < n < N_1$. Then, for $N_1 \leq n \leq N_1 + N$, you push vertically, until the vertical coordinate of $q_{N_1+N}(\lambda)$ agrees with the vertical coordinate of p_{N_1+N}. Due to approximate linearity, the equality of the horizontal coordinates, achieved after N steps, is not significantly lost for the next steps, so

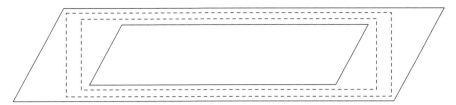

Fig. 46.12 A pair of concentric, thin, uniformly scaled parallelograms can be packed in a peanut shaped free region so that the inner one contains the good pair (p, q). Then a pair of uniformly scaled rectangles can be packed between the parallelograms. The support of the perturbation should be contained in the outer parallelogram. An ϵ-perturbation parallel to the rectangles' short edges results in proportionally effective motion of points in the inner parallelogram when judged along the its short side, but little proportional motion judged along its long edge.

after a little jiggling, both coordinates agree and there is a closed orbit through p. Figure 46.12 shows the parallelogram geometry I dealt with.

Upshot *"Pushing q_n toward p_n in the thin direction" does the trick.*

46.7 Hartman's One Word Letter

In the spring of 1964 I was finishing my thesis work at Johns Hopkins, or so it seemed, and I was looking for a job starting in the fall. Hartman had agreed that a proof of the Closing Lemma on manifolds of dimensions two and three would do.

Times were rather better than they are today. I took the Greyhound bus from Baltimore to New York City and visited Steve Smale at Columbia. After explaining to him my version of the Closing Lemma he asked what I was doing the next year, and I replied that I hoped to get a one year post-doc at RIAS. He then proposed he ask Murray Protter if they had a spot open at Berkeley. One phone call from Steve to Murray was all it took. Times were rather better then.

Berkeley gave me a position as "Acting Assistant Instructor," which is only slightly above a TA, but it was a real job and it doubled my income to $6600/year. In 1964 Berkeley was in full flower – the Free Speech Movement, marches, demonstrations, etc. It was like heaven after four isolated years in Baltimore. Steve ran a dynamics seminar where he combined differential topology and differential equations. That's where I met and became friends with Mike Shub. We followed Steve's dynamics and politics with the passion special to the era.

I had left Johns Hopkins in 1964 without *actually* turning in the thesis, promising Hartman it would soon arrive. Some time that fall, I got a letter from him. A nice Johns Hopkins envelope, nice Johns Hopkins stationery, and a one word letter from Hartman – "Well?" I got busy and was done in the spring. Thesis defense completed, degree in the mail, I was a happy man.

By the fall of 1965 I was pretty sure how to prove the Closing Lemma for recurrent orbits in higher dimensions. The idea was the same as the three dimensional case, but the geometry was much harder. The Poincaré maps of the flow along the recurrent orbit are C^1 maps $\Pi_n : P_n \to P$ where P is a fixed, smooth transversal to the flow at the recurrent point p, P_n is a small neighborhood of p on P, and $\Pi_n(y)$ is the nth point in P along the orbit through $y \in P_n$. (Π_n is the composition of n first return Poincaré maps $\pi_1, \ldots, \pi_n$.) Each Π_n is a diffeomorphism from P_n to its image in P. The derivative of Π_n at p is an isomorphism

$$T_n : T_p P \to T_{\Pi_n(p)} P.$$

The T_n incorporate all the obstacles to making the proof work, but you can't say much about them in general. Or so I thought. See Sect. 46.9. So I tried to make the parallelogram thinking ("thin directions are good") apply in higher dimensions. When parallelotopes replace parallelograms there are a great many interior altitudes to keep track of, but I eventually got a dimensional induction argument to work.

Hartman had asked that I submit the paper to the American Journal of Math, of which he was the editor, and it appeared in 1967.

46.8 The General Density Theorem

While at Johns Hopkins, I'd hardly thought beyond recurrent orbits. They were complicated enough. It was Steve Smale who introduced me to the more natural nonwandering concept. The set of nonwandering orbits, Ω, includes the recurrent orbits and is compact. Smale thought of Ω as the home of nontrivial dynamics. He later came to appreciate chain recurrence, but at the time Ω ruled. It was not much harder to close nonwandering orbits than to close recurrent orbits, and it had a nice payoff, the General Density Theorem, which is an extension of the Kupka–Smale Theorem. It states that for the generic C^1 vector field X in the space $\mathcal{X}$ of all C^1 vector fields,

$$\Gamma(X) = \Omega(X)$$

where $\Gamma(X)$ is the closure of the set of closed orbits of X and $\Omega(X)$ is the set of nonwandering orbits. In a sense, for the generic X, closing is unnecessary – near every nonwandering orbit there already pass closed orbits.

Involved in the deduction of the General Density Theorem from the Closing Lemma was the concept of a semi-continuous set valued function. I thought of $X \in \mathcal{X}$ as a variable, and the set $\Gamma_n(X)$ of hyperbolic closed orbits of period $\leq n$ as a function of X. The value $\Gamma_n(X)$ lies in the space $\mathcal{K}$ of compact subsets of the manifold. Since hyperbolic orbits are structurally stable, $\Gamma_n(X)$ is fairly indestructible under small changes of X. (Orbits of period n can relax and have period $n + \epsilon$, but that's not a serious discontinuity of $\Gamma_n(X)$.) With a little work, it follows that the closure of the set of X-orbits, $\Gamma(X)$, is a lower semi-continuous function of X as X varies in the subset KS $\subset \mathcal{X}$ of Kupka–Smale vector fields. (All the closed orbits of a Kupka–Smale vector field are hyperbolic.) If Y C^1-approximates a Kupka–Smale vector field X, $\Gamma(Y)$ can be quite a bit larger than $\Gamma(X)$ (new periodic orbits of high period can suddenly appear), but the majority of the closed orbits of X persist

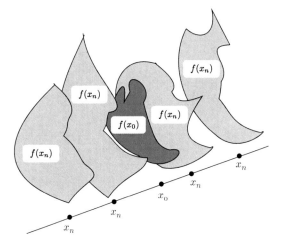

Fig. 46.13 Lower semi-continuity of a set valued function $f : x \mapsto f(x)$ at x_0. As $x_n \to x_0$ the sets $f(x_n)$ can be bigger than $f(x)$ but not much smaller in the sense that for each $\epsilon > 0$, the ϵ-neighborhood of $f(x_n)$ contains $f(x_0)$ for all large n.

for Y. They may budge a little, but not much. That's lower semi-continuity. See Fig. 46.13.

The beauty of semi-continuity is that a semi-continuous function is always continuous[6] at points in a residual subset of its domain. Since the space of C^1 vector fields $\mathscr{X}$ is a complete metric space and KS is a residual subset of $\mathscr{X}$, this is good news. The function $X \mapsto \Gamma(X)$ is continuous at the generic X. It follows that $\Gamma(X) = \Omega(X)$ generically. For if $\Gamma(X) \subsetneq \Omega(X)$, you fix a point $p \in \Omega(X) \setminus \Gamma(X)$ and apply the Closing Lemma at p. This gives a closed orbit γ that passes near p. A second perturbation makes γ hyperbolic, and a third perturbation makes γ a closed orbit of a Kupka–Smale vector field that C^1-approximates the original vector field. All together this gives a sequence of vector fields $X_n \to X$ such that p is in the limit of $\Gamma(X_n)$ as $n \to \infty$, which shows that Γ is discontinuous at X. Thus, continuity at X implies $\Gamma(X) = \Omega(X)$, so this is a property of the generic C^1 vector field.

A subtlety here is that the Closing Lemma must be applied to a C^1 vector field, not to a smoother vector field. For the C^2 vector fields are a meager subset of $\mathscr{X}$. They almost surely lie outside generic subsets of $\mathscr{X}$. See Sect. 46.10 on pseudo-flowboxes.

Anyway, the Closing Lemma for nonwandering orbits and the General Density Theorem got published as a separate paper in the American Journal of Math, also in 1967.

---

[6] A set valued function $f : x \mapsto f(x)$ is continuous at x_0 if $f(x_n) \to f(x_0)$ as $x_n \to x_0$ in the sense that for each $\epsilon > 0$, the ϵ-neighborhood of $f(x_n)$ contains $f(x_0)$ and the ϵ-neighborhood of $f(x_0)$ contains $f(x_n)$ for all large n. That is, f is both lower and upper semi-continuous at x_0.

736   C. Pugh

46.9 Anosov's Visit

In 1968 Dimitry Anosov visited Berkeley. We talked about many things including his work on geodesic flows. I understood almost nothing about that, but Anosov had some superb ideas about simplifying the geometry in the Closing Lemma. I had a weak background in linear algebra – to say the least – and was unaware of the Polar Factorization Theorem.

Given an isomorphism $T : \mathbb{R}^n \to \mathbb{R}^n$ there is a unique way to factor it as

$$T = OP$$

where O is orthogonal and P is positive definite symmetric. (This is like $z = re^{i\theta}$ where r is P and $e^{i\theta}$ is O.)

The use of this factorization is the following. I had a sequence of isomorphisms $T_n : T_p \Sigma \to T_{\Pi_n(p)} \Sigma$ (the derivatives of the Poincaré maps along the recurrent orbit) about which I knew nothing. They were pretty much arbitrary. But if you express them as $T_n = O_n P_n$ then P_n carries all the geometry. For O_n is an isometry. O_n changes nothing intrinsic to a parallelotope, only its placement in space. The sequence of positive definite symmetric matrices P_n is much less daunting than the sequence T_n. You have eigenvalues and eigenvectors. The eigenvectors are orthogonal, so P_n has a subsequence for which the normalized eigenbases converge to an orthogonal basis. To understand "thin directions" you just have to compare eigenvalues.

Anosov led me to understand my own proof better.

46.10 Pseudo-Flowboxes

It is dangerous to construct a perturbing vector field Δ in an X-flowbox if the construction uses the flowbox coordinates. For if X is C^r then the flowbox coordinates are C^r and Δ is only C^{r-1}. This apparently implies that for vector fields, all my careful perturbation work was worthless, at least to get the General Density Theorem.[7] When I realized this, I had many anguished days and nights. You can imagine that!

Lack of differentiability can actually happen. Specifically, in dimension two the C^∞ vector field $\Delta = \Delta(x, y)$ is transformed under a C^1 change of variables $G(x, y) = (u, v)$ to the vector field on the surface

---

[7] This is a technical issue only for vector fields. It is not present for diffeomorphisms. Perturbing a diffeomorphism in a C^1 coordinate system is fine. There's no loss of differentiability. The same is true for C^1 flows. Note that a C^1 vector field generates a C^1 flow, but a C^1 flow can be generated by a C^0 vector field!

46 The Closing Lemma in Retrospect

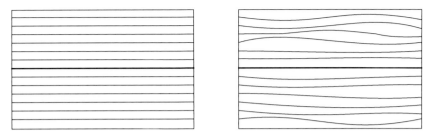

Fig. 46.14 The flow represented in a flowbox is rigid horizontal translation at all points of the flowbox, while in a pseudoflowbox the flow need only be rigid horizontal translation (to first order) at the points along the pseudo-flowbox's central orbit.

$$(G_*\Delta)(u,v) = (DG)_{G^{-1}(u,v)} \circ \Delta \circ G^{-1}(u,v).$$

(Here, (u, v) are smooth coordinates on the surface.) Thus, if $G(x, y) = (x, g(y)) = (u, v)$ and $g(y)$ is the integral of a positive, continuous, nowhere differentiable function then g is C^1, G is a C^1 diffeomorphism, and the vector field $\Delta = \epsilon\lambda\beta(x, y)(\partial/\partial y)$ becomes the nowhere differentiable vector field

$$(G_*\Delta)(u,v) = \begin{bmatrix} 1 & 0 \\ 0 & g'(g^{-1}(v)) \end{bmatrix} \begin{bmatrix} 0 \\ \epsilon\lambda\beta(u, g^{-1}(v)) \end{bmatrix}$$

$$= \begin{bmatrix} 0 \\ \epsilon\lambda\beta(u, g^{-1}(v))g'(g^{-1}(v)) \end{bmatrix}.$$

Fixing this problem required a kludge. The orbits of a C^1 vector field are actually C^2, so at least one of the flowbox coordinates is C^2. You want to get C^2 coordinates along a C^2 orbit γ, and you want the coordinates to have the same first derivative properties along γ that the flowbox coordinates have. The Whitney Extension Theorem is relevant here. It lets you do that.[8] See Fig. 46.14.

Upshot *The perturbation Δ constructed in the pseudo-flowbox is a C^1 vector field, and it has the same first order effect in the pseudo-flowbox that the old perturbation Δ had in the flowbox. And "thus" it is no loss of generality to calculate proportional lift in flowboxes, because it will all work out right in pseudo-flowboxes.*

---

[8] It's much like finding a tubular neighborhood of a curve. It you just exponentiate the normals to a C^r curve, the coordinate system loses one degree of differentiability. For the field of normals is once less differentiable than the curve. (When the curve is C^1 then exponentiation doesn't even give a local bijection. It's locally onto but not locally 1:1.) Instead, you exponentiate a C^r field of approximate normals, and you get a tubular neighborhood. But maybe you want the fibers of the tubular neighborhood to be normal to the curve, not just transverse to it, and you don't want to sacrifice a derivative. The Whitney Extension Theorem is set up to do just that.

46.11 Hamiltonianisms

My first PhD student was Clark Robinson. He got his degree in 1969. Clark generalized much of the qualitative genericity theory (such as the Kupka–Smale Theorem) to conservative dynamics – dynamical systems that preserve a smooth measure, a symplectic form, or have other special properties.[9] A natural candidate was a Hamiltonian Closing Lemma and General Density Theorem. After all, Poincaré himself speculated about the ubiquity and importance of periodic orbits – "La seule brèche ..." in the quote at the beginning of this article.

Given a C^2 real valued function H on a symplectic manifold, such as a cotangent bundle, then all the orbits of the corresponding Hamiltonian vector field X_H are nonwandering. Can you C^2 perturb H so the new Hamiltonian vector field has a dense set of closed orbits?

I worked out a Hamiltonian pseudo-flowbox proof of this and in the summer of 1966 I presented it at the International Congress of Mathematicians in Moscow. Later, Clark and I refined the proof, axiomatized the hypotheses, and published the C^1 Closing Lemma for a great many classes of conservative systems in the Journal of Ergodic Theory and Dynamical Systems in 1973. We also incorporated Anosov's improvements to the geometric linear algebra.

One class for which we had no luck was geodesic flows. Given a Riemann structure on a manifold there is a geodesic flow on its unit tangent bundle. The vector field generating the geodesic flow is a special kind of Hamiltonian vector field, so all the orbits are nonwandering. The orbits are the unit tangent fields along the geodesics, so closed orbits in the unit tangent bundle correspond to closed geodesics on the manifold. The question is: *For the generic Riemann structure, is the set of closed geodesics dense in the manifold?* The question remains unanswered.

46.12 Subsequent History

Since the 1970s a great deal of Closing Lemma related work has been done. There are several areas.

(a) Simpler proofs of the C^1 Closing Lemma.
(b) Attempted counterexamples to the general C^r Closing Lemma, $r \geq 2$.
(c) Attempted proofs to modifications of the C^r Closing Lemma, $r \geq 2$.
(d) C^1 Closing Lemmas for recurrence that is more general than nonwandering.
(e) Ergodic Closing Lemmas.

---

[9] Genericity of a property for vector fields in $\mathscr{X}$ does not automatically imply its genericity for vector fields in some $\mathscr{Y} \subset \mathscr{X}$. For if $\mathscr{Y}$ is a closed, nowhere dense subset of $\mathscr{X}$ then the generic member of $\mathscr{X}$ never lies in $\mathscr{Y}$. Thus, Clark's results were not at all straight forward to prove.

46 The Closing Lemma in Retrospect 739

Here is an extremely brief account of some of these areas.

(a) It was Jiehua Mai who got rid of the parallelogram/rectangle packing shown in Fig. 46.12. He essentially solved the linearization of the perturbation problem, imagining all the Poincaré maps are linear, and interpolated a sequence of overlapping ellipsoids between a good pair of points. This gives a great simplification of the Closing Lemma geometry, and it also led to extensions of the Closing Lemma to more general recurrence. Marie-Claude Arnaud, Christian Bonatti, and Sylvain Crovisier took this approach further.

(b) I think that Carlos Gutierrez has the best result in the second area. Carlos was a student of Jorge Sotomayor who was a student of Peixoto. In 1986 Carlos showed that while the C^1 Closing Lemma has a proof which is essentially local, the same is not true for the C^2 Closing Lemma. His example is a smooth flow on the 2-torus having a very degenerate fixed point q. The flow has a recurrent orbit through $p \neq q$, but every C^2 small perturbation whose support does not include q fails to close the recurrence.

Before Carlos made his counter-example to local C^2 closing I found a counter-example to a kind of double C^2 closing. You have two recurrent points and you want to close both at once. In both Carlos' and my examples Denjoy estimates are crucial.

Michael Herman worked with Hamiltonian vector fields and constructed an example of recurrence that cannot be C^∞ closed in the Hamiltonian category. His Hamiltonian flow lives on a smooth, compact symplectic manifold. There is a band B of compact energy hypersurfaces containing no closed orbits, and C^∞-small Hamiltonian perturbations never produce closed orbits in B. However, the symplectic structure is non-standard in that the symplectic form is not exact near B. In particular, the C^r Hamiltonian Closing Lemma is an open question on the cotangent bundle with its standard symplectic form.

(c) In 1965 I had a student, David Wallwork, who read Mauricio's structural stability paper assiduously, especially the Closing Lemma part. Mauricio asserts that a recurrent orbit of a flow on a surface can be closed or a new saddle connection can be produced by a twist perturbation along a transversal through a point p on the recurrent orbit. This is Mauricio's version of the Closing Lemma – "twist perturbations either produce a new closed orbit or a new saddle connection." (In fact, although Mauricio did not realize it at the time, twist perturbations for recurrent orbits always produce new saddle connections.) Its proof is essentially a monotonicity argument, and it is claimed to work for flows on all compact surfaces, orientable or not. Wallwork found what seemed to be a flaw in the argument for the non-orientable case, and after some initial disbelief on my part, he convinced me of his assertion. There is indeed a flaw. As Carlos Gutierrez and others have shown, not only does the argument have a gap, but twist perturbations sometimes do fail to produce closed orbits in the non-orientable case. Up to the present day, the C^r Closing Lemma and Mauricio's structural stability theorem remain unsolved problems for flows on the general non-orientable surface. It is right to name this as the *Peixoto-Wallwork problem*.

740 C. Pugh

On the positive side of things, using the C^1 Closing Lemma, I showed that the C^1 version of Mauricio's structural stability theorem does hold on non-orientable surfaces.

Further work on how well the twist perturbation on the torus succeeds in closing recurrence has been done by Catherine Carroll, Carlos Gutierrez, Benito Pires, and others.

The Closing Lemma arises in holomorphic dynamics too. It has been proved there by John Erik Fornaess and Nessim Sibony.

Finally there is the *Anosov Closing Lemma*. It is an important feature of Anosov's ground breaking work on ergodicity of geodesic flows, and it asserts that in the presence of the correct type of hyperbolicity, the closed orbits are dense in the non-wandering set. But in my opinion it is not a Closing Lemma. It is an "Already Closed Lemma." No perturbation of the flow is made to produce a closed orbit. The closed orbits are already present, so producing closed orbits is a question of perturbing the initial nonwandering orbit rather than perturbing the flow. Also, there is no issue about C^1 versus C^r.

(d) Christian Bonatti and Sylvain Crovisier have gone the furthest in closing recurrence more general than nonwandering. Their work is closely related to the *Connecting Lemma* which attempts (by a C^1-small perturbation) to weld together a pair of orbits such that the ω-limit of the first orbit meets the α-limit of the second. Building on work of Clark Robinson, Dennis Pixton, Jiehua Mai, Lan Wen, Jeff Xia, Shuhei Hayashi, Enrique Pujals, and many others, they show that if all the periodic orbits are hyperbolic and p is chain recurrent then there is a C^1-small perturbation that makes p periodic. As a result there is an improved General Density Theorem – generically the chain recurrent set equals the nonwandering set equals the closure of the periodic orbits.

(e) By perturbing a vector field X, the C^1 Closing Lemma produces a periodic orbit $\gamma(t)$ such that $\gamma(0)$ approximates a given nonwandering point p of X, but nothing is said about how well $\gamma(t)$ approximates the rest of the X-orbit through p. The *Ergodic Closing Lemma* of Ricardo Mañé for diffeomorphisms (proved for flows by Lan Wen) addresses this. A point p is said to be *strongly closable* if, given $\epsilon > 0$ and a C^1-neighborhood $\mathcal{U}$ of X in $\mathcal{X}$, there is a $Y \in \mathcal{U}$ and a closed orbit γ of Y such that

$$0 \le t \le \text{period}(\gamma) \quad \Rightarrow \quad d(\gamma(t), \varphi_t(p)) < \epsilon$$

where φ is the X-flow. For the generic X with respect to the C^1 topology, almost all points are strongly closable points in the following sense. There is a residual set of $X \in \mathcal{X}^1$ such that for every Borel probability measure μ that is invariant with respect to the X-flow, we have

$$\mu(\text{SC}(X) \cup \text{Sing}(X)) = 1$$

where $\text{SC}(X)$ is the set of strongly closable points of X and $\text{Sing}(X)$ is the set of singular points of X, i.e., its zeros.

46.13 Flute Music

Exactly why Chris Dench decided to a write piece of modern music for the flute and title it "Closing Lemma" I don't know. Nevertheless he did. Perhaps he had some kind of idea that a "Closing Lemma" in math is similar to a closing act or closing musical movement.

References

Any bibliography I compile will be far from complete or exhaustive. Googling is a better choice. Just enter "closing lemma". Perhaps the best printed source is the book *Dynamics Beyond Uniform Hyperbolicity* by Christian Bonatti, Lourenço Diaz, and Marcelo Viana. It has a bibliography with 466 references.

Chapter 47
From Peixoto's Theorem to Palis's Conjecture

Enrique R. Pujals

47.1 Survey

Roughly speaking, Peixoto's foundational works in the global theory of ordinary differential equations corresponds to the papers [17–19] which are nowadays referred to as Peixoto's Theorem.

In few words, his theorem is fundamental in putting the qualitative theory of flows on differentiable manifolds on a solid set-theoretical basis with well defined goals and problems exhibiting a certain unity. In few words his contribution here are: (a) the introduction of the space of all flows; (b) recasting the notion of structural stability; and maybe what is most important, (c) providing a paradigmatic picture of what it can be considered "*a nice and complete description of a dynamical system*".

Moreover, it could be said, that his result made sense of many aspects of the qualitative theory proposed by Poincaré and Birkhoff, and solved the ambiguities due to the lack of formal topological structure and no precise definition.

But Peixoto's theorem can not be reduced to clarifying the subject. The qualitative picture that follows from his theorem, can be understood as a hallmark of a proposed paradigm of what should be "a nice description of the global dynamic of a system". In particular, he showed that this nice description can be obtained for typical (in the Baire category's sense) flows acting on orientable surfaces. A useful way to recast Peixoto's theorem is the following:

Among all smooth flows on surfaces, (compact two-dimensional boundaryless manifolds), there is an open and dense set in the space of all flows endowed with the C^1 topology such that the set of non-wandering points consists only of finite hyperbolic periodic orbits and fixed points. Moreover, those systems are structurally stable.

The described achievement of Peixoto is paradigmatic of the view that "non pathological" systems behaves in a very simple form: the nonwandering set

---

E.R. Pujals
Instituto de Matemática Pura e Aplicada (IMPA), Estrada Dona Castorina, 110, 22460-320
Rio de Janeiro, RJ, Brazil
e-mail: enrique@impa.br

M.M. Peixoto et al. (eds.), *Dynamics, Games and Science I*, Springer Proceedings
in Mathematics 1, DOI 10.1007/978-3-642-11456-4_47,
© Springer-Verlag Berlin Heidelberg 2011

consisting of *finitely many periodic elements*. However, in the early sixties (by Anosov and Smale and following Birkhoff, Cartwright and Littlewood, etc) it was shown that "chaotic behavior" may exist within stable systems and this was the starting point of the *hyperbolic theory* and the modern nonconservative dynamical systems theory. A major result in this theory is the fact that for these systems (nowadays called *hyperbolics*), the nonwandering set can be decomposed *into finitely many compact, disjoint and transitive pieces*. Although this pieces could exhibit a chaotic behavior there are just finitely many of them and this recover the old vision by replacing finitely many periodic elements by these finitely many "nontrivial elementary" pieces. In this sense, the nicely described systems of Peixoto's theorem, gets a new form in the now called Spectral Decomposition Theorem proved by Smale.

However, it was soon realized that hyperbolic systems were not as universal as was initially thought: there were given examples of open sets of diffeomorphism were none of them are hyperbolic (see [25]). Nevertheless in all these new examples (nowadays called partially hyperbolic) the nonwandering set *still decompose into finitely many compact, disjoint and transitive pieces*. Moreover, this phenomena holds in a *robust* way: *Any perturbation of the initial system still has a only a finite number of transitive pieces*. In certain sense, Peixoto's picture is still recovered in this context.

In other words, the picture proposed by Peixoto still is presented if the requirement that the set of non-wandering points *consists only of a finite number of periodic orbits and fixed points* is replaced by requiring that the non-wandering set consists of *a finite number of isolated transitive pieces*. This kind of description, it is what we called *"generalized Peixoto's picture"*.

It was through the seminal work of Newhouse (see [14–16]) where a new phenomena was shown: *the existence of infinitely many periodic attractors (today called Newhouse's phenomena) for residual subsets in the space of C^r diffeomorphisms ($r \geq 2$) of compact surfaces*. These non-hyperbolic systems can not fitted in the type of nice descriptions inspired by Peixoto's result: they have *infinitely many isolated transitive attracting pieces*. The underlying *mechanism* here was the presence of a homoclinic bifurcation named *homoclinic tangency*: non-transversal intersection of the stable and unstable manifold of a periodic point. After the works of Newhouse, many other results were obtained in the direction of understanding the dynamics induced by unfolding homoclinic tangencies, especially in the case of one-parameter families. Many fundamental dynamical prototypes were found in the context of this bifurcation, namely the so called cascade of bifurcations, the Hénon-like strange attractor [5, 13] and infinitely many coexisting ones. Other well understood obstruction to hyperbolicity is the so-called *heterodimensional cycle* introduced in [9, 10] and used, for instances in [6], to construct examples of non-hyperbolic robust transitive systems.

These results naturally suggested the following question: *Is it possible to identify the dynamical mechanism underlying any generic nonhyperbolic behavior?*

In the early 80's Palis conjectured (see [20, 24]) that homoclinic bifurcations are very common in the complement of the hyperbolic systems: *Every C^r*

47 From Peixoto's Theorem to Palis's Conjecture

diffeomorphism of a compact manifold M can be C^r approximated by one which is hyperbolic or by one exhibiting a heterodimensional cycle or by one exhibiting a homoclinic tangency.

In other words, *Palis conjectured that avoiding homoclinic bifurcation, the generalized Peixoto's picture can be recovered.*

For the case of surfaces and the C^1 topology, the theorem A in [22] proves the above conjecture (see also [23]). When the manifold has dimension greater than two, the main result in this direction is the following one proved in [8]:

Any $f \in Diff^1(M)$ can be C^1-approximated by another diffeomorphism such that either

1. *It has a homoclinic tangency or*
2. *It has a heterodimensional cycle or*
3. *It is essentially hyperbolic*

Given $f \in Diff^1(M)$, it is said that f is *essentially hyperbolic if there exists a finite number of transitive hyperbolic attractors such that the union of their basins of attraction are open and dense.*

In certain sense, the above result, shows that a weak version of the generalized Peixoto's picture holds far from systems exhibiting homoclinic bifurcations.

So, what happens when we consider the space of all dynamics?

In this direction, Palis has proposed a probabilistic and subtle conjecture, that provides a global scenario for dissipative or, more precisely, non-conservative dynamics. Roughly speaking, it can be said that Palis's conjecture about the finiteness of attractors, states that generalized Peixoto's picture holds for a dense set of systems. In fact, the main focus of the conjecture, asserts that *there is dense set of systems in the C^r topology, having only finitely many transitive attractors, such that union of their basins of attraction have total Lebesgue probability.*

To be accurate, the conjecture is more broad and we refer to [21] for a precise statement. Actually, Palis conjecture can splitted in two parts: one part involving the typical behavior for a dense set of systems and the second part related to the type of dynamics that follows for perturbations of the initial system.

Palis's conjecture about finiteness of attractors has been fully proved for one-dimensional maps displaying only one critical point: by Lyubich [12] for quadratic families, Avila, de Melo and Lyubich [1] in the analytic case under the hypothesis of negative Schwarz derivative and, finally, by Avila and Moreira [2,3] for C^r families, $r \geq 2$. Such a perspective has been enriched by the recently published results in [11] showing the density of hyperbolicity for multimodal maps.

In any case Palis's approach certainly led for quite a while to a vacuum with respect to the possibility of formulating a global scenario for dynamics, providing key properties of a typical dynamical system. Moreover, recovers in a new framework the idealized picture that follows from Peixoto's theorem, proposing that *"a nice, and completely describable system"* are very common in the space of all dynamics.

References

1. Avila A., de Melo, W., Lyubich, M.: Regular or stochastic dynamics in real analytic families of unimodal maps. Invent. Math. **154**, 451–550 (2003)
2. Avila, A., Moreira, C.: Phase-parameter relation and sharp statistical properties for general families of unimodal maps. Contemp. Math. **389**, 1–42 (2005)
3. Avila, A., Moreira, C.: Statistical properties of unimodal maps: the quadratic family. Ann. Math. **161**, 831–881 (2005)
4. Anosov, D.V.: Geodesic flows on closed riemannian manifolds with negative curvature. Proc. Steklov Inst. of Math. (translated by the AMS) 90 (1967)
5. Benedicks, M., Carleson, L.: The dynamics of the Hénon map. Ann. Math. **133**, 73–169 (1991)
6. Bonatti, C., Diaz, L.J.: Persistence of transitive diffeomorphisms. Ann. Math. **143**, 367–396 (1995)
7. Colli, E.: Infinitely many coexisting strange attractors. Annales de l'Inst. Henri Poincaré, Analyse Nonlinéaire Annales de l'I. H. P., section C, tome 15, no 5, 539–579 (1998)
8. Crovisier, S., Pujals, E.R.: Essential hyperbolicity and homoclinic bifurcations: a dichotomy phenomenon/mechanism for diffeomorphisms. ArXiv:1011.3836
9. Diaz, L.J.: Robust nonhyperbolic dynamics at heterodimensional cycles. Ergod.Theory Dyn. Syst. **15**, 291–315 (1995)
10. Diaz, L.J.: Persistence of cycles and nonhyperbolic dynamics at heteroclinic bifurcations. Nonlinearity **8**, 693–715 (1995)
11. Kozlovski, O., Shen, W., van Strien, S.: Density of hyperbolicity in dimension one. Ann. Math. **166**, 145–182 (2007)
12. Lyubich, M.: Almost every real quadratic map is either regular or stochastic. Ann. Math. **156**, 1–78 (2002)
13. Mora, L., Viana, M.: Abundance of strange attractors. Acta Math. **171**, 1–71 (1993)
14. Newhouse, S.: Non-density of Axiom A(a) on S^2.Proc. A.M.S. Symp. Pure Math. **14**, 191–202 (1970)
15. Newhouse, S.: Diffeomorphism with infinitely many sinks. Topology **13**, 9–18 (1974)
16. Newhouse, S.: The abundance of wild hyperbolic sets and nonsmooth stable sets for diffeomorphisms. Publ. Math. I.H.E.S. **50**, 101–151 (1979)
17. Peixoto, M.: On structural stability. Ann. Math. **69**, 199–222 (1959)
18. Peixoto, M.: Structural stability in the plane with enlarged boundary conditions. Anais da Academia Brasileira de Ciencias **31**, 135–160 (1959). (in collaboration with M.C. Peixoto)
19. Peixoto, M.: Structural stability on two-dimensional manifolds. Topology **1** (1962) pp. 101–120.
20. Palis, J.: Homoclinic orbits, hyperbolic dynamics and dimension of Cantor sets. The Lefschetz centennial conference. Contemp. Math., 58, III, (1984) 203–216
21. Palis, J.: A global view of dynamics and a conjecture on the denseness of finitude of attractors. Gemetrie complexe et systemes dynamiques (Orsay, 1995). Asterisque **261**, 335–347 (2000)
22. Pujals, E.R., Sambarino, M.: Homoclinic tangencies and hyperbolicity for surface diffeomorphisms. Ann. Math. **151**, 961–1023 (2000)
23. Pujals, E.R., Sambarino, M.: On the dynamic of dominated splitting. Ann. Math. **169**, 675–740 (2009)
24. J.Palis and F.Takens *Hyperbolicity and sensitive-chaotic dynamics at homoclinic bifurcations* Cambridge University Press, 1993.
25. Shub, M.: Topologically transitive diffeomorphism of T^4. In: Symposium on Differential Equations and Dynamical Systems (University of Warwick, 1968/69), pp. 39–40. Lecture Notes in Math., vol. 206. Springer, Berlin (1971)

Chapter 48
Dynamics Associated to Games (Fictitious Play) with Chaotic Behavior

Colin Sparrow and Sebastian van Strien

Abstract In this survey we will discuss some recent results on a certain class of dynamical systems, called *fictitious play* which are associated to game theory. Here we simply aim to show that the dynamics one encounters in these systems is unusually rich and interesting. This paper does not require a background in game theory.

48.1 Introduction

Consider games with two players A and B which both can play, randomised, n strategies. So the *state space* of the players is described by two probability vectors

$$p^A \in \Sigma_A \text{ and } p^B \in \Sigma_B$$

where Σ_A and Σ_B are the space of probability vectors in $\mathbb{R}^n$. By convention, p^A is a row vector and p^B a column vector. We assume that player A has *utility* (i.e. *payoff*) $p^A A p^B$ and player B has payoff $p^A B p^B$ where A and B are $n \times n$ matrices.

At a given moment in time, player A can best improve her utility by choosing the unit vector $\mathscr{B}R_A(p^B)$ which corresponds to the largest component of Ap^B. This is the *best response* of player A to position p^B. Formally,

$$\mathscr{B}R_A(p^B) := argmax_{p^A} p^A A p^B. \tag{48.1}$$

Of course, $\mathscr{B}R_A(p^B)$ can be a whole collection of vectors. However, when A is a non-degenerate matrix, this happens only for p^B in certain hyperplanes. (More precisely, in one of the $n \times (n-1)/2$ hyperplanes in Σ_B corresponding to $p^B \in \Sigma_B$ where Ap^B has two or more equal components.) Outside these hyperplanes,

---

C. Sparrow (✉) and S. van Strien
Maths Department, University of Warwick, Coventry CV4 7AL, UK
e-mail: C.Sparrow@warwick.ac.uk, S.J.van-Strien@warwick.ac.uk

M.M. Peixoto et al. (eds.), *Dynamics, Games and Science I*, Springer Proceedings in Mathematics 1, DOI 10.1007/978-3-642-11456-4_48,
© Springer-Verlag Berlin Heidelberg 2011

$\mathscr{B}R_A(p^B)$ is a unit basis vector. Similarly denote the best response of player B by $\mathscr{B}R_B(p^A)$.

A *Nash equilibrium* is a choice of strategies from which no unilateral deviation by an individual player is profitable for that player. That is, (p_*^A, p_*^B) is a Nash equilibrium if

$$p_*^A \in \mathscr{B}R_A(p_*^B) \text{ and } p_*^B \in \mathscr{B}R_B(p_*^A).$$

The Nash equilibrium is *unique* if A and B are invertible and if, moreover, there exists a purely mixed Nash equilibrium (E^A, E^B) (see, for example, [17, Theorem 1.5]). In this case $E^B \in \Sigma_B$ is the vector so that all components of AE^B are the same (so player A is indifferent to all different strategies). The vector E^A can be found similarly.

In the 1950s Brown [4] proposed *fictitious play* as a way in which players are able to naturally find the Nash equilibrium by flowing according to the following differential equation:

$$
\begin{aligned}
dp^A/dt &= \mathscr{B}R_A(p^B) - p^A \\
dp^B/dt &= \mathscr{B}R_B(p^A) - p^B
\end{aligned}
\tag{48.2}
$$

where $\mathscr{B}R_A(p^B) \in \Sigma_A$ is the best response of player A to player's B's position, and similarly for $\mathscr{B}R_B(p^A) \in \Sigma_B$. So each player's tendency is to adjust his or her strategy in a straight line from his/her (current) strategy towards their (current) best response.

There is an interpretation of this game as a mechanism by which the players *learn* from the other player's previous actions and then one often writes

$$
\begin{aligned}
dp^A/ds &= \left(\mathscr{B}R_A(p^B) - p^A\right)/s \\
dp^B/ds &= \left(\mathscr{B}R_B(p^A) - p^B\right)/s,
\end{aligned}
\tag{48.3}
$$

see for example the monograph [6]. The dynamics of this system and the previous are the same up to time-parametrisation $s = e^t$. Since $\mathscr{B}R_A$ and $\mathscr{B}R_B$ are not necessarily single-valued, (48.2) and (48.3) are really differential inclusions, rather than differential equations. Since the right hand side of (48.3) is upper-semi continuous see [1], these differential inclusions have solutions. In actual fact, as we shall see, in many examples the solutions are still unique.

This survey will describe some results on the dynamics of these games and pose some conjectures and open questions.

48.2 A Short Introduction into Game Theory and Some Simple 2 × 2 Examples

Let us discuss first the simplest (and essentially trivial) case, where both players have only two strategies to choose from, i.e. when Σ_A and Σ_B both correspond to the one-dimensional simplex $\{(p_1, p_2) \in \mathbb{R}^2; p_i \geq 0, p_1 + p_2 = 1\}$. Often instead of writing $A = \begin{pmatrix} a_1 & a_2 \\ a_3 & a_4 \end{pmatrix}$ and $B = \begin{pmatrix} b_1 & b_2 \\ b_3 & b_4 \end{pmatrix}$, these two matrices are denoted using the following notation $\begin{pmatrix} (a_1,b_1) & (a_2,b_2) \\ (a_3,b_3) & (a_4,b_4) \end{pmatrix}$. Equivalently, these matrices are encoded in the following way:

$$\begin{pmatrix} \text{Payoff's} & \begin{array}{c|cc} & \text{Player B} & \text{Player B} \\ & \text{chooses left} & \text{chooses right} \\ \hline \text{Player A chooses top} & (a_1,b_1) & (a_2,b_2) \\ \text{Player A chooses bottom} & (a_3,b_3) & (a_4,b_4) \end{array} \end{pmatrix},$$

where the 2nd part of each entry corresponds to the payoff to player B (the column player). As mentioned, $\Sigma_A \times \Sigma_B$ can be thought of as $[0, 1] \times [0, 1]$ and because of this notation it is traditional (and convenient) to identify the vertical side with the position of player A (the player with the row vector p^A) and the top left corner of $[0, 1] \times [0, 1]$ with $(1\ 0)$, $\begin{pmatrix} 1 \\ 0 \end{pmatrix} \in \Sigma_A \times \Sigma_B$. When we use this identification, payoffs at each of the corners of the square is the corresponding entry of the matrix. (We note that p^A is a row vector, even though p^A represents the position of player A and is displayed on the vertical side of the square.)

Of course, the evolution described by fictitious play, i.e. the differential inclusion (defined on the unit square, see Fig 48.1)

$$dp^A/dt = \mathscr{B}R_A(p^B) - p^A$$
$$dp^B/dt = \mathscr{B}R_B(p^A) - p^B$$
(48.4)

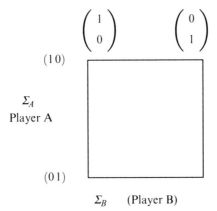

Fig. 48.1 The identification for 2 × 2 games: the horizontal side corresponds to player B and the top left corner corresponds to the first unit base vector for both players (i.e. the first strategy)

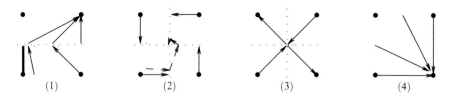

Fig. 48.2 The possible motions in 2 × 2 games (up to relabeling, and shifting the indifference lines (drawn in *dotted lines*))

is completely determined by the (multivalued) functions $\mathscr{B}R_A(p^B)$ and $\mathscr{B}R_B(p^A)$. Note that $(\mathscr{B}R_A(p^B), \mathscr{B}R_B(p^A))$ corresponds to one of the corners of $[0, 1] \times [0, 1]$, except where it is multivalued. So where it is not multivalued, the motion of (48.4) is towards one of the corners. There is a vertical line which determines where $\mathscr{B}R_B(p^A)$ changes, i.e. where the motion (48.4) switches direction (moving towards one of the top corners on one side and to one of the bottom corners on the other side), and a horizontal line where player A switches direction (moving towards one of the left corners on one side and to one of the right corners on the other side). In Fig. 48.2 we have drawn a few cases. Case (1) corresponds to a situation where the player A always prefers to move up (except on the segment marked in the figure on the left side, where she is indifferent). There are many matrices which would correspond to this situation, for example when $\begin{pmatrix} (0, -1) & (0, 0) \\ (0, 0) & (-1, -1) \end{pmatrix}$. Case (2) corresponds to $\begin{pmatrix} (-1, 1) & (0, 0) \\ (0, 0) & (-1, 1) \end{pmatrix}$. Here both players have opposite interests (the sum of the payoff's is always zero). Player B is copying A's behavior (because the largest component of $p^A B$ is then equal to the largest component of p^A), whereas player A is doing the opposite to what player B is doing. Finally, Case (3) corresponds to $\begin{pmatrix} (1, 1) & (0, 0) \\ (0, 0) & (1, 1) \end{pmatrix}$ where both players agree to choose the same strategy. In the prisoner dilemma $\begin{pmatrix} (3, 3) & (0, 5) \\ (5, 0) & (1, 1) \end{pmatrix}$ the players always move to the bottom right corner, see Case (4) and for both players the best response is always the 2nd strategy ($\mathscr{B}R_A, \mathscr{B}R_B$ are both constant in this case), even though they both would receive higher payoffs playing the first strategy.

In fact, it is easy to see that the dynamics in any 2 × 2 game is topologically of one of these types. Therefore, from this point of view, the next interesting case is that of a 3 × 3 game. (The dynamics of 2 × 3 games can be essentially reduced to that of a 2 × 2 game, with a normal direction added, see [17, Theorem 1.5].) In a later section we will review some results on 3 × 3 games, and see that these are much more complicated than 2 × 2 games.

48 Dynamics Associated to Games (Fictitious Play) with Chaotic Behavior 751

48.3 Convergence to Nash Equilibria in the Zero-Sum Case

If $B + A = 0$ then we have a so-called *zero-sum* game. It was shown in the 1950s by Robinson [13] that then the differential inclusion (48.4) converges (albeit slowly) to the set of Nash equilibria. This situation corresponds to Case (2) in Fig. 48.2.

Of course, matrices A, B for which $A + B \neq 0$, could have the same best responses $\mathscr{B}R_A$ and $\mathscr{B}R_B$ as matrices $\tilde{A}, \tilde{B}$ for which $\tilde{A} + \tilde{B} = 0$. For example, $(A, B) = \begin{pmatrix} (2,-1) & (1,0) \\ (1,0) & (2,-1) \end{pmatrix}$ and $(\tilde{A}, \tilde{B}) = \begin{pmatrix} (1,-1) & (0,0) \\ (0,0) & (1,-1) \end{pmatrix}$ have the same best responses. (Indeed, since p^B is a probability vector, $Ap^B = p^B + \begin{pmatrix} 1 \\ 1 \end{pmatrix} = \tilde{A}p^B + \begin{pmatrix} 1 \\ 1 \end{pmatrix}$ and hence player A has for both games the same best-response; for player B the same holds because his matrix is the same for both games.) Because of this, one calls two matrices A, B zero-sum if and only if they induce the same best reponses as two matrices $\tilde{A}, \tilde{B}$ for which $\tilde{A} + \tilde{B} = 0$.

In the zero-sum case, it is easy to see that the motion (48.4) converges. Indeed, take

$$H(p^A, p^B) = \mathscr{B}R_A(p^B) A p^B - p^A A \mathscr{B}R_B(p^A).$$

Note that $\mathscr{B}R_A(p^B) A p^B \geq p^A A p^B \geq p^A A \mathscr{B}R_B(p^A)$. That is, $H \geq 0$ and $H(p^A, p^B) = 0$ iff (p^A, p^B) is a Nash equilibrium. Since $\mathscr{B}R_A$ and $\mathscr{B}R_B$ are piecewise constant, (48.2) implies

$$\frac{dH}{dt} = \mathscr{B}R_A(p^B) A \frac{dp^B}{dt} - \frac{dp^A}{dt} A \mathscr{B}R_B(p^A)$$

$$= \mathscr{B}R_A(p^B) A (\mathscr{B}R_B(p^A) - p^B) - (\mathscr{B}R_A(p^B) - p^A) A \mathscr{B}R_B(p^A) = -H.$$

It follows that solutions go to the zero-set of H, i.e. to the set of Nash equilibria.

There are other examples for which it is shown that the game converges (for example in $2 \times n$ games see [2], and games with some other special properties, see for example [3,7,9,11,12]). However, for all those other cases the Nash equilibrium is on the boundary of the state space $\Sigma_A \times \Sigma_B$, and usually in those cases the Nash equilibrium is not unique and the flow does not have unique attractor. Therefore, following [8], we would like to pose the following:

Conjecture 48.1. Assume that all solutions of (48.2) converge to a unique equilibrium. Then (48.2) is associated to a zero-sum game.

We would like to mention here that we have shown in [16] that for any zero sum game (with some non-degeneracy conditions), the motion (48.4) can be viewed as the product of the *Hamiltonian motion* $\dfrac{d\bar{p}}{dt} = \dfrac{\partial H}{\partial \bar{q}}, \dfrac{d\bar{q}}{dt} = -\dfrac{\partial H}{\partial \bar{p}}$ on $\Sigma_A \times \Sigma_B$ associated to the (Hamilton) function H and a motion towards the Nash equilibrium.

752 C. Sparrow and S. van Strien

More precisely, because of the non-degeneracy conditions, the game has a unique Nash equilibrium $E = (E^A, E^B) \in \Sigma := \Sigma_A \times \Sigma_B$ and $H^{-1}(1)$ is the boundary of a ball around E. Moreover, there exists a continuous map $\pi: \Sigma \backslash \{E\} \to H^{-1}(1) \cap \Sigma$ so that $\pi(p^A, p^B) = \lambda(p^A, p^B) \cdot (E^A, E^B) + (1 - \lambda(p^A, p^B)) \cdot (p^A, p^B)$ for some scalar $\lambda(x) > 0$ and $\pi(x) \in H^{-1}(1)$. (Take $\lambda(x) = 1 - 1/H(p^A, p^B)$.) So

$$(p^A, p^B) \mapsto (\pi(p^A, p^B), \lambda(p^A, p^B)) \in (H^{-1}(1) \cap \Sigma) \times \mathbb{R}^+$$

can be viewed as (higher dimensional) spherical coordinates around E. The dynamics (48.3) in these spherical coordinates $(\bar{p}^A, \bar{p}^B) = \pi(p^A, p^B), \lambda = \lambda(p^A, p^B)$ becomes

$$\frac{d\bar{p}}{dt} = \frac{\partial H}{\partial \bar{q}}, \frac{d\bar{q}}{dt} = -\frac{\partial H}{\partial \bar{p}} \quad on \quad \Sigma_A \times \Sigma_B,$$

$$\frac{d\lambda}{dt} = -\lambda.$$

(48.5)

Of course, the Hamiltonian is not smooth. It is continuous and piecewise affine, and locally the flow is just a translation flow. However, as is shown in [16], the associated Hamiltonian flow is *unique and continuous*.

48.4 A Family of (Not Necessarily Non-Zero) Sum Games Containing Shapley's Example Displaying a Periodic Orbit

In the case of non-zero sum games, one certainly does not always convergence. Indeed, there is a famous example due to Shapley [14] from the 1960s which shows that in general the evolution does NOT converge to a Nash equilibrium of the game. The Shapley example exhibits periodic behavior.

Indeed, take the family of 3×3 games

$$A_\beta = \begin{pmatrix} 1 & 0 & \beta \\ \beta & 1 & 0 \\ 0 & \beta & 1 \end{pmatrix} \quad B_\beta = \begin{pmatrix} -\beta & 1 & 0 \\ 0 & -\beta & 1 \\ 1 & 0 & -\beta \end{pmatrix}, \quad (48.6)$$

which depend on a parameter $\beta \in \mathbb{R}$.

This family of examples was chosen in [15] because it contains Shapley's example (when $\beta = 0$) for which he had shown the existence of a periodic attractor. For $\beta = \sigma$, where $\sigma := (\sqrt{5} - 1)/2 \approx 0.618$ is the golden mean, the game is equivalent to a zero-sume game (rescaling B to $\tilde{B} = \sigma(B - 1)$ gives $A + \tilde{B} = 0$), so then play always converges to the interior equilibrium E^A, E^B as we have seen in the previous section. So varying $\beta \in [0, 1)$ should reveal how this periodic orbit disappears. In fact, it reveals a lot more interesting behavior!

48 Dynamics Associated to Games (Fictitious Play) with Chaotic Behavior

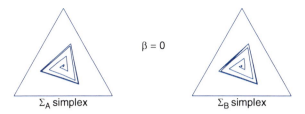

Fig. 48.3 Shapley's periodic orbit

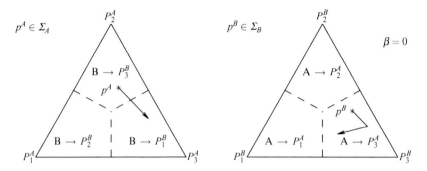

Fig. 48.4 The preferences of the players when $\beta = 0$

In the case of 3×3 games, Σ_A, Σ_B are both the set of probability vectors in $\mathbb{R}^3$ and so they are both a triangular simplex. So the state space is the product of two such triangular simplices (i.e. topologically a ball in $\mathbb{R}^4$). Shapley's periodic orbit is drawn Fig. 48.3. The periodic orbit, which lives in $\Sigma_A \times \Sigma_B$, spirals in a clockwise fashion when projected on each of the triangles Σ_A and Σ_B (where the corners are labelled as in Fig. 48.4).

For the family of games (48.6), when $\beta = 0$, the sets where the players are indifferent to two or more strategies are marked in dotted lines in Fig. 48.4. In [15] it was proven that for $\beta \in (0, \sigma)$ where $\sigma = (\sqrt{5} - 1)/2$ there still exists a periodic attractor.

Throughout the remaing part of this survey we will only consider games coming from this family (48.6).

48.5 Analysis of Stationary Points of Flow: Stable Sets of the Stationary Point are Extremely Complicated

In this section we will give a more detailed description of some results about the dynamics associated to the above family of games and show that the situation is rather different than that of smooth dynamical systems.

754   C. Sparrow and S. van Strien

Let $E^A = (E^B)^T = (1/3, 1/3, 1/3)$ and $E = (E^A, E^B)$. At this point, the players are indifferent between all three strategies (i.e. $\mathscr{B}R_A(E^B) = \Sigma_A$ and $\mathscr{B}R_B(E^A) = \Sigma_B$). So the right hand side of (48.2) includes the zero vectors, and E can be thought of as a stationary point of (48.2). (In fact, E is the only point in $\Sigma_A \times \Sigma_B$ where the right hand side of (48.2) can be zero.)

It seems reasonable to expect that we should be able to determine the local behavior near the Nash equilibrium E. It turns out that this question is rather subtle.

Theorem 48.1 (*Continuity of flow*).

- *For $\beta \in (0, 1)$, the differential inclusion (48.2) has a unique continuous flow outside E.*
- *For $\beta \in (-1, 0]$, the differential inclusion (48.2) has a flow which is not continuous (in many places).*

This theorem is proved in [17] and, as it turns out, when $\beta \neq \sigma$, orbits which start in E can choose to remain there or can leave this set. So E is not genuinely a stationary point.

Theorem 48.2 (*Stable manifold of equilibrium is extremely complicated*). *Consider $\beta \in (0, \sigma)$ and let ϕ_t be the flow of the fictitious play (48.2). Then the stable manifold*

$$W^s(E) = \{x \in E; \phi_t(x) \to E \text{ as } t \to \infty\}$$

of E is extremely complicated:

- *There exists a countable infinite number of polygons in $\Sigma \setminus E$ so that the cone with apex E over all these polygons is contained in the stable manifold of E.*
- *There exists an attracting periodic orbit γ in $\Sigma \setminus E$ (the continuation of Shapley's periodic orbit) and which also attracts points arbitrarily close to E.*

So the (local) stable manifold of the Nash equilibrium E is definitely not a full neighbourhood of E, but does contain a countable union of codimension-one sets. We believe that the stable manifold of E is equal to this set:

Conjecture 48.2. The stable manifold of E is a union of codimension-one sets (cones over certain polygons with apex E).

Question 48.1. Is the (local) stable manifold of E a closed set?

Usually, in smooth dynamical systems a stable manifold of a singular point is a manifold. Here the situation is rather more complicated, regardless of whether the above conjecture is true or not.

The stable manifold of the attracting periodic orbit γ contains a neighbourhood of γ and points arbitrarily close to E, but definitely not a countable union of cones with apex E. So we would like to ask the following:

Question 48.2. Determine the global topology of the stable manifold of γ.

48 Dynamics Associated to Games (Fictitious Play) with Chaotic Behavior 755

48.6 Bifurcations of Periodic Orbits of this Family

Let us concentrate on some simple periodic orbits of the fictitious play (48.3) associated to the matrices (48.6). One way of describing an orbit is by a symbolic sequence, indicating the sequence of corners of $\Sigma_A \times \Sigma_B$ the solution is successively heading for. Indeed, note that the best response of A to any $p^B \neq E^B$ is either an integer $i \in \{1, 2, 3\}$ or a mixed strategy set $\bar{i}$ where $\bar{i} := \{1, 2, 3\} \setminus \{i\}$ corresponding to where player A is indifferent between two strategies but will *not* play i. Similarly for B. Hence one can associate to any orbit $(p^A(t), p^B(t))$ outside E, a sequence of times $t_0 := 0 < t_1 < t_2 < \dots$ and a sequence of best-response strategies $(i_0, j_0), (i_1, j_i), (i_2, j_2), \dots$ where

$$(i_n, j_n) = (BR^A(p^B(t)), BR^B(p^A(t)))) \text{ for } t \in (t_n, t_{n+1})$$

with i_n and j_n equal to 1, 2, 3, $\bar{1}$, $\bar{2}$ or $\bar{3}$ for each $n = 0, 1, 2, \dots$ and so that

$$(i_n, j_n) \neq (i_{n+1}, j_{n+1}) \text{ for all } n \geq 0$$

(i.e. the players really do switch strategy at time t_n).

So $(\bar{1}, \bar{1}), (\bar{1}, \bar{2}), (\bar{2}, \bar{2}), (\bar{2}, \bar{3}), (\bar{3}, \bar{3}), (\bar{3}, \bar{1})$ means that during the first leg of the orbit both players initially do *not* play strategy 1, and so the leg of the orbit lies in the set where the players are indifferent to strategy 2 and 3. During the 2nd leg of the orbit player A is still indifferent between 2 and 3, and player B between 1 and 3. Such an orbit lies on the dashed lines indicated in Fig. 48.4 for the case when $\beta = 0$. For $\beta \in (0, 1)$ the corresponding dashed lines will be tilted clockwise.

Shapley's orbit is of the following type $(1, 2), (2, 2), (2, 3), (3, 3), (3, 1), (1, 1)$. So this periodic orbit, heads successively in six directions, and indeed is a hexagon. In the theorem below we describe the periodic orbits which form hexagons.

Theorem 48.3 (*The existence and stability of simple periodic orbits*). *There exists $\tau \in (\sigma, 1)$ (here $\tau \approx 0.915$ is a root of some polynomial of degree 6) with the following property.*

- *For $\beta \in (0, \sigma)$ the clockwise periodic Shapley orbit, which has symbolic sequence* $(1, 2)$, $(2, 2)$, $(2, 3)$, $(3, 3)$, $(3, 1)$, $(1, 1)$, *exists and is (locally) attracting.*
- *For $\beta \in (\sigma, 1)$ there exists another periodic orbit. We call this the anti-Shapley orbit, because it goes anticlockwise around the triangles and has symbolic sequence* $(1, 3), (1, 2), (3, 2), (3, 1)$, $(2, 1), (2, 3)$. *This orbit is of saddle-type when $\beta \in (\sigma, \tau)$ and attracting when $\beta \in (\tau, 1)$.*
- *For $\beta \in (\sigma, 1)$ there exists a third periodic orbit, called Γ, where both players choose mixed strategies* $(\bar{1}, \bar{1}), (\bar{1}, \bar{2}), (\bar{2}, \bar{2}), (\bar{2}, \bar{3}), (\bar{3}, \bar{3}), (\bar{3}, \bar{1})$.
- *This sequence of strategies corresponds to a fully-invariant set $C(\Gamma)$ (so an orbit starting in this set remains in this set, and an orbit starting outside this*

756 C. Sparrow and S. van Strien

set remains outside this set); this fully invariant set exists for each $\beta \in (0, 1)$ and contains a periodic orbit when $\beta \in (\sigma, 1)$.

There are no other periodic orbits with a symbolic sequence of length at most six.

We would like to state the following:

Conjecture 48.3. There are no periodic orbits other than the Shapley orbit when $\beta \in (0, \sigma)$.

Conjecture 48.4. There are no attracting periodic orbits when $\beta \in (\sigma, \tau)$.

The bifurcation which occurs when $\beta = \sigma$ is somewhat reminiscent of a Hopf bifurcation, except that one has immediately complicated dynamics right after the bifurcation.

Theorem 48.4 (*The bifurcation at $\beta = \sigma$*). *At the bifurcation $\beta = \sigma$ the following happens:*

- *As $\beta \uparrow \sigma$, the Shapley orbit shrinks to E;*
- *When $\beta = \sigma$ the Nash equilibrium E is a global attractor;*
- *When $\beta > \sigma$ there exist infinitely many periodic orbits.*
- *When $\beta \downarrow \sigma$ all periodic orbits, including the anti-Shapley orbit and Γ shrink to E.*

At $\beta = \tau$ the anti-Shapley periodic orbit undergoes a non-generic periodic doubling bifurcation: at $\beta = \tau$ there exists a whole continuum of periodic orbits.

48.7 Random Walk Behavior

The dynamics is indeed much more complicated than one normally encounters.

Theorem 48.5 (*The Hamiltonian flow acts like a 'random walk':*). *There exists a periodic orbit Γ (described in Theorem 48.3) with the following property: If one takes the first return map F to a section Z transversal to Γ (through some point $x \in \Gamma$), then for each $k \in \mathbb{N}$*

- *There exists a sequence of periodic points $x_n \in Z$ of exactly period k of the first return to Z accumulating to x*
- *The first return map F to Z has infinite topological entropy*
- *The dynamics acts as a random-walk. More precisely, there exist annuli A_n in Z (around $\Gamma \cap Z$ so that $\cup A_n \cup \{x\}$ is a neighbourhood of x in Z) shrinking geometrically to $\Gamma \cap Z$, so that for each sequence $n(i) \geq 0$ with $|n(i + 1) - n(i)| \leq 1$ there exists a point $z \in Z$ so that $F^i(z) \in A_{n(i)}$ for all $i \geq 0$.*

48 Dynamics Associated to Games (Fictitious Play) with Chaotic Behavior   757

One obvious consequence of the random walking described in the theorem, is the following unusual behavior. Take $\epsilon > 0$ small and define the local and unstable stable set corresponding to rate τ as

$$W_\epsilon^{s,\tau}(\Gamma) := \{x; \text{dist}(\phi_t(x), \Gamma) \le \epsilon \text{ for all } t \ge 0 \text{ and}$$

$$\lim_{t \to \infty} \frac{1}{|t|} \log(\text{dist}(\phi_t(x), \Gamma)) \to \tau\}$$

$$W_\epsilon^{u,\tau}(\Gamma) := \{x; \text{dist}(\phi_t(x), \Gamma) \le \epsilon \text{ for all } t \le 0 \text{ and}$$

$$\lim_{t \to -\infty} \frac{1}{|t|} \log(\text{dist}(\phi_t(x), \Gamma)) \to \tau\}.$$

Then the above system has for each $\epsilon > 0$ and *each* $\tau \ge 0$ close enough to zero, that both $W_\epsilon^{s,\tau}$ and $W_\epsilon^{s,\tau}$ are non-empty in any neighbourhood of Γ.

The reason why one has such strange dynamics is that the first return map $P: Z \to Z$ near to Γ has a very special form. If we identify Z with $\mathbb{R}^2$ and $\Gamma \cap Z$ with $0 \in \mathbb{R}^2$ (by projecting using the projection π introduced in Sect. 48.3), then P is essentially a composition of maps of the form

$$P(x) = A \circ R_{1/\|x\|}(x).$$

Here $\|(x_1, x_2)\| = |x_1| + |x_2|$ is the l_1 norm on $\mathbb{R}^2$, R_t is a rotation through angle t leaving the 'circles' in the l_1 norm invariant (i.e. $\|R_t(x)\| = \|x\|$) and A is a matrix of the form $A = \begin{pmatrix} \lambda_1 & 0 \\ 0 & \lambda_2 \end{pmatrix}$ with $0 < \lambda_1 < 1 < \lambda_2$.

48.8 Robustness

The above results do not require the matrices A and B to be of a special form, and hold for games corresponding to an open set of matrices:

Theorem 48.6 (*Robustness*). *For each* $\beta \in (0, 1)$ *with* $\beta \ne \sigma$, *there exists* $\epsilon > 0$ *so that for all* 3×3 *matrices* A *and* B *with*

$$\|A - A_\beta\|, \|B - B_\beta\| < \epsilon$$

the previous theorems also hold.

48.9 Connection with Other Results on Non-Smooth Dynamical Systems

The transition maps of (48.3) between hyperplanes are piecewise projective maps. In fact, as we show in [16], taking the appropriate induced flow, we get that the transition map is piecewise a translation. This connects this paper with an exciting body of work on piecewise isometries (with papers by R. Adler, P. Ashwin, M. Boshernitzan, A. Goetz, B. Kichens, T. Nowicki, A. Quas, C. Tresser and many others). Most of these paper deal with piecewise continuous maps, while the maps we encounter are continuous. Another loose connection of our work is to that of the huge and very active field of translation flows (associated to interval exchange transformations, translation surfaces and Teichmüller flows) (with recent papers by A. Avila, Y. Cheung, A. Eskin, G. Forni, P. Hubert, H. Masur, C. McMulen, M. Viana, J-C. Yoccoz, A. Zorich and many many others). But of course our flow does not act on a surface with a hyperbolic metric, and so this connection seems also rather remote. Finally, there is a growing literature on bifurcations on nonsmooth dynamical systems, mainly motivated by mechanical systems with 'dry friction', 'sliding', 'impact' and so on. As the number of workers in this field is enormous, we just refer to the recent survey of M. di Binardo et al. [5] and the monograph by M. Kunze [10]. Of course our paper is very much related to this work, although the motivation and the result seem to be of a different nature from what can be found in those papers.

48.10 Conclusion

We have seen that there is a lot of complicated behavior associated to fictitious play. Of course an economist can say: people behave rationally and if they do not converge then they will notice this. So periodic behaviour and chaos is a mathematical curiosity. Perhaps this is not so clear though. But this reminds us how chemists and other scientists have changed their approach. They used to be only interested in stationary processes. But now they realize that non-stationary processes are often more efficient, and certainly that they occur in a wide-range of important situations. Moreover, perhaps it is to be expected that learning behaviour does not converge to equilibria except in somewhat exceptional cases, such as zero-sum games?

References

1. Jean-Pierre Aubin and Arrigo Cellina. Differential inclusions, volume 264 of Grundlehren der Mathematischen Wissenschaften [Fundamental Principles of Mathematical Sciences]. Springer, Berlin, (1984). Set-valued maps and viability theory.
2. Ulrich Berger. Fictitious play in 2×n games. J. Econom. Theory **120**(2), 139–154 (2005)
3. Berger, U.: Two more classes of games with the continuous-time fictitious play property. Games Econom. Behav. **60**(2), 247–261 (2007)
4. Brown, G.W.: Iterative solution of games by fictitious play. In Activity Analysis of Production and Allocation, Cowles Commission Monograph No. 13, pp. 374–376. Wiley, New York, N.Y. (1951)
5. di Bernardo, M., Budd, C.J., Champneys, A.R., Kowalczyk, P., Nordmark, A.B., Tost, G.O., Piiroinen, P.T.: Bifurcations in nonsmooth dynamical systems. SIAM Rev. **50**(4), 629–701 (2008)

48 Dynamics Associated to Games (Fictitious Play) with Chaotic Behavior 759

6. Fudenberg, D., Levine, D.K.: The theory of learning in games, volume 2 of MIT Press Series on Economic Learning and Social Evolution. MIT , Cambridge, MA (1998)
7. Hahn, S.: The convergence of fictitious play in games with strategic complementarities. Econom. Lett. **99**(2), 304–306 (2008)
8. Hofbauer, J.: Stability for the Best Response Dynamics. Preprint, University of Vienna (1995)
9. Krishna, V.: Learning in Games with Strategic Complementarities. Preprint, Harvard University (1992)
10. Kunze, M.: Non-Smooth Dynamical Systems, volume 1744 of Lecture Notes in Mathematics. Springer, Berlin (2000)
11. Milgrom, P., Roberts, J.: Adaptive and sophisticated learning in normal form games. Games Econom. Behav. **3**(1), 82–100 (1991)
12. Monderer, D., Shapley, L.S.: Fictitious play property for games with identical interests. J. Econom. Theory **68**(1), 258–265 (1996)
13. Robinson, J.: An iterative method of solving a game. Ann. Math. (2) **54**, 296–301 (1951)
14. Shapley, L.S.: Some topics in two-person games. In Advances in Game Theory, pp. 1–28. Princeton University Press, Princeton, N.J. (1964)
15. Sparrow, C., van Strien, S., Harris, C.: Fictitious play in 3×3 games: the transition between periodic and chaotic behaviour. Games Econom. Behav. **63**(1), 259–291 (2008)
16. van Strien, S.: Hamiltonian flows with random-walk behavior originating from zero-sum games and fictitious play. preprint 2009. http://arxiv.org/abs/0906.2058. Submitted for publication (2009)
17. van Strien, S., Sparrow, C.: Fictitious play in 3×3 games: chaos and dithering behaviour. To appear in Games Econom. Behav.

Chapter 49
A Finite Time Blowup Result
for Quadratic ODE's

Dennis Sullivan

Dedicated to Mauricio Peixoto who abstracted the practical theory of ODE's and to David Rand who applied the subsequent powerful abstract theory to practical problems.

Abstract We show that typical ODE's sharing the obvious algebraic property of Euler's equation exhibit finite time blowup. In a subsequent paper (Sullivan [1]) we study other finite dimensional analogues of Euler's equation which have no finite time blowup.

49.1 Finite Time Blow Up

The famous Euler ODE of incompressible frictionless fluid dynamics expressed in terms of the variable X = vorticity has the following algebraic form: The underlying space can be viewed as the (infinite dimensional) vector space V of exact two forms on a closed Riemannian manifold. The evolution of the exact two form = vorticity is described by an ODE $dX/dt = Q(X)$ where Q is a homogeneus quadratic mapping from V to V, namely one whose deviation from being linear $Q(X+Y) - Q(X) - Q(Y)$ is a symmetric bilinear form on V and which is homogeneous of degree two. We make a few comments about finite time blowup with given initial conditions for such algebraic, specifically homogeneous quadratic, ODE's.

1. Of course the most simple example on the real line $dx/dt = x.x$ with initial condition a at $t = 0$ has solution $-1/(t - 1/a)$. This blows up at the critical time $t = 1/a$.
2. The same calculation works for $dX/dt = X.X$ where X is a linear operator, so $V = End(W)$ for some other linear space W. If A is the desired operator at time zero and A is invertible, then $X(t) = -Id/(t(Id) - Id/A)$ is the solution. So

---

D. Sullivan
Mathematics Department, CUNY Graduate Center, New York, USA
and
Mathematics Department, SUNY, Stony Brook, NY, USA
e-mail: dennis@math.sunysb.edu

M.M. Peixoto et al. (eds.), *Dynamics, Games and Science I*, Springer Proceedings
in Mathematics 1, DOI 10.1007/978-3-642-11456-4_49,
© Springer-Verlag Berlin Heidelberg 2011

$X(t)$ blows up at a finite time iff the spectrum of A contains a real number. So unless W is an odd dimensional vector space for an open set of initial conditions there is a solution for all time. And for another open set of initial conditions there is finite time blowup.

3. One may hope to find some structure like this in the Euler fluid equation referred to above that would prevent finite time blowup for a large set of initial conditions.

Thus we ask the following questions:

1. For a finite dimensional vector space V how likely is it in the variable Q for the quadratic ODE $dX/dt = Q(X)$ to have finite time blow up for some initial condition A.

2. Fixing Q how likely is it in the variable A to have finite time blowup.

The following Theorem answers question (1):

Theorem 49.1. *If V is any finite dimensional vector space, then for each Q outside a proper algebraic subvariety of quadratic mappings from V to V, the ODE $dX/dt = Q(X)$ exhibits finite time blowup for some initial condition.*

Proof (Offer Gabber). The condition that there exists a non zero vector Y so that $Q(Y) = 0$ defines a proper algebraic subvariety in the space of quadratic mappings from V to V. Outside this subvariety Q defines by rescaling a map from S, the sphere of directions in V, to itself. This mapping agrees on antipodal points because Q is quadratic. Thus this mapping from S to S has even topological degree. Then the Lefschetz number of this mapping is non zero and this mapping has a fixed point. This means the original Q keeps a line invariant. By a linear change of variable the ODE restricted to this line becomes example (1) which has finite time blowup. $\square$

We have not studied question (2) in general which in example (2) is interesting.

Acknowledgements The Theorem came out of a two day discussion at IHES in the early 1990s with Ofer Gabber where the author made up the question (1) and Ofer came up with the proof of the Theorem.

Reference

1. Sullivan, D.: The Topology, Algebra and Algebraic Topology of 3D Incompressible Fluid Flow. To appear in "Low-Dimensional, Contact, and Symplectic Topology" (Athens, GA, 2009), published by the Proceedings and Symposia in Pure Mathematics series by the AMS

Chapter 50
Relating Material and Space-Time Metrics Within Relativistic Elasticity: A Dynamical Example

E.G.L.R. Vaz, Irene Brito, and J. Carot

Abstract Given a space-time and a continuous medium with elastic properties described by a 3-dimensional material space, one can ask whether they are compatible in the context of relativistic elasticity. Here a non-static, spherically symmetric spacetime metric is considered and we investigate the conditions for that metric to correspond to different 3-dimensional material metrics.

50.1 General Results

Let (M, g) be a spacetime. The *material space* X is a 3-dimensional manifold endowed with a Riemannian metric γ, the *material metric*; points in X can then be thought of as the particles of which the material is made of. Coordinates in M will be denoted as x^a for $a = 0, 1, 2, 3$, and coordinates in X as y^A, $A = 1, 2, 3$. The material metric γ is not a dynamical quantity of the theory and it roughly describes distances between neighboring particles in the relaxed state of the material.

The spacetime configuration of the material is said to be completely specified whenever a submersion $\psi : M \rightarrow X$ is given; if one chooses coordinate charts in M and X as above, then $y^A = y^A(x^b)$ and the physical laws describing the mechanical properties of the material can then be expressed in terms of a hyperbolic second order system of PDE. The differential map $\psi_* : T_p M \rightarrow T_{\psi(p)} X$ is then represented in the above charts by the rank 3 matrix $\left(y^A_b \right)_p$, $y^A_b = \partial y^A / \partial x^b$ which is sometimes called *relativistic deformation gradient*. The kernel of ψ_* is spanned by a single timelike vector which we take as $\mathbf{u} = u^a \partial_a$, satisfying

---

E.G.L.R. Vaz (✉) and I. Brito
Departamento de Matemática para a Ciência e Tecnologia, Universidade do Minho, 4800 058 Guimarães, Portugal
e-mail: evaz@mct.uminho.pt, ireneb@mct.uminho.pt

J. Carot
Departament de Fsica, Universitat de les Illes Balears, Cra Valldemossa pk 7.5, 07122 Palma de Mallorca, Spain
e-mail: jcarot@uib.es

M.M. Peixoto et al. (eds.), *Dynamics, Games and Science I*, Springer Proceedings in Mathematics 1, DOI 10.1007/978-3-642-11456-4_50,
© Springer-Verlag Berlin Heidelberg 2011

$y^A_b u^b = 0$, $u^a u_a = -1$, $u^0 > 0$. $\mathbf{u}$ is called the *velocity field of the matter*, and in the above picture in which the points in X are material points, the spacetime manifold M is then made up by the worldlines of the material particles.

The material space is said to be in a *locally relaxed state* at an event $p \in M$ if, at p, it holds $k_{ab} \equiv (\psi^* \gamma)_{ab} = h_{ab}$ where $h_{ab} = g_{ab} + u_a u_b$. Otherwise, it is said to be *strained*, and a measurement of the difference between k_{ab} and h_{ab} is the *strain*, whose definition varies in the literature; thus, while it can be defined simply as $S_{ab} = -\frac{1}{2}(k_{ab} - h_{ab}) = -\frac{1}{2}(k_{ab} - u_a u_b - g_{ab})$, we shall follow instead the convention in [1] and use $K_{ab} \equiv k_{ab} - u_a u_b$. The strain tensor determines the elastic energy stored in an infinitesimal volume element of the material space (or energy per particle), hence that energy will be a scalar function of K_{ab}. This function is called *constitutive equation* of the material, and its specification amounts to the specification of the material. We shall represent it as $v = v(I_1, I_2, I_3)$, where I_1, I_2, I_3 are any suitably chosen set of scalar invariants[1] associated with and characterizing K_{ab} completely. Following [1] we shall choose

$$I_1 = \frac{1}{2}(\text{Tr}K - 4) \qquad I_2 = \frac{1}{4}\left[\text{Tr}K^2 - (\text{Tr}K)^2\right] + 3 \qquad I_3 = \frac{1}{2}(\det K - 1).$$
(50.1)

Notice that for $K_{ab} = g_{ab}$ (equivalently $k_{ab} = h_{ab}$) the strain tensor S_{ab} is zero, in which case one has $I_1 = I_2 = I_3 = 0$.

The energy density ρ will then be the particle number density ϵ times the constitutive equation, that is $\rho = \epsilon v(I_1, I_2, I_3) = \epsilon_0 \sqrt{\det K}\, v(I_1, I_2, I_3)$ where ϵ_0 is the particle number density as measured in the material space, or rather, with respect to the volume form associated with $k_{ab} = (\psi^* \gamma)_{ab}$, and ϵ is that with respect to h_{ab}.

In the case of elastic matter, it can be seen using the standard variational principle for the Lagrangian density $\Lambda = \sqrt{-g}\rho$ (see for instance [2] or [3]) that decomposing the energy-momentum with respect to $\mathbf{u}$ (the velocity of the matter) yields $T_{ab} = \rho u_a u_b + p h_{ab} + P_{ab}$, where $h_{ab} = g_{ab} + u_a u_b$, $P_{ab} = h^m_a h^n_b (T_{mn} - 3ph_{mn})$, $\rho = T_{ab} u^a u^b$, $p = \frac{1}{3}h^{ab}T_{ab}$ which satisfy $h_{ab}u^b = 0$, $P_{ab}u^b = g^{ab}P_{ab} = 0$. The above energy-momentum tensor is of the diagonal Segre type $\{1, 111\}$ or any of its degeneracies so that an orthonormal tetrad exists $\{u_a, x_a, y_a, z_a\}$ (with $u_a u^a = -1$, $x^a x_a = y^a y_a = z^a z_a = +1$ and the mixed products zero) such that:

$$T_{ab} = \rho u_a u_b + p_1 x_a x_b + p_2 y_a y_b + p_3 z_a z_b, \qquad p = \frac{1}{3}(p_1 + p_2 + p_3),$$

$$h_{ab} = x_a x_b + y_a y_b + z_a z_b, \qquad \text{etc.}$$
(50.2)

One can show that the Dominant Energy Condition (DEC) is fulfilled if and only if $\rho \geq 0, |p_A| \leq \rho, A = 1, 2, 3$.

---

[1] Recall that one of the eigenvalues is 1, therefore, there exist three other scalars (in particular they could be chosen as the remaining eigenvalues) characterizing K^a_b completely along with its eigenvectors.

50 Relating Material and Space-Time Metrics Within Relativistic Elasticity

50.2 Spherical Symmetry and Material Metrics

For a spherically symmetric spacetime, coordinates $x^a = t, r, \theta, \phi$ exist (and are non-unique) such that the line element can be written as

$$ds^2 = -a(r,t)dt^2 + b(r,t)dr^2 + r^2 d\theta^2 + r^2 \sin^2\theta d\phi^2 \qquad (50.3)$$

with a and b positive. This metric possesses three Killing vectors, namely $\xi_1 = -\cos\phi \, \partial_\theta + \cot\theta \sin\phi \, \partial_\phi, \xi_2 = \partial_\phi$ and $\xi_3 = -\sin\phi \, \partial_\theta - \cot\theta \cos\phi \, \partial_\phi$ which generate the 3-dimensional Lie algebra $so(3)$.

The existence of symmetries has some important consequences on physics (see [4]). For example, matter 4-velocity, pressure, density, anisotropic tensor all stay invariant along the above KVs, together with the projection tensor $h_{ab} = g_{ab} + u_a u_b$. One can also show that any timelike vector field $\mathbf{v}$ that remains invariant along the three Killing vectors is necessarily of the form $\mathbf{v} = v^t(t,r) \, \partial_t + v^r(t,r) \, \partial_r$.

Let us now consider in more detail the problem of elasticity in a spherically symmetric spacetime $(M, \bar{g})$ with associated material space $(X, \bar{\gamma})$. We shall demand that the submersion $\psi : M \longrightarrow X$ mentioned in Sect. 50.1 preserves the KVs, that is $\psi_*(\xi_A) = \eta_A$ are also KVs on X. This implies that the metric $\bar{\gamma}$ is also spherically symmetric and therefore coordinates $y^A = (y, \tilde{\theta}, \tilde{\phi})$ exist with $y = y(t,r)$, $\tilde{\theta} = \theta$ and $\tilde{\phi} = \phi$, and are such that $\eta_A = \xi_A$ are KVs of the metric $\bar{\gamma}$. Thus, the line element $d\bar{s}^2$ of $\bar{g}$ is obtained from (50.3), with a and b substituted by $\bar{a}$ and $\bar{b}$, respectively. The line element of $\bar{\gamma}$ may be written as:

$$d\bar{\Sigma}^2 = f^2(y)(dy^2 + y^2 d\theta^2 + y^2 \sin^2\theta d\phi^2), \qquad (50.4)$$

This last expression is completely general, as any 3-dimensional spherically symmetric metric is necessarily conformally flat, as it is immediate to show. Notice also that the relation between $\bar{\gamma}$ and the flat material metric γ used in [1], is simply $\bar{\gamma}_{AB} = f^2(y)\gamma_{AB}$. Writing $\bar{k} = \psi^*(\bar{\gamma})$, one has:

$$\bar{k}^a_b = \begin{pmatrix} -f^2(y)(\dot{y}^2/\bar{a}) & -f^2(y)(\dot{y}y'/\bar{a}) & 0 & 0 \\ f^2(y)(\dot{y}y'/\bar{b}) & f^2(y)(y'^2/\bar{b}) & 0 & 0 \\ 0 & 0 & f^2(y)y^2/r^2 & 0 \\ 0 & 0 & 0 & f^2(y)y^2/r^2 \end{pmatrix}, \qquad (50.5)$$

where a dot indicates a derivative with respect to t and a prime a derivative with respect to r. The velocity field of the matter, defined by the conditions $\bar{u}^a y^A_a = 0$, $\bar{g}_{ab}\bar{u}^a\bar{u}^b = -1$ and $\bar{u}^0 > 0$, can be expressed as $\bar{u}^a = \bar{\Gamma}\bar{a}^{-1/2}(1, -\dot{y}/y', 0, 0)$, with $\bar{\Gamma} \equiv [1 - (\bar{b}/\bar{a})(\dot{y}/y')^2]^{-\frac{1}{2}}$.

The projection tensor $h^a_b = \delta^a_b + \bar{u}^a\bar{u}_b$ follows now easily from these expressions.

We will use the an orthonormal tetrad $\{\bar{u}, \bar{x}, \bar{y}, \bar{z}\}$, with $\bar{u}$ given above and such that the remaining vectors are eigenvectors of the pulled back material metric $\bar{k}_b^a$: $\bar{x}^a = \left(-(\bar{a}\bar{b}^{1/2})(\dot{y}/y')\bar{\Gamma}, \bar{\Gamma}/\sqrt{\bar{b}}, 0, 0 \right)$, $\bar{y}^a = (0, 0, 1/r, 0)$, $\bar{z}^a = (0, 0, 0, 1/(r \sin \theta))$, so that $\bar{g}_{ab} = -\bar{u}_a \bar{u}_b + \bar{x}_a \bar{x}_b + \bar{y}_a \bar{y}_b + \bar{z}_a \bar{z}_b$. It is now immediate to see that the pressure tensor has the same eigenvectors as $\bar{k}_{ab}$ and can be written as $\bar{p}_{ab} = \bar{p}_1 \bar{x}_a \bar{x}_b + \bar{p}_2(\bar{y}_a \bar{y}_b + \bar{z}_a \bar{z}_b)$. Therefore, (50.2) yields $\bar{T}_{ab} = \bar{\rho} \bar{u}_a \bar{u}_b + \bar{p}_1 \bar{x}_a \bar{x}_b + \bar{p}_2(\bar{y}_a \bar{y}_b + \bar{z}_a \bar{z}_b)$, where $\bar{\rho}$ is the energy density, $\bar{p}_1$, the radial pressure and $\bar{p}_2$, the tangential pressure. These and other related issues are studies in depth in [4].

In order to know whether the spacetime metric $\bar{g}$ can be associated with different conformally related material metrics, it will be assumed that $g_{ab} = \bar{g}_{ab}$, with g and $\bar{g}$ associated, respectively, with a flat (γ) and a non flat ($\bar{\gamma}$) material metric, related by $\bar{\gamma} = f^2 \gamma$.

Therefore the expressions relating the eigenvalues of $\bar{k}$ and k are: $\bar{s} = f^2 y^2 / r^2 = f^2 s, \bar{\eta} = f^2 y'^2 / (\Gamma^2 b) = f^2 \eta$. These expressions are used to relate the invariants in (50.1), namely $\bar{I}_1$, $\bar{I}_2$, $\bar{I}_3$, with the corresponding ones I_1, I_2, I_3 through the conformal factor f, as follows:

$$\bar{I}_1 = f^2 (I_1 + 3/2) - 3/2, \quad \bar{I}_2 = f^4 (I_1 + I_2 - 3/2) - f^2 (I_1 + 3/2) + 3,$$

$$\bar{I}_3 = f^6 (I_3 + 1/2) - 1/2. \tag{50.6}$$

The above expressions for $\bar{s}$ and $\bar{\eta}$ lead to the following relations

$$\bar{\rho} = f^3 \frac{\bar{v}}{v} \rho \qquad \bar{\epsilon} = \rho_0 \bar{s} \sqrt{\bar{\eta}} = f^3 \epsilon. \tag{50.7}$$

Taking the above expressions for the invariants together with (50.7) one obtains

$$\frac{\partial \bar{\rho}}{\partial \bar{I}_1} = \frac{1}{f^2} \frac{\partial \rho}{\partial I_1} - \frac{\partial \rho}{\partial I_2} \left(\frac{1}{f^2} - \frac{1}{f^4} \right), \qquad \frac{\partial \bar{\rho}}{\partial \bar{I}_2} = \frac{1}{f^4} \frac{\partial \rho}{\partial I_2},$$

$$\frac{\partial \bar{\rho}}{\partial \bar{I}_3} = \frac{1}{f^6} \frac{\partial \rho}{\partial I_3}. \tag{50.8}$$

These expressions lead to the following relationship for the energy-momentum tensors:

$$\bar{T}_b^a = f^3 \frac{\bar{v}}{v} T_b^a. \tag{50.9}$$

However the assumption on equal metric tensors leads to equal energy-momentum tensors, so that the following relation for constitutive equations must hold:

$$\bar{v} = \frac{1}{f^3} v. \tag{50.10}$$

It is now straightforward to conclude that $\bar{\rho} = \rho$.

References

1. Magli, G.: Gen. Rel. Grav. **25**, 441–460 (1993)
2. Magli, G., Kijowski, J.: Gen. Rel. Grav. **24**, 139–158 (1992)
3. Karlovini, M., Samuelsson, L.: Class. Quant. Grav. **20**, 3613–3648 (2003)
4. Brito, I., Carot, J., Vaz, E.G.L.R.: Gen. Rel. Grav. **42**, 2357–2382 (2010)

Chapter 51
Strategic Information Revelation Through Real Options in Investment Games

Takahiro Watanabe

Abstract An investment game with an incumbent and an entrant is examined. The profit flows involve two uncertain factors: (1) the basic level of demand of the market observed only by the incumbent and (2) the fluctuation of the demand described by a geometric Brownian motion which is common to both firms. In our model, the incumbent enters into the market earlier than the entrant. The high demand type of the incumbent can invest earlier than the low demand type. This earlier investment, however, reveals the information, so that the entrant would accelerates the timing of the investment by observing the incumbent's timing of the entry and it reduces the monopolistic profit of the incumbent. Thus, the incumbent who knows the high demand may delay the timing of the investment to hide the information strategically. I characterize this signaling effect and investigate the real option values of both firms.

51.1 Introduction

Recently, many studies in the field of real options incorporate the analysis of decisions about investments under uncertainty to game theory which examines strategic interactions of firms under competition. Most typical results are shown in a duopolistic market with symmetric firms by integrating real options and optimal stopping games, e.g. [1, 5, 8, 11, 12]. Kong and Kwok [7] and Pawlina and Kort [10] develop the results to two asymmetric firms. These studies obtains implications about strategic interactions for the investment timings under incomplete information, but they do not refer the effect of information.

On the other hand, information effects and strategic interactions of firms also investigated by games under incomplete information. The strategic investments

---

T. Watanabe
Department of Business Administration, Tokyo Metropolitan University, Minamiosawa 1-1, Hachiouji, Tokyo, Japan
e-mail: contact_nabe08@nabenavi.net

M.M. Peixoto et al. (eds.), *Dynamics, Games and Science I*, Springer Proceedings in Mathematics 1, DOI 10.1007/978-3-642-11456-4_51,
© Springer-Verlag Berlin Heidelberg 2011

770   T. Watanabe

under uncertainty and the asymmetry of information has been investigated by [3] and [9]. Grenadier and Wang [2] investigates a conflict between managers and owners taking into the account of the asymmetric information and contract theory.

These past studies, however, do not concern about the information revelation and consider only static situations about the private information of the firms. Watanabe [13] examines strategic information revelation of the investments under uncertainty by integrating dynamic games known as signaling games and real options. This paper is a summary of [13] which is my work in progress.

I consider two asymmetric firms, an incumbent and an entrant, attempting the entry into a new market of a product. The demand of the market has two uncertainty factors. One factor is the fundamental size of the market which is determined at the beginning of the game and private information of the incumbent. Hence, there are two types of the incumbents, high-demand type and low-demand type. The entrant cannot obtain the information and the other factor of uncertainty, the fluctuation of the demand given by a stochastic process, is common to both firms.

In the model, the incumbent assumed to be invest earlier than the entrant for any demand. If the timing of investment by a high-demand type of the incumbents would be earlier than a low-demand type, the information is revealed. Then, the entrant would accelerates the timing of the investment if the incumbent's earlier investment is observed. Since this reduces the monopolistic profit of the high-demand type of the incumbents, the high-demand type may strategically delay the timing of the investment to hide the information.

In this paper, I characterize whether the incumbent reveals the information truthfully or not by using the concept of a weak perfect Bayesian equilibrium. I specify a condition for a high-demand type invests strategically in the equilibrium, and show that it is necessary for the incumbent to use a mixed strategy in the equilibrium under some condition.

In some numerical examples, the equilibria and the values of the incumbent are calculated. In the examples I show that if a duopoly profit for the high demand type of the incumbent is small, the incumbent invests strategically while the incumbent does truthfully if this duopoly profit is sufficiently large. The incumbent also invests strategically, if the volatility is large, or the entrant's cost is small.

Section 51.2 provides the notation and description of the model. Section 51.3 gives a value of the entrant and non-strategic values of the incumbents which gives a benchmark of the analysis. In Sect. 51.4, I define the solution of the game by Perfect Bayesian equilibrium and give two candidates of the solution, Truthful Revelation and Strategic Revelation, which corresponds to a separating and a pooling equilibrium, respectively. I also present conditions which specify either of the two candidates to the equilibrium. These conditions characterize an equilibrium in pure strategies, but in some cases there are no equilibrium in pure strategies. Section 51.5 deals with the mixed strategies and gives conditions of the equilibrium. Since these equilibria in mixed strategies include the case of the equilibria in pure strategies examined in the previous section, the condition characterize the equilibrium comprehensively. Section 51.6 shows numerical examples and Sect. 51.7 gives conclusions and the further research.

51.2 The Model

Two asymmetric firms, an incumbent and an entrant, consider the optimal timing of the entrance into the market of a new product. The incumbent and the entrant are denoted by firm I and firm E, respectively. The investments for the entry of both firms are assumed to be irreversible and the sunk cost of firm i's investment is denoted by K_i for $i = I, E$. The revenue flow of each firm by the entrance depends on the market structure, monopoly or duopoly, and the following two uncertain factors of the demand.

One uncertain factor of the demand represents a stochastic process, denoted by X_t, as a standard real option setting. X_t is interpreted as the unsystematic shocks of the demand over the time and it is common to both firms. X_t is assumed to follow a geometric Brownian motion:

$$dX_t = \mu X_t dt + \sigma X_t dz$$

where μ is the drift parameter, σ is the volatility parameter and dz_t is the increment of a standard Winner process. Both firms are assumed to be risk neutral, with the risk free rate r. As usual assumption of real option approach for convergence, I assume $r > \mu$.

The other uncertain factor of the demand represents a systematic risk determined at the beginning of the game and it is assumed to be a constant over the time. I denote the factor by θ where $\theta = H$ and $\theta = L$ means that the demand is high and low, respectively. The prior probability of drawing $\theta = H$ and $\theta = L$ are denoted by p and $1 - p$, respectively.

When only firm i enters in the market, the profit flow of firm i becomes $\pi_{i1}^{\theta} X_t$. On the other hand, when both firms enter in the market, the profit flow of firm i becomes $\pi_{i2}^{\theta} X_t$. The profit flow of the firm which has not entered in the market is assumed to be zero. I assume that $\pi_{i1}^{\theta} > \pi_{i2}^{\theta} > 0$ for $i = 1, 2$ and $\theta = H, L$.

I assume that the incumbent has two advantages stated as follows. First, the uncertain factor θ can be observed only by the incumbent, i.e., it is the private information of the incumbent. Secondly, I assume that the profit in monopoly of the incumbent is sufficiently larger than that of the entrant and/or the cost of the investment of the incumbent is sufficiently smaller than that of the entrant.

51.3 Value Functions of a Benchmark Case

A number of studies under the joint framework of real options and game theory, such as [1, 4–6, 8, 11, 12] consider the symmetric firms in order to examine preemptive behavior of competition. In these models, if the value of the leader's optimal entry is greater than the value of the best reply of the follower, then both firms want to become a leader. In this case, the leader's optimal threshold is solved by equations of an equilibrium and the value of the leader is not determined by maximizing the expected profit of either firm.

772   T. Watanabe

However, I assume that the incumbent's profit is sufficiently large and the incumbent's cost is sufficiently small. Moreover, I also assume that X_0 is sufficiently small to wait the investment for both firms. Under these assumptions, the incumbent must be the dominant leader and the entrant must be the follower (see, [7] and [10]), so that both values of the leader and the follower are solved backward. In the next subsection, first, I consider the value of the entrant as the follower. Then, the value of the incumbent as the leader will be discussed.

51.3.1 The Value of the Entrant

The value of the entrant is a function of the entrant's belief for the demand level θ. Suppose that the entrant believes the high demand $\theta = H$ occurring with probability q. Let $u_E^*(q)$ be the value function of the entrant with the entrant's belief q. The value function $u_E^*(q)$ is given by

$$u_E^*(q) = \max_{t_E} E^x [\int_{t_E}^{\infty} e^{-rs} (q\pi_{E2}^H + (1-q)\pi_{E2}^L) X_s ds - e^{-rt_E} K_E]$$

where E^x denotes the conditional expectation on $X_0 = x$. Let $x_E^*(q)$ be the optimal threshold for the belief q, i.e., $x_E^*(q) = \inf\{t \geq 0 | X_t \geq x_E^*(q)\}$. The usual calculation of real option analysis implies

$$x_E^*(q) = \frac{\beta}{\beta - 1} \frac{r - \mu}{q\pi_{E2}^H + (1-q)\pi_{E2}^L} K_E$$

where β is defined by

$$\beta = \frac{1}{2} \left(1 - \frac{2\mu}{\sigma^2} + \sqrt{(1 - \frac{2\mu}{\sigma^2})^2 + \frac{8r}{\sigma^2}} \right).$$

Let $x_E^H = x_E^*(1)$, $x_E^L = x_E^*(0)$ and $x_E^M = x_E^*(p)$. x_E^H and x_E^L are the thresholds when the entrant believes that the demand are high and low, respectively. x_E^M is the threshold when the entrant predicts the high demand with prior probability p.

We easily find that

$$x_E^H \leq x_E^M \leq x_E^L. \tag{51.1}$$

51.3.2 The Value of the Incumbent

Let $u_I(x_I, x_E, \theta)$ be the expected profit of the incumbent with his private information of the demand θ when the incumbent invests at the threshold x_I and the entrant invests at x_E under the condition $x_I < x_E$. $u_I(x_I, x_E, \theta)$ is given by

$$u_I(x_I, x_E, \theta) = E^* \left[\int_{t_I}^{t_E} e^{-rs} \pi_{I1}^\theta X_s ds - e^{-rt_I} K_E + \int_{t_E}^\infty e^{-rs} \pi_{I2}^\theta X_s ds \right],$$

where t_i is the first passage time at threshold x_i for $i = I, E$, i.e., $t_i = \inf\{t \geq 0 | X_t \geq x_i\}$. This equation can be written as

$$u_I(x_I, x_E, \theta) = E^* \left[\int_{t_I}^\infty e^{-rs} \pi_{I1}^\theta X_s ds - e^{-rt_I} K_E - \int_{t_E}^\infty e^{-rs} (\pi_{I1}^\theta - \pi_{I2}^\theta) X_s ds. \right]$$

Let the first term and the second term be denoted by $v_I(x_I, \theta)$ and $\Delta v_I(x_E, \theta)$, i.e.,

$$v_I(x_I, \theta) = E^* \left[\int_{t_I}^\infty e^{-rs} \pi_{I1}^\theta X_s ds - e^{-rt_I} K_E \right]$$

and

$$\Delta v_I(x_E, \theta) = E^* \left[\int_{t_E}^\infty e^{-rs} (\pi_{I1}^\theta - \pi_{I2}^\theta) X_s ds \right].$$

Then, $u_I(x_I, x_E, \theta)$ is given by

$$u_I(x_I, x_E, \theta) = v_I(x_I, \theta) - \Delta v_I(x_E, \theta). \tag{51.2}$$

where $v_I(x_I, \theta)$ is explicitly described as

$$v_I(x_I, \theta) = \left(\frac{\pi_{I1}^\theta}{r - \mu} x_I - K_I \right) \left(\frac{x}{x_I} \right)^\beta.$$

Note that

$$\Delta v_I(x_E^H, \theta) \geq \Delta v_I(x_E^M, \theta) \geq \Delta v_I(x_E^H, \theta) \tag{51.3}$$

because $\Delta v_I(x_E, \theta)$ is decrease in threshold x_E and (51.1).

If x_E is independent of the incumbent decision x_I, the second term $\Delta v_I(x_E, \theta)$ is independent of the incumbent decision x_I. In this case, hence, the incumbent maximizes the expected profit by maximizing $v_I(x_I, \theta)$. Taking into the account of the signaling effect, however, the optimal threshold of the entrant x_E depends on the threshold of the incumbent x_I.

This signaling equilibrium is examined in the next section. In this section, I consider the case in which x_E is independent of x_I. Let $x_I^*(\theta)$ be the optimal threshold of the incumbent with the private information θ under the condition that x_E is independent of x_I. Then, $v_I(x_I^*(\theta), \theta)$ is given by

$$v_I(x_I^*(\theta), \theta) = \max_{x_I} v_I(x_I, \theta) = \max_{x_I} E^* \left[\int_{t_I}^\infty e^{-r(s - t_I)} \pi_{I1}^\theta X_s ds - e^{-r(s - t_I)} K_E \right].$$

774   T. Watanabe

Note that x is assumed to be sufficiently small, so that $x \leq x_I^*(\theta)$. The usual calculation of real option analysis implies

$$x_I^*(\theta) = \frac{\beta}{\beta-1} \frac{r-\mu}{\pi_{I1}^\theta} K_I,$$

and

$$v_I(x_I^*(\theta), \theta) = \frac{K_I}{\beta-1} \left(\frac{x}{x_I^*(\theta)}\right)^\beta.$$

$\Delta v_I(x_E, \theta)$ is given by

$$\Delta v_I(x_E, \theta) = \frac{\pi_{I1}^\theta - \pi_{I2}^\theta}{r-\mu} x_E \left(\frac{x}{x_E}\right)^\beta$$

Let $x_I^H = x_I^*(H)$ and $x_I^L = x_I^*(L)$.
$x_I^H = x_I^*(H)$ and $x_I^L = x_I^*(L)$ express the optimal threshold when the incumbent knows that the demand is high and low, respectively, if the incumbent's decision is independent of the entrant's decision.

51.4 Equilibrium Analysis

51.4.1 Definitions of the Solution

For the analysis of the signaling effect, a Perfect Bayesian Equilibrium (PBE) is considered. PBE is an *assessment* which consists of three elements: (1) the incumbent's timing for each type, (2) the entrant's timing for each observation of the incumbent timing and (3) the entrant's belief for each observation of the incumbent timing. Formally, three components $\{(a_I(H), a_I(L)), a_E(\cdot), q(\cdot)\}$ is called an assessment if

- $a_I(H)$ and $a_I(L)$ are incumbent's threshold for private information H and L, respectively.
- $a_E(x_I)$ is the entrant's threshold for observed incumbent's threshold x_I.
- $q(x_I)$ is the entrant's belief for observed incumbent's threshold x_I.

A PBE is an assessment $\{(a_I^*(H), a_I^*(L)), a_E^*(\cdot), q^*(\cdot)\}$ satisfying the following three conditions.

First, $a_I^*(\theta)$ is the optimal threshold of the incumbent for $\theta = H, L$ such that

$$u_I(a_I^*(\theta), a_E^*(a_I^*(\theta)), \theta) = \max_{x_I} u_I(x_I, a_E^*(x_I), \theta). \tag{51.4}$$

Secondly, $a_E^*(\cdot)$ provides the optimal threshold of the entrant observing the entry of the incumbent at any threshold x_I with the belief $q^*(\cdot)$ such that

$$a_E^*(x_I) = x_E^*(q^*(x_I)). \tag{51.5}$$

51 Strategic Information Revelation Through Real Options in Investment Games 775

Finally, $q^*(\cdot)$ is the belief of the entrant for the high demand which is consistent to any threshold of the incumbent x_I observed by the entrant in the sense of Bayes rule stated as follows.

Let X_I be a random variable with respect to a threshold of the incumbent. Then $q^*(x_I) = Prob[\theta = H | X_I = x_I]$. Bayes rule implies

$$Prob[\theta = H | X_I = x_I]$$

$$= \frac{Prob[X_I = x_I | \theta = H] Prob[\theta = H]}{Prob[\theta = H] Prob[X_I = x_I | \theta = H] + Prob[\theta = L] Prob[X_I = x_I | \theta = L]}.$$

By $Prob[\theta = H] = p$ and $Prob[\theta = L] = 1 - p$, the condition of the consistent belief is expressed by

$$q^*(x_I) = \frac{p Prob[X_I = x_I | \theta = H]}{p Prob[X_I = x_I | \theta = H] + (1 - p) Prob[X_I = x_I | \theta = L]}. \tag{51.6}$$

$Prob[X_I = x_I | \theta = H]$ and $Prob[X_I = x_I | \theta = L]$ would follow the probability distributions according to a mixed strategy of the incumbent. In Sect. 51.5, I investigate the mixed strategies of the incumbent. However, in this section, I restrict the analysis to the pure strategies, so $Prob[X_I = x_I | \theta = H]$ and $Prob[X_I = x_I | \theta = L]$ can be explicitly written as

$$Prob[X_I = x_I | \theta = H] = \begin{cases} 1 & x_I = a_I^*(H) \\ 0 & x_I \neq a_I^*(H), \end{cases} \quad Prob[X_I = x_I | \theta = L] = \begin{cases} 1 & x_I = a_I^*(L) \\ 0 & x_I \neq a_I^*(L). \end{cases} \tag{51.7}$$

(51.6) and (51.7) imply that

$$q^*(x_I) = \begin{cases} p & x_I = a_I^*(H) \text{ and } x_I = a_I^*(L), \\ 1 & x_I = a_I^*(H) \text{ and } x_I \neq a_I^*(L), \\ 1 - p & x_I \neq a_I^*(H) \text{ and } x_I = a_I^*(L). \end{cases} \tag{51.8}$$

If $a_I^*(H) \neq x_I$ and $a_I^*(L) \neq x_I$, any belief $q^*(x_I)$ is consistent.

Thus, a PBE in pure strategies is formally defined as follows.

Definition 51.1. An assesment is said to be a (weak) perfect Bayesian equilibrium in pure strategies (PBEP) if it satisfies (51.4), (51.5) and (51.8).

51.4.2 Candidates of the Solution

The following two assessments are considered as candidates of the solution in this section. The first assessment is called *Truthful Revelation* defined by

$$a_I^*(H) = x_I^H, \ a_I^*(L) = x_I^L$$

$$a_E^*(x_I) = \begin{cases} x_E^H & x_I \neq x_I^L, \\ x_E^L & x_I = x_I^L, \end{cases}$$

$$q^*(x_I) = \begin{cases} 1 & x_I \neq x_I^L, \\ 0 & x_I = x_I^L. \end{cases}$$

In Truthful Revelation, any type of the incumbents truthfully enters to the market at the optimal threshold with respect to the demand. This truthful behavior reveals the information of the demand that the incumbent has. The entrant obtains the information about the demand by observing the incumbent's behavior and enters to the market optimally with full information. If the entrant observes that the incumbent enters to the market at neither x_I^H nor x_I^L, any belief of the entrant is consistent. In other words, the entrant's belief is assigned arbitrarily in the entrant's observation in off-equilibrium path. For this unexpected deviation of the equilibrium for the incumbent, the entrant is assumed to believe high demand.

Second assessment is called *Strategic Revelation* defined by

$$a_I^*(H) = a_I^*(L) = x_I^L$$

$$a_E^*(x_I) = \begin{cases} x_E^H & x_I \neq x_I^L, \\ x_E^M & x_I = x_I^L, \end{cases}$$

$$q^*(x_I) = \begin{cases} 1 & x_I \neq x_I^L, \\ p & x_I = x_I^L \end{cases}$$

In Strategic Revelation, the high-demand type of the incumbents does not enters at the optimal threshold but invests at the threshold of the low demand. This delay of the investment hides the information of high demand and the entrant cannot distinguish the type of the incumbents by observing the behavior. Thus, the entrant expects the level of the demand according to the prior probability p and enters at the threshold for the prior expectation of the demand. In off-equilibrium path, the entrant is assumed to believe the high demand, as well as Truthful Revelation.

51.4.3 The Equilibrium Strategies

In this subsection, I analyze conditions where either of candidates, Truthful Revelation or Strategic Revelation, is a solution. Since both candidates are constructed by satisfying both the optimality of the entrant and the consistency of the entrant's belief, it remains to consider the optimality of the incumbent for given entrant's strategy $a_E^*(\cdot)$ and belief $q^*(\cdot)$. Moreover, the low demand type of the incumbent does not have the incentive to deviate the optimal timing x_I^L because pretending the high demand type only accelerates the timing of the entrant's investment and reduce the incumbent's value. Hence, only the timing of the high type of the incumbent should be focused on.

51 Strategic Information Revelation Through Real Options in Investment Games 777

First, suppose that Truthful Revelation is a PBE. In Truthful Revelation, the entrant believes that the later investment of the incumbent at x_I^L reveals truthfully the information of low demand. If the incumbent with the high demand does not have the incentive for hiding the information to delay the entrant's investment, the following condition holds,

$$u_I(x_I^H, x_E^H, H) \geq u_I(x_I^L, x_E^L, H). \tag{51.9}$$

Secondly, suppose that Strategic Revelation is a PBE. In Strategic Revelation, the incumbent with information of the high demand strategically delays the investment to the optimal timing for the low demand, and the entrant cannot obtain the information about the demand. Then, the entrant observing the incumbent's investment at x_L predicts the level of the demand by prior probability p, so that the expectation of the profit is π_{E2}^M. Then the entrant enters to the market at x_E^M which is optimal threshold for π_{E2}^M. The high type of the incumbents does not have an incentive to hide information if the expected value for this delayed entrance at x_I^L is greater than that of the optimal entrance at the threshold of the high demand x_I^H. This condition is expressed by

$$u_I(x_I^H, x_E^H, H) \leq u_I(x_I^L, x_E^M, H). \tag{51.10}$$

Above arguments are summarized and proved formally in the following proposition.

Proposition 51.1. *(1) (51.9) holds if and only if Truthful Revelation is a PBE.*
(2) (51.10) holds if and only if Strategic Revelation is a PBE.

Proof. First, I show (1). Suppose that (51.9) holds. I show that if assessment $\{(a_I^*(H), a_I^*(L)), a_E^*(\cdot), q^*(\cdot)\}$ is Truthful Revelation, then it is a PBE. To prove this, it is sufficient to show that Truthful Revelation satisfies three conditions: the incumbent's optimality (51.4), the entrant's optimality (51.5) and the consistency of the entrant's belief (51.8). By the definition of Truthful Revelation, it always satisfies the entrant's optimality (51.5) and the consistency of the belief (51.8), it remains to show that it satisfies the incumbent's optimality (51.4), i.e., for $\theta = H, L$,

$$u_I(a_I^*(\theta), a_E^*(a_I^*(\theta)), \theta) \geq u_I(x_I, a_E^*(x_I), \theta), \tag{51.11}$$

for any $x_I \neq a_I^*(\theta)$.

First, let $\theta = L$ and choose any $x_I \neq a_I^*(\theta)$. Since $a_I^*(L) = x_I^L$, $a_E^*(x_I^L) = x_E^L$ and $a_E^*(x_I) = x_E^H$ for any $x_I \neq x_I^L$ in Truthful Revelation, (51.11) can be expressed as $u_I(x_I^L, x_E^L, L) \geq u_I(x_I, x_E^H, L)$ for any $x_I \neq x_I^L$. Note that $\Delta v_I(x_E^H, \theta) \geq \Delta v_I(x_E^L, \theta)$ by (51.3). $u_I(x_I, x_E, L) = v_I(x_I, \theta) - \Delta(x_E, \theta)$ implies that $u_I(x_I, x_E^L, L) \geq u_I(x_I, x_E^H, L)$ because the payoff of the incumbent increases in later investment of the entrant. Since x_I^L is the optimal threshold of

778   T. Watanabe

the incumbent, i.e. $v_I(x_I^L, \theta) = \max_{\hat{x}_I} v_I(\hat{x}_I, \theta)$, $u_I(x_I^L, x_E^L, L) \geq u_I(x_I, x_E^L, L)$. Hence, (51.11) hold for $\theta = L$.

Secondly, let $\theta = H$ and choose any $x_I \neq a_I^*(\theta)$. If $x_I \neq x_I^L$, then (51.11) can be expressed as $u_I(x_I^H, x_E^H, H) \geq u_I(x_I, x_E^H, H)$. This holds because x_I^H is the optimal threshold of the incumbent. Suppose that $x_I = x_I^L$, then (51.9) yields $u_I(x_I^H, x_E^H, H) \geq u_I(x_I^L, x_E^L, H)$. Then, Truthful Revelation is a PBE.

Conversely, suppose that (51.9) does not hold, i.e., $u_I(x_I^H, x_E^H, H) < u_I(x_I^L, x_E^L, H)$. Then, the high demand type of the incumbent strictly increases the payoff by deviating x_I^L from $a_I^*(H) = x_I^H$ in Truthful Revelation and this means that Truthful Revelation is not a PBE. Hence, Truthful Revelation is a PBE, only if $\pi_{I1}^H - \pi_{I2}^H \leq \xi_H(K_I, K_E, \beta)$.

The proof of (2) is similar. $\qquad\qquad\qquad\qquad\qquad\qquad\qquad\qquad\qquad\qquad\square$

Since $u_I(x_I^L, x_E^M, H) \leq u_I(x_I^L, x_E^L, H)$ neither Truthful Revelation nor Strategic Revelation is PBEP for $u_I(x_I^L, x_E^M, H) \leq u_I(x_I^H, x_E^H, H) \leq u_I(x_I^L, x_E^M, H)$. In this interval, the mixed strategy of the incumbent should be considered to ensure the existence of the equilibrium, which is analyzed in the next section.

The following lemma shows that (51.9) and (51.10) can be solved for difference of the incumbent's profits between monopoly and duopoly.

Lemma 51.1. *(1) (51.9) holds if and only if*

$$\pi_{I1}^H - \pi_{I2}^H \leq \frac{1}{\beta}\left(\frac{K_E}{K_I}\right)^{\beta-1}\left\{\frac{(\pi_{I1}^H)^\beta - \phi(\pi_{I1}^L)^\beta}{(\pi_{E2}^H)^{\beta-1} - (\pi_{E2}^L)^{\beta-1}}\right\}, \qquad (51.12)$$

and
(2) (51.10) holds if and only if

$$\pi_{I1}^H - \pi_{I2}^H \geq \frac{1}{\beta}\left(\frac{K_E}{K_I}\right)^{\beta-1}\left\{\frac{(\pi_{I1}^H)^\beta - \phi(\pi_{I1}^L)^\beta}{(\pi_{E2}^H)^{\beta-1} - (\pi_{E2}^M)^{\beta-1}}\right\}, \qquad (51.13)$$

where

$$\phi = \frac{\beta\pi_{I1}^H - (\beta - 1)\pi_{I1}^L}{\pi_{I1}^L}.$$

Let the right hand side of (51.12) and (51.13) be $\xi_H(K_I, K_E, \beta)$ and $\xi_M(K_I, K_E, \beta)$.

By above arguments, the equilibrium strategies are characterized by $\xi_H(K_I, K_E, \beta)$ and $\xi_M(K_I, K_E, \beta)$. Proposition 51.2 summarizes equilibrium strategies.

Proposition 51.2. *(1) $\pi_{I1}^H - \pi_{I2}^H \leq \xi_H(K_I, K_E, \beta)$ if and only if Truthful Revelation is a PBE.*
(2) $\pi_{I1}^H - \pi_{I2}^M \geq \xi_M(K_I, K_E, \beta)$ if and only if Strategic Revelation is a PBE.

51 Strategic Information Revelation Through Real Options in Investment Games 779

51.5 Equilibria in Mixed Strategies

As the discussion of the previous section, it is found that an equilibrium in the pure strategies does not exists for $u_I(x_I^L, x_E^M, H) \le u_I(x_I^H, x_E^H, H) \le u_I(x_I^L, x_E^L, H)$. Hence, I extend the model to the case where the high-demand type of the incumbents uses the mixed strategies. Let $x_I(\lambda)$ be a mixed strategy of the incumbent where the incumbent chooses x_I^H with probability λ and x_I^L with probability $1 - \lambda$ for $0 \le \lambda \le 1$. Moreover, u_I is extended to the set of mixed strategies $x_I(\lambda)$ for $0 \le \lambda \le 1$ and entrant strategies $a_E(\cdot)$, defined by

$$u_I(x_I(\lambda), a_E(\cdot), \theta) = \lambda u_I(x_I^H, a_E(x_I^H), \theta) + (1 - \lambda)u_I(x_I^L, a_E(x_I^L), \theta)$$

for any x_E and $\theta = H, L$.

The consistent belief of the entrant for $a_I^*(H) = x_I(\lambda)$ and $a_I^*(L) = x_I^L$ is solved by Bayes rule (51.6). $Prob[X_I = x_I | \theta = H]$ and $Prob[X_I = x_I | \theta = L]$ are given by

$$Prob[X_I = x_I | \theta = H] = \begin{cases} \lambda & x_I = x_I^H \\ 1 - \lambda & x_I = x_I^L \\ 0 & x_I \ne x_I^H, x_I^L, \end{cases} \qquad Prob[X_I = x_I | \theta = L] = \begin{cases} 1 & x_I = x_I^L \\ 0 & x_I \ne x_I^L, \end{cases}$$

$$(51.14)$$

(51.6) and (51.14) imply the consistent belief $q^*(\cdot)$ as

$$q^*(x_I^H) = \frac{p\lambda}{p\lambda + (1 - p) \times 0} = 1$$

and

$$q^*(x_I^L) = \frac{p(1 - \lambda)}{p(1 - \lambda) + (1 - p) \times 1} = \frac{p(1 - \lambda)}{1 - p\lambda}.$$

If $x_I \ne x_I^H, x_I^L$, any belief $q^*(x_I)$ is consistent.

This consistent belief indicates that the entrant observing the incumbent's investment at x_I^H completely learns that the level of the demand is high, because only the high-demand type of the incumbents invests at x_I^H. Hence, the optimal timing of investment of the entrant observing the incumbent's investment at x_I^H is x_E^H. In contrast, since both types of the incumbents have the possibility of the investment at x_I^L, the entrant predicts the high demand according to the probability $q^*(x_I^L)$ when the entrant observes the incumbent's investment at x_I^L. The optimal timing of the investment of the entrant observing the incumbent's investment at x_I^L is $x_E^*(q^*(x_I^L))$. For simplify notation, $q^*(x_I^L)$ is denoted by q^λ and let $x_E^*(q^\lambda)$ be x_E^λ.

By above arguments, the following assessment $\{(a_I^*(H), a_I^*(L)), a_E^*(\cdot), q^*(\cdot)\}$, called λ-*Hybrid Revelation*, is a general candidate of the solution, which satisfies the optimality of the entrant and the consistence of the belief.

$$a_I^*(H) = x_I(\lambda), \, a_I^*(L) = x_I^L$$

$$a_E^*(x_I) = \begin{cases} x_E^H & x_I \neq x_I^L, \\ x_E^\lambda & x_I = x_I^L, \end{cases}$$

$$q^*(x_I) = \begin{cases} 1 & x_I \neq x_I^L, \\ q^\lambda & x_I = x_I^L, \end{cases}$$

Note that λ-Hybrid Revelation for $\lambda = 1$ is identical to Truthful Revelation while $\lambda = 0$ is to Strategic Revelation. Hence, by solving a condition on λ where λ-Hybrid Revelation is an equilibrium for $u_I(x_I^L, x_E^M, H) \leq u_I(x_I^H, x_E^H, H) \leq u_I(x_I^L, x_E^\lambda, H)$, an equilibrium for any case can be characterized comprehensively.

The probability λ can be solved as follows. Similarly to Strategic Revelation and Truthful Revelation, the low-demand type of the incumbents also does not have incentive to deviate the optimal timing for low demand. It remains to examine the equilibrium strategies of the high-demand type of the incumbents. Let $\{(a_I^*(H), a_I^*(L)), a_E^*(\cdot), q^*(\cdot)\}$ be λ-Hybrid Revelation. If $u_I(x_I^H, x_E^H, H) > u_I(x_I^L, x_E^\lambda, H)$, then,

$$u_I(x_I^H, a_E^H(\cdot), H) = u_I(x_I^H, x_E^H, H) > \lambda u_I(x_I^H, x_E^H, H) + (1 - \lambda) u_I(x_I^L, x_E^\lambda, H)$$
$$= u_I(x_I(\lambda), a_E^H(\cdot), H).$$

Hence, the incumbent has incentive to deviate from mixed strategy $x_I(\lambda)$ to pure strategy x_I^H. Otherwise, if $u_I(x_I^H, x_E^H, H) < u_I(x_I^L, x_E^\lambda, H)$, then the incumbent similarly has incentive to deviate from mixed strategy $x_I(\lambda)$ to pure strategy x_I^L.

Hence, the incumbent's mixed strategy $x_I(\lambda)$ is an equilibrium if and only if it satisfies $u_I(x_I^H, x_E^H, H) = u_I(x_I^L, x_E^\lambda, H)$. The results can be summarized as the following proposition.

Proposition 51.3. (1) $u_I(x_I^H, x_E^H, H) \geq u_I(x_I^L, x_E^\lambda, H)$ if and only if λ-Hybrid Revelation for $\lambda = 1$, which is identical to Truthful Revelation, is a PBE.
(2) $u_I(x_I^H, x_E^H, H) \leq u_I(x_I^L, x_E^M, H)$ if and only if λ-Hybrid Revelation for $\lambda = 0$, which is identical to Strategic Revelation, is a PBE.
(3) $u_I(x_I^L, x_E^M, H) \leq u_I(x_I^H, x_E^H, H) \leq u_I(x_I^L, x_E^\lambda, H)$ if and only if λ-Hybrid Revelation for λ satisfying $u_I(x_I^H, x_E^H, H) = u_I(x_I^L, x_E^\lambda, H)$ is a PBE.

51.6 Numerical Examples

In this section, I show some results of comparative statics about equilibrium strategies and values of the incumbent by numerical examples. Parameters in examples are basically set as $\mu = 0.03$, $r = 0.07$, $p = 0.5$, $\sigma = 0.2$, $x = 0.05$, $\pi_{I1}^H = 12$, $\pi_{I1}^L = 7$, $\pi_{I2}^H = 4$, $\pi_{I2}^L = 4$, $\pi_{E2}^H = 4$, $\pi_{E2}^L = 1$, $K_I = 50$ and $K_E = 100$.

First, I examine a relation between $u_I(\cdot, \cdot, H)$ and π_{I2}^H, that are values and the duopoly profits, respectively, for the high-demand type of the incumbents. Figure 51.1 illustrates the values $u_I(x_I^H, x_E^H, H)$, $u_I(x_I^L, x_E^M, H)$, $u_I(x_I^L, x_E^L, H)$.

51 Strategic Information Revelation Through Real Options in Investment Games 781

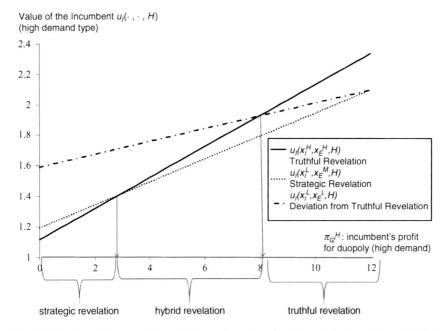

Fig. 51.1 Values of the incumbent with high demand $u_I(\cdot, \cdot, H)$ and the duopoly profit of the incumbent with the high demand π_{I2}^H

For $\pi_{I2}^H \geq 8.0$, $u_I(x_I^H, x_E^H, H)$ is greater than $u_I(x_I^L, x_E^L, H)$. In this range, the high-demand type does not deviate the truthful optimal timing of the investment, because the duopoly profit of the incumbent is sufficiently large and the incumbent does not have strong incentive to make the entrant's investment delay. Hence, the high-demand type of the incumbents enters to the market at the optimal timing of the investment for the high demand and reveals his information truthfully.

In contrast, for $\pi_{I2}^H \leq 2.9$, $u_I(x_I^H, x_E^H, H)$ is less than $u_I(x_I^L, x_E^M, H)$. In this range, the incumbent with high type invests at the optimal timing for the low demand to hide information for high demand because the duopoly profit of the incumbent is small and the decrement of the incumbent's profit by the investment of the entrant is critical. The incumbent enters to the market at the optimal timing of the investment for the low demand and does not have incentive to deviate to the optimal timing of the high demand in this range. For $2.9 \leq \pi_{I2}^H \leq 8.0$, $u_I(x_I^L, x_E^M, H) \leq u_I(x_I^H, x_E^H, H) \leq u_I(x_I^L, x_E^L, H)$, the incumbent uses a mixed strategy as λ– Hybrid Revelation. In this interval, the value of the high-demand type of the incumbents is same as $u_I(x_I^H, x_E^H, H)$ because the mixed strategy should satisfy condition $u_I(x_I^H, x_E^H, H) = u_I(x_I^L, x_E^\lambda, H)$. Therefore, the value of hight type of the incumbent in the equilibrium strategy is identical to $u_I(x_I^H, x_E^H, H)$ for $\pi_{I2}^H \leq 2.9$ while it is $u_I(x_I^L, x_E^M, H)$ for $\pi_{I2}^H \geq 2.9$.

Secondly, the effect of volatility is examined. Figure 51.2 illustrates relation between values of the high-demand type of the incumbents and the volatility.

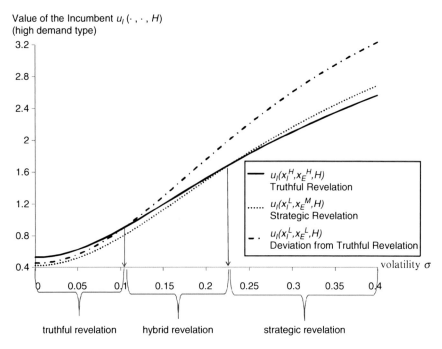

Fig. 51.2 Values of the incumbent with high demand $u_I(\cdot, \cdot, H)$ and volatility σ

If the volatility is small, the incumbent invests truthfully while if the volatility is large, the incumbent invests strategically. In medium range, the incumbent uses the mixed strategy.

Finally, the impact of the entrant's cost on an incumbent's value is investigated. It is interesting that the incumbent's value is affected not only by the incumbent's own cost, but also by the rival's cost because smaller entrant's cost pushing forward the entrant's investment reduces the incumbent's value. Figure 51.3 depicts relation between the values of the high-demand type of the incumbent and cost of the entrant. If the entrant's cost is large, the timing of the entrant's investment is late. Since the entrant's investment is negligible effect on the incumbent's value, the incumbent with high type invests truthfully. On the other hand, the incumbent invests strategically for small entrant's cost. For the medium interval of the entrant's cost, the incumbent uses the mixed strategy.

51.7 Conclusion

This paper examines investment game for an incumbent and an entrant for optimal entries into a new market in which the incumbent only has information of demand, high or low, and the entrant predict the demand by observing the incumbent's timing

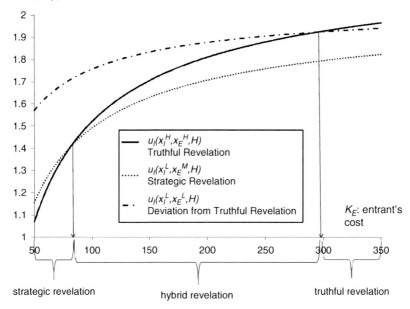

Fig. 51.3 Values of the incumbent with high demand $u_I(\cdot, \cdot, H)$ and cost of the incumbent K_E

of the investment I investigate whether the incumbent reveals the information truthfully or not taking into account signaling effect by using the concept of a weak perfect Bayesian equilibrium. I characterize a condition for the incumbent with information of high demand invests strategically in the equilibrium, and show that it is necessary for the incumbent to use a mixed strategy in the equilibrium under some condition.

Further research is needed to obtain the above results analytically. We will obtain the results by differentiating the values with respect to profit flows, costs and volatility. Some extensions of the model would be interesting. First, preemptive behavior should be considered by eliminating the assumption where the incumbent is leader and the entrant is follower. Second, other stochastic processes could be considered.

Acknowledgements I would like to thank Alberto Pinto for giving me the opportunity to write this summary of my work. I also thank participants of 14th Annual International Conference of Real Option and EURO 2010 for helpful comments.

References

1. Grenadier, S.R.: The strategic exercise of options: Development cascades and overbuilding in real estate markets. J. Finan. **51**(5), 1653–1679 (1996)
2. Grenadier, S.R., Wang, N.: Investment timing, agency and information. J. Financ. Econ. **75**, 493–533 (2005)

3. Hsu, Y.W., Lambrecht, B.M.: Preemptive patenting under uncertainty and asymmetric information. Ann. Operat. Res. **151**, 5–28 (2007)
4. Huisman, K.J.M.: Technology and Investment: A Game Theoretic Real Options Approach. Kluwer Academic, Dordrecht (2001)
5. Huisman, K.J.M., Kort, P.M.: Effect of Strategic Interactions on the Option Value of Waiting. Tilburg University, Netherlands (1999)
6. Kijima, M., Shibata, T.: Real options in a duopoly market with general volatility structure 2002. (2002)
7. Kong, J.J., Kwok, Y.K.: Real options in strategic investment games between two asymmetric firms. Eur. J. Oper. Res. **181**, 967–985 (2007)
8. Kulatilaka, N., Perotti, E.C.: Strategic growth options. Manage. Sci. **44**(8), 1021–1031 (1998)
9. Lambrecht, B., Perraudin, W.: Real options and preemption under incomplete information. J. Econ. Dyn. Control **27**, 619–643 (2003)
10. Pawlina, G., Kort, P.: Real options in an asymmetric duopoly: Who benefits from your competitive disadvantage? J. Econ. Manag. **15**, 1–35 (2006)
11. Smets, F.: Exporting versus fdi: The effect of uncertainty, irreversibilities and strategic interactions. Working Paper, Yale University (1991)
12. Smit, H.T.J., Trigeorgis, L.: Strategic delay in a real options model of r & d competition. Rev. Econ. Stud. **69**, 729–747 (2002)
13. Watanabe, T.: Real options and signaling in strategic investment games. Discussion Paper (2010)

Chapter 52
On Consumption Indivisibilities, the Demand for Durables, and Income Distribution

David Zilberman and Jenny Hsing-I Liu

Abstract This paper presents a framework to assess demand of durable products recognizing their indivisibility. The paper builds on the household production framework recognizing consumers demand for product characteristics and this demand can be satisfied by purchasing durables that combined with variable inputs to generate these characteristics or renting services that provide these characteristics. One example is buying a washer and dryer or going to a laundromat. Our analysis recognizes heterogeneity among consumers and suggests that some segments of the population will buy the durables while others will rent. We derived demand for the durables and associated variable inputs by aggregating over population. We show the demands are affected by prices and income distribution parameters.

52.1 Introduction

The limited range of issues addressed by traditional theory motivated Becker [1] and Lancaster [7] to introduce new approaches to consumer theory. The new approaches are capable of analyzing issues such as quality changes among goods and consumers' reaction to new goods. These approaches depart from traditional theory by rejecting the assumption that consumers derive utility from goods and services purchased in the market per se. Instead, they assume that utility is obtained from entities, which are produced by the family itself with purchased market goods, services, and the time of some members of the family.

D. Zilberman (✉) and J. Hsing-I Liu
Department of Agricultural and Resource Economics, UC Berkeley, Berkeley, CA, USA
e-mail: zilber11@berkeley.edu, jenny.liu@berkeley.edu

M.M. Peixoto et al. (eds.), *Dynamics, Games and Science I*, Springer Proceedings in Mathematics 1, DOI 10.1007/978-3-642-11456-4_52,
© Springer-Verlag Berlin Heidelberg 2011

Becker and Lancaster use different specifications of the family-production technologies. Lancaster's model (see [7]) assumes multiproduct-linear technologies (each activity generates several characteristics keeping a fixed input–output relationship). In Becker's work [1], each activity produces one output following a neoclassical production function.

This paper uses the family production-function approach to explain the differences and interdependencies between the demand for durable and nondurable goods. As in Becker, each production activity is assumed to generate one commodity. However, the family-production technology is assumed to have putty-clay properties. Namely, each commodity can be produced via several processes. Each process has its own fixed proportion between variable inputs and the commodity and may require the use of a specific capital good. In the analysis durables play the role of capital goods, while nondurable goods are the variable inputs. The choice of durable goods determines the fixed nondurables-output ratios, while the actual amount of the commodity consumed is determined by the amounts of nondurables used. Unlike the traditional putty-clay model, however, there are a relatively small number of durables for producing each commodity, and the amount of commodities produced by each durable is assumed to be unconstrained (or, alternatively, the productive capacities of the durable goods are above the range of practical levels of consumption). Thus, the selection of durable goods is the result of discrete choices.

Home appliances, such as washers, dryers, stoves, dishwashers, furnaces, etc. are examples of durables that suggest the model developed here.[1] Washing machines and dryers, for example, are usually utilized by households only for relatively small periods of time, and their variable costs are approximately constant per load of clothes washed, dried, etc. In most cases, a family has to make a discrete choice whether to purchase a washer and dryer to clean its clothes incurring a fixed cost and relatively low variable cost, or to use a laundromat not making any investment but paying a higher variable cost (in terms of time and operation cost).

Using this model, the first part of the paper analyzes the individual consumer's demand functions for durables, commodities, and goods as functions of prices and income.[2] The model is developed for the simple case of two commodities–one a composite commodity and the other a specific commodity (such as clean clothes). Two processes can be used to produce the specific commodity; one requires a

---

[1] One element, which is not included in this work and should be incorporated in future research, is quality changes among commodities. Many times the durable good is the source of differences in quality. The commodity transformation service has different qualities when one moves in a new Cadillac or an old Pinto. Moreover, in many cases capacity can be analyzed as an additional quality characteristic.

[2] Small and Rosen [10] and Hanemann [4] have analyzed the demand for variables? selected by a discrete choice as part of their welfare impact analysis of quantal choice models. Their analysis does not allow changes in other variables simultaneously with the discrete choice (less time is required for washing when washers and dryers are bought than when a laundromat is used), and they do not analyze extensively income effects. The aggregate relations they consider assume constant income level (or constant marginal utility of income), while we aggregate here over income.

52 On Consumption Indivisibilities, the Demand for Durables, Income Distribution   787

purchase of a durable, and the other does not require such a purchase. This simplified model allows graphical presentations and proofs using some of the traditional tools of graphical analysis of consumer behavior (income consumption curves, price consumption curves, etc.) and thus is useful for instructional purposes.

The second part of the paper presents a tractable approach for generating durable demand relationships given income distributions and utility function specifications. The results suggest that several income distribution parameters, rather than simply average income level, are essential in deriving demand for durables. With the wide use of discrete choice model in adoption studies (see Sunding and Zilberman [11]), as well as studies of consumer behavior (see Train [12]), and the increased availability of disaggregated data on consumer characteristics and income distribution, the approach introduced here becomes more applicable.

52.2 The Model

A consumer derives his utility from consuming y_0 units of a numeraire commodity and y_1 units of a specific commodity each period. The consumer's utility function is traditionally increasing in both commodities, concave, and twice differentiable, and it is denoted by $U(y_0, y_1)$.[3] Commodity 1 can be produced using $K + 1$ alternative processes denoted by $k = 0, 1, \ldots, K$. The consumer has to purchase a specific durable in order to use each of the processes with $k > 0$. The zero process does not require purchase of a durable. (It may use an already-owned durable.) Let δ_k be a dichotomous variable taking the value one when the kth durable is used to produce commodity 1 and zero otherwise. Assuming that a family is employing only one of the processes to produce the specific commodity, the dichotomous variable, δ_k, is constrained by

$$\sum_{k=0}^{K} \delta_k = 1. \tag{52.1}$$

There are m nondurable goods (including time), which are used in the production of commodity 1. The amount of good j required to produce one unit of commodity 1 using process k is denoted by β_{jk}. The price of the jth nondurable is P_j. Thus, the average variable cost of consuming commodity 1 using technology k is

$$\pi_k = \sum_{j=1}^{m} P_j \beta_{jk}, \tag{52.2}$$

where $k = 0, \ldots, K$. π_k will be referred to as the price of commodity 1 under technology k. The fixed annual cost associated with the use of process k is denoted

---

[3] It is assumed that $U'_{y_i}(y_0, y_1) = \infty$ when $y_i = 0$ for all i. This assumption ensures that all commodities are consumed.

by rI_k, where I_k is the purchase price of the kth durable, and r is the sum of the interest and amortization rate. Since process 0 does not involve purchase of a durable, $I_0 = 0$.

The permanent income of the family is denoted by R. Income is spent on periodical payments for the durable purchase, the purchase of nondurables used to produce commodity 1, and purchases associated with consumption of commodity 0 (numeraire commodity which price is 1).[4]

Thus, the budget constraint of a family is given by

$$\sum_{k=o}^{K} \pi_k y_1 \delta_k + y_0 + r \sum_{k=o}^{K} \delta_k I_k = R. \tag{52.3}$$

The consumer choice problem is

$$\max_{\delta_0,\delta_1,\dots,\delta_k,y_0,y_1} U(y_0, y_1) \tag{52.4}$$

subject to (52.1) and (52.3), where δ_k is either 0 or 1.

The consumer problem can be solved in two steps. First, compute the optimal consumption pattern and the resulting utility under each process and then select the process (and the durable) that maximizes utility. To simplify the graphical analysis, consider the case where commodity 1 can be produced only by two processes ($K = 1$). One is a technology that does not require the purchase of a new durable, and the associated price of the commodity is π_0. The alternative technology involves the purchase of a durable and requires fixed cost of rI_1 dollar per period and average variable cost of π_1 dollar per unit of commodity consumed.

The optimal consumption choice will be determined by comparing V_0, the maximum utility derived under technology 0 with V_1, the maximum utility derived when durable 1 is installed where

$$V_0 = \max_{y_0,y_1} U(y_0, y_1) \tag{52.5}$$
$$\text{subject to } y_1 \pi_0 + y_0 = R$$

and

$$V_1 = \max_{y_0,y_1} U(y_0, y_1) \tag{52.6}$$
$$\text{subject to } y_1 \pi_1 + y_0 = R - rI_1.$$

---

[4] The income R can also be interpreted as Becker's "full income", i.e., income from profit and all potential income from labor. In this case the model can be extended to include leisure as one of the commodities in the utility function, and it is produced only by labor.

52 On Consumption Indivisibilities, the Demand for Durables, Income Distribution

Fig. 52.1 The choice problem

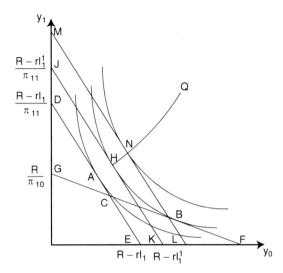

The choice problem is illustrated graphically in Fig. 52.1. The budget constraint, when one is restricted to operate without the durable, is GF. The budget constraint, when the durable is used, is DE. The two budget lines intersect at C.

To find the optimal solution, one has to compare the utility index at the tangency points of the indifference curves with the budget lines. In Fig. 52.1 these intersection points are A and B and, since the indifference curve at B represents a higher utility level, point B will denote the optimal utility choice.

Figure 52.1 represents only a certain type of outcome for the consumption choice problem. Different budget lines or preference orderings may result in other types of outcomes. That is, the budget constraint of the consumer generates a set of efficient points. This set includes all the feasible commodity bundles where the consumption of one commodity cannot be increased without a reduction in the consumption of the other commodity. In the case illustrated in Fig. 52.1, the set of efficient points is denoted by the broken line, DCF. The globally optimal consumption choices belong to the set of efficient points. All of the efficient consumption bundles can be optimal choices except the point of the intersection of the two budget lines. This point, while efficient, cannot be optimal since, even when it is the best choice under one of the processes, the budget line of the other process will be tangent to a higher indifference curve, and that tangency point will be the global optimum.

A unique solution to the consumer choice problem always exists in cases where the optimal consumption pattern under one of the processes belongs to the set of efficient consumption choices, but the optimal choice under the other process does not belong to the set. This happens when the optimal consumption of y_1 under both processes is greater or smaller than at the intersection point of the budget line. When it is greater, process 1 is superior and, in terms of Fig. 52.1, the global optimum belongs to DC. When it is less, the process without the fixed cost is more desirable, and the optimum belongs to CF in Fig. 52.1.

If the optimal consumption choices under both processes belong to the efficient set, one may have a multiple solution when the same indifference curve is tangent to both budget lines. This, however, will not be the usual case and, when both processes belong to the efficient set, one has a unique solution almost always. Figure 52.1 demonstrates this point. Given the consumption technology, nondurable prices, and the consumer's income, only a certain price of the durable goods will result in a multiple solution. This happens when the price of the durable is I_1^1; the budget line, when the durable is used, is JK; and both points H and B are optimal choices. When the price of the durable is greater than I_1^1 (as in the case of Fig. 52.1 when the durable is associated with DE), the consumer will not use the durable, and the optimal consumption point will be B. However, in cases where the price of the durable is smaller than the critical price I_1^1, the consumer will prefer the process, which uses the durable, and the final consumption pattern will be determined accordingly. In these cases a reduction in the price of the durable has the effect of an increase in income in a traditional consumption choice problem. Thus, the consumption points, when durable prices are below I_1^1, belong to curve HNQ, which has all the properties of an income consumption curve.

This analysis can be extended to derive the properties of several interesting relationships. They include the individual's demand for the durable as a function of its price, the commodities' prices (good prices) and income, the demand for commodity 1 as a function of its prices under the different technologies and income, the demand for the nondurables, and the indirect utility function. The following sections will analyze the effects various parametric changes have on these relationships.

52.3 The Effects of Changes in the Durable and Commodity 1 Prices on the Demand for the Durable and Commodity 1

Let $\tilde{V}(\pi_1, R)$ be the indirect utility function associated with a traditional choice problem

$$\tilde{V}(\pi_1, R) = \max_{y_0, y_1} U(y_0, y_1) \qquad (52.7)$$
$$\text{subject to } y_0 + y_1 \pi_1 = R.$$

The analysis associated with Fig. 52.1 suggests a qualitative choice model for durable demand. Specifically, using (7) to combine (5) and (6) yields the formulation of the demand for durable

$$\delta_1 = \delta_1^D(\pi_{10}, \pi_{11}, I_1, R) = \begin{cases} 1 & \text{if } \tilde{V}(\pi_{11}, R - rI_1) > \tilde{V}(\pi_{10}, R); \\ 0 & \text{if } \tilde{V}(\pi_{11}, R - rI_1) < \tilde{V}(\pi_{10}, R). \end{cases} \qquad (52.8)$$

When equality holds, we do not have a unique solution; and the consumer is indifferent to purchasing or not purchasing the durable. To define the demand for

52 On Consumption Indivisibilities, the Demand for Durables, Income Distribution

commodity 1, let $D(\pi_1, R)$ be the traditional demand for y_1 derived from solving the optimization problem in (7). The demand for y_1 in our case becomes

$$y_1 = y_1^D(\pi_{10}, \pi_{11}, I_1, R) = \delta_1 D(\pi_{11}, R - rI_1) + (1 - \delta_1) D(\pi_{10}, R). \quad (52.9)$$

As (52.8) and Fig. 52.1 indicate, the demand curve for the durable (given income, R, and commodity prices, π_{10}, π_{11}) is a step function which equals one when the durable price is smaller than some critical level (I_1^1 in Fig. 52.1) and zero when the price of the durable is greater than I_1^1. The demand for the durable at price, I_1^1, is indeterminate.

Changes in the durable goods price will affect the consumption of commodity 1 only when the durable is used. In this case (52.9) suggests that a reduction in the durable goods price has the same effect as increasing income by t times the amount of the reduction. This and Fig. 52.1 yield a graphical presentation that relates the consumption of commodity 1 to the price of the durable. Such a graph is derived in Fig. 52.2. The relationship consists of two disconnected parts. For prices that are greater than I_1^1, the consumption of commodity 1 is constant and equal to y_1^B (the consumption at point B in Fig. 52.1). The shape of the second part of the graph, which corresponds to prices smaller than I_1^1, is determined by the income elasticity of the traditional demand for commodity 1 (when income is $R - rI_1^1$). Four possible shapes are depicted in Fig. 52.2. A luxury commodity, with income elasticity greater than one, has a negatively sloped and convex curve relating consumption of y_1 to durable price (given durable price smaller or equal to I_1^1). This case is depicted by BF in Fig. 52.2. The negatively sloped linear curve, BE, corresponds to cases of unitary income elasticity. Normal commodities (with income elasticities

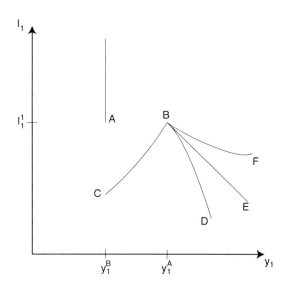

Fig. 52.2 The relation between the consumption of commodity 1 and the price of the durable

Fig. 52.3 The relation between the price of commodity 1 and the quantity of the commodity demanded

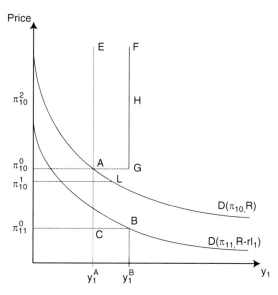

between zero and one) have negatively sloped concave curves like BD. Finally, inferior commodities have positively sloped curves (BC).

Suppose that commodity 1 prices under both processes are independent (i.e., the processes use different nondurables) and consider how changes in these prices affect the choice of the durable and quantity demanded of commodity 1.[5] Here again, Fig. 52.1 and (52.8) suggest that the demand for durable is a step function of π_{10} given R, π_{11}, and I_1^1. The durable will be purchased for all levels above the critical level and will not be purchased for lower prices. Similarly, the demand for the durable is a step function of π_{11} (given π_{10}, R, I_1^1).

Using (52.9), the relationship between the price of commodity 1 under each of the processes and the quantity of the commodity demanded is investigated in Fig. 52.3.

The starting point for the analysis is commodity 1 of price combinations (π_{10}^0, π_{11}^0), which results in a multiple solution given R and I_1. It is also assumed that commodity 1 is a normal commodity; thus, the demand curve, $D(\pi_{10}, R)$, is above $D(\pi_{11}, R - rI_1)$. From (52.9), it is concluded that, when the price of commodity 1 under process 0 is π_{10}^0, the graph of the relationship between quantity demanded of commodity 1 and its price under process 1 consists of two disconnected parts. The first, for prices greater than π_{11}^0, is denoted by the line, CE. For these π_{11} levels, the durable is not used; instead, process 0 is used and commodity 1 price under this process determines y_1 demand level to be $y_1^D = D(\pi_{10}^0, R)$. When π_{11} is smaller than the critical level, π_{11}^0, the durable good is used, the consumption

---
[5] Of course, the prices of commodity 1 under the two processes are interdependent. Some inputs are used in both processes (on different proportions), and change in these inputs will affect the prices of commodity 1 under both processes. These cases will be investigated later using the results derived here.

of y_1 is determined according to $D(\pi_{11}, R - rI_1)$, and the second part of the demand curve for y_1 (when π_{11} varies) consists of point B and all the points of $D(\pi_{11}, R - rI_1)$ to the right of B. When $\pi_{11} = \pi_{11}^0$, we have a multiple solution, and the demand for y_1 can be either y_1^A or y_1^B. The demand for y_1 as a function of π_{10} gives R, I_1, and $\pi_{11} = \pi_{11}^0$ is determined similarly. When π_{10} is greater than the critical level, π_{10}^0, the durable is used, commodity 1 demand is equal to y_1^B, and the corresponding part of the demand curve is denoted by FG. When commodity 1 prices under process 0 are smaller than π_{10}, this process is used, and the corresponding segment of the demand curve for π_{11} consists of A and all the points, $D(\pi_{10}, R)$, to the right of A. Note that an *increase* in the price of commodity 1 under process 0 from, let's say, π_{10}^1 to π_{11}^2 will generate a move from L to H, and actual demand will *increase*. The availability of a durable with lower commodity 1 prices allows a switch that increases consumption of commodity 1 when the price of commodity 1, under the traditional technology, rises above a critical level.

The results do not change essentially when commodity 1 is an inferior commodity. The only difference is that the curve, $D(\pi_{11}, R - rI_1)$, is above $D(\pi_{10}, R)$ in this case. Both demand relationships of y_{11}, however, have two discontinuous segments. The demand for y_{11} cannot increase when π_{11} increases ($\pi_{10} = \pi_{10}^0$), while it may increase when π_{10} increases and π_{11} is kept constant at π_{11}^0.

52.4 The Effects of Changes in Income on the Demand for the Durable and Commodity 1

It is of interest to find how changes in income affect the consumption of commodities and the use of durables. This subject can be analyzed using Fig. 52.4. In Fig. 52.4 the budget lines are drawn for two cases with the same prices of commodities and durables, but different income levels. Note that, in both cases, the budget line for the process using the durable and the budget line for consumption without the durable intersect always on the same level of y_1 denoted by $\bar{y}_1 = rI_1/(\pi_{10} - \pi_{11})$. This happens because an increase in income causes the same horizontal shift in both budget lines, i.e., the efficient set is homogeneous of degree 1 in income. This result and the fact that the process using the durable (without the durable) is superior when the optimal outcomes under both processes are above (below) the intersection points of the budget lines lead to some interesting insights.

These results indicate, for example, that poor people will not use the durable. This is the case when the budget lines do not intersect (thus, both budget lines are below the line $y_1 = \bar{y}_1$), and the budget line for the process without the durable dominates the one with the durable. It can also be deduced that the durable will always be purchased above a certain income if the commodity 1 is not inferior and its income elasticity is always positive.[6] Under these conditions, there must be an

---

[6] The notion of an inferior commodity used here is the traditional one. A commodity is inferior under a given consumption technology if its consumption is reduced when income is increased

Fig. 52.4 The relation between the consumption of commodities and the use of durables

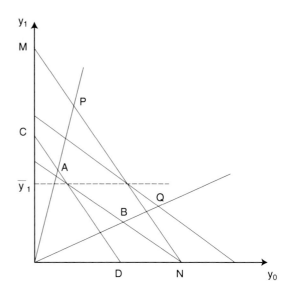

income level above, which the optimal solutions under both processes are above the line $y_1 = \bar{y}_1$. This condition also implies that, for given prices, there is at least one critical income level when the two budget lines are tangent to the same indifference curve. Moreover, if preferences are homothetic, then there is *only one* critical point, which separates lower income levels for which it is optimal not to use the durable and higher levels for which the use of the durable is optimal.[7] This property of homothetic preferences results from the fact that income consumption curves are rays from the origin and that, if the consumer is indifferent to two consumption combinations, he/she is also indifferent to the combination generated by multiplying both of them by the same scalar. Let the points A and B in Fig. 52.4 be two optimal outcomes for a given income R_0, and let them lie on the same indifference curve. The points P and Q will be the outcomes associated with increasing the income by ΔR dollars. The transformation from A to P is a scalar multiplication of A by $1 + \Delta R/(R_0 - rI_1)$, and the transformation from B to Q is a scalar multiplication of B by $1 + \Delta R/R_0$. Since the first scalar is larger, P is preferred over Q. Hence, the consumer will prefer to use the durable at all income levels above the one at which the consumer is indifferent between the two processes. Similarly, one can prove that a reduction of income below the critical income level (R_0) will cause the consumer to prefer the process without the durable.

---

and prices do not change. Note that one can define here a new notion of inferiority and say that a commodity is inferior if its consumption is reduced when income is changed. A commodity might be normal in the traditional way and inferior under the new notion.

[7] Actually, this result is more general; here it will be proven only for homothetic utilities.

52.5 The Effects in Changes in Nondurable Price

This section will analyze the effects of changes in nondurable good prices on the choice of the durable and their own demand curves. First, distinguish between goods that are used only in one of the processes (specialized goods) and goods that are used in both processes. For specialized goods, the analysis is rather simple. When their prices are very high, they make their processes less desirable and the consumer will not use the processes. For each specialized good, there is a critical level; and, once it is below the level (other variables kept constant), its process is adopted. The demand curve for a specialized good has two disconnected segments. It is 0 for all prices above the critical level when the specialized good process is not used. For all prices below the critical level, the demand is negatively sloped (assuming commodity 1 is not a Giffen commodity).

The slope of these demand curves can be derived from the demand for commodity 1. Let good j be a specialized input in process 1 with input–output coefficient δ_{j1}. When process 1 dominates, the price elasticity of good j is simply the price elasticity of the demand for commodity 1 times the share of the expenses on good j in the variable cost of producing commodity 1. Let η_{x_j} be the price elasticity of good j and $\eta_{y_{11}}$ the price elasticity of the demand for commodity 1 under process 1, then[8]

$$\eta_{x_{11}} = \frac{P_j}{\pi_{11}} \beta_{j1} \eta_{y_{11}}. \tag{52.10}$$

Equation (52.10) transfers to consumption theory a familiar condition from production theory stating that elasticity of derived demand of a good is lower as the share of this good in variable cost is lower.

The analysis is much more complicated for goods which participate in both processes. To simplify somewhat, consider first the case when the good's input commodity coefficients under the two processes are proportional to the price of commodity 1 under both processes $\left(\text{i.e, for good } j, \frac{\beta_{j0}}{\beta_{j1}} = \frac{\pi_{10}}{\pi_{11}}\right)$.

This will be the case, for example, when only one nondurable is used in producing commodity 1 in both processes. In this case the change of the price of good j has

---

[8] Let $y_{11} = D(\pi_{11}, R - rI_1)$. Then using (52.9),

$$x_1 1 = \delta_1 \beta_{j1}^2 y_{11},$$

since $\frac{d\pi_1}{dP_j} = \beta_{j1} \frac{dx_j}{dP_j} = \delta_1 \beta_{j1} \frac{dy_{11}}{d\pi_{11}}$. Using

$$\eta y_{11} = \frac{dy_{11}}{d\pi_{11}} \frac{\pi_{11}}{y_{11}},$$

one derives

$$\eta_{x_j} = \frac{dx_j}{dP_j} \frac{P_j}{x_j} = \delta_1 \frac{\beta_{j1}^2 y_{11} P_j}{x_j \pi_{11}} \eta_{y_{11}} = \eta_{y_{11}} \frac{P_j}{\pi_{11}} \beta_{j1}.$$

Fig. 52.5 The effects in changes in the price of a nondurable used in producing commodity 1

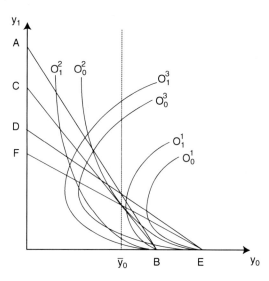

the same proportional effect on the prices of commodity 1 under both processes, and the new commodity prices keep the same proportions as the technical coefficients. Therefore, the effects of changes in good j prices in this case can be derived by analyzing the optimal choices when both prices of commodity 1 are changed in the same proportion. The lines AB and DE in Fig. 52.5 are budget constraints for one set of commodity 1 prices. The budget lines CB and FE are derived by proportional changes in the initial prices. Note that, in both cases, the intersection points of the budget lines have the same y_0 value denoted by $\bar{y}_0$. Thus, in both cases, process 1 is preferred if both optimal outcomes are left to the line $y_0 = \bar{y}_0$; process 0 (the durable is not used) is preferred when both outcomes are to the right of the critical y_0 level

$$\bar{y}_0 = rI_1 \left[\frac{\beta_{j0}}{\beta_{j0} - \beta_{j1}} - \frac{R}{rI_1} \right] \quad (52.11)$$

where $\frac{\beta_{j1}}{\beta_{j0}} = \frac{\pi_{11}}{\pi_{10}}$.

More insight into the behavior of the optimal outcomes can be gained by constructing the price possibilities curves (offer curves for both processes).[9] Two points, one on each curve, correspond to each pair of commodity 1 prices associated with the given proportions. The offer curves are convex to the origin. They are also negatively sloped when the price of the commodity is high but may reverse slopes for lower prices (for normal goods).[10] The relationship between these offer curves

---

[9] Each curve is the locus of all optimal y_0, y_1 combinations under each process resulting from change in commodity 1 price under the process given the other parameters.

[10] A special behavior pattern occurs when the elasticity of demand for commodity 1 is always unitary. In this case the PCC lines are parallel to the y_1.

52 On Consumption Indivisibilities, the Demand for Durables, Income Distribution

and the line $y_0 = \bar{y}_0$ determines the choice of optimal technology when prices are changed, but the ratio of the two prices is kept constant.

One possibility is that, under all prices, the use of the process without the durable will be preferred. This is the case, for example, where the offer curves are O_0^1 and O_1^1 in Fig. 52.5. A second possibility is that the process without the durable is preferred when the prices of commodity 1 are very high. But when the prices of commodity 1 become lower, use of a durable to increase the consumption of commodity 1 is preferable. This is the case when the offer curves are O_0^2 and O_1^2 in Fig. 52.5. Both intersect the line $y_1 = \bar{y}_0$ only once. Such is typically the case with luxury goods or normal goods with high-income elasticity. Another possibility is that low prices of commodities result in the use of process 0; higher prices cause a switch to the process without the durable, but very high prices will result in a reversal of technology and the use of the process without the durable. This is the case when the offer curves are O_0^3 and O_1^3, and it may occur for some normal commodities or in cases where the commodity is inferior for some income levels.[11] In these cases high prices of commodity 1 result in a low level of consumption of the commodity, and the use of the durable is not justified. When prices become lower, the substitution effect will increase the consumption of commodity 1 and encourage use of the durable. When prices become very low, however, the income effect will cause an increase in the demand for other goods, and the money spent to pay the fixed cost of the durable good can yield higher utility in other uses.

The relationship between the quantity consumed and the price of commodity 1 under process 1 when prices keep fixed proportions is especially interesting for the later case O_0^3 and O_1^3. This relationship is described in Fig. 52.6 where the segments AB and EF correspond to prices for which process 0 is preferred, while the segment CD corresponds to prices for which the use of the durable is preferred. This demand relationship is peculiar since it results in situations where an increase in both prices of commodity 1 (they keep fixed proportions) implies an increase in its consumption (i.e., the movement from M to N).

The results for cases when both prices of commodity 1 keep a fixed proportion suggest several patterns of durable choice when a single same good, x_j, is the variable input in both processes. The critical value, $\bar{y}_0$, is an indicator, independent of the consumer taste, of a durable choice pattern a consumer may have. Consumers with negative $\bar{y}_0$ will not purchase the durable at any price (their $\delta_1 = 0$ for all P_j). From (52.11), this condition applies when the share of the fixed durable cost in total income is larger than the relative efficiency gain from the durable purchase (i.e., $rI_1/R > \frac{\beta_{j0}-\beta_{j1}}{\beta_{j0}}$). Thus, we should not expect consumers with less than the critical income level, i.e., with $R < rI_1 \frac{\beta_{j0}}{\beta_{j0}-\beta_{j1}}$ to purchase the more efficient durable even if the price of the nondurable used to produce commodity 1 is rising very much. The purchase of the durable is also unlikely for all P_j by consumers

---

[11] Another special case is when the elasticity of demand is unitary under both processes. In these cases the optimal process choice is not affected by a price change given that both prices of commodity 1 follow fixed proportions.

Fig. 52.6 The relation between the quantity consumed and the price of commodity 1 under process 1

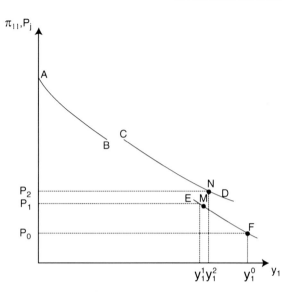

with small positive $\bar{y}_0$. For consumers with a higher $\bar{y}_0$, the durable choice will be a step function of the nondurable prices. As Fig. 52.5 suggests, the durable will not be purchased for high nondurable prices. There will be a segment of lower nondurable prices, which will result in the purchase of the nondurable, and a third segment of reswitching might occur if commodity 1 is not a luxury commodity (its elasticity is less than 1) in the relevant segment.

From (52.9), the demand for commodity 1 as a function of the nondurable price is given by

$$y_1^D = \delta_1 D(\beta_{j1}, P_j, R - rL_1) + (1 - \delta_1) D(\beta_{j0}, P_j, R).$$

The demand curve for commodity 1 as a function of the nondurable good price will be well behaved and negatively sloped for consumers who do not switch to use the durable for any P_j (low-income consumers with negative y_0). It will have a point of discontinuity but will continue to suggest that a reduction in a good price will increase the demand for the commodity it produces for consumers who switch to the durable good when P_j is smaller than a certain critical level. For consumers who reswitch, this demand curve will behave like the one in Fig. 52.6, and there will be a range of prices when increase in the nondurable price may increase commodity 1 consumption. This type of consumer behavior is not unrealistic. For example, suppose commodity 1 is home heating, both technologies use natural gas, and the durable is a new more efficient furnace. The initial price of natural gas may be very low (like P_0), and the consumer will be at F. An increase of natural gas from P_0 to P_1 will reduce natural gas consumption to y_1^1, however, an additional increase will cause a purchase of the more efficient furnace and increase in heating consumption (from y_1^1 to y_1^2).

52 On Consumption Indivisibilities, the Demand for Durables, Income Distribution 799

The demand for the nondurable good, in our case, is derived from (52.2) and (52.9) to be

$$x_j = \delta_1 \beta_{j1} D(P_j \beta_{j1}, R - rI_1) + (1 - \delta_1)\beta_j D(P_{j0}\beta_j, R). \tag{52.12}$$

Using (52.12), one can see that the price elasticity of the demand for the nondurable, x_j, is equal to the price elasticity of the demand for commodity 1 with respect to the price of commodity 1 for the process used, i.e.,

$$\eta_{x_j} = \delta_1 \eta_{y_{11}} + (1 - \delta_1)\eta_{y_{10}},$$

where

$$\eta_{y_{10}} = \frac{\partial(\pi_{10}, R)}{\partial D \pi_{10}} \frac{\pi_{10}}{y_{10}}. \tag{52.13}$$

Like other demand curves, the demand for the nondurable as a function of its price is likely to have discontinuity points at critical prices as one switches from one technology to another. However, unlike the demand for y_1 as a function of P_j, the demand for x_j will not necessarily increase at the critical price as one switches from technology 0 to 1. In many cases the switch to the use of a durable may reduce the nondurable consumption; and, in some rare cases, the consumption of x_j at the critical price will not be affected by the technological switch (the demand for x_j at this point will be continuous but not differentiable).

To illustrate and comprehend better how a technological switch affects the consumption of x_j at a given level, consider the case where the consumer has a CES utility function

$$U(y_0, y_1) = A(e_0 y_0^{-\rho} + e_1 y_1^{-\rho})^{-\frac{1}{\rho}} \tag{52.14}$$

where A is a scale parameter, e_0 and e_1 share a coefficient, and $\sigma = \frac{1}{1+\rho}$ elasticity of substitution between commodity 0 and 1 of the utility function. Using (52.12) for this case, the demand for x_j for the CES utility function will be

$$x_j^D = \delta_1 \frac{(R - rI_1)}{P_j \left[(\frac{e_0}{e_1})^\sigma \beta_{j1}^{\sigma-1} P^{\sigma-1} + 1 \right]} + \frac{(1 - \delta_1)R}{P_j \left[(\frac{e_0}{e_1})^\sigma \beta_{j0}^{\sigma-1} P^{\sigma-1} + 1 \right]}. \tag{52.15}$$

From (52.15), one derives that, given prices and income, the demand for good j under technology 0 (x_{j0}) is greater than under technology 1 if the income share of the fixed cost associated with technology one exceeds the income share of commodity 1 under technology 0 times $\left[1 - (\frac{\beta_{1j}}{\beta_{0j}})^{\sigma-1} \right]$, i.e.,

$$x_{j0} \gtrless x_{j1} \text{ if } \frac{rI_1}{R} \gtrless \left[1 - \left(\frac{\beta_{j1}}{\beta_{j0}} \right)^{\sigma-1} \right] \left[1 + \left(\frac{e_1}{e_0} \right)^\sigma \beta_{j0}^{1-\sigma} P_j^{1-\sigma} \right]. \tag{52.16}$$

Thus, for the important special case of the Cobb–Douglas utility function ($\sigma = 1$), (52.16) suggests that consumption of good j will decline as one switches from technology 0 to technology 1. Condition (52.16) also indicates that this behavior will always occur when the elasticity of substitution is smaller than one and for cases of higher elasticity of substitution when the initial income share of commodity zero is small.

The lesson from these results is that, when the elasticity of substitution in consumption between commodities 0 and 1 is small, a reduction in the price of commodity 1, resulting from purchasing the durable, will reduce the demand for x_j although the demand for y_1 will increase. The effect of the lower variable input requirement per unit associated with the new technology will over the effect of the increase in y_1 associated with the lower π_1. Condition (52.16) also suggests that, when the elasticity of substitution is large and a large share of income was spent initially on commodity 0, the increase in demand for y_1 associated with a switch from technology 0 to 1 is large enough to increase the demand for x_j in spite of the reduction of the input per commodity unit. The likelihood of increase in x_j with a switch to technology 1 is higher as the fixed cost associated with technology 1 and its relative input requirement $\frac{\beta_{j1}}{\beta_{j0}}$ are smaller.

The generalization suggested by the CES results is that, when the substitution effect is not strong enough to overcome the increase in efficiency of the nondurable associated with the new technology, the demand for x_j may have the shape depicted in Fig. 52.7. In this case the consumer buys less of the nondurable while consuming more of commodity 1 as the nondurable price is declining and the consumer purchases the durable good (movement from P_{j0} to P_{j1} will result in a switch from A to B in Fig. 52.7). If Fig. 52.7 describes the demand for energy used for heating (following an earlier example), an increase in energy price from P_{j3} to P_{j2} will reduce energy use; but, as Fig. 52.6 indicates, it may increase consumption of heat because of a switch to more energy-efficient technology.

However, note that when the substitution effect is very strong, the demand curve for x_j can have the same general shape as the demand for y_1 as P_j changes. Namely, a reduction in x_j price, causing a switch from technology 0 to 1, will increase consumption of y_1 and demand for x_j; and an additional reduction in prices, which results in a reswitch, may reduce the demand for x_j as well as y_j.

The effects of a change in the price of a good that participates in both processes should be analyzed according to a sum of two changes. The first is a proportional change in both prices of commodity 1, and the second is a change in the price of commodity 1 under the process when the good's shares in variable cost are higher. For example, if the price of good j is increased by ΔP_j and the initial ratio $\frac{\beta_{j0}}{\pi_{10}} > \frac{\beta_{j1}}{\pi_{11}}$, then the effects of the change are the sum of: (1) an increase of

$$\Delta P_j \frac{\beta_{j1}}{\pi_{11}}$$

Fig. 52.7 The demand for energy used for heating

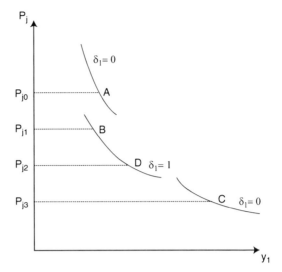

in the price of commodity 1 under both processes and (2) an increase of

$$\Delta P_j \frac{\beta_{j0}\beta_{j1}}{\pi_{10} + \beta_{j1}\Delta P_j}$$

in the price of commodity 1 under process 0. The total effect of the price change can be analyzed using the results of this and the previous sections.

52.6 The Behavior of the Indirect Utility Function

Let $V(\pi_{10}, \pi_{11}, I_1, R)$ be the indirect utility function of the durable choice problem considered here. For each price income combination, it will be equal to the indirect utility function under the selected technology, i.e.,

$$V(\pi_{10}, \pi_{11}, I_1, R) = \delta_1 \tilde{V}(\pi_{11}, R - rI) + (1 - \delta_1)\tilde{V}(\pi_{10}, R) \quad (52.17)$$

The indirect utility function is continuous since each of the functions generating it is continuous, and they are equal at the switching point. However, it is not always differentiable since, at the switch points, the indirect utility functions under each of the technologies have different gradients. Moreover, the indirect utility function is not concave in income and is not convex in prices. Again, the behavior of the function near the switch point is the reason for the irregularities.

To better understand the function's behavior, consider Fig. 52.8. The curve ABC depicts indirect utility as a function of income for the likely case of one switch. The segment, AB, corresponds to technology 0 and segment, BC, to the use of

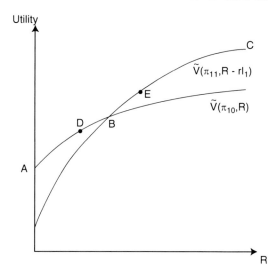

Fig. 52.8 The indirect utility function $V(\pi_{10}, \pi_{11}, I_1, R)$

the durable. At the switching point, B, the marginal utility of income, when the durable is used, is greater than when it is not used (otherwise, there is no reason to switch). Therefore, marginal utility is not declining with income; and, in some cases, marginal utility may be larger at higher income levels (as compared to the marginal utility that points D and E demonstrate). Thus, assuming identical preferences, it might happen that an individual with higher income (above the critical switching level, R_B) will enjoy more from a given increase in income than a poorer individual who does not own the durable and uses a less efficient consumption technology.

52.7 Aggregation of Demand

Thus far, this paper has analyzed the behavior of an individual consumer. This section builds on the previous results to derive aggregate demand relationships. It is assumed that all consumers have identical tastes but differ in their income. It is also assumed that their preferences can be expressed by homothetic utility functions.

Most of the analysis will be for the case where commodity 1 can be produced with only two processes – one of which requires the use of a durable good. Some of the results will be extended to a case where several durables producing commodity 1 are available.

Four aggregate relationships will be derived: the desired stock of durable good k at time t, $S_k(t)$; the aggregate demand for durable good k at time t, $Q_k(t)$; the aggregate demand for commodity 1 at time t, $y_1(t)$; and the aggregate demand for nondurable j for the production of commodity 1 in period t, $X_j(t)$.

52.8 The Demand for Durable Good Stock

Consider the case where only one durable is available. Recall that, when utilities are homothetic, there is a cutoff income level that separates lower incomes, which do not lead to use of the durable and higher income levels, causing consumers to prefer the use of the durable good. This cutoff income level is a function of commodity 1 and the durable good prices. Let this cutoff level be denoted by $R_1 = R_C(\pi_{10}, \pi_{11}, r I_1)$. From (52.8), the critical income is determined by solving

$$\tilde{V}(\pi_{10}, R_C) = \tilde{V}(\pi_{11}, R_C - r I_1). \tag{52.18}$$

Thus, given prices, the desired quantity of the durable good in the economy is equal to the number of consumers with income above the cutoff level. Therefore, knowledge of the income distribution is necessary for deriving the demand for durables.

Let the function, $f_t(R)$, be the income density function at time t such that $f_t(R)\Delta R$ is the fraction of households in the total population with income between R and $R + \Delta R$. This density function is defined on positive income levels with a minimum level, R_m. The function $F_t(R)$ denotes the share of households whose income does not exceed R and is derived by integration of $f_t(R)$. Let the total population size be denoted by T, thus $T F_t(R)$ is the number of households whose income does not exceed R.

Using these definitions and the critical income defined by (52.18), the total quantity of the durable good desired at time t is a function of population size, income distribution parameters, and prices and is given by

$$S_1(t) = T \int_{R_C(\pi_{10},\pi_{11},I_1)}^{\infty} f(R) dR = T \left[1 - F_t \left[R_C(\pi_{10}, \pi_{11}, I_1) \right] \right]. \tag{52.19}$$

To demonstrate the suggested approach, consider the case where income has Pareto distribution and consumers have Cobb–Douglas utility functions. The Pareto distribution has been found to fit empirical data rather well (Champernowne [3]). The density function is given by

$$f(R) = \begin{cases} \gamma R_m^{\gamma} R^{-\gamma-1} & \text{for } 0 \le R_m \le R \text{ and } \gamma > 0, \\ 0 & \text{otherwise.} \end{cases} \tag{52.20}$$

This income distribution assumes that the minimal income in the economy is R_m; this income level has the highest population density, and population density declines as income rises.[12] Moreover, this income distribution approximates the share of population with income above a critical level, R, to be equal to $\frac{R_m^{\gamma}}{R}$. The parameter, γ, is

---

[12] Actually, the Pareto distribution is approximating very well the behavior of the tail end of income distribution, with income above the mode level. When it is used empirically, one has to accommodate the exclusion of the very low-income groups from the analysis.

804   D. Zilberman and J. Liu

the elasticity of the population share of the high income group, and it approximates
the percentage change in the fraction of population with income above a critical
level when this level is reduced by 1%.

By introducing (52.20) into (52.19), one finds

$$S_1(t) = T R_m^\gamma [R_C(\pi_{10}, \pi_{11}, I_1)]^{-\gamma}. \tag{52.21}$$

Consider the Cobb–Douglas utility function

$$U(y_0, y_1) = y_0^{\alpha_0} y_1^{\alpha_1}, \tag{52.22}$$

where $\alpha_0 + \alpha_1 \le 1$ and $\alpha_0, \alpha_1 \ge 0$.

The indirect utility function associated with (52.21) is

$$\tilde{V}(\pi, R) = (R\alpha_1)^{\alpha_1} (R\alpha_0)^{\alpha_0} (\alpha_0 + \alpha_1)^{-(\alpha_0 + \alpha_1)} \pi^{-\alpha_1}. \tag{52.23}$$

From (52.18), the critical income is

$$R_C(\pi_{10}, \pi_{11}, I_1) = r I_1 \left[1 - \left(\frac{\pi_{11}}{\pi_{10}} \right)^{\frac{\alpha_1}{\alpha_1 + \alpha_0}} \right]^{-1}. \tag{52.24}$$

To simplify the expression later, assume $\alpha_0 + \alpha_1 = 1$ and denote $\alpha_1 = \alpha$.
Introducing (52.24) into (52.21) obtains the optimal stock of durable 1 at time t

$$S_1(t) = T \left(R_m \left\{ 1 - \left(\frac{\pi_{11}}{\pi_{10}} \right)^\alpha \right\}^{-\gamma} \right) (r I_1)^{-\gamma}. \tag{52.25}$$

Both the price elasticity (in absolute value) and the mode income elasticity of the
stock demand for the durable are equal to the elasticity of population share of the
high – income group. The increase in stock demand resulting from a decline in
the variable cost of the modern technology relative to the old technology $\left(\frac{\pi_{11}}{\pi_{10}} \right)$ is
higher as the elasticity of population share of high income group and the share of
commodity 1 in current expenses (α) increase. Increase in the elasticity of population share of the high-income group will reduce aggregate stock demand, and the
impact will be greater as the critical income that results in durable purchase is lower.
When $\gamma \ge 1$, the average income in the economy is given by $\bar{R} = \frac{\gamma R_m}{\gamma - 1}$.[13] Thus,
the demand for the durable good stock can be rewritten as

$$S_1(t) = T \left\{ \frac{\gamma - 1}{\gamma} \left(1 - \left(\frac{\pi_{11}}{\pi_{10}} \right)^\alpha \right) \bar{R} \right\}^\gamma (r I_1)^{-\gamma}. \tag{52.26}$$

Again, the average income elasticity of stock demand of the durable is equal to γ.

---

[13] If R is bound from above, one can express $S_1(t)$ as a function of $\bar{R}$ for $\gamma < 1$.

52.9 The Aggregate Flow Demand of the Durable Good

While $S_1(t)$ denotes the desired stock of capital good 1 at time t, the actual demand for new durables at time t is the difference between the demanded stock and the stock accumulated prior to time t. For simplicity, let us exclude the possibility of physical deterioration of the durable and thus replacement demand. Let us also assume that the actual stock of the durable is adjusting instantaneously to increase in stock demand. Thus, the flow demand for the durable good is equal to the change in the stock demand for the durable over time. If changes in parameters (prices, income distribution, etc.) cause reduction in the desired stock level of the durable good, the stock demand will be equal to zero.

Considering the cases when stock demand for the durable is rising over time and assuming differentiability of the relevant functions with respect to time, the flow demand for capital is derived by differentiating (52.19) to yield

$$Q_1(t) = \dot{S}_1(t) = T \left\{ S_1(t) \frac{\dot{T}}{T} - \frac{\partial F_t}{\partial t} - \frac{\partial F_t}{\partial R_C} \left[\frac{\partial R_C}{\partial \pi_{11}} \dot{\pi}_{11} - \frac{\partial R_C}{\partial \pi_{10}} \dot{\pi}_{10} + \frac{\partial R_C}{\partial r I} (r \dot{I}) \right] \right\}.$$

(52.27)

where the upper dot denotes differentiation with respect to time. The flow demand for the durable good is an increasing function of the rate of population growth, the change over time in the fraction of population with income above critical level, the decline over time of the variable cost associated with the modern technology compared to the old technology, and the decline in the price of the durable good.

Using the example when income is Pareto distributed and the utility function is Cobb–Douglas, the flow demand for the capital good is derived (assuming $\gamma > 1$) to be

$$Q_1(t) = S_1(t) \left\{ \frac{\dot{T}}{T} + \gamma \left[\frac{\dot{R}_m}{R_m} - \left(\frac{r \dot{I}_1}{r I_1} \right) - \alpha \frac{(\pi_{11}/\pi_{10})^\alpha}{1 - (\pi_{11}/\pi_{10})^\alpha} \left(\frac{\dot{\pi}_{11}}{\pi_{10}} \right) \right] \right\}.$$

(52.28)

The flow demand for the durable is a product of the stock demand and the rate of change in the desired stock of the durable good (the expression in the brackets). The rate of change in the desired stock of the durable is linearly dependent on the rate of decline in the price of the durable and the interest rate, and the rate of increase in the minimum (average) income elasticity of the population share of the high-income group in the linear coefficient. The effect of the rate of reduction in the variable cost associated with technology 1 relative to technology 0 on the rate of change in desired durable stock is proportional to the product of the elasticity of the population share of the high-income group, the share of coefficient of commodity 1 in current expenses, and an increasing function of $(\frac{\pi_{11}}{\pi_{10}})^\alpha$.

52.10 The Aggregate Demand for Commodity 1 and Nondurables

The aggregate demand for commodity for cases where there is one critical income above which consumers purchase the durable (i.e., homothetic utility functions) is derived from (52.9) to yield

$$
Y_1(t) = \left\{ T \int_0^{R_C(\pi_{10},\pi_{11},rI_1)} D(\pi_{10}, R) f(R) dR \right.
$$
$$
\left. + \int_{R_C(\pi_{10},\pi_{11},rI_1)}^{\infty} D(\pi_{11}, R - rI_1) f(R) dR \right\} \qquad (52.29)
$$

For the case of Cobb–Douglas utility function,

$$
D(\pi, R) = \frac{\alpha R}{R}. \qquad (52.30)
$$

Introducing (52.24) and (52.30)–(52.29) for the case of Pareto income distribution defined in (52.20) with $\gamma > 1$ yields

$$
Y_1(t) = T \frac{\alpha \bar{R}}{\pi_{10}} \left\{ 1 + \left(\frac{R_m}{rI_1} \left(1 - \left(\frac{\pi_{11}}{\pi_{10}} \right)^\alpha \right) \right)^{\gamma-1} \right.
$$
$$
\left. \times \left(\left(\frac{\pi_{10}}{\pi_{11}} \right)^{1-\alpha} + \frac{1}{\gamma} \left(1 - (\frac{\pi_{11}}{\pi_{10}}) \right)^\alpha (\frac{\pi_{10}}{\pi_{11}}) - 1 \right) \right\}. \qquad (52.31)
$$

The aggregate demand for commodity 1 in (52.31) is equal to aggregate demand for commodity 1 under technology 0, $(T \frac{\alpha \bar{R}}{\pi_{10}})$, plus an additional element that expresses the effect of introducing technology 1 on the demand of commodity 1. While average income is the only income distribution parameter required to compute aggregate demand for commodity 1 when only technology 0 is available (and we have a "traditional" demand model), the availability of durable good 1 and the discrete choice it implies requires more than one income distribution parameter to compute aggregate demand. As (52.31) indicates, the aggregate demand for commodity 1 is an increasing function of the minimum income, population size, and the share of commodity 1 in current expenses (α). It is a decreasing function of the elasticity of the population share of the high-income group, the price of commodity 1 under technology 1 relative to its price under technology 0, and the price of commodity 1 under technology 0 (when $\frac{\pi_{11}}{\pi_{10}}$ is kept constant).

The derivation of aggregate demand for nondurable j used in producing commodity 1 is similar to the derivation of the demand for the commodity. Using (52.12) and (52.28), the aggregate demand for nondurable j is

52 On Consumption Indivisibilities, the Demand for Durables, Income Distribution   807

$$X_j(t) = T \int_0^{R_C(\pi_{10},\pi_{11},rI_1)} \beta_{10} D(\pi_{10}, R) f(R) dR$$

$$+ \int_{R_C(\pi_{10},\pi_{11},rI_1)}^{\infty} \beta_{11} D(\pi_{11}, R - rI_1) f(R) dR. \qquad (52.32)$$

Consider the case where commodity 1 is the only nondurable used in both technologies where the utility function is Cobb–Douglas and income is Pareto distributed. In this case (52.32) becomes

$$X_j(t) = \frac{\alpha T \bar{R}}{P_j} \left\{ 1 + \left(\frac{R_m}{rI_1} \right)^{\gamma-1} \frac{\gamma - 1}{\gamma} \left(1 - \left(\frac{\pi_{11}}{\pi_{10}} \right)^{\alpha} \right)^{\gamma} \right\}. \qquad (52.33)$$

The aggregate demand for nondurable good j is equal to the demand for good j when only technology 0 is available $\frac{\alpha T \bar{R}}{P_j}$ minus an additional element reflecting the contribution of technology 1 in saving good j. Note that in our examples, the availability of the new durable will increase the aggregate demand for commodity 1 while reducing the aggregate demand of the nondurable.

52.11   The Case of More than One Durable

The analysis thus far can be easily extended to the case where more than one durable exists for generating commodity 1. When there are two durables, for example, it can be concluded that, if the second durable is more expensive than the first but requires lower variable cost (i.e., $I_2 > I_1$ but $\pi_{12} < \pi_{11}$), then the consumer with homothetic preferences will have two critical income levels, R_C^1 and R_C^2. The level of income for which the consumer will switch from no durable to durable 1 is R_C^1 and the level where the consumer switches from durable 1 to 2 is R_C^2 and R_C^2 will be larger than R_C^1. Thus, population with identical tastes will be segmented according to income into three groups. The low-income group $R < R_C^1$ will not use any durable; the middle-income group $(R_C^1 < R < R_C^2)$ will use durable good 1; and the high-income group will use durable good 2. Assuming a two-parameter Pareto income distribution $S_1(t)$, the aggregate stock demand for durable 1 at time t, for example, is given by

$$S_1(t) = TR_n^{\gamma} \left[(R_C^1)^{-\gamma} - (R_C^2) - \gamma \right]. \qquad (52.34)$$

Using (52.23), the critical values for a Cobb–Douglas utility function are

$$R_C^1 = \frac{rI_1}{1 - \left(\dfrac{\pi_{11}}{\pi_{10}} \right)^{\alpha}}$$

$$R_C^2 = \frac{rI_2\left[1-\left(\dfrac{\pi_{12}}{\pi_{11}}\right)^{\alpha}\right]I_1}{1-\left(\dfrac{\pi_{12}}{\pi_{11}}\right)^{\alpha}}. \tag{52.35}$$

Thus, the aggregate stock demand for durable good 1 becomes

$$S_1(t) = T\frac{R_m}{r}\left\{\left[\frac{1-\left(\frac{\pi_{11}}{\pi_{10}}\right)^{\alpha}}{I_1}\right]^{\gamma} - \left[\frac{1-\left(\frac{\pi_{12}}{\pi_{11}}\right)^{\alpha}}{I_2-\left(\frac{\pi_{12}}{\pi_{11}}\right)I_1}\right]^{\gamma}\right\}. \tag{52.36}$$

Similarly, the aggregate stock demand for durable 2 becomes

$$S_2(t) = T\frac{R_m}{r}\left[\frac{1-(\pi_{12}/\pi_{11})^{\alpha}}{I_2-(\pi_{12}/\pi_{11})I_1}\right]^{\gamma}. \tag{52.37}$$

One can extend the analysis to derive aggregate flow demands for the durables, and the aggregate demand functions for the nondurables and commodity 1.

52.12 Conclusions

Explicit consideration of the indivisibilities caused by the availabilities of consumer durables modifies significantly the behavioral pattern predicted by consumer theory and the nature of its basic relationship. Individual consumers' demand relationships will have points of noncontinuity reflecting technological switches, and changes in product prices may result in total realignment of the consumers' durable mix, which will drastically change the nature of demand for all goods.

Recent developments in econometrics allow estimation of simultaneous discrete and continuous choices made by consumers (see, for example, Heckman [6]). Indeed, frameworks similar to the one presented here have been applied to estimate demand relationships to durables like air conditioners (see Hausman [5]) and refrigerators (see Brownstone [2]). The results of individual demand estimations can be used for estimating aggregate demand for durables when estimators of joint distribution of key variables that affect consumer choices (i.e., income) are available. These probability measures will be used as weights in generating the aggregate demand relationships following an aggregation procedure similar to the one taken in the latest part of this chapter. These later reactions have developed aggregate demand relationships for durables and nondurables analytically. These aggregates depend on more than one income distribution parameter even when consumers have identical homothetic preferences (such preferences allow aggregation with the knowledge of mean income only when traditional consumer behavior models are used).

While the framework introduced here is useful to understanding durable and nondurable choices in many cases, it is still limited and requires improvements and

52 On Consumption Indivisibilities, the Demand for Durables, Income Distribution

generalizations. Two particular elements that have to be incorporated in the analysis are product quality differences[14] and labor leisure choices, in particular, the effects of different sources of income (wage earning and returns from assets) on durable choices.

References

1. Becker, G.S.: A theory of allocation of time. Econ. J. **75**, 493–517 (1965)
2. Brownstone, D.: Econometric models of choice and utilization of energy using durables. EFRI Workshop on the Choice and Utilization of Energy Using Durables, Boston, Massachusetts, November 1–2, (1979)
3. Champernowne, D.G.: The Distribution of Income Between Persons. Cambridge: Cambridge University Press (1973)
4. Hanemann, W.M.: Welfare evaluations in contingent valuation experiments with discrete responses. Am. J. Agr. Econ. **66**(3), 332–341 (1984)
5. Hausman, J.A.: Individual discount rates and the purchase and utilization of energy-using variables. Bell J. Econ. **10**, 33–54 (1979)
6. Heckman, J.J.: Dummy endogenous variables in a simultaneous equations systems. Econometrica **46**, 679–694 (1978)
7. Lancaster, K.J.: A new approach to consumer theory. J. Polit. Econ. 74, 132–157 (1966)
8. Novshek, W., Sonnenschein, H.: Marginal consumers and neoclassical demand theory. J. Polit. Econ. 87(6), 1368–1376 (1979)
9. Rosen S.: Hedonic prices and implicit markets: Product differentiation in pure competition. J. Polit. Econ. **82**(1), 34–55 (1976)
10. Small, K.A., Rosen, H.S.: Applied welfare economics with discrete choice. Econometrica **49**, 105–130 (1981)
11. Sunding, D., Zilberman, D.: The Agricultural Innovation Process: Research and Technology Adoption in a Changing Agricultural Sector. In: Bruce, L.G., Gordon, C.R. (eds.) Handbook of Agricultural Economics, Volume 1A Agricultural Production, pp. 207–261. Elsevier Science B. V., Amsterdam, the Netherlands (2001)
12. Train, K.E. Discrete Choice Methods with Simulation. Cambridge University Press, Cambridge (2003)

---

[14] The approach taken by Rosen [9] and Novshek and Sonnenschein [8] seems to be useful for extending the analysis to consider differences in product qualities. This seems especially appropriate in situations where there is a large variety of durables with small differences in production coefficients and quality properties.